To Glen Lucas
(more weight for your
bookshelves!)
Greetings, Jack Hawkes

ANNALS OF BOTANY MEMOIRS

No. 3

THE POTATOES OF ARGENTINA, BRAZIL, PARAGUAY AND URUGUAY

A BIOSYSTEMATIC STUDY

THE POTATOES OF ARGENTINA, BRAZIL PARAGUAY, AND URUGUAY

A BIOSYSTEMATIC STUDY

BY

J. G. HAWKES

AND

J. P. HJERTING

OXFORD
AT THE CLARENDON PRESS
1969

Oxford University Press, Ely House, London W. 1

GLASGOW NEW YORK TORONTO MELBOURNE WELLINGTON
CAPE TOWN SALISBURY IBADAN NAIROBI LUSAKA ADDIS ABABA
BOMBAY CALCUTTA MADRAS KARACHI LAHORE DACCA
KUALA LUMPUR SINGAPORE HONG KONG TOKYO

PRINTED IN GREAT BRITAIN
AT THE UNIVERSITY PRESS, OXFORD
BY VIVIAN RIDLER
PRINTER TO THE UNIVERSITY

PREFACE

THIS monograph on the tuber-bearing *Solanum* species of Argentina, Brazil, Paraguay, and Uruguay grew out of our first attempts to identify herbarium and living collections made by Petersen and Hjerting in Argentina during the years 1948 to 1953. We were also faced with the problem of identifying the rich and interesting material sent on loan by the Instituto Miguel Lillo, Tucumán, and by the herbaria in and around Buenos Aires, which we shall detail presently.

It soon became evident that in Argentina, and more especially in the north-western provinces, a highly complex pattern of potato variability existed that could not be dealt with satisfactorily by reference to the existing taxonomic works on that area. Fortunately, the region has been very well studied botanically, and indeed more collections of wild potatoes have been made (most of them by local botanists and collectors) in that region than in any other area in South America of comparable size.

Although we began by thinking in terms of a monographic treatment of the wild potatoes of Argentina we soon came to the conclusion that it would be necessary to extend our terms of reference to include the adjacent countries of Paraguay, Uruguay, and south-east Brazil, since there are no well-marked natural boundaries to the spread of species into those countries from Argentina. On the west, however, the high ranges of the Andes constitute a boundary that has not made it necessary for us to consider the Chilean species also, except in a very few cases. To the north-west we have unfortunately had to draw an artificial boundary from a phytogeographical point of view, and to exclude the Bolivian species from this monograph, since they are on the whole rather poorly known in comparison with the Argentine species. Nevertheless, when we have had the necessary information we have included Bolivian data for Argentine species that were to be found north of the border. We have also quoted synonyms or variants from Bolivia when we considered this to be necessary for the understanding of the taxonomy of the Argentine or Paraguayan species under discussion.

In a group of plants such as the tuber-bearing Solanums the use of classical taxonomic methods alone is often doomed to failure. This is especially so with the Argentine species, the complexities of which can, in our view, be unravelled only by combining the classical with the experimental approach.

Indeed, our work which began as a simple study of identification finally developed into a complex biosystematic survey of the species as living and evolving continua. We have attempted to treat species as dynamic systems of great complexity, composed of a wide range of populations adapted to a correspondingly wide range of geo-ecological conditions. In this book we have tried to assess the pattern of variability and to interpret this pattern in terms of environmental and habitat factors, the reproductive biology of the material studied, breeding mechanisms, pollination and dispersal phenomena, and barriers to gene exchange. By controlled hybridization and field observations, as well as by other experimental methods and graphic analyses, we have attempted to show the relationship between species and series, and to examine patterns of natural hybridization, introgression, and gene flow.

This has involved, on the one hand, the examination of some 1500 collections of dried material from herbaria in South America, Europe, and the U.S.A.; and, on the other hand, the cultivation of some 180 accessions of living material gathered by ourselves and by other collectors. In this way we were able to form a concept of the range of phenotypic variability expressed by different growth forms of the same species under very variable natural conditions, and to compare these with the phenotypes of identical genetic lines grown under controlled conditions in the experimental field and glasshouse in Europe.

We also raised some 86 families of experimental hybrids, including 62 F_1 and 24 F_2 generations, with a limited number of backcrosses also, to assess the sterility barriers, if they existed, between species and to be able to verify by comparison the hybrid nature of certain natural variants observed in herbarium specimens and in living collections. Furthermore, we made many hundreds of chromosome counts and were also able to link the taxonomic results to the extensive literature of cytological research and crossability studies that exist for several of the more important species that have formed the subject of studies by geneticists and plant-breeders.

During the course of the work we have assessed critically 96 specific and 64 infraspecific taxa described previously, and have reduced them with a reasonable degree of certainty to 22 species and 12 subspecies in the light of the studies to be described in this monograph. The task has been a slow one, spreading over a period of some ten years, and we have purposely delayed its completion (apart from other delays caused by our own slowness and our other commitments) in order to take advantage of several collecting expeditions made by ourselves and by other botanists and plant-breeders who have kindly allowed us to use their data. The raising of families of hybrids is also a slow process, which cannot be hurried. We began with the immense advantage of being able to work with the material obtained by Petersen and Hjerting in the period 1948 to 1953 and then to return for further collecting trips to key areas to help solve outstanding problems, where previous lack of living material made it impossible to arrive at a correct conclusion. As a result, our understanding of the group of species under survey has enabled us to solve many problems, and at least to formulate those which we have not been able to solve in terms that indicate how they might be studied. Indeed, we hope that this work will act as a stimulus to other botanists, and especially to those who live and work in Argentina and adjacent countries, to take a renewed interest in the potatoes of that area.

We have been fortunate enough to collect at various times, both jointly and individually, in most of the area studied. Such a first-hand knowledge is an essential prerequisite to any floristic or monographic work. Thus, one of us (J. P. H.) has collected extensively in the south, west, north-west, north, and north-east provinces of Argentina, and also in Bolivia. The other (J. G. H.) has collected in north-west and south-east Argentina, in Bolivia, Uruguay, and south Brazil. In our recent joint expedition to Argentina we made extensive collections in most of the western and north-western provinces. Unfortunately, neither of us has been able to collect in Paraguay, south-east Brazil (Sta. Catarina, Paraná, Minas Gerais), or in the eastern provinces of Argentina. Apart from this, we have been able to observe wild potatoes in their natural habitats over a very wide area, and have had the additional advantage of being able to talk to Argentine, Brazilian, and European collectors who have visited regions not personally known to us; we have also been able to cultivate material generously donated by them.

This book is arranged in two parts. The first is of more general interest to the non-specialist, especially Chapter 4, which summarizes the findings described in greater detail in Part II. In that chapter, also, the evolutionary and biological relationships of the species are discussed in general terms. The first three chapters deal with historical and phytogeographical matters relating to *Solanum* taxonomy in the area studied. The fifth chapter shows the importance to the plant-breeder of the indigenous potatoes of this area in relation to problems of disease resistance and adaptation to different conditions.

Information on the exact treatment for each species and the experimental techniques adopted is given in Chapter 6. The taxonomy of the group within the genus, and dichotomous keys for the identification of series and species appear in Chapter 7. Part II describes the series and species in detail.

When we had finished the text of this book we decided to make yet another expedition to Argentina, to try and solve some of the remaining outstanding problems of potato taxonomy and distribution. As a result, some of our hypotheses have been strengthened, whilst others have had to be discarded; and we have certainly been able to solve a few more problems and add fresh data on distribution and ecological adaptation of the species concerned. The general results of this 1965–6 expedition are given in Appendix III, whilst some of the detailed information has been incorporated into the text. All the specimens collected on this expedition are listed in the Appendix. Data from these records have been added to the species distribution maps only in those cases where they are from areas not previously recorded.

We have received so much help from so many people during the course of this work that it is difficult to know whom to thank first.

We should like to record our gratitude to Mr. T. J. Williams, late Director of the Williams Foundation, Buenos Aires, who made possible the collecting trips of 1955–6 and 1965–6; also to Dr. Gunnar Hiorth, Director of the Instituto de Genética of the Facultad de Agronomía, Tucumán, under whose auspices the collections of 1948–53 were made. Thanks are due also to the Commonwealth Agricultural Bureaux for providing funds for the 1939 visit to north-west Argentina, and to the Government of Uruguay for making possible the collecting trips in that country and in south Brazil. We also wish to thank the Carlsberg Foundation of Denmark and the Royal Society of London for generously providing funds for the 1965–6 expedition; to the Kaiser Company of Argentina for kindly lending us an 'Estanciera' station wagon and to the J. Lauritzen Shipping Co. for providing free passages for the Danish members of the two expeditions of 1955–6 and 1965–6. We should also like to thank the following Directors of Institutes for providing hospitality and transport at various times: Ing. Agr. D. Pascuale, Director of the Estación Experimental Agropecuaria de Balcarce, Argentina; Ing. Agr. U. C. García, Director of the Instituto Nacional de Tecnología Agropecuaria (INTA), Argentina; Dr. G. Fischer, late Director of the Instituto Fitotécnico y Semillero Nacional 'Alberto Boerger', La Estanzuela, Uruguay; and the Director of the Instituto Agronômico do Sul, Pelotas, Brazil. We owe a special debt of gratitude to Dr. A. Digilio and Dr. T. Meyer of the Instituto Miguel Lillo, Tucumán, to Professor A. T. Hunziker of the Department of Botany, Córdoba, to Professor O. Boelcke of the Fac. de Agron. y Vet. Buenos Aires, to Professor A. Ruiz Leal of Mendoza, and to Dr. J. A. Hunziker of Buenos Aires, for their help in 1965–6; our thanks are also due to Lic. A. K. Okada, who travelled with us for a month in 1966 and helped us collect potatoes and many other plants.

The fact that one of us lives in Denmark and the other in England has necessitated a certain amount of travel between the two countries, in addition to a vast amount of correspondence. We are very much indebted to the Friderichsen Fund in Denmark and to the University of Birmingham for providing money to make these visits possible; to Professor Th. Sørensen, Director of the Botanic Gardens and Institute, Copenhagen, for providing growing facilities and leave of absence; and also to the Landbrugets Kartoffelfond, Foraedlingsstationen at Vandel, Denmark, and especially to the Director of this station, Lic. Agron. B. Jacobsen, for kindly providing growing facilities, hospitality, and accommodation to enable part of this co-operative work to be carried out at his station.

Our thanks are also due to the following institutes and museums in Argentina and Uruguay for so kindly sending us very extensive herbarium material on loan and allowing us to keep it for such long periods.

Instituto Miguel Lillo, Tucumán, Argentina.

Museo Argentino de Ciencias Naturales 'Bernardino Rivadavia', Buenos Aires, Argentina.

Instituto de Botánica Agrícola del I.N.T.A., Castelar, Argentina.

Museo de La Plata, División de Plantas Vasculares, La Plata, Argentina.

Instituto de Botánica 'C. Spegazzini', La Plata, Argentina.

Instituto de Botánica Darwinion, San Isidro, Argentina.
Museo Botánico, Facultad de Ciencias E. F. y Naturales, Córdoba, Argentina.
Laboratorio de Botánica, Facultad de Agronomía, Montevideo, Uruguay.
Museo de Historia Natural, Montevideo, Uruguay.

We are also greatly indebted to many institutions in Europe and the United States of America for the loan of herbarium material or for allowing us to study it in the institutes concerned. We ask their forgiveness for not listing them here, but we give a full list, together with the standard abbreviations, in Appendix I, p. 446.

We wish to thank the following for kind permission to reproduce certain illustrations.

1. The Council of the Royal Horticultural Society (Fig. 9).
2. The editor of the *Annals and Magazine of Natural History* (Fig. 29).
3. *Schweizerbart'sche Verlagsbuchhandlung* (Engler's *bot. Jb.*) (Fig. 41).
4. The Council of the Linnean Society (Fig. 45).
5. The editor of *Euphytica* (Figs. 19, 20; Tables 9, 10; Map 6).
6. Professor S. M. Bukasov, Leningrad (Fig. 4(*a*) (modified); Figs. 17, 42).

It is a pleasure also to record our thanks to the following collectors who have most kindly sent us their material for identification or who have made special gatherings of wild potato material at our request. Many also have sent us very useful detailed notes on habitats, distribution, and plant communities, which has been of the greatest value in presenting as complete a picture as possible of the biology of the wild potatoes of the areas concerned.

Dr. H. Sleumer, Rijksherbarium, Leiden, Netherlands. (Many interesting dried collections from key areas in Argentina, as well as several living collections; much advice and information on habitats, natural variation, and plant communities.)

Dr. F. Vervoorst, Instituto Miguel Lillo, Tucumán, Argentina. (Dried and living collections from Catamarca province and special information on the plant communities of the north-west provinces of Argentina.)

Mr. E. Petersen, Bariloche, Nahuel Huapi, Argentina. (Invaluable help from this excellent collector with 'a good eye' for wild potatoes, who helped so much with his enthusiasm to make the 1948–53 and 1956 collecting trips a success.)

Mr. K. Rahn, Botanic Garden, Copenhagen. (Excellent help with collecting during the 1955–6 and 1965–6 expeditions in north-west Argentina.)

Professor H. Ross, Max-Planck-Institut für Züchtungsforschung, Cologne, Germany. (Loan of interesting collections from his South American Expedition of 1959; also corolla preparations, information on plant communities, chromosome numbers, and growth of material in Germany.)

Sr. A. Castronovo, Centro Nacional de Investigaciones Agropecuarias, Castelar, Argentina. (Living collections of *S. chacoense* and other species.)

Dr. A. Ruiz Leal and Sr. F. Roig, Facultad de Ciencias Agrarias, University of Mendoza, Argentina. (Loan of herbarium collections and information on the ecology and distribution of species in Mendoza.)

Sr. J. Diem, Bariloche, Neuquén, Argentina. (Living material, herbarium specimens, and information on the wild potatoes of the Nahuel Huapi region.)

Dr. R. Millán, Batán, Mar del Plata, Argentina. (Living collections from various parts of Argentina.)

Dr. E. Grondona, formerly at the Estación Experimental Agropecuaria, Balcarce, Argentina. (Herbarium collections and information on collectors.)

Sr. J. Reche, Córdoba, Argentina. (Living material from Catamarca and La Rioja provinces.)

Sr. E. Viirsoo, Tucumán, Argentina. (Herbarium specimens, information on cultivated varieties, and enthusiastic help at the Estación Experimental 'El Suncho', Catamarca, Argentina.)

Sr. A. Schulz, Colonia Benítez, prov. Chaco, Argentina. (Seed and tuber collections of wild potatoes.)

Mr. F. Kemp, Primavera, Paraguay. (Seeds and herbarium specimens of wild potatoes.)

Professor B. Rambo, Colegio Anchieta, Porto Alegre, Brazil. (Herbarium and living collections from south Brazil.)

Sr. D. Mota da Costa, Instituto Agronômico do Sul, Pelotas, Brazil. (Living and herbarium collections from south Brazil.)

Professor F. G. Brieger, Instituto de Genetica, São Paulo, Brazil. (Living collections from Minas Gerais and information on habitats, etc.)

Dr. D. Dedecca, Department of Botany, Instituto Agronômico do Estado de São Paulo, Brazil. (Tubers of wild potatoes from south-east Brazil.)

Dr. Floriano Guimaraes, Estacão Experimental de Horticultura 'Domingos Petrolini', Rio Grande do Sul, Brazil. (Tubers of wild potatoes from Rio Grande region.)

Professor M. Cárdenas, Departamento de Botánica Aplicada, Universidad Mayor de San Simón, Cochabamba, Bolivia. (Herbarium and living collections of Bolivian potatoes that are also to be found in Argentina or Paraguay; also reprints of his interesting publications.)

Our grateful thanks are due to Dr. D. S. Correll for sending us a copy of his scholarly monograph of 1962, *The Potato and its Wild Relatives*, and for stimulating discussions.

We should also like to extend our most sincere thanks to the following for advice and criticism, especially in relation to ecological and habitat information, distribution maps, phytogeographical maps, vegetation profiles, and locality and place-names.

Professor L. A. Cabrera, División Botánica, Facultad de Ciencias Naturales, Universidad de La Plata, Argentina.

Dr. T. Meyer, Universidad Nacional de Tucumán, Argentina.

Dr. F. Vervoorst, Instituto Miguel Lillo, Tucumán, Argentina.

We also wish to thank K. Jakobsen, P. Jacobsen, O. Mattsson, and R. Tarn for making some of the chromosome counts and Miss S. Branford for making the photographs of all the herbarium specimens shown in the plates. Our thanks are also due to Miss S. Branford and Mrs. M. Cotton for mounting the herbarium material in the Birmingham collection and to Mrs. M. White and Mrs. J. Alabaster for typing the manuscript of this book.

Some of the happiest memories in our lives are of our travels and collecting experiences in the magnificent landscapes of Argentina, Uruguay, and Brazil; we should like to express our most sincere thanks to all our friends from those regions, whether we have mentioned them above or not, who by their warm-hearted generosity and kindness have made our travels so pleasant.

We dedicate this book to our wives and families who endured our absences and preoccupations with nobility and restraint. Without their understanding, help, and sympathy it would certainly never have been written.

Birmingham J.G.H.

Copenhagen J.P.H.

CONTENTS

PART II. DETAILED TAXONOMY

LIST OF PLATES

(at end)

(*Note:* Unless otherwise stated the specimens photographed are in the herbarium of J. G. Hawkes, Birmingham.)

'*S. guaraniticum* Hassl. var. *angustisectum*' Hassl., collected by Hassler (*No. 2805*—herb. Geneva) at Santa Elisa, Paraguay.

36. *S. chacoense* Bitt. Type for '*S. guaraniticum* Hassl. var. *latisectum* Hassl. f. *glabrescens*' Hassl. Collected by Hassler (*No. 326*—herb. Geneva) in the Cordilleras de Altos, Paraguay.

37. *S. chacoense* Bitt. A collection from Villa Elisa, near Asunción, Paraguay, collected by T. Myndel Pedersen (*No. 5144*).

38. *S. chacoense* Bitt. Type specimen of '*S. boergeri*' Buk., grown at Leningrad from material collected by T. Henry (Accession K. 1070—herb. Leningrad, WIR) near Río Uruguay, Uruguay.

39. *S. chacoense* Bitt. Type specimen of '*S. laplaticum*' Buk., grown at Leningrad from material collected by R. Millán (*No. G–972*. Accession K. 597—herb. Leningrad, WIR) from the neighbourhood of the city of Buenos Aires, Argentina.

40. *S. chacoense* Bitt. Authenticated specimen of '*S. gibberulosum*' Juz. & Buk., grown at Leningrad (herb. J.G.H.), presumably from material of the type collection of Millán (*G–1183*) from Manfredi, Córdoba, Argentina.

41. *S. chacoense* Bitt. Authenticated specimen of '*S. garciae*' Juz. & Buk., grown at Leningrad (herb P.G.S.), presumably from the type collection of García (*Millán 1201*), collected near the city of Córdoba, Argentina.

42. *S. chacoense* Bitt. Type specimen of '*S. parodii*' Juz. & Buk., grown at Leningrad from material collected by R. Millán (*No. G–980*. Accession K. 603—herb. Leningrad, WIR) at Río Chico, Tucumán, Argentina.

43. *S. chacoense* Bitt. Authenticated specimen of '*S. schickii*' Juz. & Buk., grown at Leningrad (herb. Leningrad, WIR), presumably from the type material of Schick from Siambón, prov. Tucumán, Argentina.

44. *S. chacoense* Bitt. Type specimen of '*S. knappei*' Juz. & Buk., grown at Leningrad from material sent by P. T. Knappe (herb. Leningrad, WIR), and said to have been collected at Salta, Argentina.

45. *S. chacoense* Bitt. Type specimen of '*S. jujuyense*' Hawkes. Collected by Balls, Gourlay, and Hawkes (*Balls 5921*—herb. P.G.S.) from San Antonio, near Jujuy, Argentina.

46. *S. chacoense* Bitt. Paratype specimen of '*S. saltense*' Hawkes, grown at Cambridge from material collected by Balls, Gourlay, and Hawkes (*Balls 5936* p.p. CPC: 51 b—herb. P.G.S.) near Salta, Argentina.

47. *S. chacoense* Bitt. Type of '*S. emmeae*' Juz. & Buk., grown at Leningrad from material sent by P. T. Knappe (herb. Leningrad, LE), possibly collected near Tucumán, Argentina.

48. *S. chacoense* Bitt. subsp. *muelleri* (Bitt.) Hawkes & Hjerting. Collected by Petersen and Hjerting (*No. P–66*; CPC: 2426) in prov. Misiones, Argentina.

49. *S. chacoense* Bitt. subsp. *muelleri* (Bitt.) Hawkes & Hjerting. Type specimen of '*S. muelleri* Bitt. f. *densipilosum*' Corr. Collected by Smith, Reitz, and Caldato (*No. 9516*—herb. US) at Chapecó, Santa Catarina, Brazil.

Petersen and Hjerting (*No. P–332a*) at Infiernillo, dept. Tafí, prov. Tucumán, Argentina.

118. Natural hybrid of *S. vernei* Bitt. & Wittm. × *S. spegazzinii* Bitt. One of the seedlings raised at Birmingham from a natural berry of *Petersen & Hjerting P–332a.*

119. *S. gourlayi* Hawkes. Isotype specimen of *Balls 5979* collected by Balls, Gourlay, and Hawkes at San Gregorio, Tilcara, prov. Jujuy, Argentina.

120. *S. gourlayi* Hawkes. Collected at Rodero Colorado, dept. Iruya, prov. Salta, Argentina by Petersen and Hjerting (*No. P–365*).

121. *S. gourlayi* Hawkes. Material of *Petersen & Hjerting P–365* (CPC: 2480/1) grown at Birmingham.

122. *S. gourlayi* Hawkes. Collected by Hjerting and Rahn (*No. 190*) at Cuesta de San José, dept. Sta. Victoria, prov. Salta, Argentina.

123. Probable natural hybrid of *S. gourlayi* Hawkes × *S. vidaurrei* Cárd. collected by Hjerting and Rahn (*No. 254*) near Lizoite, dept. Santa Victoria, prov. Salta, Argentina.

124. *S. oplocense* Hawkes. Lectotype. Collected by Balls, Gourlay, and Hawkes (*Balls 6127*—herb. P.G.S.) between Oploca and Oro Ingenio, dept. Potosí, Bolivia.

125. *S. oplocense* Hawkes. Topotype material collected by M. Cárdenas (CPC: 2503) and grown in Birmingham in 1954.

126. *S. oplocense* Hawkes. Material collected by O. Braun (CPC: 2116) from the Salo district near Tupiza, dept. Potosí, Bolivia, and grown in Birmingham in 1954.

127. *S. oplocense* Hawkes. Isotype specimen of '*S. subandigenum* Hawkes var. *camarguense* Cárd.', collected by Cárdenas (*No. 5078*) near Camargo, dept. Chuquisaca, Bolivia.

128. *S. oplocense* Hawkes. Specimen of '*S. camarguense*' or '*S. subandigenum* var. *camarguense*' collected by Alandia and Ross (*No. 123*—herb. Max) in Yavi, prov. Jujuy, Argentina.

129. *S. oplocense* Hawkes. Collected by Petersen and Hjerting (*No. 962*) between Iturbe and Tres Cruces, prov. Jujuy, Argentina.

130. *S. oplocense* Hawkes. Tetraploid material of *Petersen & Hjerting 962* grown in the experimental field at Birmingham.

131. *S. vidaurrei* Cárdenas. Isotype specimen of *Cárdenas 5075*, collected near Muyuquira, prov. Nor Cinti, dept. Chuquisaca, Bolivia.

132. *S. vidaurrei* Cárdenas. Collection made by Sleumer (*No. 4059*) from Lizoite, dept. Sta. Victoria, prov. Salta, Argentina.

133. *S. spegazzinii* Bitt. Lectotype and isotype from Amblayo to Cachipampa, prov. Salta, Argentina, collected by C. Spegazzini (s.n.—herb. La Plata, Spegazzini 14211).

134. *S. spegazzinii* Bitt. Syntype specimen from Cerro de Cachi, prov. Salta, Argentina, collected by C. Spegazzini (s.n.—herb. La Plata, Spegazzini 14210).

135. *S. spegazzinii* Bitt. Duplicate of one of the original specimens on which Bitter and Wittmack's description of '*S. famatinae*' was made; collected by Kurtz

(*No. 13445b*—herb. Córdoba) between Los Corrales and Escalera de Famatina, Sierra Famatina, prov. La Rioja, Argentina.

136. *S. spegazzinii* Bitt. Specimen collected by Ross and Vervoorst (*No. 36*—herb. Max Planck) at Zanja de los Cardones, dept. Tafí, prov. Tucumán, Argentina, in a fairly dry situation.

137. *S. spegazzinii* Bitt. Specimen collected by Petersen and Hjerting (*No. 1893*) at Corrales, Sierra Famatina, prov. La Rioja, Argentina.

138. *S. spegazzinii* Bitt. Material grown under glass at Vandel, Denmark of *P. & H. 1893*.

139. *S. spegazzinii* Bitt. Specimens collected by Ruiz Leal and F. A. Roig (*No. 19369*) at Chepical, prov. San Juan, Argentina.

140. *S. spegazzinii* Bitt. Collected by Hjerting, Petersen, and Rahn (*No. 334*) from dept. Cachi, prov. Salta, Argentina.

141. *S. spegazzinii* Bitt. Material grown from seeds of *H.P. & R. 334* (Accession H. 135) which contrasts strongly with the original specimen.

142. *S. kurtzianum* Bitt. & Wittm. Isotype specimen of *Kurtz 15422* from herb. Córdoba, now designated by us as Lectotype, collected in Sierra Velasco, prov. La Rioja, Argentina.

143. *S. kurtzianum* Bitt. & Wittm. Duplicate of the specimen on which Bitter and Wittmack based their description of '*S. velascanum*'. Collected by F. Kurtz (*No. 15380*—herb. Córdoba) in Sierra Velasco, prov. La Rioja, Argentina.

144. *S. kurtzianum* Bitt. & Wittm. Duplicate specimen from herb. Córdoba of '*S. juanense*' Wittm. *nom. nud.*, collected by F. Kurtz (*No. 9828*) in prov. San Juan, Argentina.

145. *S. kurtzianum* Bitt. & Wittm. Collected by Petersen and Hjerting (*No. P–40*) at Río Pantanillos, dept. Andalgalá, prov. Catamarca, Argentina.

146. *S. kurtzianum* Bitt. & Wittm. Flower dissection of material grown at Cambridge (CPC: 1296) of the probable type collection of '*S. macolae*' Buk.

147. *S. kurtzianum* Bitt. & Wittm. Collected by Ruiz Leal (*No. 11906*—herb. Ruiz Leal) at El Portezuelo, dept. Las Heras, prov. Mendoza, Argentina.

148. *S. kurtzianum* Bitt. & Wittm. Collected by Ruiz Leal and Roig (*No. 18390*—herb. Ruiz Leal) near El Portezuelo, dept. Las Heras, prov. Mendoza, Argentina.

149. *S. maglia* Schlechtd. Plant grown in Argentina from tubers collected by Ruiz Leal (*No. 10470*—herb. Lillo) at Quebrada de Alvarado, dept. San Carlos, prov. Mendoza, Argentina.

150. *S. maglia* Schlechtd. Grown in Cambridge and Birmingham from tubers collected by Schnack (s.n.—CPC: 2057) at Quebrada de Alvarado, dept. San Carlos, prov. Mendoza, Argentina.

LIST OF MAPS

PART I

GENERAL

1

HISTORICAL SURVEY

THE interest to botanists and plant-breeders of the tuber-bearing species of *Solanum* has largely been a phenomenon of the present century. A few species were certainly described in the nineteenth century, and some search was also made for new living material to obtain more resistant forms in the period of the disastrous blight attacks and potato famines in Europe. The earlier botanists in general considered most wild potatoes to be ancestral or escaped forms of the cultivated potato, *S. tuberosum*, and they were therefore slow to pay proper attention to the great specific diversity of this group. Having once done so, in the second decade of the twentieth century, the flood-gates were opened, and almost every new variant discovered was honoured with a new specific or infraspecific name. We have now arrived at the third, or synthetic phase, when very few new species are being described for the area of Argentina, Paraguay, Uruguay, and south Brazil, and a great majority of the names once proposed have now been reduced to synonymy.

It is of some interest to follow the historical progress of work with the wild tuber-bearing species of this area, especially since the first species to be collected and the first to be described[1] in the whole group was *S. commersonii* from Montevideo. This was gathered by the French explorer Philibert Commerson in May 1767, and was described by Dunal in the supplement to Poiret's *Encyclopédie methodique* in 1813, and again in the *Synopsis* of 1816.

Sabine (1824) took up the question of the wild nature of *S. commersonii*, quoting a statement by Baldwin, an American botanist, to the effect that the plant had been seen commonly in the vicinities of Montevideo and of Buenos Aires, and was probably a wild form of *S. tuberosum*. Understandably enough, Sabine was not able to accept such a statement without an examination of Commerson's material, which he obtained on loan from Paris. After careful study he concluded that Commerson's potato was not a wild form of

[1] Leaving out of consideration the *nomen dubium*, *S. papa*, from Colombia, described in 1809.

the cultivated potato, though he considered that plants of *S. maglia* sent him from Chile by a Mr. Caldcleugh were none other than the cultivated species.

Sendtner's studies of Sellow's collections from eastern Brazil, published in Martius's *Flora Brasiliensis* in 1846, added a new species, *S. tenue*, and a new variety, *S. commersonii* var. *pubescens*. Unfortunately, no exact localities were given, and we can only assume that the material was obtained somewhere in the southern or south-eastern parts of that country. We have classed both these taxa of Sendtner as *S. commersonii*.

In the same year, 1846, the Solanums of J. D. Hooker's *Flora Antarctica* were published. In this work Hooker placed various entities from Chile, the Andes of Mendoza, and the region of Buenos Aires, under *S. commersonii*, though he separated them into four distinct varieties. We now know that the Chilean material belongs to *S. maglia*, that from Buenos Aires to *S. commersonii* (possibly also to *S. chacoense*) and those from the Andes of Mendoza to *S. kurtzianum* (see later, pp. 160, 203, and 415).

A few years earlier Schlechtendal (1841) had described *S. maglia* from Chile, on the basis of material collected by Pöppig, though the presence of that species in Argentina was not noted until very recently. In the same work Schlechtendal described a *S. renggeri*, merely on the basis of a short description given by Rengger of a wild potato seen by him on his travels in Paraguay in 1835. This might have been *S. chacoense* or *S. commersonii* subsp. *malmeanum*, but since no specimens were collected we cannot be certain of the true nature of this potato and we must therefore regard *S. renggeri* as a *nomen dubium*.

Dunal (1852) in de Candolle's *Prodromus* redescribed the three species, *S. commersonii*, *S. tenue*, and *S. maglia*; he added also *S. collinum*, which we now think of as a variant of *S. maglia*, and the new Bolivian species, *S. boliviense*, which we have shown in the present monograph to be found also in Argentina.

Some twenty years elapsed between the publication of de Candolle's *Prodromus* and the appearance in 1873 of a work by the Chilean botanist and collector, R. A. Philippi, in which a non-tuberiferous species, *S. brevidens*, was described. The morphological similarities between this and the tuber-bearing species are great enough for it to be included with them in recent taxonomic treatments. Philippi's species was described from a collection made by Fonck in Llanquihue, only about 125 km from the regions of the Argentine provinces of Rio Negro and Neuquén in which it was discovered many years later.

Work on the wild species of the interior of Argentina was slow to develop, possibly because the older collectors gathered material in regions of easy access near the coasts of South America rather than in the interior, where travel was difficult. Grisebach's *Symbolae ad floram Argentinam*, published in Göttingen in 1879, is perhaps the first attempt to list those known for

Argentina, and shows clearly how little was known about the wild potatoes of that country by then. In addition to *S. tuberosum*, under which Grisebach also groups a collection of Lorentz and Hieronymus which we now classify as *S. vernei*, only three other species are mentioned.

1. *S. commersonii*, which he thinks has white flowers and correctly notes as occurring also in Uruguay and south Brazil.

2. '*S. caldasii* Kth. var. *glabrescens* Dun.' This taxon has caused perhaps more confusion than any other tuber-bearing species. The original *S. caldasii* was described by Dunal in the *Synopsis* of 1816, and again by the same author in Humboldt, Bonpland, and Kunth's *Nova genera et species plantarum*, vol. 3, 1818. It was again described, also by Dunal, in de Candolle's *Prodromus* of 1852, but was then assigned to Kunth, even though this latter author expressly stated that all the Solanums in Humboldt, Bonpland, and Kunth were monographed by Dunal. *S. caldasii* is an Ecuadorean species belonging to series Juglandifolia (yellow-flowered, non-tuberiferous) and was shown later to be synonymous with *S. ochranthum* from the same region. In the *Prodromus* of 1852 Dunal further described a so-called variety '*glabrescens*' of *S. caldasii*, but from Chile, and based on Schlechtendal's '*S. palustre* var. *parviflora glabrescens*' which Walpers (1844–5) renamed '*S. palustre* var. *glabrescens*'. All three names were based in fact on a single specimen collected in Chile by Pöppig. This Chilean collection has nothing to do with *S. ochranthum* in series Juglandifolia, but should in fact be classified with *S. brevidens* in series Etuberosa (see Hawkes 1956*b*). Grisebach evidently mistook some mountain forms of *S. chacoense* (see *Lorentz & Hieronymus 1050*, det. Grisebach) for the so-called *S. caldasii* var. *glabrescens*, and this seems to be the reason also why many of the living collections of *S. chacoense* in breeding stations and botanic gardens were until comparatively recently given that name.

3. Grisebach also mentioned the occurrence in Argentina of '*S. collinum* Dun.' (a taxonomic synonym of *S. maglia* Schlechtd.) but this information was based on erroneous determinations of *S. sanctae-rosae* (*Lorentz & Hieronymus 664*) and *S. acaule* (*L. & H. 662*). Of course Grisebach cannot be held to blame for such errors, since the state of knowledge of this group at that time would have made it impossible for him to determine the material with greater accuracy.

Grisebach's *Symbolae* was followed in 1882 by a work of Hieronymus entitled *Plantae diaphoricae florae Argentinae*. The same four species were dealt with, and little of value was added to Grisebach's account, though the two works differed somewhat in detail.

The year 1883 saw the publication of Carrière's description of *S. ohrondii* based on a collection made on the Isle of Gorriti, off the coast of Uruguay, by a doctor of the French fleet, M. Ohrond. This was undoubtedly a variant of the species *S. commersonii*, though so far we have not been able to discover any type or authenticated material for study, and possibly none was preserved.

In the last decade of the nineteenth century the Chilean botanist Philippi, whom we have already had cause to mention, described in 1891 a new species of wild potato from the north of that country in a region that is now part of Bolivia. This species is *S. infundibuliforme*, which was subsequently noted by Robert Fries in the province of Jujuy. An account of its occurrence in Argentina appears in his work on the alpine flora of that country, published in 1905.

The period from 1910 to 1920 was the most productive decade in the study of the plant group under review, since it was then that the illuminating works of Hassler, Bitter, and Wittmack appeared.

Hassler, a Swiss botanist, who lived and collected for many years in Paraguay, published the first description of the well-known species *S. chacoense* in 1911, under the name *S. guaraniticum*. This was, unfortunately, a homonym and was therefore changed by Bitter in 1912 to *S. chacoense*. Hassler also described in 1911, and again in 1917, several varieties and forms of '*S. guaraniticum*', together with a variety of *S. commersonii*, namely, var. *pseudostipulatum*, which we now regard as part of subsp. *malmeanum*. It seems indeed unfortunate that despite his careful work on the Paraguayan species none of Hassler's names remain in current use.

Bitter's detailed and classic works were, in reality, the first attempt to solve the taxonomic problems of the wild potatoes of Argentina. This was made possible by the presence of the fine collections of Lillo, Carlos Spegazzini, Kurtz, Lorenz and Hieronymus, Hieronymus and Niederlein, Schickendantz, Fiebrig, and others that had been accumulating in the large European and Argentine herbaria over the previous fifty years.

Bitter's first descriptions of Argentine species appeared in 1912. In that year he published *S. microdontum* and *S. bijugum* from what was then south Bolivia, and *S. gigantophyllum* and *S. simplicifolium* from Tucumán and Salta respectively. In 1913 he examined more material of this species complex and erected several new varieties of *S. simplicifolium*. All these species and varieties have now been classed by us as *S. microdontum* and its subspecies *gigantophyllum*, since we have been able to satisfy ourselves, with the benefit of more material, that no good species boundaries exist within this complex. In 1912, also, Bitter described the two interesting high Andean species, *S. acaule* and *S. megistacrolobum*. In 1913 the species *S. alticola* was described, which we now consider to be a variant of *S. megistacrolobum*. All three descriptions were based on Bolivian material, but the species also spread into north-west Argentina, as was pointed out by Brücher, Hawkes, and others, at a later date.

Turning his attention to the lowlands of Brazil, Bitter then considered the complex of *S. commersonii*, out of which, in 1912 and 1913, he distinguished *S. acroleucum*, *S. malmeanum*, and *S. calvescens*. The first-named species we now think of as a variant of *S. commersonii*, the second as a subspecies of that species, and the third has been retained as still worthy of specific rank. In

1913 also, Bitter described *S. muelleri* from plants that had been grown in the Vienna botanic garden from tubers collected by Müller in south Brazil. On the basis of a wider range of living and herbarium material we now class this as a subspecies of *S. chacoense*. Bitter also described a variety of *S. maglia* from Chubut, south Argentina, but this was shown by Hawkes in 1956 to be no more than an escaped variety of *S. tuberosum*. Finally, in the same year, came the descriptions of the mountain species, *S. spegazzinii* and *S. setulosistylum*, from prov. Salta in north-west Argentina. The latter species is regarded by us as a hybrid of *S. chacoense* × *S. spegazzinii*, though other authors (Correll 1962) retain it as a distinct entity of specific status and Brücher (1964*a*) considers it to be no more than *S. chacoense* itself.

In 1914, together with Wittmack (see Wittmack 1914) Bitter described five other mountain species from Argentina, namely, *S. kurtzianum*, *S. velascanum*, *S. vernei*, *S. famatinae*, and *S. aemulans*. We have retained two of these (*S. kurtzianum* and *S. vernei*) in our treatment, reduced two more (*S. velascanum* and *S. famatinae*) to taxonomic synonyms, and ranked the last one (*S. aemulans*) as a southern subspecies of *S. acaule*.

Taxonomic work on the tuber-bearing Solanums seems to develop in waves, and probably the work of one author has stimulated an interest among other workers to continue along the same lines. After the 1914 publication just mentioned there were no other works (apart from the 1917 paper of Hassler, already noted) of interest on this group for the area with which we are concerned until Schroeder's monograph on *S. commersonii* appeared. This little-known but interesting work, published in 1928 at Buenos Aires, contains much valuable information on morphological variation, chemical content, and growth and development under experimental conditions of Uruguayan plants of this species.

We now come to the period of the intensive studies on wild and cultivated potatoes by the Russian taxonomists and plant-breeders, Juzepczuk, Bukasov, and Lechnovicz. During the period 1935 to 1941 they published descriptions of seventeen new species, as well as several new varieties, from Argentina and Uruguay. In addition, they arranged the species then known to them in a number of groups or series, according to their supposed evolutionary relationships. This system of series grouping has been a valuable one when not carried to excess, and has been used by all subsequent workers with this group. In our opinion, and also those of Correll and of Brücher, the species described by the Russian workers should not be retained, since we consider that they represent only small variants of species already described. We have accordingly placed nine of them as synonyms of *S. chacoense*, six as synonyms of *S. commersonii*, one of *S. acaule*, and one of *S. kurtzianum*.

The value of the Russian work was perhaps more marked in the field of plant breeding, since most of the material that formed the basis of their taxonomic descriptions was studied in the living state. It was from this work that interest in the breeding potentialities of the species of the tuber-bearing section of *Solanum* was awoken, since plant breeders in other countries also became anxious to study living material with a view to using it for breeding better yielding and more resistant potato varieties.

The bulletins of Hawkes (1941, 1944) on the British potato-collecting expedition in 1939 dealt, amongst other things, with Bolivia and with the small part of north Argentina visited during this expedition by Balls, Gourlay, and Hawkes in provs. Jujuy and Salta. Of the ten Argentine and Bolivian species described, following at that time the species concept of the Russian school, only three have 'survived' to be included by us in the present treatment, namely *S. tarijense*, *S. gourlayi*, and *S. oplocense*. We have grouped *S. ballsii* as a subspecies of *S. vernei*, whilst the other six species have been reduced to taxonomic synonyms of *S. chacoense*, *S. infundibuliforme*, and *S. megistacrolobum*. Later papers dealt with the breeding value of the indigenous wild and cultivated species (1945), the creation of a new series, Megistacroloba (1945) in co-operation with Martín Cárdenas, indigenous potato names (1947), and several papers concerned with the origin of the cultivated potato. A new Argentine species, *S. sanctae-rosae*, was described in 1954, and a detailed survey of the taxonomy of *S. tuberosum* and its tetraploid relatives appeared in 1956. Also in 1956 a completely new treatment (*Revision of the Tuber-bearing* Solanum *Species*) of the wild and cultivated potatoes was published, in which a drastic reduction in the number of species was proposed. A further edition of this work, published in 1963, was based, so far as the Argentine, Uruguayan, Paraguayan, and Brazilian species are concerned, on work undertaken by ourselves in the preparation of the present monograph. A work largely devoted to introgression in *S. chacoense*, based on results obtained by Hawkes and Hjerting, was published by Hawkes in 1962.

An interesting series of papers was published from 1951 onwards by H. Brücher (1951; 1953; 1954; 1956*a*, *b*; 1957*a*, *b*; 1958; 1959*a*, *b*, *c*; 1960*a*, *b*; etc.; Brücher and Ross 1953). In these, besides giving his views on the origin of the cultivated potato from the wild Argentine species, *S. vernei*, Brücher reviews the reproductive biology and pattern of variability of the species known to him in north-west Argentina and elsewhere. In spite of several inaccuracies and some lack of knowledge of taxonomic techniques, Brücher's contribution has been considerable. He has laid stress on the species as living populations of biotypes, with useful resistance to disease and ability to withstand adverse climatic conditions, often forming hybrids in certain areas of geographical or ecological overlap. In his later works (1960, 1961, 1964) he discussed *Heterodera* in Argentina and described two new species, *S. diemii* and *S. sanmartiniense*, which are, in our opinion, no more than escaped forms of *S. tuberosum*. He also (1962) described two further species, *S. ruiz-lealii* and *S. improvidum*, whose status we shall discuss in a later section.

Correll's valuable monograph on the potato and its wild relatives, published in 1962, deals with all the species in the area which we are reviewing and mentions three new ones, *S. sleumeri*, *S. bruecheri*, and *S. puberulofructum*, all previously described by the same author in 1961. This detailed and painstaking monograph contains much excellent work; the figures, short descriptions, and collection citations are very good, but in our view the classification of species into series is over-complex, and in some cases misleading. We consider *S. sleumeri* to bc synonymous with *S. spegazzinii*; *S. puberulofructum* seems to be part of a hybrid complex of *S. spegazzinii* and *S. chacoense*; whilst *S. bruecheri*, or possibly only part of it, seems to be a hybrid of *S. acaule* × *S. megistacrolobum*.

It would certainly be appropriate at this point to mention the work of the well-known Bolivian botanist, Martín Cárdenas. Although he did not describe species based on material collected in the area under review, he nevertheless contributed descriptions (1956) of Bolivian species, such as *S. vidaurrei*, which are known on the Argentine side of the border. A number of his other Bolivian species are reduced by us to synonyms because we believe that they are northern forms of species already described for Paraguay and Argentina. Cárdenas also collaborated with Hawkes in the formulation of the series Megistacroloba, which extends from Argentina northwards into Bolivia and Peru.

In the course of preparing the present work we have described two species, *S. venturii* (Hawkes & Hjerting 1960) and *S. rechei* (in Hawkes 1963). In the first-mentioned paper, also, we dealt with the taxonomy of *S. microdontum* and its two subspecies, *microdontum* and *gigantophyllum*.

It would not be feasible in the present chapter to mention all the genetical, cytological, and plant-breeding work that has been published on the wild and cultivated species of the area with which we are concerned. Much work has come from Holland, Russia, Germany, U.S.A., Britain, and elsewhere, and this will be referred to later in the present monograph, where it throws light on the taxonomy and evolutionary relationships of the species concerned.

2

WILD POTATO COLLECTIONS

IN the present chapter we shall sketch in some of the historical background to the more important potato collections that have been made in Argentina, Brazil, Paraguay, and Uruguay. Space does not permit us to mention all the potato collectors in those countries, but a more complete list will be found in Appendix II on p. 449, where all those who made more than one gathering of wild potatoes are listed alphabetically. Some single, historically important collections are also mentioned.

The historical section is followed by information on areas of particular interest for the potato collector and also those regions where more exploration is necessary. Some information on modern collecting methods is given, where the material gathered is intended for genetical or plant-breeding research.

(*a*) *Past collections of historical or taxonomic importance*

We have drawn our information for this section partly from the data given on herbarium labels and partly from biographical information given by the following authors: J. Britten and G. S. Boulger (1931), A. Davy de Virville (1954), C. C. Hosseus (1921), L. R. Parodi (1961), F. W. Pennell (1945), W. B. Turrill (1920), and K. F. P. von Martius (1906, *Vitae collectorum*). We are also grateful to the many living Argentine botanists who have kindly furnished additional biographical data verbally and by letter. Their names appear on pp. ix–xi of the preface.

Potato collecting in Argentina, Brazil, Paraguay, and Uruguay may be conveniently divided into three historical periods, linked with the development of botanical exploration in the countries concerned. The first period, lasting from 1767 to about 1870, is one of pioneer plant exploration by European botanists and explorers, most of whom made only one or a very few gatherings of potatoes as a part of their general botanical collections. The second period may be conveniently defined as lasting from the 1870s until roughly the second or third decade of the present century. During this period serious attempts were being made to assess the botanical resources of more than just the coastal strip near the Atlantic ocean, and although no special collections of potatoes were made the numbers gathered increased greatly as the attempts to collect the whole flora became more effective. During this period many European botanists settled in the area we are discussing, and together with Argentine, Uruguayan, Paraguayan, and Brazilian botanists, began much more detailed surveys of the vegetation of the plains, forests, and mountains.

The third period of potato collecting overlaps the second to some extent.

This is the period of specialized collecting expeditions and smaller trips in which the gathering of wild potatoes is the prime or sole reason for the expedition. This period stems from the work of the Russian botanist and plant-breeder, N. I. Vavilov, though he himself probably made no potato collections here. Although herbarium collections are made in most cases, the chief need is for living tubers and seeds of species that may be of value in breeding hardier and more disease-resistant cultivars of the domestic potato, *S. tuberosum*.

To return now to a discussion of the important collections made during the first period, we begin with the historically significant collection of Philibert Commerson in May 1767. This wild potato from Montevideo, which was later named by Dunal in Commerson's honour, but which Commerson thought might have been a wild tomato, must have been one of his first collections in the area. He arrived at the Río de La Plata region on 28 May (Pasfield Oliver 1909) and made his now famous collection 'on the beach at the foot of the mountain' (*plage au pie du morne*) before the end of the month. This was the first wild potato to be collected, not only in this area but in the whole of the Americas, as we have stated in an earlier section.

The botanical explorer, F. Sellow (1789–1831), who came to Brazil in 1814 and from 1820 to 1830 collected in the south of that country, made at least five collections of wild potatoes. Unfortunately, none were given exact locality data or dates of collection, so far as we have been able to ascertain, but all belong to the species *S. commersonii*. The French botanist, A. P. de Saint-Hilaire (1779–1853), who collected many plants in Brazil, made two potato collections in the State of Rio Grande do Sul. This was during the period 1816–21 and both collections were also *S. commersonii*, since very few other species occur in Brazil (Saint-Hilaire 1822).

The same species was also gathered in Buenos Aires in 1820 by the Scottish collector John Gillies (1747–1836). In 1823 he went to Mendoza where, in 1824, he gathered two interesting examples of *S. kurtzianum*. These were the first wild potatoes to be collected in the Andes of Argentina. All Gillies's material was classified by J. D. Hooker (1846) as variants of *S. commersonii*.

Captain P. P. King also collected this species, *S. commersonii*, in Uruguay during the voyage of H.M.S. *Adventure* and *Beagle* in 1826 to 1836 (King 1839). This was not of course the more famous voyage in which Darwin took part. It was yet again obtained, in 1836, by the French botanist Charles Gaudichaud-Beaupré (1789–1864) during his third voyage on *La Bonite*, when he visited Uruguay in 1836–7.

The Scottish landscape gardener and collector, John Tweedie (1775–1862), went to Buenos Aires in 1825 and collected many thousands of plants, which he sent to Hooker at Kew. Among these were *S. commersonii* and also *S. chacoense*, obtained on the pampas of Buenos Aires province and elsewhere in 1837. A detailed account of Tweedie's travels and collections is given by Parodi (loc. cit.).

The collections of the Swedish physician, A. F. Regnell (1807–84), must also be mentioned. Regnell lived in the Brazilian State of Minas Gerais during the period 1841–74, chiefly in the town of Caldas. Many collectors obtained plants for him in various parts of Brazil and Argentina, amongst whom we should mention Mosén, Malme, R. E. Fries, Widgren, Lindman, and Henschen, all of whom made potato collections (some after his death). Besides *S. commersonii*, the first collections of *S. calvescens*, '*S. malmeanum*', *S. infundibuliforme*, and *S. acaule* were obtained, though these were not to be described until much later, and then often on the basis of other collections.

Fritz Müller (1822–97) made two collections of '*S. muelleri*' from Brazil in the States of Rio Grande and Santa Catarina in 1868, and these must be included in our first period. They link on to later periods, however, since Müller was among the first to send back living material for cultivation, in this case to the botanic garden of Vienna.

The second historical period of potato collecting begins with the well-known German bryologist, P. G. Lorentz (1835–81), whose career and travels in Argentina are described in some detail by Parodi (loc. cit.). Lorentz came to Argentina from Munich in 1870, collecting until 1872 alone and from then to 1874 with his assistant, G. Hieronymus (1846–1921). Hieronymus, also born in Germany, was appointed to the Chair of Botany in 1874 at Córdoba in place of Lorentz. The latter also collected in Entre Rios province in 1877, whilst the Hieronymus and Niederlein collections in prov. La Rioja, made in 1879, are also deservedly famous. The route of this latter expedition is given by Hosseus (1921). Grisebach's and Hieronymus's works, in which are described some of the potatoes collected by the latter author, have already been noted in a previous section. The extensive collections of these distinguished German savants from Uruguay, Entre Rios, La Rioja, Catamarca, Tucumán, and Salta provided material that was later to be of the greatest value to Bitter in describing such species as *S. vernei*, '*S. aemulans*', '*S. gigantophyllum*', '*S. simplicifolium*', etc. They also collected many other wild potato species (see Appendix II) and very materially increased our knowledge of the plants of the plains and mountains of Argentina.

The interesting collections of the German Professor of Chemistry at Tucumán University, F. Schickendantz, which were made in the provinces of Catamarca and Tucumán, are also worthy of note. Several collections of *S. chacoense*, *S. spegazzinii*, and *S. microdontum* were made by him during the period 1873 to 1893. No less interesting are those of the German amateur collector Cornelius Osten (1863–1936), who collected in Buenos Aires province and in Uruguay during the period 1886 to 1926. All his potato material was *S. commersonii*.

We now come to the valuable collections of the Italian botanist, Carlos Spegazzini (1858–1926), who arrived in Argentina and settled in Buenos Aires in 1879, moving to La Plata in 1885. Although a mycologist by profession he extended his interests to phanerogams and made an extensive herbarium of Argentine plants. A great deal of his *Solanum* material (collected from 1887 to 1921) was sent to Bitter for determination and formed the basis for Bitter's species *S. spegazzinii* and *S. setulosistylum*, as well as for a number of varieties of '*S. simplicifolium*'. Spegazzini's potato collections were made in Córdoba, Salta, and Buenos Aires provinces, and included material also of *S. chacoense*, *S. commersonii*, and hybrids of *S. chacoense* × *S. microdontum*.

Miguel Lillo (1862–1931), who was born and died in Tucumán, was a self-taught botanist and was instructed in chemistry by Schickendantz. According to Parodi (loc. cit.), from whose account these details are taken, Lillo was a professor of chemistry but not of botany, though the latter subject must have been his chief interest. His deservedly famous collections form the basis of the Instituto Miguel Lillo in Tucumán. The potatoes in these were gathered between 1888 and 1917 in Salta, Tucumán, Santiago del Estero, and Jujuy provinces, one of them forming the basis for Bitter's species '*S. subtilius*'. Much valuable material of *S. chacoense*, *S. microdontum* subsp. *gigantophyllum*, *S. venturii*, and several other species has enabled later botanists to derive a more exact concept of the altitude range and pattern of variability of these species than would otherwise have been possible.

C. M. Hicken (1875–1933), who was born at Buenos Aires and died at Mar del Plata, travelled and collected over the whole of Argentina. In 1911 he founded the Instituto Darwinion, adding numerous collections and books to this world-famous herbarium. His potato collections, made between 1899 and 1932 are to be found in the Darwinion and include good material of *S. chacoense*, *S. commersonii*, and *S. kurtzianum*.

The Chair of Botany at the University of Córdoba, left vacant by Hieronymus on his return to Germany in 1884, was occupied from then until 1915 by another German botanist, F. Kurtz (1854–1920). Kurtz's collections of some 16 000 numbers are now incorporated in the University of Córdoba herbarium. Kurtz's Jujuy collections were apparently made by Fritz Claren in 1900–1, but those from the provinces of Mendoza, San Juan, and La Rioja were made by Kurtz himself in 1906–8 and are of considerable interest. His routes are given by Hosseus (1921) in some detail, as well as those of Hieronymus and Niederlein and a number of other botanists who collected in those provinces. The potatoes collected by Kurtz were of particular interest in that they were obtained in the drier north-western regions of Argentina. Certain of these were used by Bitter and by Bitter and Wittmack as types for '*S. famatinae*', *S. kurtzianum*, and '*S. velascanum*' as well as the named but not described '*S. juanense*'. The high-altitude species *S. infundibuliforme* and *S. acaule* (collected by F. Claren) were also included in Kurtz's material. Many of Kurtz's specimens were sent to Berlin and were destroyed during the Second World War; fortunately, duplicates and isotypes of most of these exist in the Kurtz type collection at Córdoba.

The Argentine botanist, Santiago Venturi, lived in Tucumán and was administrator of the Parque 9 de Julio for many years. He made the first collection of Argentine woods in 1910, and gathered some particularly interesting potato material from 1902 to 1929, of which one duplicate was always deposited in the Miguel Lillo herbarium. One of his collections was used by Hawkes as the type for *S. sanctae-rosae*, and Venturi also made several gatherings of the species *S. venturii*, which we have named in his honour in a previous publication. Several collections of *S. chacoense*, *S. microdontum* subsp. *gigantophyllum*, and the hybrid between them were also made, as well as the hybrid of *S. sanctae-rosae* and *S. venturii*. In prov. Jujuy, amongst other collections from Volcán in dept. Tumbaya, is an intriguing possible hybrid of *S. megistacrolobum* × *S. microdontum*, as well as the two putative parents. In fact, many of Venturi's collections are unique, and give a clear indication of the care and thought that this remarkable collector gave to his work. Several other species were collected also, and Venturi's material has helped enormously in promoting an understanding of the relationships between the potato species of northern Argentina.

Paraguay was almost unknown botanically until the work of the eminent Swiss botanist and medical doctor, Emile Hassler (1861–1937), whose extensive collections from that country were reported on by Chodat between the years 1898 and 1905. Hassler's interesting potato collections were made in 1903–16 (some of them possibly having been gathered by his pupil, T. Rojas). They provide ample material for us to obtain a reasonably good concept of the variability of *S. chacoense* and *S. commersonii* subsp. *malmeanum* in the general region of Asunción. Hassler's descriptions of *S. guaraniticum* (*S. chacoense*) and its varieties and forms have already been referred to in a previous chapter; his carefully prepared specimens are to be seen in Geneva and in several other European and American herbaria.

The Belgian botanist Lucien Hauman (1880–1965) collected in Argentina during the period 1904–25 (potatoes from 1906 to 1923), contracted by what is now the Facultad de Agronomía y Veterinaria of the University of Buenos Aires. In 1912 he occupied the Chair of Microbiology in the same University. He made several gatherings of *S. chacoense* and *S. commersonii* in Buenos Aires province from 1906 to 1907 and again in 1922–3. Material of *S. infundibuliforme* and *S. acaule* was obtained in prov. Jujuy, and of *S. commersonii* and its subspecies *malmeanum* in Entre Ríos and Misiones respectively from 1916 to 1922. He returned to Brussels in 1925.

The collector León Castillón was a priest and held the Chair of Botany in the National College of Tucumán for various years. His potato collections are deposited in the herbarium of the Instituto Miguel Lillo and were gathered during the period 1906–27 in the provinces of Tucumán, Jujuy, and Catamarca. They include a fair range of species and one or two interesting hybrids (*S. chacoense* × *S. spegazzinii* at Rio Churqui; *S. vernei* × *S. spegazzinii* at Infiernillo). Some interesting collections made at Volcán, prov. Jujuy, in 1927, are also worthy of note.

The collections of W. G. Herter (b. 1884), which formed the basis for his floristic studies of Uruguay, also included eight collections of potatoes, made in the years 1907–47. As might be expected, all these were of *S. commersonii*.

The self-taught German naturalist, R. Schreiter (d. 1939 or 1940), worked with Miguel Lillo and after Lillo's death was the first director of the Instituto Miguel Lillo. His interesting potato collections, made during the period 1917 to 1940(?), are housed in the Institute and were obtained mostly in the provinces of Tucumán and Catamarca. He, also (see under Venturi), collected the very rare hybrid of *S. megistacrolobum* × *S. microdontum* (Jujuy, Lomal de Tambo). Other interesting material included hybrids of *S. vernei* × *S. spegazzinii*, *S. microdontum* subsp. *gigantophyllum* × *S. vernei*, and *S. sanctae-rosae* × *S. venturii*, as well as useful collections of *S. spegazzinii*, *S. sanctae-rosae*, *S. chacoense*, and other species. His three collections of *S. sanctae-rosae* from dept. Tinogasta, prov. Catamarca, are the most westerly yet known for that species.

The Argentine botanist A. Castellanos (b. 1896), who replaced L. Hauman in the Museo Nacional at Buenos Aires and at the death of Hicken occupied his Chair in the Faculty of Sciences, collected potatoes amongst his other material during the period 1917–43. His collections come from various parts of Argentina, but those from the north-western provinces are of greatest interest. Especially worthy of mention are the various gatherings of *S. infundibuliforme* and *S. megistacrolobum* from Jujuy province. Also particularly valuable are the gatherings from Sierra de los Quilmes (prov. Salta) of *S. acaule*, *S. spegazzinii*, and the probable hybrid between them; also hybrids of *S. infundibuliforme* and *S. megistacrolobum* at San Antonio de los Cobres and various collections of *S. spegazzinii*. From La Rioja province the material classified by us as *S. microdontum* subsp. *gigantophyllum* × *S. kurtzianum* is of obvious importance, as well as a probable *S. spegazzinii* × *S. kurtzianum* hybrid from Sierra Famatina. Interesting collections of a distinct form of *S. chacoense* were made in provs. Córdoba and San Luis. To sum up, the collections of Castellanos are particularly valuable as dried material, and it is greatly to be hoped that he or some other collector will retrace his steps, so as to obtain living material from the areas mentioned.

During the extensive collecting trips of A. E. Burkart (b. 1906), Director of the Instituto Darwinion at San Isidro, many wild potatoes were gathered between the years 1927–65 in various parts of the country. *S. commersonii*, *S. chacoense*, *S. acaule*, *S. brevidens*, and other species were collected. Similar extensive collecting expeditions have been made by L. A. Cabrera (b. 1908), Curator of the Museo de La Plata. Wild potatoes were gathered in many parts of Argentina and Uruguay between 1930 and 1955, including *S. acaule*, *S. chacoense*, *S. commersonii*, *S. megistacrolobum*, etc. Both Burkart and Cabrera gathered an interesting possible hybrid of *S. commersonii* and *S. chacoense*, though not at the same locality. Cabrera's collections in Salta and Jujuy provinces from 1940 onwards are especially valuable.

The present Curator of the phanerogamic herbarium of the Instituto Miguel Lillo, T. Meyer (b. 1910), is another eminent Argentine botanist who has made extensive collecting expeditions to many parts of Argentina. We are indebted to him for the loan of much material and for certain biographical information embodied in the present chapter. Meyer's collections of potatoes, made from 1933 onwards, include a range of ten species and several hybrids from various provinces. A particularly interesting sequence made on 9–10 March 1949 (Nos. 14868–15075) in prov. Tucumán (dept. Chicligasta), from El Bolsón through La Cascada to Las Cuevasi ncludes gatherings of *S. venturii* (2100 m), *S. microdontum* subsp. *gigantophyllum* (2500 m), *S. venturii* × *S. vernei* and *S. venturii* × *S. sanctae-rosae* (2800 m), and *S. venturii* × *S. sanctae-rosae* and *S. sanctae-rosae* at 3000 m.

No account of the botanists and collectors of Argentina would be complete without a mention of the work of L. R. Parodi (1895–1966), former Professor of Botany and Director of the herbarium of the Facultad de Agronomía y Veterinaria of the University of Buenos Aires. During his many expeditions he has collected numerous samples of wild potatoes, grasses, and many other plants. We are indebted to him for much of the material in the present sketch, and would refer the reader again to his valuable biographical work on Argentine botanists, published in 1961.

Mention should also be made of the Director of the Botanical Museum of the Facultad de Ciencias in Córdoba, Professor A. T. Hunziker (b. 1919). He has made a special study of the family Solanaceae and has also obtained material of various species of wild potatoes during his extensive collecting trips in various parts of Argentina.

The Swedish botanist Benkt Sparre, who was contracted by the Instituto Miguel Lillo from about 1947 to 1952, made a number of special collections of wild potatoes during that time, all from the province of Tucumán. Most of the species known in that region were gathered, including hybrids of *S. venturii* × *S. vernei* and *S. venturii* × *S. sanctae-rosae*, from dept. Tafí.

We now come to the third phase of potato collecting, which overlaps in time with the second to quite an extent. This is the period of specialized collecting trips and expeditions, in which specific attention is paid to wild and cultivated potatoes, and every effort is made to obtain living material in the form of seeds or tubers for subsequent cytogenetical and plant-breeding research.

We begin, appropriately enough, with the Argentine agronomist Roberto A. Millán, who collected potatoes in the living state in various provinces from 1926 to 1941. These collections (mostly of *S. commersonii*) were grown in the experimental field in Argentina and many were also sent to the Russian geneticists S. M. Bukasov and S. W. Juzepczuk. In addition, Millán gathered herbarium material of potatoes, both by himself and with E. A. Pujals (in 1939). Much of the living material provided the basis for studies on the disease resistance and breeding behaviour of the wild Argentine potatoes described by the Russians in numerous publications. Millán's collections of escaped forms of *S. tuberosum* from Chubut, grown in Buenos Aires for several years, are also of considerable theoretical importance (see below, p. 435).

One of the Russian botanists, S. W. Juzepczuk, visited Argentina in October 1927, calling at Tucumán, Buenos Aires, and various other cities, but he apparently made no collections of wild potatoes there, so far as we know. Cultivated material was probably obtained in the northern provinces. The German plant-breeders Bauer and Schick also visited Argentina in 1930 and made collections of cultivated varieties in Tucumán and La Quiaca on their way north to Bolivia and Peru. The living collections of T. Henry in Uruguay, which were sent to Russia in 1935, formed the basis of several of Bukasov's microspecies, now included in *S. commersonii* and *S. chacoense*.

Particularly valuable are the collections of the Argentine botanist, A. Ruiz Leal (b. 1898), made between the years 1935 to 1957 chiefly in Mendoza province. Many of these gatherings were made in collaboration with his colleague, F. A. Roig, especially those of *S. kurtzianum*. Other interesting collections of Ruiz Leal are *S. chacoense* × *S. kurtzianum* hybrids, *S. maglia* (from Quebrada de Alvarado—so far the only known locality of this species for Argentina, first discovered there by Ing. Agr. Schnack in 1945); also *S. chacoense* from Córdoba and *S. brevidens* from Neuquén. Ruiz Leal and Roig were also responsible for two curious collections of *S. spegazzinii* from San Juan, the most southern locality known for that species.

The British Commonwealth potato-collecting expeditions of 1938–9 travelled both in Mexico and the Andes of South America. The members visiting Argentina in 1939 were E. K. Balls, W. B. Gourlay, and J. G. Hawkes, and all the Balls collections mentioned in the species locality lists (Nos. 5935–6081) were in fact gathered by these three collectors. Only the northern provinces of Jujuy and Salta were visited.

Introgressed forms of *S. chacoense* were obtained in the vicinity of Jujuy, whilst in prov. Salta at San Lorenzo the type locality of '*S. simplicifolium*' was visited, and this species was collected, together with introgressed forms of *S. chacoense*. In prov. Jujuy interesting collections were made in the mountains and valleys east of Tilcara (Balls Nos. 5956–6056). *S. infundibuliforme* was found in the lower part of the Quebrada de Huasamayo at c. 2450 m, and higher up at San Gregorio. It continued up to the cooler wetter regions at 3700 m, where *S. gourlayi* and *S. megistacrolobum* began to appear. At 3900 to 4000 m *S. infundibuliforme* had disappeared entirely, but *S. megistacrolobum* and *S. gourlayi* became quite frequent and were joined by *S. acaule*. The latter species was seen at 4100 m at Laguna Colorada and down into the upper reaches of the Chorro Valley, where subsp. *aemulans* appears. In this east-facing much wetter valley of the Montano district at lower altitudes *S. acaule* disappeared and was replaced by *S. vernei* subsp. *ballsii*, *S. microdontum*, and the hybrids between the two. At the higher, west-facing elevations of San Gregorio, the low rosette form of *S. megistacrolobum*, was of particular interest, since it differs in phenotype very strikingly from the type material of Fiebrig, collected earlier in south Bolivia, in more protected localities.

Hawkes made further living and dried collections in Uruguay, in March to April 1958 (Nos. 701–25), accompanied by G. J. Fischer, then Director of the plant-breeding station of La Estanzuela. Collections were also made in Rio Grande do Sul (Pelotas region) in south Brazil, and in prov. Buenos Aires (Balcarce region), Argentina, in company with various Argentine and Brazilian agronomists, notably Mota da Costa, Pascuale, and Grondona. The collections made during the survey of this area were of diploid and triploid *S. commersonii* (typical subspecies), together with a few accessions of subsp. *malmeanum*. Much useful ecological and other information was gained on that occasion, which enabled us to form a clearer picture of the variation pattern of *S. commersonii* than had hitherto been possible.

Other living collections of *S. commersonii* were sent to us by the late Padre B. Rambo (1906–61) from Porto Alegre, Rio Grande do Sul, and elsewhere. He also made a number of special gatherings of dried material of that species.

The Danish botanists E. Petersen and J. P. Hjerting collected during the years 1948 to 1953 for the University of Tucumán, Argentina, where they were contracted as plant collectors by the Instituto de Genética (later known as the Instituto de Fitotecnia and finally the Facultad de Agronomía). The Instituto de Genética was founded by the Norwegian geneticist G. Hiorth who directed the work of Petersen and Hjerting. The main purpose of this work was to collect wild and cultivated potatoes, as well as other native cultivated plants and their wild relatives.

In the summer of 1948–9 collections of wild material were made in the provinces of Tucumán and Catamarca (dept. Andalgalá) (Nos. P–1 to P–58[1]). In April and May 1949 visits were made to the provinces of Santiago del Estero, Chaco, Corrientes, Misiones, and Tucumán. Several collections of *S. chacoense* (both subspecies) and *S. commersonii* subsp. *malmeanum* were made from all over the area, and much living material was obtained (Nos. P–60 to P–80).

Later on, in the same year, some wild and much cultivated potato material was obtained in Catamarca, Tucumán, Salta, and Jujuy provinces (see pp. 428–9) (Nos. P–82 to P–319).

In January and February 1952 the region of Quebrada Honda (Infiernillo) of dept. Tafí, prov. Tucumán, was investigated; in addition to several species collections a series of hybrids between *S. sanctae-rosae* and *S. venturii* was found (Nos. P–324 to P–336). Later in 1952 (March and April) visits were made to Jujuy (depts. Tumbaya, Tilcara, and Humahuaca) and Salta (dept.

[1] The Petersen and Hjerting potato collections from 1 to 369 are marked with a 'P' in front of all the numbers, to distinguish them from a corresponding series of collections of other plants with the same numbers.

The Petersen and Hjerting numbers from 856 to 1901, as well as the Hjerting, Petersen, and Rahn, and the Hjerting and Rahn collections consist of a single series in which all plant collections as well as potatoes are included.

Iruya) provinces, and a number of high-mountain species were collected. The classic potato locality, Quebrada del Chorro, near Tilcara, was included in this trip (Nos. P–337 to P–369).

Finally, in February to June 1953, an expedition to north-west Argentina, Bolivia, and Peru was made, and material was gathered in prov. Jujuy from Tumbaya northwards to the Bolivian border, mainly in the puna region (Nos. 958–75).

Several subsequent trips were made by J. P. Hjerting, either in company with Petersen, or with the Danish botanist K. Rahn, or with both of them. We shall deal with each separately.

In 1956, from January to March, J. P. Hjerting took part in a Danish botanical expedition sent out from the University of Copenhagen and headed by Professor T. W. Böcher. This expedition was made possible by the generous support of the Williams Foundation of Buenos Aires, and consisted of two parts. The first (October to December 1955) was concerned with an ecological survey of the Atuel Valley area of Mendoza province. The second was dedicated to the collection of wild potatoes in Jujuy, Salta, Tucumán, and La Rioja provinces. A plan for visiting certain key localities had been prepared by the present authors, in an attempt to solve some of the chief problems of potato taxonomy still outstanding in those areas. Special attention was paid to the collection of living material wherever possible, since it was hoped to arrive at a better understanding of phenotypic variation in certain species by growing a wide range of material under constant conditions in Europe. Visits to type localities were also included in this plan, and material from these was collected in the following localities.

1. Toldos (dept. Santa Victoria, prov. Salta) for *S. microdontum* and '*S. bijugum*'.
2. Dept. Cachi (prov. Salta) for *S. spegazzinii*.
3. Estancia Pampa Grande (dept. Guachipas, prov. Salta) for *S. setulosistylum*, *S. microdontum* subsp. *gigantophyllum* and its hybrids with *S. chacoense*.
4. Sierra Famatina (dept. Famatina, prov. La Rioja) for '*S. famatinae*' and '*S. aemulans*'.

Especially interesting was the region of Santa Victoria to Toldos (prov. Salta), where much material of *S. microdontum* was collected, enabling us to solve the problem of the relationships of this species with '*S. bijugum*' and '*S. simplicifolium*'. Without the living material and the mass collections obtained in this region, such a solution would have been virtually impossible. Here also *S. boliviense* was found for the first time in Argentina. An interesting zone of species hybridization in the mountain region between Santa Victoria and Yavi was discovered, and much additional material was collected.

Another very interesting area was Sierra Famatina, where '*S. aemulans*' (*S. acaule* subsp. *aemulans*) was re-collected, probably for the first time since the collection of Kurtz in 1907. In this region certain hybridization seems to be taking place between this tetraploid species and the diploid *S. spegazzinii*, where the two species grow together.

The sequences of numbers (including other plants also) were as follows:

Hjerting, Petersen, and Rahn, Nos. 1–114 and 264–399. Provs. Jujuy, Salta, and Tucumán.

H. & R. Nos. 151–255. Prov. Salta.

P. & H. Nos. 1892–901. Prov. La Rioja.

Unfortunately, we have been able to see very few of the apparently extensive collections made by the German geneticist H. Brücher from 1948 to 1963. We know from his published works, however, that he made many collections of *S. vernei*, *S. chacoense*, *S. microdontum* subsp. *gigantophyllum*, *S. venturii*, and *S. sanctae-rosae* from the Aconquija mountains of Tucumán and Catamarca provinces. Brücher also collected in provs. Salta, Jujuy, La Rioja, Córdoba, Mendoza, and Neuquén, as well as in the neighbouring country of Chile. Working first at Tucumán and later at Mendoza, he made collections, chromosome counts, and other observations on living material both in the field and in experimental plots. His series of papers has already been mentioned in a previous chapter.

Particularly noteworthy are the excellent sequences of collections, both of dried and, to a limited extent, of living material, made by the German botanist H. Sleumer (b. 1906), sometimes in company with the Argentine botanist and phytogeographer F. Vervoorst. Many of the type localities and several areas of great interest that had not previously been examined for wild potatoes were investigated by Sleumer, either alone or with Vervoorst.

In 1950 to 1952 Sleumer collected in Tucumán, dept. Tafí, and Catamarca, dept. Andalgalá, finding most of the species known from the Aconquija Sierra area (included in his general numbers from 14 to 2751). Amongst his many collections during this period, two particularly interesting series stand out:

1. Prov. Catamarca, dept. Andalgalá, Río Potrero Superior (Nos. 1830–40): *S. vernei* at 2750 m; *S. sanctae-rosae* at 2700–2900 m; *S. sanctae-rosae* × *S. venturii* at 2700–2900 m; *S. acaule* at 3600 m.

2. Prov. Catamarca, dept. Andalgalá, Mina Capillitas through Cuesta de la Negrilla to Cerro Yutuyaco (Nos. 2686–2751): *S. sanctae-rosae*, *S. sanctae-rosae* × *S. spegazzinii*, *S. spegazzinii*, *S. acaule* all at 3000–3100 m; *S. sanctae-rosae* at 2900 m; *S. acaule* at 3600–3800 m; *S. spegazzinii* at 3700 m. The zone of 3000 to 3100 m seems to be one of overlap between the lower growing *S. sanctae-rosae* and the higher-growing *S. acaule* and *S. spegazzinii* in this particular area.

In January to February 1953 Sleumer travelled through the Quebrada de Humahuaca in prov. Jujuy, exploring the surrounding mountains in depts. Tumbaya, Tilcara, and Humahuaca, and collecting a series of interesting species (Nos. 3198–590 and 4105–19). Finally, from La Quiaca on the Bolivian border he crossed over the Sierra Santa Victoria to Santa Victoria and Toldos, near the Bermejo river at 1500 m. Here he also visited the type locality of *S. microdontum*, where it had been collected by Fiebrig in 1903. Between here and Lizoite, west of Santa Victoria, a long series of most interesting collections were made, including many hybrids (Nos. 3608–4059). Sleumer also collected the Bolivian species *S. vidaurrei* for the first time in Argentina. In this region the sequence collected in Santa Victoria department at Cuesta de Mesón is of interest (Nos. 3998–4003): *S. vidaurrei* at 2700 m; *S. tarijense* and its introgressed hybrids with some other species (*S. vernei*, *S.vidaurrei*, or *S. gourlayi*?) at 2600 m, the hybrid also at 2500 m, and *S. megistacrolobum* at 2600 to 3000 m. On the road from La Huerta to Rodeopampa (Nos. 4037–48) were collected *S. microdontum* at 2700–2900 m and *S. megistacrolobum* × *S. boliviense* at 3000 m.

Sleumer's collections of *S. brevidens* and *S. tuberosum* (escaped form) in Neuquén province are also of interest.

In 1952 Sleumer and Vervoorst collected in prov. Catamarca, dept. Belén, Quebrada de Belén, and the mountain valleys just west of it. Within a week (*c.* 1 February) some twenty-five interesting collections were made (within Nos. 2326 to 2645), and the following species were found: *S. chacoense* (1300–1700 m); *S. spegazzinii* (1900–2800 m); *S. kurtzianum* (1400–1900 m); *S. microdontum* subsp. *gigantophyllum* (2500 m); *S. sanctae-rosae* (2500–2800 m); to *S. acaule* (2800 m). Hybrids of *S. chacoense* × *S. spegazzinii* (1700–2000 m) and *S. sanctae-rosae* × *S. spegazzinii* (2600–2800 m) were also gathered. Collections of this type, made throughout a specific altitudinal zone, are extremely valuable for understanding the pattern of species variation and areas of hybridization amongst wild potato species. Later on, in the same year, Sleumer and Vervoorst collected in the little-known area of Quebrada del Río Potrero (del Castillo) in dept. Caldera of prov. Salta, obtaining some interesting material.

Vervoorst's own collections, made in 1951–2, are no less interesting than those cited above. In the little-collected region of Sierra de Ambato, depts. Pomán and Capayán, prov. Catamarca, Vervoorst gathered the following species: *S. chacoense* (1600–2000 m); *S. vernei* (2400 m); *S. sanctae-rosae* (2500–3000 m). Hybrids of *S. kurtzianum* × *S. chacoense* (2000 m), *S. sanctae-rosae* × *S. spegazzinii* (2500 m), and *S. chacoense* × *S. spegazzinii* (1600 m) were also discovered (Nos. 3427–526). Vervoorst made many collections with Professor Hans Ross in 1959, which we shall mention in more detail below.

During the period 1952–63 many botanists, plant breeders, and others have made living and dried collections for us that have been of considerable value in the species-crossing programme which we mention later in this work. These collaborators, to whom we have drawn attention in the preface, and whose names are listed in Appendix II, are as follows: Professor F. G. Brieger (*S. calvescens* from Minas Gerais, Brazil—the first living collections of this species to be made); Sr. José Diem (*S. brevidens* and *S. tuberosum* from Nahuel Huapi, Argentina); Ing. Agr. E. Grondona (*S. commersonii* from Buenos Aires province); Mr. F. J. Kemp (*S. chacoense* from Paraguay and *S. commersonii* from Uruguay); Sr. Delorge Mota da Costa (both subspecies of *S. commersonii* from South Brazil); Sr. A. G. Schulz (*S. chacoense* and *S. commersonii* subsp. *malmeanum* from prov. Chaco, Argentina). Many of these collections have been keys to the understanding of variability and crossing relationships of the species concerned, and our grateful thanks are due to the collectors.

In 1958 the well-known Russian breeder and authority on the origins of cultivated plants Professor P. M. Zhukovsky visited Argentina in the course of an expedition to study the centres of origin of cultivated plants in Latin America (see Zhukovsky 1959). He was accompanied for part of the time by the German geneticist Heinz Brücher, to whom we have already referred. Unfortunately we have not been able to examine any of Zhukovsky's material, but from the account of his travels we gather that visits were made to the Argentine provinces of Buenos Aires, Jujuy, Salta, Tucumán, Catamarca, the Chaco, and elsewhere. Apparently a wide range of material was collected.

The German 'Botanisch-Landwirtschaftlichen Anden Expedition' of 1959, led by Professor Hans Ross, visited Argentina and made many interesting collections in the north-western provinces, before moving north into Bolivia and Peru. Professor Ross was accompanied by R. Rimpau and the Argentine collector and phytogeographer F. Vervoorst during his travels in Argentina, collecting in the provinces of Tucumán, Catamarca, Salta, and Jujuy. Most of the wild potato species of those provinces were obtained, and much living material was sent back to Germany for further study. We are most indebted to Professor Ross for allowing us to examine all his herbarium material and corolla preparations, and for inviting one of us (J. G. H.) to view the living material in his experimental plots at Cologne. Amongst others, the collections of *S. spegazzinii* (Nos. 16–23) were particularly interesting. These were gathered in prov. Catamarca, dept. Andalgalá, Ingenio del Arenal at 2500 m. Another set from prov. Tucumán, dept. Tafí, Zanja de los Cardones (2700 m) (Nos. 30–6) is also worthy of mention; also the *S. sanctae-rosae* and *S. spegazzinii* collections with the hybrids between these two species from prov. Tucumán, dept. Tafí (2900–3300 m). Several good collections of *S. chacoense* and *S. microdontum* subsp. *gigantophyllum* were made in Jujuy and Salta provinces, as well as of other species. Of interest also were the collections of diploid and tetraploid forms of *S. oplocense*, gathered just over the border, in Bolivia.

The 1960 collections of the American potato taxonomist, D. S. Correll (b. 1908), are of considerable importance. In company with the British geneticists K. S. Dodds and G. J. Paxman and the German geneticist

H. Brücher, Dr. Correll made several gatherings of wild species in the provinces of Jujuy and Tucumán. The species of the puna region in Jujuy, *S. infundibuliforme*, *S. megistacrolobum*, *S. acaule*, *S. oplocense*, and *S. gourlayi* were found, as well as the hybrid species *S.×bruecheri*, which has apparently been derived as a cross of *S. acaule*×*S. megistacrolobum*, or some other wild species. In Tucumán *S. sanctae-rosae*, *S. vernei*, and *S. microdontum* subsp. *gigantophyllum* were collected. Correll also travelled in provs. Salta, Catamarca, San Juan, La Rioja, and Mendoza, but we have not seen his collections from those regions. He states, furthermore, that he visited Buenos Aires province, but he does not cite collections from there, and it is possible that he travelled there to examine herbaria or visit research stations.

J. O. Reche's collections in La Rioja province in 1960, although few, are of great interest, since he was able to gather living material of the interesting hybrid *S. microdontum* subsp. *gigantophyllum*×*S. kurtzianum*, which had been first collected as herbarium specimens by Castellanos in 1928. Reche also obtained *S. spegazzinii* from the same province.

Potato collections were made also by the Esthonian agronomist E. Viirsoo in 1961 in La Rioja province. These were nearly all of *S. spegazzinii*. Viirsoo, who arrived and settled in Argentina in 1950, maintained the Petersen and Hjerting collections of wild and cultivated potatoes to which we have already referred, and conducted tests for frost resistance and other qualities on the material.

With this we come to the end of the long list of distinguished potato collectors, even though space has not permitted us to mention everyone. Others will be found in Appendix II. It is largely due to the untiring efforts of these field-workers that it has been possible for us to write the present work. Unfortunately, space has not been sufficient for us to mention much cytogenetical and plant-breeding work carried out by potato breeders in the areas concerned. Thus D. Pascuale, C. J. Induni, E. L. Ratera, G. Weiseth, and others in Argentina; and Delorge Mota da Costa and F. Guimarães in Brazil have also made important contributions in potato breeding, but their work is nevertheless outside the scope of the present book.

(*b*) *Areas of special interest to the potato collector*

The many *quebradas* or valleys of the eastern slopes of the Andes which face the lowland plains are of great interest to the potato collector. They occur from Jujuy in the north to the eastern parts of Catamarca province, and are occupied mainly by the western subtropical phytogeographical province. On ascending these valleys one is struck by the abundance of wild potato species, as well as by the generally exuberant vegetation, especially in the region from about 2500 to 3500 m. Although wild potatoes are normally rather rare components of the natural or semi-natural vegetation of the Americas as a whole, in the region of which we are speaking they become quite abundant. Indeed, the potatoes of certain valleys are so frequent and present so many points of general interest to the botanist that these places can be considered as 'classic areas', well worth a visit even by the non-specialist in the group. Accordingly, we show in Fig. 1 in schematic form the altitudinal ranges for wild potatoes and their hybrids in six different localities in provs. Jujuy, Salta, Tucumán, and Catamarca. Details of the routes are as follows:

1. *Toldos* (Fig. 1 (*a*))

This is a two days' trip each way by mule to Toldos from Santa Victoria. No hotels are available and a tent should be taken. The journey to Santa Victoria from La Quiaca (prov. Jujuy) is mentioned in the second list of 'classic areas' on p. 25.

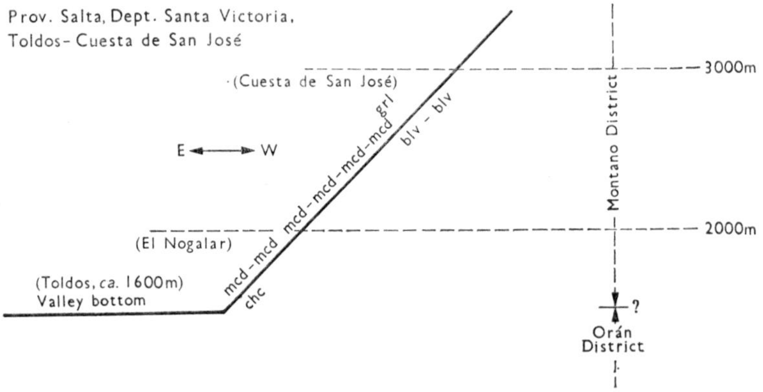

Fig. 1(*a*)

FIG. 1. Transect diagrams of six mountainous areas in north-west Argentina where wild potato species and hybrids occur with some frequency.

(*a*) Toldos; (*b*) Quebrada del Chorro; (*c*) Cuesta del Obispo; (*d*) Trancas; (*e*) Yunka Suma; (*f*) Quebrada de Belén.

Key to species and subspecies abbreviations: see Appendix II, p. 448.

Key to hybrids:

1. *S. sanctae-rosae* × *S. spegazzinii.*
2. *S. microdontum* subsp. *microdontum* × *S. vernei* subsp. *ballsii.*
3. *S. chacoense* × *S. tarijense.*
4. *S. chacoense* × *S. microdontum* subsp. *gigantophyllum.*
5. *S. microdontum* subsp. *gigantophyllum* × *S. vernei* subsp. *vernei.*
6. *S. vernei* subsp. *vernei* × *S. spegazzinii.*†
7. *S. sanctae-rosae* × *S. venturii.*
8. *S. chacoense* × *S. spegazzinii.*

* Forms intermediate between the two sub-species are shown thus: vrn-bal. See Fig. 1 (*c.*)

† The altitude for this hybrid, 1500 m, is anomalous, and is due probably to a seed or tuber having been washed down by the river from somewhere higher up, where the two parents are to be found.

2. *Quebrada del Chorro* (Fig. 1 (*b*))

By mule from Tilcara (prov. Jujuy), taking one or two days to get to San Antonio and the same for the return journey. No hotels. A tent should be taken.

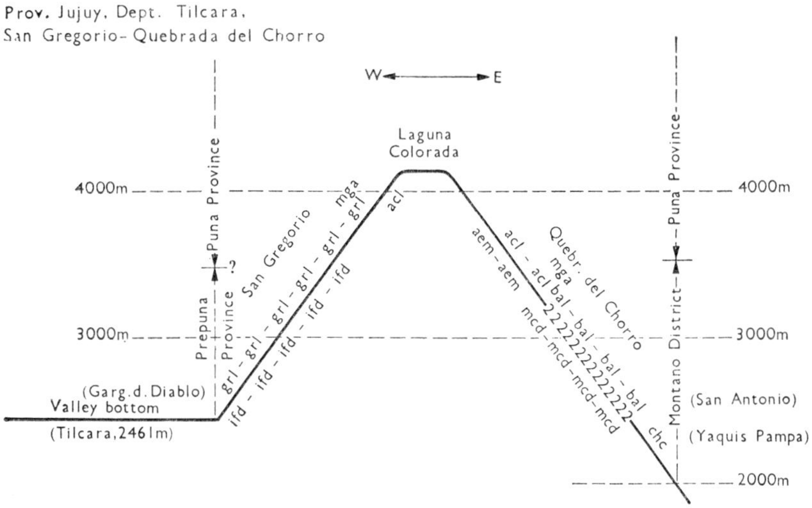

Fig. 1(*b*) (For explanation see facing page)

3. *Quebrada de Escoipe and Cuesta del Obispo* (Fig. 1 (*c*))

There is a good motor road from Salta, continuing to Cachi, where hotels are available. This is a one-day trip each way. A detour can be made to Amblayo on a recently constructed motor road.

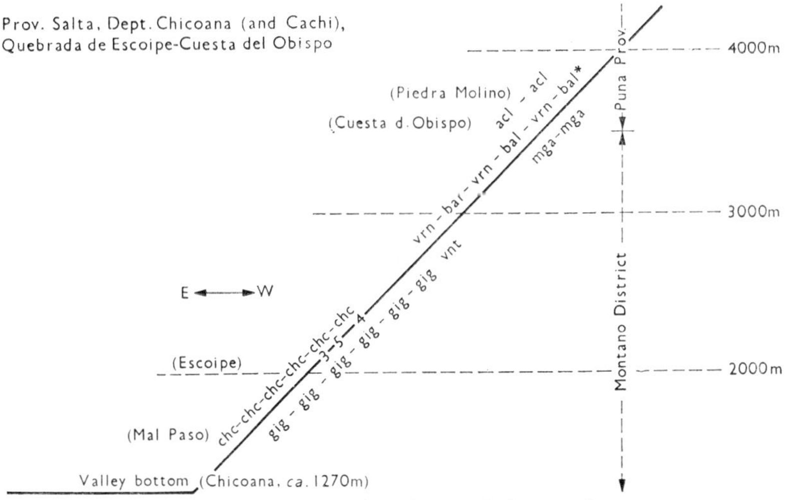

Fig. 1(*c*) (For explanation see facing page)

4. *Trancas* (Fig. 1 (*d*))

A road goes almost to Tacanas (prov. Tucumán); from there it takes one or two days by mule to Cuesta Grande and back. There is a good hotel in San Pedro de Colalao.

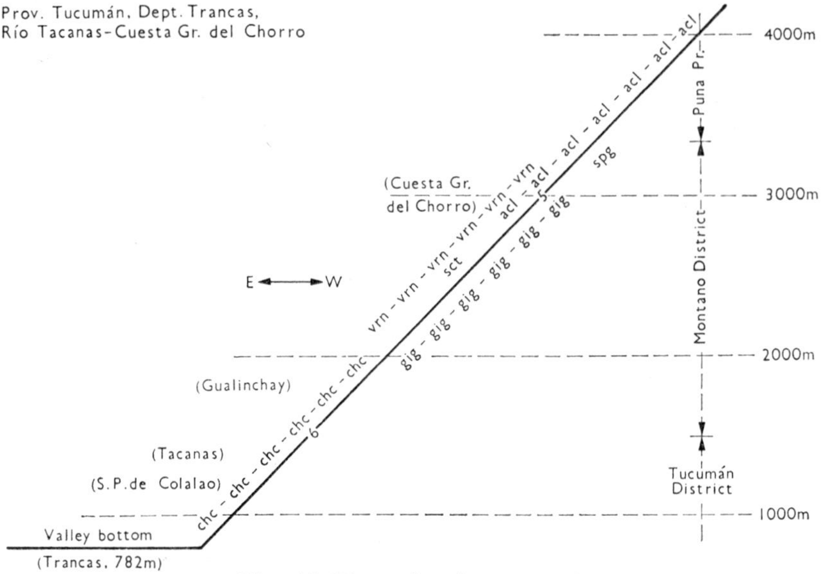

Fig. 1(*d*) (For explanation see p. 22)

5. *Yunka Suma* (Fig. 1 (*e*))

The entrance to this Estancia lies on the road between Concepción and Andalgalá, in Catamarca province. The collector should allow one day up and

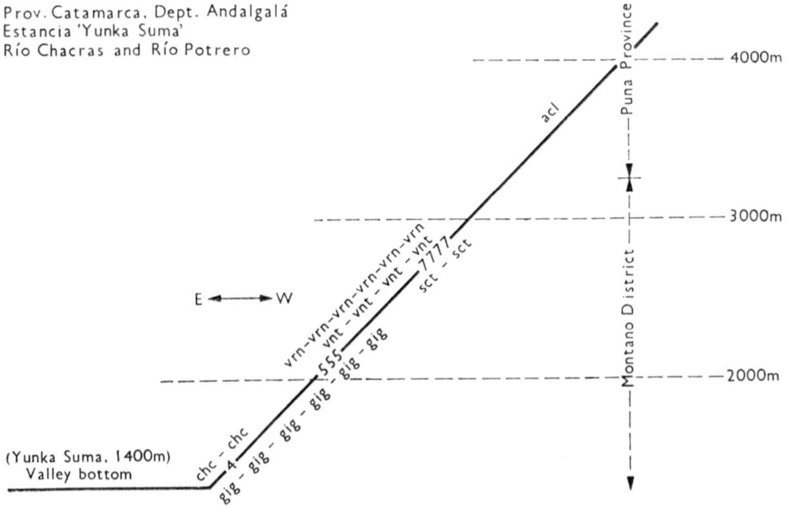

Fig. 1(*e*) (For explanation see p. 22)

one day down by mule or on foot, though it could be accomplished in a single day. No hotels. A tent should be taken for the two-day journey. Very humid.

6. *Quebrada de Belén and Pozo de Piedra* (Fig. 1 (*f*))

There is a fairly good motor road from Belén (prov. Catamarca), through Puerta de Belén (San José) to Pozo de Piedra. The return journey to Pozo de Piedra can be accomplished in less than a day.

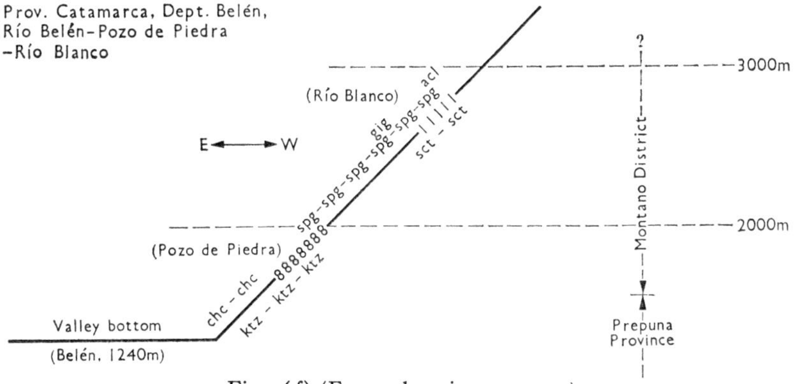

Fig. 1(*f*) (For explanation see p. 22)

The routes given could well be called 'nature trails' for wild potatoes, and are based on the actual collections mentioned in this book. They have all been studied personally by one or both of us, though unfortunately not in as great a detail as we would have liked. The picture would emerge even more clearly, of course, if more collections were available or if a detailed study were carried out with this special aim in mind.

A number of other places are also of considerable interest, and we give a list of some of these below, together with a few notes on the transport conditions up to 1966 or thereabouts, according to our own experience, though conditions may have changed since then:

1. *Prov. Salta, dept. Santa Victoria.* Route from La Quiaca through Yavi, Rincón de Cajas and Abra de Lizoite to Santa Victoria. This is a rough but most interesting journey (*c.* 120 km) by car at very high altitude; the road runs for 52 km at heights of over 4000 m, at times reaching 4600 m, through wild and beautiful scenery. No hotels are available. The most interesting potato zone is between Santa Victoria, La Huerta, Rodeopampa, and Patahuasi, where the following species and hybrids are found: tar, acl, mga, mcd, vdr, mcd × mga, mcd × bal, mga × blv.[1]

2. *Prov. Salta, dept. Guachipas, Estancia Pampa Grande.* This Estancia lies on the road from Ruiz de los Llanos to Guachipas in a wide upland valley

[1] See Appendix II, p. 448, for key to abbreviations.

at 1700 m, where chc and gig, as well as their hybrids, have been found. Higher up, in the region of Cerro Cristal and Cerro Pirgua, the species mga, vrn, and spg are seen, as well as several hybrids. Further exploration in the mountains surrounding the main valley would be rewarding. No hotels are available.

3. *Prov. Tucumán, Quebrada de Río de las Sosas to Tafí del Valle*, and thence further north to Quebrada Honda and Infiernillo. A good road runs all the way, and a wide range of species with interesting zones of hybridization can be seen. Good hotels in Tafí del Valle. The most interesting part of this route is the region of Infiernillo, Río Tafí, and Quebrada Honda, where hybrids of vrn × spg and sct × vnt, as well as several species (acl, sct, gig, vrn, spg), can be found in abundance.

4. *Prov. Tucumán, dept. Chicligasta, Alpachiri to Quebrada de Las Pavas* and thence up to El Bolsón and La Cascada. This is mostly a very bad humid trail where horses and tents are indispensable, taking at least one day to ascend and another to return. Wild potatoes are very common: chc, sct, gig, vnt, vrn, and the hybrids of sct × vnt and vnt × vrn. These occur mostly in the zone 2000–3000 m.

5. *Prov. Catamarca, Sierra de Ambato*, from Pomán to El Simbolar, Los Cienagos, and Alto de la Cruz. Details of this route are known to Dr. F. Vervoorst and no motor road is available. Potatoes found here are: chc, vrn, sct, and the hybrids chc × ktz, chc × spg, and sct × spg.

6. *Prov. La Rioja, Sierra Famatina.* An interesting trip is by road from Chilecito, through Carrizal to Corrales and then on foot or by mule to La Encrucijada, passing Cuevas de Noroño. The trip from Corrales can be accomplished in one day, returning on the following day, but it would be better to spend one or two days in the higher regions, thus making three or four days in all. On this route *S. spegazzinii* is very common, especially at about 3000 m where it occurs also with both subspecies of *S. acaule* and hybrids of *S. spegazzinii* × *S. acaule*, subsp. *aemulans*. This is the classic zone for '*S. famatinae*' and '*S. aemulans*'. Good hotels are available at Chilecito.

(c) *Important areas for future collections*

Despite the numerous potato-collecting expeditions in Argentina and elsewhere, there are still many places that have been insufficiently studied or not even visited at all. This is generally due to their difficult accessibility. Many localities are so remote from the motor roads that one, two, or even more weeks are needed to reach the right places and return with satisfactory results. The best way of travelling is of course with horses or mules, but often the problem of obtaining animals is so difficult that, after having waited in vain for several days, one finally gives up and starts out on foot. A local guide with a few animals to carry the baggage is a useful compromise.

It may still be possible to find new wild potato species in Argentina, but the probability is rather low in an area so well investigated botanically. Rather than trying to count a new species or two to its credit, every collecting expedition can contribute to the more important task of improving our knowledge and understanding of wild potatoes and can bring the solution of some of the many difficult problems a step nearer.

We shall now mention some places that are insufficiently known or completely untouched, so far as wild potatoes are concerned, in the hope that collectors in the future will turn their attention to these areas.

ARGENTINA

1. Prov. Jujuy, dept. Valle Grande and Sierra de Calilegua.
2. Prov. Jujuy, Sierra Santa Bárbara, Sierra del Centinela, and Sierra del Maíz Gordo, all east of San Pedro de Jujuy.
3. Prov. Salta, depts. Santa Victoria and Iruya, between Santa Victoria and Orán.
4. Prov. Salta, Cerros de la Cresta del Gallo, east of Grl. M. M. de Güemes.
5. Prov. Salta, Quebrada del Toro. A few collections have been made here by Cabrera and others, and further exploration of this rather dry valley region would be of great interest.
6. Prov. Catamarca, dept. Santa María, Sierra de Quilmes o del Cajón (northern part is in prov. Salta, dept. Cafayate).
7. Prov. Catamarca, Sierra de Ancasti o del Alto, south-east of Catamarca.
8. Prov. La Rioja, Sierra de Velasco and less well-known parts of Sierra de Famatina.
9. Prov. San Juan. Very little has been collected in this dry province, but a search in the more humid valleys would be most rewarding, as also in the western parts of prov. La Rioja, adjacent to prov. San Juan.
10. Prov. Mendoza, to study the distribution of *S. maglia*.
11. Provs. Santa Fé, Entre Ríos, and Corrientes. A special search in these provinces for the two subspecies of *S. commersonii*, to elucidate the nature of the transitional zone between them would be most valuable. A search for *S. chacoense* and for its possible hybrids with *S. commersonii* would also be of great interest.

PARAGUAY

12. Very few living potato collections have been made in that country. A detailed search in the mountainous area east of Asunción would be most rewarding.

BRAZIL

13. Further collections from the southern States (especially their western regions) of Rio Grande do Sul, Santa Catarina, and Paraná would be valuable in elucidating the pattern of variability in *S. chacoense* subsp.

muelleri. More study of the pubescent form of this subspecies described by Correll is needed.

14. Further search for *S. calvescens* from Minas Gerais is needed, elsewhere than at the type locality of Caldas.

(*d*) *Modern collecting methods*

We very much hope that collectors will explore the areas mentioned above, as well as other remote valleys and mountain regions that have hitherto been overlooked.

Herbarium collections can still be of value, especially if they are well arranged, quickly dried, and the flowers well pressed. Several different collections within a small area with careful explanatory notes and small sketch maps can tell us a great deal, whereas a single specimen may not be of much value. When there seems to be a great deal of variation in the populations of wild potatoes or when there appears to be a gradual trend or cline, linked in some way with altitudinal or other differences, then a random series of samples (say, every fifth or tenth plant, or one taken every 50 m, according to the frequency with which they occur), can be very instructive, even though no living collections might be made.

Adequate collection data are of the very greatest importance. Not only should as exact a locality as possible be given, but an accurate altitude measurement should be taken wherever possible. A note on the plant habit and height and the colour of the flower should also be made. It is also very important to add as much information as possible on habitat, light conditions, soil (e.g. 'dry, sandy', 'humus-rich', etc.) and a note of the other plants associated with it, especially the dominant species.

Whenever possible, it is clearly an enormous advantage for living material to be collected. Tubers can be valuable, but tend to dry up or lose their power to sprout if not dug when mature. On the other hand, they are indispensable if one needs to analyse a population where suspected hybridization between species of different chromosome number and consequent hybrid sterility is taking place. In such a case seed collections will merely confirm the presence of the parents and will leave one with no knowledge about the hybrids. If it is suspected that diploid and triploid forms of a species are occurring in the same area then tuber collections are again essential, since the triploids will not normally be fertile, and seed collections will be obtainable only from diploid plants.

In general, apart from the cases just mentioned, seed collections are much better than tuber collections, since if the berries are more or less mature the seeds will retain their viability for many years. Seeds also possess the additional advantage of providing a wider sample of the genetic variability of the population of plants from which they were collected, even if taken from a single plant only, since most potato species are out-breeding.

If seeds are not available and tuber collections are made, these may be taken from a single clone without this fact being realized at the time of collection. This means that, if the species is a diploid and possesses the usual self-incompatibility mechanisms of wild potatoes, no further sexual generations can be obtained from that collection and it will sooner or later die out. If tubers must be collected it is advisable, therefore, to make several distinct tuber collections from plants separated by a distance of about 10 to 20 m. In this way it will be reasonably certain that more than one clone has been obtained, and interclonal crosses can be effected in the experimental field between these plants, which will represent at least some of the variability present in the original population. This fact is of great importance, as plant breeders now realize, but it was completely overlooked by collectors until comparatively recently.

Finally, we should reiterate that population samples and mass collections of tuber or seed material, or even of dried plants, are very much more valuable, and can give us very much more information than the old style haphazard collecting methods. A small number of well-collected and pressed collections from a small area with complete collection data and additional notes is of much greater value than a very large collection of single plants taken in a hurry, with inadequate or poorly remembered data, not noted down at the time of collection. This may seem to be an obvious truism to some, but it still needs saying for the benefit of beginners, who can make a real contribution by collecting in the right way.

Collections of this sort will help to deepen our understanding of the evolutionary relationships and variability of wild potato species, and the living collections may be of immense value to the plant breeder in providing new genetic material with economically desirable characters.

3

DISTRIBUTION AND ECOLOGY OF THE WILD SPECIES

(*a*) *General distribution and altitude ranges*

THE distribution of the species and subspecies of wild potatoes in Brazil, Paraguay, Uruguay, and twenty of the provinces of Argentina is given in Table 1. From this one can see clearly that the north-western provinces of Argentina are by far the most important regions for wild potatoes. This is, of course, due to the enormous variation in climatic conditions, as well as the mountain barriers that restrict species distribution in many regions. A great part of the provinces of Jujuy, Salta, Tucumán, Catamarca, La Rioja, and Mendoza is occupied by mountain ranges, ascending here and there to peaks of over 6000 m (19 700 ft). Thus, in the province of Salta, where fifteen out of the twenty-four wild species and subspecies are found, the climate varies from subtropical in the eastern lowland to the cold alpine conditions of the high mountains to the west.

We show the altitude ranges of the wild species, arranged in taxonomic order of series, in Fig. 2. The ranges are based on the altitudes given in the collections studied by ourselves, but we should like to emphasize that the picture for each species can only be approximate, since many collections bear no altitude data, whilst in others the indications given are doubtful and are apparently based on very rough estimates. Some collectors give a general altitude range for everything collected on one trip or perhaps during one day. Thus, for instance, Kurtz indicates for his collection No. 14466 of *S. spegazzinii* an altitude of 1400–4525 m! One can hardly believe that the same species could occur through an altitude range of more than 3000 m, even though some species (*S. chacoense*, *S. microdontum*) possess a range of at least 2000 m.

For some species we have been able to see only a few collections, and the ranges for such species will certainly need to be extended considerably when more material becomes available. On the other hand, for the majority we have been able to examine so many collections that we consider the general picture to be reasonably sound. At the upper and lower limits of distribution we have generally seen one or two collections of plants that appear to be outside the normal range of the species concerned. In such cases we have indicated the abnormal range by a dotted line. A further point emerging from Fig. 2 that seems to be of some interest is the generally quite wide altitude ranges for wild potato species in Argentina and adjacent countries. Some

TABLE 1

Geographical Distribution of Wild Potato Species in Brazil, Paraguay, Uruguay, and Argentina

				Argentina																			
	Brazil	Paraguay	Uruguay	Jujuy	Salta	Tucumán	Catamarca	La Rioja	San Juan	Mendoza	San Luis	Cordoba	Neuquén	Río Negro	La Pampa	Buenos Aires	Entre Ríos	Sante Fé	Corrientes	Misiones	Chaco	Formosa	Sant. del Estero
S. brevidens													×	×									
S. calvescens	×																						
S. commersonii																							
subsp. *commersonii*	×		×													×	×						
subsp. *malmeanum*	×	×	×														×	×	×	×	×	×	
S. chacoense																							
subsp. *chacoense*		×	×	×	×	×	×	×		×	×	×			×	×	×	×	×	×	×	×	×
subsp. *muelleri*	×																			×			
S. tarijense				×	×																		
S. acaule																							
subsp. *acaule*				×	×	×	×	×															
subsp. *aemulans*				×				×															
S. infundibuliforme				×	×																		
S. megistacrolobum				×	×																		
S. boliviense					×																		
S. sanctae-rosae					×	×	×																
S. microdontum																							
subsp. *microdontum*				×	×																		
subsp. *gigantophyllum*				×	×	×	×	×															
S. venturii				×	×	×	×																
S. vernei																							
subsp. *vernei*					×	×	×																
subsp. *ballsii*				×	×																		
S. gourlayi				×	×																		
S. oplocense				×																			
S. vidaurrei				?	×																		
S. spegazzinii				(?)	×	×	×	×	×														
S. kurtzianum							×	×	×	×													
S. maglia										×													
Total numbers of species and sub-species	4	2	3	12	15	7	8	6	2	3	1	1	1	1	1	2	3	2	2	3	2	2	1

fourteen out of a total of twenty-four species and subspecies have an altitude range of 1000 m or over, whilst of these, five have an altitude range of 2000 m or over (total ranges considered in every case). The species with the greatest total range is *S. chacoense*, extending from sea level to 2300 m, but this is

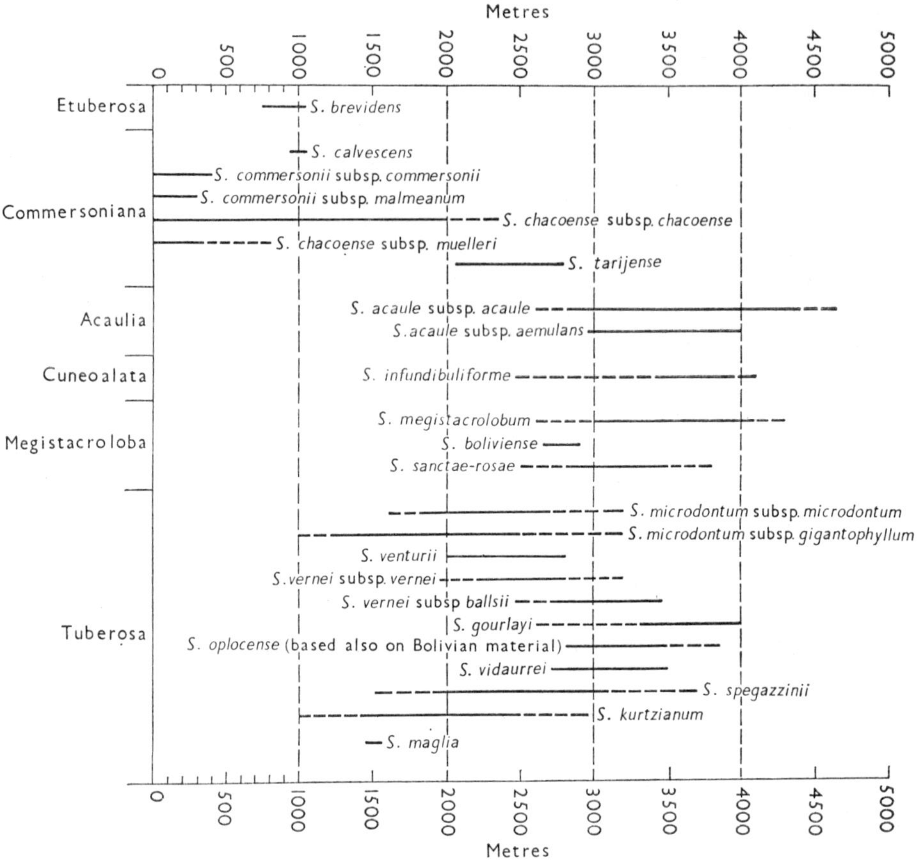

FIG. 2. Altitude ranges of wild potato species.

perhaps a special case due to introgression of genes from a high-altitude species, which has enabled it to extend its range in this spectacular way (see pp. 212–18).

In Fig. 3 we show the frequency of wild potato species and subspecies at different altitudes, using the same data as for Fig. 2. It would appear that 2500–2950 m is the optimum altitude belt for wild potatoes in Argentina, with no less than seventeen species and subspecies out of the total of twenty-four. A generally preferred range could be given as (2000–) 2500–3500 (–3700) m. This range falls within the altitude range of the Montano district of the Western Subtropical phytogeographical province (see below, p. 39).

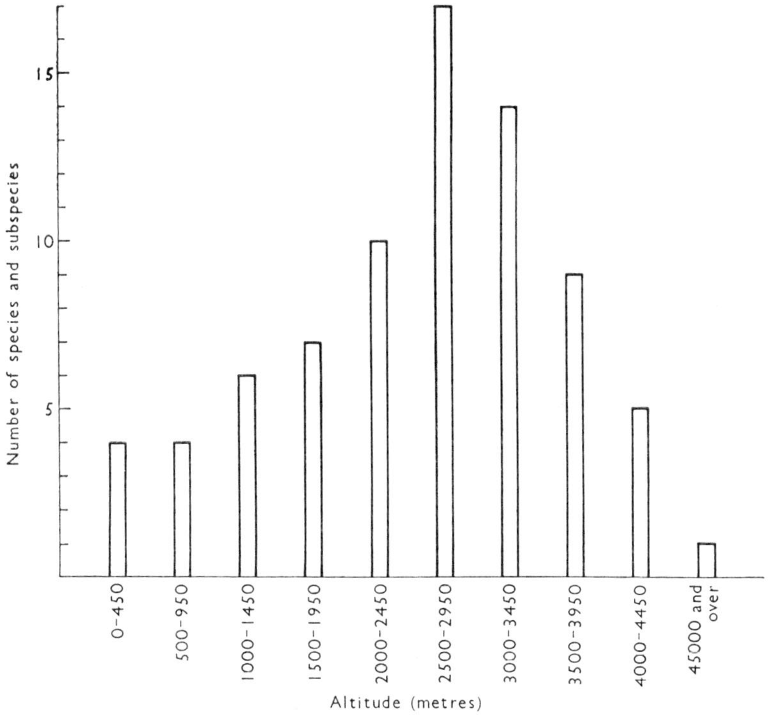

FIG. 3. Frequency of wild potato species at different altitudes.

(*b*) *Phytogeographical aspects*

In the section of this monograph dealing with the detailed taxonomy of each species we have attempted to give as complete a picture as possible of their distribution and ecological requirements. This information is based partly on our own observations in the field and partly on the (often sparse) notes given on the herbarium labels by many collectors. We have also used data from the literature when they were available. It is to be regretted that so few collectors give details of the vegetation type and climatic conditions for the material they collect. We feel, however, that any modern broad treatment of a group of plants should include a study of their habitat relationships, and we have therefore been at pains to give as much information for each species as was available to us. In this study the interesting work by Bukasov (1941*a*) on the geography of the endemic potatoes of South America has been consulted.

The vegetation of Argentina has been studied in some detail by many workers, and several attempts have been made to construct phytogeographical maps. These are compared and discussed in detail by Cabrera in an excellent work published in 1953, and Cabrera's own phytogeographical scheme has been followed by us in the present work. A standard list of English translations of the Spanish names of Cabrera's phytogeographical provinces and some of the districts is given in Table 2. This has been adhered to throughout. We are extremely grateful to Cabrera for allowing us to base our Map 1 on his map (loc. cit., Fig. 12, p. 107), and for generously giving us much help and advice.

TABLE 2

Phytogeographical territories in Argentina (after Cabrera 1953)

I. NEOTROPICAL REGION	(Región Neotropical)
A. AMERICAN SUBTROPICAL DOMINION	(Dominio de la América Subtropical)
1. Western Subtropical province	(Provincia Subtropical Occidental)
a. Orán district	(Distrito Oranense)
b. Tucumán district	(Distrito Tucumanense)
c. Montano district	(Distrito Montano)
2. Eastern Subtropical province	(Provincia Subtropical Oriental)
a. Araucaria Forest district	(Distrito de los Pinares)
b. Mixed Forest district	(Distrito de las Selvas Mixtas)
c. Field or Savanna district	(Distrito de los Campos)
B. CHACO DOMINION	(Dominio Chaqueño)
1. Chaco province	(Provincia Chaqueña)
a. Eastern Chaco district	(Distrito Chaqueño Oriental)
b. Western Chaco district	(Distrito Chaqueño Occidental)
c. Chaco Serrano district	(Distrito Chaqueño Serrano)
2. Espinal province	(Provincia del Espinal)
a. Ñandubay district	(Distrito del Ñandubay)
b.	(Distrito del Algarrobo)
c.	(Distrito del Caldén)
d. Tala district	(Distrito del Tala)
3. Prepuna province	(Provincia Prepuneña)
4. Monte province	(Provincia del Monte)
5. Pampa province	(Provincia Pampeana)
a. Uruguay district	(Distrito Uruguayense)
b. Eastern Pampa district	(Distrito Pampeano Oriental)
c. Western Pampa district	(Distrito Pampeano Occidental)
d. Southern Pampa district	(Distrito Pampeano Austral)
C. ANDEAN DOMINION	(Dominio Andino)
1. High Andean province	(Provincia Altoandina)
a.	(Distrito Altoandino Septentrional
b.	(Distrito Altoandino Central)
c.	(Distrito Altoandino Austral)
2. Puna province	(Provincia Puneña)
3. Patagonian province	(Provincia Patagónica)
a.	(Distrito Patagónico Subandino)
b.	(Distrito Patagónico Occidental)
c.	(Distrito Patagónico Central)
d.	(Distrito del Golfo de San Jorge)

II. AUSTRAL REGION	(Región Austral)
A. Subantarctic Dominion	(Dominio Subantártico)
1. Subantarctic province	(Provincia Subantártica)
a. Valdivian district	(Distrito Valdiviano)
b.	(Distrito Magellanico)
c. Pehuén district	(Distrito del Pehuén)
d. District of deciduous Woodland	(Distrito del Bosque Caducifolio)
2.	(Provincia Insular)
B. Antarctic Dominion	(Dominio Antártico)
1.	(Provincia Antártica)

The English translation has been omitted for the provinces and districts that are not dealt with in the present monograph.

In Cabrera's scheme five dominions, divided into thirteen phytogeographical provinces, are distinguished in Argentina, of which we include ten in our map. The extrapolations made by us into Paraguay, Uruguay, and south Brazil are based largely on sketch maps kindly supplied by Dr. Cabrera, though we ourselves must be held responsible for any errors our own maps might contain. Other alterations have been our inclusion of Cabrera's High Andean province with the Puna province so as not to make the map too confusing, and because the High Andean province is of little interest with respect to wild potatoes. Only *S. acaule* sometimes apparently ascends into it from the Puna province. We have also shown the strips of gallery forest along the Paraná and Uruguay rivers (Eastern Subtropical province) and the Tala district of the Espinal province along the north-east coast of Buenos Aires province as far too broad. However, conventional exaggerations of this sort are unavoidable in small-scale maps when very narrow geographical features need to be portrayed.

A more detailed representation of the phytogeographical provinces of north-west Argentina is given in Map 2. This map has been kindly prepared for us by Dr. Federico Vervoorst from Tucumán, to whom we are much indebted. His names for the phytogeographical provinces differ somewhat from those of Cabrera, and for this reason we give the corresponding names in the key to Map 2. Here also, for the reasons mentioned above, the High Andean province has been included with the Puna province. The so-called 'selva palo-blanco' (*Calycophyllum multiflorum* forest), which has been included with the Orán district of the Western Subtropical province, is, according to Vervoorst, more related in its vegetation type to the Eastern Subtropical province and the Brazilian forests. Since the zone is probably of very little interest with respect to wild potatoes, even though it has not been explored very well from this point of view, we have followed the generally accepted custom of including the 'selva palo-blanco' as part of the Orán district, but have shown it separately on Map 2.

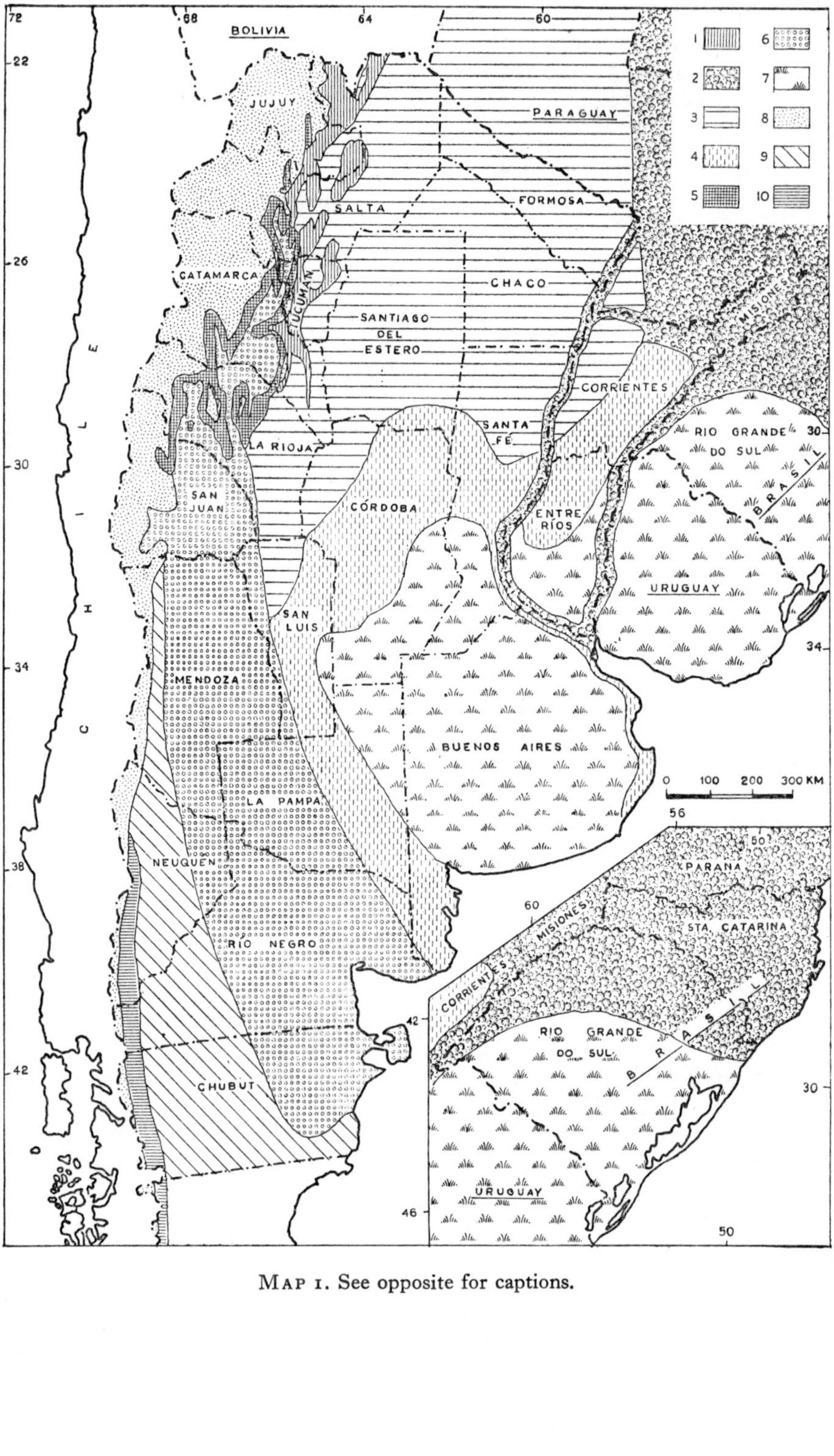

Map 1. See opposite for captions.

The special region of 'pajonales' in the Montano district of the Western Subtropical province has also been treated in different ways by various authorities. It will be discussed in detail later, but we have followed Vervoorst in showing it as a distinct zone on Map 2. In this connexion the interesting paper by Czajka and Vervoorst (1956) should be referred to. The authors show fourteen sections or vegetation profiles in north-west Argentina, running roughly west to east, in which the vegetation types are accurately portrayed. This has helped us considerably to clarify our concept of the phytogeographical zones in that region, and we are additionally indebted to Dr. Vervoorst for his great help and advice with detailed information on special problems.

In the present treatment we have, as far as possible, referred all the wild potato species (with the exception of the Brazilian *S. calvescens*, for which we have insufficient data) to Cabrera's phytogeographical provinces. The distribution of the various species within provinces may be seen in Table 3, where we have separated the Western Subtropical province into its three component districts to give a clearer picture. It is quite evident that the most important region for wild potatoes is the Montano district of the Western Subtropical province, where fifteen out of the total of twenty-three species and subspecies occur. The distribution of *S. chacoense* is of interest here, since it has been collected in at least eight and possibly nine of the phytogeographical provinces. Due to introgression from at least one other species (*S. microdontum*) it has become adapted to the more humid cooler upper part of the Western Subtropical province, and this would no doubt account for the very great altitude range of *S. chacoense* to which we have already referred.

It is important to note that, in general, wild potato species do not form part of the dominant vegetation of each province in which they occur. Indeed, they are often found in specialized or in man-made habitats. Too much stress must therefore not be laid on the occurrence of species in certain phytogeographical provinces or districts; this point is taken up again in the detailed information on each province which follows.

MAP 1. Phytogeographical provinces in Argentina, Paraguay, Uruguay, and South Brazil. (Re-drawn from Cabrera (1953) and with additional details from other sources. Area of north-west Argentina adapted from Map 2, supplied by Dr. F. Vervoorst. For further details, see p. 35.)

1. Western Subtropical province (Provincia Subtropical Occidental)
2. Eastern Subtropical province (Provincia Subtropical Oriental)
3. Chaco province (Provincia Chaqueña)
4. Espinal province (Provincia del Espinal)
5. Prepuna province (Provincia Prepuneña)
6. Monte province (Provincia del Monte)
7. Pampa province (Provincia Pampeana)
8. Puna province (including High Andean province) (Provincia Puneña, incluyendo Prov. Altoandina)
9. Patagonian province (Provincia Patagónica)
10. Subantarctic province (Provincia Subantártica).

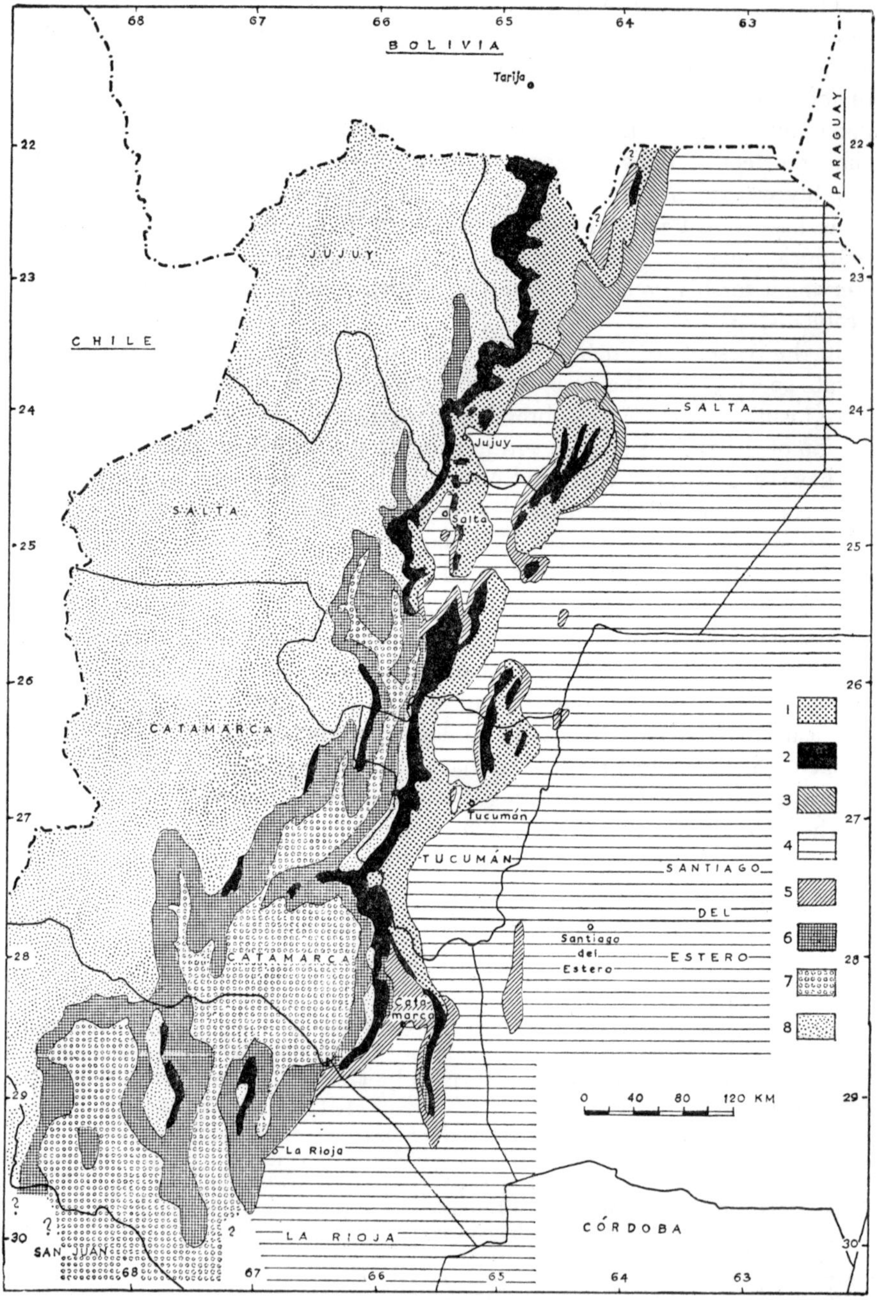

MAP 2. See opposite for captions.

Western Subtropical Province (*Provincia Subtropical Occidental*)

This province extends in a narrow band along the valley and mountain slopes of the north-west provinces of Argentina, running down from Bolivia through the provinces of Salta, Jujuy, Tucumán, and Catamarca. Together with the Eastern Subtropical province it forms what Cabrera terms the American Subtropical Dominion. On its eastern side the Western Subtropical province grades into the Chaco province, and on its west it grades into the Prepuna and Puna provinces. In general, the climate of this province is warm and damp, with rainfall chiefly in the summer (September–March) and frosts during the winter mainly in the higher parts. Rainfall is high, reaching 2000 mm annually, or perhaps more.

The province has been divided by Cabrera into three districts: Orán, Tucumán, and Montano. The two lower districts, Orán and Tucumán, are situated on the lower east-facing slopes of the Andes from the Bolivian border southwards to Catamarca between 500 and 1200 (–1500) m. They consist of dense summer-green forests, the so-called 'selva tucumano-boliviana', which extend southwards from Bolivia. Here are found *Phoebe porphyria* (laurel), *Juglans australis* (nogal), several species of Myrtaceae, and many other trees, shrubs, and herbs. Amongst the wild potatoes only *S. chacoense* and *S. microdontum* are found in these districts, though not very commonly, and generally in disturbed or cultivated places along the roads and rivers. Roughly speaking, the Orán district extends through the departments of San Martín, Orán, Santa Victoria, and Iruya, in the north of Salta province, and also in the east of Jujuy province. The Tucumán district spreads southwards from the south of Salta province through the centre of Tucumán province and into the east of Catamarca province.

The Montano district is situated above the two other districts just mentioned, or in places is found above the Chaco and Prepuna provinces. Its altitude range is from about (1200–) 1500 m up to probably about 3500 (–4000) m. It extends along the eastern slopes of the cordilleras in the provinces of Jujuy, Salta, Tucumán, Catamarca, and La Rioja, and, because of the greater altitude, possesses a cooler climate than that of the Orán and Tucumán districts. The Montano district normally passes into the humid

---

MAP 2. Phytogeographical provinces in north-west Argentina. (From an original map prepared by Dr. F. Vervoorst, Tucumán. Names in brackets are those supplied by Dr. Vervoorst. A more detailed explanation is given in the text, pp. 35 and 37.)

1. Western Subtropical province, Orán and Tucumán districts, excluding Selva palo-blanco (Selva tucumano-boliviana).
2. Western Subtropical province, Montano district (Pajonales y (en parte) bosque de Aliso y Queñoa).
3. Western Subtropical province, Orán district (in part) (Selva palo-blanco).
4. Chaco province, Western Chaco district (Chaco).
5. Chaco province, Chaco Serrano district (Quebrachal serrano).
6. Prepuna province (Prepuna).
7. Monte province (Monte).
8. Puna province, including High Andean province (Puna).

TABLE 3. *Distribution of wild potato species in Cabrera's phytogeographical provinces and districts*

Species	Phytogeographical provinces and districts											
	Western Subtropical province (Provincia Subtropical Occidental) Orán District (Distrito Oranense)	Western Subtropical province (Provincia Subtropical Occidental) Tucumán District (Distrito Tucumanense)	Western Subtropical province (Provincia Subtropical Occidental) Montano District (Distrito Montano)	Eastern Subtropical province (Provincia Subtropical Oriental)	Chaco province (Provincia Chaqueña)	Espinal province (Provincia del Espinal)	Prepuna province (Provincia Prepuneña)	Monte province (Provincia del Monte)	Pampa province (Provincia Pampeana)	High Andean province (Provincia Altoandina)	Puna province (Provincia Puneña)	Subantarctic province (Provincia Subantártica)
S. brevidens												×
S. commersonii												
subsp. *commersonii*						×			×			
subsp. *malmeanum*				×	×	×						
S. chacoense												
subsp. *chacoense*	×	×	×	×	×	×	(×)	×	×			
subsp. *muelleri*				×								
S. tarijense			×				×					
S. acaule												
subsp. *acaule*			×							×	×	
subsp. *aemulans*			×								×	
S. infundibuliforme							(×)				×	
S. megistacrolobum			×				(×)				×	
S. boliviense			×									
S. sanctae-rosae			×								×	
S. microdontum												
subsp. *microdontum*	(×)		×									
subsp. *gigantophyllum*		×	×									
S. venturii			×									
S. vernei subsp. *vernei*			×									
subsp. *ballsii*			×									
S. gourlayi			×				×				×	
S. oplocense							×				×	
S. vidaurrei			×									
S. spegazzinii			×				×	×			×	
S. kurtzianum							×	×		?	?	
S. maglia								†				

×: Normally occurs in province or district indicated. (×): Rarely found. ?: Uncertain occurrence.
†: Occurs in transition zone between Monte and High Andean province according to Ruiz Leal (verbal communication).

part of the Puna province in its upper limit. In its lower part it is dominated by woods of *Alnus jorullensis* (aliso), *Podocarpus parlatorei* (pino), *Sambucus peruviana* (sauco), and others. These woods are normally found between 1200 and 1800 m, but the *Alnus* ascends in some places to about 2500 m.

Higher up the woods are replaced by groups or single trees of *Polylepis australis* (queñoa) and grassland with many shrubs and herbs (see Plates 1 and 2). This upper grassland zone is called 'prados alpinos' (alpine meadows) or 'pajonales'. The flora here is very luxuriant and the climate is cool and humid during the growing season (September–March), though normally dry during the winter (April–August). In those places where the Montano district is found above the Prepuna or the Serrano district of the Chaco province (generally in Catamarca and La Rioja; see Map 2), it does not contain woods of *Alnus* and *Polylepis*, etc., but consists of pure grassland with herbs and small shrubs ('pajonales').

The Montano district is by far the most important zone for wild potatoes in Argentina (see Plates 1–4), and at least fifteen out of the twenty-three species and subspecies are to be found here. Of these fifteen, about five seem to be found exclusively in the Montano district, namely, *S. boliviense*, *S. venturii*, *S. vernei* (both subspecies), and *S. vidaurrei*. *S. microdontum* subsp. *gigantophyllum* is found in the Montano and the Tucumán districts, whilst subsp. *microdontum* occurs in the Montano and probably in the Orán districts. *S. tarijense* occurs in the Montano district and Prepuna province so far as our rather scanty evidence shows. It also seems to occur in the Prepuna province of southern Bolivia. Other species with a higher altitude range occur in the Montano district and the Puna province, as follows: *S. sanctae-rosae*, *S. megistacrolobum* (perhaps also in the Prepuna province), *S. acaule* (both subspecies, and also in the High Andean province). Finally, *S. chacoense*, *S. gourlayi*, and *S. spegazzinii* are, in addition to the Montano district, known also from several other provinces.

None of these species could be said to form part of the dominant vegetation, but such species as *S. microdontum* and *S. vernei* can be quite abundant in places where they find optimum conditions for growth. Others, such as *S. venturii* and *S. boliviense*, seem to prefer the competition of various grass species, whilst ones like *S. acaule*, *S. megistacrolobum*, *S. sanctae-rosae*, and *S. gourlayi*, are mostly found in open places, often on bare soil, though occasionally amongst short grass and stones. Again, other species, such as *S. chacoense*, *S. tarijense*, *S. acaule*, *S. spegazzinii*, and also sometimes *S. microdontum* very often invade cultivated fields and sometimes act as troublesome weeds, or occupy disturbed places along roads and around houses in this province.

Eastern Subtropical Province (*Provincia Subtropical Oriental*)

This province occupies only a small part of Argentina, but extends over a large area of eastern Paraguay and southern Brazil, where it is known also as the

'South Brazilian Forest and Savanna zone'. In Argentina it spreads through the whole of prov. Misiones and into north-east Corrientes, where it follows the Paraná and Uruguay rivers southwards to the head of the River Plate, forming narrow strips of gallery forest along the rivers (shown conventionally too broad on Map 1). To the south and the west the Eastern Subtropical province borders on the Chaco Dominion.

The climate of the Eastern Subtropical province is warm and humid, though with night frosts during the winter. The vegetation consists of dense forests and savannas. One of the three component districts (Distrito de los Pinares) is named after the famous *Araucaria angustifolia* or Paraná Pine forests, and is chiefly found in the Brazilian states of Paraná and Santa Catarina, though it extends also into Rio Grande do Sul, and into the north-eastern tip of the Argentine province of Misiones. The other two districts are the Savanna district and the Mixed Forest district (see Plate 5), but for the purposes of the present work we have not considered it worth while to deal with these three separately, since only a very few wild potatoes are known in this province. These are *S. commersonii* subsp. *malmeanum* and *S. chacoense* (both subspecies). Of these, only *S. chacoense* subsp. *muelleri* is found exclusively in this province (mostly in Brazil). The typical subspecies of *S. chacoense* ranges through a large number of phytogeographical provinces, whilst *S. commersonii* subsp. *malmeanum* is also met with in the Chaco and Espinal provinces. It is doubtful whether this latter subspecies ever forms a real part of the natural vegetation, since it seems to prefer disturbed places along roads and rivers, and often grows as a weed of cultivated fields. Despite its being the wild potato species from the area under survey that exists in the hottest climate (apart perhaps from *S. chacoense* subsp. *muelleri* and the as yet comparatively unknown *S. calvescens*), *S. commersonii* subsp. *malmeanum* is nevertheless fairly frost-resistant, probably due to the fact that it often grows and flowers in the coldest time of the year, when slight night frosts are not uncommon.

The typical subspecies of *S. chacoense* is also quite often met with in Misiones and Paraguay, but is in addition very common in several other phytogeographical provinces. In the Eastern Subtropical province, although it is chiefly found in disturbed vegetation and as a weed of roadsides and cultivated fields, *S. chacoense* occurs as a component of the natural forest vegetation in a few localities.

Chaco Province (*Provincia Chaqueña*)

This widespread province occupies the centre of north Argentina (Formosa, Chaco, Santiago del Estero, eastern Salta, Tucumán, and Catamarca, western Corrientes, northern Santa Fé, northern Córdoba, northern San Luis, and south-eastern La Rioja provinces), western and northern Paraguay, and south-eastern Bolivia. The climate is warm and dry, and the vegetation consists of xerophilous shrubs and trees (e.g. *Schinopsis*, *Aspidosperma*,

Prosopis, etc.). The rainfall is restricted to the summer and there are light frosts from May to November. In the west the Chaco province grades into the Western Subtropical province and to the east it borders the Eastern Subtropical province. To the south-west it grades into the Prepuna and Monte provinces and to the south into the Espinal province.

Wild potatoes are mainly confined to the more humid eastern and western parts of the province, and in the drier central region are probably found only in irrigated fields, along ditches, railroads, and waysides, and perhaps also on river banks. Only two species are known from here: *S. commersonii* subsp. *malmeanum* in the eastern part, which has probably migrated from the wetter Eastern Subtropical province; and *S. chacoense*, which is found all over the province but most abundantly in the more humid eastern and western parts. *S. chacoense* seems to be quite common in the Chaco-Serrano district, which occupies the low foothills of the eastern promontories of the Cordilleras, in places up to about 2000 m. From the Chaco province it has migrated into the Western Sub-tropical province, possibly by means of rivers, roadsides, and cultivated land. It is rather doubtful whether the wild potatoes of this province form part of the natural vegetation except for *S. chacoense* in the Chaco Serrano district of Córdoba and San Luis (see Plate 6).

Espinal Province (*Provincia del Espinal*)

This province forms an arc round the Pampa province from the centre of prov. Corrientes through Entre Ríos, Santa Fé, Córdoba, San Luis, La Pampa, and the extreme south of prov. Buenos Aires. It is bounded on the south by the Pampa province, on the west by the Monte province, and on the north by the Chaco and the Eastern Subtropical provinces. Indeed, the Espinal province is closely related to the Chaco and has been considered as a type of poorer Chaco, dominated by species of *Acacia*, *Prosopis*, etc. The climate is warm and humid in the northern districts and dry and warm in the southern. A special subdivision, the Tala district, follows the valley of the Río Paraná southwards from Santa Fé in a narrow band to the Río de La Plata coast and continues along the Atlantic coast of prov. Buenos Aires to the town of Mar del Plata.

S. chacoense has been found as a part of the natural vegetation of the Espinal province of Córdoba and San Luis and also occurs as a colonizer of irrigated fields and other places, as has been described above for the Chaco province. In the Tala district, which forms a narrow strip, often only 1–2 km broad, along the shore of prov. Buenos Aires, *S. commersonii* and *S. chacoense* (typical subspecies in both cases) are in parts quite common and might be a component of the natural vegetation. This is a problem that would repay further study.

Finally, *S. commersonii* subsp. *malmeanum* is found here and there in the northern part of the Espinal province (Ñandubay district) in Corrientes, Entre Ríos, and Santa Fé.

Prepuna Province (*Provincia Prepuneña*)

This phytogeographical province is to be found on the dry slopes and valleys of the mountains of north-west Argentina from Jujuy, through Salta, Tucumán, Catamarca, and La Rioja to San Juan province. The formation is most typically developed in Salta, Catamarca, and La Rioja, and is quite characteristic for parts of these provinces. It ranges from (800–) 1500 to 3000 (–3400) m, and is generally found on the slopes and quebradas in the more western regions, which are in the rain shadow of the eastern mountain chains. As a rule the Prepuna province is situated above the Monte province and sometimes above the Chaco province, with the Puna province nearly always at its upper limit. In some places on east-facing slopes, however, a belt of 'pajonales' may be found between the Prepuna and Puna provinces. The climate of the Prepuna province is dry and warm, with rains confined to the summer, and total precipitation rather low (302–354 mm, according to Cabrera 1953).

The Prepuna province is related to the Monte province, which is also very dry. They are both dominated by low, often spiny shrubs of the families Zygophyllaceae, Compositae, Leguminosae, etc. Giant columnar cacti, such as *Trichocereus terscheckii*, *Tr. pasacana*, *Tr. poco*, and others are characteristic for the Prepuna province (see Plates 7 and 8). Species of *Abromeitiella*, *Tillandsia*, *Dyckia*, and other Bromeliaceae are frequent on rocks and dry slopes (see Plates 7–9).

The wild potato species characteristic of the Prepuna province are *S. spegazzinii*, *S. kurtzianum*, *S. gourlayi*, and also probably *S. tarijense*. Occasionally also *S. chacoense*, *S. infundibuliforme*, *S. megistacrolobum*, and *S. oplocense* occur here.

Some of the species are no doubt fairly drought resistant, but even these seem to have difficulties in the dry years and then develop very poorly, if at all. Thus the wild potato collector can easily be very unlucky in looking for species typical of the Prepuna province in a dry year. Species such as *S. gourlayi* and *S. oplocense*, for instance, are very poorly represented in most herbaria. The most typical species, perhaps, for this phytogeographical province is *S. spegazzinii*, which is quite common in many places. In good wet years it develops well and shows a luxuriance of growth in strong contrast with its development in dry years, so that the two contrasting phenotypes have received two distinct names, *S. spegazzinii* (dry form) and *S. famatinae* (damp form). *S. chacoense* is seen only occasionally in this province, and then probably always in connexion with cultivation.

Apparently only *S. spegazzinii* and *S. kurtzianum* occur with such frequency as to form a perhaps limited part of the typical vegetation of the Prepuna province in some places and in certain years. This needs further verification, however. All the other species mentioned seem to be fairly rare in this province, though *S. gourlayi*, *S. oplocense*, and *S. tarijense* are still not well known enough for us to be certain of the frequency with which they occur.

Monte Province (*Provincia del Monte*)

The Monte province occupies some of the interior valleys in the drier parts of provs. Salta, Catamarca, and La Rioja and continues southwards in a wide band through San Juan, Mendoza, Neuquén, La Pampa, Río Negro, and Chubut. In the northern valleys it grades into the Prepuna province on its upper margin and spreads over the bottoms of the valleys and intermont basins. Further southwards its western border is formed by the Puna and Patagonian provinces, whilst to the east it grades into the Chaco and the Espinal provinces. The climate is very warm and dry in the northern parts, gradually becoming cooler in the south. Rainfall is low, 86–323 mm, according to Cabrera (1953).

The vegetation is a bush steppe formed mainly by species of *Larrea*, *Prosopis*, *Bougainvillea*, *Cassia*, *Bulnesia*, and other genera (see Plate 10). Several wild potato species are found within this area, but possibly only *S. kurtzianum* could be said to be truly native to the Monte province, since it seems in places to form part of the natural vegetation (see p. 410). *S. chacoense* and *S. spegazzinii* have also been recorded for the Monte province but are much more frequent in other provinces, spreading into the Monte here and there where they probably occur in special places where the humidity exceeds the normal for this very dry zone. Finally, *S. maglia*, which is known only from one locality in Argentina, prov. Mendoza, seems to grow in the transition zone between the Monte and High Andean provinces.

Pampa Province (*Provincia Pampeana*)

This phytogeographical province occupies most of prov. Buenos Aires, Uruguay, and the south of Rio Grande do Sul State, in Brazil. It also spreads into the south of Entre Ríos, Santa Fé, and Córdoba and the north of La Pampa provinces. Most of the Pampa province is composed of flat grassy plains, with isolated ranges of hills and occasional marshes and thickets. The climate is warm-temperate, and humid, with precipitation all the year round. On its north and west sides the Pampa province grades into Espinal, whilst on the north-east it is replaced by the Eastern Subtropical province. The Pampa rainfall ranges from about 700 to 950 mm per annum.

Two potato species are found here, namely, *S. chacoense* and *S. commersonii* (typical subspecies in each case). The former does not seem to be a component of the natural vegetation, but is found only in a few places where it is confined to special habitats such as ditches, roadsides, grassy places along rivers and streams, and other disturbed habitats. *S. commersonii*, although to some extent found in ruderal habitats, is also to be found in 'natural' ones as well. It prefers damper and more marshy places than *S. chacoense* and is far more common in this province than the latter species, especially in the eastern parts (Uruguay and Brazil). In the 'matorrales (= thickets) de curro' (*Colletia paradoxa*), *S. commersonii* seems quite definitely to form part of the natural vegetation.

High Andean Province (*Provincia Altoandina*)

For reasons of clarity we have included this province with the Puna province in Maps 1 and 2. It is situated in western Argentina, on the highest peaks and upland plateaux of the Andes from Jujuy southwards to Tierra del Fuego. In Jujuy and Salta it usually begins at about 4500 m and goes up to the permanent snows. The climate is very cold and fairly dry, with night frosts all the year round. Further south, in Mendoza, the High Andean province descends to about 2300 m (according to Roig 1960). The vegetation is often sparse, consisting mostly of grass steppe. The only potato species to occur in the High Andean province is *S. acaule*, which ascends here and there from the Puna province. In Mendoza, at the transition of the High Andean and Monte provinces, *S. maglia*, as well as *S. kurtzianum* are to be found, though rarely.

Puna Province (*Provincia Puneña*)

This high cold province is not differentiated on our maps from the High Andean province, for reasons already stated. It is situated on the high mountains and plateaux of north-west Argentina, from Jujuy southwards to San Juan, from about 3400 to 4500 m. Above it is the High Andean province, whilst below are found the Montano district of the Western Subtropical province in wetter parts, the Prepuna province in the drier parts, and the Monte province in the south. The eastern parts of the Puna, especially above the Montano district, are comparatively humid, and this is where wild potatoes are most common (see Plates 11–14). The dominant vegetation of the Puna is a bush steppe, consisting of species of *Mulinum*, *Lepidophyllum*, *Fabiana*, *Chuquiraga*, *Azorella*, etc., with the grass genera *Festuca*, *Calamagrostis*, and *Stipa* well represented. The climate is cold, and night frosts are common the whole year round. The rainfall is said to be usually very low, only 100 to 300 mm a year, but is no doubt higher in certain regions.

Of wild potatoes, *S. infundibuliforme* is probably the most typical for the province and is no doubt highly drought resistant. This species is also found occasionally in the Prepuna province. *S. acaule* is also quite typical for the Puna province, but is in addition found in the Montano district and also enters the lower part of the High Andean province. It is highly frost-resistant and withstands 8–10° C below zero; it seems to grow mostly in humid places by springs and in damp meadows, in corrals (= stock yards) and around houses and cultivated fields. It can also form part of the natural vegetation, just as *S. infundibuliforme*. Another species often to be met with in the Puna province is *S. megistacrolobum*, but only in the damper places, and it is common in the Montano district as well. *S. sanctae-rosae* is mostly a plant of the Montano district but ascends in places into the Puna province. Finally the following species have also been recorded, though not very frequently: *S. gourlayi*, *S. oplocense*, and occasionally *S. spegazzinii*.

Subantarctic Province (*Provincia Subantártica*)

This narrow province runs along the east side of the southern Andes close to the Chilean border and to the west of the Patagonian province. The climate is cold and damp, with snow in the winter and frosts almost the whole year round. The dominant vegetation is *Nothofagus* forest (see Plate 15). Here is found the most southern representative of the wild 'potatoes', namely, the non-tuberiferous *S. brevidens*, belonging to the mainly Chilean series, Etuberosa. It has been collected frequently in the well-known 'Parque Nacional de Nahuel Huapi', where it is sometimes associated with an escaped form of *S. tuberosum*, the cultivated potato.

S. brevidens seems to be found in the Valdivian as well as in the Deciduous Woodland districts, but whether it is also found in the northern Pehuén district (named after the local word for *Araucaria araucana*) is not yet clear. *S. brevidens* prefers humid and not too densely shaded habitats, and is apparently a true component of the natural vegetation.

4

TAXONOMY AND EVOLUTIONARY RELATIONSHIPS

A DETAILED survey of any group of plant species, especially if experimental as well as observational data are included, ends by revealing perhaps as many problems as it solves. The present work is no exception. Yet in the process of answering some questions and asking others we have been able to look more deeply into the micro-evolutionary processes at work in this group of species and to understand some of the reasons for the variation patterns of the group of tuber-bearing species as a whole. At every step we have tried to keep firmly in touch with the results of exact observations and experiments, and to interpret our facts in the light of the selective processes of the environment, acting on the variables of mutation and gene recombination.

In the present chapter we have attempted to draw together the information given in Part II of the book, in order to give the reader who is not a specialist in the group a general synthesis of the evolutionary relationships of the species so far as they are known. We present in this way a summary of the main observations and conclusions for the individual species, of which the full documentation appears later. We also have tried to give our views on the species concept, and past and present evolutionary trends in this very fascinating group of plants.

(*a*) *The species concept in* Solanum

Ideas of what constitutes a species have varied from age to age, from group to group, and from taxonomist to taxonomist. Although it is currently rather unfashionable to hold to the concept of the 'biological species', one in which the species boundaries on paper relate directly to reproductive barriers of some sort in nature, nevertheless, in this group of plants it will be seen that this concept is by and large a very satisfactory one, providing that we confine our attentions to the normally sexually reproducing outbreeding forms. Exceptions occur, as we shall see later, but these only serve to emphasize that the modern biological species concept in *Solanum* is useful and also satisfactory from a theoretical point of view.

Historically, the views of taxonomists who have specialized on this group have varied considerably. Dunal (1852, etc.) described so few tuber-bearing species that one might be excused for saying that he probably had no concept of the species at all, except that common to his age of considering all species to be divinely created and to be distinguishable from each other by well-defined combinations of characters which were immutable and fixed for all time. It will be noted in passing that nearly every species was described from a single specimen.

The German taxonomist, Bitter, working almost exclusively with dried herbarium material, thought of *Solanum* species as very narrow units with little variability. Even the slightest variants were given specific, subspecific, or varietal rank, and his taxa also, were based on one or at the most a very few specimens. Bitter realized that some grouping of these species was needed below the level of section, and considered that certain groups of species should be associated under the concept of 'collective species'. This concept is today taxonomically inadmissible. Examples of Bitter's use of collective species are *S. commersonii* sp. coll., under which he included *S. commersonii* sensu stricto, *S. acroleucum* and *S. calvescens*; and *S. maglia*, sp. coll., under which were included *S. maglia* s. str., *S. weberbaueri*, *S. medians*, and *S. hypacrarthrum*. The latter three species are from Peru and certainly are not closely related to *S. maglia*. Bitter also established the concept of the 'series' or slightly higher grouping of species. He described one such series, the conical-berried Conicibaccata, and mentioned others, such as 'the series of *S. maglia*' (including the collective species, *S. maglia*, and what we now call *S. microdontum* subsp. *gigantophyllum*) as well as a number of other groups of species that it was left to other authors to name and describe much later.

Bitter was undoubtedly feeling his way towards a more completely formulated grouping within Hyperbasarthrum, one which he had attained with the establishment of the Conicibaccata, but not or only imperfectly with the other groups of species. This is not surprising in view of the fact that he was actively engaged in the pioneer work of species delimitation and description, that he cultivated very few of the species, made no crosses or transplant experiments and had not collected or observed material in the wild. Unfortunately he did not live to complete his long-awaited monographic treatment of the genus as a whole. Despite the fact that his species concept was rather a narrow one by present-day standards he made a fundamental contribution to our knowledge of the group by his meticulous examination of specimens and his wonderfully clear and detailed descriptions.

The Russian botanists and plant-breeders, Juzepczuk and Bukasov, followed Bitter's methods fairly closely, though they based their descriptions for the most part on living material grown at Leningrad. Any collection that was found to differ significantly from the descriptions of Bitter or other taxonomists was described as a new species, variety, or form. Their 'micro-species' were generally grouped together into collective species and these were in turn classified into series, many of which were based on Bitter's groupings that he himself had not named. Although Bitter had made a few subspecies the Russian botanists, following current practice in their country, gave specific rank to any taxon that might by botanists from other countries have been considered as subspecies. Since they were interested in the breeding value of the native wild and cultivated potatoes they paid much more attention to the latter than botanists had done in the past and split *S. tuberosum* up into a number of species with different chromosome numbers and different ecological preferences. Bukasov's collective species are roughly equivalent to the Linnean species of the present work, whilst his ordinary species, which we would rank at the most as microspecies, are not generally recognized by us of sufficient value to be retained as distinct taxa. We suspect that in some cases they are little more than clones and we regard them as ecotypes, local variants or at best as subspecies (subsp. *punae* of *S. acaule*, for instance, named by Juzepczuk as *S. punae* Juz.). Although Bukasov showed that the differences between these microspecies were genotypically controlled there seem to be no sharp discontinuities between them, so far as we can judge, and their characters intergrade considerably. In our view, also, too much attention was generally paid by Bukasov and Juzepczuk to geographical considerations in establishing new species, without a sufficient basis of fact to demonstrate the presence of well-marked morphological differences. In some instances even the series proposed by Bukasov and Juzepczuk (see, for instance, the latest views on Commersoniana quoted by us on p. 137) have been equated by us with our species.

J. G. Hawkes, working during the war on the collection of potatoes brought back by the Empire Potato Collecting Expedition and without a complete range of type specimens available to him, continued with the taxonomic methods of Bitter, Juzepczuk, and Bukasov until about 1945. It was then decided to begin a revision of the whole subsection Hyperbasarthrum by reviewing type and other material from European and American herbaria. It became evident during the course of that study that our concept of the species in *Solanum* must be drastically revised if we were not to arrive at a state where nearly every new collection was given a distinct varietal or specific name. Even from herbarium studies it was possible to see that the species must be regarded as larger units, comprising a wide range of infra-specific variability which could only rarely be fitted into the conventional pattern of varieties and forms, though geographical variation enabled one to see that subspecific divisions were often useful and necessary.

Working quite independently, the same conclusion was reached by Correll in 1952 and also by Brücher from 1953 onwards, though these two workers have not used the concept of geographical subspecies developed by Hawkes for this group and retained by us here.

In the present work we have tried to take experimental results into consideration when fixing species boundaries and have regarded potato species as being both highly variable genotypically and of extreme phenotypic plasticity. Environmental differences are sufficient to cause great phenotypic differences, sometimes overriding or obscuring the genotypic ones. Ideally we should have liked to have been able to grow every collection of *Solanum* examined by us in the experimental field to assess the extent of this kind of variation. This was obviously impossible, but even so, one needs to possess a yardstick with which to assess the vast quantities of herbarium material for which no seed or tuber collections were made. As a compromise, if one can cultivate at least some accessions of most species under one or more series of uniform conditions it will then be possible to compare these phenotypes with those of identical genotypes growing in nature. This we have attempted to do, and have thus gradually built up a picture of the range of phenotypic variation of most of the species discussed in this book, under the widely varying environmental conditions encountered by them in nature, and have contrasted this with the phenotypic ranges of closely allied species.

We have seen that sometimes two apparently distinct species (as judged from the nomenclatural types and other specimens gathered in the wild, e.g. *S. spegazzinii* and *S. famatinae*; *S. microdontum* and *S. bijugum*) are one and the same entity and that their apparent distinctness was due to widely varying habitat, and no more. In other examples the phenotypic variation *in nature* was so intense as to obscure almost completely the expression of essential genotypic variation that would have pointed to differences between species (e.g. some growth forms of *S. microdontum* and *S. venturii*). In this way it is possible to understand which of the characters are of little taxonomic value in the separation of two taxa, because these characters vary in a parallel manner in two or more species according to environmental changes (e.g. stem development in *S. acaule*, *S. megistacrolobum*, and *S. sanctae-rosae*). The 'essential' characters that are of value in delimiting the species can easily be overlooked until one has had such an opportunity of cultivating material under various types of controlled conditions.

(*b*) *Breeding barriers and gene flow*

As we shall mention several times in this book, most species of tuber-bearing *Solanum* in the area we have studied are evolving under geo-ecological isolation barriers. Each is adapted to a particular set of habitat conditions which it does not (or, rather, did not in the past) transgress. Therefore, there are generally no very well-marked sterility barriers between the species, since no selection for this sort of barrier has taken place (except, apparently, between *S. commersonii* and *S. chacoense*).

It is therefore not surprising that hybrids may be obtained with ease between most of these species and that the hybrids themselves are moderately to completely fertile. The ease with which artificial hybrids can be formed does not mean that they will necessarily also be found in nature, or that if found they will be very common. The situation is a very propitious one for the plant-breeder since transference of genes from most of the South American wild species to *S. tuberosum* can be effected without too much difficulty.

This ease of hybridization in *Solanum* and other groups similar in this respect has often been used as evidence in the argument that such species are not 'good' or 'valid' ones since they are not separated by sterility barriers.

The geo-ecological barriers seem to have been very effective in restricting gene flow, so that the species, although often difficult to identify because of the phenotypic plasticity to which we have already referred, do not in fact lose their identity in nature. We shall see in a later section that the number of hybrids identified by us is very small compared with the number of non-hybrid collections examined (9·5 per cent).

It is possible that occasional F_1 hybrids have always been formed in nature to a very limited extent between most species whose distribution areas overlapped. It is probable, however, that since each of the parental species is adapted to different environmental conditions the hybrids between them will not survive for long because they are not genotypically fitted to the 'adaptive peaks' of their parents. Only perhaps in a very narrow zone of intergradation between one habitat and the next will the hybrids of such species be able to survive. And if the F_1 hybrids do not survive then the F_2 or backcross segregates will not be formed and the species boundaries will remain fairly sharp.

This completely natural situation has in many areas long since ceased to exist, since much of the natural ecological balance has been changed by the activities of man. Especially in recent historical times and perhaps even in pre-conquest days man has altered and destroyed the natural plant communities by means of farming, settlement, and other processes, so producing various intermediate or 'hybrid' habitats where hybrids between species can now exist and reproduce. Here the F_1 hybrids will be frequent and well adapted, and so will be in a position to give rise to F_2 and backcross generations, resulting in gene flow or introgression from one species to another. We shall later mention one clear case and several other probable ones where the boundaries between species that were once presumably clear cut are now becoming blurred and obscured by this process.

It would seem also that the flow of genes from one species into another (those of *S. microdontum* into *S. chacoense*, for example) has resulted in renewed vigour and a greater range of adaptation to a wider variety of habitats than was possible in the non-introgressed species. We tend to observe most easily the morphological evidence for this, but undoubtedly the physiological adaptation is increased and we also have some evidence for the flow of resistance genes between certain species. Introgressive hybridization has been claimed by Edgar Anderson as a positive evolutionary factor; in some of the tuber-bearing species of *Solanum* there seems to be considerable evidence to support this claim.

(*c*) *Breeding systems*

The pattern of variability in the tuber-bearing *Solanum* species cannot be adequately understood without a realization of the importance of the two alternative means of reproduction available to them—sexual and vegetative. Although the cultivated potatoes commonly reproduce vegetatively the reverse is largely true for the wild diploid and even-numbered polyploid species. Even amongst the South American cultivated diploid and tetraploid species sexual reproduction probably plays a more important role than has generally been admitted.

On the one hand, the tuber-bearing species are able to maintain themselves indefinitely by vegetative or tuber reproduction, thus building up a large population of individuals which are the ramets of a single clone and are therefore genetically identical. On the other hand, most wild species are highly fertile and are capable of producing abundant seeds from cross-pollination, resulting in new ranges of biotypes, each differing genetically from its fellow.

If the environment is favourable and moderately stable, with not too intense a level of competition from other vegetation, then vegetative reproduction will be at a premium; if it is changing, or if competition is more intense, the clonally propagated plants will die out and new plants will arise from true seed in favourable positions elsewhere. A delicate balance is thus preserved. By sexual reproduction potato species are able to adapt themselves to changing environments by means of natural selection acting on a wide range of genotypes. When a successful genotype becomes established in a favourable situation it can then maintain itself indefinitely by clonal propagation. This combination of variation on the one hand with stability on the other may perhaps be likened to the processes of facultative apomixis and of inbreeding combined with occasional outbreeding that seem to have occurred with such frequency and success in other flowering-plant genera.

In the sexually reproducing diploid species of the area we are surveying in this book, allogamy or outbreeding is the dominant breeding system. All the diploids, with the exception of the self-fertile *S. brevidens*, are self-incompatible, the mechanism involved being the simple oppositional factor type, with multiple *S* alleles at the same locus. This outbreeding system ensures that gene flow will take place through large populations and seems to have been responsible for maintaining the high degree of infraspecific variability that we have already commented upon.

In the two tetraploid species known in the area under discussion, the wild *S. acaule* is not only self-compatible but is normally self-pollinated, or so it appears. One would therefore expect to be able to see a reflection of this in the different variation pattern of this species as contrasted with the outbreeding diploids. *S. acaule* should exist as a series of pure lines, of which at least some should be identifiable morphologically. Such pure lines have not been distinguishable up to the present. They may of course exist and have not been seen by reason of insufficiently detailed observations; or perhaps under natural conditions some degree of cross-pollination takes place and a mixture of inbreeding and outbreeding results. This merits further investigation.

The other tetraploid, *S. tuberosum*, is also self-compatible, but it seems to need an insect pollinizing agent before seed can be set, due probably to the difference in position of the anther pores and the stigmatic surface. In *S. acaule* the style is short and the anthers dehisce directly on to the stigma.

No data exist for *S. tuberosum* as to the extent of self-pollination under natural conditions, though certainly very little occurs on plants grown in insect-proof glasshouses. Probably in nature both self and cross pollination occur. We do not possess any information whatever about the breeding behaviour of the tetraploid forms of *S. oplocense*.

Forms with thirty-six chromosomes occur and are of two types:

1. The simple hybrids between the tetraploid species *S. acaule* and various diploid species, such as the triploid wild *S.*×*bruecheri* (*S. acaule*×*S. megistacrolobum*) and the triploid cultivated *S.*×*juzepczukii* (*S. acaule*×*S. stenotomum*). These are highly sterile and reproduce wholly vegetatively.

2. Apparently autotriploid forms of diploid species. Examples of this type in Argentina and adjacent countries are *S. commersonii* (both subspecies), *S. calvescens* (the only collections so far known in the living state are triploid, but we suspect the presence of diploids also), *S. chacoense* (this species is normally diploid, but one triploid clone has been examined), and *S. maglia* (triploid forms known so far only in Chile). All these triploid forms seem to be rather sterile, though they will function as female parents when pollinated by fully fertile diploid or tetraploid species. They are chiefly to be found in the environs of large cities or in cultivated fields where the disturbance of the natural vegetation probably enables them to grow vegetatively under the protection of man for perhaps indefinite periods of time. *S. commersonii* and *S. maglia* seem also to do well on the sea coast for perhaps the same reason.

Such sterile hybridogenic or autotriploid forms are obviously quite distinct in their formation and reproduction from the sexually fertile ones. The hybridogenic pentaploid species, *S.*×*curtilobum*, is of the same nature. The advisability of giving the hybrids or indeed any naturally or artificially occurring hybrids specific names may be questioned. If a hybrid of this sort occurs frequently in nature then it seems useful to have a name for it, but if it is of rare occurrence or has only been made artificially, then it seems better not to give it a name but to characterize it by the hybrid combination of the two parental names.

(*d*) *Chromosome numbers and polyploidy*

The basic chromosome number in *Solanum* is generally considered to be 12. Analysis of secondary associations of the meiotic chromosomes at first and second metaphase has been interpreted by certain workers, however, as evidence of remoter chromosome affinities, indicative of a basic number of 6, instead of 12 (Müntzing 1933, von Olah 1938, Choudhuri 1943, etc.). In view of the fact that cytological analysis of odd-numbered polyploids has failed to show any secondary balance and that genetical analysis of diploid progenies has not revealed evidence of tetrasomic inheritance we can assume for all practical purposes that the basic number is 12.

In the wild species of the area dealt with in this book, chromosome numbers of diploid ($2n = 24$), triploid ($2n = 36$), and tetraploid ($2n = 48$) levels have been established, whilst the cultivated species range from triploid to pentaploid ($2n = 60$). Yet polyploidy is not very common amongst the South American potato species, and by far the greatest number of species is diploid. Thus seventeen of the nineteen wild species described in the present book are diploid. If we conclude that diploid clones of *S. calvescens* will also be found in time, then the number can be raised to eighteen. Two of these diploid species are also known as triploid clones, and one as tetraploid (*S. oplocense*). There is only one exclusively tetraploid species, *S. acaule*, and even this varies its chromosome number in north Peru, where a hexaploid subspecies has been recorded.

Thus, unlike the Mexican potatoes, amongst which polyploidy is a frequent phenomenon, the wild potatoes of Argentina and adjacent countries conform to the general pattern of diploidy in South America. We have no explanation to offer for the occurrence of triploid clones amongst the diploid ones, but the phenomenon seems to be widespread, and is known for other countries in South America, as well as in Mexico.

The cultivated South American species show a range of ploidy from diploid to pentaploid. The two diploid species, *S. stenotomum* and *S. phureja*, have not yet been seen in north-west Argentina.[1] The triploid species, *S.×juzepczukii*, which has been formed as a natural hybrid between *S. acaule* and *S. stenotomum*, is known from our area. Both subspecies of the tetraploid *S. tuberosum* are known, as well as the pentaploid species, *S.×curtilobum*. This latter seems to have arisen as a hybrid between *S.×juzepczukii* and *S. tuberosum*.

(*e*) *Chromosome homology and genome analysis*

From the results of analysis of meiotic configurations in the F_1 of interspecific hybrids it would seem that the twelve chromosomes of the haploid set of every diploid species in the area under discussion are exactly homologous with those of every other diploid species in this region. No meiotic analysis of this sort has shown any substantial difference of pairing behaviour from that of the species themselves. Occasional irregularities, in the form of anaphase bridges and fragments, have been reported from time to time, but these do not seem to be more frequent in the hybrids than in the parents. From this point of view, therefore, the chromosomes of all the diploid species from this region so far analysed are homologous.

Artificial autotetraploids of a diploid species such as *S. chacoense* (Howard & Swaminathan 1952) show multivalent frequencies similar to those of other autotetraploids, both in *Solanum* and *Lycopersicon* (3·57 in *S. chacoense*; 3·7 and 4·0 in *S. phureja*, Lamm 1945; 3·8 in *Lycopersicon esculentum*, Upcott 1935). Swaminathan (1953) analysed quadrivalent frequences in artificial autotetraploid lines of diploid species, comparing them with the amphidiploids

[1] See, however, Brücher's report on diploid cultivated varieties (p. 428).

derived from diploid × diploid crosses and subsequent colchicine doubling. Interestingly enough, although the quadrivalent frequencies in the autotetraploids varied from 3·45 to 4·56 they were very much lower, ranging from 0 to 1·19, in the amphidiploids of species that were considered by other criteria to be distinct. Those amphidiploids of species that were not generally considered to be distinct showed quadrivalent frequencies similar to the autotetraploids.

These results may be interpreted by assuming that the chromosomes of different species possess cryptic structural differences that are so small as not to interfere with pairing in the simple F_1 hybrids (Stebbins 1945). A chromosome in the presence of its exact partner or homologue (in an amphidiploid) will pair with it in preference to a similar chromosome (homeologue) derived from another species. This preferential pairing of homologues cannot take place in the diploid F_1 hybrid, and where the homeologue only is present, pairing will then take place with that. This appears to be evidence of genome differentiation between species which is sufficiently slight not to be visible in simple F_1 hybrids but is large enough to manifest itself in amphidiploid hybrids where differential pairing between homologues at the expense of homeologues is clearly visible. Unfortunately, not very much has yet been accomplished along these lines, and a great deal remains to be done.

The evidence of genome, and hence of species, evolution by means of cryptic structural differences is by and large corroborated by the results of genetical studies of F_2 progenies. If the chromosomes are essentially similar in two taxa, then the F_2 generation of hybrids between them will be thrifty and vigorous, since the chromosomes or portions of them will be able to substitute for each other in the F_2 segregates, giving physiologically well-balanced plants. If, on the other hand, the species have evolved apart and their chromosomes are no longer similar, differing either by minute structural differences or perhaps by multiple gene substitution (Harland 1933), then these chromosomes will no longer be able to substitute for each other in the F_2 segregates. The exact gene combinations necessary for normal growth and development will be lacking, and many plants will die at an early stage or remain small and unthrifty. A complete range from normal plants to inviable seeds should be observable.

This technique of F_2 analysis has been of great value as an indicator of the boundaries between potato species in the present studies, and correlates reasonably well with results from genome analysis mentioned above, and with morphological and ecological data. It is, in fact, a particularly interesting biosystematic tool for judging the degree of genetic relationship between species, and is thus another criterion in the task of making a final assessment on how the populations of wild potatoes should be grouped and classified. Details of these studies appear on pp. 116–18.

Although a start has been made, we are still very far from being able to present a complete genome picture for potato species, and any attempts that have been made in the past suffer from being over-simplifications. A tentative series of genome symbols was set out by Hawkes (1958, p. 29) in which Series Tuberosa, Commersoniana, and the Mexican species *S. verrucosum* were assigned the symbol A_1, whilst the tetraploid Series Acaulia was given the symbols A_2A_3. Evidently the types A_1, A_2, and A_3 differ from each other to an almost equal degree. In the future it will certainly be necessary to divide A_1 into a number of other symbols when the results of further work along the lines mentioned above become available. What seems abundantly clear at present is that our knowledge of the process of genome differentiation in the South American species is in its infancy, and much more work must be done before we can obtain a really clear picture of evolutionary trends in this group.

(*f*) *Affinities of series and species*

In this section we shall attempt to summarize the main observations and conclusions on variation and relationship in the species and series that are treated in more detail in the second part of this book.

Wild potatoes are known to occur over a very large part of the American continent, from the United States southwards through Central America into South America, as far as latitude 42° S. in Argentina and Chile. Their wide range of morphological and physiological variability has induced monographers to subdivide them into numerous groups or series, reflecting the assumed phylogeny to some extent, on the one hand, and reflecting a purely utilitarian need for a workable classification on the other. In a complex group of plants such as that with which we are dealing, where the affinities are not always very clearly understood, the views of different authors as to how the groups should be defined do not always coincide. We have followed the rather conservative classification of Hawkes (1963), by grouping the species studied here into a small number of series of moderately great amplitude.

1. ETUBEROSA (p. 131)

This is one of the two series of non-tuber-bearing species that are grouped with the tuber-bearers in subsection Hyperbasarthrum because of their general morphological similarity. Series Etuberosa is mainly confined to Chile, and only one species, *S. brevidens*, penetrates into Argentina. The series is evidently deeply separated from the tuberiferous species since no hybrids have been formed between them, despite numerous attempts. The lack of stolons and tubers seems to be evidence of an ancient separation from the ancestral stocks of the true potatoes, which were in any case probably formed in North America in the first place.

The species of this series included in the present treatment, *S. brevidens* (p. 132), has been hybridized with the type species, *S. etuberosum* (Hawkes, unpublished) and they are sufficiently close genetically for no genetic breakdown to occur in the F_2 generation. The Argentine forms of *S. brevidens* have been collected only in provs. Neuquén and Río Negro, in the humid *Nothofagus* forest zone and possibly elsewhere. The material varies somewhat in leaflet width and degree of pubescence, from densely to sparsely pubescent. The Argentine material that we have cultivated does not differ essentially from the Chilean collections grown by us. A diploid chromosome number of $2n = 24$ has been recorded by various authors, including ourselves, on material from both countries. It is of interest to note that the species is completely self-compatible, so far as our present observations indicate.

2. COMMERSONIANA (p. 137)

In this series we have included all the South American species with stellate corollas and round to ovate berries, except for the Bolivian series Circaeifolia and some forms of the widespread mountain series Conicibaccata, though in any case these two latter series are outside the scope of the present work.

In the wide sense in which we have defined series Commersoniana we include the two widespread species, *S. chacoense* and *S. commersonii*, each with two geographical subspecies, and the much more restricted species, *S. tarijense* to the north-west and *S. calvescens* to the north-east. There seems little doubt that *S. chacoense* and *S. tarijense* are more closely related to series Tuberosa and other South American series than are the species *S. commersonii* and *S. calvescens*, since hybrids can be formed with ease between the two former species and other series, and the serological spectrum is basically similar (see p. 138). In contrast, hybrids between *S. commersonii* and species from other South American series are more difficult to make. Even hybrids between *S. commersonii* and *S. chacoense* are somewhat difficult, so that it may be necessary to subdivide the series in the future, as Bukasov has done. *S. commersonii* possesses a distinct serological spectrum, which separates it from *S. chacoense*, *S. tarijense*, and all other South American species. Not enough is known of *S. calvescens* in the living state to make a division of the series advisable at the present time.

S. calvescens (p. 138) is a little-known species from Minas Gerais State in eastern Brazil, separated geographically from the nearest related species by some 400 km. So far, it has been collected only in the vicinity of the town of Caldas, and at Alfenas, some 100 km distant (see Appendix III), though it is possibly more widespread. Both collections at present known in the living state are triploid, but diploids almost certainly occur. It seems morphologically closest to *S. commersonii* subsp. *malmeanum*, but possesses curious stamens similar to those of *S. maglia* in series Tuberosa, in which the anthers and filaments grade into each other. Some features of *S. chacoense* are also possessed by *S. calvescens*, and in view of the combination of characters that distinguish it from other species in Commersoniana we think that this species is a valid one, though its affinities are still rather imperfectly known.

S. commersonii (p. 142) is a very widespread species, which seems most closely allied to *S. calvescens* but nevertheless has several features in common with *S. chacoense*. It occurs as two geographical subspecies, *commersonii* and *malmeanum*.

Subsp. *commersonii* occurs chiefly in the region of the Río de La Plata estuary of Argentina and Uruguay, spreading northwards along the coast into south-east Brazil. For some as yet unknown reason it has rarely been seen more than 300 km from the coast and is most abundant in a fairly narrow coastal and estuarine strip. It is frequently to be seen in marshes and dune slacks, though it also spreads into a wide range of habitats along river banks and in a large variety of grassy, wayside, and waste places, even at times amongst bushes and in forests. It seems to occur naturally also in thickets of *Colletia paradoxa*. Its frost resistance enables it to grow and flower in the cooler winter season as well as in the summer. Subsp. *commersonii* has a wide morphological range, both genetically and environmentally determined. It has been known since 1767, when it was collected for the first time near Montevideo by Commerson. Altogether, it has received nine specific epithets in addition to the original one, as well as various infraspecific names. We have been able to show that the variation of the subspecies as a whole is continuous, encompassing the type material of these 'species' and thus rendering them taxonomically invalid. Continuous variation is seen in habit, from the short rosettes of open ground to the long trailing stems of plants growing amongst tall grass and other vegetation. Even so, genotypically controlled differences in stem length also exist. Pubescence varies from biotype to biotype; some are genetically glabrescent, and there is a range right through to plants that possess quite dense pubescence. Genotypic differences are again modified by environmental influences, the drier habitats provoking a more pubescent and the wetter a less pubescent expression of the genotype. Leaf dissection also, varies from simple to 4- to 5-paired laterals, though 2–4 pairs are by far the commonest. Interjected leaflets vary also, from 0 to *very rarely* 2 pairs. The peduncle forks are typically rather short, though they may be lengthened in certain forms. The pedicel articulation fluctuates at about the centre, whilst the calyx lobes vary from short triangular to ovate or ovate-lanceolate. There is also some variation in the intensity of corolla pigmentation. All this variation is undoubtedly due to the dual effect of changing environment and genotype.

Subsp. *commersonii* has received some attention from the cytologists but most of the investigations have been carried out on the triploid forms, to be mentioned later. Meiosis is regular in diploid forms and also in hybrids of these with other species outside the series. On the whole, both subspecies of *S. commersonii* are difficult to hybridize with other species and the hybrids are often weak and unthrifty, either in the F_1 or the F_2 generations. Thus the F_2 generation of *S. chacoense* × *S. commersonii* was weak and unthrifty, but an F_2 of *S. chacoense* × subspecies *malmeanum* could not be obtained as only one F_1 seed germinated.

Natural hybrids between the two sympatric species *S. chacoense* and *S. commersonii*, which are also found in roughly the same habitat range, are rare, and we have seen only two herbarium collections and one living one of probable hybrids. Interestingly enough, the putative hybrid studied in the living state possesses a serological spectrum identical with that of *S. chacoense* and is a triploid, though in its morphology it is obviously intermediate between the two species.

Subsp. *malmeanum* is morphologically rather more similar to *S. chacoense* and *S. calvescens* than the typical subspecies, and its distribution area lies more to the interior. It has been recorded in Argentina, Paraguay, and southern Brazil, as well as in the north-western regions of Uruguay. On the whole its habitat range is rather similar to that of the typical subspecies, though it is not, of course, to be found in the dune slacks and estuarine marshes, and it also tends to have a greater preference for shade. It, also, is frost resistant, and it is often to be found in the mixed forest and savanna areas as well as in the damper parts of the Chaco.

Subsp. *malmeanum* was placed by Bitter as a distinct species, and has also been confused with *S. chacoense*, with which species it is to be found growing in association. A frequency histogram of leaflet indices (breadth/length × 100) for all specimens examined by us showed a more or less normal curve, with the known triploid forms grouped toward one end and including the triploid 'species' *S. millanii* of Bukasov and Lechnovicz. The rosette form of the species is shown in plants growing in the open but not in those growing in the shade of trees and bushes. The lateral leaflet number is on average rather higher than that of subsp. *commersonii*, 4–6 pairs being most frequent. Interjected leaflets are more common than in the typical subspecies, also. The typical subsessile to shortly petiolulate leaflets distinguish subsp. *malmeanum* from subsp. *commersonii* (which generally has sessile leaflets), as well as the consistently white flowers. The pubescence range is similar in the two subspecies, as well as the peduncle and pedicel form, calyx lobes and acumen shape, and the stellate corolla. The peduncle branches are generally not so contracted as in the typical subspecies.

Cytologically, subsp. *malmeanum* has not been so well examined as subsp. *commersonii*, but diploid and triploid clones seem to occur without any discernible geographical pattern, just as in the typical subspecies.

More information is required on the possibility of hybridization between subsp. *malmeanum* and *S. chacoense* in Paraguay and eastern Argentina (see also under *S. chacoense*, p. 218), and also the pattern of variation of the two subspecies of *S. commersonii* in the boundary areas where they merge.

Anomalous triploid forms. In the course of this study we have come across certain anomalous variants of *S. commersonii* that are not placeable with certainty in either subspecies. Some of them come closer to subsp. *commersonii*, but differ in the longer stem, well-developed petiolules, and very short peduncle branches. Some of these have been examined cytologically and are found to be triploids. Possibly the others may have been triploids also, but living material was not available of most of them. These forms are very similar to, if not identical with, triploid material known in plant-breeding stations, which we have called 'conventional' *S. commersonii* and which Bukasov has regarded as *S. commersonii sensu stricto*. They may have been grown in Europe since the introduction made from Uruguay by Saint-Quentin in 1896. Even Commerson's original collections seem to have been mixed, and at least one specimen seems to be morphologically similar to these 'conventional' forms.

Other anomalous forms are not so close to subsp. *commersonii*, and some of them approach subsp. *malmeanum* in a number of features, especially the gatherings from Montevideo by Felippone. Some may perhaps have been escapes from the Botanic Gardens at the Prado Park of introduced material or plants of hybrid origin, since subsp. *malmeanum* does not occur naturally in the region of Montevideo.

Cytologically the triploid forms so far investigated act as though they were autotriploids of diploid subsp. *commersonii*, though none of the *malmeanum*-like material has yet been thoroughly investigated. Therefore, although these triploid forms might conceivably be hybrids it seems more likely that they are autotriploids, derived directly from each of the two subspecies by fusion of a reduced and a non-reduced gamete. Why this should take place with such frequency and what the evolutionary or adaptive significance is of this phenomenon are questions that cannot yet be answered. Clearly, the matter would well repay further investigation.

S. chacoense (p. 180). This is undoubtedly by far the most interesting species in the whole area. It occurs in Argentina, Paraguay, Uruguay, and eastern Brazil, as well as spreading northwards into central Bolivia. Thus its north–south range is about 1800 km and its east–west range about 1700–1800 km. It is also the commonest, most aggressive, and most adaptable species, with an immense range of morphological and physiological variation. Such variability has induced some authors to subdivide it into a number of distinct species, varieties, and forms. It has thus received no less than twenty-two specific epithets, as well as a large number of infraspecific names.

On the basis of our comprehensive studies of this species we have been able to fit the variation into two geographical subspecies, *chacoense*, which is the more widespread of the two, and *muelleri*, with much longer and narrower leaflets, confined to the forests and savannas of north-east Argentina and south Brazil.

Subsp. *chacoense* possesses a wide altitude range, from sea level to 2350 m or more, though it is primarily a species of the plains and low foothills (to 1000 m). It has evidently acquired a secondary adaptation to higher altitudes by means of introgression, as we shall explain later. It occurs in a very wide range of habitats, often as a weed of fields, forest borders, waysides, river banks, waste places, etc., both in shade and sun. It spreads into seven different phytogeographical provinces, though it appears to be native perhaps only to the eastern and western borders of the Chaco, Espinal, and portions of the eastern and western subtropical forests. Elsewhere, it grows as an adventive in subclimax vegetation and perhaps was once much more limited than it is today.

In the course of our study of some hundreds of collections of *S. chacoense*, both as living and dried material, and of a comparison of these with the various microspecies into which *S. chacoense* had been divided, we were unable to discover any reliable and constant discontinuities between these collections; the microspecies seemed merely to represent various character combinations in a highly variable species of great complexity. In spite of this, however, there seemed to be a certain amount of correlation between certain characters and mountain areas, chiefly concerned with a shortening of petiolules, lack of basal leaf rosetting, greater pubescence, wider stem wings, higher pedicel articulation, and longer calyx acumens. We scored as many specimens as possible for these characters, as well as a number of others, arranging the collections geographically and by altitude. From this it became quite clear that *S. chacoense* is fairly uniform from 0 to 450 m, but after that begins to show variation, particularly over the 1000 m level, in a constant direction. In the north-west provinces we were able to show that *S. chacoense* is varying towards the mountain species, *S. microdontum*, with which it undoubtedly forms hybrids. Hybrid index figures underline this, showing very high values for the provinces of Jujuy (77 per cent) and Salta (68 per cent). A flow or introgression of *S. microdontum* genes seems to be taking place, that has enabled *S. chacoense* to form a whole series of well-adapted mountain biotypes in areas where it was probably originally not able to grow. There is also some evidence for flow of *S. microdontum* genes down into the plains in Santiago del Estero province. Some anomalies in the hybrid index figures for Paraguay suggest, also, that there may have been hybridization there with *S. commersonii*, subsp. *malmeanum*, though more evidence is needed before we can be certain of this.

Comparatively little is known about subsp. *muelleri*, though it has been in cultivation once or twice. We have felt that it should be placed as a subspecies of *S. chacoense* because F_2 hybrids between the two taxa are vigorous and fertile, because the two taxa replace each other geographically, and because the differences between them involve only the dimensions of the leaflets (those of subsp. *muelleri* being longer and narrower) and the constant lack of a rosette in subsp. *muelleri*. Some forms of this subspecies are known with rather broader leaflets and one, from Santa Catarina State, Brazil, with long, spreading hairs (*f. densipilosum*).

Our studies of living material of *S. chacoense* (for the most part of subsp. *chacoense*) in the experimental field have confirmed that the differences between mountain and plains forms are genetically determined. We raised progenies of plants from each of the two areas and were able to show that the plains seedling plants possessed long petiolules, narrow stem wings, absence of pubescence, and very short calyx acumen. In contrast, the mountain plants were more variable and with shorter petiolule, wider wing, wider range of pubescence, and longer calyx acumens.

Basal rosettes appear in culture on all plains plants of subsp. *chacoense*, but not on the mountain forms, nor in subsp. *muelleri*.

Cytologically, *S. chacoense* is almost entirely diploid, only one triploid clone ever having been reported. It shows a good capacity to hybridize with most other diploid species, except *S. commersonii*, and the hybrids are fertile, with regular bivalent pairing. Triploid hybrids of *S. chacoense* crossed with tetraploid species behave as autotriploids. Details of the crosses made by ourselves and by other authors, both F_1 and F_2, appear in the chapter on *S. chacoense* and need not be repeated here. Natural hybrids are also common, as would be expected, with all species that occur in the same area as *S. chacoense*, except for *S. commersonii*. Those generally found are with *S. tarijense*, *S. microdontum* (both subspecies), *S. spegazzinii*, and *S. kurtzianum*. The problem of *S. setulosistylum* is an interesting one, since we consider that it is no more than a hybrid of *S. chacoense* with *S. spegazzinii*; the original collections were made at Pampa Grande in Salta province, where two specimens of *S. chacoense* and one of *S. microdontum* have been collected with the curious style setae that Bitter used to characterize *S. setulosistylum*. Another species, *S. puberulofructum*, is also, in our view, a hybrid of the same type, but taken from an area in which the style seta gene or genes do not occur. The puberulent fruit is a curious feature, however, which needs re-examination. Altogether, *S. chacoense* is a species that is full of interest to the experimental taxonomist, as well as the plant-breeder, and there are many fascinating problems connected with it that should be investigated further.

S. tarijense (p. 239). This is mainly a species of the Bolivian Andes, which only just penetrates into Argentina in provs. Jujuy and Salta. In Argentina it occurs mainly between 2400 and 2600 m but can also be seen lower and higher than this altitude belt. It is a plant of rather dry regions, in the drier part of the 'pajonales' (Montano vegetational district) and also in the Prepuna province of Argentina and Bolivia.

The white stellate flowers, long calyx acumens, and heavy glistening glandular pubescence are particularly striking in cultivation, and even visible enough on dried material. They distinguish it at once from all other Commersoniana species. *S. tarijense* is entirely diploid so far as we know. Meiosis is fairly regular and hybrids with other diploids are easy to obtain, showing regular bivalent pairing also. We have made hybrids with *S. chacoense*, *S. vernei* subsp. *ballsii*, and *S. kurtzianum*. Natural hybrids seem to be formed with some frequency, and it would appear that the Bolivian species *S.* × *berthaultii* is a series of hybrid and introgressed forms of *S. tarijense* with an as yet unidentified blue-flowered mountain species. Many of the Argentine forms show evidence of introgression and others are obvious hybrids with a blue-flowered species, though we are not yet certain whether *S. vernei*, *S. vidaurrei*, or *S. gourlayi* is the blue-flowered parent. The area of Santa

Victoria in Salta province where these hybrids occur is a complex mosaic of climates with different species and hybrids, and should be investigated further. A hybrid of *S. tarijense* with *S. microdontum* has also been identified, as well as ones with *S. chacoense* and with *S. spegazzinii*. These latter two are from depts. Chicoana and La Viña respectively, also in prov. Salta, but considerably further south than *S. tarijense* had been recorded. Further search in those regions has revealed its presence in the Chicoana locality and may well do so in the La Viña one after further investigation.

3. ACAULIA (p. 247)

This series consists of one rather polymorphic species, *S. acaule*, which extends from northern Peru to north-west Argentina, as far south as prov. La Rioja, always at high altitudes. By some authors the single species is split into a number of distinct specific units, and certain hybridogenic species, such as the cultivated *S.×juzepczukii* and *S.×curtilobum*, are added. The natural affinities of the series probably lie somewhere in the region of the other high mountain series, Megistacroloba, and probably not very far removed from *S. megistacrolobum* itself.

S. acaule (p. 248) in Argentina occurs as two subspecies, *acaule* and *aemulans*, though the latter is accorded specific rank by some authors. The typical short stems and rosettes of leaves of *S. acaule*, together with the very short peduncle, high pedicel articulation that is never well marked and is sometimes completely lacking, and the small blue-violet (occasionally paler, almost white) corolla, with very short lobes are all features that serve to distinguish it from other species.

Subsp. *acaule* is by far the commoner of the two subspecies known in Argentina and is distinguished from subsp. *aemulans* by the virtual lack of leaflet decurrency, as well as the larger number of leaflet pairs. Its normal altitude range is 2800 to 4400 m, but it has been recorded as low as 2600 m and as high as 4650 m. Both subspecies seem to be highly frost resistant and occur in the wetter parts of the Puna phytogeographical province, the Montano district of the Western Subtropical province, and the lower wetter parts of the High Andean province. Both seem to be plants of high Andean grassland, stony places, waysides, cultivated fields, irrigation ditch sides, and waste places.

Variation in subsp. *acaule* is not very intense when grown under standard conditions, though there are genotypic differences in stem development, interjected leaflet number, intensity of pedicel articulation, and flower colour. None of these is exactly geographically correlated, except that there is a very rough indication of a better-marked pedicel and a longer stem in the Argentine forms as compared with the Bolivian and Peruvian ones. No matter how well marked the articulation, the fruits never drop off at maturity in either subspecies, as they do in other wild potatoes.

A somatic chromosome number of $2n = 48$, with four instead of the usual two satellite chromosomes, has been established for subsp. *acaule*. Meiotic analysis of hybrids has suggested that this species is a segmental allotetraploid whose two genomes are sufficiently well differentiated not to form multivalents in the tetraploid and to form no higher associations than quadrivalent in the artificial octoploid; but they will become associated in triploid hybrids (and probably would do so in polyhaploids, if these were yet formed) when their exact homologues are absent. *S. acaule* is fully self-fertile and autogamous, a condition that probably arose through lack of pollinating insects in the high bleak altitudes at which it normally grows. We have already indicated, however, that we have some doubts as to whether autogamy is obligatory under natural conditions, because of the apparent lack of the 'pure line' type of variation in this species.

In our crossing programme with *S. acaule* we began by crossing both subspecies together and found the progeny, both F_1 and F_2, to be fully fertile and vigorous. *S. acaule* forms hybrids with ease with diploid species from South America, but, strangely enough at first sight, not with tetraploids. We have crossed it with *S. sanctae-rosae* and *S. megistacrolobum* in the present programme. The resulting triploid hybrids are sterile or almost so, though colchicine hexaploids reported in the literature may often possess restored fertility. Hexaploid hybrids of reasonable fertility result often from *S. acaule* × tetraploid species crosses, due to the functioning of non-reduced egg cells from the former parent.

Spontaneous hybrids of subsp. *acaule* × wild diploid species are not very frequent. We have seen probable hybrids with *S. megistacrolobum* and with *S. spegazzinii*. The former were first reported by Brücher, and later named *S. bruecheri* in his honour. They are said to occur in high windswept sandy places where neither of the parents can grow.

Subsp. *aemulans* possesses a tendency to leaflet decurrency, larger terminal leaflets and fewer laterals than in the typical subspecies. The pedicel articulation is also much better marked. This subspecies is comparatively rare and has so far been seen only in the Tilcara region of prov. Jujuy and in Sierra Famatina in prov. La Rioja. This disjunct distribution is curious and needs further investigation. The known altitude range of this subspecies is less than that of subsp. *acaule*, only 2950 to 4000 m. The habitat range is also not as wide as that of subsp. *acaule*, but the apparently restricted altitude and habitat may possibly be due to the fact that not so much material of subsp. *aemulans* has been available for study. Only two living collections of the subspecies have been grown by us. There seem to be some genetically determined differences between them and there is also some variation in the seedling progenies of one of them. The longer stems, decurrent leaflets, and well-marked pedicel articulation are all fairly constant in cultivation.

Little cytological work has been carried out on subsp. *aemulans*. According to our counts it is tetraploid, though Brücher claims that it is diploid. The results of crossing with other species are essentially similar to those already reported for subsp. *acaule*. We have crossed it with *S. kurtzianum*, though so far no natural hybrids of this combination have been seen. We have, however, observed what appear to be natural hybrids of subsp. *aemulans* × *S. spegazzinii*, but have not been able to verify this yet by the study of living material.

The well-marked pedicel articulation of subsp. *aemulans* might well be indicative of its primitive nature, and it is significant that both this feature and the decurrent leaflets render subsp. *aemulans* vegetatively similar to *S. megistacrolobum*. The fewer, smaller laterals and larger terminal leaflet also increase this similarity. The difference in corolla form between them is of course well marked, but this might partly be explained by the difference of chromosome number, since petals as well as leaves seem to increase in width with increase in the level of ploidy.

If this hypothesis is correct, then we could envisage subsp. *aemulans* as being closest to the ancestral form of *S. acaule*, produced by natural chromosome doubling from *S. megistacrolobum* or an allied form, or as an amphiploid hybrid of *S. megistacrolobum* and some other, as yet unidentified, species. From this primitive *S. acaule* type would have been produced in succession subsp. *acaule* in Argentina, Bolivia, and Peru; then subsp. *punae* in central Peru, and finally the curious hexaploid subspecies *albicans* from north Peru.

4. CUNEOALATA (p. 271)

Series Cuneoalata is, like Acaulia, monotypic, with one widespread variable species, *S. infundibuliforme*. This species had previously been divided into several taxa on the basis of extreme habitat variants. The series is probably most closely related to series Megistacroloba and Tuberosa.

S. infundibuliforme (p. 271) is easily recognized by its pinnatisect leaves with narrow segments. Interjected segments are absent. It occurs in Bolivia, northern Chile, and north-western Argentina. In the latter country it is known in provs. Jujuy and Salta at high altitudes, commonly between 3400 and 3900 m. It grows in dry sandy and stony regions in open communities with cacti and other xerophytic vegetation of the Puna and occasionally the Prepuna phytogeographic provinces.

The width of leaf segment and rhachis wing seem to vary greatly with habitat, being very narrow in plants from dry regions. The pubescence is denser, also, in plants from drier habitats. There is also a considerable difference in these characters, except perhaps in pubescence, at different growth stages of the individual plant, and the upper leaves seem on the whole to have much narrower leaf segments and wings than the lower ones. The corolla exhibits a range of mainly genetically controlled differences, varying from rotate to semi-stellate, and the colour also varies from white, mauve and white, to uniformly mauve. Plants grown in the experimental plots and glasshouse vary according to conditions, as noted above for plants in the wild.

The chromosome number is always $2n = 24$. *S. infundibuliforme* can be crossed easily with diploid species such as *S. chacoense* and *S. megistacrolobum*, and various species in series Tuberosa. Meiotic irregularities occur in the *S. chacoense* hybrids, however.

Natural hybrids of *S. infundibuliforme* with *S. megistacrolobum* and with *S. gourlayi* have been identified.

It seems on the whole that *S. infundibuliforme* may be most closely related to species in series Megistacroloba and Tuberosa, especially to *S. spegazzinii*, *S. leptophyes*, and *S. vidaurrei*, judging by the morphological similarities. Its apparent drought resistance indicates that it might be useful initial material for breeding with that end in mind.

5. MEGISTACROLOBA (p. 283)

This series of generally high-altitude species is widely distributed from north Peru to north-west Argentina. The chief distinguishing feature is the pinnatisect leaf with very large terminal lobe and small laterals; these lateral lobes are often completely absent. The generally well-marked wide rhachis wings and semi-stellate corolla are also useful characters which help in the identification of plants in this series. Megistacroloba seems to be in some way ancestral to series Acaulia, though the details of this process are not yet known. In Argentina the series is represented by three species.

S. megistacrolobum (p. 284) is a widespread species with rounded obtuse leaf lobes and typical semi-stellate corolla. The peduncles are short and the pedicels long, with high articulation. It is distributed from south Peru through Bolivia into the provinces of Jujuy and Salta in north-west Argentina. Further south it is replaced by the vicarious species *S. sanctae-rosae*. The altitude range of *S. megistacrolobum* is commonly 3000 to 4100 m, and it is found in the wetter part of the Puna, occasionally in the upper Montano, and very rarely in the upper Prepuna vegetational provinces. It is a frequent plant of Andean meadows, fields, waysides, and waste places. In open habitats at high altitudes the stems are very short, giving it a very close vegetative similarity to *S. acaule* subsp. *aemulans*. Plants in shade or growing amongst tall vegetation, as well as those grown in Europe, produce longer stems and more exuberant growth. Variation in lateral leaf lobe number varies both through environmental and genetical differences. Flower colour varies genotypically, but is generally light blue-purple.

Very little cytological work has been carried out on *S. megistacrolobum*, apart from routine chromosome counts, which indicate a constant diploid number ($2n = 24$). Hybrids with *S. acaule*, *S. infundibuliforme*, and *S. raphanifolium* (a Peruvian species in the same series) have been formed, but little success so far has been obtained with *S. microdontum* and none with *S. gourlayi*. Natural hybrids with *S. acaule*, *S. infundibuliforme*, *S. boliviense*, *S. microdontum* subsp. intermediate, and either *S. gourlayi* or *S. oplocense* have been identified.

S. boliviense (p. 297) is a little-known Bolivian species, allied to *S. megistacrolobum*, which just penetrates into northern Argentina. It is distinguished by its simple (rarely lobed) leaves, stem bent from node to node and pentagonal to rotate deep purple corolla with well-defined lobes. Only three collections have been made in Argentina, all from prov. Salta, dept. Santa Victoria. It seems to grow at 2650 to 2900 m in the humid Montano district 'pajonales' of the Western Subtropical phytogeographical province in long grass and scrub. No Argentine material has been examined in the living state, but Bolivian material had been grown previously by one of us and had a chromosome number of $2n = 24$. No crossability data are available but probable hybrids with *S. megistacrolobum* have been examined and also forms of *S. microdontum* that seem to be contaminated with *S. boliviense* germ plasm. Further investigations are needed in the area of these probably introgressed forms.

S. sanctae-rosae (p. 301) is endemic to Argentina and is known only from the provinces of Tucumán and Catamarca, with the exception of one record from south Salta province, near the Tucumán border. It occurs from 2500 to 3800 m in the moderate to high rainfall zones of the Puna province and the upper part of the Montano district ('pajonales') of the Western Subtropical phytogeographical province. It is chiefly a plant of open ground, sandy and stony places, field borders, enclosures, grassland, etc., and occasionally grows in the shade of low shrubs. *S. sanctae-rosae* was probably derived from *S. megistacrolobum* and can be considered as a southern vicariad of it. *S. sanctae-rosae* is distinguished by the often rhomboid acute terminal leaf lobe and invariably acuminate laterals, which in *S. megistacrolobum* are obtuse to acute.

Variation in *S. sanctae-rosae* is considerable. What may be considered as the typical form occurs mostly from 2800 m upwards, growing in open situations or on very poor soil, with much exposure. It has a short stem and narrow leaf lobes, the laterals being especially narrow. Another variant is characterized by simple or sub-simple leaves with the lateral lobes almost entirely absent and the terminal rather longer and broader than that of the first variant mentioned. A range of intermediates occurs, and although the differences may be partly due to environment the general control of the degree of leaf dissection is probably largely genotypic, depending on several genes. There is some evidence to indicate that forms with simple leaves may be derived by a process of introgression with the almost simple leaved *S. venturii*, since hybrids between these two species have been recorded with some frequency and generally in the altitude zone where the simple-leaved forms of *S. sanctae-rosae* occur. A third type of variation is evidently due to no more than shade or rich soil, acting on the genotypes mentioned above. Under these conditions the plants of either leaf type produce long, often branched stems and large leaves with broader lobes. This variant is not found higher than 3500 m whilst types 1 and 2 ascend to 3800 m. Type 3 with luxuriant growth but simple to sub-simple leaves is restricted to the zone of 2550 to 2900 m, and some of these collections are difficult to separate from the hybrid of

S. sanctae-rosae × *S. venturii*. Several collections show a gradated range from type 1 to 3, and we have merely picked out the extreme variants as a means of scoring the collections from different regions.

Plants cultivated in Europe showed the more luxuriant growth form of the third or shade form mentioned above. This was probably largely due to the richer soil in which they were grown. The semi-stellate blue-purple corolla with broadly triangular lobes and acuminate lateral leaf lobes remained constant, however.

Root-tip examinations showed a somatic chromosome number of $2n = 24$, but little meiotic chromosome study has been carried out. *S. sanctae-rosae* has been crossed with *S. acaule*, *S. toralapanum* (a Bolivian species related to *S. megistacrolobum*), *S. microdontum*, and *S. vernei* subsp. *ballsii*. Natural hybrids of *S. sanctae-rosae* × *S. venturii* have been recorded with some frequency, especially in the zone of overlap between the two species (2500–2800 m). Curiously, hybrids have also been recorded up to 3200 m in a zone in which *S. venturii* has not yet been recorded. This might be due to upward migration of hybrid forms, or gene introgression from a lower to a higher zone, or even to the disappearance of *S. venturii* by hybridization or other means from a zone where it once occurred. Detailed collections are needed from such regions.

Hybrids of *S. sanctae-rosae* with *S. spegazzinii* have also been recorded, chiefly from the rather drier zones of Catamarca province and within the altitude range of 2500 to 3100 m. A progeny grown in Europe from seed collected from a population of hybrid plants showed some variation but was on the whole closer to *S. spegazzinii*. The plants also exhibited aborted anthers and in some cases no stamens whatsoever. Mass collections are needed from several areas where hybrids have been recorded to clarify the situation and to arrive at a better understanding of the processes at work.

6. TUBEROSA (p. 313)

Series Tuberosa contains the rest of the species dealt with in this book. It is distinguished by a well-developed bifurcate peduncle, rotate to semi-stellate corolla and more or less globular berry. The series has been treated in various ways and has by some authors been, in our opinion, excessively subdivided. We have retained a rather broad concept, and although we recognize various circles of affinity within it we have not considered subdivisions advisable either from a theoretical or a practical viewpoint.

We first describe the two closely related species *S. microdontum* and *S. venturii*, and the not very far removed *S. vernei*. These three species are found chiefly in the humid mountain forests and link phylogenetically to similar species further north. The other wild Tuberosa species occur chiefly in the drier and/or colder regions of north-west and western Argentina; these are: *S. gourlayi*, *S. oplocense*, *S. vidaurrei*, *S. spegazzinii*, *S. kurtzianum*, and *S. maglia*. The first three species, *S. gourlayi*, *S. oplocense*, and *S. vidaurrei*, are not well known, and the latter two of these are probably more common in Bolivia than in Argentina. The more widespread species *S. spegazzinii* and *S. kurtzianum* are found in the drier valleys and mountain slopes of the western regions and may possess some drought resistance. *S. maglia* is very poorly known in Argentina and appears to be rare, though the areas where it might be found are not well studied.

Three cultivated species in the same series are known in Argentina, namely, *S.×juzepczukii*, *S. tuberosum*, and *S.×curtilobum*. Escaped forms of *S. tuberosum* are to be found in the south of Argentina also.

S. microdontum (p. 316) has proved a particularly difficult species to resolve, and it was only by means of repeated collections from certain key areas and the cultivation of much material in Europe that we were able to solve the taxonomic problems involved, especially those concerned with its delimitation from the closely allied species, *S. venturii*. *S. microdontum* is characterized by the rather large simple leaves or leaves with a few pairs of lateral leaflets that are much smaller than the terminals, coarse shining pubescence and white pentagonal to semi-stellate corolla. The unequal lengths of the calyx acumens are also worthy of note.

We have been able to differentiate two geographical subspecies, the northern one, subsp. *microdontum*, in provs. Jujuy and Salta (northern part), extending also into Bolivia, where it is known under the names of *S. cevallos-tovari* and *S. higueranum*; and the southern one, subsp. *gigantophyllum*, which is abundant in southern Salta and Tucumán provinces, but also extends northwards into south Bolivia and south-westwards into Catamarca and La Rioja provinces.

Subsp. *microdontum* is more slender than subsp. *gigantophyllum* with narrow stem and stem wings, the latter being generally straight-edged, and smaller leaves with terminal leaflets of 8×5 cm or less; subsp. *gigantophyllum* is more robust, with thick stems and wide wings that are generally marginally undulate; the terminal leaflets vary from 8 to 18 cm long×4 to 9 cm wide. The flowers of both subspecies are more or less identical. The altitude range of subsp. *microdontum* is mainly 1800–2700 m and that of subsp. *gigantophyllum* 1200–2500 m, but collections of both have been found rather lower and rather higher than these limits. Both are plants of the rain and cloud forest of the Western Subtropical province, growing often in dense shade but also in more open habitats in clearings and along paths, also in the upper more open communities of grass and shrubs above the dense forest zones. Both are found very rarely in dry open places and then show atypical stunted growth.

S. microdontum was known for many years as a single collection from Toldos, prov. Salta (formerly in Bolivia), and another species from the same region, *S. bijugum*, was considered to be distinct. Mass collections by Hjerting and Rahn demonstrated that *S. microdontum* and *S. bijugum* were no more than two biotypes, with a different degree of leaf dissection, of continuously varying populations, and that the two species could not be maintained.

Up to 1960 subsp. *gigantophyllum* material was known under the name *S. simplicifolium*, and occasionally as *S. gigantophyllum*, the latter name being used for plants with the largest leaf dimensions. It was later given subspecific rank under *S. simplicifolium*, but no real subspecific differences exist between the two taxa, since when the whole range of variation is viewed together there are numerous linking forms and no discontinuity whatsoever, with regard to leaf dissection and size, which were the only characters that had been used to separate them. Similarly, four varieties of *S. simplicifolium* could not be maintained for the same reasons. Until 1956 we had thought that *S. simplicifolium* was a good species, closely related in some way to the northern *S. microdontum* and to the southern *S. venturii*. These last two species looked superficially similar, but when we grew material of northern *S. microdontum* it became clear that it was quite distinct from *S. venturii* but that it was so similar to *S. simplicifolium* that the two must be regarded as the same species. Only stem diameter, stem winging, and leaf size differed, whilst the floral characters were very close indeed. Since *S. microdontum* was the earlier name the other must be suppressed, though there were certainly differences of a subspecific magnitude between the two taxa. Since the name *gigantophyllum* existed at the subspecific level this had to be used in place of *S. simplicifolium*, which thus became invalid at any level.

This example illustrates very clearly the need for a comparison of living material in the tuber-bearing Solanums, since the original *S. microdontum* material had been collected from dense shade, which increased the apparent distinctness between it and *S. simplicifolium*. The distinct pubescence was less marked and the inflorescences were smaller, with fewer flowers, but when collections were grown in less-shaded conditions the true affinities were revealed. The typical subspecies varies in the degree of leaf dissection from 0 to 3-jugate, though plants with 0 to 1 pair of laterals are most common, as was shown in a population sample taken at Toldos. Evidently there are slight differences of genotype which, in combination with environmental differences, bring about the range of leaf dissection observed. A certain population in Salta province, dept. Santa Victoria, gives evidence of introgression of genes from *S. boliviense*, and differs from all other forms of subsp. *microdontum* yet seen.

Certain forms from provs. Jujuy and Salta seem intermediate between the two subspecies and retained this difference when grown in Europe; hybrids between them did not show any marked variation from the parental types, and it is therefore probable that these are true intermediates between the subspecies and not natural hybrids, as was once thought. One intermediate gave a rather wide segregation to both subspecies, but there is evidence that the situation was a more complex one, involving past hybridization with another species, *S. vernei*.

Subsp. *gigantophyllum* also shows a range of leaf dissection from simple to 3-jugate. On the whole, a greater proportion of plants may be seen further south with well-dissected leaves, and they are very abundant around Tucumán (72 per cent), less so further north at Pampa Grande (47 per cent), and still less near Salta (29 per cent) and Jujuy (33 per cent). We have no suggestion to explain this clinal variation in leaf dissection, unless the more dissected forms are due to introgression from *S. chacoense*. On the whole this does not seem likely, in view of the known range of hybrids between the two species. The growth and general habit of subsp. *gigantophyllum* under cultivation in Europe is very similar to that in Argentina. Stem and wing pigmentation is present in some samples, which might be due to introgression from *S. vernei*.

All lines so far investigated cytologically show a diploid chromosome number and regular meiosis. *S. microdontum* crosses with a wide range of species (mostly subsp. *gigantophyllum* has been used for these crosses) in series Commersoniana, Acaulia, and Tuberosa. We have made crosses with *S. chacoense*, *S. sanctae-rosae*, *S. vernei* (both subspecies), and *S. maglia*, as well as crosses within and between the subspecies of *S. microdontum* itself. Curiously, the F_2 between *S. microdontum* and *S. vernei* shows no genetical breakdown, thus indicating that the two species are very closely related, despite their morphological differences. The crosses with *S. maglia* demonstrate that the peculiar form of the stamens in this latter species is dominant.

Natural hybrids with *S. chacoense*, *S. tarijense*, and *S. megistacrolobum*, and introgressed forms probably influenced by *S. boliviense* have already been noted. Natural hybrids with *S. vernei* are not very frequent (ten recorded so far), even though the two species overlap altitudinally. However, the latter species seems to be adapted to open places amongst shrubs and long grass, and is never found in dense forest. It is of interest that six out of the nine hybrids recorded with altitude data seen between these two species are from the zone 2700–3200 m where the forests give way to shrubs and grassland and *S. microdontum* is thus forced into contact more frequently with *S. vernei*. Probably many hybrids and introgressed populations of the two species exist in this zone, which would repay further study.

A hybrid of *S. microdontum* subsp. *gigantophyllum* × *S. kurtzianum* has been collected several times in prov. La Rioja, Sierra Famatina, which is apparently the only area where the two species come into contact. Since one is adapted to conditions of high humidity and the other to dry Monte vegetation the fact that they coincide at all in their distribution is of great interest. A hybrid of subsp. *gigantophyllum* with either *S. vernei* or *S. spegazzinii* has also been recorded.

Altogether, *S. microdontum* is a species of very great interest, and there are many problems in connexion with it that would repay further investigation.

S. venturii (p. 349) is a little-known species related to *S. microdontum*, of very restricted distribution. Until 1966 we thought that it was endemic to the Aconquija mountains of provs. Tucumán and Catamarca, Argentina, but in our recent expedition we found it also in prov. Salta and there is one record even from prov. Jujuy. It differs from *S. microdontum* in the weak decumbent habit, thin unwinged stem, sparse pubescence of triangular 3-celled hairs on all green parts, shorter peduncle below the inflorescence, and other characters. It occurs only in the very humid grassland ('pajonales') from 2000 to 2800 m of the upper part of the Montano district of the Western Subtropical province, and virtually nowhere else. We thought at one time that *S. venturii* was barely distinct from *S. microdontum* subsp. *microdontum*, but there is no doubt now that it is a good, well-defined species.

Material has been cultivated in Argentina, Germany, and possibly elsewhere, and we have grown a hybrid of *S. venturii* × *S. sanctae-rosae*. Information from these sources confirms its distinctness from *S. microdontum* but emphasizes that the two species are certainly closely related. The chromosome number is $2n = 24$ (36). Little information is available on artificial hybrids, but natural ones have been recorded with *S. sanctae-rosae*, and also probably with *S. vernei*.

S. vernei (p. 356) is a distinctive species with dark purple stems and large dark purple flowers. It occurs as two geographical subspecies, one, the typical, in Salta, Tucumán, and Catamarca, and the other, subsp. *ballsii*, in Jujuy and north-west Salta. They differ in the presence or absence of petiolules and the form of the leaflet base and apex. Although *S. vernei* has not yet been recorded for Bolivia it very probably occurs in that country also.

The altitude range of the species as a whole is 1950 to 3450 m, but subsp. *vernei* is mostly found at 2200 to 2800 m, whilst subsp. *ballsii* occurs at 2700 to 3450 m. Both subspecies grow in the cloud forest region of the Montano district, generally in open habitats on valley bottoms amongst stones and bushes or in old fields. They are often found in partial shade but not in dense forest, as is *S. microdontum*. We do not agree with Brücher's hypothesis that *S. vernei* is the ancestral diploid species from which the cultivated potatoes were derived, for various reasons connected with their morphology, ecological adaptation, and distribution.

Material of both subspecies cultivated by us showed some variation in leaf dissection; on the whole the leaf of subsp. *ballsii* seemed to be better dissected and the flower seemed smaller than that of subsp. *vernei*. The deep purple anthocyanin pigmentation over the whole plant was very characteristic, and also the large tubers.

Various counts have shown that this species is a diploid. Artificial hybrids of *S. vernei* with several species in series Commersoniana, Demissa, Megistacroloba, and Tuberosa are easy to make, and quite fertile. In our own hybrids we observed no F_2 breakdown in the combination of *S. vernei* × *S. microdontum*, indicating that the two species are rather closely related. On the other hand, sterility was observed in hybrids with *S. kurtzianum*, and the seeds of crosses with *S. oplocense* were inviable.

Natural hybrids with *S. chacoense*, *S. tarijense*, *S. microdontum*, and possibly *S. venturii* have been recorded, and also several occurrences of

S. vernei × *S. spegazzinii*. A seedling family of one of these was raised by us and showed typical F_2 segregation.

S. gourlayi (p. 371) is one of the little-known high mountain species from provs. Jujuy and Salta, which probably occurs with frequency also in south Bolivia, though so far only one collection has been identified from that country by ourselves. *S. gourlayi* was placed by Correll as a form of the Bolivian species, *S. leptophyes*, and it is certainly very closely related to it on morphological grounds. We have very meagre evidence so far, however, about these two species, and hence we think it better to retain *S. gourlayi* as a distinct species until the Bolivian species are better studied. *S. gourlayi* is also probably related to *S. spegazzinii* from further south. The 3–4-jugate leaf with 0–2 pairs of interjected leaflets, as well as the purple pentagonal corolla, serve to distinguish *S. gourlayi* from other related species, as well as other characters. It grows at very high altitudes, mainly between 3300 and 4000 m, in humid regions of the upper Prepuna province and the Montano district, as well as in the Puna province. Exposed places amongst grass and small shrubs in sandy or stony soil are frequent habitats for *S. gourlayi*.

Some considerable difference in phenotype occurs in the different collections. The one line that we were able to study in the living state was morphologically very similar to the type but came from a small stunted plant very different from the type; we have therefore added a number of similar variants to *S. gourlayi* assuming that they were from drier habitats or poorer soil, whilst the phenotype of the type specimen was probably indicative of wetter conditions. Certain forms were so stunted, however, that we did not feel able to identify them beyond saying that they were 'probably *S. gourlayi*'. Wild potatoes growing under such extremely dry or inclement conditions are practically impossible to identify unless one can grow tubers of them in the experimental field subsequently.

The chromosome number of *S. gourlayi* is $2n = 24$, but no meiotic analysis has been carried out. Artificial hybrids of various types have been made (e.g. with *S. chacoense*, *S. infundibuliforme*, and *S. vernei* subsp. *ballsii*), but no success was obtained with *S. megistacrolobum*. Various natural hybrids have been recorded, some of which are probable, such as the hybrids with *S. vidaurrei* and *S. infundibuliforme*, whilst others are possible but not perhaps very probable. Undoubtedly, much remains to be learned about this little-known high-mountain species.

S. oplocense (p. 377) is known chiefly in southern Bolivia, and only three collections have been made in Argentina. These were from prov. Jujuy, not very far from the Bolivian border. *S. oplocense* is probably related to *S. gourlayi*, as well as to *S. spegazzinii* and *S. vidaurrei*. Its chief distinguishing features are the coarse pubescence, attenuate decurrent leaflets, often 4-lobed anther base, purple semi-stellate corolla, and very minute stigma. It grows at lower altitudes than *S. gourlayi*, from 2800 to 3500 m and in rather drier regions, chiefly in the upper Prepuna and the Puna phytogeographical provinces.

A wide and variable range of phenotype exists in this species. From an examination of South Bolivian specimens one gains the impression that there are two species, *S. oplocense* and *S. camarguense*. Material of these two phenotypes when grown in Europe under similar conditions all proves to be the same thing, with a phenotype rather similar to *S. camarguense*. Since *S. oplocense* was described first, this takes priority. The situation is further complicated by the fact that *S. oplocense* is found as two cytological races, diploid ($2n = 24$) and tetraploid ($2n = 48$). Diploids and tetraploids are known in Bolivia, but only tetraploids so far in Argentina. The diploid collections we have studied from Bolivia are rather more robust than the single tetraploid from Argentina; but tetraploid collections made by Professor Ross in Bolivia seem just as vigorous as the diploids. Possibly some or all of the tetraploids might be hybrids with other species, but at the moment we have no certain information on that point.

Attempts at crossing diploid *S. oplocense* with other species have so far resulted in failure, with no more than a few inviable seeds per berry. There seems, therefore, to be some sort of sterility barrier, but obviously the matter needs looking into further. A few hybrid plants of the cross *S. tuberosum* × tetraploid *S. oplocense* have been produced and seem fully intermediate between the parents. No authentic records of natural hybrids of *S. oplocense* with any other species have yet been obtained.

S. vidaurrei (p. 384) is another species that seems to be quite common in south Bolivia and only just penetrates into northern Argentina, where it is known from two localities in prov. Salta, dept. Santa Victoria. A single doubtful collection is known from prov. Jujuy, and several apparent intermediates further south in that province seem to connect it with *S. spegazzinii*. *S. vidaurrei* is no doubt related to *S. oplocense* and *S. gourlayi*, as well as to *S. leptophyes* and *S. spegazzinii*, but the precise details of this are obscure. Its distinctive features are the delicate habit, yellow-green colour, generally 3-jugate leaf with narrow lanceolate sessile leaflets and absence of interjected leaflets, as well as several other features. It occurs at approximately 2700 to 3500 m, and has thus virtually the same altitude range as *S. oplocense*. The two species have not been observed together, and *S. vidaurrei* seems to prefer wetter regions than *S. oplocense*. It has been recorded in Argentina in the pajonales of the Montano district of the Western Subtropical province, and in Bolivia in what are probably alpine meadow formations.

We have not been able to grow living material of this species, but Professor Ross reports that the yellow-green colour is constant when grown in Germany, as well as the other features we mention above. He also states that the species is diploid ($2n = 24$). No artificial hybrids are known to us, though possibly some may have been made elsewhere. The only reasonably authentic natural hybrid we have seen is *S. vidaurrei* × *S. gourlayi*, which we have already mentioned under the latter species.

S. spegazzinii (p. 387) is a species wholly confined to Argentina, but with close morphological connexions with *S. pachytrichum* and *S. leptophyes* in Bolivia. Further study of these two Bolivian species may make it necessary in the future to link all three together, perhaps as subspecies of a widely distributed species, but at the moment it seems better to keep them apart because the Bolivian species are not well enough known. *S. spegazzinii* is characterized by long elliptic leaflets, narrowing towards each end, the upper laterals tending towards decurrency except when the petiolules are well developed, though even then some decurrency can often be observed. The interjected leaflets are narrow, the pubescence is coarse, though not so markedly so as in *S. kurtzianum*, and the corolla is semi-stellate to pentagonal, and medium purple in colour.

The species is widely distributed in north-west Argentina, through Salta, Tucumán, Catamarca, La Rioja, and into San Juan. In Jujuy forms possibly linking it with *S. vidaurrei* have been recorded. The general altitude range is from 1900 to 3100 m, with occasional records down to 1500 m and up to 3700 m.

S. spegazzinii is a characteristic plant of dry places in the Prepuna and Monte provinces but it also penetrates into the lower parts of the Puna province and the wetter pajonales of the Montano district. It thus ranges from quite dry to rather wet and cold regions, though it never extends into the mountain rain forests. It seems to possess a wide range of tolerance and acts as a ruderal and weed species as well as growing in open stony areas with thorn scrub and cacti, whilst at higher levels it often grows in mountain grassland.

S. spegazzinii is yet another species of wild potato that presents a very wide range of phenotype, due to the wide range of soil and rainfall conditions under which it will grow. This has given rise to past misunderstandings of the identity of certain material collected in Argentina, so resulting in the erection of two apparently distinct species, which are in fact no more than ecads, based on soil moisture and exposure differences. The resolution of the problem has been possible only by cultivation of material under standard conditions, taken from plants of very different original phenotype. It has also been possible to vary the growth conditions experimentally, so as to reproduce a part of the phenotypic range as seen in nature.

From this work we have been able to deduce that the dry habitat form of this species conforms to the material described originally as *S. spegazzinii*, whilst the wet habitat form conforms to the material of the later-described *S. famatinae*. These we have named the 'spegazzinii' type and 'famatinae' type, respectively. The collections known as *S.*×*setulosistylum*, although vegetatively similar to *S. spegazzinii*, are shown to possess a stellate corolla and to be hybrids of that species with *S. chacoense*. The setose style character is not species specific but is evidently due to a rare mutation, found only in one area, and then in several different species and hybrids.

The growth form of the 'famatinae' type from sheltered damp conditions is of large plants with large, well-dissected leaves, leaflets gradually attenuated into the well-developed petiolules but hardly decurrent, and numerous long oval interjected leaflets springing from both rhachis and petiolules. The other extreme of the range from open dry habitats, the 'spegazzinii' type, is of small plants with smaller leaves and fewer leaflets which are sessile, narrow, and generally decurrent; the interjected leaflets are few and may often be absent. Occasionally there is a range from 'famatinae' to 'spegazzinii' leaf type on a single plant, probably due to a change of soil moisture during the growing season. Without such intermediates, either on the one plant or within a single population, one would be justified in assuming that there were two distinct species involved.

We grew twelve different collections from distinct parts of the range of *S. spegazzinii*, and of which the original South American material was widely disparate. Under standard conditions in Europe the uniformity was high, except for some genotypically distinct collections from Salta. Accessions varied slightly in pubescence, leaf dissection, and leaflet width, but there was some intra-plant and intra-family variation, also. In the good soil of our experimental plots the leaflets were, on the whole, wider than in the Argentine plants. Glasshouse-grown material matched very well indeed the 'famatinae' type, and it seems evident that this type was taken from a moist situation, shaded by other vegetation or by rocks. Material grown in small pots with low moisture and soil fertility grew into phenotypes very similar to the 'spegazzinii' type, with sessile decurrent leaflets and few interjected leaflets. We have thus been able to match the 'famatinae' and 'spegazzinii' material from Argentina by appropriate régimes of soil and moisture, but we have not been able to reproduce the variants with widely decurrent leaflets, seen in both Salta and San Juan. The Salta material when grown in Europe lost completely the broad decurrency, even though on the original specimens it was so striking that we thought they were hybrids with *S. infundibuliforme* or *S. megistacrolobum*. Such are the pitfalls of *Solanum* taxonomy! Similar broad decurrency was seen on the San Juan collections, but we had no living material of these to test in Europe. Evidently certain unknown combinations of soil, light, and temperature cause such extreme variants.

There seems to be evidence of a cline in the paling of the corolla petal centres and acumens from north to south, and there are also genotypic differences in leaflet width and style length. In general, the collections made by Hjerting, Petersen, and Rahn from prov. Salta are rather distinct, with their long coarse pubescence and white central petal streak, from all other material we have seen of this species, and they may merit distinct taxonomic status after further study.

A chromosome number of $2n = 24$ has been verified from various counts. Hybrids within the species are uniform and fertile. Hybrids have also been made with *S. chacoense*, *S. infundibuliforme*, *S. vernei* (both subspp.), and attempted with *S. kurtzianum*, but only a few inviable seeds resulted from this last cross. Natural hybrids are fairly common, due possibly to the wide ecological range of *S. spegazzinii*, and we have recorded ones with *S. chacoense*, *S. tarijense*, *S. acaule* (both subspp.), *S. sanctae-rosae*, and *S. vernei*, and possible ones with *S. microdontum* subsp. *gigantophyllum* and *S. kurtzianum*. The rarity of hybrids with *S. kurtzianum* is perhaps strange, since the two species overlap geographically and altitudinally.[1] This may possibly be due to some kind of sterility barrier, judging from the results of the artificial crosses, but confirmatory evidence is needed. It is worthy of note that *S. spegazzinii* forms hybrids with even more species than does *S. chacoense*.

S. kurtzianum (p. 406) is another endemic Argentine species from the drier valleys and hillsides of the western and north-western regions. Its chief distinguishing features are the 3–5-jugate leaf with 0–2 pairs of interjected leaflets, conspicuous coarse marginal leaflet hairs, pedicels articulated near the base, and pentagonal to substellate corolla which frequently bears a rich magenta purple central petal streak on the external surface only. *S. kurtzianum* is possibly related to *S. chacoense*, *S. spegazzinii*, *S. oplocense*, and *S. maglia*. It has a rather more southerly distribution than *S. spegazzinii*, and occurs from Catamarca province, through La Rioja and San Juan to Mendoza. This species seems better able to withstand drought than any other species in Argentina, with the possible exception of *S. infundibuliforme*. Its altitude range is (1000–)1400 to 2500(–3000) m. It grows in dry soil on bushy slopes and amongst stones and rocks as well as in dry woods in Mendoza province. It is the only potato to be found chiefly in the Monte phytogeographical province, though it also occurs in others, such as the Prepuna.

Young plants frequently form semi-rosettes with short internodes, and these are seen on plants grown in the experimental field but not in the glasshouse. The juvenile foliage is marked by much wider leaflets than the adult foliage. The glandular pubescence on peduncles and pedicels was mentioned many years ago by Hooker for plants collected in Mendoza, though not of course identified as *S. kurtzianum* at that time. Glandular pubescence is not noted for the type specimen from La Rioja province, but on scoring our own material for the glandular as well as non-glandular pubescence we noted that glandular hairs were moderate to frequent in Mendoza, but moderate to absent further north in Catamarca. Coarse non-glandular hairs tended to be absent in some Mendoza material and very frequent on others. Some indication of a north–south cline for increase in glandular pubescence of the inflorescence seems therefore to be present. More material is needed from San Juan and La Rioja, as well as northern Mendoza, to fill in the details. Differences in corolla colour can also be seen, and the white-flowered forms which sometimes appear seem to be more frequent towards the north.

Material grown in Europe showed evidence of small genotypic differences in degree of leaf dissection, leaflet width, and other characters. Material in the field showed semi-rosette growth. The coarse marginal leaf hairs, often set on

[1] See Appendix III, however.

small teeth, were characteristic, as also the below-centre articulation and the showy corolla. The chromosome number of all material examined is $2n = 24$. Colchicine-tetraploids of the hybrid *S. kurtzianum* × *S. microdontum* subsp. *gigantophyllum* showed a very high frequency of bivalents at meiosis, indicating the marked degree of genome differentiation between these two species (Swaminathan 1956). Genetic analysis also bore this out, with most characters examined inherited disomically; some tetrasomic inheritance was reported, however, evidently due to genes borne on chromosomes which normally form quadrivalents at meiosis in these hybrids.

Hybrids of *S. kurtzianum* with many other species are reported in the literature, and we have raised hybrids with *S. chacoense*, *S. tarijense*, *S. acaule* subsp. *aemulans*, *S. vernei*, and *S. maglia*. An F_2 family was difficult to obtain from this latter cross, and when grown exhibited extreme genetic breakdown.

Natural hybrids with *S. chacoense* and with *S. microdontum* have already been noted. Three probable hybrids with *S. spegazzinii* have also been observed, but none so far with *S. maglia*.

S. maglia (p. 420) is very rare in Argentina, although quite common in some parts of Chile. It has so far been observed in Argentina in a single locality, namely, Quebrada de Alvarado in prov. Mendoza.

It is a curious species in many ways, easily distinguished from all other species but *S. calvescens* by the lack of a clear delimitation between anthers and filaments. The anthers are very loose, not tightly pressed together as in other species. The coarse hairs, set on teeth or cushions of tissue, separate it from many other species, but link it to *S. kurtzianum*.

In Argentina *S. maglia* has been collected at 1500 m, but in Chile it is known at sea level and somewhat above. It grows in Argentina in a region that is probably transitional between Monte and High Andean provinces, and amongst bushes of various types. Possibly further search would show a wider distribution area for *S. maglia*. We regard it as a relic of a southward migration of white-flowered forms which eventually crossed the Andes into central Chile.

The living material of Mendozan origin was grown by us and shown to have reddish-brown tubers which sprouted slowly. The fully developed leaves were similar to those of Chilean forms of this species grown by us, as also were other vegetative and floral characters. Until recently *S. maglia* had been assumed to be wholly triploid, but we have shown that this Argentine form, as well as a Chilean biotype, were diploid. Thus this species probably conforms to the diploid-triploid pattern of species already mentioned, such as *S. commersonii*. Not only the triploid, but also the diploid forms of *S. maglia* seem to be rather male sterile, with scanty pollen. Nevertheless, hybrids between diploid Argentine and Chilean collections have been made, giving a uniform and vigorous F_1. We have crossed *S. maglia* with *S. chacoense* and *S. microdontum*, as well as with *S. kurtzianum* and *S. medians* from Peru. No natural hybrids are reported, however.

Cultivated species (p. 428). We must finally say a few words about the indigenous cultivated potato species of Argentina, the majority of which have been brought into that country from Bolivia. Some forms, however, probably moved eastwards from Chile into southern Argentina. Most indigenous potato cultivation takes place in the 2500–3500 m zone, but the frost-resistant species, *S.* × *juzepczukii* extends up to 4000 m.

No cultivated diploid species have been recorded by us from Argentina, though these are mentioned by Brücher. The triploid *S.* × *juzepczukii* has been recorded several times. It is a natural hybrid of *S. acaule* × *S. stenotomum* and is highly frost-resistant. *S. stenotomum* is a primitive diploid species from Bolivia and Peru, and it seems rather unlikely that *S.* × *juzepczukii* could have been formed in Argentina, though it might have been cultivated there for a long time. The triploid species, *S.* × *chaucha*, has also been recorded once by us (see Appendix III).

The tetraploid species, *S. tuberosum*, occurs as two subspecies, of which the typical one is cultivated in temperate and subtropical regions of Argentina, Brazil, and Uruguay, up to altitudes of about 2000 m. We shall mention only the semi-indigenous cultivations of 'papas criollas' in the north-western provinces, whose origin is uncertain and which have been cultivated there for a long period of time. Certain escaped forms from provs. Chubut and Neuquén are also described. These we consider to be escapes or relics of cultivation left by semi-nomadic tribes. Certain authors regard such potatoes as native to the region of southern Argentina and southern Chile, and possibly ancestral to *S. tuberosum*, not derived from it. We consider this explanation to be most unlikely, since there are no diploid wild species in that region from which the tetraploids could have been derived. Other data also point towards the theory of these wild forms being relics of old cultivations, such as the gene for red pigmentation which is not found in any wild species but is not uncommon in cultivated forms.

Subsp. *andigena* is cultivated at high altitudes in provs. Jujuy and Salta, and has also been recorded here and there in certain other provinces. We have classified the varieties into cultivar groups and have applied a standard name to each. The few varieties from further south may perhaps be relics of the ancient migration route of peoples who took the potato from its centre of origin in Peru/Bolivia southwards to southern Chile.

The pentaploid species, *S.* × *curtilobum*, was derived from a natural cross between the triploid *S.* × *juzepczukii*, which furnished an unreduced gamete, and *S. tuberosum* subsp. *andigena*. It is less frost resistant than *S.* × *juzepczukii* and much less than *S. acaule*. *S.* × *curtilobum* has been recorded only twice in Argentina, and only in the far north near the Bolivian border. It is doubtful, therefore, whether it could be considered indigenous, and has probably been brought into the country in very recent times.

This brings us to the end of these notes on the wild and cultivated species of Argentina, Brazil, Paraguay, and Uruguay. We can see that the wild species especially are a highly diverse group, some with very wide distribution area and large variation range, and others restricted to one or a few valleys or mountains. Some are highly aggressive, with a wide range of tolerance, able to colonize new man-made habitats and to hybridize and exchange genes with other species, so evolving rapidly and extending their range of adaptation to an ever-increasing series of habitats. From these there is a continuous range to the species at the other end of the scale that possess only a very narrow tolerance range, and are perhaps highly restricted to a single habitat, such as *S. venturii* in the pajonales of the Aconquija and other ranges. Yet even this species is possibly not on the way to extinction, for it too can exchange genetic material with other species and so perhaps enlarge its range of biotypes little by little. Undoubtedly the group is in a state of flux and change, and possesses a great evolutionary potential.

One fact in particular in series Tuberosa strikes us of great interest: the curious uniformity in corolla shape of members of this series in Argentina. This semi-stellate to pentagonal corolla of the Argentine Tuberosa is not seen in Peru, nor perhaps to any great extent in Bolivia. The same corolla form occurs in Cuneoalata partly, and the corolla is even more semi-stellate in series Megistacroloba. In Commersoniana it is of course completely stellate. Only series Acaulia remains apart from this general trend. It is difficult to understand this fact or even to think of a hypothesis to explain it; for it seems to be taking the theory of introgression too far to try to explain the trend in corolla shape entirely by reference to some past series of hybridizations with series Commersoniana. Perhaps we have here the remains of the more primitive vanguard of the migration of potato species in Eocene times from Mexico to South America. Since the diploid Mexican species all possess a stellate or semi-stellate corolla, and since there is some evidence to support the theory that all potatoes were originally derived from a centre of origin in Mexico, possibly these Argentine potatoes may be showing a remnant of the primitive stellate corolla of their remote Mexican ancestors.

5

BREEDING VALUE

THE value to plant-breeders of the potato species of Argentina, Brazil, Paraguay, and Uruguay is considerable, perhaps only to be matched by that of the potato species of Mexico. Amongst the four countries we are discussing here, Argentina is outstanding as a source of genes for plant-breeders, though we should add that less work has been carried out on the species of the other three countries and that many of the valuable Argentine species extend northwards into Bolivia and even also into Peru in some instances. Indeed, some of the results we shall quote below have been obtained on non-Argentine forms of species which nevertheless are also to be found in that country.

Whilst Mexico can be considered as an area of prime importance for resistance to late blight (*Phytophthora infestans*), Argentina and Bolivia are of equal value in presenting valuable genes for resistance to eelworm (*Heterodera rostochiensis*). Both areas, Mexico and Argentina, possess valuable resistance to insects and insect-transmitted viruses, as well as to a number of other pathogens and to frost damage. Several useful biochemical qualities are also found in species of both regions. Even now it is very probable that their genetic richness has not yet been fully exploited, since each year brings fresh evidence of their value to the breeder in the improvement of our domestic potato varieties.

In the present chapter we shall outline some of the results obtained in the search for disease resistance and other qualities, grouping our data according to the topic under discussion rather than in a systematic sequence of host species. The literature references are not a complete list, though we have attempted to cite the more important papers known to us that deal with disease resistance of species in the area under survey. Much disease resistance is known outside this area, but we felt that it would have led us too far away from our terms of reference to deal with a complete range of resistance known in all potato species. On the whole, also, we have not included any mention of resistance known in domestic varieties of *S. tuberosum*, unless the work had been carried out specifically on species known to occur in Argentina or adjacent countries.

A. FUNGUS RESISTANCE

1. *Phytophthora infestans* (Mont.) de Bary — Late blight

As we have pointed out above, Argentina and the adjacent countries do not seem to be of great importance in comparison with Mexico as a major source of genes conferring resistance to *Phytophthora infestans*.

Thus, Castronovo (1950) found no late blight resistance in the Argentine wild species, whilst Schechaj (1954) stated that all the Argentine cultivars were susceptible in spring and summer plantings in Catamarca province; however, he noted some resistance to the local races of blight in the native variety Blanca or Imilla, as well as in certain varieties introduced from Europe. Bazán de Segura (1953) showed that *S. chacoense* was susceptible to blight when grown in Peru. This was confirmed by Brücher (1954), who stated that he saw it attacked by blight in the field in Argentina, as well as *S. microdontum* subsp. *gigantophyllum*. On the other hand, Brücher observed field resistance to blight in *S. vernei*, and this was verified by Black (in Anon. 1958) who stated that this species showed more resistance to race 1, 2, 3, 4 than did *S. tuberosum* subsp. *andigena*. Black later (in Anon. 1963) detected new R genes in seedlings derived from *S. vernei*, which conferred hypersensitivity to the pathogen of the usual type. Ross and Rowe (1965) claimed *Phytophthora* resistance for three lines of *S. microdontum* subsp. *gigantophyllum*, though this seems not to be in accordance with the results of other workers, except Kameraz (1964) who noted that some lines of *S. microdontum* subsp. *gigantophyllum* and of *S. vernei* were resistant to local (Russian) races of blight. He observes, however, that they were not so resistant to these blight races as the Mexican species.

Some resistance to blight was found by Black and Gallegly (1957) in one progeny of *S. kurtzianum*, which, when tested against race 1, 3, 4, gave 13 susceptible to 62 resistant seedlings. All other progenies tested of this species were susceptible.

A certain degree of resistance to *Phytophthora* has also, according to Ross (1958*b*), been reported in *S. commersonii*, *S. commersonii* subsp. *malmeanum*, *S. tarijense*, *S. acaule*, *S. spegazzinii*, *S. curtilobum*, and *S. tuberosum* subsp. *andigena*, but not of a very high order.

Apart, therefore, from the work of Black and of Black and Gallegly in showing that R genes occur in *S. vernei* and in *S. kurtzianum*, respectively, it would seem that a certain fairly low level of field or polygenic non-specific resistance is to be found in some species, but not in others. On present evidence, then, Mexican potatoes seem to offer the best prospects in blight breeding, though possibly some of the field resistance in the Argentine potatoes may prove to be of value in building up a useful level of resistance in breeding lines derived in part from the indigenous potatoes of this area.

2. *Synchytrium endobioticum* (Schilb.) Perc. Wart

Wart resistance is well known in varieties of the common domestic potato, *S. tuberosum*; it has also been found in the Andean subspecies *andigena* (Hawkes 1945). Many wart-resistant clones have been noted in nearly all the species of the area we are considering, e.g. *S. commersonii*, *S. chacoense*, *S. tarijense*, *S. acaule*, *S. microdontum* subsp. *gigantophyllum*, *S. vernei*, *S. maglia*, *S. juzepczukii*, *S. tuberosum* subsp. *andigena*, and *S. curtilobum* (Diehl 1934,

Bukasov 1940*a*, 1955*a*, Bukasov & Naumov 1938, Braun 1942, Hawkes 1945, Anselmi Laukin 1952, Kameraz 1957*a*). Often both resistant and susceptible lines are found in the same species, though Novák (1962) claimed that all the forms he tested of *S. acaule* and *S. microdontum* subsp. *gigantophyllum* were resistant. On the other hand, resistant and susceptible forms of '*S. punae*' were noted—a species that we class with *S. acaule* as a central Peruvian subspecies.

Apart from the common biotype of wart to which the species mentioned above show resistance, other biotypes were recognized by various workers in the early 1940s (Braun 1942). These biotypes were characterized by the initial letter of the locality in which they were first discovered, the common one being labelled biotype 1 or D (= Dahlem). Rothacker (1957*b*, 1961, Rothacker & Müller 1960) observed resistance to biotype G (= Giessübel) in the species *S. chacoense*, *S. tarijense*, *S. microdontum* subsp. *gigantophyllum*, *S. vernei*, *S. spegazzinii*, *S. maglia*, *S. tuberosum* subsp. *andigena*, and *S. curtilobum*.

Wart resistance was noted by Ross and Rowe (1965) for one line of *S. acaule* and one of *S. chacoense*, whilst immunity was recorded for one line of *S. microdontum* subsp. *gigantophyllum*. Unfortunately the biotype used was not mentioned, but presumably the results refer to the common biotype 1. Probably resistance to other biotypes of wart occurs in the Argentine species.

3. *Streptomyces scabies* (Thaxt.) Waksman & Henrici Common scab

No true immunity is known to common scab, though a high degree of resistance is found in some cultivars of *S. tuberosum* and its subspecies *andigena*, as well as in several wild species. Some twenty-five years ago, Reddick (1939) reported apparent immunity to scab in *S. commersonii* and *S. chacoense*, both in field and glasshouse tests, but later workers have regarded these results as indicating a high degree of resistance rather than immunity.

Hoffmann (1958) reported that a few individual plants of *S. chacoense* showed only slight infection, but others were susceptible, as also were all lines tested of the species *S. commersonii*, *S. acaule*, *S. microdontum* subsp. *gigantophyllum*, and *S. vernei*. Dionne and Lawrence (1961) found promising diploid selections resistant to scab, obtained by back-crossing the F_2 of (*S. chacoense* × *S. phureja*) × *S. phureja*. These were tetraploidized by colchicine treatment as a preliminary step to crossing with *S. tuberosum*.

Rothacker (1961) lists ten species that have shown resistance to scab, including, besides the ones we have noted above, *S. acaule*, *S. microdontum* subsp. *gigantophyllum*, and *S. vernei*, as well as three Mexican and one Bolivian species.

Calderoni and Induni (1952) found that the Argentine variety Runa was moderately scab resistant; this would certainly be a form of *S. tuberosum* subsp. *andigena*. There are many varieties of *S. tuberosum* with very high resistance to scab, so that the wild species may not at the present be of very great interest.

4. *Spongospora subterranea* (Wallr.) Johns Powdery scab

Powdery scab seems to be native to the mountain regions of Argentina, and it also occurs frequently on cultivated potatoes in the highlands of Bolivia and Peru, generally as the 'canker' form. It was recorded by Hawkes (1945) on plants of *S. megistacrolobum* in Jujuy province, and has been seen elsewhere on wild plants. (See also Appendix III.)

Most Chilean varieties of potato (*S. tuberosum* subsp. *tuberosum*) are susceptible to *Spongospora*, according to Mujica (1942) and Montaldo (1951*b*), though a few varieties were classed as resistant. It is possible that this might also be so for Argentine varieties.

Bukasov (1938) reports that Dorozhkin in 1936 found *S. curtilobum* to be immune to *Spongospora*, but Boyd (1951 and in Anon. 1951*b*) pointed out that certain lines were susceptible, at least in the roots and stolons.

Zadina (1958) reported resistance in certain lines of *S. chacoense*, *S. commersonii*, *S. acaule*, and *S. microdontum* subsp. *gigantophyllum*, as well as various Andean cultivated species further north, including *S. tuberosum* subsp. *andigena*.

5. *Alternaria solani* (Ell. & Mart.) Jones & Grant Early blight

The resistance of potatoes to early blight has been little investigated, though most forms so far screened for resistance have proved to be susceptible. Charles (1956) tested sixty species of the Commonwealth Potato Collection for resistance and found that only *S. chacoense* and *S. toralapanum* (related to *S. megistacrolobum*) were highly resistant, though not immune. However, the resistance was recessive and was inherited in a complex manner.

Resistance was claimed by Vargas (1950) for *S. juzepczukii*, *S. tuberosum* subsp. *andigena*, and *S. curtilobum*, but these observations were made on Peruvian and not Argentine varieties.

Bukasov and Kameraz (1959) report early blight resistance in several lines of *S. chacoense*, as well as in *S. tarijense*, *S. toralapanum*, *S. acaule*, *S. juzepczukii*, *S. tuberosum* subsp. *andigena*, and *S. curtilobum*. Early blight resistance is also reported by Montaldo and Sanz (1962) for a number of Chilean varieties of *S. tuberosum*.

6. *Other fungal diseases*

Mastenbroek (1955) reported certain species and hybrids that were not attacked when exposed to infection by *Oidium solani* (powdery mildew). These were *S. microdontum* subsp. *gigantophyllum*, *S. kurtzianum*, *S. microdontum* subsp. *gigantophyllum* × *S. kurtzianum* hybrids, and hybrids of subsp. *gigantophyllum* with two varieties of *S. tuberosum*. Certain species from other countries were also more or less immune.

Immunity to *Macrophomina phaseoli* (charcoal rot) was found by Thirumalachar and Pushkarnath (1953) in interspecific hybrids within series

Commersoniana, as well as in certain clones of *S. tuberosum* subsp. *andigena*. The resistance was of a hypersensitivity type. Later, Paharia, Pushkarnath, and Deshmukh (1962) reported that fifteen clones of *S. chacoense* were found to be highly resistant to the disease and that they were being used in a breeding programme.

Tolerance to *Verticillium albo-atrum* Rke. & Btd. (*Verticillium* wilt) was established by Webb and Buck (1955) in *S. chacoense*, *S. chacoense* × *S. commersonii*, and *S. chacoense* × *S. tuberosum* var. Menominee. Bazán de Segura and Ochoa (1959) found eleven resistant and five moderately resistant varieties of *S. tuberosum* subsp. *andigena*, amongst some 100 varieties tested, though these were, of course, all Peruvian ones. Webb and Hougas (1959) and Ross and Rowe (1965) recorded resistance in most lines tested of *S. chacoense*, one of *S. tarijense*, one of *S. acaule*, three of *S. microdontum* subsp. *gigantophyllum*, one of *S. vernei*, two of *S. spegazzinii*, one of *S. kurtzianum*, and one of *S. maglia*, as well as in several previously unidentified lines, two of which have later been shown to be *S. spegazzinii* and one of *S. tarijense*.

Hrobry (1958) and Bukasov and Kameraz (1959) show that, amongst a range of *Solanum* species tested for resistance to *Oospora pustulans* Owen and Wakefield (skin spot), the least susceptible were *S. chacoense*, *S. vernei*, and *S. tuberosum* subsp. *andigena*.

Boyd (in Anon. 1951*b*) gave some data on resistance to *Fusarium* wilt in certain samples of *S. tuberosum* subsp. *andigena*, *S. curtilobum*, and other Andean cultivated species from the Commonwealth Potato Collection.

B. BACTERIAL RESISTANCE

1. *Erwinia phytophthora* Appel — Blackleg

Some resistance to blackleg has been discovered in the species growing in the area we are discussing. Thus, Zaiczeva (1950) claimed resistance for *S. chacoense*, whilst Stapp and Spicher (1955) and Stapp and Hartwich (1956) also showed resistance to blackleg in various forms of *S. chacoense*, as well as in *S. microdontum* subsp. *gigantophyllum* and in various Mexican species. Novakova (1958) also found *S. chacoense* to show the highest resistance, together with the Mexican species, *S. stoloniferum*.

2. *Corynebacterium sepodonicum* (Spiek. & Kotth.) Skapt. & Burkh. — Ring rot

Knorr (1948) found resistance to this disease in *S. acaule* and in various strains of Mexican species; there was also a limited degree of resistance in one form of *S. chacoense*. Zaiczeva (1950) confirmed resistance in *S. chacoense*, as well as in several Mexican species.

3. *Pseudomonas solanacearum* E. F. Smith — Brown rot; Bacterial wilt

High resistance to this disease was found in *S. chacoense* by Thung (1947), as well as in *S. tuberosum* subsp. *andigena* and certain other species. *S. commersonii* may also have been resistant. Ross and Rowe (1965) mention resistance to bacterial wilt (presumably *Pseudomonas*) in several species, though strangely enough, not in *S. chacoense*, even though many lines were tested. Resistance was noted in two lines of *S. brevidens*, three lines of *S. acaule*, and two of *S. curtilobum*, but many other Argentine species tested were wholly susceptible.

C. NEMATODE RESISTANCE

1. *Heterodera rostochiensis* Woll. Potato-root eelworm; Golden nematode

During the process of screening the Commonwealth Potato Collection for resistance to eelworm Ellenby (1948) discovered true high-grade resistance in one line of *S. vernei* subsp. *ballsii*, and later (1952, 1954) in the typical subspecies of *S. vernei* as well as in several lines of *S. tuberosum* subsp. *andigena* from Bolivia. These results were confirmed by Toxopeus and Huijsman (1952), Mai and Peterson (1952), Howard (1955), and several other workers.

Breeding work with this material was quickly taken up in various countries and the genetic basis for resistance was elucidated. Toxopeus and Huijsman (1953) and Huijsman (1955) showed that inheritance in the subsp. *andigena* material was tetrasomic, depending on a single dominant gene, *H*. Rothacker and Stelter (1957) confirmed these results and pointed out that the resistance of *S. vernei* was polygenic. Later, Rothacker (1959) suggested that there might be two main genes in *S. vernei* which conferred resistance, but that modifiers were present also. Focke (1959) asserted that two independent genes, *N* and *R*, conferred resistance, so that if only one were present as a dominant the resistance was low. Goffart and Ross (1954) showed that of all the species tested, *S. vernei*, *S. venturii*, and *S. spegazzinii* were the most constant in their resistance over a three-year period, and that, whilst the resistance of *S. vernei* was polygenically controlled, that of *S. spegazzinii* was monogenic. Colchicine treatment of *S. vernei* produced tetraploid plants, which were to be used for future crosses, and hybrids of diploid *S. vernei* with domestic varieties showed considerable promise. Huijsman (1956) demonstrated resistance also in *S. kurtzianum*, whilst Rothacker (1957*a*) mentioned a relatively low degree of attack in forms of *S. acaule* and various other species.

Studies on the nature of resistance to eelworm have been carried out by various workers (Toxopeus & Huijsman 1952, Jones 1954, Williams 1956, 1958, etc.). It appears that the hatching power of the root diffusates of the resistant species (though not always that of *S. vernei*) was about equal to that of susceptible species and that their roots were also invaded by just as many larvae. However, in the resistant species the larvae bring about a cell necrosis of the roots and the females are not able to develop into mature individuals or cysts, though a certain number of mature males may be formed.

Soon after the work of breeding for resistance with *S. tuberosum* subsp.

andigena had begun Dunnett (1957) showed that an 'aggressive' race of eelworm occurred in Great Britain which completely broke down the resistance derived from subsp. *andigena*. On the other hand, *S. vernei* was shown to resist both the common (biotype A) and the 'aggressive' (biotype B) races. A similar 'aggressive' race was discovered in Holland by Huijsman (1959) and by workers in Germany (Goffart 1957, etc.). Dunnett (in Anon. 1957 to 1963, Dunnett 1960*a, b*) showed that *S. gourlayi*, *S. spegazzinii*, *S. kurtzianum*, and *S. maglia* were similar to *S. tuberosum* subsp. *andigena* in possessing resistance to the common race (biotype A). On the other hand, *S. megistacrolobum* and the Chilean-Bolivian-Argentine species *S. infundibuliforme* resisted the 'aggressive' race (biotype B) and were susceptible to the common one, whilst *S. sanctae-rosae* and *S. vernei* were resistant to both. He confirmed that the resistance of *S. vernei* was polygenic, and thought that that of *S. sanctae-rosae* appeared to be inherited in a fairly simple manner. A further disadvantage of *S. vernei*, according to Dunnett, is that the resistant plants do not stimulate a high level of larval emergence; thus this species is of no use as a trap crop, even though it may be of value in contributing useful polygenic field resistance, as a background to combine with single-gene resistance derived from other species. The two genes of *S. sanctae-rosae* were thought to be dominant, the one indistinguishable in its effect from the H_2 gene of the Peruvian *S. multidissectum* (resistant to biotype B), whilst the other conferred resistance of the subsp. *andigena* type, though to a lower degree.

Cole and Howard (1962) confirmed resistance to biotype B and susceptibility to A in *S. multidissectum* and showed that a similar reaction was to be found in *S. juzepczukii*, whilst subsp. *andigena* showed the reverse pattern.

Rothacker and Stelter (1961) demonstrated the presence of dominant heterozygous resistance to biotype A in *S. tarijense*, whilst Huijsman (1956, 1959, 1960) showed that *S. kurtzianum* resisted not only the A biotype, but also several Dutch strains of the B biotype. Two dominant genes were postulated. He later (Huijsman 1962) reported that *S. vernei* and *S. kurtzianum* were being currently used as sources of resistance to biotype B in his breeding programme. Goffart (1962) confirmed the results mentioned above for *S. kurtzianum* and *S. multidissectum*. Resistance was also found in *S. oplocense* (as *S. camarguense*, from Bolivia) by Dunnett (in Anon. 1965).

Jones and Pawelska (1963) tested forty-seven eelworm populations for ability to form cysts on plants with resistance derived from various resistant tuber-bearing and non-tuberiferous *Solanum* species. They concluded that many biotypes occurred and that *S. vernei* resisted more of them than any other tuberiferous species, but that a Rothamstead race of *S. nigrum* was completely resistant to all races. The results of Ross (1962) indicated that another promising species was *S. spegazzinii*, which was resistant to all the tested populations of both biotypes. Resistance appeared to be determined by two independent genes, designated F_a and F_b.

Stelter and Rothacker (1965) claimed resistance to race A in *S. kurtzianum* and *S. multidissectum*, and to races *A* and *B* in *S. juzepczukii*. These results, it will be noted, do not conform exactly with those of other workers.

In the list of tests on the material at Sturgeon Bay, Ross and Rowe (1965) add another species with resistance, namely, *S. boliviense*. However, of the two forms tested, one was resistant and the other susceptible; neither was from Argentina. One further point should be noted: many species marked on the list as susceptible have been tested for only one accession, whilst other species recorded in the literature elsewhere as resistant (e.g. *S. gourlayi*, *S. megistacrolobum*, etc.) are also seen to possess susceptible variants.

In summing up these results there is perhaps little need to emphasize the immense importance of the Argentine potatoes as sources of eelworm resistant material for plant-breeders (see also Brücher & Ross 1953, Brücher 1959*c*, 1963*c*), though it must not be forgotten that certain Bolivian, Peruvian, and Mexican species also exhibit some resistance. Brücher (1961) has pointed out that *Heterodera* is to be found on wild potatoes in Argentina, and it is no doubt owing to the presence of eelworm in the whole mountain area that resistance genes are so commonly found in such a large number of species.

Of the twenty-two species recognized in the present work about two-thirds have been shown to bear resistance to *Heterodera rostochiensis*. There seems to be no resistance, so far as can be seen at present, in the plains species, *S. calvescens*, *S. commersonii*, and *S. chacoense*, though one might have expected to find resistance in some of the introgressed mountain forms of the last-named species. Resistance in the mountain species is quite common, and only *S. brevidens*, *S. microdontum*,[1] *S. vidaurrei*, and *S. curtilobum* have shown no signs of resistance so far. It is possible that most of these have not been subjected to tests, or if they have, then only a very small number of accessions may have been screened. Further work might well add other species to the list of those so far known to exhibit resistance, so opening up new perspectives to breeders in this field.

2. *Meloidogyne* spp. (= *Heterodera marioni*) Root knot nematode

Some resistance to this eelworm has been recorded by Mujica (1939) and Montaldo (1950, 1951*a*) in Chilean and European varieties of *S. tuberosum* subsp. *tuberosum*. Ross and Rowe record susceptibility in *S. venturii* and *S. vernei*, as well as in *S. tuberosum* subsp. *andigena* from Argentina. At present, therefore, so far as the present writers are aware, no resistance to this nematode has been seen in Argentina or the other countries we are now discussing.

3. *Other nematodes*

Resistance to horsenettle cyst nematode and Osborne's cyst nematode is noted by Ross and Rowe (1965) in several Argentine species, namely,

[1] Reports in the literature on the resistance of *S. microdontum* (e.g. Goffart & Ross 1954) refer to *S. venturii*.

S. chacoense, *S. tarijense*, *S. acaule*, and *S. maglia*. Resistance is also noted for certain Bolivian and Mexican species, including lines of *S. tuberosum* subsp. *andigena* resistant to the common eelworm, *Heterodera rostochiensis*.

D. INSECT RESISTANCE

1. *Leptinotarsa decemlineata* Say. Colorado beetle

An enormous amount of work has been carried out on resistance to this insect, from which it seems clear that the species from Mexico, on the one hand, and Argentina and adjacent countries, on the other, are of greatest importance as sources of resistance. The most promising species appears to be *S. chacoense*, which was first shown to be resistant by Trouvelot (1938), Schwartz and Müller-Böhme (1938), and Schaper (1939). A high grade of resistance was observable in *S. chacoense* by all these workers, and this was reaffirmed by Bukasov (1940*a*), Kameraz (1940), and Müller and Sellke (1941). Other resistant forms were discovered by Schwartz and Müller-Böhme (1938), namely, *S. commersonii*, *S. acaule*, and the hybrid of *S. chacoense* × Pepo.

Further work by Stelzner (1943*a*), Stelzner and Torka (1948), and Torka (1943, 1949, 1950, 1954) indicated that not all biotypes of *S. chacoense* were resistant and that none of the resistant biotypes were homozygous resisters. Four grades of resistance could be established for *S. chacoense* and *S. chacoense* × *S. tuberosum* hybrids, the resistance showing clear evidence of being polygenically inherited. Another species, *S. kurtzianum*, as well as hybrids of *S. chacoense* × *S. kurtzianum* and *S. chacoense* hybrids with species from other regions were said to be promising. According to Ross and Rowe (1965), one line of *S. boliviense* from Bolivia and one of *S. gourlayi* were resistant to Colorado beetle, whilst Bukasov and Lebedeva (1955) claim that the solacaulin in some forms of *S. acaule* also protects this species from attack. Several Mexican species are also resistant, especially promising being *S. demissum*. However, it is not generally regarded as being of equal value to *S. chacoense*, since the hybrids with *S. tuberosum* do not retain a high level of resistance after several generations. Breeding is not easy, even with *S. chacoense*, owing to the polygenic nature of the resistance, but Schaper (1954) reports hybrids of *S. chacoense* × *S. tuberosum* which were not attacked and yielded to 60 per cent of the standard varieties. Promising hybrids were also reported by Kameraz (1957*b*, 1961), which were derived from *S. chacoense* and *S. kurtzianum* crosses with *S. tuberosum*.

The nature of the resistance in *S. chacoense* and other species seems to be due to unpalatability (Boczkowska 1951) and to the partial or complete suppression of egg-laying (Schaper 1953). Kuhn and Löw (1954) point out the particularly high alkaloid content and exceptionally high proportion of chaconins, but conclude that the total amount of these is insufficient to account for its resistance. Lipinsky (1957) thinks that chaconin plays an important part in *S. chacoense* resistance. Schreiber (1955) described a glycoalkaloid in *S. chacoense* and *S. commersonii*, identical with solanin from *S. tuberosum*, and concludes therefore that this could not be responsible for the beetle resistance of these two species. Langenbuch (1952) described tests from which it could be concluded that resistance was due to a repellent principle in *S. chacoense*. Evidently the unpalatibility of *S. chacoense* and other resistant species, and the poisonous nature of the alkaloids they contain are sufficient to explain their resistance.

2. *Other insects*

Resistance to the east-Asiatic *Epilachna vigintioctomaculata* Motsch. is found in the same potato species that are resistant to Colorado beetle, according to Bukasov and Kameraz (1948). Antipova (1950) claims a high resistance in various forms of *S. chacoense*, though not in all accessions. Starceva (1960) also found resistance in *S. chacoense*, as well as in *S. tarijense*.

Leaf hopper (*Empoasca fabae* Harr.) and Flea beetle (*Epitrix cucumeris* L.) resistance was demonstrated by Sleesman (1940, 1960, 1963) in *S. chacoense*, *S.×setulosistylum*, and *S. tarijense*, as well as in certain Mexican species. *S. microdontum* subsp. *gigantophyllum* was somewhat less resistant. None of the F_1 hybrids between *S. tuberosum* and the above-mentioned species was resistant to either insect. Ross and Rowe (1965) note resistance to flea beetle in certain lines of *S. chacoense* and *S. acaule*, though other lines were susceptible. Resistance in one line each of *S. tarijense*, *S. sanctae-rosae*, *S. microdontum* subsp. *gigantophyllum*, *S. vernei*, *S. gourlayi*, *S. maglia*, and *S. tuberosum* subsp. *andigena* was also recorded. Resistance to leaf hopper was also recorded by the same authors in most lines of *S. chacoense*, *S. tarijense* (one line), *S. microdontum* subsp. *gigantophyllum* (one line), *S. kurtzianum* (two lines), and *S. maglia* (one line).

Resistance to aphid attack was demonstrated for *S. chacoense* by Adams (1946). Ross and Rowe (1965) mention one resistant and two tolerant lines of this species, as well as one tolerant line each of *S. sanctae-rosae* and *S. microdontum* subsp. *gigantophyllum*.

E. VIRUS RESISTANCE

1. *Virus X*

Immunity to this virus was first discovered in the United States Department of Agriculture breeding line, 41956 (Schultz *et al.* 1937), and also, much later, in *S. acaule* (Stelzner 1950, Ross & Baerecke 1950, 1951). This species was shown to owe its immunity to virus X to a single gene in the homozygous dominant condition. Later, Ross (1954*a*, *b*) considered that the resistance represented an extreme form of hypersensitivity, and Rudorf (1958) showed that the R^x allele determined extreme resistance to all strains of the virus. Cockerham (in Anon. 1951*a*, Cockerham 1957) also confirmed Stelzner's report on the immunity of *S. acaule*, and established three alleles *Xi*, *Xn*, and *x*, which, in descending order of dominance, determined immunity, necrotic, and tolerant responses respectively. Cockerham contrasted this single gene resistance to the resistance of 41956, which was due to the effects of two complementary genes. Both Ross (1954*a*, *b*) and Cockerham (1957) showed that the *S. acaule* immunity gene showed disomic inheritance, as would be expected in an allotetraploid, functionally diploid species.

Before the discovery of immunity to virus X in *S. acaule* Cockerham (1943) showed that *S. chacoense*, *S. juzepczukii*, and *S. curtilobum* exhibited top necrosis to virus X, for which the gene N_x was postulated. The immune reaction to virus X was also reported to exist in *S. curtilobum* (Toxopeus 1958), who considered that the gene concerned was inherited from the *S. acaule* ancestor of this hybridogenic species. The frost tolerance and ease of crossing with *S. tuberosum*, together with the virus X immunity of *S. curtilobum*, would render it of great value in breeding varieties for these qualities.

Ross (1958*b*) noted resistance to virus X in *S. kurtzianum* also, whilst Brücher (1954) postulated X resistance genes in *S. microdontum* subsp. *gigantophyllum*. In addition, Cockerham (in Anon. 1956) stated that a gene in *S. tuberosum* subsp. *andigena* controlling necrotic response to virus X was allelic with the gene for immunity in *S. acaule*, but that a gene controlling *immunity* to virus X in subsp. *andigena* was independent of the *S. acaule* X-immunity gene. He suggested that the presence of the same locus in *S. acaule* and *S. tuberosum* subsp. *andigena* might very possibly be due to introgression of *S. acaule* germplasm into subsp. *andigena*. This most interesting piece of work gives added support to the hypothesis of the origin of *S. curtilobum* as a two-stage hybrid, involving the three species *S. acaule*, *S. stenotomum*, and *S. tuberosum* subsp. *andigena* and indicates the possibility of natural backcrosses of *S. curtilobum* (which can be regarded as having the composition: [*S. acaule* × *S. stenotomum*] × *S. tuberosum* subsp. *andigena*) to the tetraploid parent. On this basis we should also expect *S. juzepczukii* to exhibit X resistance of some kind. It is interesting to note that this species was recorded by Cockerham (1943) and in the Commonwealth Potato Collection 1946 Report (unpublished) as necrotic or field immune to virus X.

A mixture of immune, resistant, and susceptible lines was noted by Ross and Rowe (1965) for *S. acaule*; some resistant but most susceptible lines of *S. chacoense* and *S. curtilobum*; hypersensitive, susceptible, and resistant lines of *S. kurtzianum*; one resistant line of *S. maglia* and one resistant line of *S. tarijense* (unidentified on list: P.I. 208881). Evidently this resistance is of the necrotic response type.

2. *Virus A*

Necrotic response to virus A is moderately frequent in wild species, and Cockerham (1943) noted it in *S. chacoense*, *S. infundibuliforme*, *S. microdontum* subsp. *gigantophyllum*, *S. tuberosum* subsp. *andigena*, and *S. curtilobum*. He postulated at that time the presence of a gene *Na*, but considers now, at least for certain species (in Anon. 1964 and personal communication) that 'there are genes which condition necrotic reactions to all strains of virus Y, including virus A and virus C'. This is corroborated by Ross (1960*b*) for *S. stoloniferum*, a Mexican species.

Ross and Baerecke (1950) considered that virus A immunity was as common as virus Y immunity. Ross (1958*b*) found virus A resistance in *S. chacoense*, *S. acaule*, and *S. vernei*. Hawkes (1947) noted that *S. brevidens* was shown by W. R. Wortley to be apparently immune to all the usual sap-transmissible potato viruses. Much later Bukasov and Kameraz (1959) stated that *S. brevidens* was immune to virus A. Webb and Hougas (1959) and Ross and Rowe (1965) indicate both immunity and resistance (necrotic response?) to virus A in certain lines of *S. chacoense*, *S. microdontum* subsp. *gigantophyllum*, *S. kurtzianum*, *S. maglia*, and *S. tuberosum* subsp. *andigena*.

3. *Virus Y*

Hypersensitivity to virus Y was observed by Cockerham (in Anon. 1945, 1946, 1947, etc.) in *S. microdontum* subsp. *gigantophyllum*, and it was thought that this conferred immunity on older plants at least. Later, Cockerham (in Anon. 1949) postulated a single gene, *Ny*, conditioning the basic necrotic response on which hypersensitivity depended, but other genes were thought to be needed to enable the host to localize the virus at the point of infection.

Ross and Baerecke (1950, 1951) and Stelzner (1950) found immunity to virus Y in *S. chacoense*, *S. vernei*, and *S. kurtzianum*. Brücher and Ross (1953) also, emphasized that resistance to viruses Y and C occurred in some lines of *S. microdontum* subsp. *gigantophyllum* and *S. vernei*. Similar results for subsp. *gigantophyllum* were also obtained by Easton (1957) and Rothacker and Witt (1959). Bukasov and Kameraz (1959) note that *S. brevidens* is immune to this virus, and this confirms the results of Wortley (Hawkes 1947) noted above for virus A. It is of considerable interest to note that the same gene for the necrotic reaction to virus Y has been identified by Cockerham (in Anon. 1962) in *S. chacoense* and *S. microdontum* subsp. *gigantophyllum*. This has led him to postulate (in Anon. 1965), that the character of necrotic reaction to virus Y might be useful in introgression studies between the two species (see later, in this work, p. 219).

Apparently, resistance in the form of immunity is rather uncommon. In South America it is chiefly to be seen in *S. chacoense*, which, in addition, exhibits resistant, hypersensitive, and susceptible responses, according to the line tested (Commonwealth Potato Collection Report, 1946, unpublished; Ross & Rowe 1965). It appears that *S. chacoense* has a far wider range of resistance genes than any other species, with the exception of *S. stoloniferum* from Mexico, and it would seem that more use might be made of it in the future in programmes involving the breeding for resistance to virus Y.

4. *Leaf roll*

Apparent resistance to leaf-roll virus in potatoes generally appears to be due to some kind of resistance to infection, and up to now no true immunity has been found. Thus, infection resistance was seen by Ross and Baerecke (1950) in *S. chacoense* and *S. tuberosum* subsp. *andigena*, as well as tolerance in other lines. Baerecke (1956) considered that the inheritance of leaf-roll resistance was polygenic, and that some genes are inherited as dominants; this was based on crosses of *S. acaule*, *S. tuberosum* subsp. *andigena*, and *S. demissum*.

Ross (1958*a*) mentioned greater resistance to leaf-roll infection in hybrids derived from *S. chacoense*, *S. acaule*, *S. tuberosum* subsp. *andigena*, and *S. demissum* than could be found in domestic varieties, but the resistance was not found in the wild species themselves. This is thought to be due possibly to the action of complementary genes from each parental species, and it is suggested that further improvement of resistance could be effected by combining the most resistant hybrids.

Clark (1962, 1963) reported that no infection with leaf-roll virus took place in the glasshouse under controlled aphid inoculations of certain lines of *S. chacoense*, *S. tuberosum* × subsp. *andigena* hybrids, and a few species from Peru and Mexico. Ross and Rowe (1965) mention one tolerant and two resistant lines of *S. acaule*, two tolerant lines of *S. brevidens*, several of *S. chacoense* as resistant, two of *S. kurtzianum* as tolerant and one as resistant, one of *S. megistacrolobum* and one of *S. sanctae-rosae* as resistant and susceptible (presumably in different seedlings of the one accession), and two of *S. microdontum* subsp. *gigantophyllum* of the same type. Unfortunately, no further information is given as to the possible nature of the tolerant and the resistant material.

5. *Other viruses*

Ross and Rowe (1965) record that all lines tested of *S. chacoense* were resistant to virus F (see also, Bagnall 1960). One line of *S. kurtzianum* was resistant to spindle tuber.

F. PHYSIOLOGICAL AND BIOCHEMICAL PROPERTIES

1. *Frost resistance*

Perhaps the first mention of frost resistance in potatoes was made by Reddick (1930), who reported tolerance to frost in *S. commersonii*. The most resistant species so far known to us is *S. acaule*. Reports on its resistance came from Bukasov and his co-workers from the early 1930s onwards (Bukasov 1932*a*, *b*, 1933*a*, *b*, 1936, 1937*a*, etc., Kovalenko 1932).

S. acaule was generally stated to withstand −8° or even −10° C, whilst for the related frost-resistant cultivated species, *S. juzepczukii* and *S. curtilobum*, a resistance to −5° C was claimed. Bukasov also pointed out that a lower grade of frost resistance was to be found in *S. commersonii* (including subsp. *malmeanum*—both of these split into various micro-species). Kameraz (1940) mentioned that a tolerance to −3° or even −5° C was to be found in *S. commersonii.*

Razumov (1935) showed that resistance to frost in *S. acaule* diminished with advancing season, and whilst at the beginning it showed no damage at −6° and very little at −7°, at the end of the season it withstood only −3·5° C. Later work by various authors has tended to confirm these results.

The techniques of testing material accurately for frost resistance have been dealt with by various workers, especially Rzhavitin (1936), Stelzner (1938), Mastenbroek (1956), and Hudson (1961). Whilst uniform conditions of test are important, it is equally necessary to subject material to uniform pre-treatment and even to a uniform period of 'hardening' for several weeks previously; the age of the leaves in cut leaf tests is also most important, in ensuring uniformity of test.

The use of *S. acaule* in breeding has been mentioned by numerous workers (Dremliug 1937, Traczenko 1938, 1941, Stelzner 1943*b*, Zvereva 1945, 1951, etc.), but the results were not immediately successful, owing to the difficulty in crossing *S. acaule* with *S. tuberosum*. Often non-reduced gametes function, giving a not very fertile hexaploid F_1. Dremliug showed that hybrids of *S. acaule* × *S. tuberosum*, backcrossed to *S. tuberosum* withstood −5·5° C. Pissarev (1933) considered that cold resistance behaved as a dominant in all crosses of cold-resistant wild species with *S. tuberosum*, and this was confirmed by Mastenbroek (1954, 1956), who thought that the resistance was controlled by several independent genes.

Ratera (1947*a*) exposed a series of wild Argentine species to low temperatures near Buenos Aires and found that the most resistant were *S. commersonii* subsp. *commersonii* and subsp. *malmeanum*. Lechnovicz (1940) claimed that *S. rionegrinum* (= *S. commersonii* subsp. *commersonii*) from Uruguay was the most resistant wild species known, not excepting *S. acaule*, and that it withstood temperatures of −11° C, when *S. demissum* was completely killed.

Viirsoo (1954*a*), observing a collection of Argentine potatoes in the valley of Humahuaca (prov. Jujuy) after a night of frost of −5° C, found that only *S. vernei*, *S. commersonii* subsp. *malmeanum*, *S. juzepczukii*, and *S. curtilobum* were not destroyed. A certain degree of resistance was also seen in *S. tuberosum* subsp. *andigena* cv. Collareja. He does not state, unfortunately, what species were damaged. In further experiments (1963) he mentions that two varieties of subsp. *andigena* showed resistance to −2° C and two others possessed some tolerance of low temperatures.

The frost resistance of the Argentine potatoes was also discussed by Weisaeth (1956*a*, *b*) in which the value is mentioned of the species *S. commersonii* subsp. *malmeanum*, *S. acaule*, *S. sanctae-rosae*, *S. vernei*, *S. curtilobum*, and several varieties of *S. tuberosum* subsp. *andigena*, as well as *S. stenotomum* and *S. demissum*, in breeding frost-resistant varieties. The resistance of *S. vernei* is also mentioned by Brücher (1954, 1959*c*) who claims that it resists −4° C, whilst *S. microdontum* subsp. *gigantophyllum* is thought to withstand −5° C. Brücher (1959*b*, *c*) also mentions that *S. megistacrolobum* and *S. sanctae-rosae* (as '*S. catamarcae*') are frost-resistant, as well as some cultivars from the high valleys of prov. Jujuy.

Firbas and Ross (1961) and Firbas (1962*a*, *b*) reconfirm the high degree of frost resistance in forms of *S. acaule*, *S. juzepczukii*, and *S. curtilobum* subjected to a temperature of −2·5° C for 3 hours in a controlled-temperature chamber. These species were included in 'resistance group 1', showing least damage. Second-grade resisters, in group 2, were *S. commersonii*, *S. megistacrolobum*, and *S. boliviense*. Resistance group 3 included *S. sanctae-rosae*, whilst *S. camarguense* (= *S. oplocense* in this work), and *S. vernei* were classed in group 4. Group 5 included species which were completely damaged and showed no resistance; in this group were classed *S. chacoense*, *S. tarijense*, *S. infundibuliforme*, *S. microdontum*—both subspecies, *S. gourlayi*, *S. spegazzinii*, *S. kurtzianum*, and *S. maglia*. Firbas and Ross were not able to find any close correlation between the altitude of origin of a species or collection and its frost resistance, though there is of course a loose connexion here. Most frost-resistant species extended into areas where there was no danger of frost and the authors do not consider that there are any anatomical characters which can be correlated with frost resistance, either. This is contrary to the conclusions of Nassonov and Yager (1933), who claimed that all frost-resistant species and hybrids showed little secondary wood, an unusually large number of stomata and other features.

Frost resistance was recorded by us on plants growing in the field at Vandel, Denmark, in *S. commersonii*, most lines of *S. acaule* and all seedlings of *S. infundibuliforme*. A certain degree of resistance was also noted in *S. megistacrolobum*, *S. sanctae-rosae*, and *S. oplocense*.

Recent work on breeding for frost resistance still utilizes *S. acaule* as source material, though *S. curtilobum* might be better from the point of view of fertility. Thompson and Olien (1964) claim tolerance to −4·5° C in *S. acaule* hybrids with *S. tuberosum*, and similar promising hybrids have also been described by Filippov (1954) and Vavilova (1960). Finally, Veselovsky and Kuznecova (1963) report a hybrid from octoploid *S. acaule* crossed with *S. demissum* and then with *S. tuberosum* cv. Kameraz No. 1. The hybrids yielded well, were resistant to various races of *Phytophthora*, resisted −3·5° C of frost, and the tubers had high starch and protein content (27 and 2·87 per cent, respectively).

2. *Drought resistance*

Problems of resistance to drought have not been dealt with to such an extent as those concerned with frost resistance or the major potato diseases and insect pests. It seems to us very probable that useful drought and/or heat resistance could be bred into our potato varieties, using basic material from the wild and primitive cultivated species from selected areas. We have observed species of wild potatoes growing under very warm conditions, not only in Argentina and adjacent countries, but also in Peru, Mexico, and the U.S.A. The following account deals mainly with material from Argentina, Brazil, Paraguay, and Uruguay.

A drought-resistant wild potato is mentioned by Ratera (1938*a*) from La Pampa; presumably he is here referring to a form of *S. chacoense*. Hawkes (1945) pointed out the possible drought resistance of forms of *S. infundibuliforme*, judging from the areas in which it is to be found (Prepuna province), and Perlova (1951) mentioned that *S. commersonii* and *S. maglia* were drought resistant, according to her observations of material cultivated under experimental conditions at Moscow. From the observations on distribution and adaptation to different phytogeographical provinces given in the present work (see Chapter 3) it would seem very probable that species such as *S. chacoense*, *S. commersonii*, *S. calvescens*, *S. tarijense*, *S. infundibuliforme*, *S. oplocense*, *S. spegazzinii*, *S. kurtzianum*, and *S. maglia* might be used as initial material for screening, hybridization, and selection. All these grow in fairly dry and/or warm regions and might therefore be expected to possess drought-resistant ecotypes. Ross and Rowe (1965) mention drought-resistant lines of *S. tuberosum* from Chile and *S. sparsipilum*[1] and *S. gandarillasii* from Bolivia, though no Argentine drought-resisters are mentioned.

As was pointed out earlier (Hawkes 1945), it is essential that drought-resistant varieties should incorporate resistance to insect-transmitted viruses, and as we have seen, many of the Argentine species possess such qualities. At the moment, so far as we know, potato breeding for drought resistance has been carried out in Sweden and Russia (see Demidovicz 1938, Josefsson 1944, Kruzhilin 1946), as well as in Argentina, where the variety Huinkul shows promise. It seems very likely, therefore, that a considerable advance in the tolerance of potatoes to heat and drought could be made, on the basis of wild and primitive indigenous potatoes crossed with selected breeding lines or domestic varieties.

3. *High dry matter and starch content*

Species in series Commersoniana seem to be of promise in breeding for high dry matter, and in fact this series, together with Demissa and Longipedicellata from Mexico, shows the highest dry matter of all the wild species.

Thus, Bukasov and Naumov (1938) report dry matter of 21 to 28·4 per cent in *S. commersonii*, whilst Prokoshev and Mattison (1940) reaffirm this, indicating a starch content of up to 33·4 per cent in *S. chacoense* and 27·7 per cent in *S. commersonii*. In contrast, the dry matter of domestic potato varieties rarely exceeds 23 per cent. Undoubtedly there are enormous intra-specific differences in dry matter, and just a single result for a species is quite meaningless. Thus, Kameraz (1940) records dry matter of up to 32·5 per cent for

[1] One of these may have been *S. tarijense*.

S. chacoense and starch varying from 12 per cent in some forms to 26 per cent in others. The starch content of *S. commersonii* subsp. *malmeanum* varied from 15·4 to 23 per cent in different forms, whilst that of *S. acaule* reached 21·9 per cent. Ross and Rowe (1965) record many lines of *S. chacoense* with high starch content and one of *S. curtilobum* with high dry matter.

On the whole, the primitive cultivated species also show great promise. Commonwealth Potato Collection results (unpublished) record dry matter of over 30 per cent in certain clones of the cultivated Andean species, *S. juzepczukii* and *S. tuberosum* subsp. *andigena*. However, Bukasov (1940*d*) finds that the non-Argentine cultivated species *S. ajanhuiri* and *S. phureja* are most promising, and mentions that very wide ranges of dry-matter content are found in both subspecies of *S. tuberosum*. An estimation of the starch content of Bolivian and Peruvian varieties of *S. tuberosum* subsp. *andigena* and wild species has been carried out by Viirsoo (1956). The average starch content for all the samples was 25·47 per cent, 8·9 per cent of them containing more than 30 per cent starch.

As was pointed out by Akeley and Stevenson (1943) dry matter depends on the interaction of genotype and growing conditions, the latter being especially important, so that results must be obtained on material grown under strictly comparable conditions if they are to be of any value.

4. *High protein content*

So far as the wild species are concerned high protein is again to be found chiefly in those from Mexico on the one hand, and Argentina on the other. Thus, Bukasov and Naumov (1938) note a crude protein content in percentage of the fresh weight of 3·75 for *S. commersonii*. Prokoshev (1939) gave a figure of 3·66 per cent for *S. commersonii* and 3·44 per cent for *S. chacoense*. This may be compared with figures of from 1·89 to 3·26 per cent in European varieties of *S. tuberosum*.

Considering the analyses of protein in some of the more promising cultivated species the results compare very favourably with the best of the wild species. Hawkes (1945) mentions certain *S. tuberosum* subsp. *andigena* forms with over 4 per cent and one with over 6 per cent raw protein per fresh weight. Even better results were obtained by Evgrafova (1955—quoted by Rothacker 1961) who analysed protein in a range of wild and cultivated species, finding the best values in certain forms of *S. tuberosum* subsp. *andigena* (to 7 per cent) and *S. phureja* (to 7·65 per cent), which was higher than the highest wild species, *S. chacoense* (to 6·06 per cent).

It must be noted that, as pointed out by Bukasov and Kameraz (1959), the protein content varies greatly according to the climatic, geographical, and soil conditions obtaining during growth.

From this rather brief survey of the value to plant-breeders of the potato species of Argentina, Brazil, Paraguay, and Uruguay one can see quite clearly that there is a strong parallelism between these species and the wild species from Mexico. This may possibly be due in part to the similar range of climatic conditions under which the species in each area grow. Whatever the reasons, however, there seems no doubt that the area under consideration here is of prime importance as a source of germ plasm in breeding resistance of various types into our domestic varieties. This has already been noted by Brücher (1959*c*).

In comparing the value of the different species as sources of genetic resistance to the diseases and pests mentioned above or as possessors of desirable chemical characteristics, *S. chacoense* is outstanding, since it seems to embody no less than 25 qualities out of a possible total of 32. No other species from this area possess such a large number, though *S. tuberosum* subsp. *andigena* with 17, *S. acaule* with 16, *S. microdontum* with 13, *S. kurtzianum* with 11, and *S. tarijense* and *S. vernei* with 10 each are still of considerable value. It should not be forgotten that some of these species have been known for a long time (e.g. *S. chacoense*) and have been repeatedly subjected to tests of all kinds, whilst others, such as *S. calvescens*, *S. venturii*, *S. oplocense*, and *S. vidaurrei* are almost unknown in the living state. In any case, even though a species is of value in only one respect, this would be ample justification for its use.

It is to be hoped that further evaluation of the species of this area will be carried out in the future and that breeders will take advantage of the valuable opportunities offered by making even more use of them in their breeding programmes than at present.

6

TAXONOMIC METHODS USED IN THE PRESENT WORK

THE present chapter outlines the methods used in the second part of this monograph, where a detailed account of the taxonomy of the wild and cultivated species of Argentina, Brazil, Paraguay, and Uruguay is given. We shall here explain the descriptive terms and deal with the treatment accorded to each species, paying particular attention to the question of artificial and natural hybrids.

The twenty-two species that we have recognized are arranged in taxonomic sequence, grouped into six series, namely: Etuberosa, Commersoniana, Acaulia, Cuneoalata, Megistacroloba, and Tuberosa. A dichotomous key for the identification of series and wild species is given on p. 127 and a key for the identification of cultivated species appears on p. 429.

We have taken a conservative and rather broad view of the taxonomy of the series, and have tried not to split them up any more than was necessary. For each series we have given a list of synonyms (where applicable), followed by a short description and a note on its geographical distribution in South America and approximate altitude range. A note on the chromosome number or numbers is also given, followed by a general discussion of the taxonomic treatment and boundaries given by other authors as well as ourselves. Finally, we give a list of species included in the series for the area under survey.

A. WILD SPECIES

In dealing with the wild species we have used a standard format, grouping the information under the same headings for each species, except when very little knowledge was available, when a certain 'telescoping' of some of the sections has been effected.

We have attempted, on the whole, to give a broad synthetic treatment to each species, starting with formal descriptions and the presentation of data of a non-experimental nature, and moving to a discussion of the living plants and the results of experimental studies in the later sections. Distribution maps, line drawings, and photographs are also included.

The general layout for each wild species is given below, followed by descriptions of the methods employed and explanatory notes, where necessary. Notes on the cultivated species appear on p. 123.

1. Name and synonyms (at species level).
2. Description.
3. Affinities and methods of distinguishing from allied species.
4. Key to subspecies (if any).
5. Names and synonyms of subspecies (if any).
6. Descriptions of subspecies (if any).
7. Specimens examined.
8. Geographical distribution.
9. Habitat.
10. Taxonomic (and nomenclatural) notes.
11. Notes on living plants.
12. Cytology.
13. Fertility and crossability.
14. Natural hybrids.
15. Localities of natural hybrids.

1. *Name and synonyms*

The name of the species, its author, and a full literature reference is followed by a list of synonyms, where appropriate, with their respective literature references. These are given in chronological sequence. Where the species has been divided into two subspecies the synonyms for the typical subspecies are given here, under the species name, whilst those of the non-typical subspecies follow later, just before the description of that subspecies. All varietal names have been listed as synonyms, since we do not consider that these are worth retaining as separate taxa in the group with which we are dealing.

2. *Description*

The description of each species is based on actual specimens, both dried and living, of all material available to us. We have attempted to make the descriptions more easily readable by providing separate paragraphs for: (*a*) habit and underground parts, (*b*) stem, (*c*) leaves, (*d*) inflorescence, peduncles, and pedicels, (*e*) calyx, (*f*) corolla, (*g*) anthers and filaments, (*h*) style, stigma, etc., (*i*) berries. The use of thicker, bold face type for **stem, leaves, inflorescence, calyx, corolla,** etc., is also intended to help the reader in picking out the relevant data without too much difficulty. Features of special significance in helping to characterize a species or to identify it easily are written in italics.

Although we have attempted to show as complete a range of variation as was available to us it is impossible to be certain that every future collection will fit into the range of measurements given. Wild potatoes are notoriously plastic, that is, they respond to changes in the environment by large changes in phenotype. Hence, some future collections may enlarge our concept of the variation ranges of the species concerned.

When we have seen a very few specimens of a species we have given a single number or a simple range of size or number (e.g. *S. calvescens*, five collections seen: 'leaves 22–27 cm long, 3–4-jugate'), since the data are insufficient for any more complex treatment. On the other hand, where a large number of specimens have been examined it is possible to give a range of measurements into which the majority of specimens fit, together with other figures in brackets representing extreme values that are rarely met with (e.g. *S. commersonii*, about 250 collections seen: 'leaves with (0–)2–4(–6) pairs of lateral leaflets'. This indicates that the normal range is 2–4 pairs, but specimens have also been seen with 0, 1, 5, or 6 pairs.)

The usual form of descriptions of *Solanum* species in section Tuberarium has been adhered to. The plants are all herbaceous, with tubers borne on more or less horizontal or obliquely descending underground stems or *stolons* in all species described here except *S. brevidens*. In this species, both stolons and tubers are absent. Tuber skin patterning is absent in the wild species, though the tubers may be flushed with anthocyanin occasionally. The tuber form or depth of eyes is not so characteristic or bizarre in the wild as in the cultivated species, and in most cases does not call for a great deal of comment. Occasionally the stolons grow up through the soil to the surface some distance from the mother plant and there form daughter plants. They are then called *sobols* instead of stolons since they rarely bear tubers under these circumstances.

In giving the length of the stem we have measured it from ground level to the apical bud, but have excluded the inflorescence. The stems in the tuber-bearing species are normally angled, these angles running down from the bases of the leaves; in most of the species described here the angles bear *wings*, again running down from the leaf bases and hence termed *decurrent*. The wings are straight or crinkled (*crisped*), or *undulate* and are very often marginally toothed (*crenulate*), and sometimes bear a few coarse hairs on the apices of the teeth. These hairs may also be wide based (*gibbous*, *gibberulous*), or at any rate set on a cushion of tissue. Wing pigmentation is not infrequent.

The leaf in most potato species is compound, made up of a number of lateral leaflets and a terminal one, all set on a main stalk or *rhachis*, which itself is known as the *petiole* in the region below where the leaflets are given off (see Fig. 4(*a*)). Such compound leaves are spoken of as *imparipinnate*, but since this condition is normal for all potato species (they are in fact *interruptedly imparipinnate*, as we shall see below) other than those with simple leaves, the term imparipinnate is not used here. The individual leaflets may not be stalked (*sessile*) or may be set on stalks or *petiolules*. In describing the leaf it is customary to say that the laterals are, for instance, 3-paired or 3-*jugate*; frequently the phrase is shortened to 'leaf 3-jugate'. This of course refers only to the main or primary lateral leaflets which gradually diminish in size from apex to base of leaf; in some species, however, the penultimate leaflet pair is as large or larger than the uppermost pair (see Fig. 4(*b*)).

All smaller leaflets that interrupt the regular sequence of the primary lateral leaflets, that are, as it were, thrown in between them, are termed by us *interjected leaflets*. We feel that this term is less confusing than 'intermediate leaflets' or 'folioles', since the former gives an erroneous meaning of size, whilst the term 'foliole' could well be applied to any leaflet. Since some

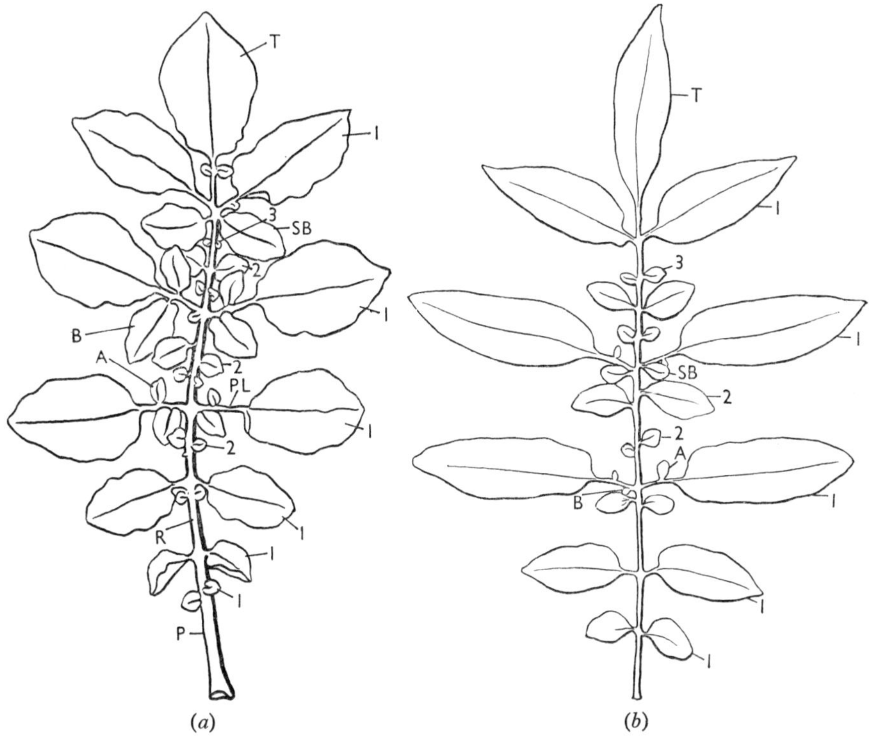

FIG. 4. (*a*) Diagrammatic form of potato leaf (adapted from Bukasov 1933*b*). (*b*) Leaf of *S. spegazzinii*. Note that the penultimate leaflet pair is larger than the ultimate one.

T	Terminal leaflet	B	Basiscopic leaflet
1	Primary lateral leaflet	SB	Semi-basiscopic leaflet
2	Secondary lateral } Interjected leaflet	P	Petiole
3	Tertiary lateral } Interjected leaflet	PL	Petiolule
A	Acroscopic leaflet	R	Rhachis

interjected leaflets near the apex of the leaf may be larger than the primary laterals near the base one must define these interjected leaflets by position rather than size. They can be recognized easily enough on one size criterion, however, since they are always smaller than the primary laterals *between which they occur*. Between two pairs of primary laterals there can occur a sequence of secondary or interjected leaflets, also decreasing in size from above to below; and in some well-dissected leaves of cultivated varieties there may even be tertiary series of leaflets between those of higher order.

This is a very rare occurrence in wild species and can be disregarded in general. This scientific description of the leaflet system in potatoes was proposed by Bukasov (1933*b*).

Although the interjected leaflets are normally to be found on the leaf rhachis they may occur also on the petiolules of the primary lateral leaflets. Such interjected leaflets must then be characterized as *acroscopic* (pointing towards the leaf apex) or *basiscopic* (pointing towards the leaf base); very occasionally they are *semi-basiscopic* (pointing diagonally from the lower angle between the petiolule and the rhachis). Although one can theoretically speak of the number of interjected leaflets *per series* (i.e. between each two pairs of primary lateral leaflets) it is in practice easier to give the total number of pairs of interjected leaflets *per leaf*, and this is what has been done in the present treatment. In the leaves of certain species the interjecteds do not occur in completely regular pairs. To maintain conformity of treatment we have then counted the total number of interjected leaflets and divided by two, expressing the result correct to the next higher number where a fraction occurs in the answer.

With sessile leaflets the tissue at the leaflet base may often run down from the base along the rhachis. This condition is known as *decurrency*, and is especially well marked in the leaves of series Megistacroloba and Cuneoalata. In these two series the decurrent portion forms a triangular or wedge-shaped wing respectively. The decurrent leaflet bases of *S. acaule*, and especially of subsp. *aemulans*, are also noteworthy. In series Cuneoalata and Megistacroloba the leaflets themselves seem to be no more than lobes or segments which project from the rhachis wings. Interjected leaflets are not normally developed in Megistacroloba and Cuneoalata, possibly due to some inhibitory effect from the wing-tissue development. In certain other species, where a petiolule is developed and the leaflets are not sessile the leaflet decurrency can descend along the basiscopic side of the petiolule on to the rhachis. This curious phenomenon is seen in some growth forms of *S. spegazzinii* and allied species.

The length of the *petiole* (i.e. the distance from the leaf base to the lowest pair of laterals) is on the whole a rather unreliable character, and we have not given it in our descriptions; nor have we mentioned the length of the petiolule of the terminal leaflet, since this is in reality merely the upper part of the rhachis, and its length does not seem to be of any particular significance taxonomically.

The size and form of the *pseudostipular leaves* is of no great taxonomic value in the species here dealt with. These leaves are really the terminal leaflets of two small modified leaves with generally no laterals, borne on a very reduced axilliary shoot. They are always approximately *semilunar* and more or less clasp the stem. In some Mexican species (e.g. *S. jamesii*) these leaves are of great taxonomic importance.

The leaflet shape in this group of *Solanum* species lies somewhere between *ovate* and *lanceolate*, sometimes as narrow as *narrowly lanceolate* or *narrowly elliptic*, etc., and sometimes as broad as *broadly ovate* or *broadly elliptic*. We have used the terms *ovate* and *lanceolate* to express a shape that is broader below the centre and the term *elliptic* to mean broadest at the centre. The word *oblong*, meaning that the leaflet is not very tapered at either end, has also been used, as well as *rhomboid* (roughly diamond shaped). *Obovate*, etc., means that the leaflet is broadest above the centre.

The leaflet apex may be *obtuse* (blunt), *acute* (sharp), *attenuate* (drawn out), or *acuminate* (with a suddenly drawn out sharp end). The base may be *cuneate* (wedge-shaped), *truncate* (cut off at right angles to the stalk), *cordate* (heart-shaped at the base), or *auriculate* (with very well-marked lobes). It is often *oblique* (tissue is not equally developed on both sides of the petiolule).

The normal type of leaf and stem hair covering (*pubescence*) is of simple multicellular weak hairs of various lengths which bend gradually over about a third to a half of their length, so becoming rather *adpressed*, and pointing towards the apex of the leaf or the plant. In some species (e.g. *S. venturii*) the hairs are very short and few-celled, but normally they are longer, several to many-celled, and curved in the way described above. In one or two species (*S. maglia*, *S. kurtzianum*, some forms of *S. chacoense*) the hairs are raised on swellings or cushions, besides being bent over, and are then spoken of as *gibbous* or *gibberulous*. In some species the pubescence is very sparse, and these are spoken of as *glabrescent*. In a few organs the hairs are absent and the organs concerned are then spoken of as *glabrous*. Turbinate glands borne on single-celled stalks are to be met with in most species, but are very sparse except in species such as *S. tarijense*, or in certain parts of others (upper part of inflorescence in *S. kurtzianum*—some forms). Very occasional adpressed glands of a multicellular type can be seen in most species examined, but these do not seem to be of much taxonomic significance in the species dealt with here and have consequently not been mentioned.

The margins of the leaflets are in many species minutely toothed in a rather irregular manner (e.g. *S. microdontum*), and on the apices of these teeth or *crenulations* one or more hairs are usually developed. The coarse marginal hairs of *S. kurtzianum* and other species allied to it are particularly noticeable.

The general inflorescence stalk is known as the *peduncle*. It is normally *dichotomously branched* or forked (divided into two equal branches), and the lengths of the divided and undivided portions are very useful diagnostic characters. In *S. brevidens* and *S. chacoense* the peduncle has long branches, which are themselves again branched once or twice. In *S. venturii* the branches are also very long, though not always branched again. In most species the branches are much shorter than the unforked part, and in some triploid clones of *S. commersonii* they are very short indeed. In species such as *S. acaule* and the Argentine and Bolivian species of series Megistacroloba the peduncle is very short and is normally unforked. The peduncle pubescence is normally rather similar to that of the stem, though one occasionally finds that the hairs increase or decrease in density, often become smaller, and are frequently mixed with a greater percentage of glands.

The flower stalks or *pedicels* are developed on the branches of the peduncles in *monochasial ebracteate cymes* (in which the main axis ends in a flower and its growth is continued by a side branch which does not come from the axil of a small leaf or bract). They bear a joint or *articulation* at a point where an abcission layer develops, so causing flower or fruit drop. No abcission layer is developed in series Acaulia, where the articulation is generally not very well marked and may often be completely absent. The position of the articulation is a useful taxonomic character, very low in *S. brevidens* and *S. kurtzianum*, in the middle third in most species, and very high up in *S. acaule* and most members of series Megistacroloba. The pubescence is often similar to that of leaf and stem, but often of much smaller hairs. As with the peduncle, the hair number may gradually increase or decrease in the pedicel, as compared with the lower part of the stem; there is certainly a tendency for more glandular hairs to be developed on it, especially above the articulation. If the calyx pubescense differs markedly from that of the stem and peduncle this same difference is also developed on the upper part of the pedicel, whilst the lower part, below the articulation, shows the pubescence type of peduncles and stems.

The calyx is generally *campanulate* (bell-shaped), with free *lobes* and *acumens*, but tubular below (see Fig. 5(*a*)). The lobe length is not of great diagnostic value, but the acumen length is, on the contrary, generally of great value. The calyx is normally green but becomes pigmented with purple in those species with deeply pigmented corollas (e.g. *S. vernei*). Often the pigment is darker in the centre midribs or *nerves* of the sepals, and paler towards the margins.

The corolla is normally showy, varying from whitish-cream to various shades of blue-violet or purple in different species. Occasionally the pigment is developed only on the lower or outer surface (*S. commersonii*, *S. kurtzianum*), but then generally shows through slightly to the upper or inner surface. A colour gene for reddish pigmentation occurs sporadically in cultivated potatoes and in what we consider to be escaped forms of *S. tuberosum* (see p. 435) but never in genuinely wild species. A rather reddish-purple colouration is seen in *S. kurtzianum* due to the presence of the anthocyanin pigment malvidin (Harborne, in Dodds 1956). Often the pigments are not developed evenly throughout the flower. Darker central petal streaks are seen in some forms of *S. spegazzinii*, whilst others have paler central areas and acumens. This can also be seen in some forms of *S. infundibuliforme*. *S. kurtzianum* is unique in showing deeply pigmented central petal streaks and acumens, with white areas between. The whole corolla of *S. vernei* is very deeply pigmented, and this pigmentation often extends to the calyx and other parts of the plant, too.

The shape of the corolla varies from truly *stellate* to very rotate (see Fig. 5(*b*)). In the genus *Solanum* there is a very short *tube* and a flat or saucer-shaped *limb*, whose margin is shaped into five *lobes*, corresponding to the petals; each lobe has a triangular point or *acumen*. In the stellate (star-shaped) flowers the acumens are hardly distinguishable, and the free parts of the petals or lobes are longer than their breadth at the base. If the length/breadth ratio varies around one, and the acumens are plainly visible, we speak of the corolla as *sub-stellate* or *semi-stellate*. This is a frequent condition in Argentine potatoes (e.g. series Megistacrolaba, and many Tuberosa). The *pentagonal* corolla has 'shoulders' to the lobes, which give it a pentagonal outline. The fully *rotate* (wheel-shaped) corolla has even better-developed petal shoulders and is rounded in outline. The *very rotate* corolla of *S. acaule* possesses very short lobes that are often more than three or four times as broad as long. Often the lobes are so flat that a ten-angled appearance is presented. In all measurements given we have taken the whole length of the lobe, including the acumens. The pentagonal or rotate corolla may be considered as theoretically derived from an originally stellate one in which the spaces between the lobes have been filled in by the growth of interpetalar tissues. Such tissues are often spoken of as *interpetalar membranes*.

The *star*, composed of tongues of vascular tissue, one in the lower central part of each petal, may be sometimes pigmented but is generally green and of little diagnostic value in the wild species. Short hairs or papillae are normally present on the edges and outer surface of the corolla acumens, often extending down the lobe centres. These hairs are often slightly branched. They are found in all species and are therefore of little taxonomic value.

The androecium or male part of the flower consists of five epipetalous stamens, the *filaments* being joined at their bases to the corolla tube (see Fig. 5(*c*)). They are glabrous and white in our species, except for *S. maglia* and *S. calvescens*, in which the yellow pigment from the anthers suffuses down on to them and the point of junction of anthers and filaments is not clear. In other species this point of junction is well marked. The *anthers* vary from pale lemon yellow to deep yolk yellow and are pressed together into a conical to barrel-shaped column except in *S. maglia*, where they spread apart loosely. The base of the anthers is generally 2-lobed but occasionally may be 4-lobed (as in *S. oplocense*). The apex is 2-lobed and possesses two terminal pores when just dehisced; later these elongate into splits running down the inner-facing side of the anthers.

The gynoecium has a number of important diagnostic features, though the *ovary* in the species dealt with here is of no particular taxonomic value. The *style* is normally covered in its lower $\frac{1}{3}$ to $\frac{1}{2}$ with very minute obtuse-ended papillae, only visible under a strong lens. In certain species longer hairs are interspersed, which are sharp-ended (especially in the so-called *S. setulosistylum*—see pp. 233–5). The style may protrude (be *exserted*) quite some way beyond the top of the anther column.

The *stigma* is frequently capitate (more or less globular), or *clavate* (elongated, club-shaped), and is often grooved on its upper surface in one direction and lobed below at right angles to this (see Fig. 5(*d*)). Occasionally it is bi-lobed above (as in *S. maglia*).

The berries are in this group of species either approximately spherical to globular, or ovoid to conical (see *S. commersonii*). They are often streaked

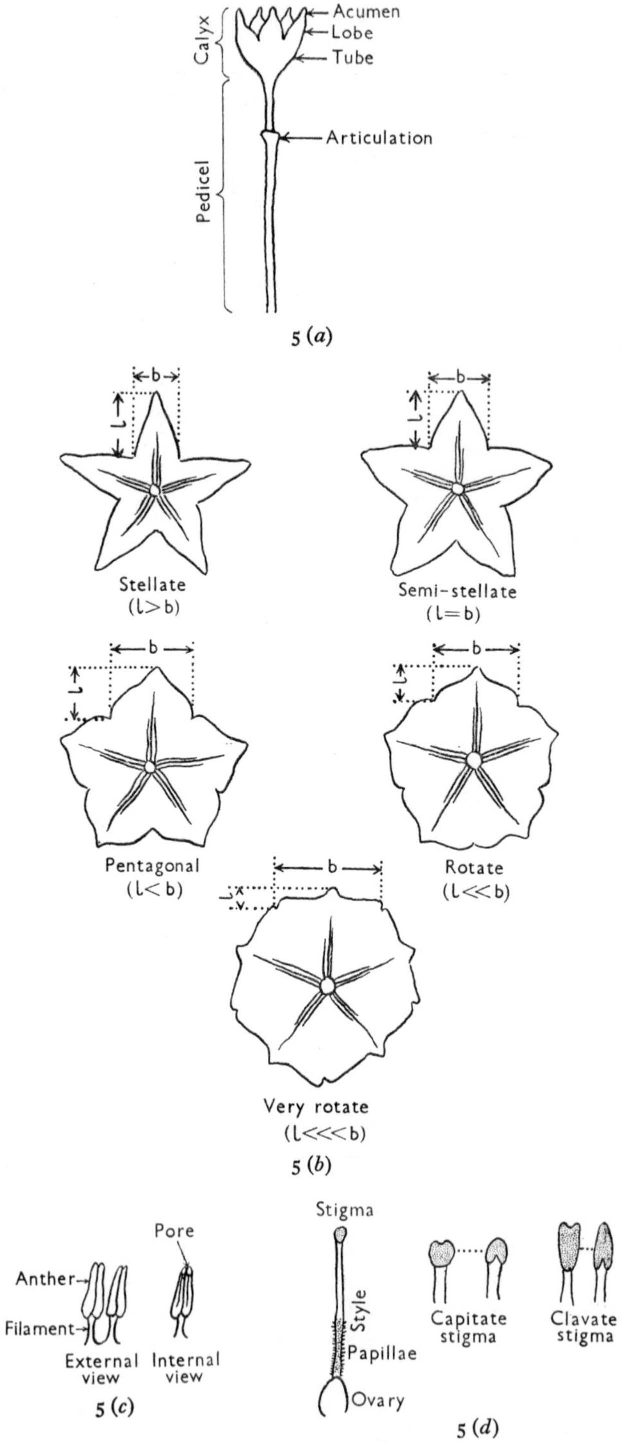

FIG. 5. (*a*) Typical calyx and pedicel. (*b*) Various corolla forms seen in Argentine wild potato species. (*c*) Details of potato stamen structure. (*d*) Style and stigma structure.

with white on a green ground, whilst purple pigmentation is sometimes present in the form of blotches or streaks. Very occasionally two shallow and poorly marked grooves are also present.

At the end of each species description we give the somatic chromosome number. Most species are diploid ($2n = 24$), but where occasional triploid or other chromosome numbers occur these data are given in parentheses.

3. *Affinities*

In this section we have given for each species our ideas of its affinities, based chiefly on morphological similarities, but also, where the data were available, on the results of crossing experiments and other investigations.

We have also added here some notes on how to distinguish each species from other closely similar ones with which it might easily be confused. Here we have given what are often spoken of as 'spot characters', those which are of particular value in distinguishing a species from its confrères because of their great constancy. These are of special value for identification, since they vary less under changes of environment and do not change so much because of genotypic variations as do many characters in the descriptions.

4. *Key to subspecies*

Where a particular species is distinguishable as two fairly well-marked subspecies a key is inserted here to aid in their identification. It is generally impossible to separate subspecies on a single character, and for this reason the key really takes the form of two contrasting sets of 'spot characters', of which most from each set will be possessed by each subspecies.

5. *Names and synonyms of subspecies*

The synonyms for the typical subspecies are given in Section 1, immediately following the name of the species itself. Those of the non-typical subspecies are listed chronologically and include varietal names, as was mentioned before (see p. 101).

6. *Descriptions of subspecies*

These descriptions follow the same lines as the species descriptions, except that only the features which serve to separate the two subspecies are mentioned.

7. *Specimens examined*

In this section we give details of all the specimens examined by us, arranged in the following geographical sequence.

I. BRAZIL

States of: Rio Grande do Sul
Santa Catarina
Paraná
Minas Gerais

II. PARAGUAY

III. URUGUAY. Departments arranged in alphabetical sequence.

(BOLIVIA. Records sometimes added where necessary for elucidation of distribution of Argentine species.)

IV. ARGENTINA. Provinces arranged in an anti-clockwise sequence, as follows:

1. Jujuy
2. Salta
3. Tucumán
4. Catamarca
5. La Rioja
6. San Juan
7. Mendoza
8. San Luis
9. Córdoba
10. Neuquén
11. Río Negro
12. Chubut
13. La Pampa
14. Buenos Aires (including Capital Federal)
15. Entre Ríos
16. Santa Fé
17. Corrientes
18. Misiones
19. Chaco
20. Formosa
21. Santiago del Estero

Within each Argentine province the collections are arranged alphabetically by Department where this has been possible.

Within each State of Brazil, each Department of Uruguay, Bolivia, or Argentina, and in Paraguay (without further subdivisions) the collections are arranged alphabetically by the names of the collectors. Insufficiently localized collections or ones in which the collector's name is missing are listed at the end of each section.

We have tried as far as possible to standardize the entries by arranging the data in the following sequence:

(*a*) Locality, in order of increasing exactness.
(*b*) Altitude (always in metres), where known.
(*c*) Date of collection.
(*d*) Collector's notes on flower colour, habitat, etc.
(*e*) Collector and number (see also Appendix II, p. 449).
(*f*) Herbaria from which the material was obtained on loan or where it was examined (see Appendix I, p. 446).
(*g*) Our own notes, where appropriate.

The standard herbarium abbreviations are given and one or two special collections are also quoted. All these are listed in Appendix I.

Where the species has been divided into two subspecies the collection data are given in two separate sections.

8. *Geographical distribution*

The locality data given in Section 7 have been synthesized into a more simple form here, so that one can see at a glance the main outlines of the distribution as known at present. Besides noting the provinces, departments, etc., in which the species has been recorded, we have also discussed its distribution outside the area under survey for those species such as *S. brevidens* and *S. acaule* that spread beyond it.

Distribution data are given separately for each subspecies when these occur, and possible gaps in knowledge are indicated. The information is presented in tabular form in Table 1, p. 31, and also graphically in the distribution maps.

It is realized that the area of our present survey has not been studied in the same detail all over, and that more collections have been made in the more easily accessible areas in the region of Montevideo, Buenos Aires, Córdoba, Tucumán, etc., than in the more remote areas.

9. *Habitat*

We have in this section for each species attempted to bring together information on altitude range and habitat from herbarium labels, literature sources, our own observations and those of other collectors. The altitude ranges are given when possible to illustrate normal and extreme ranges (e.g. *S. spegazzinii*: (1500–)1900–3100(–3700) m, which indicates that the normal range is from 1900 to 3100 m but plants have been seen as low down as 1500 m occasionally and as high up as 3700 m). It is of course possible that some records are not very accurate, and when an obvious inaccuracy occurred we have not included it. We may well have missed other inaccuracies, but have made every attempt to check the altitudes against geographical features on maps and by referring to our own memories of places visited (see Figs. 2 and 3).

The flowering season is also given, largely taken from herbarium labels; we have also added whatever information we could obtain from other sources. Notes on species found frequently with the wild potato species under discussion are given, as well as the habitats in which they are found, with, wherever possible, information on soil type and degree of humidity. Rainfall data are added from literature sources where known.

The species has also been placed in its main phytogeographical dominions, provinces and districts, based on the scheme put forward by Cabrera (1953). (See Maps 1 and 2 and Tables 2 and 3.) Its frequency and status in each of these phytogeographical regions has been assessed where possible, that is to say, we have attempted to see whether it forms a part of the natural vegetation or whether it grows as a weed, a ruderal or viatical plant, or as an adventive in special man-made habitats. A general discussion of this is also given in Chapter 3.

10. *Taxonomic (and nomenclatural) notes*

This is perhaps the most interesting section for each species, since it incorporates much information on variability and relationships that is given only in bare outline in Sections 1, 2, and 3.

A discussion of the history of the naming and description of the species is given here, which is also discussed in the chapters on the historical survey and potato collections (1 and 2), though there only in very general terms. Here we go into more detail and attempt to explain the reasons underlying the changes made in the naming and classification of the species, as well as such other taxa that were at one time considered to be distinct and are now regarded as synonyms. We also mention here certain names given that were not validated by descriptions (*nomina nuda*). The nomenclatural discussion is linked to a discussion of the pattern of variability of the species in question, thus giving the reasons for the large number of synonyms where these occur. Where cytological races (cytotypes) occur, which can be linked to morphological differences, these also are discussed. A general description of the range of variability of the species, organ by organ, is given in most cases. Geographical clines or variations from one part of its distribution area to another are also mentioned, as well as possible introgression and its effect on the nomenclature. If great variations occur, due to differences of habitat (altitude, exposure, insolation, soil type, moisture) these are discussed. We have also presented the views of other *Solanum* taxonomists, chiefly Bitter, Bukasov, Brücher, and Correll, on the species in question.

Finally, we have noted areas of doubt and uncertainty in our knowledge of these species and indicated a few lines of interest along which investigation should be directed in the future.

11. *Notes on living plants*

One of the points to which we have paid a great deal of attention in the present work has been the study of living material under experimental conditions, and to some extent also, when opportunity has permitted it, to a study of the living plants in their natural environment. The present section deals with our work on such material, which totals 179 accessions of wild potatoes from Argentina, Brazil, Paraguay, and Uruguay, as well as a few from Bolivia and Chile. This material has been grown in England, at Cambridge and at Birmingham; and/or in Denmark, at Vandel and Copenhagen. Table 4 indicates the number of collections grown for each species and subspecies. The material was collected and grown in the form of tubers or seeds, and has been maintained whenever possible in the form of seeds.

In addition to our own observational and experimental work we have noted the results of other colleagues where appropriate, taken either from personal communications or from literature sources. We are greatly indebted to several of them for allowing us to use their unpublished data.

TABLE 4

Living collections of wild potatoes from Argentina, Brazil, Paraguay, and Uruguay grown in England and/or Denmark

S. brevidens	1+(1)
S. calvescens	1
S. commersonii subsp. *commersonii*	28
,, *malmeanum*	12
S. chacoense subsp. *chacoense*	59
,, *muelleri*	2
S. tarijense	1+(3)
S. acaule subsp. *acaule*	9
,, *aemulans*	3
S. infundibuliforme	5
S. megistacrolobum	8
S. sanctae-rosae	1
S. microdontum subsp. *microdontum*	1
,, intermediate	3
,, *gigantophyllum*	11
S. vernei subsp. *vernei*	3
,, *ballsii*	2
S. gourlayi	2
S. oplocense	1+(2)
S. spegazzinii	11
S. kurtzianum	5
S. maglia	1+(3)
Total	179

Note. Numbers in parentheses indicate accessions of Chilean or Bolivian material.

One of the main features of this section for each species has been the noting of differences between field- and glasshouse-grown material, on the one hand, and South American material on the other (the latter based on observations made on herbarium specimens). The chief observable differences have been: tuber shape and size, as well as tuber pigmentation, where known; habit; height of plant; branching; leaf dissection; pubescence type (including presence or absence of glands); rosetting and internode length; pigmentation of stems and leaves, etc.; corolla colour, size, and form; berry shape and size, as well as the markings. Some of these features are not always discernible from herbarium specimens or may be considerably modified owing to indifferent methods of preservation; such are corolla colour and form, berry shape and markings, tuber shape and colour. Other features are not normally available from the accompanying information on the label: maximum tuber size, tuber weight, total yield (this has been recorded by us under artificial short days at Vandel), frost and blight resistance. We have noted such features wherever possible, in order to make the descriptions of the species more complete.

Unpublished notes kindly provided by Professor Ross have also been added where we ourselves had little or no information, as with the species

S. venturii, *S. oplocense* (2*x*), and *S. vidaurrei*. These notes were based on his collections grown at Cologne in Germany.

Notes from the literature on plants observed under cultivation in Europe, and especially those provided by Bukasov and his colleagues, have also been included. We have noted in many places the presence of resistance to blight, frost, and eelworm, together with virus resistance or content; for a full account of this work reference should be made to Chapter 5.

Observations made in the field by ourselves in South America appear partly in this section, though much of this information is scattered throughout the whole book. We have added references to published work on field observations in South America, chiefly that of Brücher (various publications), Schroeder (1928), and many others.

In the discussions of our own results we have attempted to make exact comparisons of clonal material grown in Europe with the corresponding phenotypes of the same material in South America. This has helped to show us the range of plasticity, induced by differences of environment, on a standard genotype. It has also been valuable to grow families from seeds collected in South America, so as to assess the normal range of genotypic variation within a population and to compare this with that of plants collected from the same populations in their native countries. Unfortunately, we have been able to make only a small number of comparisons of this sort through lack of material, but we are convinced that more work of this nature would be most valuable.

Where a species or subspecies is known to occur as two different cytological races we have tried to compare the morphology of the different cytotypes (2*x* and 3*x*; 2*x* and 4*x*) in an attempt to throw more light on the possible origins of the non-diploid forms.

We have also tried to assess the proportions of genotypic and environmental control of phenotype by growing accessions of previously differing phenotype under uniform conditions. When this has been done we have generally been able to observe that variation was due largely to environmental differences, but not completely so. A part of this variation was genetically controlled but nevertheless could not be considered as sufficient to divide the material into more than one species, since no obvious and constant discontinuities could be observed.

Conversely, a comparison of glasshouse and experimental field material of the same clones or accessions, when grown under the different conditions of Birmingham and Vandel has also given us some idea of the amount of variability induced by environmental differences, much of which could be matched closely with variants collected in South America. It has thus been possible for us to understand at least some of the variation of our South American material, in terms of how much of the variation could be assigned to environmental causes and how much to differences of genotype. As a case in point, the glasshouse-grown material from England and Denmark matches naturally occurring shade forms very closely (i.e. with '*S. famatinae*').

Such studies have enabled us to obtain a clearer idea of the characters which remain relatively constant, both in the wild and under cultivation, and are therefore of greatest value in identification.

Finally, by comparing material grown under the various conditions mentioned we have been able to solve problems posed by extreme ranges of morphological variation (as with *S. spegazzinii*), of introgression (as with *S. chacoense*), and of specific and subspecific delimitation (as with *S. chacoense*, *S. commersonii*, *S. acaule*, *S. microdontum*, and *S. vernei*). It has also been possible to speak with more confidence of the inadvisability of dividing the species into formal varieties and forms, and to show in many cases what sort of variation is due to the natural interplay of environmental and genetical variation within the species and how much to the effects of natural hybridization (see pp. 119–22).

12. *Cytology*

In this section for each species we report on the somatic chromosome number as established by ourselves or by our colleagues, K. Jakobsen, P. Jacobsen, O. Mattsson, and R. Tarn, or by Professor H. Ross in written communications. Literature references are also given, where appropriate.

With one exception (*S. infundibuliforme*) meiotic studies have not been carried out, but are reported from the literature. Meiosis in the species itself is followed by descriptions (where such exist) of this process in artificially produced hybrids between the species in question and other species, generally from the area that we are considering. Artificially produced polyploids are also mentioned, and some reference is made to the pachytene studies of Gottschalk and his co-workers, as well as to the criticisms of these methods by von Wangenheim and others.

Finally, where cytological races are known we have discussed their occurrence in relation to morphological differences and to their geographical distribution. The general results so far obtained indicate that the non-diploid forms do not occur with any type of geographical or other regularity.

13. *Fertility and Crossability*

In this section we have discussed the fertility of each species within itself, and have followed with an account of the successes and failures resulting from attempts to hybridize it with other species, for the most part from the countries dealt with here, but occasionally with other species from further afield.

In each of these two items we have given the data from the literature first, followed by our own results.

Most diploid species of *Solanum* are self-incompatible but fully fertile when biotypes with different sterility factors are crossed. *S. brevidens* is one of the rare exceptions to this rule and is fully self-compatible. Several other species have been shown to be partially self-compatible (e.g. *S. chacoense*,

S. tarijense, etc.) when enough attempts at selfing have been made. In other *Solanum* species outside the region we are now dealing with, certain self-compatible clones have been discovered in species that are normally thought to be self-incompatible, but such a phenomenon has not been seen in the group of species at present being discussed.

The triploid forms of all the diploid species so far studied are highly self-sterile, due to defective pollen and low female fertility. Most will form hybrids with some difficulty when acting as female parents if pollinated by fertile diploid or tetraploid male parents.

The tetraploid species, *S. acaule*, is not only fully self-compatible but is habitually self-pollinating.

Little information is available on pollinating agents, though it is generally supposed that Hymenoptera are the usual pollinators. Observations on *S. chacoense* and *S. commersonii* are given, but further information is needed for other species.

The reasons for our attempts to produce hybrids between the various species studied in this work fall into two categories. In the first place we wished to have available a series of voucher specimens and living families of known parentage, in order to be able to make valid deductions on the nature of the suspected hybrids collected or observed in South America.

Secondly, we wished to arrive at an understanding of the relationships between species by observing the fertility of F_1 hybrids and the degree of genetic breakdown observable in F_2 families, to measure the similarity or otherwise of the genetic 'architecture' of the pairs of species so tested. We wished to be able to see whether gene flow could occur between species, given the necessary conditions in nature, though we fully realized that care must always be taken in extrapolating the results of crosses made in the experimental field or glasshouse to the situation as it occurs in the wild. We shall not deal with the full significance of the F_2 families here, since that topic has already been discussed in Chapter 4.

Of the 86 families raised, most (62) were F_1, and unfortunately we were only able to raise 24 F_2 families, partly through lack of time and space, or of suitable initial material, and partly through a lack of fertility of some of the F_1 families, which effectively prevented us from taking the crosses any further.

The details of the F_1 crosses made by us and reported in the literature are given schematically in Table 5, where the natural hybrids identified by us and discussed below in Section 14 are also presented. It will be seen that only 36 cross combinations made by ourselves are noted in the table, and not 62, as stated above. This is because we made a certain number of intra-specific crosses, which are not shown on the table, and also because we often tried several combinations of accessions, as well as reciprocals, between a given pair of species.

Details of the F_2 progenies are given in Table 6 (p. 117). They show that there is a complete genome compatibility between the subspecies of the

TABLE 5. *Natural and artificial hybrids between the wild species*

Species	*S. brevidens*	*S. calvescens*	*S. commersonii* subsp. *commersonii*	,, subsp. *malmeanum*	*S. chacoense* subsp. *chacoense*	,, subsp. *muelleri*	*S. tarijense*	*S. acaule* subsp. *acaule*	,, subsp. *aemulans*	*S. infundibuliforme*	*S. megistacrolobum*	*S. boliviense*	*S. sanctae-rosae*	*S. microdontum* subsp. *microdontum*	,, subsp. (intermediate)	,, subsp. *gigantophyllum*	*S. venturii*	*S. vernei* subsp. *vernei*	,, subsp. *ballsii*	*S. gourlayi*	*S. oplocense*	*S. vidaurrei*	*S. spegazzinii*	*S. kurtzianum*	*S. maglia*
S. brevidens																									
S. calvescens																									
S. commersonii subsp. *commersonii*					X		—																	[×]	
,, subsp. *malmeanum*					X																				
S. chacoense subsp. *chacoense*			×			X	X	[×]		X					X	[×]			X	X			X	X	X
,, subsp. *muelleri*																									
S. tarijense					×					[×]						[×]			X					X	
S. acaule subsp. *acaule*									X		X		X			[×]									
,, subsp. *aemulans*																								X	
S. infundibuliforme											X					[×]		X		X			X		
S. megistacrolobum								×		×				—		—				—					
S. boliviense											×														
S. sanctae-rosae														X	X	X			X						
S. microdontum subsp. *microdontum*					×							?			X										
,, subsp. (intermediate)							×				×					X		X	X						X
,, subsp. *gigantophyllum*					×													X			—			[×]	[×]
S. venturii													×												
S. vernei subsp. *vernei*																×	×				—		X	X	
,, subsp. *ballsii*					?		?							×	×					X			X		
S. gourlayi							?			×	?														
S. oplocense											?									?					—
S. vidaurrei							?													×					
S. spegazzinii					×		×	×	×				×			?		×						—	
S. kurtzianum					×											×							×		X
S. maglia																									

Artificial hybrids (above the diagonal)

Natural hybrids (below the diagonal)

× Natural hybrids observed and/or cultivated by us.
X Artificial hybrids made by us (to F_2 in some cases).
[×] Hybrids reported in the literature.
— No hybrids obtained.

Note. This table does not convey information on the *direction* of the cross; see appropriate text pages for details of crosses made by ourselves.

TABLE 6

Genome compatibility within and between species from Argentina and surrounding countries as shown by F_2 progeny tests (typical subspecies unless otherwise stated)

F_1 Family No.	F_1 Combination	F_2 Family Nos.	F_2 Combinations	Results
274	*S. chacoense* (pl. 5) × *S. commersonii* (H. 185)*	286, 287	274/3 × /1; 274/2 × /1	Breakdown in both families
275	*S. chacoense* (pl. 5) × *S. commersonii* (H. 184)*	292, 293	275/1 × /5; 275/4 × /2	Breakdown in both families
120	*S. chacoense* subsp. *muelleri* (2426/3) × subsp. *chacoense* (2062)	218, 219	120/13 × 120/17; 120/17 × 120/14	No breakdown; all seedlings vigorous
273	*S. chacoense* (H. 137) × *S. tarijense* (H. 122)	294, 295	273/3 × /4; 273/5 × /1	Breakdown in both families
289	*S. gourlayi* (2480) × *S. chacoense* (2435/1)	337, 338	289/2 × /1; 289/3 × /1	Breakdown in both families
61	*S. acaule* (H. 129/1) × subsp. *aemulans* (82/1)	174, 175	61/23 × 61/19; 61/3 × 61/1	No breakdown; all seedlings vigorous
130	*S. sanctae-rosae* (2483/4) × *S. toralapanum*† (1773)	185, 221	130/7 × 130/3; 130/1 × 130/2	Breakdown in 185, not in 221
131	*S. toralapanum*† (1773) × *S. microdontum* subsp. *gigantophyllum* (51a)	186, 222	131/2 × 131/1; 131/1 × 131/2	Breakdown in 186; poor germination in 222
297	*S. infundibuliforme* (2477) × *S. vernei* (2413/1)	339	297/2 × 297/1	Breakdown
140	*S. microdontum* subsp. *gigantophyllum* (2417/3) × subsp. *gigantophyllum* (51a)	188	140/6 × 140/4	No breakdown
13	*S. vernei* (2414/3) × *S. microdontum* subsp. *gigantophyllum* (51a)	159, 178	13/2 × 13/1; 13/11 × 13/1	No breakdown; all seedlings vigorous
143	*S. vernei* (2414/3) × *S. spegazzinii* (H. 68)	190	143/7 × 143/9	Breakdown
290	*S. gourlayi* (2480/1) × *S. sparsipilum* (533)‡	320, 321	290/3 × /2; 290/4 × /2	Breakdown in both families
121	*S. maglia* (2057) × *S. kurtzianum* (H. 62)	220	121/1 × 121/2 [121/8 × 121/6: no germination]	Breakdown

* Hybrid seed sent by Dr. Toxopeus, Wageningen. † A Bolivian species allied to *S. megistacrolobum*.
‡ A Peruvian and Bolivian weed species.

species tested (*S. chacoense*, *S. acaule*), and also within one subspecies (*S. microdontum* subsp. *gigantophyllum*). All seedlings from this sort of cross are uniform and thrifty, as would be expected, though unfortunately not enough combinations were made for us to be able to speak with certainty about the complete group of species. It seems, however, reasonable to suppose that genome evolution will not have taken place within each species.

Clear incompatibility of genomes is shown in the combinations: *S. chacoense* × *S. commersonii*, *S. chacoense* × *S. tarijense*, *S. toralapanum* (a species allied to *S. megistacrolobum*) × *S. microdontum* subsp. *gigantophyllum*, *S. gourlayi* × *S. chacoense*, *S. infundibuliforme* × *S. vernei*, *S. vernei* × *S. spegazzinii*, *S. gourlayi* × *S. sparsipilum* (a Peruvian and Bolivian weed species), and *S. maglia* × *S. kurtzianum*.

Curious and apparently anomalous results are shown in the cross *S. sanctae-rosae* × *S. toralapanum*, where one F_2 family showed breakdown and the other did not; and in the cross *S. vernei* × *S. microdontum* subsp. *gigantophyllum*, where both F_2 families were thrifty. Although *S. vernei* and *S. microdontum* are very distinct morphologically and seem to remain so in nature, with very few natural hybrids between them observable (a total of ten seen by us), they seem to be genomically identical. Further work is needed to verify these results, and it would also be valuable for them to be checked by Swaminathan's comparison of multivalent frequencies in the autotetraploids and amphidiploids mentioned in Chapter 4.

Each hybrid family is described in the text in some detail, and information is also given on the ease with which the cross was made and its apparent fertility. Similar details are also given from the literature, when reported, as well as any reciprocal differences observed. Looking back on this work we have some considerable regrets that we were not able to try more combinations between each pair of species in those cases where some failure in cross-ability or fertility ensued. It is becoming increasingly known that a single negative result is not to be relied upon absolutely, and that many attempts with different accessions should be made before an investigator can say with reasonable certainty that two species do not cross. Unfortunately, in many instances we had insufficient material to try enough crosses, but we have given the results here, so that other workers will be able to attempt the crosses again, using different accessions.

Our results have been arranged here in taxonomic sequence under each species, but are described fully only under the first species in this sequence, to avoid unnecessary duplication. For example, the crosses of *S. chacoense* × *S. microdontum* are described under *S. chacoense* and are only mentioned briefly under *S. microdontum*, though they are always cross-referenced back to the page where they are described fully. This method carries the possible disadvantage that large numbers of hybrids are described under species such as *S. chacoense* that appear early in the taxonomic sequence, and very few under those that appear late, such as *S. spegazzinii* and *S. kurtzianum*.

We have also mentioned the crosses made between lines of the same species and those between the two subspecies, when these occur. On the whole, we made very few intra-specific crosses, however.

It will be realized that many artificial hybrids have been made that could not occur in nature, at any rate at present, since the species concerned do not overlap, either geographically or altitudinally. However, as we have already mentioned, this work was partly concerned to show the relationships between species as exhibited by their fertility and the nature of the hybrids, if any, that could be formed between them. For those sympatric species that seem to preserve their identities by different ecological preferences the production of artificial hybrids is, on the other hand, of considerable interest, since such species frequently seem to form natural hybrids in certain regions, and these artificial hybrids can usefully be compared with the natural hybrids described in the following section.

14. *Natural hybrids*

The occurrence of natural hybrids in the tuber-bearing Solanums adds considerably to the difficulties of classification and identification of this group of plants and makes it even more necessary to study living material whenever possible.

No doubt at one time natural hybrids were very rare, but since most species have been evolving under barriers of an ecological nature any interference by man with the natural environment will tend to render such barriers less absolute. This will result in the formation of intermediate or 'hybrid' habitats, as Anderson has termed them, into which species can migrate which were once kept apart by their adaptation to distinct natural habitats. Under these conditions the species will form hybrids and the hybrids themselves will find a series of suitable ecological niches which would not have been available to them in undisturbed conditions. As we have seen in the previous section on artificial hybrids, there seem to be a rather limited number of genetical barriers to intercrossing, at least among the diploid species under discussion. Successfully established hybrids with a reasonable amount of fertility may be the means to a flow of genes from one species into another, and this phenomenon can certainly be discerned in a number of cases. The classic example is that of *S. chacoense* and *S. microdontum* which we have discussed fully under the former species (pp. 213–18). There is also some evidence of hybridization and introgression with *S. sanctae-rosae* and *S. venturii*, with *S. tarijense* and a blue-flowered species that we have not been able to identify with certainty from a study of the introgressed forms, with *S. megistacrolobum* and *S. boliviense*, and with *S. microdontum* and *S. boliviense*. More observational and experimental data are needed on this topic, and we undoubtedly have here an ample field for interesting future research.

Perhaps at first sight it will be thought that we have seen hybrids and introgressed forms everywhere! We can assure the reader that we have tried to resist this temptation as far as possible, and have attempted not to fall into the common trap of labelling everything as a hybrid that does not fit into a preconceived narrow concept of what each species should conform to. As we have seen, there is a wide range of variability in each species which has nothing to do with hybridization or introgression, though we are not at all certain in some cases whether further study might not reveal the evidence of introgression which is at the moment obscured through lack of material and inadequate knowledge.

Returning to the problem of the actual amount of hybridization noted by us, it will be of interest to see that we have designated 123 collections of possible or definite hybrids out of a total number of collections examined of about 1300.[1] This is not more than about 9·5 per cent, which is a fairly low figure when one considers the amount of success obtained in the production of artificial hybrids. It seems to point towards the fact that under natural conditions there must be certain sterility or other barriers, besides the ecological ones, to prevent the formation and establishment of hybrids. Perhaps the hybrids are not really so well adapted, even to the 'hybrid habitats' as their parents, and may not survive for long, even though they could, if well-adapted, reproduce indefinitely by vegetative means. Another possibility is that collectors may not have made so many gatherings of hybrids as of species, in the belief that the hybrids were 'atypical' and therefore to be avoided. On the whole we do not think this explanation is very likely, since very few collectors have known what the species were in any case, and would therefore collect species and hybrids at random.

The problem of the identification of hybrids is not an easy one, and it is obviously necessary to steer a middle course between the Scylla of seeing a hybrid in every slight variant, on the one hand, and the Charybdis of classing all variants as new species, varieties, or forms. We have gradually formulated the following methods and criteria for hybrid identification:

(*a*) *Morphological.* By studying the morphology of the collection in question it may be possible to decide whether it is intermediate in character expression between two previously known species. It need not, of course, lie necessarily half-way between such species, even if it is a simple F_1 hybrid; and if it is some sort of back-cross or F_2 segregate then it is unlikely to be exactly intermediate on that score either. Flower colour is often of considerable help, since if hybridization is taking place between white and purple-flowered species the flower will generally possess a 'washed out' or pale purple colour, with variation in intensity between the petal centres and the interpetalar membranes.

(*b*) *Variation.* If a reasonable number (sometimes as few as four or five may be sufficient) of specimens has been taken in a collection the variation of these may be a good indication of hybridity, since the population from which they were taken is in all probability a series of segregants ranging in their

[1] This does not include our 1966 collections. See Appendix III.

morphology from one parent to the other, though rarely approaching the parental character combinations completely. Very often the collection will have been made in the form of duplicate sheets that were distributed to several different herbaria in various parts of the world. It is helpful then to be able to reassemble these duplicates by obtaining them on loan, so as to make a considered comparison of all the specimens at one time. This method presupposes that the collector took all plants at random; unfortunately this cannot always be assumed, but since, as we have just said, knowledge of the species in this group of plants is not very great, most collectors will in fact be making a fairly random collection of the material of both species and hybrids.

(*c*) *Locality study*. A useful corroborative method is to make a careful examination of all material collected at the locality and altitude where the probable hybrid occurred to see whether the parents were also gathered or recorded. It is often necessary to compare collections made by different collectors at different times, and for that reason we have grouped such data together in the form of a table, for each species. In any case it is useful to collate and compare all the material from any one location, since it is sometimes possible to discern the influence of more than two species in a hybrid population.

(*d*) *Habitat study*. Ideally it would also be valuable to estimate the degree of 'habitat hybridization', the influence of man in changing the environment, so providing conditions in which hybridization could take place. This is difficult to do unless herbarium labels bear more habitat information than is usually the case; and generally a concept of habitat changes of this sort could be adequately obtained only by means of a careful study in the field. On a simpler level, it is useful to know the approximate habitat ranges and preferences of the two putative parental species so as to decide whether they could possibly have grown under similar conditions together and so formed a hybrid.

(*e*) *Chromosome numbers*. If living material is available, and if the two probable parents possess different chromosome numbers, then the identification of a hybrid on the basis of chromosome number is of comparitive simplicity. Unfortunately, this method has not been available to us, since we have had no living material of natural crosses with the tetraploid species, *S. acaule*, though Brücher (1959*a*) used the method with some success. All the natural species hybrids examined by us in the living state have been diploids.

(*f*) *Comparison with artificial hybrids*. Another very useful method of hybrid identification is to compare the naturally occurring hybrids with artificially produced ones of known parentage. As we explained in Section 13 this has perhaps been one of the most valuable methods available to us for the identification of hybrids. If the artificial and natural hybrids can be grown together under closely similar conditions in the experimental garden an exact comparison can be made, resulting in a high degree of probability as to whether the hybrid is genuine or not. Even where living material of the natural hybrids is not available, artificial ones grown in the field and glasshouse can provide very good material for comparison with the herbarium specimens of natural hybrids.

(*g*) *Experimental study of hybrids.* It is of the greatest value to possess living material of the possible hybrid collections for study. A single tuber line is the least valuable material, but even this can be grown next to artificial hybrids, as mentioned above. It can also be backcrossed to either parent, to observe whether segregation occurs if the line is a hybrid. Serological tests (as with the Camañ collection of *S. commersonii* × *S. chacoense*) can also provide a key to the hybrid nature of the line in question.

If natural seed is available then the problem becomes ideal, since it is then necessary only to grow a small family to observe the segregation of characters that should occur if the original collection was a hybrid.

We have been fortunate enough to have been able to grow a total of eleven seed and tuber lines of naturally occurring hybrids during the present work; and we have also grown a certain amount of material which we thought, from a preliminary examination of herbarium specimens, to have been of hybrid origin, only to find that it was not of this nature when we examined the progeny. For instance, we had assumed that certain collections of *S. spegazzinii* were hybrids with *S. infundibuliforme* because of the strong leaflet decurrency in the original specimens. The seedlings revealed, on the contrary, no rhachis wings or leaflet decurrency, and a considerable uniformity!

The natural hybrids grown by us are noted in Table 7.

TABLE 7

Living collections of natural hybrids

	Number	Tuber or seed
S. commersonii × *S. chacoense*	1	T
S. chacoense × *S. microdontum* subsp. *gigantophyllum*	1	T
S. chacoense × *S. spegazzinii*	1	S
S. chacoense × *S. kurtzianum*	1	S
S. tarijense × *S. sp.* (blue-purple fls.)	4	T & S
S. sanctae-rosae × *S. venturii*	1	T
S. sanctae-rosae × *S. spegazzinii*	1	S
S. vernei × *S. spegazzinii*	1	T & S
	11	

We should perhaps remark at this stage that we have certainly not been able to apply all seven of these methods to every suspected hybrid, but have used some of them where possible in each case.

To return to the presentation of the data in the chapters that follow, it should be mentioned that we have followed the same sequence of taxonomic arrangement as with the artificial hybrids; thus the natural hybrids are fully described under the species in the combination to appear first in the taxonomic sequence. They are also mentioned under the second parent, and cross-referenced to the first. A table of natural hybrids (Table 5, p. 116), as well as artificial ones, has already been referred to. The hybrids are described in terms of the combination of parental characters exhibited by them, and compared with the artificial hybrids when these have been made. The data in the tables of occurrence of hybrids and parents are arranged by locality and altitude, the collections being referred to by means of the collector's name and number. Where one parent is missing the hybrids have still been noted, but not when neither parent has been found in the same area.

15. *Localities of natural hybrids*

These are arranged in the same way as the descriptive matter referred to above. Within any one hybrid combination the localities are arranged in the sequence of species localities referred to in Section 7: *Specimens Examined.*

B. CULTIVATED SPECIES

The cultivated species are dealt with after the wild species. For these the data are presented in a completely different way from the wild species, chiefly because our method of work with them was different and the information available was not so complete. A key for their identification appears on p. 429.

We have arranged the species in ascending order of ploidy, from triploid (*S.*×*juzepczukii*) through tetraploid (*S. tuberosum*) to pentaploid (*S.*× *curtilobum*).

The sequence for each species is basically as follows:

1. *Name and synonyms.* Treated in the same way as in the wild species.
2. *Short description.* This attempts to give only as much information as is necessary to distinguish each species from the others.
3. *Herbarium specimens examined.* This section is similar to the *Specimens Examined* sections of the wild species.
4. *Tuber collections grown in Argentina.* Here are listed the tuber collections made and grown by J. P. Hjerting and his associates in Argentina. The native name and a short description of the tubers is given in each case. Fuller details of the classification of these is given in the introductory section to the cultivated species on p. 428.
5. *Distribution.* The data given in the previous two sections are summarized in this section.
6. *Status in Argentina.* A general discussion of the origin and present status of the species in Argentina is given here, together with a discussion of taxonomic points of interest, in a similar way to the Taxonomic Notes section in each wild species.

Since *S. tuberosum* occurs as two distinct subspecies the treatment for this is a little more complex. Subsp. *tuberosum* has been dealt with in two sections:

(*a*) *Cultivated semi-indigenous forms.* It should be noted that the commonly grown cultivated potatoes of the lowland plains of Argentina, Uruguay, and Brazil are not dealt with in this book.

(*b*) *Escaped forms.* These have been found in southern Argentina, and are considered to be truly wild by some authors, but we believe them to be relics of ancient cultivations left by nomadic Indian tribes.

Subsp. *andigena* is dealt with in the usual way for cultivated species.

7

CLASSIFICATION OF THE GENUS *SOLANUM*

THE family Solanaceae was divided by Bentham and Hooker (1876) and by von Wettstein (in Engler 1895) into five tribes: Solaneae, Atropeae, Hyoscyameae, Cestrineae (or Cestreae), and Salpiglossideae. The genus *Solanum* was grouped by von Wettstein in subtribe Solaninae of the tribe Solaneae, together with such other genera as *Capsicum*, *Physalis*, *Saracha*, and *Bassovia*. *Lycopersicon* was included in *Solanum* in that treatment but has generally been regarded as distinct.

Solanum is by far the largest of the genera in the family and contains over 2000 species. Most are herbs or small shrubs, often clothed with thorns, and less than one-tenth of the genus is tuber-bearing.

Dunal (in de Candolle 1852), who was the first specialist in the genus, divided it into two sections, *Pachystemonum* and *Leptostemonum*. The former section included plants with thick anthers and simple, non-stellate hairs, whilst the latter contained species with long narrow anthers and stellate hairs; in addition, some species were thorny. Linnaeus included *Lycopersicon* in *Solanum*, as also did von Wettstein, as we have seen above. The possession of elongated anthers which dehisce by slits along their whole length and whose narrow apical portions are sterile, serves to distinguish *Lycopersicon* from *Solanum*. In contrast, the anthers of *Solanum* dehisce by apical pores which extend downwards as slits with age. There is no doubt, however, that the two genera are very closely related.

The tuber-bearing species belong to section *Pachystemonum*, which Dunal further divided into five subsections: (1) Tuberarium; (2) Morella; (3) Dulcamara; (4) Micranthes; and (5) Lycianthes. The potatoes were, of course, placed in subsection Tuberarium.

The German *Solanum* specialist, George Bitter (1912, etc.), elevated sections *Pachystemonum* and *Leptostemonum* to subgenera and the subsections to sections. *Lycianthes* was separated off into a distinct genus. Bitter formed many new sections but unfortunately was not able to complete his monographic work of the genus before his death. Thus the classification of the complete genus was left unresolved, since by Bitter's time Dunal's monograph, which even to begin with was unsatisfactory in places, had become very out of date.

The classification of the genus has been considerably improved recently, with the publication of the work of Almut Seithe (1962). This author, on the basis of Bitter's unfinished work and her own studies on hair type, has set out a new classification of the genus, to which the work of Dunal has also been related.

Seithe established six sub-genera, as follows:

1. *Solanum* (Sect. Pachystemonum *pro parte*)
2. *Lycopersicum* (Genus *Lycopersicon*)
3. *Bassovia* (Genus *Bassovia pro parte*)
4. *Archaesolanum* (Sect. Pachystemonum *pro parte*)
5. *Lyciosolanum* (Sect. Pachystemonum *pro parte*)
6. *Stellatipilum* (Sect. Leptostemonum)

Twenty-four sections were recognized by Seithe within the subgenus Solanum, and Correll (1962) added a further one, Neolycopersicon.

Section Tuberarium was retained by Bitter and by Seithe and was formerly divided by Bitter (1912) into two subsections, Basarthrum and Hyperbasarthrum. Bitter later (1917) suggested that the name Hyperbasarthrum should be replaced by Tuberarium, and the two subsections elevated to sections, but this has never been generally accepted. Seithe has retained the name Tuberarium for the subsection in which the tuberiferous species are included. In these the pedicel articulation is normally not situated right at the base, whilst in subsection Basarthrum the pedicel articulation lies at the base, and the plant is provided with characteristic 2-celled bayonet hairs. None of the species in subsection Basarthrum is tuber-bearing.

Although most species in subsection Tuberarium are tuber-bearing a few do not develop tubers or even stolons. One or two seem to produce stolons but no tubers.

The subsection has been variously divided by different authors into a number of groups or series. Thus Bukasov (in Bukasov and Kameraz 1959) recognized twenty-two series, whilst Correll (1962) added several more, bringing the total to twenty-six. In the present work we shall follow the fairly conservative scheme of Hawkes (1963), in which seventeen series are recognized. It is considered possible that further division of series Tuberosa may be necessary when certain of the wild species in this series become better known. Some future subdivisions of certain other series might also become necessary with further increase of knowledge.

We give below a list of these series, with notes on the distribution of each. We have also shown those that occur in Argentina, Brazil, Paraguay, and Uruguay by means of bold type for ease of reference. It will be seen that only six series occur in the area under survey:

1. JUGLANDIFOLIA Rydb. Central America and north-west South America as far south as the coastal ranges of Peru and north Chile. Non tuber-bearing.
2. **Etuberosa** Juz. South Argentina, central to south Chile, Juan Fernández Isles. Non tuber-bearing.

3. MORELLIFORMIA Hawkes. Guatemala, Mexico.
4. BULBOCASTANA Rydb. Guatemala, Mexico.
5. PINNATISECTA Rydb. Mexico, U.S.A.
6. **Commersoniana** Buk. Argentina, Bolivia, Brazil, Paraguay, Uruguay.
7. CIRCAEIFOLIA Hawkes. Bolivia.
8. CONICIBACCATA Bitt. Bolivia, northwards to Mexico.
9. PIURANA Hawkes. Colombia, Ecuador, Peru.
10. **Acaulia** Juz. Argentina, Bolivia, Peru.
11. DEMISSA Buk. Guatemala, Mexico.
12. LONGIPEDICELLATA Buk. Mexico, U.S.A.
13. POLYADENIA Buk. Mexico.
14. **Cuneoalata** Hawkes. Argentina, Bolivia, Chile.
15. **Megistacroloba** Cárd. & Hawkes. Argentina, Bolivia, Peru.
16. INGAEFOLIA Ochoa. Peru.
17. **Tuberosa** Rydb. Andes of South America, including Argentina. Coasts of Chile and Peru. This series includes the cultivated species.

KEY TO THE SERIES AND WILD SPECIES[1] OCCURRING IN ARGENTINA, SOUTH BRAZIL, PARAGUAY, AND URUGUAY

1. Plant never bearing stolons or tubers; pedicel articulation very low, generally 2–3 mm above base; flowers pale blue (ETUBEROSA) — **1. S. brevidens** (p. 132)

   Plant with stolons and tubers; pedicel articulation generally in the middle to upper third (or if low, flowers white or white and purple) — 2

2. Corolla stellate, i.e. lobes longer than their breadth at the base (very rarely substellate, i.e. as long as broad) (COMMERSONIANA) — 3

   Corolla substellate, pentagonal or rotate, i.e. lobes as long as or shorter than their breadth at the base — 6

3. Plant covered with fairly dense indumentum of stalked glands (visible under lens) —easily seen on pedicels and peduncles even in dried state. Flower normally white, or in introgressed forms flushed with pale blue-violet — **5. S. tarijense** (p. 239)

   Plant without glandular indumentum — 4

4. Lateral leaflets acute, generally acuminate, petiolulate (except in introgressed mountain forms); peduncle frequently twice-forked; flowers white — **4. S. chacoense** (p. 180)

[1] A key to the cultivated species will be found on p. 429.

Lateral leaflets not acuminate, normally obtuse, subsessile to sessile and often decurrent; peduncle not or only once-forked; flowers white or purple 5

5. Calyx acumens 1–3 mm long; habit rosette-like or decumbent; anthers clearly delimited from filaments; flowers white or purple **3. S. commersonii** (p. 142)

Calyx acumens minute, less than 1 mm long; habit upright; anthers not clearly delimited from filaments; flowers always white (Brazil: Minas Gerais) **2. S. calvescens** (p. 138)

6. Habit rosette-like or decumbent, or stem upright but short; peduncles generally unforked; pedicels long, often with high articulation 7

Habit not rosette-like (except in very poorly developed plants); peduncles normally forked; pedicel articulation low to high 10

7. Corolla very rotate, with very short, flat lobes; peduncle very short; articulation of pedicel very high, often almost or completely invisible (ACAULIA) **6. S. acaule** (p. 248)

Corolla substellate to pentagonal or (very occasionally) rotate; peduncle short to long; articulation of pedicel high but always clearly visible (MEGISTACROLOBA) 8

8. Leaves normally simple; flowers rotate to pentagonal; habit rosette-like or with decumbent zigzag stem **9. S. boliviense** (p. 297)

Leaves normally with lateral lobes; flowers semi-stellate; habit rosette-like to decumbent, but stem not zigzag 9

9. Lateral leaf lobes acuminate, rather narrow (occasionally absent); articulation 2–3(–6) mm below calyx base **10. S. sanctae-rosae** (p. 301)

Lateral leaf lobes obtuse, rarely acute, and never acuminate (occasionally absent); articulation 5–10 mm below calyx base **8. S. megistacrolobum** (p. 284)

10. Leaf with well-developed wedge-shaped wings on the rhachis; lateral leaf lobes or segments linear to linear-lanceolate; interjected leaflets absent (CUNEOALATA) **7. S. infundibuliforme** (p. 271)

Leaf without well-developed wedge-shaped wings on the rhachis (or if some are visible, then interjected leaflets also present); lateral leaf lobes or leaflets varying from linear-lanceolate to broad ovate, etc.,

	according to species; interjected leaflets present or not (TUBEROSA)		11
11.	Terminal leaflet normally larger than laterals which are few-paired, 0–3(–4), and decrease in size rapidly towards base of leaf; flowers white		12
	Terminal leaflet not or only slightly larger than the (3–)4 or more paired laterals which normally do not decrease in size rapidly towards base of leaf; flowers generally not white (except sometimes in *S. kurtzianum*)		14
12.	Filaments and anthers confluent (or if distinct in form, then at least with similar yellow colour); hairs on leaflet margins well-marked, thick	**19. S. maglia** (p. 420)	
	Filaments and anthers completely distinct both in form and colour; hairs on leaflet edges (if present) not or only moderately conspicuous		13
13.	Pubescence of leaf, stem, peduncle, pedicel and calyx very sparse and composed of short triangular 3-celled hairs; habit weak, decumbent	**12. S. venturii** (p. 349)	
	Pubescence of leaf, stem and calyx frequent, easily visible, often shining; hairs many-celled; habit vigorous	**11. S. microdontum** (p. 316)	
14.	Flowers white or white with magenta streak on centre of each petal externally; pedicel articulation often very low	**18. S. kurtzianum** (p. 406)	
	Flowers mauve to deep violet, never white or white with magenta petal streak; pedicel articulation median to high		15
15.	Lower surface of leaflets (especially in young leaves) covered with conspicuous whitish-grey tomentum; corolla large (generally more than 3·5 cm diam.), showy, dark purple; calyx very dark greenish-purple	**13. S. vernei** (p. 356)	
	Lower surface of leaflets with normal pubescence; corolla medium-sized, lavender or medium purple; calyx green or only slightly pigmented		16
16.	Lateral leaflets at least (2·5–)3 times as long as broad, narrowly elliptic to narrowly lanceolate		17
	Lateral leaflets about 2–2·5 times as long as broad, ovate to ovate-lanceolate		18

17. Leaf 3(–5)-jugate, rather pale yellowish-green; interjected leaflets normally absent; marginal hairs well-marked — **16. S. vidaurrei** (p. 384)

Leaf (3–)4–5(–6)-jugate; interjected leaflets normally quite frequent, 2–7(–14)-jugate; marginal hairs absent or inconspicuous — **17. S. spegazzinii** (p. 387)

18. Leaflets subsessile to shortly petiolulate, not decurrent, the margins not denticulate; filaments 0·5–1·5 mm long; stigma larger than style apex — **14. S. gourlayi** (p. 371)

Leaflets decurrent on to the well-marked petiolule, the margins conspicuously denticulate; filaments 0–0·5 mm long; stigma minute, not or barely thicker than style apex — **15. S. oplocense** (p. 377)

PART II

DETAILED TAXONOMY

SERIES ETUBEROSA JUZ.

(*Izv. Akad. Nauk SSSR* **2**, 301, 1937, *nomen nudum; ex* Buk. & Kameraz, *Bases of Potato Breeding* 18, 1959)

ESTOLONIFERA Buk. (*Problemý Bot.* **2**, 1955, *nomen nudum*)

PLANTS herbaceous, without stolons and tubers. Leaves interruptedly imparipinnate. Pedicels articulated near or almost at the base. The plant does not, however, possess the 'bayonet hairs' typical of the subsection Basarthrum, where the articulation is always completely basal. Corolla rotate, purplish or bluish. Berries globose, deciduous.

General distribution: Central Chile, as far south as the Island of Chiloé; Isles of Juan Fernández. Nahuel Huapi region of southern Argentina.

Chromosome number of species so far investigated: $2n = 24$.

This series was named by Juzepczuk in 1937 but no Latin description was published until that of Bukasov in 1959. The series was mentioned in the literature many times by Bukasov and others, but neither Juzepczuk nor Bukasov gave reasoned descriptions in Russian so far as we are aware. Hawkes (1956*b*, 1958, 1963) and Correll (1962) have described Etuberosa in English, and Bitter (1912) gave a description in German to the group of Chilean species which comprise it, though he gave it no name.

The basis for the inclusion of Etuberosa in Tuberarium/Hyperbasarthrum, along with the tuber-bearing species, is rather slender. None of the species so far studied as living plants (*S. etuberosum*, *S. fernandezianum*, and *S. brevidens*) can be induced to form stolons or tubers under any conditions to which they have so far been subjected. Even the grafting of tuber-bearing species on to stocks of *S. brevidens* (Hawkes 1947) has been completely unsuccessful in this respect. On the other hand, the flower and leaf are very similar to those of many tuber-bearing species, as well as the general pubescence type, and in any case the pedicel articulation is not completely basal, except perhaps in *S. fernandezianum*.

Correll (1962) states that 'the most natural position for this series would be to consider it as immediately following Canensa, the last series in subsection Basarthrum'. But, of course, the lack of bayonet hairs would prevent its inclusion in that group. We shall therefore continue to consider Etuberosa as a rather anomalous series in subsection Hyperbasarthrum until fresh evidence for its affinities is forthcoming. So far, no one has succeeded in crossing species of series Etuberosa with those of any of the tuber-bearing series, but it is not improbable that in the near future this might be accomplished.

Species of this series included in the present treatment:

S. BREVIDENS Phil.

1. SOLANUM BREVIDENS Philippi

(*An. Univ. Chile* **43,** 521, 1873)

S. palustre Poepp. var. *parviflora glabrescens* Schlechtd. (*Hort. Hal.* **1,** 5, 1841).
S. palustre Poepp. var. *glabrescens* Walpers (*Rep. bot. Syst.* **3,** 39, 1844/5).
S. caldasii Dun. var. *glabrescens* Dun. (in *DC. Prodr.* **13,** 1, 37–38, 1852).
S. bridgesii A.DC. (*Archs. Sci. phys. nat.* **5,** 437, 1886; *non S. bridgesii* Phil. in *Linnaea* **33,** 203, 1864).
S. pearcei Phil. (*An. Univ. Chile* **91,** 5–6, 1895; *non S. pearcei* Britton ex Rusby in *Mem. Torrey bot. Club* **4,** 227, 1895).
S. tuberosum subsp. *brevidens* Reiche (*Flora Chile* **5,** 352, 1910).
S. tuberosum subsp. *pearcei* Reiche (ibid. 353, 1910).

Herbaceous, *non-tuberiferous*, probably generally perennating by means of a thick underground rootstock or rhizome.

Stem generally tall, 1–2(–3) m or perhaps taller, to 1 cm thick or more at base, normally with frequent weak spreading hairs.

Leaf large, 14–30 cm long, (3–)4–6(–8)-jugate; larger lateral leaflets paired or alternating, *broadly lanceolate*, oblong-lanceolate or long ovate, not as large as terminal, 4–7·5(–10) × 1–2·5(–3·8) cm, sessile or with up to 1–2 mm long petiolules, apically acuminate; terminal leaflet to 10·5 × 4 cm; interjected leaflets generally numerous, (2–)4–12(–15)-jugate, sessile, orbicular to ovate-cordate, apically obtuse; leaf provided with sparse to quite frequent *silky adpressed hairs* (in the Argentine forms); pseudostipular leaflets well developed, semi-lunate, to 20 mm long.

Inflorescence to 40-flowered (perhaps even more?); common peduncle (2–)7·5–14 cm long, forked one to three times above; pedicels 10–20 mm long, with *articulation only (1–)2–3(–6) mm above the base*; peduncles and pedicels clothed with sparse to quite frequent white adpressed to spreading hairs and some glands; pedicel above articulation more sparsely pubescent or even glabrescent.

Calyx campanulate, about 4 mm diam., with very short triangular lobes and *0–1 mm long acumens*; pubescence as for upper part of pedicels.

Corolla *sky-blue* (possibly bluish-violet in some plants?) *paling to white at edges*, rotate, 18–30 mm diam.; corolla lobes *very short*, 8–18 mm broad × 3–5 mm long, including the 2–4 mm acumen.

Filaments 1–2 mm long; anthers 4–6 mm long.

Style thin, about (5–)8–10 mm long, with minute papillae on its lower $\frac{1}{4}$–$\frac{1}{2}$; stigma globose, slightly larger than style apex.

Berries frequent, globose, to *c.* 1·5 cm diam., *dark purple or nearly black, shining*.
Chromosome number: $2n = 24$.

AFFINITIES

S. brevidens may be recognized by the absence of stolons and tubers, highly dissected softly pubescent leaves, many-flowered inflorescence, low pedicel articulation, very short calyx acumens and pale blue corolla paling towards the edge with very short lobes. It is only very distantly related to the other species described in the present work, and the absence of stolons and tubers, together with the low pedicel articulation are sufficient to indicate its remoteness from the tuber-bearing species.

SPECIMENS EXAMINED

ARGENTINA

Prov. NEUQUÉN:

Parque Nacional de Nahuel Huapi, Hotel Correntoso. *R. de Barba 1319* (LIL).

Parque Nacional de Nahuel Huapi, Desfiladero Huemul. 17 Dec. 1946. *R. de Barba 1329* (LIL).

Parque Nacional de Nahuel Huapi, entre Espejo y Correntoso. 16 Jan. 1947. *R. de Barba 1498* (LIL).

Parque Nacional de Nahuel Huapi, Lago Espejo. 17 Jan. 1947. *R. de Barba 1561* (LIL).

Parque Nacional de Nahuel Huapi, Puerto Baratta. 18 Jan. 1947. *R. de Barba 1622* (LIL).

Hotel Ruca Malén, clearing in the forest. 15 Dec. 1955. *Böcher, Hjerting & Rahn 1711* (J.G.H., J.P.H.).

Nahuel Huapi, Puerto Manzano. 13 Feb. 1934. *Burkart 6505* (BA).

Near Lago Huechulafquen. Alt. 1050 m. 31 Dec. 1926. *H. F. Comber 942* (E, K, LP, J.G.H.). See Plate 17.

Region of Lago Nahuel Huapi, entre Puertos Tigre y Sin Nombre. 12 Feb. 1927. 'Entre las piedras. Altura orilla lago.' *Cordini 13* (S, US).

Nahuel Huapi, Quetrihué. Alt. 800 m. 29 Jan. 1952. 'En bosque abierto. Flor violácea; baya púrpura. Alt. 0·6–1 m. Forma matas grandes en tierra negra con humus.' *Diem 2034* (J.G.H.).

Dept. Los Lagos, near Villa La Angostura. 15 Feb. 1952. 'A vigorous herb, growing to a height of 3–4 ft, with lilac coloured flowers. Along water courses or in similar moist places in the evergreen forest, not particularly common.' *Myndel Pedersen 1516* (C, J.G.H., J.P.H.).

Nahuel Huapi, Península Quetrihué. Alt. 800 m. Seeds sent from Sr. José Diem in 1952 to J. P. Hjerting (*P & H 1890*), and to J. G. Hawkes (CPC: 2451). Herbarium specimen made in Birmingham (H. 610 = *P & H 1890*; J.G.H.). See Plate 16.

Dept. Lacar (Parque Lanín), Lago Lacar, Pucará. 9 Nov. 1956. 'En pantanos. Estéril. No se encontraron tubérculos.' Leg. Ruiz Leal and F. Roig. *Ruiz Leal 18248* (Hb. Ruiz Leal).

Dept. Los Lagos, Parque Nacional Nahuel Huapi, Los Arrayanes. 13 Nov. 1956. 'En pantanos; no abundante. No se encontraron tubérculos. Fl. azulada.' Leg. Ruiz Leal and F. Roig. *Ruiz Leal 18296* (Hb. Ruiz Leal).

Nahuel Huapi, Península Quetrihué, Estancia Quetrihué. Alt. 750 m. 16 Jan. 1951. 'Fl. viol.' *Sleumer 1520* (p.p.) (LIL, US, J.G.H.).

Lago Nahuel Huapi, Correntoso (camino aserradero). Alt. 850 m. 23 Mar. 1934. 'Flor azul viol. Fruto: color violaceo. Encontrado en un desmonte.' *Rutile A. Spegazzini 172* (BAB).

Prov. RÍO NEGRO:

Lago Nahuel Huapi, Llao-Llao. Alt. 770 m. 25 Dec. 1933. *E. Ljungner 591* (NY).

Nahuel Huapi, Llao-Llao. 16 Feb. 1941. 'Hab. bosque higrófilo. Flor violácea. Alt. de la planta: 0·70 m.' *R. Maldonado 592* (LP).

Nahuel Huapi. 1936. 'Flor lila azulada.' *Julia B. de Saint 12824* (SI).

DISTRIBUTION

S. brevidens is known in Argentina only in the south-western parts, namely in the provinces of Neuquén and Río Negro[1] (see Map 13). In Chile the species occurs in provinces Bío-Bío, Malleco, Cautín, Valdivia, Osorno, Llanquihue, and Chiloé (also Arauco, according to Montaldo and Sanz 1962). No doubt it has migrated from Chile into Argentina, since it is related to several other species in the former country.

In Argentina *S. brevidens* has so far been seen only in the regions of Lago Nahuel-Huapi, Lago Lacar, and Lago Huechulafquen but it is possible that it may be much more widespread in the Argentine Lake region than the present records indicate.

HABITAT

In Argentina *S. brevidens* occurs at altitudes of 750–1050 m, whilst in Chile it is well known from sea level to 950 m. It grows in forest and scrub, always in deep rich soil, and sometimes in marshes, in the humid *Nothofagus* forest-zone. On Cabrera's (1953) classification it would apparently occur in the Valdivian district and also in the Deciduous Woodland district (Bosque Caducifolio) of the Subantarctic phytogeographical province. It is not clear whether it also occurs in the Pehuén (*Araucaria araucana* forest) district, belonging to the same province. It is occasionally found associated with escaped forms of *S. tuberosum* (see p. 434).

TAXONOMIC NOTES

S. brevidens varies considerably in the shape, size, and number of leaflets as well as in the flower size. It can always be distinguished from the other Argentine *Solanum* species in Hyperbasarthrum by the low pedicel articulation, very short sepals, minute sepal acumens, and sky-blue (? to violet) flowers with very short petal lobes.

Philippi's type material was collected in 1865 in the region of the Chilean Lake Llanquihue by Dr. Fonck, only about 100 km from Nahuel Huapi where most of the Argentine collections were made. The isotype (see Fig. 6) possesses rather narrower lateral leaflets (1·2×4·8 cm) than the Argentine ones (2·5×5·5 cm, etc.), though many Chilean collections are of the broader leaflet form also. The biotype described by Philippi under the name *S. pearcei* has even wider leaflets (3·8×7·5 cm, etc.) which overlap each other. It is undoubtedly referable to *S. brevidens*, as also is *S. bridgesii* of Alphonse de Candolle.

The pubescence varies quite markedly in different collections of *S. brevidens*. In all the Nahuel Huapi collections it is quite dense but rather shorter in Comber's Huechulafquen gathering (see Plate 17), and in this latter specimen the pedicels are glabrescent. An especially glabrescent form from Chile, which has been given various specific names but has always received the varietal name '*glabrescens*' was placed by Hawkes as *S. brevidens* var. *glabrescens* (*ined.*, see Correll, 1962, p. 121). It does not occur in Argentina.[2]

Previous collectors in south-west Argentina considered that *S. brevidens* might be a form of *S. etuberosum*. This latter species, however, occurs only in Chile, where it is to be found in the Mediterranean scrub forest much further north. It almost certainly does not spread into Argentina.[1] Montaldo and Sanz (1962) are inclined to think, however, that it may be necessary to group the two species together after further study.

[1] Correll (1962) mentions that *S. brevidens* has been reported for prov. Mendoza and that *S. etuberosum* probably occurs there. We have seen no specimens of the former species in that province, and only one collection of the latter (see Appendix III). The climate of Mendoza would appear to be too dry for *S. brevidens*, whilst there is some doubt as to the Argentine provenance of the *S. etuberosum* collection.

[2] Although Correll (1962) cites the Argentine collection *Comber 942* as belonging to var. *glabrescens* we consider that this collection should be placed as the type variety and *not* as var. *glabrescens*.

FIG. 6. *S. brevidens* Phil. Isotype (herb. Vienna). Gathered by Fonck (s.n.) in the region of Lake Llanquihue, Chile, about 100 km from Nahuel Huapi, Argentina, where most of the Argentine material of this species was obtained. Leaflets rather narrower than those of most collections of this species, both from Argentina and Chile.

NOTES ON LIVING PLANTS

Seeds of *S. brevidens* from Argentina were sent to England through the kindness of Sr. J. Diem from Villa La Angostura, Península Quetrihué in 1952 (grown under CPC: 2451; *P. & H. 1890*; H. 610).

The seedlings raised from them agree very well with typical *S. brevidens* in all respects (see Plate 16). The plants are vigorous, with up to seven pairs of lateral leaflets and eighteen pairs of interjecteds, including acroscopics and basiscopics. The inflorescences are very large and the flowers are deep blue in the centre, paling to almost white at the edges. *S. brevidens* is apparently very resistant to insect-transmitted viruses (Hawkes 1947) and to bacterial wilt (see Chapter 5).

CYTOLOGY

Chromosome counts on both Argentine and Chilean material of *S. brevidens* showed a somatic number of $2n = 24$ (Rybin 1933, Avanzi 1949, Sanz de Cortázar 1955, Knud Jakobsen unpublished (*P. & H. 1890*), J.G.H. unpublished) and a haploid number of $n = 12$ (Sanz de Cortázar 1955).

FERTILITY AND ARTIFICIAL AND NATURAL HYBRIDS

The species is of interest in that all collections so far examined are perfectly self-fertile and set abundant seed. This is an exception to the general rule of self-incompatibility in diploid *Solanum* species.

So far, it has only been possible to hybridize this species with *S. etuberosum*, and no success whatsoever has been obtained in pollinations with *S. demissum*, *S. phureja*, and *S. fendleri*. No natural hybrids between *S. brevidens* and any other species of *Solanum* have been reported for Argentina.

SERIES COMMERSONIANA BUK.
(*sensu* Hawkes 1956*b*)

(*Izv. Akad. Nauk SSSR* 714, 1938, *nomen nudum; ex* Buk. & Kameraz, *Bases of Potato Breeding* 19, 1959)

GLABRESCENTIA Buk. (*Problemý Bot.* **2**, 1955, *nomen nudum; ex* Buk. & Kameraz, *Bases of Potato Breeding* 19, 1959)
TARIJENSA Corr. (*Tex. Res. Found. Contrib.* **4**, 233, 1962)
YUNGASENSA Corr. (ibid. 220–2, 1962)

TUBER-BEARING herbs with imparipinnate leaves and stellate corolla. Corolla lobes generally (except in the Bolivian *S. yungasense*) less than twice as long as broad. Flowers white or mauve tinted.

General distribution: Bolivia, Argentina, Paraguay, Uruguay, Brazil, generally at low altitudes but occasionally ascending to 1500 m or even exceptionally to 2350 m or over.

Chromosome numbers: $2n = 24, 36$.

The species that we have here included in Series Commersoniana have been moved about to quite an extent by Bukasov, and also latterly by Correll. In his first paper that dealt with the systematics of *S. commersonii* and its allies Bukasov placed them in Rydberg's North American Series Pinnatisecta, to the members of which they certainly bear a close superficial resemblance. In 1938, however, Bukasov realized that this arrangement was artificial, and he therefore created the series Commersoniana for the South American group of species with stellate corollas. No description was given, however, though Hawkes published a short one in English in 1956.

In 1955 Bukasov proposed that all the species from the mountainous regions of Argentina that we have classed as *S. chacoense* in the present work should be separated off into a new series, Glabrescentia. It was not until 1959 that Glabrescentia and Commersoniana were validated by Latin descriptions, and by that time the series Commersoniana was obviously much more restricted in its connotations than it had been in 1938 when it was first proposed. In Bukasov's 1959 publication (Bukasov & Kameraz 1959) all the other species that we have grouped here in Commersoniana, namely, *S. tarijense*, *S. yungasense*, and *S. calvescens*, were placed in Glabrescentia, even though several of them were far from being glabrescent. In his series Commersoniana as defined in 1959 only what we have grouped here as the two subspecies of *S. commersonii* are included.

Brücher (1956*a*) retained the wider connotation of this series, but mistakenly added two species from series Tuberosa, namely, *S. kurtzianum* from north-west Argentina, and *S. pampasense* from Río Pampas in Peru, in the erroneous belief that it came from the pampas of Argentina.

The anomaly effected by Bukasov in placing the very glandular-pubescent species *S. tarijense* in the glabrescent series Glabrescentia is of course not a nomenclatural error, though it may be an error of taxonomic judgement. It was evaded by Correll (1962), who placed *S. tarijense* in a distinct series, Tarijensa. To this he also added the glandular Bolivian species, *S. berthaultii*, which is probably a hybrid of *S. tarijense* and a species in series Tuberosa. Another series, Yungasensa, included, besides the type species, *S. yungasense,* various discordant elements of other affinities. Some of these we ourselves class as forms of *S. chacoense* and its subsp. *muelleri,* whilst another, *S. trigalense,* is undoubtedly a synonym of *S. tarijense.* It seems to us that both these series of Correll are somewhat artificial and serve to confuse the taxonomic situation rather than to clarify it. If any division of the series Commersoniana is needed we are inclined to think that Bukasov's scheme may have a little more validity, since it is supported to some extent by sero-taxonomic data, as follows:

When an antiserum is raised to tuber extracts of *S. commersonii* and tested against a wide range of clones of that species, both the typical subspecies and subsp. *malmeanum,* a five to six-line gel-diffusion spectrum results. On the other hand, both the other species in series Commersoniana so far tested (*S. chacoense* and *S. tarijense*) give a four-line spectrum, as also do *S. tuberosum* and all other South American species investigated (see Lester 1965). No results are yet forthcoming for the little-known *S. calvescens* from Brazil, since this has only just been received for study in the living state, and without this information it would be difficult in our view to follow Bukasov's division. We also need to know a great deal more about what takes place in the zones of natural overlap between *S. chacoense* and *S. commersonii* subsp. *malmeanum* in eastern Argentina and Paraguay before coming to a conclusion on this matter, since the literature is already full of fractionated series and species. If *S. calvescens* shows the same series of precipitin bands as *S. commersonii,* then there would be a great deal to be said for separating all the other species in Commersoniana into a distinct series, for which the name Glabrescentia would have to be used.

For the moment, however, we consider it wiser, both from a practical and a theoretical point of view, to retain the series Commersoniana in the original sense of Bukasov (1938), and in line with Hawkes's original English description of 1956.

Species of this series included in the present treatment:

S. CALVESCENS Bitt.
S. COMMERSONII Dun.
S. CHACOENSE Bitt.
S. TARIJENSE Hawkes

2. SOLANUM CALVESCENS Bitter

(In *Reprium nov. Spec. Regni veg.* **11**, 436–7, 1912)

Herbaceous, tuber-bearing.

Stem up to 50 cm high, not very pubescent, with narrow decurrent wings according to Hj. Mosén.

Leaf up to 22–27 cm long, 3–4-jugate, set at a wide angle to the stem and gradually curved upwards (though not in all specimens), without interjected leaflets (or rarely with a solitary minute one, according to Bitter); lateral leaflets elliptic to ovate-oblong, narrowing at both ends, rather remote from each other, apically obtuse or acutish, sessile or with short-winged petiolules, slightly basally decurrent; uppermost lateral leaflets to about 6·5 × 2·5 cm; terminal leaflet more broadly elliptic or ovate than laterals, to 8 × 3·5 cm; pubescence fairly short and sparse; pseudostipular leaves curved, clasping the stem.

Inflorescence 7–15-flowered, lax; peduncle 4·5–10·5 cm long, often once-forked; pedicels rather long, up to 3 cm, articulated slightly above the middle; peduncle and pedicels clothed with sparse short hairs and minute glands.

Calyx about 4–7 mm long, with short triangular to ovate lobes and 1 mm long acumen, very sparsely pubescent.

Corolla stellate, white, large, 3–3·5(–4) cm diam.; lobes 8–14 mm long, 8–12 mm broad.

Filaments about 1·5–2·5 mm long; anthers lanceolate, ±5–6×1·5 mm, *gradually narrowing into the filaments*, the boundary between them not well marked, either in form or colour (cf. *S. maglia*). See Plate 19.

Style protruding over stamens, 10–12 mm long, entirely glabrous. Stigma obconical, slightly notched at the apex or entire.

Berries ovate-conical, dark green, spotted with white, about 2·5×2·0×1·75 cm.

AFFINITIES

S. calvescens is obviously related to *S. commersonii* and especially to subsp. *malmeanum*, agreeing with it in the rather obtuse lateral leaflets which are sub-sessile to shortly petiolulate, the larger terminal leaflets, often once-branched peduncles and large white corolla. It may be distinguished from subsp. *malmeanum* by the more upright growth, no interjected leaflets, and especially by the curious stamens in which the anthers are not delimited from the filaments either in form or colour. The ovate-conical berries are also rather similar to those of *S. commersonii*, though the white spotting reminds one rather of *S. chacoense*; the very short calyx acumens are also similar to those of the plains forms of that species.

SPECIMENS EXAMINED

BRAZIL

Estado de MINAS GERAIS:

Caldas, damp shady banks of Rio Verde. 30 Oct. 1873. Leg. Hj. Mosén 4294. *Regnell 658* (LD, S, UPS-Lectotype). See Fig. 7.

Caldas. 27 Oct. 1868. Leg. S. E. Henschen. *Regnell III 1623* (UPS-Syntype). See Plate 18.

Caldas. 18 Nov. 1871. 'In cultis dictu tapeira.' Leg. Regnell. *Regnell III 1623* (P, UPS-Syntype, US).

No exact locality. 1845. Leg. Widgren 1171. *Regnell* (s.n.) (BR, S).

Caldas, Pocinho do Rio Verde. Alt. 1100 m. 16 Mar. 1963. 'Em terreno cultivado à beira de rio.' Leg. Alaa de Oliveira. *Brieger* (s.n.) (J.G.H.—living material grown at Birmingham, Accession No. H. 653). See Plate 19.

(See also Appendix III.)

DISTRIBUTION

Brazil, State of Minas Gerais, Caldas. (See Map 4.)

HABITAT

On damp shady river banks and in fallow fields (*Cultis dictu tapeira* = abandoned farmland covered again by weeds and natural flora—according to *Caldas-Aulete*, Brazilian edition by Hamílcar García, 1958), at altitudes of about 1200 m. According to Brieger (in a written communication) this plant is also a weed of maize fields. From the data given above it seems to flower in October and November. Brieger states: 'Farmers in the area know the "wild potato" and inform us that it is flowering at the time when they are working the maize fields. Thus the plants appear at the beginning of the rainy season, are fast growing, and disappear entirely from the surface around Christmas.'

TAXONOMIC NOTES

Bitter's description of *S. calvescens* was based on three collections of Regnell in Herb. Uppsala. As will be seen from the above list, two of these were actually collected by other botanists and not by Regnell himself. We have designated *Regnell 658* (in Herb. Uppsala) as a lectotype of this species and the two collections of *Regnell III 1623* have been left as syntypes. (See Fig. 7 and Plate 18.)

Until recently very little was known about *S. calvescens*. Bitter considered that it should be included in the group of *S. commersonii*, whilst Correll (1962) thought that *S. calvescens* showed characteristics of a possible natural hybrid of *S. chacoense* × *S. commersonii*, possessing a calyx typical of the former species and a terminal leaflet typical of the latter. We doubt whether this explanation could be correct, not only because of its geographical isolation from the two species just mentioned (at least 400 km north-east of the nearest known locality of either of them), but also because of the curious stamens. These are rather similar to the stamens of the west Argentine and Chilean species, *S. maglia*, though they seem to be longer and paler. It would seem to us more likely that *S. calvescens* is a species of equal status to *S. commersonii* and *S. chacoense*, and has originated with them from the primitive ancestral stock of series Commersoniana a long time ago.

According to our present knowledge, *S. calvescens* seems to be curiously restricted in its distribution. The three Regnell collections and that sent by Brieger are all from Caldas in Minas Gerais State. Correll (1962) mentions another collection from Santa Catarina State, which we have not yet seen (*Smith & Reitz 10210*).

The epithet '*calvescens*' indicates the almost hairless nature of this species. Even so, some pubescence can generally be seen quite clearly with a hand lens, especially on living plants (see below).

NOTES ON LIVING PLANTS

Until quite recently *S. calvescens* was quite unknown in the living state. We are accordingly very much indebted to Professor F. G. Brieger of the Institute of Genetics, São Paulo University, for sending tubers to us in December 1963 and for providing us with information on the plants found growing at Caldas. We were able to raise three plants from thirteen tubers kindly sent us by Professor Brieger. These plants were fairly tall growing and developed 2–3-jugate leaves with an enlarged terminal leaflet and obtuse to sub-acute laterals. These possessed 1–2 mm long petiolules. The leaves were not curved upwards in the manner developed in some of the material seen by Bitter, and were, on the whole, rather short. The pubescence of stem and leaf was moderately dense, but the hairs were rather short and seen with ease only by using a hand lens. The peduncles were always once-forked and often bore reduced leaves about half-way along the lower unbranched portion. The pedicels were about 2 cm long and articulated a little above the centre; they were provided with frequent short hairs and plentiful stalked glands. The calyx form and pubescence, and the corolla form and colour agree well with those of the lecto- and syntype specimens; the corolla is creamy white, darkening slightly on drying, and the lobes are generally broadly triangular, only slightly longer than broad (*c.* 13 × 12 mm). The characteristic stamens were well shown in these plants but the anthers were smaller, paler and more distorted than in Regnell's specimens. The glabrous style and entire obconical stigma match well those of the lecto- and syntype material. See Plate 19.

Although these plants were rather sterile, some berries were formed (see below). These were ovoid-conical (cf. *S. commersonii*), attaining a size of 2·5 × 2·0 × 1·75 cm, and were dark green, with abundant white spots.

On the whole, the living plants show rather strong similarities to *S. commersonii*

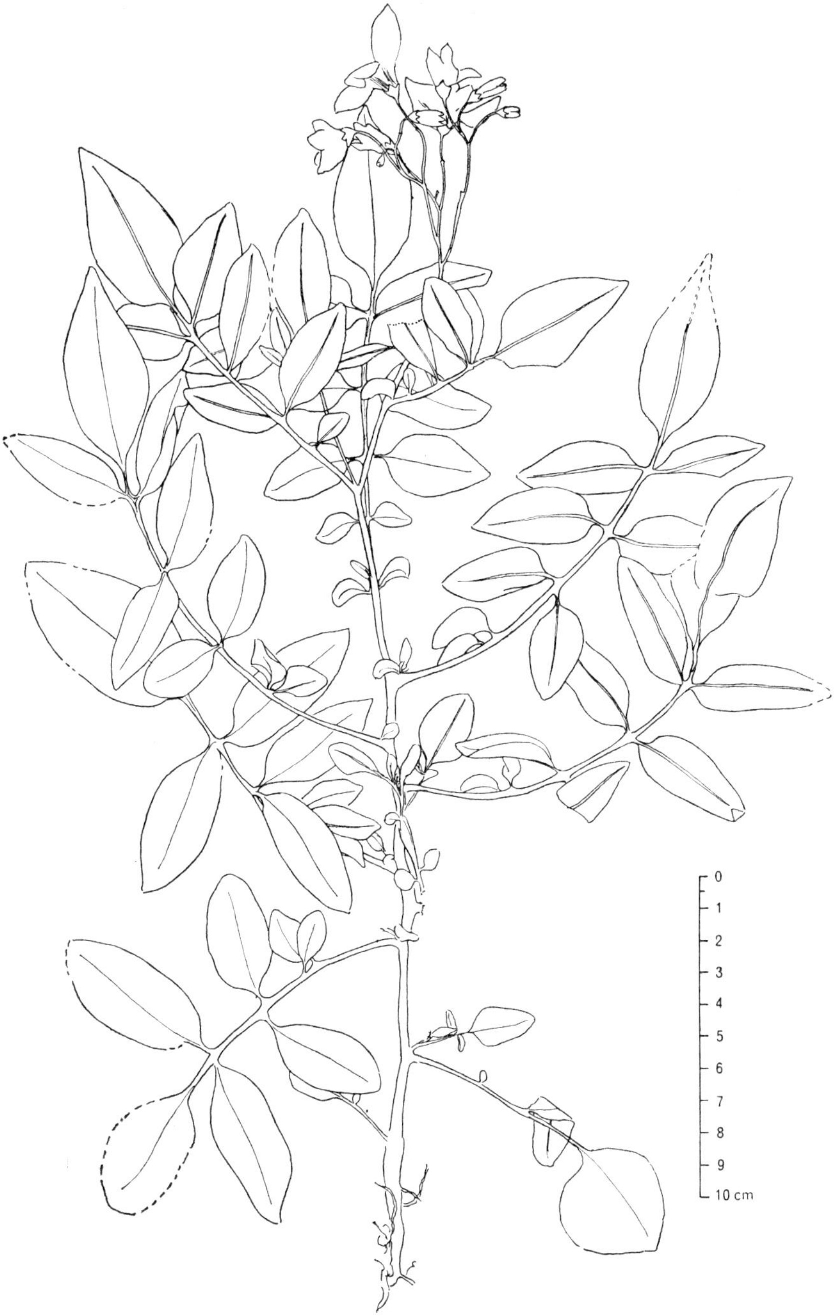

FIG. 7. *S. calvescens* Bitt. Lectotype. Collected by Hj. Mosén (*Regnell 658*—Herb. Uppsala) on the banks of the Rio Verde at Caldas, Minas Gerais, Brazil. Note the large curved leaves with no interjected leaflets and the very large corolla.

subsp. *malmeanum*, but there seem to be enough differences between this subspecies and *S. calvescens* to warrant the retention of the latter as a good species (see section 'Affinities', p. 139).

CYTOLOGY

Chromosome counts by Mr. R. Tarn revealed that all three plants were triploid, with $2n = 36$ chromosomes. No doubt further search would reveal the presence of diploid plants of *S. calvescens* in the Caldas area, and they might well occur in mixed populations with the triploids, as has been demonstrated in *S. commersonii*, *S. maglia*, *S. bulbocastanum*, *S. cardiophyllum*, etc., in other regions of South America and Mexico, No meiotic studies have so far been carried out on *S. calvescens*.

FERTILITY

The fertility of the triploid plants was very low, but some berries were obtained by Mr. R. Tarn, using growth hormones, both in selfing and by artificial pollinations with tetraploid experimental lines of *S. commersonii* subsp. *malmeanum*. The work is still in progress at the moment of writing but it seems likely that all or most of the embryos become aborted at a fairly early stage of development. Diploid material of *S. calvescens* is needed for us to be able to make an adequate assessment of the fertility and crossability of this species. No natural hybrids have been observed by us or reported in the literature. This is, of course, only to be expected, since *S. calvescens* does not overlap in distribution with any other species, so far as we are aware at present.

3. SOLANUM COMMERSONII Dunal, *emend.* Hawkes et Hjerting

(In *Poiret, Encyclopédie méthodique botanique Suppl.* **3**, 746, 1813)

S. commersonii Dun. var. *glabriusculum*[1] Hook. f., p. pte. (*Flora Antarct.* **2**, 330, 1846).
S. commersonii Dun. var. *glabratum*[2] Hook. f. (ibid.).
S. commersonii Dun. var. *pubescens* Sendtn. (in Martius, *Flora Brasil.* **10**, 12, 1846).
S. tenue Sendtn. (in ibid. 13, 1846).
S. ohrondii Carrière (*Revue hort.* **55**, 496–500, 1883).
S. acroleucum Bitt. (in *Reprium nov. Spec. Regni veg.* **11**, 435–6, 1912).
S. henryi Buk. et Lechn. (*Revta argent. Agron.* **2**, 182–3, 1935).
S. henryi f. *pubescens* Lechn. and f. *laticalyx* Lechn. (ibid., p. 183).
S. mechonguense Buk. (ibid. **7**, 363, 1940).
S. mercedense Buk. (*Vest. sots. Rasteniev.* No. 4, 3–12, 1940).
S. sorianum Buk. (ibid.).
S. compactum Buk. *nomen nudum* (≡ *S. mercedense* Buk.).
S. rionegrinum Lechn. *nomen nudum*[3] (*Vest. sots. Rasteniev.* No. 4, 31–32, 1940).
S. commersonii Dun. f. *mechonguense* (Buk.) Corr. (*Wrightia* **2**, 173, 1961).

Small herb with long stolons or sobols and medium-sized tubers 2–5(–8) cm long.

Stem branched or unbranched, short or sometimes slender and elongated, (1–)10–30(–80) cm long, glabrescent to densely pilose, slightly angled but *unwinged*.

Leaves *in a rosette* or more or less widely spaced according to stem development, entire to pinnate, *sub-pinnatifid* or *sub-lyrate*, up to 25 cm long, with (0–)2–4(–6) pairs

[1] See notes on pp. 160 and 203; also footnote to p. 160.
[2] See notes on pp. 160 and 168.
[3] See notes on pp. 148 and 167.

of lateral leaflets and 0–2(–5) pairs of minute interjected leaflets; terminal leaflet *larger than the laterals* which often decrease rapidly in size from apex to base of leaf; lateral leaflets or lobes *sessile or shortly petiolulate,* basally decurrent on to the rhachis, *obtuse or rounded at the apex,* ovate-elliptic to oblong-ovate; terminal leaflet generally broadly ovate, elliptic, suborbicular or elliptic-rhomboid, apically rounded or obtuse, basally cuneate; leaf surface varying from glabrous to densely but shortly pubescent; pseudostipular leaflets generally absent, sometimes well developed, to 8×5 mm.

Inflorescence often apparently terminal, *rather compact,* often in the form of a cymose umbel, (2–)4–10–(–16)-flowered; peduncle (1·5–)5–13 cm long, *not or only once-forked,* the branches of the fork generally *reduced in length* (in subsp. *commersonii*); pedicels 15–25(–30) mm long, articulated at the centre or slightly below it or even two-thirds of the length above the base according to the subspecies under consideration; pubescence of peduncle and pedicels variable but generally sparser than that of the leaves.

Calyx campanulate, 4–9 mm long, the lobes typically *ovate to ovate-lanceolate* (frequently shorter) with a well-marked short to medium-lengthed triangular acumen (1–3 mm); pubescence as for pedicels.

Corolla *stellate,* white to purple, the pigment when present generally situated on the outer surface of the corolla; diam. 15–35 mm; lobes lanceolate or occasionally triangular, reflexed in flower, (7–)10–15 mm long×(3–)5–7(–10) mm broad at the base.

Anthers 4·5–7·5 mm long; filaments 1–2·5 mm long.

Style 8–12 mm long, furnished with papillae in its lower $\frac{1}{3}$–$\frac{1}{2}$; stigma *generally bifid,* capitate to clavate, markedly thicker than the style apex; ovary ovate to conical.

Berries *frequently cordate, ovoid or conical,* rarely±globose.

Chromosome number: $2n = 24, 36$.

AFFINITIES

S. commersonii appears to be most closely related to *S. chacoense,* from which it differs in the sessile to sub-sessile apically obtuse leaflets, often decurrent on to the rhachis, once-forked peduncle, quite well-developed calyx acumens, often purple-tinted corolla, ovate to cordate berries, and the resistance to frost. *S. commersonii* is a more delicate-growing species than *S. chacoense,* with shorter leaves, few interjected leaflets and often reduced peduncle forks. Subsp. *malmeanum* would seem perhaps to be closer to *S. chacoense* than subsp. *commersonii,* but more experimental work is needed, together with more collections from Paraguay and south Brazil, before we can be quite certain of the relationship of the two subspecies of *S. commersonii* to *S. chacoense.* Subsp. *malmeanum* also appears to bear some relationship to *S. calvescens* but differs in possessing normal stamens and interjected leaflets, as well as non-spotted berries.

Some difficulty may be encountered in distinguishing *S. chacoense* from *S. commersonii* subsp. *malmeanum,* since both taxa possess petiolulate leaflets. However, the petiolules of *S. chacoense* are generally much longer (2–8 mm) than those of subsp. *malmeanum* (2–3 mm), and the leaflets are obtuse at the apex in the latter taxon. Furthermore, the sepal acumen in *S. chacoense* is normally not much longer than 0·5 mm, whilst that of subsp. *malmeanum* is 1–3 mm long; if the acumen of *S. chacoense* is lengthened to 1 mm or a little more in exceptional cases it still remains narrow and linear, not more than 0·5 mm wide, whilst that of subsp. *malmeanum* is broadly triangular, about 1 mm wide, or even more, at the base.

The anomalous (probably always) triploid forms of *S. commersonii* mentioned below (pp. 154–6) possess petiolules at times reaching even to 5 mm; here again, they may be readily differentiated from *S. chacoense* by the obtuse leaflet apices, and especially by the very reduced peduncle branches, which are, on the other hand, long and spreading in *S. chacoense*.

KEY TO SUBSPECIES

Lateral leaflets decreasing rapidly to base of leaf, often markedly decurrent, normally sessile; peduncle branches somewhat contracted; pedicel articulation low to ±median; corolla generally purple, the lobes about 1½ times as long as broad or even longer subsp. ***commersonii***

Lateral leaflets generally decreasing gradually to base of leaf, narrowly decurrent, normally petiolulate; peduncle unbranched or the branches not markedly contracted; pedicel articulation ±median to high; corolla always white, the lobes about as long as broad subsp. ***malmeanum***

1. Subspecies ***commersonii***

(For synonyms see p. 142)

Leaves short, 5–10(–15) cm, the terminal leaflet always much larger than the (0–)2–4(–5)-paired laterals which decrease in size rapidly towards the base of leaf; lateral leaflets sessile or with very short petiolules (1–2 mm), generally decurrent on to the rhachis; interjected leaflets absent or very rarely 1–2-paired; inflorescence branches generally contracted; pedicels generally articulated near the centre or slightly below it; corolla generally purple, the lobes longer and narrower than in subsp. *malmeanum*, about 1½ times as long as broad, or longer.

2. Subspecies ***malmeanum*** (Bitter) Hawkes et Hjerting, ***comb. nov.***

(*Scott. Pl. Breed. Stn Rec.* 105, 1963)

S. commersonii β pubescens Chodat (*Bull. Herb. Boissier* Ser. II, **2**, 812, 1902).
S. guaraniticum Hassl. var. *latisectum* Hassl. f. *pilosulum* Hassl. (in *Reprium nov. Spec. Regni veg.* **9**, 116, 1911).
S. commersonii Dun. var. *pseudostipulatum* Hassl. (in ibid. **9**, 116, 1911).
S. malmeanum Bitt. (in ibid. **12**, 447–8, 1913).
S. chacoense Bitt. var. *latisectum* Hassl. f. *pilosulum* Hassl. (equivalent to f. *puberulum* Hassl. *nom. nud.*) (*Annls. Conserv. Jard. Bot., Genève* **20**, 186–7, 1917).
S. pseudostipulatum (Hassl.) Buk. (*Revta argent. Agron.* **2**, 180, 1935).
S. millanii Buk. et Lechn. (ibid. 180–2, 1935).
S. hassleri Wittm., *nomen nudum; non S. hasslerianum* Chodat (*Bull. Herb. Boissier* Ser. II, **2**, 747, 1902).
S. commersonii f. *malmeanum* (Bitt.) Corr. (*Wrightia* **2**, 173, 1961).

Leaves longer than in subsp. *commersonii*, 8–17(–24) cm long, the terminal leaflet broader but generally barely longer than the (2–)4–6-paired laterals which decrease in size more slowly towards the base of the leaf than in the typical subspecies; lateral leaflets shortly petiolulate (rarely completely sessile) and only very narrowly decurrent on to the rhachis; petiolules (1–)2–3(–4) mm long; interjected leaflets sometimes absent but frequently 2–5-paired; peduncle often unbranched, the branches when present not so contracted as in subsp. *commersonii*; pedicel articulation about two thirds or more above the base, sometimes ±median; corolla always white, the lobes only slightly longer than broad.

SPECIMENS EXAMINED

1. Subspecies *commersonii*

BRAZIL

Estado do RIO GRANDE DO SUL:

Pedro Sola, 60 km from Pelotas on highway to Rio Branco. Alt. ±200 m. 30 Mar. 1958. 'Grassy field by patch of natural *Araucaria brasiliensis*. Flowering.' *Hawkes 713* (J.G.H.). (Tubers from natural population grown in Birmingham, No. 185; showed variation between plants in pubescence and corolla size.)

Est. Exp. Hort. 'Domingos Petrolini.' Tubers given by Dr. F. Guimarães 31 Mar. 1958. *Hawkes 714* (Accession H. 186).

Municipio General Camara. Tubers given by Dr. F. Guimarães 31 Mar. 1958. *Hawkes 715* (Accession H. 187. $2n = 24$).

Instituto Agronómico do Sul, Pelotas. Collected 1 Apr. 1958 in fields. *Hawkes 716* (Accession H. 188. $2n = 24$).

Instituto Agronómico do Sul, 13 km south of Pelotas. Alt. 10–20 m. 2 Apr. 1958. 'Weed of ploughed ground in front of new Institute building. Flowers and immature fruits.' *Hawkes 717* (J.G.H.). (Grown in Birmingham from tubers, No. 189; fairly uniform.) See Plate 21.

Pelotas. 14 and 20 Mar. 1955. *Mota da Costa 1 & 2* (K).

Est. Exp. do Pelotas. 1958. *Mota da Costa* (s.n.) (Living material only. Accession H. 203. $2n = 24$.)

Encruzilhada do Sul. Seeds collected by Dr. Delorge Mota da Costa and sent 23 July 1959. (Accession H. 558. $2n = 24$.)

Piratini. Seeds collected by Dr. Delorge Mota da Costa and sent 23 July 1959. Identical with H. 558. (Accession H. 559. $2n = 24$.)

Instituto Agronómico do Sul. Seed collected by Dr. Delorge Mota da Costa, and sent 23 July 1959. (Accession H. 561. $2n = 24$.)

Prov. Porto Alegre, Canôas. 28 Jan. 1948. *Palacios & Cuezzo 326* (LIL).

Quaraí. 18 Mar. 1948. *Palacios & Cuezzo 1976* (LIL).

Porto Alegre. *B. Rambo 1173* (LIL).

Montenegro, Parecí. 19 Sept. 1945. Leg. Henz, *B. Rambo 29699* (LIL).

Ilha das Flores, pr. P. Alegre. 22 Apr. 1949. 'In subpaludosis.' *B. Rambo 41209* (LIL).

Esteio, pr. P. Alegre. 1 June 1949. 'In campestribus graminosis.' *B. Rambo 41790* (LIL).

Esteio, pr. S. Leopoldo. 11 Nov. 1955. 'In arenosis subhumidis herbosis; florens.' *B. Rambo 57308* (J.G.H.).

Cristo Rei, pr. S. Leopoldo. 11 May 1957. 'In eucalypto aperto; florens.' *B. Rambo 60585* (J.G.H.).

Arroio Grande. 1 Mar. 1893. 'Ad marginem silvae primaevae.' Leg. C.A.M. Lindman, *Regnell A. 1439* (S).

1816–21. *Saint-Hilaire 1835* (P).

1816–21. *Saint-Hilaire 2779 bis* (P).

Estado de SANTA CATARINA:

Tubarão. May 1889. 'Nos campos cultivados perto de Tubarão.' *E. Ule 1201* (CORD, F, P). (Type collection for *S. acroleucum* Bitt.) See Fig. 12.

Exact locality unknown:

Bande orientale (Bresil). *Bâcle 1834* (G–DC).

From Instituto Agronómico do Sul, Pelotas, Rio Grande do Sul. Dec. 1947. *Deslandes* (s.n.) (LIL ex Herb. Jard. Botânico do Rio de Janeiro, 61781).

Sellow 276 (K, P). See Fig. 11.
Sellow 1971 (BR, K, LE). (Designated as *S. tenue* Sendtn.)
Sellow 3499 (BR, K, LE).
Sellow (s.n.) (BR, CGE, K, LE, P, W). (Designated as *S. commersonii* var. *pubescens* Sendtn.)

URUGUAY

Dept. CANELONES:

Bañados de Toledo. Apr. 1907. 'En tierras bajas, húmedas.' *Arechavaleta* (s.n.) (MVM).

Carrasco, fixed dunes 500 m from the sea. 21 Mar. 1958. 'Damp sandy places amongst grass. Flowering.' *Hawkes 704* (J.G.H., grown in Birmingham as No. 181. $2n = 24$). See Plate 20.

Independencia. 3 May 1908. 'In saxosis humidis. Fl. violaceis.' *Osten 5228* (G, MVM).

Olmos. 31 Jan. 1915. *Osten 7057* (MVM).

Dept. COLONIA:

Colonia, recinto ciudad. Apr. 1913. 'Fl. blanca, algo violácea.' *M.B. Berro 6776* (MVFA).

Riachuelo. Apr. 1935. 'Suelo arenoso.' *Cabrera 3344* (LP).

Arroyo Rosario. 11 July 1936. 'Médanos. Flor azulada.' *Cabrera 3723* (LP).

Riachuelo. 22 Mar. 1937. 'Dunas. Corola blanca.' *Cabrera 4028* (LP).

Isla San Gabriel. 10 May 1854. *Courbon 181* (G).

San Gabriel (Montevideo, Plata). 10 May 1854. *Courbon 185* (P).

La Estanzuela, just outside the plant-breeding station. Alt. 80 m. 23 Mar. 1958. 'Rough grass, damp places.' *Hawkes 705, 706* (J.G.H. Accession 182, 183. Both $2n = 24$).

Estanzuela. *T. Henry* (s.n.) (WIR, No. K. 189) (Type for *S. henryi* Buk. & Lechn.) See Plate 25.

Arroyo Pintos. Alt. 5–20 m. 23 May 1947. 'Silv. Eucalypt. Flor viol. In dunis aren. sicc. umbras.' *Herter 1337b (61319)* (US) (var. *violaceum* Herter, *nom. nud.*).

Nueva Palmira. Apr. 1926. 'Ufer des Uruguay. Fl. alb. Galeriewald.' Leg. Herter, *Osten 19022* (MVM).

Caño. Apr. 1926. 'Fl. alb.' Leg. Herter, *Osten 19024* (MVM).

Dept. DURAZNO:

Maestro Campo. 21 Nov.–24 Dec. 1931. *Dauber 159* (BA).

Dept. FLORIDA:

Florida. '25 de Agosto.' Apr. 1907. 'Fl. azulada.' *M.B. Berro 4792* (MVFA).

Dept. LAVALLEJA:

Paraje de Aguas Blancas. Apr. 1958. 'Ploughed field. Berries and tubers.' *Lucía Koch de Brotas* (s.n.). (J.G.H. 724, grown as No. 195; no herbarium specimen. $2n = 24$.)

Dept. MALDONADO:

Puente del Rey. 'En lugar húmedo arenoso. Fl. violeta.' *M.B. Berro 3864* (MVFA).

Solis Grande. 25 Mar. 1907. 'Fl. violácea.' *M.B. Berro 4085* (MVFA).

Isla Gorriti, east side, 500 m from landing stage. 5 Apr. 1958. 'Grassy glade between pine trees. No fls.' (type locality for *S. ohrondii* Carr.) *Hawkes 721* (J.G.H.). (Grown in Birmingham as No. 192. Different tuber lines showed different development of sepals, from short triangular to ovate, and different petiolule lengths. $2n = 36$. Rather similar to 'anomalous forms'—see p. 154.)

Punta del Este. Mar. 1945. 'En veredas cerca de la costa.' *Diego Legrand 3845* (MVM).

Punta del Este. 4 Mar. 1922. *Osten 16560* (MVM).

Maldonado. 8 Apr. 1925. 'In arenosis.' Leg. Herter, *Osten 17653* (MVM).

Isla Gorriti (plants propagated in Leningrad as *S. ohrondii* Carr.). Coll. unknown (WIR).

Dept. MONTEVIDEO:

La plage du pied du morne de Montevideo. May 1767. 'Flores sunt palliduli.' *Commerson 47* (P—Lectotype, C, GOET (drawing), MPU). See Figs. 8, 9, and 10 and Plate 22. *Note*: The Copenhagen sheet does not bear the number *47*, but it is assumed to be part of Commerson's original collection.

Alrededores de Montevideo. Apr. 1909. 'En tierras húmedas sueltas, grietas de peñascos. Tallos acostados, pétalos grandes.' *Arechavaleta* (s.n.) (MVM).

Alrededores de Montevideo. Apr. 1909. 'En tierras húmedas sueltas, grietas de peñascos. Tallos acostados, pétalos medianos.' *Arechavaleta* (s.n.) (MVM).

Punta Gorda (nr. Carrasco). Apr. 1885. 'Terrenos arenosos de la costa.' *Arechavaleta* (s.n.) (MVM).

Manga. 10 May 1906. *M.B. Berro 4084* (MVFA).

Malvín. 20 Apr. 1912. 'Fl. violácea.' *M.B. Berro 6697* (MVFA).

Montevideo, Carrasco. 2 Jan. 1914 and 25 Dec. 1913. *M.B. Berro 7203, 7204* (MVFA), also *M.B. Berro 7529* from same locality, 24 Dec. 1914 (MVFA).

Montevideo, Malvín. Apr. 1914. 'Fl. caerulea.' *M.B. Berro 7527, 7528* (MVFA).

Playa del Cerro. 'Forma litófita.' *Chebataroff 5229* (MVM) (Topotype).

Lago Buceo. 1852. *Courbon (1856)* (P).

Montevideo, Plata, Miguelete. 20 May 1854. *Courbon 181* (P).

Punta Brava (Montevideo). 26 Sept. 1876. *Fruchard* (s.n.) (P).

Montevideo. Apr. 1836. *Gaudichaud* (s.n.) (G, P).

Montevideo. 1839. *Gaudichaud* (s.n.) (FI).

Montevideo. Mar. 1858. *Gibert 54* (K).

Montevideo. 1859. *Gibert 110* (K).

Montevideo. May 1868. 'Inter rupes maritimas.' *Gibert 962* (GOET, K, LP, MVM, S).

Montevideo. 29 May 1827. Mus. Henslow. (CGE). Note: This may have been collected by Capt. P. P. King.

Miguelete. May 1925. 'In graminosis argil. humid umbros. Plantae 30 cm fl. alb.' *Herter 276 (78033)* (G, LE, S, Z).

Entre Sayago y Colón, N. of Montevideo. 14 Apr. 1907. *Herter 276a (2504a)* (F, LIL, NY, SI, UC).

Apr. 1907/10. *Herter 276b (10382)* (U).

Pajas Blancas. Alt. 0–10 m. 1935. 'Camp., 25 cm high; fl. viol.' *Herter 1337b (95791)* (U) (var. *violaceum* Herter, *nom. nud.*).

Montevideo. 1838. *Arsène Isabelle 23* (FI, K, P).

Montevideo. 1826–30. *Capt. King* (s.n.) (K, TCD).

Carrasco. Mar. 1934. 'Fl. blanco-morada.' *Diego Legrand 186* (MVM).

Carrasco. 19 Mar. 1936. 'En arena cerca de caminos. Fl. blanca por arriba, morada por abajo.' *Diego Legrand 829* (F, MVM).

Dunas de Carrasco. Apr. 1922. *L.H.* (s.n.) (BA 2337) [coll. L. Hauman?].

Colón. 12–25 Mar. 1899. 'In graminosis. fl. albis, pallide violascentibus, exterius violaceis.' *Osten 3670* (G, MVM) (var. *raphanistrum* Bitt., *nom. nud.*).
Carrasco. 11 Feb. 1912. *Osten 5777* (MVM) (var. *indigoticascens* Bitt., *nom. nud.*).
Carrasco. 6 Apr. 1912. 'In dunis.' *Osten 6450* (MVM).
Buceo. July 1922. Leg. J. Schröder, *Osten 17108* (MVM).
Montevideo. Received 24 July 1930. Grown by Vilmorin at Verrières (s.n.) (P).
Montevideo. Nov. 1855? (Collector unknown) *55* (TCD).
Montevideo area. Exact locality unknown. 1960. *Kemp* (s.n.). (Living material only. Accession H. 598. $2n = 24$.)

Dept. RÍO NEGRO (or SORIANO):

Río Negro, Island of Naranjo. *Rómulo Rubbo* (s.n.) (WIR No. K. 1216, 2884). (Type for *S. rionegrinum* Lechn. *nom. nud.*) See Plate 29.

Dept. RIVERA:

Tranqueras. Nov. 1899. *Arechavaleta* (s.n.) (LY).
Rivera. *Arechavaleta* (s.n.) (MVM).

Dept. ROCHA:

Road no. 9, between Rocha and Castillos, 250·5 km from Montevideo. 4 Apr. 1958. 'Marsh by stream passing under road. Flowers purple.' *Hawkes 719* (J.G.H.).
Balneario Aguas Dulces, cerca de Castillos. 29 Apr. 1943. 'En lugares húmedos y abrigados. Fl. de pétalos violetas con punta blanca.' *Diego Legrand 3291* (MVM).

Dept. SAN JOSÉ:

Barra Santa Lucía. 30 Sept. 1956. *Arrillaga 361* (MVFA).
Barra Santa Lucía. 28 Oct. 1956. *Arrillaga 484* (MVFA).
La Barra. *Chebataroff 5228* (MVM, US).
San José. May 1867. 'Quinta y campos.' *Christison* (s.n.) (E).
Near Río Santa Lucía and Puente La Barra, 21·5 km from Montevideo. 20 Mar. 1958. 'Among coarse grass in damp sandy places. Flowering.' *Hawkes 702* (J.G.H. No. 180).
San José, Barra Santa Lucía. 30 Apr. 1917. 'Fl. viol. 20 cm high.' *Herter 1337* (*85714*) (F, G, LIL, SI, UC, Z) (var. *violaceum* Herter, *nom. nud.*).
Eifler. Alt. 5 m. Apr. 1933. 'Subhumid. 30 cm high. Fl. purple-viol.' *Herter 1337 a* (*91303*) (U) (var. *violaceum* Herter, *nom. nud.*).
Eifler. 24 Nov. 1935. *Herter 1337c* (*96596*) (G, LA, M, UC, Z) (var. *violaceum* Herter, *nom. nud.*).
La Barra. 4 Apr. 1947. 'Fl. violácea y blanco-violácea; pajonales de bañado en suelos sub-arenosos de preferencia.' *Diego Legrand 1919* (MVM).
Barra Santa Lucía. 27 Nov. 1944. *H. Osorio 6* (MVM).
Inter Santa Lucía et Libertad. 28 Feb. 1918. 'In pratis, locis humosis umbrosis. Fl. violaceis albo-variegatis.' *Osten 12031* (G, MVM).

Dept. SORIANO:

Vera. 30 Mar. 1898. 'Fleur violacée.' *M.B. Berro 124* (G, MVFA).
Bords du Río Negro, près de Mercedes. Apr. 1867. 'Sables. Très commune.' *Fruchard 263* (K, P).
Juan Jackson, Estancia Monzón-Heber. 17 Apr. 1940. 'Pratense, flor morada.' *Gallinal, Aragone, Bergalli, Campal & Rosengurtt PE—4323* (LIL, MVM).
Near Mercedes, Arroyo Cololó, Esta. S. Joaquín. *T. Henry* (s.n.) (WIR, No. K. 1071) (Type for *S. mercedense* Buk., formerly named *S. compactum* Buk., *nom. nud.*). See Plate 27.

Cerca Mercedes, Arroyo Bequeló. *T. Henry* (s.n.) (WIR, No. K. 1073. Type for *S. sorianum* Buk.). See Plate 28.

Mercedes. 11 Nov. 1892. *Osten 2915* (BREM, MVM) (subsp. *ellipticans* Bitt. *nom. nud.*).

Uruguay, locality unknown. 20 May 1854. 'Lieux sablonneux. Pantanoso.' *Courbon* (s.n.) (G, P). (var. *rosulans* Bitt. *nom. nud.*).

Uruguay, locality unknown. *Felippone 5502* (SI).

Dept. unknown. In silvulis ad ripas Sta. Lucía. Mar. 1869. *Gibert 906* (K).

Uruguay, locality unknown. Sent by T. Henry. Grown at Leningrad. *T. Henry* (s.n.) (WIR, No. K. 1071).

72 km from Montevideo on road to Colonia (probably Dept. San José). 27 Jan. 1948. 'Among grasses at roadside in heavy black soil; locally common.' *Stephens & Hartley S.H. 88* (K).

ARGENTINA

Prov. BUENOS AIRES (including CAPITAL FEDERAL):

La Plata, Bosque. Apr. 1943. *Delia Abbiatti* (s.n.) (LP 052523, 052524).
Delta del Paraná, Arroyo Caraguatá. 23 Mar. 1927. 'Flor violeta. En frutales.' *Burkart 1230* (SI).
Punta Lara. 10 Feb. 1929. 'Flor lila. Monte.' *Burkart 3170* (SI).
Delta del Paraná, Espera. 26 Dec. 1931. 'Hab. monte de durazneros; sin flores, tubérculos que se prolongan en estolones.' *Burkart 4535* (GH, SI).
Punta Piedra. 2 May 1931. 'Praderas de *Spartina*.' *Cabrera 1728* (LP).
Delta del Paraná. Ile Brunet. 10 Apr. 1876. *Calot 65* (P).
Balcarce, Sierra Bachicha. Apr. 1925. *Castellanos* (s.n.) (BA 25/429).
Cinco Cerros. Apr. 1925. *Castellanos* (s.n.) (BA 25/1428).
El Toro, Rosas, F.C.S. 10 Mar. 1926. 'Papa cimarrona.' *Daguerre 326* (BA).
Capital Federal, Tiro [Federal]. Feb. 1911. 'Papa Cimarrona.' *Dieckmann* (s.n.) (SI 15942).
Capital Federal, Paseo Colón. Mar. 1910. *Dieckmann* (s.n.) (SI 15943).
Cabo San Antonio. 10 Feb. 1961. *Fabris & Cullen 2571* (LP).
Partido Balcarce, Sierra Brava. 20 Apr. 1950. *J. M. Gallardo* (s.n.) (BA 52185).
Buenos Aires. May 1820. 'Hedges.' *Gillies* (s.n.) (K—Syntype for *S. commersonii* var. 4, *glabratum* Hook. f.).
Buenos Aires. 'Planta parva inferior ad *S. commersonii* Dun. pertinet—Bitter.' *Gillies* (s.n.) (K—Syntype for *S. commersonii*, var. 4, *glabratum* Hook. f.).
Balcarce, Estación Experimental de Balcarce. 'Cerros.' 20 Mar. 1958. *Grondona 6300* (BAL).
Balcarce, Estancia San Simón, Cerro de la Virgen. 29 Mar. 1958. *Grondona 6316* (BAL).
Balcarce, Cerro 'La Peregrina', cara N.E., 500 m del camino. Alt. 120 m. 30 Mar. 1958. 'Entre rocas de cuarzo, suelo humífero.' *Grondona 6343* (BAL, J.G.H.).
Balcarce, Cerro 'Cota 320', 1500 m N. del Cerro 'El Sombrero'. 31 Mar. 1958. *Grondona 6372* (BAL, J.G.H.).
Balcarce, Cerro 'El Cristo'. 31 Mar. 1958. 'Suelo humífero.' *Grondona 6375* (BAL).
Balcarce, Cerro del Morro. 31 Mar. 1958. *Grondona 6376* (BAL).
Balcarce, Sierra Larga de punta N.—oriental. 2 Apr. 1958. *Grondona 6392* (BAL).
Balcarce, Sierra 'El Sombrero'. 3 Apr. 1958. *Grondona 6407* (BAL).
Balcarce, 'Cinco Cerros'. 23 Apr. 1958. 'En suelo humífero, entre rocas.' *Grondona 6414* (BAL, J.G.H.).
Balcarce, Cerro cerca del Tanibo. 23 Apr. 1958. *Grondona 6416* (BAL).

Buenos Aires, prairies humides au bord de Río de la Plata. Mar. 1906. *Hauman* (s.n.) (BRLU).

Sierra Balcarce, Yráizoz. 30 Dec. 1922. *Hauman* (s.n.) (BA 2417).

Capital Federal, alrededores. *Hauman* (s.n.) (SI 13767).

Balcarce, Estación Experimental de Balcarce. Cerro 'La Paja'. 26 Mar. 1958. Alt. 200 m. 'In black shallow soil on granite rocks amongst short grass. Plants with flowers, berries and mother tubers.' *Hawkes 708* (J.G.H.) (H. 184. $2n = 24$).

Mar del Plata, Cerro Copelina. 29 Feb. 1932. 'Frecuente entre los cerros E.' *Hicken 380* (SI).

Mar del Plata, La Copelina. 1 Mar. 1932. *Hicken* (s.n.) (SI 409).

Olavarría, Sierra Dos Hermanas. 5 Apr. 1899. 'Muy abundante.' *Hicken* (s.n.) (SI 13766).

Capital Federal, Palermo. 19 Mar. 1902. *Hicken* (s.n.) (SI 15946).

Mar del Plata. 1932. *Hicken* (s.n.) (SI 15957).

Delta del Paraná, Sec. Primera. Feb. 1938. 'En un albardón cerca del Arroyo Gálvez.' *A. T. Hunziker 897* (SI).

Buenos Aires, Palermo, en el fondo desecado del lago. 28 Apr. 1945. 'Corola blanco-azulada.' *J. H. Hunziker 644* (CORD, LIL).

Zárate, en barranca del Río Paraná de las Palmas. 19 May 1946. 'Geófita tuberosa. Corola blanca.' *J. H. Hunziker 1611* (CORD, LIL).

Partido Olavarría, La Providencia, inmediaciones del Arroyo San Jacinto. 18 Apr. 1947. *Krapovickas 3356* (LIL).

'Seis de Sept.' (Morón), Castelar. Apr. 1938. 'Muy raro.' *Lourteig 154* (LIL).

Delta del Paraná. Oct. 1942. 'Flor blanca.' *Martínez Crovetto 1365* (LIL, SI).

General Alvarado, Las Piedritas, F.C.S., Mechongué. 'Crece entre los curros (*Colletia cruciata*).' *Millán M.A. 31* (WIR No. K. 1088, Type for *S. mechonguense* Buk.). See Plate 26.

Same locality. 'Bajo matas de curro.' 26 Feb. 1941. *Millán* (s.n.) (BAL, J.G.H.).

Mechongué, Est. Las Piedritas, entre los cerros. 2 May 1939. 'Campo virgen.' *Millán & Pujals 8* (SI).

Balcarce, Sierra Larga. 4 May 1939. *Millán & Pujals 25, 27, 29, 31, 33, 35, 37* (SI).

Tandil. 26 Apr. 1949. 'Humid places, amongst rocks.' *Myndel Pedersen 364* (C, US).

Delta del Paraná, Arroyo La Espera. 'Común. Flor blanco-azulada; fruto alargado.' *Nicora 4472* (SI).

Azul, Est. La Armonía. 6 Apr. 1945. 'Corola violeta; hojas verde clara.' *J. A. Núñez 383* (BA).

Azul, Boca de las Sierras. Alt. 200–400 m. 21 Mar. 1886. 'In montibus saxosis.' *Osten 2628* (MVM) (var. *depauperatum* Bitt. *nom. nud.*).

Tigre. 27 Feb. 1934. 'Flor violácea. Hab. adventicia en los cultivos en el Delta del Río Paraná.' *L. R. Parodi 11882* (WIR).

Río de La Plata, Isla Martín García. Apr. 1935. 'Médano. Al margen, cerca de los arbustos.' *Ada Pastore 225* (LP, SI).

Balcarce, en el cerro al S.O. de la laguna 'La Brava.' 10 May 1928. *Pelosi 64* (SI).

La Plata, Punta Lara. 6 May 1937. *C. Pujals 37* (LIL).

Lobería. Feb. 1918. *A. Scala* (s.n.) (LP 010996).

La Plata, Ferrocarril Francés. Jan. 1921. 'Fl. violac.' *C. Spegazzini* (s.n.) (LP 14180).

La Plata. Mar. 1914. *C. Spegazzini* (s.n.) (LP 14181).

Buenos Aires (see J. D. Hooker, 1846). 1837. 'This var. of the Pampas is finely sorted.' *Tweedie* (s.n.) (K—possibly type for *S. commersonii* var. 1, *glabriusculum* Hook f. p. pte.).

Barracas al Sud. 23 Feb. 1902. *Venturi 2* (SI).

Partido de Lobos, Lobos, 3 km a Beguesia. 3 Apr. 1958. 'En el pastizal.' *Vervoorst 5892* (LP).
Capital Federal, Palermo, Tiro Federal. Dec. 1916. 'Escasa; tubérculos comestibles.' Coll. unknown (SI 15952).

Prov. ENTRE RÍOS:

Médanos. 3 Dec. 1930. *Burkart 3568* (SI).
Delta del Paraná, Arroyo Martínez. 12 Oct. 1944. 'Fl. violeta, baya blanquecina.' *Burkart 15127* (SI) and *15129* (GH, SI).
Palermo. June 1933. *Friedman 85* (BAB).
Ibicuy. Nov. 1918. *Hauman* (s.n.) (BA 2336).
Partido Gualeguaychú, Estancia Rincón de Landa. 19 Apr. 1943. 'Con un pequeño tubérculo. Se presenta formando manchones.' *Krapovickas 189* (LIL).
Concepción del Uruguay. 2 May 1877. *Lorentz 497* (M, Z).
Concepción del Uruguay. Apr. 1876. 'Bl. weiss.' *Lorentz 553* (CORD, GOET, JE, W).
Concepción del Uruguay. 2 May 1877. *Lorentz 966* (BREM, FI, G, GH, GOET, JE, K, L, LE, LY, P, UPS).
Locality unknown. Nov. 1877. *Lorentz & Grisebach* (s.n.) (K).
Gualeguaychú, Holt. May 1943. 'Hab. en los médanos. Flores violáceas.' *Martínez Crovetto 2230* (LIL).
Gualeguaychú, Pehuajó Norte. Mar. 1926. *Serié & Migoya* (s.n.) (BA 26/678).

Country and province uncertain:

Río Paraná (not Brazil). *Tweedie* (s.n.) (TCD).
Pampas. *Tweedie* (s.n.) (LE).

Origin unknown:

Cultivated at Hedisburgum, Batavia. 3 July 1905. *Buysman* (s.n.) (LY).
24 Aug. 1893. '*S. ohrondii*.' Herb. Groningen (s.n.) (GRO 5194 and 5195).
Bot. Garten Wien. 27 July 1909. '(Angeblich die Stammpflanze der Kartoffel).' *Schiffner* (s.n.) (WU).
Cult. en hort. bot. Monspel. June 1902. *Daveau* (s.n.) (MPU).
Material grown in Leningrad, 1938, and stated by Bukasov to be *S. ohrondii*. (P.G.S. 0012.)
Material grown in Leningrad, 1938, and stated by Bukasov to be genuine *S. mechonguense*. (P.G.S. 0015.)

2. Subspecies *malmeanum*

BRAZIL

Estado do RIO GRANDE DO SUL:

Ijuí. 5 Apr. 1893. 'In ripa arenosa rivuli; loco aperto terra fere denudata.' Leg. G. A. Malme, *Regnell 756* (S—lectotype (lower plant), UPS, US). See Plate 31.
Ijuí. Seed collected by Dr. Delorge Mota da Costa and sent 23 July 1959. (Accession H. 560. $2n = 24$.)

PARAGUAY

Dept. Encarnación, Ihú. Alt. 123 m. 16 June 1949. 'Fl. blanca.' *Bertoni 4274* (LIL).
San Bernardino, in campos. Dec. 'Herba 30–40 cm, pétala alba.' *Hassler 3674* (G, type for *S. guaraniticum* var. *latisectum* Hassl. forma *pilosulum* Hassl. and for *S. commersonii* β *pubescens* Chodat). See Plate 34.

Campos entre montes San Bernardino. May. 'Herba de 30–40 cm. Pétalos blancos.' *Hassler 140a* (G, SI).

San Bernardino. May 1915. 'Alt. 0·3–0·4 m. Pét. blanco; campos orillas montes.' Leg. Rojas. *Hassler 2515* (MVM). (*S. bitteri* var. *latisectum* f. *pilosulum*, det. Hassler.)

Dept. Guairá, Iturbe, próximo Monte Santa Clara. 3 May 1952. 'Hierba an? alt. 0·30–0·70 m con fl. Flores blancas, estambres amarillos. Hab. desmonte, lugar alto, escasa, con bulbo raíces. "Papa del yaguá." ' *J. E. Montes 15656* (LIL).

PARAGUAY OR ARGENTINA

In Campis altis in regione cursus inferiores fluminis Pilcomayo. May 1906. '30–50 cm. Pét. blanco.' Comisión argentino-paraguaya de límites 1906. *T. Rojas 136* (G—Type for *S. commersonii* var. *pseudostipulatum* Hassl.). See Fig. 13.

URUGUAY

Dept. MONTEVIDEO:

Montevideo, El Prado Park Botanical Garden, near Museum. 8 Apr. 1958. 'In grass under partial shade of trees, damp soil. White flowers. Probably an escape from former botanical collections.' *Hawkes 723* (J.G.H. No. 194. $2n = 36$).

Dept. PAYSANDÚ:

Paysandú. *Felippone 3226* (SI).

Escuela de Práctica y Campo Experimental de Agricultura de Paysandú. 5 May 1958. *L. Mazzei* (s.n.) (J.G.H. No. 197. $2n = 24$).

Dept. SALTO:

San Antonio, Escuela Experimental de Agronomía. 16 Apr. 1958. 'Dark humus-rich soil. White flowers.' Coll. J. A. Reyes. Comm. A. Montoro Guarchi. *Hawkes 725* (J.G.H. No. 196. $2n = 24$).

ARGENTINA

Prov. ENTRE RÍOS:

Paraná. Mar. 1941. *Báez* (s.n.) (BAL, J.G.H.).

Dept. Paraná, Gobernador Racedo. 16 May 1964. 'Invasora de rastrojo; flor blanca, pedúnculos violáceos.' *Burkart, Troncoso, Bacigalupo & Crespo 25460* (SI).

Dept. La Paz, La Paz. 17 May 1964. 'Invasora de rastrojo; flor blanca, pedúnculos violáceos.' *Burkart, Troncoso, Bacigalupo & Crespo 25461* (SI).

Dept. La Paz, entre Arroyo Alcaraz y Arroyo Feliciano, ruta 126. 6 Nov. 1965. 'Terraplén camino; estolones largos, papas chicas. Corolla blanco-crema.' *Burkart 26044* (SI).

María Grande. 24 May 1940. *Cabrera 6437* (LP).

Dept. La Paz, Colonia 'Avigdor', Bovril F.C.E. 26 May 1943. 'En monte degradado (?), cerca del Arroyo Colorado.' *Krapovickas 254* (LIL).

Dept. Federación, orillas del Mocoretá. 27 Apr. 1953. 'Flores blancas, muy comunes en la arena.' *Nicora 6379* (SI).

Prov. SANTA FÉ:

Gallardo, a 11 km al N. de la Capital. May 1939. *Millán & Pujals 39, 40, 41, 42* (SI).

Dept. Capital, San José del Rincón. 3 July 1938. 'Con tubérculos. En terreno arenoso. Pétalos blancos, anteras amarillas.' *Ragonese 3442* (US).

Prov. CORRIENTES:

Dept. Berón de Astrada, Tuyutí. 3 May 1945. 'Alt. 10 cm. Fl. blanca.' *Huidobro 2134* (LIL).

Dept. Capital, Riachuelo. 3 Apr. 1944. 'Flor blanca.' *Ybarrola 171* (LIL).

Dept. General Paz, Lomas de Vallejos, propiedad de Juan Jensen. Alt. *c.* 70 m. 29 Apr. 1949. 'En tierra arenosa y húmeda, en cultivos de algodón.' *Petersen & Hjerting P-72* (C, J.G.H., J.P.H.).

Dept. General Paz, Estancia La Flecha, la huerta. Alt. *c.* 70 m. 30 Apr. 1949. 'Entre pasto, húmedo.' *Petersen & Hjerting P-73* (J.G.H., J.P.H.).

Dept. Itatí, Ruta 12, entre Tuyutí y Ramada Paso, km 1016. Alt. *c.* 70 m. 1 May 1949. 'Planta forma roseta, con rizomas subterráneos. Fruto ovalado.' *Petersen & Hjerting P-74* (J.G.H., J.P.H.).

Dept. Ituzaingó, Estero R. Carambola y Estero Laguna Iberá. 28 Apr. 1944. 'Solanácea de flores blancas muy semejante a *S. tuberosum*.' *Ybarrola 332* (LIL, SI).

Dept. Mburucuyá, Estancia 'Santa María.' 27 May 1951. *Myndel Pedersen* (s.n.) (SI 20036).

Dept. Mburucuyá, Estancia 'Santa Teresa.' 1 May 1952. *Myndel Pedersen 1706* (C, LP, US).

Dept. Mercedes, Yofre, cercanías. 24 Mar. 1945. 'Solanácea anual de 20 cm alto; flores blancas; vegeta borde arroyuelo.' *Ybarrola 2729* (LIL).

Dept. Monte Caseros, Libertad, costa Arroyo Curuzú. 24 Feb. 1945. 'Solanácea anual de 20 cm alto; flores blancas con manchas rojas.' *Ybarrola 2515* (LIL).

Dept. Saladas (?), Laguna Soto. 22 Sept. 1944. 'Planta rara.' *G. Schwarz 34* (LIL).

Dept. San Cosme, Paso de la Patria, Costa Toledo. Alt. 55 m. May 1945. *T. Meyer 8979* (LIL).

Dept. San Cosme, Paso de la Patria. 22 June 1945. 'Flor blanca.' *A. Wurth 146* (LIL).

Locality unknown, Apr. 1925. 'Bl. weiss. Akerland.' *W. Lossen 535* (GH, LE, SI).

Prov. MISIONES:

Dept. Cainguas, Arroyo Paranay-guazú, cerca de la balsa. Alt. *c.* 100 m. 26 Apr. 1949. 'En tierra arcillosa semihúmeda a pleno sol.' *Petersen & Hjerting P-69* (J.G.H., J.P.H.).

Dept. Cainguas, Puente J. U. Martín, Arroyo Cuña-Pirú. Alt. *c.* 100 m. 26 Apr. 1949. 'Flor con pétalos separados en varias plantas; frutos acorazonados.' *Petersen & Hjerting P-70* (CPC: 2495) (LIL, J.G.H., J.P.H.) ($2n = 36$). See Plate 32.

Dept. Candelaria. Colonia Finlandesa. Four collections made between 21 May and 29 June 1929. 'Selva y campos.' *Lunnasvaana* (s.n.) (H).

Dept. Candelaria, Loreto. Leg. A. Mutinelli, *Millán 973* (WIR No. K. 223. Type for *S. millanii* Buk. et Lechn.). See Plate 33.

Dept. Candelaria, Loreto. 25 Apr.–22 May 1947. 'Papa del Monte, Papa del Indio, Papa de Bugre; herba 30–60 cm alta. Flores blancas, raíz con tubérculos; escasa.' *J. E. Montes 576, 600, 661, 667, 670, 671* (BA).

Dept. Candelaria, La Pastora. Alt. 220 m. 5 June 1946. 'Hierba. 30–50 cm, estolones amarillos. "Papa Silvestre, Papa Montes".' *J. E. Montes 2321* (LIL).

Dept. Candelaria, La Pastora. Alt. 220 m. 16 July 1946. 'Hierba. Alt. 30–60 cm.' *J. E. Montes 2449* (LIL).

Dept. Candelaria, Colonia Mártires, lote 29, la chacra de Chr. Davidsen. Alt. 125 m. 15 Apr. 1949. 'En campiñas bajas, muy seco, entre vegetación pobre.' *Petersen & Hjerting P-64* (J.G.H., J.P.H.).

Dept. Candelaria, Est. Exp. de Loreto, Colonia Loreto. Alt. 163 m. 27 Apr. 1949.

'En cultivos de mandioca. En la primavera hay muchas plantas.' *Petersen & Hjerting P–71* (CPC: 2496) (C, LIL, J.G.H., J.P.H.). ($2n = 36$).

Dept. Candelaria, Candelaria. 11 June 1944. 'Hierba. Flor blanca.' *Sesmero 93* (LIL).

Dept. Capital, Arroyo Garupá, al lado del puente, ruta 12. Alt. 125 m. 14 Apr. 1949. 'En selva; tierra húmeda y selvática, sombreada. Flor blanca, tubérculos hasta 1×3 cm.' *Petersen & Hjerting P–63* (LIL, J.G.H., J.P.H.).

Dept. Iguazú, Colonia Eldorado, Cuesta Río Piray-Guazú, la chacra de P. Beck. Alt. 70 m. 21 Apr. 1949. 'En tierra de humus, húmeda.' *Petersen & Hjerting P–67* (CPC: 2493. $2n = 24$) (C, LIL, J.G.H., J.P.H.).

Dept. San Ignacio, Yabebiry. 14 May 1944. *Bertoni 51* (LIL).

Dept. San Ignacio. 16 May 1946. 'Papa Montes. Hierba an 20–50 cm. Fl. blancas; raíz con tubérculos. Hab. monte lugar alto.' *J. E. Montes 2258* (LP).

Dept. San Ignacio, Teyucuaré. Alt. 238 m. 16 Mar. 1952. 'Hierba 30–60 cm; fls. blancas. Hab. monte, lugar alto, escasa.' *J. E. Montes 15611* (LIL).

Dept. San Ignacio, Arroyo Macaco. 9 Apr. 1948. '15 cm alto, fl. blanca.' *G. Schwarz 5716* (LIL).

Dept. San Ignacio, Santo Pipó. 28 Apr. 1948. '20 cm alto, fl. blanca.' *G. Schwarz 5898* (LIL).

Dept. San Ignacio, San Ignacio. 12 May 1948. 'Flor blanca.' *G. Schwarz 5919* (LIL).

Dept. San Ignacio, Arroyo Nacanguazú. 29 June 1949. '15 cm alto. Flor blanca.' *G. Schwarz 6099* (LIL).

Dept. San. Javier, Colonia Oberá. Alt. 310 m. 22 July 1950. 'Papa del Monte. Hierba an alt. 30–50 cm. Fl. blancas. Hab. desmonte, lugar bajo. Muy escasa.' *Schwindt 4838* (p.p.) (LIL).

Dept. San Pedro, Caraguatay (centro). Alt. 206 m. 10 May 1949. 'Hierba an 30–45 cm alto, fl. blancas. Hab. monte despejado.' *J. E. Montes 1684* (LIL).

Dept. San Pedro, Colonia Teresa, Río Piray-Guazú. Alt. *c.* 100 m. 25 Apr. 1949. 'Entre yerba en tierra arcillosa.' *Petersen & Hjerting P–68* (CPC: 2494) (C, J.G.H., J.P.H.).

Dept. unknown: Francés-Cué. July 1919. *Hauman* (s.n.) (BA No. 2421).

Prov. CHACO:

Dept. Resistencia, Resistencia. May 1883. *C. von Gülich* (s.n.) (LP 14230 p.p.).

Dept. Resistencia, Cacuí. Mar. 1933. *T. Meyer 2034* (GH, LIL).

Dept. Resistencia, Fontana. June 1937. 'Crece en terrenos humíferos en el interior de los bosques altos.' *T. Meyer 2323* (p.p.) (GH, LIL).

Dept. Resistencia, Colonia Benítez, la quinta de A. Schulz. Alt *c.* 50 m. 2 May 1949. *Petersen & Hjerting P–75* (C, J.G.H., J.P.H.).

Dept. Resistencia, Colonia Benítez, la quinta de A. Schulz. Alt. *c.* 50 m. *Schulz 9024, 9026*. Living material at Birmingham (H. 83, 84, 85. $2n = 24$).

Prov. FORMOSA:

Villa Formosa. 27 Mar. 1885. *F. Kurtz 1786* (CORD).

ANOMALOUS FORMS SIMILAR TO THE TRIPLOID CLONES GROWN IN PLANT BREEDING STATIONS

In the extensive study made by us of both subspecies of *S. commersonii*, both as dried material and living plants, certain anomalous morphological variants have stood out sharply. We have not been able to place them with confidence in either subspecies, and we mention them here in the hope that more work on them may be stimulated.

Certain collections, which one would have placed with subsp. *commersonii* on other grounds, differ from it in possessing a stronger, more upright stem, well-developed petiolules and very short peduncle branches. Since none of them has been observed with fruits it seems likely (though so far verified in only three collections of known provenance) that these are all triploids (possibly autotriploids) and that the anomalous features are due for some reason to the higher chromosome number. (See also section 'Cytology', p. 175.) The collections we have placed under this head are *Hicken* (s.n.) (SI 15945); *Hawkes 701* (see Plate 22), *718*, *722*; *B. Rambo* (s.n.) (Accession 93, see Plate 23); *Felippone 2444a*, *5494*, *5350*. When grown in Birmingham the collections of Hawkes and Rambo showed very marked similarities to the known triploids found in plant-breeding stations, which we have called 'conventional' *S. commersonii*.

Other collections, especially those of *Felippone 5434*, *5436*, *5437*, *5473*, *5476*, seem closer to subsp. *malmeanum*, which also possesses triploid forms. The collections *5473* and *5476* are said to have come from the Prado Park in Montevideo, where subsp. *malmeanum* occurs, probably as a Botanic Garden escape (see p. 152), and it is therefore likely that they may be an aberrent or triploid form of that subspecies.[1] Numbers *5434*, *5436*, and *5437* bear no locality data; from their morphology, however, one might guess that they also were collected in the Prado Park at Montevideo.

The anomalous triploid forms with long petiolules and cordate berries,[2] grown in plant-breeding stations and mentioned frequently in the literature, we have called 'conventional' *S. commersonii*. They are so similar morphologically to the anomalous forms which we have noted above, collected in Brazil and Uruguay by Hawkes (*701*, *718*, *721*, *722*) *when grown under comparable conditions in Birmingham*, that it seems fairly certain that they are one and the same thing. In other words, triploid biotypes of *S. commersonii* (probably autotriploids of subsp. *commersonii* in this case) exist in nature, not only in the region of Montevideo but in Brazil also. They were no doubt collected as triploids and have existed for many years under culture in plant-breeding stations and botanical gardens as objects of interest even though they have been of no value in the breeding of new potato varieties.

LOCALITIES OF ANOMALOUS FORMS

BRAZIL

Estado do RIO GRANDE DO SUL:

25 km N. of Porto Alegre, Esteio. Tubers only. Received February 1956. *B. Rambo* (s.n.) (J.G.H. No. 93). *Note*: Very similar to the triploid plant breeders' forms, with petiolules to 5 mm and peduncle forks in some plants only 5 mm long, in others to 3 cm. Various attempts at interclone crossing within this line failed to result in any seed or even berry production. It is a triploid clone, with $2n = 36$ chromosomes. See Plate 23.

13 km S. of Pelotas, Instituto Agronómico do Sul. 2 Apr. 1958. 'In scrub near undisturbed sub-tropical forest, amongst *Baccharis*, *Lantana*, and *Ipomoea*. Flowering.' *Hawkes 718* (J.G.H. No. 190). *Note*: Petiolules to 3 mm, peduncle branches much reduced. Probably a triploid form, closely related to or forming a part of subsp. *commersonii*.

Pelotas, Instituto Agronómico do Sul. 7 May 1954. 'Beira de estrada, terreno arenoso.' *José da Costa Sacco* (s.n.) (US 2279883). *Note*: Petiolules 1–1·5 mm; peduncle branches short, 10–12 mm.

[1] Recent studies (unpublished) by R. Tarn indicate that some of these forms may be natural hybrids of subsp. *commersonii* and subsp. *malmeanum*.

[2] Formed parthenocarpically, or occasionally with some seeds when pollinated with diploid species.

URUGUAY

Dept. MONTEVIDEO:

Rincón del Cerro, Playa Sanguinete, La Colorada. 18 Mar. 1958. 'Grassy and shrubby bank just above the sea shore. Flowers but no berries.' *Hawkes 701* (J.G.H. No. 179). *Note*: In the rounded leaflets, 1–4 mm long petiolules, larger terminal, and lack of berries this specimen shows some similarities to the triploid plant-breeding station clones. The purple flowers and general leaflet dimensions indicate a close link with subsp. *commersonii*, even though the petiolules are much longer than normal for that subspecies. $2n = 36$. See Plate 22.

El Prado Park, Arroyo Miguelete, on the banks of the stream, but out of reach of flood water. Alt. ±10 m. 8 Apr. 1958. 'In grass under shade of trees. Tall plants with mauve flowers and mother tubers.' *Hawkes 722* (J.G.H. No. 193). *Note*: Differs from normal *S. commersonii* in the strong tall stem, 3–5 mm petiolules and lack of fruit. When grown from tubers at Birmingham (No. 193) it came very close in habit, petiolules, and short inflorescence branches to the plant-breeding clones mentioned below. $2n = 36$.

Prado. *Felippone 5473* (SI, sheets *a*, *b*, *c*). *Note*: This might be a hybrid of subsp. *commersonii* × subsp. *malmeanum* (cf. *Hawkes 723*) but the petiolules are very long, up to 10 mm in some cases. It also shows some similarities with *Hawkes 722* from the same general area, especially sheet *b* and some specimens on sheet *a*.

Same locality. *Felippone 5476* (SI, sheets *a*, *b*). Both sheets show some affinities to subsp. *malmeanum*.

Carrasco. 26 June 1922. *Felippone 2444a* (SI). *Note*: Similar in habit to *Hawkes 722*, with petiolules to 10 mm. No flowers. A thorough search in Carrasco by Hawkes in 1958 failed to reveal the presence of this form in that area, though the typical form was plentiful.

Buceo. *Felippone 5494* (SI). *Note*: Petiolules 0–4 mm. Probably a triploid form, close to subsp. *commersonii*.

Uruguay, localities unknown. *Felippone 5350, 5434, 5436, 5437* (SI). *Note*: *5350*: long petiolules but rather similar to subsp. *commersonii*. *5434, 5436, 5437*: all more closely allied to subsp. *malmeanum*.

ARGENTINA

Prov. BUENOS AIRES:

Quilmes. 21 Nov. 1909. *Hicken* (s.n.) (SI 15945). *Note*: Petiolules to 3 mm, but in other features such as the enlarged terminal leaflet and mauve corolla the specimen fits closest to subsp. *commersonii*.

ANOMALOUS TRIPLOID FORMS WITH LONG PETIOLULES AND CORDATE BERRIES, GROWN IN PLANT-BREEDING STATIONS

URUGUAY, locality unknown. Collector unknown. (WIR. Type for the so-called '*S. commersonii* (Dun.) s. str. Buk.' $2n = 36$.) See Plate 24.

Localities and collectors unknown. Obtained from various plant-breeding stations. Grown at Potato Genetics Station, Cambridge (CPC: 1319, 1320, 1321, 1322, 2126. All with $2n = 36$). *Note*: Strong growing plants with petiolules to 5 mm, leaflet base truncate or even slightly cordate and peduncle branches only 0–3 mm long. Both purple and white-flowered forms occur.

Locality and collector unknown. Wittmack No. 1. 3 July 1912 (GOET).

DISTRIBUTION

1. Subsp. *commersonii* is found chiefly in the region of the Río de La Plata estuary, both near Buenos Aires (Argentina) and Montevideo (Uruguay) (see Map 3). In this area it occurs with great frequency, becoming scarcer to the north and south, and rarely found more than 300 km from the coast. The exact distribution is as follows:

Brazil: States of Rio Grande do Sul and Santa Catarina.
Uruguay: Depts. Canelones, Colonia, Durazno, Florida, Lavalleja, Maldonado, Montevideo, Río Negro, Rivera, Rocha, San José, and Soriano.
Argentina: Provs. Buenos Aires (including Capital Federal) and Entre Ríos.

2. Subsp. *malmeanum* has a more inland distribution area and has been recorded with greatest frequency in the Argentine provinces of Corrientes and Misiones (see Map 3). The exact distribution is as follows:

Brazil: State of Río Grande do Sul.
Paraguay: Region of Asunción and the lower part of the Pilcomayo river.
Uruguay: Depts. Paysandú and Salto.
Argentina: Provs. Entre Ríos, Santa Fé, Corrientes, Misiones, Chaco, and Formosa.

It is quite probable that more intensive collections in southern Paraguay, southern Brazil, northern Uruguay, and the Argentine province of Formosa would reveal that subsp. *malmeanum* occurred in many parts of those regions where it has not previously been recorded.

HABITAT

1. Subsp. *commersonii* apparently grows chiefly as a ruderal plant, often very abundantly, and especially in humid places, from sea level to about 400 m. The notes given by collectors indicate a very wide range of habitat; in fact it seems to have been collected in practically every sort of situation, as the following list indicates: meadows, prairies, grassy fields, cultivated fields, fruit-tree plantations, on the margins and in the middle of woods and forests, gallery woods, eucalyptus woods, amongst bushes, under *Colletia paradoxa* (= *cruciata*), shady places, sides of streams, hedges, paths, roadsides, marshes, dry and damp sandy places, amongst maritime rocks, and in coastal sand dunes. Whilst one could not claim that subsp. *commersonii* was a true maritime plant it seems nevertheless to be found very frequently by the sea, never at high altitudes, and frequently in damp marshy places.

The flowering season is chiefly from February to May; very occasionally it flowers also in June to July and again occasionally in September to January, with a small peak in November. Thus it is able to grow and flower during the winter season, when slight night frosts are not infrequent.

In Argentina it is to be found in the following phyto-geographical regions:[1]

PAMPA PROVINCE: Uruguay district and eastern and southern Pampa districts.
ESPINAL PROVINCE: Tala district.

In Uruguay and Brazil it has been recorded for the prairies (probably equivalent to the Argentine pampa) though most records are for coast and river banks.

On the whole it would seem that subsp. *commersonii* is not to be found at all frequently in the natural Pampa and Espinal vegetation, but is rather a colonizer of damp grassy places by streams and rivers or on the edges of marshes, in ditches and by paths and rocks, etc., or in cultivated areas where the natural climax vegetation has been destroyed.

[1] Scheme proposed by Cabrera (1953).

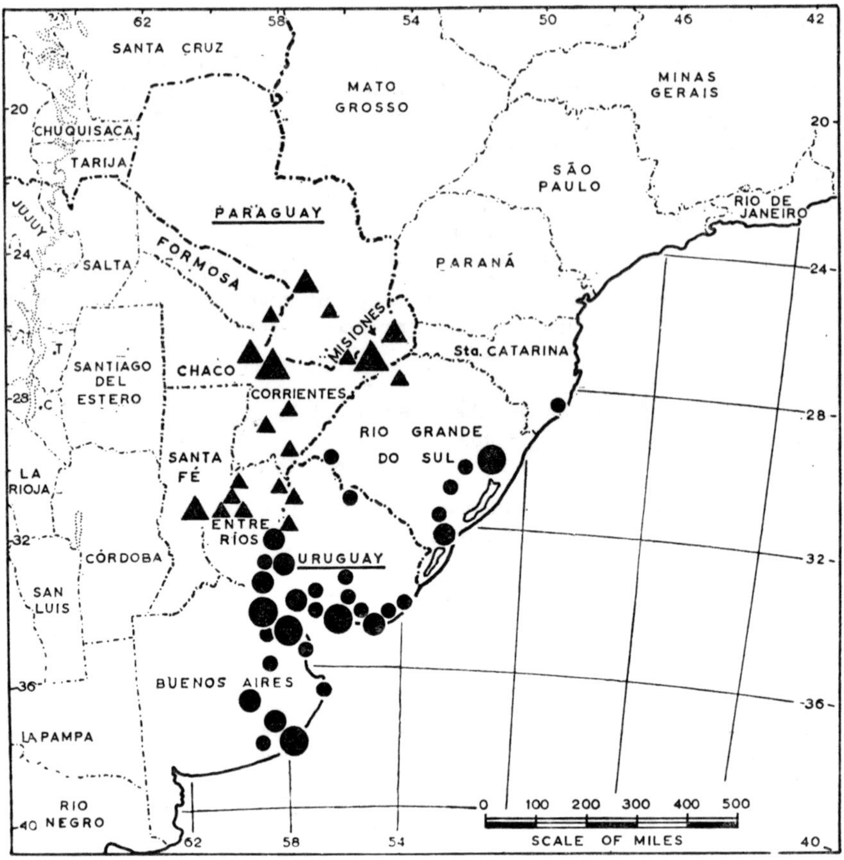

No. of Records
5 & over 3–4 1–2

● ● ● *Solanum commersonii* Dun. subsp. *commersonii*

▲ ▲ ▲ *Solanum commersonii* Dun. subsp. *malmeanum* (Bitt.) Hawkes & Hjerting

MAP 3. Distribution of *S. commersonii* Dun. subsp. *commersonii* and subsp. *malmeanum* (Bitt.) Hawkes & Hjerting.

As an exception to the above generalization there is evidence to show that subsp. *commersonii* forms a natural, though sparse, component of the 'matorrales (thickets) de curro' (*Colletia paradoxa*) described by Cabrera as being found in the Sierras of Balcarce and Mar del Plata. In the Balcarce region Millán collected it several times amongst *Colletia* bushes, whilst Grondona made frequent collections in the hills around Balcarce amongst various grasses, herbs, and small shrubs.

Records for *S. chacoense* are known from the distribution area of subsp. *commersonii*, but there is only slight evidence that the two species hybridize under natural conditions.

2. Subsp. *malmeanum* is also a ruderal plant or a weed of cultivation, found chiefly in damp places, from sea level to about 300 m. It differs from the typical subspecies, however, in exhibiting an apparent preference for greater shade, and is only occasionally seen in full sun. Nevertheless, the range of habitats in which it has been found is quite wide, since we have records of it for shady forests, grassland, damp pastures, stream sides, and cultivations of cotton, manioc, citrus fruits, etc. No soil preference is exhibited, since it grows apparently equally well in sandy river banks, clay soils and soils rich in humus. It frequently grows in disturbed soil, however, and occasionally on dry stony hills. As with other weed species it spreads easily where the soil has been disturbed or brought under cultivation.

The flowering season is from April to June, and possibly at other times as well. This is, of course, the winter season and most observers state that it grows and flowers during this period, rather than in the hotter summer season (December, January). It has even been observed in flower after a night frost of $-3°$ C, apparently quite unharmed.

In Argentina it is to be found in the following phyto-geographical regions:[1]

EASTERN SUBTROPICAL PROVINCE: Field or savanna and the Mixed Forest districts.

CHACO PROVINCE: Eastern Chaco district (more humid eastern parts).

ESPINAL PROVINCE: Ñandubay district.

In south Brazil (Río Grande) it is found in the Mixed Forest and possibly also in the Araucaria Forest districts.

In Paraguay it probably spreads through the Mixed Forest district into the wetter parts of the Chaco province, though as yet we have insufficient records to be certain of its exact distribution.

In Uruguay it seems to be found chiefly in the so-called 'Galería Uruguayense', the gallery forest along the Uruguay river, belonging to the Mixed Forest district of the Eastern Subtropical province.

As with the typical subspecies, this subspecies is chiefly a plant of open ground, or places where the natural vegetation no longer exists in an undisturbed condition.

Records for both subspecies of *S. chacoense* have been obtained from the distribution area of subsp. *malmeanum* and some collectors have taken mixtures of the two species in one gathering. Nevertheless, there is no certain evidence for the occurrence of natural hybrids between *S. chacoense* and *S. commersonii* subsp. *malmeanum*.

TAXONOMIC NOTES

The nomenclature and taxonomy of *S. commersonii* has been in some confusion up to now, due to the large range of variability in this species and to the fact that different collectors sent single isolated gatherings to Europe at various times, which were then described as distinct species or varieties. On the basis of an examination of nearly

[1] Scheme proposed by Cabrera (1953).

300 collections, belonging to both subspecies, we are convinced that *S. commersonii* cannot be divided into numerous 'microspecies', varieties, or forms in this way. The infraspecific variability is as great or perhaps greater than that of *S. chacoense* and certain other wild tuber-bearing species, but apart from the two subspecies into which we have divided it, *S. commersonii* presents a continuous variation of characters which makes impossible any attempt at further subdivision.

1. Subspecies *commersonii*

S. commersonii was described first by Dunal in 1813, from specimens collected by Commerson (No. 47 'Beach at the foot of the hill') in May 1767 at Montevideo (see Figs. 8–10). It was thus probably the first wild potato to be collected on a scientific expedition, and was also (with the exception of the little-known Colombian *S. papa* Val.) the first wild potato species to be named and described.

Dunal's description was made from the Paris sheet(s) of *Commerson No. 47*, but is inadequate on present-day standards. The existing Paris specimen, which we have designated as LECTOTYPE (Fig. 8), is a small plant (21 cm), unbranched, and with moderately frequent long white spreading hairs, (2–)3-jugate leaves to 6·5 cm long and no interjected leaflets; the laterals are sessile and decurrent, the terminal orbicular to ovato-rhomboid; inflorescence 5-flowered; pedicels 1·5 cm long, articulated near the base; calyx lobes short; style 7–8 mm long; stigma globose and notched.

Sabine (1824) reproduced a drawing made under his direction from one of Commerson's specimens in Herb. Paris which does not differ essentially from the lectotype specimen already discussed; this sheet was missing from the Paris collections when we examined them in 1947. We reproduce the drawing here (Fig. 9) as a good example of the species, though pointing out that the pedicel articulation has been omitted from the drawing. This plant is taller than the lectotype, at least 30 cm high.

Another sheet of *Commerson 47* in Herb. Montpellier (of which a drawing made by Bitter is in Herb. Goettingen; see Fig. 10) is rather larger in leaf (17 cm) with 3–4 pairs of lateral leaflets and one solitary interjected leaflet. The calyx lobes are longer also than in the Paris sheet. There is a photograph in Chicago of a further Paris sheet (also not seen by us) with 3–4-jugate leaves and one pair of interjecteds. Both this and the Montpellier sheets possess very shortly petiolulate leaflets in all lateral pairs but the uppermost ones. A collection from Copenhagen, also said to be of Commerson, is distinctly 3-jugate with petiolulate laterals and one pair of interjecteds. Thus the Montpellier sheet, the Chicago photograph, and the Copenhagen sheet differ from the Paris sheet seen by us (designated as lectotype) and from Sabine's drawing. They stand closer to the anomalous triploid forms from plant-breeding stations and may perhaps be autotriploid biotypes.

In the year 1846 J. D. Hooker (*Flora Antarctica*)[1] described four varieties of *S. commersonii*, two of which (var. 2 *pilosiusculum* and var. 3 *glanduloso-pubescens*) from Mendoza, cannot be retained as part of *S. commersonii* under modern concepts of this species (cf. *S. kurtzianum*). Hooker's first variety, *glabriusculum*, is a mixture of *S. maglia* from Valparaíso, Chile (*Bridges 401*), and a rather glabrous form of *S. commersonii* collected by Tweedie at Buenos Aires.[2] This latter comes well within the range of variability for pubescence seen by us in numerous collections and does not, in our opinion, merit distinct varietal rank. The fourth variety, named by Hooker var. *glabratum*, based on two collections of Dr. Gillies from Buenos Aires, is a glabrescent lower-growing biotype and again does not merit varietal rank when the total range of variability in *S. commersonii* is considered.

[1] Part 2, pp. 273–380, published 1846. See Wiltshire (1913).

[2] Tweedie also collected a form of *S. chacoense* in the same year and it is possible that Hooker's name may be based partly on that specimen also (see p. 203).

Up to this point only material of *S. commersonii* from Uruguay (Montevideo) and Argentina (Buenos Aires) had been examined by European botanists. However, Otto Sendtner's treatment of Solanaceae in Martius's *Flora Brasiliensis* (Solanaceae published 1 July 1846) showed that *S. commersonii* was also to be found in Brazil, though

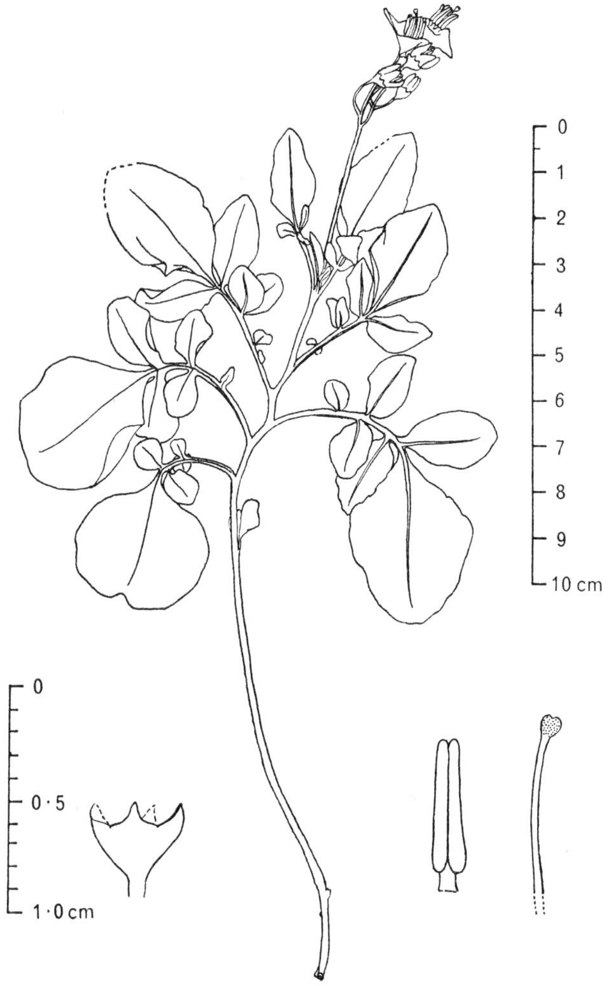

FIG. 8. *S. commersonii* Dun. Lectotype from the Paris herbarium of *Commerson 47*, collected at Montevideo, Uruguay. This was probably a diploid clone, judging from the completely sessile leaflets.

the exact place of collection is not stated. Besides *S. commersonii* itself, Sendtner describes a variety *pubescens* and a delicate tall-growing form which he designates as *S. tenue*. Although the collection numbers of the material on which these taxa were based were not published by Sendtner (whose type specimens were in Herb. Berlin and are presumably now destroyed), named material, labelled presumably by Sendtner himself, was examined from Brussels, Kew, and Leningrad. The specimens labelled *S. tenue* (*Sellow 1971*) are obviously rather etiolated specimens of subsp. *commersonii*, probably growing amongst other vegetation, the stem ±30 cm long and leaves (0–)1–2-jugate with larger terminal. *Sellow 276* (see Fig. 11), labelled as *S. commersonii*, also agrees fairly well with *Sellow 1971*, whilst *3499* is also quite close. These are 3–4-jugate with 0–2 interjected pairs. Finally, an unnumbered collection by Sellow,

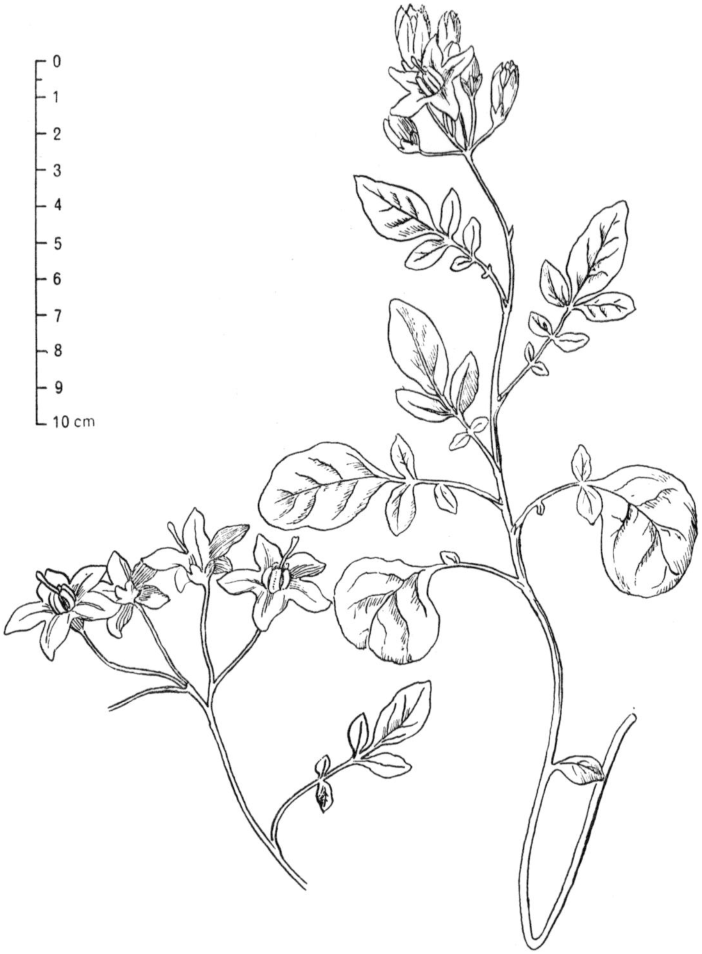

FIG. 9. *S. commersonii* Dun. Copy of Sabine's figure (*Trans. hort. Soc.* **5**, Plate 10, 1824) of one of Commerson's specimens from Herb. Paris (now lost). Note close similarity to Fig. 8.

labelled '*S. commersonii* var. *pubescens*' with rather small 3–4-jugate pubescent leaves is probably part of Sendtner's var. *pubescens* and again comes well within the range of subsp. *commersonii*. In our opinion, therefore, the taxa *S. commersonii* var. *pubescens* and *S. tenue* Sendtn. should be discontinued.

We next must consider the species *S. ohrondii*, described by Carrière (1883) from material collected by Ohrond in Uruguay. From the description and figure given by Carrière there is no doubt that this species should also be included within subsp. *commersonii*. The whitish interior and bluish-violet exterior colouration of the stellate corolla is particularly typical, though the linear-lanceolate sepals induced many people to think that *S. ohrondii* should be considered as a separate species. One should note that the calyx lobes in the type specimen of *S. commersonii* are rather short, but this character evidently varies, since in Sabine's plate and in Bitter's drawing of the Montpellier sheet (both part of Commerson's original collection) the sepals are ovate to ovate-lanceolate. This ovate or ovate-lanceolate sepal shape is in fact just as frequent

FIG. 10. *S. commersonii* Dun. Based on a drawing made by G. Bitter (Herb. Goettingen) of the Montpellier sheet of *Commerson 47* (cf. Figs. 8 and 9 and Plate 22). This was probably a triploid plant, since the petiolules are well marked and the inflorescence branches are short.

as the shortened form, as we have mentioned in the description (see also p. 143). Unfortunately no original specimen of *S. ohrondii* seems to be available in the herbaria we have examined, though a sheet from the type locality in the herbarium of the Institute of Plant Industry, authenticated by Bukasov, fits the description well. From an examination of this sheet and a careful study of Carrière's original description we are convinced that *S. ohrondii* Carr. should also be included within the typical subspecies of *S. commersonii*.

Although the German Solanologue G. Bitter stated (1912) that he had made an extensive study of *S. commersonii* he did not in fact publish any revision of the species itself, though he remarked, 'I have made elaborate investigations on the polymorphic group of *S. commersonii* but I should like to wait before I publish them until I have more material for a still more thorough elucidation of the collective species *S. commersonii*'. This work was never published, though one collection in Herb. Paris and Herb. Geneva (*Courbon* (s.n.)) was labelled by him *S. commersonii* var. *rosulans*;[1]

[1] Additional *nomina nuda* proposed by Bitter have recently come to light:

var. *depauperatum* (*Osten 2628*) (MVM)
var. *raphanistrum* (*Osten 3670*) (G, MVM)
var. *indigoticascens* (*Osten 5777*) (MVM)
subsp. *ellipticans* (*Osten 2915*) (BREM, MVM)

Another *nomen nudum* was var. *violaceum* Herter, based on Herter's collections 1337, 1337*a*, *b*, *c*. None of these names was published with descriptions, so far as we are aware, and they are certainly best left in obscurity.

Bitter continues, 'In the following I only give the definition of two East Brazilian species which have a somewhat greater difference from the southern chaos of forms' namely *S. acroleucum* from Sta. Catarina, and *S. calvescens* from Minas Gerais.

FIG. 11. *S. commersonii* Dun. Collected by Sellow in Brazil (*No. 276*—herb. Kew). This might be part of the type collection of '*S. tenue*' Sendtn., but since the collection number was not mentioned by Sendtner we cannot be certain. The material undoubtedly belongs to *S. commersonii* and was possibly a diploid.

Whilst there is no doubt that both these latter species bear some relationship to *S. commersonii* it is very doubtful if *S. calvescens* is conspecific with *S. commersonii*. *S. acroleucum* is closer, both morphologically and in its geographical distribution, to *S. commersonii*, and has been included with the typical subspecies in the present treatment. In the south Brazilian material seen by us the following specimens come closest to *S. acroleucum*: *B. Rambo 29699*, *Regnell A 1439*, *Saint-Hilaire 1835*. The

S. acroleucum type was cited for Herb. Breslau, but, although we have not been able to see this, an isotype specimen from Herb Paris (see Fig. 12) is similar in most respects to the type description, though a little larger in its leaf dimensions. It agrees with subsp. *commersonii* in the 3-jugate leaf, sessile rather decurrent leaflets, larger terminal leaflet, semi-rosette habit, contracted inflorescence branches, mauve stellate corolla, the lobes paling distally, and notched stigma. Until we have further material from south Brazil and north Uruguay for comparison we cannot yet say definitely whether *S. acroleucum* should be placed as a regional subspecies of *S. commersonii*. It seems fairly certain that it is not specifically distinct and for the time being we have classified it within subsp. *commersonii*, though reserving the right to give it separate subspecific rank if necessary, when more material becomes available for study.

Bitter's var. *rosulans* of *S. commersonii* is certainly, in our opinion, not worthy of varietal rank, since it is based on a short-growing, rosette-forming phenotype, due no doubt to the poor open sandy or stony environment in which the plants were growing. His other proposed varieties and subspecies are, in a like manner, unworthy of further consideration.

The distinguished potato-breeder and taxonomist Professor S. M. Bukasov described a number of new diploid *Solanum* species related to *S. commersonii* in the years 1935 to 1940. These were all grown by him near Leningrad under comparable conditions, and by their differences demonstrated extremely well that the variability within the species is not due entirely to environmental differences. Professor Bukasov's concept of *Solanum* species differs from ours in that he thinks of them as a number of small units (microspecies) grouped within a 'collective species'. We ourselves are treating *S. commersonii* in the wide sense, approximately equivalent to Bukasov's *S. commersonii* (*sensu latiore*). In a discussion of the taxonomy of this group Bukasov (1940*a*) points out that within the collective species are to be found both diploid and triploid forms, the latter to be obtained only within the immediate vicinity of Montevideo. Triploid *S. commersonii* has been known to potato-breeders for many years and was worked on by Heckel (1905), Labergerie (1907), Sutton (1909), and others. It differs from the diploid forms described by Bukasov and from others seen by us in its sturdier growth, shortly petiolulate leaflets, and the presence of interjected leaflets, but possesses the very characteristic reduced peduncle forks and mauve flowers of the diploid biotypes (though some strains with white flowers are also known). Bukasov claims that the original collections of Commerson were triploid and that *S. commersonii* (*sensu stricto*) should only include such biotypes (see Plate 24). With this we disagree, since although we cannot know the chromosome number of Commerson's original collections the Paris material is more similar to known diploids. The Montpellier, Chicago, and Copenhagen sheets are closer to known triploids, on the other hand, and it is quite possible that both diploid and triploid biotypes may have been included in the original Commerson collections. Although we have no actual living collections of diploids and triploids from a single population in Montevideo to substantiate this statement we can draw a parallel with *S. cardiophyllum* which in the region of Mexico City is to be found growing in populations of mixed diploid and triploid biotypes. We have also grown diploid and triploid strains of *S. commersonii* collected as weeds in and around Montevideo itself, so that the triploid form cannot, it seems, be the only one found in the vicinity of that city. We can sum up our views by saying that it is quite impossible for the chromosome number of Commerson's original collections to be known, but that very possibly these collections consisted of a mixture of diploid and triploid biotypes. Since the general view held about triploid *S. commersonii* is that it is an auto-triploid formed from the diploid biotypes by the union of reduced and non-reduced gametes (see p. 176) one would naturally expect the diploid and triploid forms to be found in juxtaposition. This point could be cleared up quite simply by small random tuber collections made at Montevideo and elsewhere

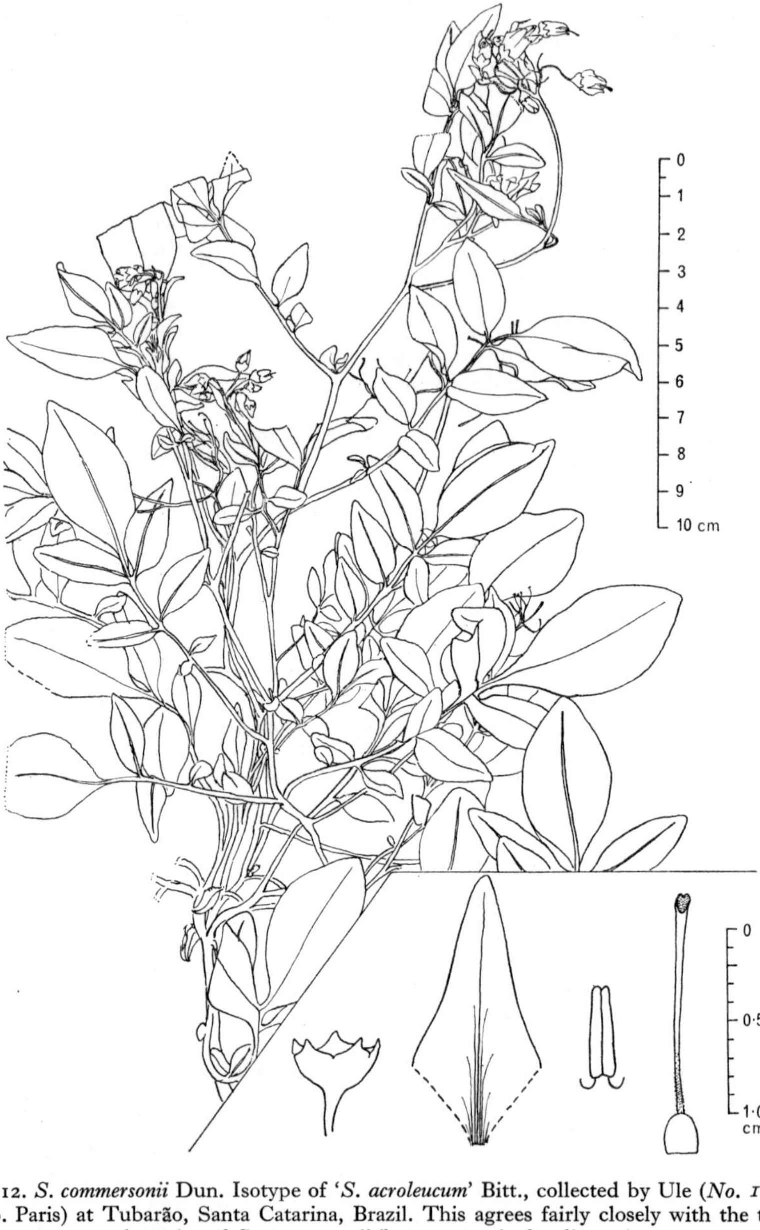

FIG. 12. *S. commersonii* Dun. Isotype of '*S. acroleucum*' Bitt., collected by Ule (*No. 1201*—Herb. Paris) at Tubarão, Santa Catarina, Brazil. This agrees fairly closely with the typical subspecies of *S. commersonii* (but see p. 165 for discussion).

on which chromosome counts and morphological analysis could then be made. Meanwhile, therefore, we shall include both diploid and triploid biotypes in our concept of *S. commersonii* and shall hope that observations on their exact distribution will one day be forthcoming.

We return now to a discussion of the diploid microspecies described by Bukasov and placed by him in *S. commersonii* as a collective species. Five such species were described, namely:

S. henryi from Uruguay, Colonia (co-author: V. Lechnovicz).
S. mechonguense from Argentina, B. Aires, Mechongué.
S. mercedense from Uruguay, Soriano, Mercedes.
S. sorianum from Uruguay, Soriano, Bequeló.
(*S. laplaticum* from Argentina near B. Aires.)

The last species in the above list, *S. laplaticum*, was placed by Hawkes (1956*b*) under *S. commersonii* on the basis of the description given by Bukasov and also because of Bukasov's own statement that it should be included in the collective species, *S. commersonii*. When, however, we were able to see the type specimen of *S. laplaticum*, kindly made available by Professor Bukasov, it became clear at once that this microspecies should be included under *S. chacoense* and not under *S. commersonii*. (See Plate 39.)

Another species, *S. rionegrinum*,[1] mentioned by Lechnovicz (1940), and which at present stands as a *nomen nudum*, should also be included within *S. commersonii*. The specimen seen by us, which Bukasov designated as the type, is of very delicate growth with narrow, shortly petiolulate lateral leaflets, typically ovate calyx lobes and possibly white corolla. The inflorescence forks are only moderately reduced, but we must remember that this specimen was in fact grown at Pushkin, near Leningrad, U.S.S.R., and cannot be compared in every detail with material gathered in Uruguay. Furthermore, it agrees with *S. commersonii* in its resistance to frost. Thus, it seems on the whole reasonable to include *S. rionegrinum* in *S. commersonii*, though it might be necessary to include it in subsp. *malmeanum* rather than in subsp. *commersonii* after further study.

We have already mentioned that the other diploid 'microspecies' of *S. commersonii* differ from each other genetically, since all the descriptions were made from material cultivated at Leningrad under roughly comparable conditions. We give below a short synopsis of the main characters in which they differ from the type specimen of *S. commersonii*, though we maintain strongly that they should be included in this species.

S. henryi differs from the type in the shorter rosette growth and the 4-jugate lateral leaflets, rather rounded at the base. It seems to agree in other features with the type though the differences listed by Bukasov presumably refer to a comparison of *S. henryi* with the triploid biotype of *S. commersonii* known to him. It is interesting to note that the fruits of this 'microspecies' are long-cordate. (See Plate 25.) *S. henryi* forma *pubescens* was described by Lechnovicz (in Bukasov and Lechnovicz 1935), and differed from the type in its greater pubescence and other characters; forma *laticalyx*, described in the same publication, differed chiefly in the larger, bilabiate calyx, as well as other features. In our opinion both these forms should be included in subsp. *commersonii*.

S. mechonguense is of less contracted habit, with 3–6-jugate laterals and 0–2 pairs of interjected leaflets. The calyx lobes are rather short, peduncle branches sometimes but not always shortened, whilst the berry is said to be rotund. (See Plate 26.)

In *S. mercedense* (also mentioned once or twice in the literature under the *nomen nudum*, *S. compactum*) the stem is said to be short, the lateral leaflets 2–3(–4)-jugate, rather short and compact and very obtuse. There are no interjected leaflets, the calyx acumens are hardly detectable and the anthers are very short, only 4·5 mm. The pubescence is, in general, very short, though some longer hairs are also to be found. (See Plate 27.)

[1] Collected in Uruguay, dept. Río Negro or Soriano, Río Negro (see Plate 29).

S. sorianum is said to be close to *S. henryi* in its rosette-like growth and 4–5-jugate lateral leaflets. The inflorescence is intermediate between the normal and the reduced type, the calyx lobes characteristically ovate-lanceolate, style notched, and berry oblong. (See Plate 28.)

In all these microspecies described by Bukasov the flowers are typically violet-purple outside and slightly paler or whitish inside.

It therefore seems quite clear to us, from a careful examination of the descriptions and type material of Bukasov's species, that they should all be included in the typical subspecies of *S. commersonii* as defined by us in the present work.

In a study of the Argentine species of series Commersoniana Brücher (1956*a*) rightly considers that many of the previously published names should be treated as synonyms. We cannot completely agree with his statement, however, that '*S. boergeri* and *S. sorianum*, together with *S. pampasense*, should be placed in this group' (i.e. Commersoniana). *S. boergeri* is in reality a part of the *S. chacoense* complex (see p. 206 and Plate 38) whilst the last mentioned, *S. pampasense*, belongs to series Tuberosa, not to Commersoniana. Evidently Brücher was confused by the name 'pampasense', which was derived (see Hawkes 1944) not from the Argentine pampas, but from the Río Pampas in Central Peru, some 3000 km distant.

Correll (1962) retained Hooker's var. *glabratum* (wrongly assigned to Walpers, however) and placed *S. acroleucum* under it as a synonym, stating that 'this rather artificial variety is maintained solely to accommodate the glabrous or glabrescent plants that are sometimes found in this species'. We are convinced (see below) that there is a complete range of variation in pubescence within *S. commersonii*, and it seems unwise, therefore, to maintain a separate taxon for part of this range.

The forms and varieties of *S. commersonii* distinguished by Correll have been thought by him to represent possible hybrid phases between *S. commersonii* and *S. chacoense*. We doubt very much whether this is so, since there seems to be little evidence that such hybridization is taking place in nature. Furthermore, both var. *glabratum* and f. *mechonguense* (Correll 1961) easily come within the continuous range of variation of *S. commersonii*. The latter form, *mechonguense*, was erected by Correll on the basis of Bukasov's *S. mechonguense*, and merely encompasses variants with more narrowly and deeply lobed corolla, of a deeper purple colour than usual. Again, we consider that it has no validity. Another form, *malmeanum*, will be dealt with when we discuss subsp. *malmeanum* on p. 172.

It has already become abundantly clear that subsp. *commersonii* is a highly variable taxon and for this reason we have tried to analyse in a little more detail the type of variation found, as seen by us in certain specimens taken in South America:

Habit. It is possible to see a complete gradation from rosette to slender trailing plants, due almost certainly to differences of habitat. Probably the rosette form is the more common.

Pubescence. Continuous variation is seen both in respect of length and density of pubescence. Six plants were quite glabrous, 28 possessed sparse to moderate pubescence whilst in 15 it was dense, at any rate on the lower surfaces of the leaves.

Leaf dissection. This varied from 0–2-jugate laterals (6 plants), 2–3-jugate (25 plants), 3–4-jugate (18 plants), 4–5-jugate (1 plant). Thus by far the largest number came within the 2–4-paired range. Of 49 collections analysed for petiolule length, 27 plants showed sessile, generally decurrent lateral leaflets, whilst in 22 plants very short petiolules could be distinguished. Interjected leaflets are often absent (37 plants) up to 1-paired (12 plants) or 1–2-paired (2 plants).

Peduncle. The usual case is for the peduncle forks to be much reduced (25 plants) but often a lengthening may take place (13 plants).

Pedicel. The articulation varies from median (12 plants) to slightly above (19 plants) or below (4 plants) the centre.

Flower. The calyx lobes are ovate to ovate-lanceolate (19 plants) or quite short (17 plants). Corolla colour varies from white (7 plants) to varying shades of mauve or violet (24 plants). In some of the south Brazilian collections (possibly to be equated with *S. acroleucum* Bitt.) the colour pales towards the tip.

Berries. In only two cases have globular berries been recorded; they are generally ovoid to conical (eight cases), or cordate in the triploid forms known to plant-breeders (see Sutton 1909, Plate 39).

The foregoing analysis is of course based on a not very large number of specimens. It serves to indicate, nevertheless, that the typical subspecies is highly variable, within certain limits, and we hope it may stimulate work on living material of this species in the region of the La Plata estuary.

2. Subspecies *malmeanum*

In the present work we have grouped a number of biotypes under subsp. *malmeanum* of *S. commersonii*, some of which have been given separate specific names. They vary amongst themselves probably as greatly as the biotypes of the typical subspecies, but seem distinct from it, differing in the generally longer leaves with 4–6 pairs of shortly petiolulate lateral leaflets, larger number of interjected leaflets, and the invariably white corolla. Some inter-grading and difficult collections from prov. Entre Ríos are hard to place and may well represent forms intermediate between this and subsp. *commersonii*.

We have used as lectotype for subsp. *malmeanum* the lower plant on the Stockholm sheet collected by Malme (*Regnell 756*) in the Brazilian state of Rio Grande do Sul (Ijuí) and named and described by Bitter as a distinct species (*S. malmeanum* Bitt.) (See Plate 31.) Recent collections by Petersen and Hjerting and others in the Argentine provinces of Misiones, Corrientes, and Chaco have shown that subsp. *malmeanum* spreads considerably westwards from the place where it was originally found. (See also Plates 32–34.)

Records from Paraguay are few but we were astonished to find on examining the type material of Hassler's varieties and forms of *S. guaraniticum* (= *S. chacoense*, see p. 201) that the type specimen of var. *latisectum* f. *pilosulum* from near Asunción (*Hassler 3674*)[1] was identical with our subsp. *malmeanum* of *S. commersonii*. (See Plate 34.) Luckily, although Hassler does not designate a type specimen for his species *S. guaraniticum* he describes under it two varieties, the first of which, var. *angustisectum*, is included in the modern concept of *S. chacoense*.[2] The second, var. *latisectum*, is divided again into two forms, namely, *pilosulum* and *glabrescens*. It is only this forma *pilosulum* of var. *latisectum* which should be removed from *S. guaraniticum* (=*S. chacoense*) and placed with subsp. *malmeanum* of *S. commersonii*. Thus, due to the fortunate circumstance of v. *angustisectum* having been placed first in Hassler's descriptions, we can remove var. *latisectum* forma *pilosulum* from *S. guaraniticum* without having to alter the name *S. guaraniticum* or *S. chacoense* for the species typified by var. *angustisectum*. Nomenclatural stability is thus ensured and the International Rules are not contravened.

Another Hassler collection from Paraguay, No. 140*a*, was separated by Hassler and Rojas from the rest of No. 140 (which was typical *S. chacoense*) and was also placed by Hassler as var. *latisectum* forma *pilosulum* (= subsp. *malmeanum*). Rojas expresses his doubts on a label: 'La presente especie es un poco diferente en las hojas un poco [mas] peludosa que las del número 140 . . .', showing that both these authors realized that this material belonged to a distinct taxon. Hassler's No. 2515, which may have

[1] This collection is also the type for Chodat's *S. commersonii* β *pubescens*.

[2] Lectotype designated from this variety by Correll. See p. 201.

been collected from the same place as his No. 140*a*, should also be placed under subsp. *malmeanum*.

A further collection of subsp. *malmeanum* from Paraguay was made by Rojas (No. 136) in the lower Pilcomayo region[1] and was named by Hassler (1911) as *S. commersonii* var. *pseudostipulatum* (see Fig. 13). In fact, the pseudostipules are rather poorly developed in this collection compared with many other potato species. *Rojas 136* differs from the type of subsp. *malmeanum* in the broader leaflets which are sessile to sub-sessile, the 3-jugate leaves without interjected leaflets, and the taller habit. If we knew only Malme's and Rojas's specimens we should be certainly justified (as were Bitter and Hassler) in assuming that the two taxa were distinct species. Since we are now in the more fortunate position of having a wide range of collections at our disposal we can see that they only represent two rather extreme types in a series of variable populations spreading over the region of southern Paraguay and adjacent countries. There is, therefore, no justification, in our view, for keeping them apart, or for linking var. *pseudostipulatum* with the typical subspecies, or for naming the Rojas collection *S. pseudostipulatum*, as was done by Bukasov (in Bukasov & Lechnovicz 1935).

The *nomen nudum*, *S. hassleri*, was a nomenclatural synonym placed by Wittmack on the Berlin sheet of *Rojas 136*. Since it was never published and the specimen itself is now destroyed this name need trouble us no further.

Bukasov and Lechnovicz (1935) and Bukasov (1940*a*) describe under the name of *S. millanii* a species from prov. Misiones, Loreto. This is of some interest as the authors state that it occurs both in diploid and triploid biotypes. The type specimen of *S. millanii* is probably triploid, judged from the greater width of the leaflets and corolla lobes, and in these features it differs a little from the type of subsp. *malmeanum*.

In order to see whether the type specimen of '*S. millanii*' (*Millán 973*, see Plate 33) was connected, in respect of leaflet width, to that of the type for subsp. *malmeanum* by a series of intergrading biotypes we obtained a 'leaflet index' (breadth/length × 100) for all specimens available. The distal leaflet on the best-developed leaf of each collection was measured and an index value obtained. The frequency histogram (Fig. 14) in which the indices are grouped into blocks indicates, to some extent, a normal curve, with the majority of plants giving an index value of 50 (i.e. the leaflet twice as long as broad). In this group are to be found the type for subsp. *malmeanum* (*Regnell 756*) and the former *S. chacoense* var. *latisectum* f. *pilosulum* (*Hassler 3674*).

The type specimen for '*S. millanii*', however, has a broader leaflet (index value of 68·9) whilst that for '*S. pseudostipulatum*' (*Rojas 136*) gives a value of 62·5. Both these specimens are not alone in the groups in which they fall, nor are they disconnected by any sudden break from the biotypes with narrower leaflets. It is of interest to note, nevertheless, that the curve represented by the frequency blocks is interrupted or 'kinked' on the right-hand side by an over-predominance of records for the 65–69·9 block, and this may be due to the influence of triploid plants with wide leaflets, to which group the triploid type specimen of '*S. millanii*' belongs.

It therefore seems fairly obvious that in respect of leaflet width '*S. millanii*' should be brought within subsp. *malmeanum*. Furthermore the two taxa agree in the leaf size, shortly petiolulate leaflet, white corolla, and other characters.

When allowance is made for the fact that the type material of *S. millanii* was grown near Leningrad and that the collections made by Petersen & Hjerting (No. P–71) from the type locality of this species, are very close morphologically to the type of subsp. *malmeanum*, then one is justified, we consider, in stating that the two taxa are synonymous. From a general knowledge of the variability of subsp. *malmeanum* there seems

[1] There is some doubt as to whether this collection came from Paraguay or Argentina; the former is more probable since most of Rojas's collections were from Paraguay.

FIG. 13. *S. commersonii* Dun. subsp. *malmeanum* (Bitt.) Hawkes & Hjerting. Type for '*S. commersonii* var. *pseudostipulatum*' Hassl., collected by T. Rojas (*No. 136*—Herb. Geneva) by the river Pilcomayo in Paraguay or Argentina. Differs from the rest of subsp. *malmeanum* in the fewer subsessile rather broader leaflets and stout erect stem.

little doubt but that *S. millanii* is only a slightly wider-leaved biotype of it, and we consider that it should not in future be retained as a good or distinct species.

Correll (1961, 1962) downgrades *S. malmeanum* to a form of *S. commersonii*, stating also that it might represent a hybridization of *S. commersonii* and *S. chacoense*. He concedes that it closely resembles his *S. chacoense* f. *pilosulum* but does not believe that they are one and the same thing. He also cites specimens under *S. chacoense* f. *pilosulum* that should certainly be classed as subsp. *malmeanum* in our view. One can see from Correll's text (1962, p. 296) that he is dealing with forms of both *S. chacoense*

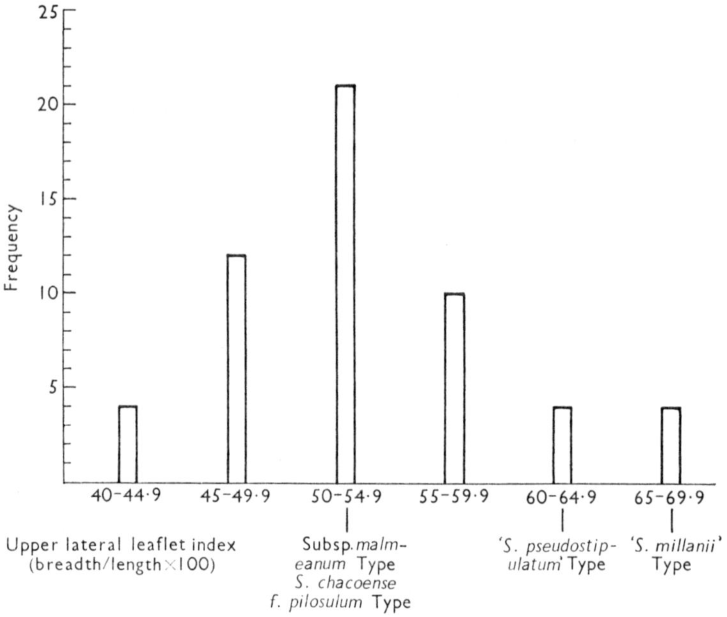

FIG. 14. Frequency histogram of leaf indices for *S. commersonii* Dun. subsp. *malmeanum* (Bitt.) Hawkes & Hjerting. For explanation see p. 170.

and *S. commersonii* subsp. *malmeanum* under his concept of *S. chacoense* f. *pilosulum*, since he says: 'Some of the plants possess the laxly branched inflorescence of typical *S. chacoense* while others, as represented by Hassler's collection *No. 11849* have the characteristically compact inflorescence of *S. commersonii*.' We ourselves have classified *Hassler 11849* under *S. chacoense*, since some sheets of it can definitely be placed with that species; other sheets seem closer to subsp. *malmeanum*, and may represent the products of natural hybridization between the two taxa (see p. 218).

We consider, therefore, that Correll is mistaken in classifying subsp. *malmeanum* as a form of *S. commersonii*, since in our view it consists of an assemblage of biotypes with a distinct geographical distribution which are adapted to a habitat range quite different from that of the typical subspecies; subsp. *malmeanum* differs from subsp. *commersonii* also in many leaf characters and in some inflorescence and corolla features. In fact, the two taxa could almost be considered to differ from each other at the species level, were it not for the presence of inter-grading forms in Corrientes and Entre Ríos provinces.

The range of variability within subsp. *malmeanum* is as follows:

Although under open conditions the plant is rosette-forming, when growing in shade or amongst other herbage it may develop a stem up to 30–60 cm tall. The lateral leaflets are mostly 3–4-paired but may be as many as 6-paired or as few as 2-paired. In some collections the interjected leaflets are absent, in others up to 4-paired, whilst in others they vary from none to several pairs in different plants of a population or even on different leaves of the same plant. Leaf and stem pubescence varies from sparse to quite dense, and in a number of collections the hairs are quite long and spreading.

The constricted or shortened type of *S. commersonii* peduncle fork is most usual in this subspecies, though a certain amount of elongation can occur. There are hardly ever more than two branches (once-forked therefore) on the peduncle of all forms of *S. commersonii*, and this character helps to distinguish it from *S. chacoense*, where the peduncles are generally forked at least twice.

NOTES ON LIVING PLANTS

S. commersonii has been known for a very long time in European potato-breeding stations but apparently only as triploid clonal material. It is thus highly sterile. Sutton (1909) describes and figures both purple and white-flowered clones which he states probably came from Uruguay. He refutes the claims made earlier by Labergerie (1907), Planchon (1909*a*, *b*), and Heckel (1905) that it was possible to cause *S. commersonii* to mutate in the direction of a cultivated potato by growing it under appropriate cultural conditions, since these so-called 'mutants' were shown by Sutton to be identical with known cultivated varieties of *S. tuberosum*.

1. Subspecies ***commersonii***

Heckel (1904, quoted by Schroeder 1928) gives detailed notes made from living plants of *S. commersonii* in Uruguay and in France. His description differs very little from the one we have given above (p. 142). In addition, he mentions that the flowers smell of jasmine and that the berries are cordate. The tubers are said to be small in the wild plant but up to 20–50 grammes in plants cultivated in rich soil; the skin is a rosy-yellowish colour and is covered with abundant lenticels. Even after boiling for a long time the flavour of the tuber is said to be still bitter and disagreeable.

Schroeder (1928) studied living plants in various parts of Uruguay and was able to furnish a number of observations on the variability within *S. commersonii*. He was particularly impressed with the range of hair density, from glabrous to pubescent with all intermediates, but considered this to be largely due to habitat, the more glabrous plants growing in damper places. He was not able, however, to link hairiness with any special soil type. All grades of flower colour from dark purple to white were found in samples growing under identical soil conditions in the same locality, though he considers that season, temperature, and sun also influence the flower colour. Although the 5-lobed corolla is usual, he speaks also of flowers with from 4 to 7 lobes, even on the same plant. The berry is said to be cordate with a groove on each side, first green and then turning whitish-yellow when mature, with a marked smell of strawberries. This strawberry smell, however, is not characteristic only of *S. commersonii*, since it is to be noted in ripe berries of nearly all species of tuber-bearing *Solanum*. Schroeder notes also that the stolons frequently grow up to the soil surface to form daughter plants up to 50 cm away from the parent. Although this species has received the name of 'papa purgante' (purging potato) Schroeder doubts whether it is used medicinally, since he has found no information from the Uruguayan peasants in support of that claim. The flowering times are said to be from October to the beginning of December and again in March to May.

However, in warm rainy years examples were found flowering in all months. The tubers, also, were found in all seasons apparently. They were yellowish, round, oblong or pyriform, with a smooth skin when young and very warty (due to lenticels) later. The flavour of some is said to be bitter but of others not bitter at all and hardly distinguishable from that of *S. tuberosum*. The flavour when cooked was more or less similar to that of the raw tuber. A chemical analysis of berries, haulms, and tubers was given by Schroeder and compared with that of *S. tuberosum*.

Observations were made on living plants in Uruguay, Buenos Aires province of Argentina, and south Brazil by Hawkes in March and April 1958. All material observed was of subsp. *commersonii* with the exception of one population from El Prado, Montevideo (see below).

The plants were seen very frequently in grassy places at the edges of marshes and streams in humus-rich soil. Other habitats were thin, humus-rich soil on damp granitic rocks, meadows, sand dune 'slacks', and sea shores just above high tide level. In one meadow habitat in south Brazil, near a grove of *Araucaria*, the plants grew especially abundantly where cattle dung had enriched the soil. Subsp. *commersonii* was also noted for arable fields in south Brazil.

No doubt the species was at one time very common all along the coast of Uruguay, but it has now been exterminated for a long region east of Montevideo due to the filling in of coastal marshes and the building of sea-front esplanades, houses, and hotels. Several searches were made on the famous hill ('Cerro' or 'Monte') from which Montevideo takes its name, to look for topotype material from Commerson's original locality. Commerson found this species on 'la plage du pied du morne de Montevideo' (the beach at the foot of the hill of Montevideo), but although this area has not been too heavily built over, the plant could not be found. Nevertheless, it has been collected by Chebataroff (No. 5229), probably not so very long ago, according to a labelled specimen in the Montevideo Natural History Museum. Possibly the time of year was not right, or *S. commersonii* may now have disappeared from the classical locality.

The different populations of subsp. *commersonii* observed and collected by Hawkes in 1958 showed some differences in growth and habit, pubescence, and intensity of corolla pigmentation, though no white-flowered clones were observed. Although variation in stem length in the naturally growing material must have been largely conditioned by habitat differences, such as the height of plants amongst which it was growing, there were also real genetical differences controlling stem length. Cultivation under comparable conditions in Birmingham showed definite height variation; differences of leaf dissection and pubescence were also seen to be due to genetical factors, though further modification in nature by environmental changes would take place.

In spite of these differences there was no doubt that all forms belonged to *S. commersonii*, and most to the typical subspecies.

What appeared to be subsp. *malmeanum* was seen growing amongst the short grass and under trees in the botanic-garden section of El Prado Park in Montevideo. Since subsp. *malmeanum* has not otherwise been recorded so far south or so close to the coast (see Map 3) it would seem reasonable to suppose that this material, which is triploid, had been at one time introduced as a botanical specimen and had later escaped from cultivation and become naturalized. It is possible that some of Felippone's collections, mentioned above, were also made here.

Material of 'conventional' triploid *S. commersonii* (CPC: 1319, 1320, 1321, 1322, 2126) grown in Cambridge, exhibited both white and purple-flowered variants. The lateral leaflets varied from 3–5 pairs with 0–2 pairs of interjected leaflets. In one clone tubers of up to 12×6 cm were obtained, though this was exceptional. Generally, however, yields are quite good.

2. Subspecies *malmeanum*

Subsp. *malmeanum* has been cultivated both by Bukasov and ourselves. It generally forms rather wide rosettes but may later in the season produce short stems. The presence of a curious yellowing and crinkling virus was noted by us in several collections, which were destroyed to prevent a spread of the infection. This is probably the same as the ring-spot strain of tobacco mosaic virus, discovered by Hansen (1960) in a collection of subsp. *malmeanum* collected by Petersen & Hjerting (No. P–67*a*), in prov. Misiones, Argentina. The symptoms were low growth, and rather rigid leaves with a diffuse interveinal yellowing. The sub-sessile or shortly petiolulate leaflets, so characteristic of this subspecies in herbarium specimens, are also always to be seen in living material. Delicate sparse spreading hairs were noted on one collection (*Schulz 9024*, seed and tuber collections). The length of the calyx acumens seems to vary in different collections, as also does the position of the pedicel articulation (generally, however, $\frac{2}{3}$–$\frac{3}{4}$ above the base).

The nodding flowers are invariably white but only a few ((1–)2–4(–7)) are produced, generally on shortened peduncle forks.

The fruits are ovate to cordate, up to 0·75 × 1 cm. Tubers vary from 1–3 × 1–5 cm in dimensions and weigh 3–15 grammes; they are brown in colour with a rough skin and raised lenticels.

Apparently some degree of frost resistance is to be found in both subspecies of *S. commersonii*. This was noted by many collectors and has been confirmed by us from field observations in South America.

Resistance to wart, scab, powdery scab, and Colorado beetle has been established for *S. commersonii*. Clones with high starch, high dry matter, and high protein have also been found in this species. (See Chapter 5 for further details.)

CYTOLOGY

The chromosome number of *S. commersonii* was first published by Vilmorin (1929) as $2n = 36$. Subsequent counts by Rybin (1930), Longley and Clark (1930), and other workers confirmed the triploid nature of *S. commersonii* as known up to that time. Vilmorin reported that his material was obtained from Heckel in 1900, and Berthault (1911) stated that these plants were derived from an introduction from Uruguay by Saint-Quentin in 1896. It seems likely that this introduction was the only original source for all the numerous lines used by investigators in France, England, Germany, Russia, and elsewhere. Even until quite recently triploid *S. commersonii* was grown at potato-breeding stations and may still be available at some of them. Since it was more or less completely sterile there is good reason to suppose that the same stock has been multiplied vegetatively for about sixty years. The fact that it has lasted so long under conditions where few precautions to prevent virus infection have been taken argues well for a high degree of resistance to virus diseases in this particular line.

Diploid *S. commersonii* was reported by Koopmans (1951), Hougas (1953), and Kawakami and Matsubayashi (1954). Unfortunately, however, these authors were in the possession of incorrectly identified material. Koopman's and Kawakami's publications included photographs of the plants they had worked on, whilst Hougas sent samples to us later for identification. In all cases the authors had been working not with *S. commersonii* but with *S. chacoense*.

Bukasov and Lechnovicz (1935) described a diploid species, *S. henryi*, which we have placed with *S. commersonii*. From later publications by Bukasov (1940*a*, etc.) we know that diploid forms[1] of this species are fairly widespread, though, as we

[1] All given specific rank by Bukasov (see p. 167).

mentioned above (p. 165), the exact distribution area of diploid and triploid clones of *S. commersonii* is not known.

Counts kindly made by Mr. R. Tarn on our living collections of *S. commersonii* have shown that in subsp. *commersonii* most of the material (twelve out of sixteen accessions) is diploid, with $2n = 24$ somatic chromosomes. All these are typical subsp. *commersonii* from Uruguay, Brazil, and Argentina, with sessile lateral leaflets and a fairly loose inflorescence. The other four specimens counted were triploid ($2n = 36$) and were collected in Brazil and Uruguay. It is of interest to record that even before the chromosome numbers were known we had grouped three of these with the 'anomalous forms, similar to the triploid clones grown at plant breeding stations' (see p. 155). The fourth triploid (*Hawkes 721* from Gorriti) was somewhat intermediate between the anomalous and typical forms, and had been placed in the latter group. Thus the evidence seems to point strongly to the fact that most, if not all, of the triploid forms of subsp. *commersonii* constitute a distinct cytotype and are fairly readily distinguishable morphologically from the diploid forms. Furthermore, they have been collected both in Uruguay and Brazil, and may perhaps be found to extend even further when more material has been examined.

Meiotic studies on triploid *S. commersonii* by Vilmorin (1929), Longley and Clark (1930), Rybin (1933), and von Oláh (1938) showed the irregular pairing that one would associate with such plants. Von Oláh's detailed results give the following average associations:

$5{\cdot}76_{III} + 7{\cdot}88_{II} + 4{\cdot}16_{I}$. Comparing this with auto-triploid *Lycopersicon esculentum* (Upcott 1935) he was able to show that the trivalent frequency was even rather higher than in the tomato, thus pointing to an autotriploid origin for *S. commersonii*. On the basis of hybrids between triploid and diploid *S. commersonii* (the latter as *S. henryi*), on the one hand, and hybrids between triploid *S. commersonii* and diploid *S. chacoense*, on the other, he demonstrated that the genome of *S. henryi* was closer to triploid *S. commersonii* than was the genome of *S. chacoense* or of other diploid species.

Thus triploid *S. commersonii* crossed with *S. henryi* gave hybrids with $2n = 24$–27(–32) chromosomes, whilst when crossed with *S. chacoense* and with *S. spegazzinii*[1] the resulting hybrids possessed 30–34 chromosomes. Emme (1937) also confirmed this in *S. commersonii* ($3x$) × *S. chacoense* (as *S. emmeae*) hybrids, which possessed about 34 chromosomes. Thus the cytological results obtained by von Oláh tend to confirm our view that '*S. henryi*' ($2x$) and *S. commersonii* ($3x$) are very closely related, and in fact may be grouped together as the same species.

Meiosis is regular in the diploid forms of *S. commersonii* (von Oláh 1938, Propach 1940, Okuno 1951, 1952) and also in hybrids between these (as *S. henryi*) and *S. verrucosum* and *S. wittmackii* respectively (see Propach 1940). Hybrids between *S. commersonii* and *S. chacoense* at the diploid level are difficult to make and if produced are often weak and unthrifty, dying before flowering and hence preventing a study of meiotic behaviour, but apparently some good combinations can be made (see below, under 'CROSSABILITY').

The above paragraphs all refer to subsp. *commersonii*. Subsp. *malmeanum* (as *S. millanii*) was analysed cytologically by Ivanovskaya (1939), who noticed 3–4 trivalents, 3 per cent bridges and fragments etc., at meiosis. This material was triploid. Bukasov and Lechnovicz (1935), when describing *S. millanii* (placed by us under subsp. *malmeanum*), mentioned that it might possibly occur in diploid as well as triploid biotypes. (See also Ratera 1938*b*, 1943*a*, 1954.) This suspicion has now been confirmed by us on living material collected by Petersen & Hjerting and others. Collections *P. & H. P–67*, *Mazzei* (s.n.), *Mota da Costa* (s.n.), *Hawkes 725*, and *Schultz 9024, 9026*

[1] Spoken of as 'Papa chusa'. From the figures given we consider that his material was probably *S. spegazzinii*.

were all diploid, whilst *P. & H. P–70*, *P. & H. P–71*, and *Hawkes 723* were triploid. It is interesting to note that *P. & H. P–71* was collected in the place where Bukasov obtained his material of *S. millanii*.

The curious triploid form of subsp. *malmeanum*, which grows as a weed in the Prado botanic gardens at Montevideo was listed on p. 152 and has been discussed above (p. 174).

Berries have been observed on collections from Argentina, prov. Corrientes (*Petersen & Hjerting P–72*, *P–73*, *P–74*; *T. Meyer 8979*) and prov. Chaco (*P. & H. P–75*), thus indicating that the plants were probably diploid. The Rojas collection from Río Pilcomayo also possesses berries.

Thus, with subsp. *malmeanum*, as with subsp. *commersonii*, both diploid and triploid clones have been found, though here again we have insufficient material available to give us any idea of their distribution. One interesting point that may be mentioned is that we see here for the first time triploid forms of a diploid species situated very far away from a large city. However, their presence may perhaps be explained by the fact that they are able to colonize fields, river-banks, or pathsides which are kept permanently at an ecological sub-climax which is suitable for wild potatoes.

FERTILITY AND CROSSABILITY

Diploid clones of *S. commersonii* are self-sterile. However, they will cross together, forming abundant fruits and seeds (Bukasov 1941*b*, and our own observations). Many workers have noted the self-sterility of triploid *S. commersonii* (Sutton 1909, Bukasov 1937*b*, 1940*b*, Kameraz 1940), which is only to be expected owing to the irregular meiosis.

Pollination of subsp. *commersonii* was observed by Hawkes in Balcarce, Buenos Aires province, in 1958. The flowers were visited by a large wild bee known as 'Mangangá' (scientific name unknown to us), which also made indiscriminate visits during the same flight to a non-tuber-bearing species with white flowers of the same shape. Since the colour of these differed from the purple-flowered subsp. *commersonii* it might be assumed that the general flower shape and scent, but not the colour, were the chief guides. The two species in question no doubt possessed strong genetic barriers to hybridization, since if they did not, hybrids would have been formed. If this pattern of behaviour in pollinating insects is general, then two sympatric species growing in close proximity could only maintain themselves as distinct entities by the development of sterility barriers of a genetical or chromosomal kind, since ethological ones appear to be absent. It is interesting to note that such genetic barriers appear to exist between *S. commersonii* and *S. chacoense* which grow in the same area under similar habitat conditions, but not between the mountain species, which for the most part seem to be maintained by virtue of their adaptation to distinct habitats.

On the whole *S. commersonii* is difficult to hybridize with other potatoes, compared with the ease with which most diploid species in this section cross together.

Rather fewer attempts at crossing the diploid forms of *S. commersonii* with other species have been made than with the triploid ones, which have been available for a longer time to plant-breeders. Propach (1940) crossed *S. commersonii* (2*x*) with *S. verrucosum*, *S. wittmackii*, *S. spegazzinii*, and *S. chacoense* respectively and also made the reciprocals of these hybrids. All the F_1 hybrids except *S. commersonii* × *S. verrucosum* were very weak and died before flowering. From this latter an F_2 generation was raised from intersib crosses which in its turn was also weak and unthrifty. Emme (1937) also records the interesting triple hybrid (*S. commersonii* × *S. phureja*) × *S. jamesii*.

Although we ourselves made several attempts at crossing *S. commersonii* with *S. chacoense* we had no success when using subsp. *commersonii* and only limited success with subsp. *malmeanum*. At one time we wondered whether subsp. *malmeanum* might have arisen as a result of hybridization between *S. chacoense* and subsp. *commersonii*, or even perhaps that it might be more closely related to *S. chacoense* than seemed at first possible, judging purely with morphological criteria. It seemed particularly necessary, therefore, to observe how the cross between *S. chacoense* and subsp. *malmeanum* behaved, both in F_1 and F_2 generations.

Seeds of a cross between *S. chacoense* (unknown provenance) and *S. commersonii* (*Hawkes 713*, from south Brazil, Pelotas) were kindly sent us by the late Dr. Toxopeus of Wageningen, Holland. Another similar cross between the same line of *S. chacoense* and *S. commersonii* (*Hawkes 708*, from Argentina, Buenos Aires province) was received from the same source. The F_1 families, 274 and 275 respectively, from these seeds were vigorous, and combined many features of both parental species. Four F_2 families (two from each F_1, Nos. 286, 287; and 292, 293) were grown. Of a total of 200 seeds sown, 120 germinated and 100 survived to the planting-out stage. Of these, many were very small and unthrifty, so that only 53 plants survived to the end of the summer. Few of these flowered and even fewer produced berries (4 plants only). The wide range of vigour and fertility indicates a clear genetic breakdown, due presumably to differences in the genetic architecture of *S. chacoense* and *S. commersonii*.

The hybridization of subsp. *malmeanum* (*P. & H. P–67a*, from Argentina, Misiones) with *S. chacoense* (*Schulz 9027*, from Argentina, Chaco) proved rather difficult, and from various attempts only one berry was produced. This contained four seeds (contrasting with 200 seeds per berry in interclonal crosses), only one of which germinated. The plant was a good hybrid, with enlarged terminal and petiolulate lateral leaflets, but since there was only one seedling it was not possible to raise an F_2 progeny.

In our attempts at crossing *S. commersonii* with *S. tarijense*, only parthenocarpic berries were obtained, with no viable seeds.

Results in the literature of crossings between *S. commersonii* and *S. chacoense* can be divided into two sections; firstly, those in which the triploid 'conventional' forms of *S. commersonii* have been used, and the other in which diploid forms of subsp. *commersonii* (named as *S. henryi* or some other synonym) were involved. Although triploid *S. commersonii* is self-sterile and has very little good pollen it can nevertheless be used as a female parent in hybridizations of the first type. For instance, Emme (1937), von Olah (1938), and Choudhuri (1948) recorded $3x \times 2x$ hybrids which were numerically unbalanced in their chromosome complement and hence sterile. The plants were also dwarfed, and showed considerable growth depression. That the sterility might be due mainly to chromosome unbalance can readily be conceded. But when we examine the results of Propach (1940) we find that the F_1 progenies of diploid *S. commersonii* crossed with *S. chacoense* were similarly very weak, and died completely before flowering. It would seem, then, that the absence of naturally occurring hybrids between *S. commersonii* and *S. chacoense* might well be explained by reference to these results.

Crosses of triploid *S. commersonii* with other diploid species such as *S. spegazzinii* (von Olah 1938) and *S. kurtzianum* (Perlova 1945*a*) have also been reported.

As female parent *S. commersonii* ($3x$) has also been crossed with *S. tuberosum* (Bukasov 1940*b*, Perlova 1945*a*), though Sutton (1909) reported a total failure in his attempts; and also with *S. stoloniferum* (Bukasov 1937*a*, Emme 1937). Broili (1921) reports *S. commersonii* ($3x$) $\times$ *S. edinense* ($5x$) and Reddick (1939) and Wilson (in Black 1945) note that success was obtained in *S. commersonii* ($3x$) $\times$ *S. demissum* ($6x$) hybridizations.

All the above-mentioned triploid and diploid *S. commersonii* hybrids were made with subsp. *commersonii*, with the exception of the work reported by ourselves, in which we used both subspecies ($2x$). Bukasov (1940*b*) states that triploid forms (as *S. millanii*) of subsp. *malmeanum* have been used, but with a complete lack of success, in crosses with other species.

As compared with *S. chacoense* it can be seen that *S. commersonii* is not only more difficult to cross with other species, but the hybrids when formed are more likely to be sterile. Despite the long time that triploid *S. commersonii* has been known to plant-breeders, practically nothing of value has come from it. The difficulty with which hybrids are obtained between *S. commersonii* and *S. chacoense* seems at first sight strange when we consider that they are both members of the series *Commersoniana*. Bukasov (in Bukasov & Kameraz 1959) indeed separates them into two series, Commersoniana and Glabrescentia. If we glance at the distribution areas of the two species we shall see that they largely overlap, and within this area of overlap they are not found in two distinct habitats. One may assume that (in contrast to most other wild species of potatoes) *S. commersonii* and *S. chacoense* are kept apart by a sterility barrier and not by a geo-ecological barrier. They may thus be closer together from an evolutionary point of view than the hybridization results indicate.

Lester (1965), working on the serological relationships of wild potato species (see also Gell, Hawkes & Wright 1960), showed that from an immunological point of view both subspecies of *S. commersonii* differed sharply from *S. chacoense*, and indeed from all other South American tuber-bearing species that he tested. Antisera raised to subsp. *malmeanum* gave at least a five-line precipitin spectrum when tested against antigens from various accessions of subsp. *malmeanum* and subsp. *commersonii*, but only a four-line spectrum with the other species. When subsp. *malmeanum* antiserum was absorbed out with *S. tuberosum* antigen, it gave 1–2-line spectra with all forms of subsp. *malmeanum* and subsp. *commersonii* (with one exception) and no lines at all with forms of *S. chacoense* and *S. tarijense*. The one exception was a morphologically intermediate form of *S. commersonii*, which possessed many features of *S. chacoense* and which might therefore have been a hybrid segregate from a natural *S. commersonii* × *S. chacoense* cross. (Accession H. 201; see below.)

It would seem, therefore, both from crossability studies and the immunological ones referred to above, that the two species *S. commersonii* and *S. chacoense* are quite distinct. We are not certain, however, that they should be separated into two distinct series, as Bukasov has suggested.

NATURAL HYBRIDS

As we have already mentioned, the distribution ranges of *S. commersonii* and *S. chacoense* partly overlap. From the evidence mentioned above in the section on artificial hybrids it would seem that there are some partial sterility barriers keeping these two species apart, and one would not expect, therefore, that they would be likely to form hybrids in nature with great frequency.

Of the 300 collections of *S. commersonii* that we have studied in the present work, there seem to be only three possible hybrids, unless some of the anomalous probable triploid forms mentioned on p. 154 are of hybrid nature. The three hybrids referred to above are as follows:

1. *Cabrera 5805*, collected in prov. Buenos Aires, partido de La Plata. This possesses the reduced peduncle branches, obtuse leaflet apex, and well-developed calyx acumen of *S. commersonii*, and the round berries, long leaves, tall habit, long petiolules, and numerous interjected leaflets of *S. chacoense* (see Plate 30).
2. *Burkart 3676* from Magdalena, also in Buenos Aires province; it is characterized by the same features as the Cabrera collection.
3. A collection by Camañ (Accession H. 201) from Buenos Aires province, General Madariaga, sent to us as living material. The lower leaves of this plant possess up to 8 mm long petiolules, the inflorescence branches are short, but twice-forked, this latter feature being typical of *S. chacoense*. Lester's serological results also indicate a rather strong affinity with *S. chacoense*, even though in its general habit and

morphology it approaches *S. commersonii*. Although *S. chacoense* has not been recorded so far south on the Argentine coast, the influence of that species in this collection is unmistakable. It is of interest that this collection is triploid.

The triploid material known to plant-breeders, with very reduced peduncle forks and broad petiolulate leaflets, together with some triploid collections made by Hawkes in Uruguay, has been placed by us in the present treatment on p. 156 as 'anomalous forms'. Also included with these are a number of collections by Felippone (*2444a, 5350, 5434, 5436, 5437, 5473, 5476, 5494*), some of which are from the Uruguay coast near Montevideo. Their appearance is very similar to that of the triploid forms. Although cytological data indicate that the diploid and triploid *S. commersonii* are conspecific we should perhaps consider as a possibility that the triploids might be hybrids between the two subspecies of *S. commersonii*, or even between *S. chacoense* and *S. commersonii*, produced as a result of the fusion of a reduced and a non-reduced gamete, since the Camañ collection mentioned above is triploid. However, we personally consider the latter explanation to be unlikely and the only probable natural hybrids of the two species we have yet seen are the collections from Buenos Aires province, Argentina (see below).

LOCALITIES OF NATURAL HYBRIDS (Probably with *S. chacoense*)

ARGENTINA

Prov. BUENOS AIRES:

Magdalena, F. C. S. Mar. 1931. 'Fl. blanca; matas abundan en vías. Frutos con aroma de ananás.' *Burkart 3676* (GH, SI).

Partido de La Plata, Bosque. Jan. 1940. *Cabrera 5805* (GH, LP). *Note*: Petiolules 5–10 mm long; leaflets rounded apically and often folded along the midrib; inflorescence branches short, to 2 cm or less; berries globular. See Plate 30.

General Madariaga. July 1958. 'Papa silvestre. Campos ondulados.' *Camañ* (s.n.) (Accession H. 201. $2n = 36$; no herbarium specimen).

4. SOLANUM CHACOENSE Bitter, *emend.* Hawkes et Hjerting

(*In Reprium nov. Spec. Regni veg.* **11**, 18, 1912)

S. tuberosum L. var. *glabriusculum* Dun. (in *DC. Prodr.* **13**, 1, 32, 1852).
S. guaraniticum Hassl. (p.pte[1]) (in *Reprium nov. Spec. Regni veg.* **9**, 115, 1911; *non S. guaraniticum* A. St. Hilaire, *Mem. Mus.* **12**, 321–2, 1825).
S. guaraniticum Hassl. var. *angustisectum* Hassl. (in *Reprium nov. Spec. Regni veg.* **9**, 115, 1911).
S. guaraniticum Hassl. var. *latisectum* Hassl. forma *glabrescens* Hassl. (in ibid.).
S. bitteri Hassl. (in ibid. **11**, 190, 1912).
S. subtilius Bitt. (in ibid. **12**, 6–7, 1913).
S. chacoense Bitt. var. *angustisectum* Hassl. (*Annls Conserv. Jard. Bot., Genève* **20**, 186–7, 1917).
S. chacoense Bitt. var. *latisectum* Hassl. forma *glabrescens* Hassl. (ibid.).
S. chacoense Bitt. var. *latisectum* Hassl. forma *plurijugum* Hassl. *nom. nud.* (ibid.).
S. gibberulosum Juz. et Buk. (*Revta argent. Agron.* **3**, 225–6, 1936).
S. parodii Juz. et Buk. (ibid. 226–7, 1936).
S. garciae Juz. et Buk. (ibid. 227–8, 1936).
S. horovitzii Buk. (ibid. **4**, 238, 1937).

[1] Part of this species, namely var. *latisectum* f. *pilosulum*, should be placed under *S. commersonii* subsp. *malmeanum*.

S. laplaticum Buk. (ibid. 238–9, 1937).
S. boergeri Buk. (ibid. 239, 1937).
S. emmeae Juz. et Buk. (*Izv. Akad. Nauk SSSR* **2**, 321–2, 1937). = *S. dolichostigma* Buk. *nomen nudum* (in Vavilov, *Theoret. Bases Pl. Breed.* **3**, 72, 1937).
S. knappei Juz. et Buk. (*Izv. Akad. Nauk SSSR* **2**, 322–3, 1937).
S. schickii Juz. et Buk. (ibid. 324–5, 1937).
S. horovitzii Buk. var. *multijugum* Hawkes (*Bull. imp. Bur. Pl. Breed. Genet., Camb.* 114, 1944).
S. horovitzii Buk. var. *glabristylum* (ibid. 19, 1944, *nomen nudum*).
S. jujuyense Hawkes (ibid. 114, 1944).
S. saltense Hawkes (ibid. 113–14, 1944, *non S. saltense* (Bitter) Morton in *Contr. U.S. natn. Herb.* **29**, 63, 1944 = *Lycianthe saltensis*).
S. chacoense subsp. *subtilius* (Bitt.) Hawkes (*Rep. Scott. Pl. Breed. Stn.* 61, 1956).
S. caipipendense Cárd. (*Boln Soc. peru. Bot.* **5**, 35–36, 1956).
S. cuevoanum Cárd. (ibid. 36–37, 1956).
S. arnezii Cárd. (ibid. 37–41, 1956).
S. chacoense forma *caipipendense* (Cárd.) Corr. (*Wrightia* **2**, 172, 1961).
S. chacoense forma *gibberulosum* (Juz. & Buk.) Corr. (ibid.).
S. limense Corr (ibid. 188–9, 1961).

The following inadequately described 'species' should also probably be included within subspecies *chacoense*. Absence of type material or reasonably detailed descriptions makes it impossible to place them with any greater degree of certainty.

S. renggeri Schlechtd. (*Hort. Hal.* **1**, 5, 1841).
S. paraguariense Dun. (in *DC. Prodr.* **13**, 1, 376, 1852 = *S. renggeri* Schlechtd.).
S. pseudo-maglia Planchon *nomen nudum* (see *Annls Fac. Sci. Marseille* **18**, 205–24, 1909).

Herbaceous, often semi-rosulate, with long underground stolons bearing rather small tubers, up to 1–3 × 1–5(–6) cm.

Stem 20–150(–200) cm tall, according to biotype and environment, 2–6 mm diam., simple or with divaricate branches especially near the base; internodes long, but generally shorter near the stem base; decurrent lines or very narrow wings generally present, much wider (up to 2–3 mm) in certain mountain biotypes, straight or marginally crisped; stem pubescence very sparse or occasionally moderately frequent, the hairs sometimes with a swollen base; pigmentation often absent but present in some forms towards the nodes or as a network of violet lines and blotches, also sometimes on the wings when these are present.

Leaf (5–)8–17(–36) cm long, frequently closely compacted into a semi-rosette at stem base; basal leaves often much larger than the stem leaves; lateral leaflets (3–)4–6(–8)-jugate, with 0–3(–10) pairs of much smaller sub-sessile interjected leaflets; petiolules of lateral leaflets (1–)2–8(–25) mm long, the leaflets themselves ovate, lanceolate, occasionally oblong, obtuse or subcordate at the base, *acute to acuminate at the apex*, 2–7·5(–10·5) × 1–2·8(–3·2) cm; terminal leaflet wider than laterals though barely longer, (1·7–)2·2–8·3(–9·5) × 1–3·5(–4·3) cm; surface of the leaf generally glabrescent or with short rather scattered hairs, occasionally with numerous hairs both on surface and margins, especially in the mountain forms.

Inflorescence typically many-flowered but a range of from 3 to 50 flowers is recorded; *peduncle typically twice or more forked*, occasionally once-forked, 2–15 cm long or perhaps longer under some conditions; pedicels (8–)12–25(–32) mm long, generally articulated in the middle third; pubescence of peduncle and pedicels absent or sometimes quite dense, composed of simple short few-celled hairs, and very occasionally some glands.

Calyx 4–6 mm long, the lobes varying from short to long oval (1·5–3·5 mm long) with typically a *very short acumen* 0·3–0·5 mm long, which occasionally in mountain forms of subsp. *chacoense* and in subsp. *muelleri* attains a length of 0·75–1 mm or more; pubescence generally absent.

Corolla *white, stellate*, from (16–)25–35 mm diam.; lobes broadly to narrowly triangular, 7–15 mm long × 4–12 mm broad.

Anthers 4·5–7·5 × 1·5–2·2 mm; filaments 0·75–2 mm long.

Style often long, varying from 9–15 mm, exserted for about 3–10 mm, glabrous or with minute papillae in its lower $\frac{1}{3}$–$\frac{2}{3}$; stigma varying from short capitate through ovoid to long ellipsoid or long obconical; apex of stigma entire or bifid; base entire or with two long sinuses.

Berries *more or less spherical* or sometimes short ovate (never as long as in *S. commersonii*) up to 2·5 cm diam., green or marked with white spots and dark purple streaks.

Chromosome number: $2n = 24$ (36).

AFFINITIES

S. chacoense may be distinguished from species in series other than Commersoniana by the white stellate corolla and sparse pubescence. From *S. commersonii* subsp. *commersonii* it differs in the larger habit, generally long-petiolulate acuminate leaflets, twice or thrice-branched peduncles, glabrous calyx with very short narrow acumens (generally about 0·5 × 0·5 mm), white corolla and globular berries. It is perhaps more difficult to distinguish from *S. commersonii* subsp. *malmeanum*, which also has a white corolla and whose leaflets are subsessile to shortly petiolulate. In this latter subspecies, however, the peduncles are unbranched or only once-forked, the calyx acumens are larger and wider (1 × 1 mm), the plant more pubescent and the berries ovate to conical.

Introgressed forms of *S. chacoense* from the north-western mountainous areas are more pubescent, with shorter petiolules and longer calyx acumens. They lose the typical rosette habit of this species on the plains and generally possess winged stems—a character not to be found in lowland *S. chacoense* or in either subspecies of *S. commersonii*.

S. chacoense is evidently closely related to series Tuberosa, with several members of which it easily forms hybrids. It is also closely related to *S. tarijense* but is easily distinguished from it by the absence of glandular pubescence. Its relationship to *S. commersonii* is perhaps not so close, judging by the difficulty with which it hybridizes with that species, both in nature and in the experimental field.

KEY TO SUBSPECIES

Lateral leaflets 2–3 times as long as broad. (Argentina, [Bolivia,] Paraguay, southern Uruguay) subsp. **chacoense**

Lateral leaflets (on well-developed leaves) 3·3–4·3 times as long as broad. (Argentina, prov. Misiones; Brazil) subsp. **muelleri**

1. Subspecies *chacoense*

(For synonyms see p. 180)

Differs from subsp. *muelleri* by the shorter leaflets, 2–3 times as long as broad, generally shorter petiolules (1–15 mm), more or less median pedicel articulation, regular calyx with very short acumens (except in the introgressed mountain forms at altitudes of over 1500 m) and generally shorter anthers.

2. Subspecies *muelleri* (Bitter) Hawkes et Hjerting, *comb. nov.*

(*Scott. Pl. Breed. Stn. Rec.* 103, 1963)

S. muelleri Bitt. (in *Reprium nov. Spec. Regni veg.* **12**, 155–6, 1913).
S. jamesii Torr. var. *grandifrons* Bitt. (in ibid. 151–2, 1913).
S. muelleri Bitt. f. *densipilosum* Corr. (*Wrightia* **2**, 173, 1961).

Distinguished from subsp. *chacoense* by the long, narrow leaflets, about 3·3–4·3 times as long as broad, the long petiolules, up to 25 mm long on the acroscopic side, the very oblique leaflet bases, the very low almost basal pedicel articulation (never more than half way above the base), well-developed calyx acumens, to 1·25 mm long, and the slightly irregularly placed calyx lobes. Plant generally glabrous, very occasionally pubescent (*S. muelleri* f. *densipilosum* Corr.).

SPECIMENS EXAMINED

1. Subspecies *chacoense*

PARAGUAY

Asunción. June 1875. 'Sur les bords humides des chemins. Spontané. Rhizomes produisant des tubercules gros comme des noix. Fleurs blanches.' *Balansa 2103* (G, K, P).

Campichuelos, orillas montes Cordilleras de Altos. Apr. *Hassler 140* (p.p.) (SI).

San Bernardino, Cordillera de Altos. Orillas montes. Mar. 1916. *Hassler 140* (p.p.) (= *1140*) (BAF, Herb. Osten 8600. MVM).

Near Cordillera de Altos. July. 'In locis arenosis. Herba 30–60 cm. Corolla alba.' *Hassler 326* (G, Type for *S. guaraniticum* Hassl. var. *latisectum* Hassl. forma *glabrescens* Hassl.). See Plate 36.

Ad marginem silvae pr. Villeta. July. 'Herba 30–60 cm, similis *326*, differt corolla lilacino caesia.' *Hassler 470* (G, Syntype for *S. guaraniticum* Hassl. var. *angustisectum* Hassl.). (It is doubtful whether this specimen could have possessed a lilac-grey corolla.) See Fig. 15.

Gran Chaco, Santa Elisa, lat. S. 23° 10′ (W. 57° 40′). Mar. 1903. *Hassler 2805* (G, Syntype for *S. guaraniticum* Hassl. var. *angustisectum* Hassl., K, P). (This collection was proposed by Correll as lectotype for *S. chacoense* Bitt.) See Plate 35.

Paraguay central, Lago Ypacaray región. Oct. 1913. *Hassler 11849* (BAF, C, E, G, GH, K, L, LY, US). (*Note*: Some specimens seem a little doubtful, and may have affinities with *S. commersonii* subsp. *malmeanum*. This collection is the type for Hassler's var. *latisectum*, forma *plurijugum* Hassler, *nom. nud.*)

Paraguay central, región de Lago Ypacaray. Jan. 1913. *Hassler 12017* (BAF, C, E, GH, L, US).

Paraguay central, región de Lago Ypacaray. Jan. and Oct. *Hassler 12607* (K, LY).

Villarrica. 13 March 1929. 'Común in capueros.' *P. Jörgensen 3681* (BA, C, F, LIL, LP, NY, S, SI, US).

Primavera, Alto Paraguay, Río Tapiracuay. Alt. *c.* 100 m. 1954. *F. J. Kemp* (s.n.) (J.G.H. No. 82).

Ascau-Cué. 16 Apr. 1885. *F. Kurtz 390* (CORD).

Dept. Cantera, Puerto Cantera. Alt. 230 m. 20 Nov. 1948. 'Pl. herbacea, an alt. 0·60 m.' *J. E. Montes 3338* (SI).

Dept. Central, Villa Elisa, *c.* 20 km south of Asunción. 16 Oct. 1959. 'Fls. white; ripe fruit green, 15–20 mm diam.; tubers frequently not formed. Sandy fields; common in cultivated fields.' *Myndel Pedersen 5144* (J.G.H.). See Plate 37.

Dept. Misiones, Santiago, Estancia 'La Soledad.' 28 Apr. 1961. 'Fls. white. Open woodland.' *Myndel Pedersen 6016* (C, J.G.H., J.P.H.).

Exact locality unknown. Paraguay Expedition 1851. *Capt. T. J. Page* (s.n.) (US).

Jardín Botánico, Trinidad, praderas sombreadas. 13 Apr. 1951. 'Flores blancas.' *T. Rojas 14522* (C, LIL).

Trinidad. Apr. Tubers sent from the Argentine Ministry of Agriculture as No. 1955. Herb. specimen from field at Cambridge 1951. CPC: 2137 (P.G.S., J.G.H.).

URUGUAY

Dept. SORIANO:

Paraje Agraciada. 1935. 'Planta ruderal, algunos metros de la orilla del Río Uruguay. Solanum de flor blanca.' Recibido por semillas. Reproducido en Leningrado. *T. Henry* (s.n.) (WIR No. K. 1070, Type for *S. boergeri* Buk.). See Plate 38.

Soriano, near the coast. Dec. Tubers sent from the Argentine Ministry of Agriculture under the name *S. boergeri*. No. 1954. Herb. specimen from field at Cambridge, 1951. CPC: 2136 (J.G.H.).

Uruguay, locality unknown. *Tweedie* (s.n.) (K, TCD).

ARGENTINA

Prov. JUJUY:

Dept. Capital, near Jujuy, Barcena. Alt. 1600 m. 8 Feb. 1939. 'Growing amongst grass and rough herbage.' *Balls 5952* (P.G.S.).

Dept. Capital, Termas de Reyes. 23 Jan. 1946. 'Flores blancas; tuberígeno.' *Cabrera 8140* (LP).

Dept. Capital, Termas de Reyes. 11 & 12 Jan. 1947. *Garolera & Romero* (s.n.) (LIL 197186, 208490).

Dept. Capital, Lagunas de Yala. Alt. 2100 m. 1 Feb. 1956. *Hjerting, Petersen & Rahn 52* (J.G.H., J.P.H.).

Dept. Capital, entre Yala y Lagunas de Yala. Alt. 1750 m. 1 Feb. 1956. *Hjerting, Petersen & Rahn 56* (C, J.G.H., J.P.H.).

Dept. Capital, Lagunas de Yala. 30 Jan. 1947. *O'Donell 4870* (LIL).

Dept. Capital, Yala. Alt. 2086 m. 11 Jan. 1941. 'Flor blanca.' *E. Romero 48* (LIL).

Dept. Capital, Reyes. 5 Feb. 1941. 'Flor blanca.' *E. Romero 58* (LIL).

Dept. Capital, Laguna de Yala. Alt. 2000 m. 17 Feb. 1959. *Ross & Vervoorst 75, 77* (Max.).

Dept. Ledesma, Quinta Yuchan vor Higueras. Alt. 640 m. 17 Feb. 1959. 'Sommergrüner Regenwald.' *Ross & Vervoorst 69* (Max.).

Dept. San Antonio, San Antonio, El Potrerillo. Alt. 1300 m. 6 Feb. 1939. *Balls 5921* (CPC: 84) (K, US, P.G.S., Types for *S. jujuyense* Hawkes, and *S. horovitzii* Buk. var. *multijugum* Hawkes, J.G.H.). See Plate 45.

Dept. Tilcara, Quebrada Yaquis Pampa, Estanque. Alt. 2300 m. 23 Mar. 1952. 'Más abajo que San Antonio, pero en la misma quebrada.' *Petersen & Hjerting P–349* (LIL, J.G.H., J.P.H.).

Dept. Tumbaya, Est. Volcán, quebradas occidentales. Alt. 2200 m. 24 Jan. 1927. 'Flor crema.' *Castillón 354* (LIL).

Dept. Tumbaya, Volcán. 11 Jan. 1947. *Garolera & Romero* (s.n.) (LIL 206246).

Dept. Tumbaya, cerca de Volcán. Alt. 2000 m. 19 Feb. 1956. *Hjerting, Petersen & Rahn 277* (J.G.H., J.P.H.).

Dept. Tumbaya, cerca de Volcán. Alt. 2000 m. 7 Feb. 1953. *Petersen & Hjerting 958* (J.G.H., J.P.H.).

Dept. Tumbaya, Volcán, campos abandonados. Alt. 2100 m. 23 Jan. 1953. *Sleumer 3566* (LIL, J.G.H.).

Prov. SALTA:

Dept. Anta, 1 Secc., Miraflores. 9 Apr. 1951. '"Sacha-Tomate"; flor blanca.' *Luna 328* (BA).

Dept. Anta, 3. Secc., Colonia de los Franciscos. 17 Mar. 1948. *Luna 777* (LIL).

Dept. Anta, 1. Secc., El Dorado. 24 Apr. 1948. *Luna 1007* (LIL).

Dept. Cafayate, Cafayate. Alt. 1650 m. 25 Feb. 1951. *Hayward 2170* (LIL).

Dept. Caldera, camino a Jujuy. 1 Feb. 1949. *Cabrera & Schwabe 209* (LP, US). (Classified by Correll, 1962, p. 224, as *S. trigalense* [≡ *S. tarijense*].)

Dept. Caldera, Quebrada de Yacones. Alt. 1650–1700 m. 20 Feb. 1956. *Hjerting, Petersen & Rahn 286* (J.G.H., J.P.H.).

Dept. Caldera, alrededores de los Potreros al pie del Nevado del Castillo. Mar. 1873. *Lorentz & Hieronymus* (s.n.) (G).

Dept. Capital, Quebrada de San Lorenzo. Alt. 1530 m. 7 Feb. 1939. 'Flowers pale cream. Loose slender plants. Growing among bushes and on sides of paths in rich leaf mould.' *Balls 5935* (P.G.S., Type for *S. saltense* Hawkes, J.G.H.).

Dept. Capital, Quebrada de San Lorenzo. Alt. 1530 m. 7 Feb. 1939. Tubers only. Herb. spec. field Cambridge. *Balls 5936* (p.p.) CPC: 51b (K, P.G.S., Paratype for *S. saltense* Hawkes, J.G.H.). See Plate 46.

Dept. Capital, Cerro 20 de Febrero. Alt. 1197 m. 23 Jan. 1949. *Filipóvich 237* (LIL).

Dept. Capital, San Luis. Alt. 1215 m. 11 Jan. 1949. *Filipóvich 381* (LIL). (Classified by Correll, 1962, p. 227, as *S. yungasense*.)

Dept. Capital, Quebrada de San Lorenzo. Alt. 1500 m. 20 Feb. 1956. 'Topotype of *S. saltense* Hawkes.' *Hjerting, Petersen & Rahn 292* (J.G.H., J.P.H.).

Dept. Capital, Salta, Mar. 1941. *Zabala 16* (LIL).

Dept. Chicoana, entre Chicoana y Escoipe, km 16·5. Alt. 1500 m. 22 Feb. 1956. 'Frutos. La planta casi 2 metros de largo.' *Hjerting, Petersen & Rahn 295* (C, J.G.H., J.P.H.).

Dept. Chicoana, entre Chicoana y Escoipe, cerca de Mal Paso. Alt. 1650 m. 22 Feb. 1956. 'Frutos.' *Hjerting, Petersen & Rahn 297* (C, J.G.H., J.P.H.).

Dept. Chicoana, 3 km arriba de Escoipe, cerca de la Hostería. Alt. 2200 m. 22 Feb. 1956. *Hjerting, Petersen & Rahn 307* (J.G.H., J.P.H.).

Dept. Chicoana, un poco arriba de la Hostería, Pie de Cuesta del Obispo. Alt. 2350 m. 22 Feb. 1956. *Hjerting, Petersen & Rahn 309* (p.p.) (J.G.H., J.P.H.).

Dept. General M. M. de Güemes, entre Caldera y Carmen, km 1650, camino de la Cornisa. Alt. 1450 m. 27 Jan. 1956. *Hjerting, Petersen & Rahn 16* (C, J.G.H., J.P.H.).

Dept. General M. M. de Güemes, entre Caldera y Carmen, km 1652, camino de La Cornisa. Alt. 1450 m. 27 Jan. 1956. *Hjerting, Petersen & Rahn 17* (C, J.G.H., J.P.H.).

Dept. Guachipas, Pampa Grande. Alt. 1700 m. Dec. 1963. 'Flor alba.' *Brücher 4101* (BA).

Dept. Guachipas, arriba de Cebilar en el camino a Pampa Grande. Alt. 1500 m. 24 Feb. 1956. *Hjerting, Petersen & Rahn 361* (J.P.H.).

Dept. Guachipas, 3 km de Ojo de Agua hacia Guachipas, camino a Pampa Grande. Alt. 1900 m. 24 Feb. 1956. *Hjerting, Petersen & Rahn 364* (J.G.H., J.P.H.).

Dept. Guachipas, Estancia Pampa Grande, cerca de la represa. Alt. 1700 m. 24 Feb. 1956. 'Frutos.' *Hjerting, Petersen & Rahn 370* (J.G.H., J.P.H.).

Dept. Guachipas, Estancia Pampa Grande, cerca de la represa. Alt. 1700 m. 25 Feb. 1956. *Hjerting, Petersen & Rahn 377a* (J.G.H., J.P.H.). (Corolla sub-rotate, due to introgression from *S. microdontum*; see *377b* classed as a hybrid between *S. chacoense* and *S. microdontum.*)

Dept. Guachipas, Ojo de Agua, entre Guachipas y Pampa Grande. Alt. 1900 m. 25 Feb. 1956. *Hjerting, Petersen & Rahn 379* (J.G.H.) and *380* (C, J.G.H., J.P.H.). (Both collections show introgression from *S. microdontum.*)

Dept. Iruya, Río Iruya. 28 Apr. 1947. *Pierotti 6622* (LIL).

Dept. Metán, El Rincón. 13 May 1949. *Luna 1480* (LIL).

Dept. Metán, Metán. 20 Apr. 1948. *R. Rocha 128* (LIL).

Dept. Rosario de Lerma, Campo Quijano. 14 Mar. 1945. 'Flor blanca.' *Abbiatti & Claps 259* (LP).

Dept. Rosario de Lerma, Campo Quijano. 14 Mar. 1945. 'Flor blanca.' *Abbiatti & Claps 484* (LIL, LP).

Dept. San Martín, Tartagal, Q. Naranjito. 15 July 1944. 'Flor blanca. Arbusto. "Papa del Monte."' *Schulz & Varela 5226* (LIL).

Dept. Santa Victoria, cerca de Condado, entre Toldos y Río Bermejo. Alt. 1350 m. 12 Feb. 1956. 'En la selva de *Alnus* y *Podocarpus*.' *Hjerting & Rahn 206* (J.G.H., J.P.H.).

Dept. Santa Victoria, Toldos, junta de Río Toldos y Río Astillero. Alt. 1600–1700 m. 13 Feb. 1956. *Hjerting & Rahn 221* (J.G.H., J.P.H.).

Dept. Santa Victoria, Toldos, dentro de cultivos. Alt. 1550 m. 8 Feb. 1953. 'Flor alba. Seguramente silvestre.' *Sleumer 3880* (LIL, J.G.H.).

Dept. La Viña, cerca de Coronel Moldes, km 1556. Alt. 1100 m. 24 Feb. 1956. '2 metros de altura.' *Hjerting, Petersen & Rahn 356* (C, J.G.H., J.P.H.).

Dept. La Viña, Ampascachi. Alt. 1200 m. 24 Feb. 1956. *Hjerting, Petersen & Rahn 357* (J.G.H., J.P.H.).

Dept. La Viña, Potrero de Díaz. 1935. 'Papa de Yuta.' 'Silvestre. Invasora en los cultivos de papa.' *Horovitz* (s.n.) (WIR No. K. 637, sent by R. Millán No. G. 1047. Type collection of '*S. horovitzii*' Buk.). See Fig. 17.

Dept. La Viña, Coronel Moldes. Alt. *c.* 1200 m. 2 Feb. 1941. 'De un cultivo en La Ollada, en tierra humífera muy buena; pétalos blancos.' *A. T. Hunziker 1281* (BA).

Dept. La Viña, La Viña, Q. de Molina. Alt. 1300 m. 15 Feb. 1959. 'Bergquebracho-wald, an feuchter Stelle, im Schatten.' *Ross & Vervoorst 56* (Max.).

Dept. La Viña, La Hoyada. Sent by Sr. A. Castronovo, Castelar, as No. Hort. 1966. Herb. spec. from field in Cambridge, 1951. CPC: 2067 (P.G.S., J.G.H.).

Dept. La Viña, La Hoyada. Aug. From the Argentine Ministry of Agriculture as No. 1959. Herb. spec. from field in Cambridge, 1951. CPC: 2138 (P.G.S., J.G.H.).

Salta, locality unknown. 'Papa del Monte.' *P. T. Knappe* (s.n.) (WIR No. 2882, tubers sent by P. T. Knappe, Type for *S. knappei* Juz. & Buk., LE). See Plate 44.

Prov. TUCUMÁN:

Dept. Burruyacú, Cerro del Campo. Alt. 1200 m. 7 Apr. 1914. 'Alt. pl. 150 cm. Bosques.' *Monetti 2223* (LIL).

Dept. Burruyacú, La Ramada (F.C.C.A.). Alt. 450 m. 2 Apr. 1933. 'Flor blanca.' *Peirano (Schreiter) 9657* (LIL).

Dept. Burruyacú, Burruyacú. 20 Mar. 1900. *Stuckert 8549* (9049) (LIL).

Dept. Burruyacú, Puestito. 22 Apr. 1944. *L. Varela* (s.n.) (LIL 101967).

Dept. Burruyacú, Quebrada Río Loro. Alt. 850 m. 21 Oct. 1951. *Vervoorst 3354* (LIL).

Dept. Capital, Parque Centenario. 11 May 1913. *Castillón 2885* (LIL).

Dept. Capital, Capital. Jan. 1923. *Hauman* (s.n.) (BA 2307).

Dept. Capital, Tucumán. 20 May 1913. 'Abundante en las afueras ' *P. Jörgensen* (s.n.) (BAB).

Probably Dept. Capital. 15 Apr. 1888. 'En los campos.' *Lillo 319* (P, Type for *S. subtilius* Bitt.). See Fig. 16.

Dept. Capital, Capital, campos cultivados. 16 Feb. 1904. *Lillo 3354* (LIL).

Dept. Capital, Capital, campos cultivados. 1 May 1913. *Lillo 12910* (LIL, SI).

Dept. Capital, alrededores de la Capital. 28 Apr. 1913. 'Común en los campos.' *Lillo* (s.n.) (SI 15940).

Dept. Capital, Tucumán, Parque 9 de Julio. Alt. 450 m. 19 Mar. 1921. 'Flor blanca.' *Schreiter 9966* (LIL).

Dept. Capital, Muñecas. Alt. 500 m. 14 Nov. 1917. 'Fl. albis.' *Schreiter* (s.n.) (Herb. Corn. Osten 12412. MVM).

Dept. Capital, Muñecas. Alt. 550 m. 16 Dec. 1923. 'Fl. bl., 50 cm, rastrera. Faldas del ferrocarril.' *Venturi 2246* (LIL, US).

Dept. Famaillá, Lules. 19 Nov. 1950. *Araque & Barkley 20Tu503* (LIL).

Dept. Famaillá, entre San Pablo y Villa Nougués, 4·5 km antes de Villa Nougués. Alt. 1000 m. 5 Dec. 1948. *Petersen & Hjerting P–21* (J.G.H.).

Dept. Famaillá, San Pablo, Ruta Nac. 38, km 1536. Alt. 400 m. 4 Apr. 1949. 'En pasto y a pleno sol; tierra arcillosa.' *Petersen & Hjerting P–58* (J.G.H., J.P.H.).

Dept. Famaillá, San Ramón a Bella Vista. Alt. 450 m. 20 Jan. 1924. 'Flor crema. Planta de 60–80 cm.' *Schreiter 9965* (LIL).

Dept. Famaillá, Quebrada de Lules, dique inferior. Alt. *c.* 650 m. 22 Oct. 1950. 'Fl. blanca.' *Sleumer 758* (LIL).

Dept. Graneros, La Madrid. Alt. 440 m. Mar. 1919. 'Flor blanca, Sacha papa!' *Bailetti 213* (LIL).

Dept. Graneros, La Cocha. 19 Mar. 1939. *M. Isabel H. Scott de Birabén & Max Birabén 1573* (LP).

Dept. Monteros (or Tafí?), el camino de Acheral a Tafí del Valle, km 33. Alt. 1000 m. 26 Feb. 1949. 'Tallo con alas violetas.' *Petersen & Hjerting P–51* (J.G.H., J.P.H.).

Dept. Monteros, el camino de Acheral a Tafí del Valle, km 17. Alt. 550 m. 26 Feb. 1949. 'En la selva húmeda. Una planta de 1 m de altura.' *Petersen & Hjerting P–52* (J.G.H.).

Dept. Monteros, Ruta Nac. 38, km 1483·7, entre Concepción y Villa Quinteros. Alt. 400 m. 9 June 1949. 'En tierra negra, entre yerbas.' *Petersen & Hjerting P–83* (J.G.H.).

Dept. Río Chico, Río Chico. 1934. *Millán G. 980* (WIR No. K. 603, Type for *S. parodii* Juz. & Buk.). See Plate 42.

Dept. Río Chico, El Potrerillo. Alt. 600 m. 28 Nov. 1913. 'Planta 80 cm. Terreno bañado. Flor crema.' *Monetti 1553* (LIL).

Dept. Río Chico, La Calera. Alt. 500 m. 30 Nov. 1913. '80 cm, matorral. Flor blanca.' *Monetti 1587* (LIL).

Dept. Tafí, Valle de Tafí. 1908. *Bruch* (s.n.) (p.p.) (LP).

Dept. Tafí, Raco, Alt. 1200 m. 23 Jan. 1933. 'Muy invasora en suelo arado; flor blanca.' *Burkart 5259* (SI).

Dept. Tafí, orilla de la vía de Tafí Viejo al Cadillal. 18 Mar. 1945. 'Flor blanca, centro amarillo; alt. 30–50 cm.' *J. Herrera 289* (LIL).

Dept. Tafí, Parque Aconquija, Villa Nougués. Alt. 1200 m. 6 May 1945. *J. Herrera 412* (LIL).

Dept. Tafí, Mollar. Alt. 1900 m. 27 Feb. 1956. *Hjerting, Petersen & Rahn 396* (J.G.H., J.P.H.).

Dept. Tafí, Mollar, cerca del puente. Alt. 1900 m. 27 Feb. 1956. *Hjerting, Petersen & Rahn 398* (J.G.H., J.P.H.).

Dept. Tafí, Parque Aconquija. Alt. 1300 m. 17 Nov. 1948. *Legname 48* (LIL).

Dept. Tafí, Siambón, campos cultivados. Nov. 1888. *Lillo 165* (LIL).

Dept. Tafí, El Lambedero. Alt. 2030 m. 3 Feb. 1908. 'Playas.' *Lillo 7515* (LIL).

Dept. Tafí, Tacanas. Alt. 2000 m. 10 Dec. 1908. 'Playa arenosa.' *Lillo 8724* (LIL).

Dept. Tafí, Siambón, Sierra de Tucumán. 9–18 Jan. 1873. '*S. caldasii* var. *glabrescens*.' *Lorentz & Hieronymus 1050* (BR, CORD, F, G, GOET, K, LE).

Dept. Tafí, Aconquija, near Tucumán, cultivated fields from village to forest. Alt. 675 m. 25 Dec. 1935. 'Herb. White flower; forms tubers. A common weed in fields. "Papa del Diablo."' *Ynes Mexia 04330* (GH, UC).

Dept. Tafí, Siambón, 1 Mar. 1945. 'Flores blancas.' *Olea 302* (LIL).

Dept. Tafí, Raco. Alt. 1000 m. 23 Jan. 1933. 'Sacha Papa. Invasora de los cultivos. Tubérculos pequeños, 2–3 cm diam.' *L. R. Parodi 10605* (WIR).

Dept. Tafí, Mollar, entre el río y el hotel. Alt. 1900 m. 26 Feb. 1949. 'En campos cultivados.' *Petersen & Hjerting P–10* (J.G.H., J.P.H.).

Dept. Tafí, Parque Aconquija, km 22. Alt. 1150 m. 13 Mar. 1949. 'En la selva grande, en tierra muy buena, en sombra.' *Petersen & Hjerting P–14* (J.G.H., J.P.H.).

Dept. Tafí, Mollar, entre los cercos de piedras. Alt. 1950 m. 25 Feb. 1949. *Petersen & Hjerting P–47* (J.G.H.).

Dept. Tafí, La Hoyada. Alt. 1500 m. 6 Jan. 1953. *Petersen & Hjerting 920* (J.G.H., J.P.H.).

Dept. Tafí, Siambón. Alt. *c.* 1300 m. 1 Feb. 1953. *Petersen & Hjerting 940* (J.G.H., J.P.H.).

Dept. Tafí, Finca Horco Molle. Alt. 600 m. Dec. 1952. (Herb. spec. made at Vandel, Denmark, 16 Aug. 1955). *Petersen & Hjerting 1889* (J.G.H., J.P.H.).

Dept. Tafí, San Javier, [Parque Aconquija]. 6 May 1949. *R. Rocha 856* (LIL).

Dept. Tafí, Siambón. Jan. 1943. *Rothe* (s.n.) (LIL 136859).

Dept. Tafí, Siambón. Tubers sent by *Schick* (s.n.). Herb. spec. from field near Leningrad, 1935 (LE, WIR 2879, 2866. Type collection of *S. schickii* Juz. & Buk.). See Plate 43.

Dept. Tafí, El Mollar. Alt. 1800 m. Jan. 1893. *Schickendantz* (s.n.) (LIL 62632).

Dept. Tafí, Parque Aconquija. Alt. 500 m. 19 May 1917. 'Flor blanca.' *Schreiter 238* (LIL).

Dept. Tafí, Yerba Buena (naranjal). Alt. 500 m. 18 Sept. 1921. 'Color de la flor blanquecino. Pl. herbácea de 40–50 cm.' *Schreiter 9964* (LIL).

Dept. Tafí, El Siambón. En la chacra del Sr. Dionisio Olea. Alt. 1300 m. 15 Jan. 1950. 'Spontanea.' *Sleumer 72* (LIL).

Dept. Tafí, Tafí del Valle. Alt. 2000 m. 13 Feb. 1949. *Sparre 5703* (LIL).

Dept. Tafí, Yerba Buena. Alt. 600 m. Feb. 1919. 'Fl. blanca; 30–60 cm. Faldas de los cerros.' *Venturi 221* (p.p.) (BA, GH, LIL, SI, US).

Dept. Tafí, camino a Taficillo. Alt. 900 m. 13 Apr. 1928. 'Fl. blanca. Falda del cerro.' *Venturi 6087* (GH, K, S, US).

Dept. Trancas, Tapia. Alt. 640 m. 10 Feb. 1959. 'Trockenwald.' *Diers 1609* (Max.).

Dept. Trancas, Estancia Riarte, Campo de Barburín. Alt. 1150 m. 28 Nov. 1948. 'En campos cultivados. Una planta con raíces violetas.' *Petersen & Hjerting P–6a* (J.G.H., J.P.H.).

Dept. Trancas, Estancia 'El Estanque'. Alt. 850 m. 30 Nov. 1948. *Petersen & Hjerting P–19* (J.G.H.).

Dept. Trancas, Tacanas, Finca Tacanas. Alt. 1350 m. 11 June 1949. 'En un campo de maíz.' *Petersen & Hjerting P–84* (J.G.H.).

Dept. Trancas, Estancia Riarte, La Higuerita. Alt. 1000 m. 13 June 1949. 'En un campo de maíz.' (Herb. spec. made at Est. Exp. 'El Suncho', Catamarca, 4 Mar. 1950.) *Petersen & Hjerting P–85* (J.G.H., J.P.H.).

Dept. Trancas, Quebrada de Tacanas, Cuevita. Alt. 1830 m. 16 June 1949. 'Papa Yuta.' (Herb. spec. made at Est. Exp. 'El Suncho', Catamarca, 4 Mar. 1950.) *Petersen & Hjerting P–86* (J.G.H.).

Dept. Trancas, Potrero Grande. Alt. 1400 m. 19 June 1949. 'Papa Yuta. En un campo de maíz.' (Herb. spec. made at Est. Exp. 'El Suncho', Catamarca, 4 Mar. 1950.) *Petersen & Hjerting P–92* (J.G.H., J.P.H.).

Dept. Trancas, San Pedro de Colalao. Alt. 1000 m. 4 Feb. 1949. *Sparre 5538* (LIL).

Prov. CATAMARCA:

Dept. Ancasti, El Potrero. 30 Mar. 1950. *J. Brizuela 1112* (C, LIL).

Dept. Ancasti, El Cajón. 18 Apr. 1950. *J. Brizuela 1251* (LIL).

Dept. Ancasti, Los Barriales. 22 Apr. 1950. *J. Brizuela 1364* (L, LIL).

Dept. Andalgalá, Quebrada del Río Pisavil. Alt. 2000 m. 26 Nov. 1948. *Filipóvich 40* (LIL).

Dept. Andalgalá, proximidades del Cerro Durazno. Jan. 1937. *M. M. Job 1345* (LP).

Dept. Andalgalá, El Candado. Alt. 2700 m. 10 Feb. 1916. 'Abundante en los cerros. Alta robusta, las flores blancas, azules o casi blancas; una papa grande, venenosa.' *P. Jörgensen 1311* (p.p.) (BA 2494).

Dept. Andalgalá, Las Mesadas. Alt. 1400 m. 13 Feb. 1949. *T. Meyer 14636* (LIL).

Dept. Andalgalá, Alto de Las Juntas. Alt. 1600 m. 14 Feb. 1949. *T. Meyer 14818* (LIL).

Dept. Andalgalá, Est. Exp. 'El Suncho', Las Mesadas. Alt. 1650 m. 4 Feb. 1949. *Petersen & Hjerting P–23* (J.G.H., J.P.H.).

Dept. Andalgalá, Estancia 'Yunka Suma'. Alt. 1450 m. 8 Jan. 1949. 'Atrás del gallinero.' *Petersen & Hjerting P–24* (J.G.H., J.P.H.).

Dept. Andalgalá, Estancia 'Yunka Suma', cerca de Ruta 65, km 61. Alt. 1400 m. 18 Jan. 1949. 'Hojas a veces muy grandes (sombra?).' *Petersen & Hjerting P–31* (J.G.H., J.P.H.).

Dept. Andalgalá, Valle del Suncho, Ruta 65, km 66·8. Alt. 1400 m. 9 June 1949. 'En un campo de maíz.' (Herb. spec. made at Est. Exp. 'El Suncho', Catamarca, 4 Mar. 1950). *Petersen & Hjerting P–82* (J.G.H., J.P.H.).

Dept. Andalgalá, Las Estancias. Alt. 1500 m. 11 Feb. 1945. *Rodriguez V. 908* (LIL).

Dept. Andalgalá, Fuerte de Andalgalá y Carrizal de Augier. Jan. and Dec. 1875. '*S. caldasii* var. *glabrescens*.' *Schickendantz 56* (CORD).

Dept. Andalgalá, Fuerte de Andalgalá. Dec. 187–. *Schickendantz 225* (CORD).

Dept. Andalgalá, Valle del Río Chacra, cerca de 'Yunka Suma'. Alt. 1600 m. 12 Feb. 1950. 'Fl. alb.' *Sleumer 16* (LIL).

Dept. Andalgalá, Estancia 'Yunka Suma'. Alt. 1350 m. 11 Feb. 1950. 'Spontanea.' *Sleumer 16 a* (LIL).

Dept. Andalgalá, Valle del Río Chacra. Alt. 1500 m. 23 Feb. 1951. 'Fl. alb.' *Sleumer 1611* (LIL).

Dept. Andalgalá, Quebrada en frente del Alto de las Juntas. Alt. *c.* 1600 m. 19 Jan. 1952. 'Fl. alb.' *Sleumer 2273* (p.p.) (LIL).

Dept. Belén, Alrededores de Belén, lomas. Alt. *c.* 1300 m. 22 Jan. 1952. 'Entre vegetación de *Dyckia velascana*, Leguminosas y *Trichocereus*.' *Sleumer & Vervoorst 2326* (LIL). (Possibly *S. kurtzianum*?)

Dept. Belén, camino entre Pozo de Piedra y El Bajo, km. 11. Alt. *c.* 1650–1700 m. 3 Feb. 1952. 'Fl. alba. Entre cercos.' *Sleumer & Vervoorst 2644* (J.G.H.). (Influenced by *S. kurtzianum*?)

Dept. Capital? (or Andalgalá), Choya. Oct. 1873. *Schickendantz 303* (GOET, SI).

Dept. El Alto, Balerzina. Alt. 1250 m. 14 Jan. 1928. 'Fl. blanca. Rastrera, borde de cerco, acequia.' *Venturi 7183* (SI).

Dept. La Paz, el Río de La Dorada. 8 Mar. 1947. *A. Brizuela 973* (LIL).

Dept. La Paz, El Moreno. 12 Apr. 1947. *A. Brizuela 1112* (LIL, TEX).

Dept. La Paz, La Huerta. 24 Apr. 1947. *A. Brizuela 1154* (LIL).

Dept. La Paz, El Potrero. 23 Apr. 1947. *A. Brizuela 1178* (LIL).
Dept. Paclín, Río Paclín. 23 Dec. 1946. *Risso 192* (LIL).
Dept. Pomán, entre El Simbolar y Los Ciénegos, donde edificio vieja escuela. Alt. *c.* 2000 m. 27 Feb. 1952. 'Faldeos secos, cerca campos cultivados, donde se observan andenes primitivos. Fl. blanco-crema.' *Vervoorst 3427* (p.p.) (J.G.H.).
Dept. Pomán, Saujil, El Potrero, quebrada Río Saujil. Alt. *c.* 1600 m. 18 Feb. 1952. 'Crece formando casi poblaciones puras en lugares húmedos, en la parte baja de la quebrada; fl. blanco-crema; se observan andenes primitivos.' *Vervoorst 3525* (J.G.H.).
Dept. Santa María, Yape. 6 Feb. 1949. *A. Reales 1630* (LIL).
Dept. unknown, [Ambato?] El Rodeo. 22 Nov. 1946. 'Flores blancas.' *Dimitri & Piccinini* (s.n.) (BAB 68337).

Prov. LA RIOJA:

Quebrada del Portezuelo, camino a la mina 'La Esperanza'. 5 Mar. 1944. 'A la sombra de *Acacia visco*, *Lithraea molleoides*, *Celtis*, *Bougainvillea*, etc. Corola blanca.' *A. T. Hunziker 5117* (LIL, SI).
Dept. Capital, 40 km antes de Patquía, en el camino de La Rioja a Chamical y en márgenes umbrosas del mismo. 11 Apr. 1955. *Ruiz Leal & Roig 18715* (*Roig 1451*) (Ruiz Leal).

Prov. MENDOZA:

Dept. San Rafael, Rama Caída. May. Tubers sent by the Argentine Ministry of Agriculture, No. 1963. Herb. spec. from field at Cambridge, 1951. CPC: 2140 (P.G.S.).

Prov. SAN LUIS:

Dept. Capital, vivero del M. A. G., como maleza. 7 May 1949. *Ruiz Leal 12423* (Ruiz Leal).
Dept. Capital, Río Trapiche. Alt. 1150 m. 3–4 Mar. 1957. 'Fl. blanca.' Leg. F. Roig 1338, *Ruiz Leal 19069* (Ruiz Leal).
Dept. Chacabuco or Junín, Larco to Merlo. 14 Nov. 1925. *Castellanos* (s.n.) (BA 25/2658).
Dept. General Pedernera, Sierra del Morro. 1913. *Franco Pastore 27* (SI).
Dept. Junín, Merlo, Piedra Blanca. Alt. 1000 m. 5 Feb. 1947. 'Corola blanca.' *Digilio & Grassi 2061* (LIL).
Dept. Junín, Bajo de los Vélez. 15 Feb. 1942. 'Flor blanca. Alt. 40 cm, terreno arenoso.' *R. Maldonado 1189* (LIL, LP).
Dept. Pringles, Pancanta. 21 Feb. 1925. *Castellanos* (s.n.) (BA 25/955).
Dept. Pringles, Trapiche. Feb. 1931. *Gez* (s.n.) (BA 31/295).
Dept. San Martín, Sierras de San Luis, Peñón Colorado. 20 Dec. 1929. *Castellanos* (s.n.) (BA 29/421).

Prov. CÓRDOBA:

Dept. Calamuchita (?), Valle de los Reartes, márgen derecha del Río de Los Molinos. 30 Mar. 1917. 'Fl. blanca; hab. terrenos arenosos o húmedos.' *Castellanos 112* (SI).
Dept. Calamuchita (?), Valle de los Reartes. Feb. 1920. *Castellanos* (s.n.) (LIL 36994).
Dept. Capital, en las cercanías de Córdoba. 1 Feb. 1878. *Galander* (s.n.) (G).
Dept. Capital, near City of Córdoba. *G. García* (s.n.) (WIR No. K. 883, tubers sent by Millán as No. 1201; type collection for *S. garciae* Juz. & Buk.). See Plate 41.
Dept. Capital, Altos de Córdoba. Leg. Bockubender, *Hicken 95* (SI).
Dept. Capital (?), Córdoba. Dec. 1887. *C. Spegazzini* (s.n.) (LPS 14212).

Dept. Capital, Puesto del Paraíso. 30 Oct. 1897. '*S. caldasii* Kth. var. *glabrescens*.' *Stuckert 3505* (CORD, G).

Dept. Capital, prope Córdoba. 30 Nov. 1897. *Stuckert 4004* (8546) (CORD, G, LIL).

Dept. Capital (?), Córdoba. 31 Mar. 1900. *Stuckert 9119* (CORD, G).

Dept. Colón, Unquillo. 1926. *Bruch* (s.n.) (LP 010990).

Dept. Colón, Ascochinga. 31 Dec. 1935. 'Flor blanca.' *Giardelli 323* (SI).

Dept. Colón, Ascochinga. 1 Apr. 1939. 'Hab. monte, escasa. Fl. blanca. Papas.' *Giardelli 1169* (SI).

Dept. Colón, Casa Bamba. 27 Nov. 1898. *Stuckert 5176* (8547) (CORD, G, LIL).

Dept. Ischilín, Sierra Copacabana, falda occidental, Morteros. 25 Jan. 1952. 'Corola blanca. Frecuente.' *A. T. Hunziker 9731* (CORD).

Dept. Juárez Celman, cerca de Pacheco de Melo, yendo desde El Rastreador, camino a Laboulaye. 1 Nov. 1956. 'En un médano fijo, poco frecuente.' *A. T. Hunziker 12770* (CORD).

Dept. Minas, cerca de San Carlos Minas, Río Jaime. 18 Feb. 1952. *A. T. Hunziker 9852* (CORD).

Dept. Punilla (?), Dique. Dec. 1890 (and 1891). '*S. caldasii* var. *glabrescens*.' *O. Kuntze* (s.n.) (LP, NY, US).

Dept. Punilla, Thea. 28 Jan. 1947. 'Flor blanca.' *Villafañe* 609 (LIL).

Dept. Río Cuarto, Río Cuarto. Leg. Maradona, *Hicken 86* (SI).

Dept. Río Cuarto, Río Cuarto. 28 Dec. 1898. *Stuckert 5803* (8548) (CORD, G, LIL).

Dept. Río Primero, Río Primero. 10 Dec. 1946. *Balegno 1023* (LIL).

Dept. Río Seco, Gutenberg. 24 Mar.–1 Apr. 1943. *H. H. Bartlett 19870* (GH, SI, UC).

Dept. Río Segundo, Manfredi. 1935. Tubers sent to Leningrad by R. Millán. *Millán G–1183* (WIR No. 2877, J.G.H., Type collection for *S. gibberulosum* Juz. & Buk.) (probably = CPC: 1157). See Plate 40.

Dept. San Javier, Yacanto. 5 Dec. 1927. *Castellanos* (s.n.) (BA 11692).

Dept. San Javier, San Javier. 3 Feb. 1939. *Castellanos* (s.n.) (BA 31225).

Dept. San Javier, Yacanto. Mar. 1921. *J. F. Molfino* (s.n.) (BA 2499).

Dept. San Javier, San Javier. 8 May 1949. *Ruiz Leal 12453* (Ruiz Leal).

Dept. Santa María, Alta Gracia, a orillas del arroyo Alta Gracia. 14 Jan. 1940. '130 cm de altura.' *A. T. Hunziker 618* (LIL).

Dept. Santa María, Alta Gracia. Alt. 600 m. 7 Dec. 1945. 'Geófita, tuberosa. Corola blanca. Rara. Borde camino.' *J. H. Hunziker 1296* (CORD, LIL, SI).

Dept. Santa María (?), camino de Alta Gracia a Córdoba. 29 Nov. 1926. 'Maleza de un trigal. Muy abundante.' *Millán 626* (BA).

Dept. Santa María, Alta Gracia. 26 Nov. 1926. 'En una zanja.' *Millán 637* (BA).

Dept. Santa María, La Cumbrecita. Alt. 1500–1600 m. 18–27 Dec. 1955. 'Muy común. Fl. blancas, subrotáceas. Pl. de 30–50 cm de altura.' Leg. F. Roig 1034, *Ruiz Leal 17765* (Ruiz Leal).

Dept. Santa María, La Paisanita. 8 Dec. 1950. *A. de la Sota 3027* (LIL).

Dept. Sobremonte, San Francisco del Chañar. 8 Dec. 1944. *Balegno 342* (LIL).

Dept. Tulumba, Cerro Sauce Punco. 2 Nov. 1945. *Cuezzo 773* (LIL).

Dept. Unión, Bell Ville. From Argentine Ministry of Agriculture, as No. 2471. Herb. spec. from field at Cambridge, 1951. CPC: 2144 (P.G.S., J.G.H.).

Dept. unknown, Sierra de Achala. 15 Dec. 1901. *Stuckert 10557* (CORD, G).

Locality unknown. CPC: 1157 (K, P.G.S.).

Prov. LA PAMPA:

La Maruja, FCO. 11 Nov. 1939. 'Vías. Flor blanca. Con papas.' *Burkart 9886* (SI).

Prov. BUENOS AIRES AND CAPITAL FEDERAL:

Bella Vista. 17 Mar. 1945. *R. Alvarez 653* (LIL) (atypical).

Buenos Aires. 1831. *Bâcle 117* (G, Z). (Type for *S. tuberosum* L. var. *glabriusculum* Dun.).

Alrededores de la Capital Federal, Palermo. 13 Feb. 1901. *Bettfreund* (s.n.) (BA 2523).

Ernestina. 27 Mar. 1938. *M. Isabel H. Scott de Birabén & Max Birabén 808* (LP).

Playa de Los Talas. Dec. 1944. *Boffa* (s.n.) (LIL 120411).

Martínez F.C.C.A. 20 Jan. 1939. 'Junto a la vía férrea.' *Bridarolli 355* (LP).

Delta del Paraná, Caraguatá. 8 Mar. 1935. 'En albardón alto, sombra.' *Burkart 7001* (GH, SI).

Alrededores de La Plata, Las Talas. 26 Mar. 1933. 'Terreno calcáreo.' *Cabrera 2782* (LP, NY).

Partido de La Plata, Villa Elisa, terraplén del ferrocarril. 9 Mar. 1940. *Cabrera 6309* (LP). (Articulation very low.)

Bolívar. 2 Apr. 1952. *C. A. Campos* (s.n.) (BAB 77393).

Alrededores de Buenos Aires. Apr. 1907. 'Campo.' *Hauman* (s.n.) (SI 15947).

Atalaya, Magdalena. 20–28 Feb. 1906. 'Monte de conchillas.' *Hauman* (s.n.) (SI 15948).

Río de la Plata, orilla, cerca de Atalaya. Feb. 1906. 'Tubercules de 5 cm diam.' *Hauman* (s.n.) (BRLU).

Alrededores de Buenos Aires, Chacarita. Apr. 1907. *Hauman* (s.n.) (BRLU).

Km 70 del F.C.C.C. 3 Feb. 1922. 'Muy abundante sobre el terraplén.' *Hicken* (s.n.) (SI 15941).

Buenos Aires. 'Horto paterno (non culta).' *Hicken* (s.n.) (SI 15954).

In campis prov. Buenos Aires. 1885. *Albert Lefebre* (s.n.) (BR).

Buenos Aires? Herb. *John Miers* (s.n.) (BM).

Near Buenos Aires (Type sheet says 'Uruguay'). 1934. *Millán G–972* (WIR No. K. 597: Type for *S. laplaticum* Buk.). See Plate 39.

Villa Ortúzar. 15 Nov. 1935; 10 Mar. 1939. *Ratera 2*; *38* (BAA, BAL).

Villa Ortúzar. 20 Apr. 1943. 'Papa cimarrona. Abundante entre las vías del Ferrocarril Central Bs. As. (Estaciones Arata y Escalantes).' *Ratera 1200* (BA).

Rincón de Viedma (Bahía de Samborombón). Jan. 1934. 'A orilla del camino de conchilla, casi abundante. Fl. blanca.' *Ringuelet 364* (LP, SI).

Buenos Aires. 1837. 'Papa Amarga of Buenos Aires. Roots large but poisonous.' *Tweedie* (s.n.) (K, possibly type for *S. commersonii* var. l. *glabriusculum* Hook. f. p. pte.).

Barracas al Sud. Feb. 1902. *Venturi 79* (BA).

Barracas al Sud. 12 Mar. and 3 May 1902. *Venturi 152* (CORD, LIL, S, SI).

Ditch sides, Buenos Aires. Jan. 1834. Collector unknown (CGE, TCD).

Buenos Aires. Mar. 1845. Collector unknown, *117* (FI).

Pergamino. From Sr. A. Castronovo, Castelar, as No. Hort. 1946. Herb. spec. from field, Cambridge, 1951. CPC: 2062 (P.G.S., J.G.H.).

Paraná, delta. Aug. From the Argentine Ministry of Agriculture as No. 1944. Herb. spec. from field, Cambridge, 1951. CPC: 2132 (P.G.S., J.G.H.).

Llavallol. Dec. From the Argentine Ministry of Agriculture as No. 1952. Herb. spec. from field, Cambridge, 1951. CPC: 2135 (J.G.H.).

Llavallol. Dec. From the Argentine Ministry of Agriculture as No. 1965. Herb. spec. from field, Cambridge, 1951. CPC: 2142 (J.G.H.).

Prov. ENTRE RÍOS:

Dept. Diamante, Diamante, en barranca. 16 May 1964. 'Flor blanca.' *Burkart, Troncoso, Bacigalupo & Crespo 25456* (SI).

Prov. SANTA FÉ:

Colonia Esperanza. 1896. 'Wild potato, grows in the fields. Flowers in November–January. Black humus soil.' *Lindner* (s.n.) (JE—herb. Haussknecht).

Kilómetro 140. F.C.N.A. Argentine Ministry of Agriculture No. 1960 (CPC: 2139).

Casilda, Escuela de Agricultura. 11 Feb. 1943. 'En abundancia en cultivo de maíz.' *Ratera 959* (BAA).

Prov. CORRIENTES:

Dept. General Paz, Paraje Angostura. 29 Nov. 1945. 'Hierba anual de 70 cm. Flores blancas, vegeta dentro monte.' *Ybarrola 3789* (LIL).

Prov. MISIONES:

Dept. Candelaria, Colonia Finlandesa. 21 May 1928. 'Campos.' *Lunnasvaana* (s.n.) (H).

Dept. Candelaria, Colonia Mártires, lote 40, chacra de Carl Davidsen. Alt. 125 m. 15 Apr. 1949. 'En selva virgen, zona pedregosa, húmeda.' *Petersen & Hjerting P–65* (C, J.G.H., J.P.H.).

Dept. Capital, Posadas. Picada. 12 Apr. 1945. 'Papa silvestre.' *Bertoni 1198* (LIL).

Dept. Capital, Posadas, Zaiman. 16 Apr. 1945. *Bertoni 1205* (LIL).

Dept. Frontera, Piñalito. 10 Feb. 1947. *T. Meyer 11951* (LIL).

Dept. Guaraní, Ruta 14, km 304, cerca de Arroyo Chafariz. Alt. 420 m. 2 Mar. 1950. 'Pl. an alt. 30–70 cm. Fl. bl. Hab. monte ralo, lugar bajo húmedo, muy escasa. "Papa del Monte."' *Schwindt 3255* (LIL).

Dept. Iguazú, Puerto Segundo, km 17. 23 Oct. 1950. *J. E. Montes 9510* (C, LIL).

Dept. Iguazú, Eldorado, centro. Alt. 160 m. 12 Oct. 1949. 'Pl. an herb. alt. 70 cm. Fl. blancas. Hab. monte ralo.' *Schwindt 2802* (LIL).

Dept. San Ignacio, San Ignacio. 1 Apr. 1914. *Horacio Quiroga 752* (BA).

Dept. San Javier, Panambí, Río Uruguay. Alt. 100 m. May 1944. *Sandeman 4826* (K).

Dept. San Javier, Colonia Oberá. Alt. 310 m. 22 July 1950. 'Papa del Monte; herba an alt. 30–50 cm. Fl. blancas. Hab. desmonte, lugar bajo. Muy escasa.' *Schwindt 4838* (p.p.) (LIL).

Dept. San Pedro, Santa Teresa. Alt. 160 m. 10 June 1949. 'Pl. herb. 45 cm alt. Fl. blancas. Hab. abierta, monte, lugar bajo. Muy escasa.' *Schwindt 1865* (LIL).

Dept. San Pedro, cercanías del Piray-Guazú. 21 July 1949. 'Pl. herbácea, 50–80 cm alto. Fl. blanca, estambres amarillas, baya verde crema. Hab. lugar bajo, escasa. "Papa de Monte."' *Schwindt 2011* (LIL).

Dept. San Pedro, San Pedro a Tobunas, Ruta 14. Alt. 570 m. 25 Apr. 1950. 'Pl. h. an 40–70 cm alto. Fl. bl. Hab. monte despejado, desmonte, lug. bajo, muy escasa. "Papa de Bugre."' *Schwindt 4069* (LIL).

Prov. CHACO:

Dept. Martínez de Hoz, Ruta 16, entre Col. Elisa y Pres. de la Plaza, *c.* km 97. Alt. *c.* 75 m. 3 May 1949. 'En un campo de algodón, tierra negra, arcillosa, semi-húmeda.' *Petersen & Hjerting P–77* (C, J.G.H., J.P.H.).

Dept. Napalpí, Campo Largo, la chacra de Jeppesen. Alt. *c.* 100 m. 24 Oct. 1949. (Herb. spec. made at Est. Exp. 'El Suncho', Catamarca, 4 Mar. 1950, tubers sent from

Chaco.) *Petersen & Hjerting P–78 a* (CPC: 2499) (J.G.H.).

Dept. Napalpí, Campo Largo, los terrenos de la represa. Alt. *c*. 100 m. 6 May 1949. 'Entre pasto, tierra arcillo-limosa. Planta hasta 1 m de altura. Flor blanca.' *Petersen & Hjerting P–79* (J.G.H., J.P.H.).

Dept. Resistencia, Resistencia. May 1883. *C. von Gülich* (s.n.) (LP 14230 p.p.).

Dept. Resistencia, Fontana. June 1937. 'Crece en terrenos humíferos en el interior de los bosques altos.' *T. Meyer 2323* (p.p.) (GH, LIL).

Dept. Resistencia, Colonia Benítez, la chacra de G. Riquelme. Alt. *c*. 50 m. 2 May 1949. 'Tallo hasta 150 cm. Tubérculos hasta 3×5–6 cm. Frutos hasta 2 cm en diam.' *Petersen & Hjerting P–76* (C, J.G.H., J.P.H.).

Locality as above. *Schulz 9027*. Living material at Birmingham (H. 86, 87).

Dept. Resistencia, Colonia Benítez. Feb. 1931. 'Fl. blanca. Hierba de 15–25 cm. Montes, orillas boscosos ríos. Abundante. Da tubérculos, aunque pequeños.' *Schulz 84* (BA).

Dept. Resistencia, Colonia Benítez. 2 Sept. 1909. *Stuckert 20264* (G).

Dept. Río Teuco, Nueva Pompeya. Apr. 1918. 'Común en el campo, 50 cm alto, robusta, extendida. Flor blanca.' *P. Jörgensen 2828* (BA, BAF, MVM, SI).

Dept. Tapenagá, Ruta k, entre Villa Angela y Samuhí. Alt. *c*. 100 m. 11 Apr. 1949. 'Entre yerbas naturales; tubérculos hasta 3×4 cm, parejas.' *Petersen & Hjerting P–61* (J.G.H., J.P.H.).

Dept. Tapenagá, Villa Berthet, chacra de J. Wammen. Alt. 65 m. 12 Apr. 1949. 'En un campo de algodón; tubérculos hasta 3×5 cm.' *Petersen & Hjerting P–62* (CPC: 2492) (J.G.H., J.P.H.).

Dept. Tobas, Fortín Nuevo, Río Bermejo. 4 Apr. 1912. *Rodríguez* (s.n.) (BA 2248).

Prov. FORMOSA:

Dept. Formosa?, Guaiculé. Mar. 1919. *P. Jörgensen 3339* (BA, GH, US).

Dept. Pilcomayo, sur de Puesto Porteño a 2 km. 24 Apr. 1947. *I. Morel 2671* (LIL).

Dept. Pilcomayo, norte de Puesto Porteño a 3 km. 26 Apr. 1947. 'Planta de tierra húmeda removida y arenosa. Crece en las chacras.' *I. Morel 2696* (LIL).

Dept. Pilcomayo, Riacho Negro. 14 May 1947. 'Planta de campo; habita en tierra húmeda.' *I. Morel 2800* (LIL).

Dept. Pilcomayo, Porteñito. 27 May 1947. 'Planta de campo, habita en tierra linda y húmeda.' *I. Morel 2886* (LIL).

Dept. Pilcomayo, Porteñito al N. a 2 km. 17 June 1947. 'Planta de campo, habita en tierra arenosa.' *I. Morel 3112* (LIL).

Dept. Pilcomayo, norte a 2 km de Riacho Negro. 11 July 1947. *I. Morel 3357* (LIL).

Dept. Pilcomayo, sur a 3 km de Riacho Negro. 4 Oct. 1947. *I. Morel 3806* (LIL).

Dept. Pilcomayo, sur a 2 km de Puente Angelito. 22 June 1949. 'Planta de orilla y de monte, habita en terreno esponjoso.' *I. Morel 7859* (LIL).

Dept. Pilcomayo, N.E. a 2 km de Puente Angelito. 24 June 1949. 'Planta de orilla de monte.' *I. Morel 7939* (C, LIL).

Prov. SANTIAGO DEL ESTERO:

Dept. Guasayán. 18 May 1940. *Argañarás 253* (BA).

Dept. Guasayán, El Cevilar. *Pierotti* (s.n.) (LIL 106118, US).

Dept. La Banda, Estación Clodomira, a los 8 km dirección N.O. 1–15 Apr. 1944. 'Flores color blanco.' *P. García* (s.n.) (LIL 109666).

Dept. La Banda, La Banda. Apr. 1907. *Lillo 6203* (LIL).

Dept. La Banda, el camino entre La Banda y Beltrán, km 1019. Alt. 200 m. 10 Apr. 1949. *Petersen & Hjerting P–60* (J.G.H.).

Dept. Ojo de Agua, Ojo de Agua. 15 Apr. 1945. 'Flor blanca.' *P. García 873* (LIL).

Dept. Pellegrini, Estancia 'El Remate'. Alt. 500 m. 21 Feb. 1928. 'Fl. blanca; 80 cm, en el campo.' *Venturi 5851* (LIL, US).

Dept. Robles, camino nacional entre Forres y Beltrán. 28 Apr. 1942. 'Flores blancas.' *Argañarás 384* (LIL).

Dept. Robles, Beltrán. 17 July 1940. 'A orillas de cultivos, terreno arenoso. Papa Silvestre. Flor blanca, estambres amarillas.' *R. Maldonado 393* (LP).

ARGENTINA, exact locality unknown:

P. T. Knappe (s.n.) Tubers sent to Leningrad by P. T. Knappe, as *S. caldasii* var· *glabrescens*. (LE, Type for *S. emmeae* Juz. et Buk; P.G.S. No. 01316, herb. spec. from field, Cambridge, 1951.) See Plate 47.

Río de la Plata. Tubers sent from Argentine Ministry of Agriculture 1945. CPC: 1720 ($2n = 36$—no herb. spec.).

Prov. Buenos Aires? L[a] P[lata]. Summer 1886. Collector unknown (LPS 14214).

Material grown in Leningrad, 1938, and stated by Bukasov to be genuine *S. knappei* (P.G.S. 0013).

Material grown in Leningrad, 1938, and stated by Bukasov to be genuine *S. gibberulosum* (P.G.S. 0016).

Material sent by Bukasov under the name of *S. gibberulosum*. (K, P.G.S.: CPC 1157).

Material grown in Leningrad, 1938, and stated by Bukasov to be genuine *S. parodii* (P.G.S. 0019).

Material sent by Bukasov under the name of *S. schickii* (P.G.S.: CPC 1155).

Material sent by Bukasov under the name of *S. garciae* (P.G.S.: CPC 1158).

Material given by Dr. R. N. Salaman under the name of *S. caldasii* var. *glabrescens* (P.G.S.: CPC 1315, 1316).

Material given by Dr. R. N. Salaman under the name of *S. chacoense* (P.G.S.: CPC 1317, 1318).

2. Subspecies *muelleri*

BRAZIL

Estado do RIO GRANDE DO SUL:

Municipio de Sarandí, ad flumen Alto Uruguay. 23 Feb.–11 Mar. 1945. 'In agris.' *B. Rambo 28190* (LIL).

(Exact locality unknown) *Fritz Müller* (s.n.) (W–Type, WU; grown in Vienna Botanic Gardens, 7 Oct. 1869, from seeds gathered by Fritz Müller and sent to Professor Hildebrand). See Fig. 18.

Estado de SANTA CATARINA:

Blumenau. 2 Nov. 1886. 'Wilde Form von Kartoffel?' *Schenck 947* (GH—photograph of destroyed Berlin specimen).

Mun. Chapecó, Campo, bog, Fazenda Campo São Vicente, 24 km west of Campo Erê. Alt. 900–1000 m. 26–28 Dec. 1956. *Smith, Reitz & Caldato 9516* (US–Type for *S. muelleri* f. *densipilosum* Corr.). See Plate 49.

An Uferabhangen des Itajahy bei Blumenau. May 1888. *E. Ule 696* (CORD, US, WU).

(Exact locality unknown) June [18]68. *Fritz Müller 154* (K).

Estado do PARANÁ:

Jacarezinho River. Seeds sent by *Dr. Olavo J. Boock* (s.n.), Oct. 1959. (Accession H. 564.)

Desvio Ribas Turma. Alt. 800 m. Oct. 1914. *Dusén 23* (S).

Ad viam ferream inter Perdices et Río des Pedros. 21 Feb. 1916. *Dusén 17725* (S).
Villa Velha. Nov. 1914. 'In margine silvae primaevae.' *Gösta (Jönsson) Ilien* (s.n.) (LD, UPS).
(Brazil, but exact locality and State unknown) *Glaziou* (s.n.) (P).

ARGENTINA

Prov. MISIONES:

Dept. Frontera, Ruta 101, km 491, cerca de Deseado. Alt. 350 m. 18 Apr. 1949. 'Tallo hasta 1 m; entre yerba.' *Petersen & Hjerting P-66* (CPC: 2426) (C, J.G.H., J.P.H.). See Plate 48.

Locality unknown:

'*S. suaveolens* Kth. et B. Mexico.[1] Hort Kew, July 1874' (K, type for *S. jamesii* Torr. var. *grandifrons* Bitt.).

DISTRIBUTION

1. Subsp. **chacoense** is widely distributed in the plains and low mountain ranges of central and northern Argentina, south-eastern Paraguay and southern Uruguay (see Map 5). In Argentina it extends for about 1500 km in a north–south direction, and records from Bolivia indicate at least another 300 km spread into that country also. Its east–west distribution is equally extensive, about 1300 km, from the foothills of the Andes to the Brazilian border. It is thus by far the most widespread species in South America (apart perhaps from *S. acaule*), and may be compared in this respect with *S. fendleri* from the northern hemisphere.

The exact distribution is as follows:

ARGENTINA: Provs. Buenos Aires (and Capital Federal), Catamarca, Chaco, Córdoba, Corrientes, Entre Ríos, Formosa, Jujuy, La Pampa, La Rioja, Mendoza, Misiones, Salta, San Luis, Santa Fé, Santiago del Estero, Tucumán.

PARAGUAY: apparently mainly confined to the region lying south-east of Asunción and perhaps rare, though few collections of wild potatoes have been made from that country.

URUGUAY: Dept. Soriano, apparently very rare.

The largest number of collections has been made in the Argentine provinces of Jujuy, Salta, Tucumán, Catamarca, Córdoba, and Buenos Aires. Whilst these regions are populated areas with good communications and hence easy to collect in, it would not be correct to say that the apparent frequency of subsp. *chacoense* in them was due only to the greater ease with which it could be collected. It seems in fact to be more frequent there than in the drier central regions of Argentina or in Uruguay and Paraguay.

2. Subsp. **muelleri** is much less common than subsp. *chacoense*, and has been recorded only from Brazil, States of Rio Grande do Sul, Paraná, and Santa Catarina, and from Argentina, prov. Misiones (see Map 4).

HABITAT

S. chacoense seems to be by far the most vigorous and adaptable of all the wild species in the region under discussion. Its altitude range is very extensive, from sea level to 2000 m and occasionally to 2350 m or even higher. It would appear that, although it was primitively a species of the plains and low foothills (to 1000 m), it acquired a secondary adaptation to higher altitudes by means of gene introgression from certain mountain species (see pp. 212–19).

[1] Both name and locality are obviously erroneous.

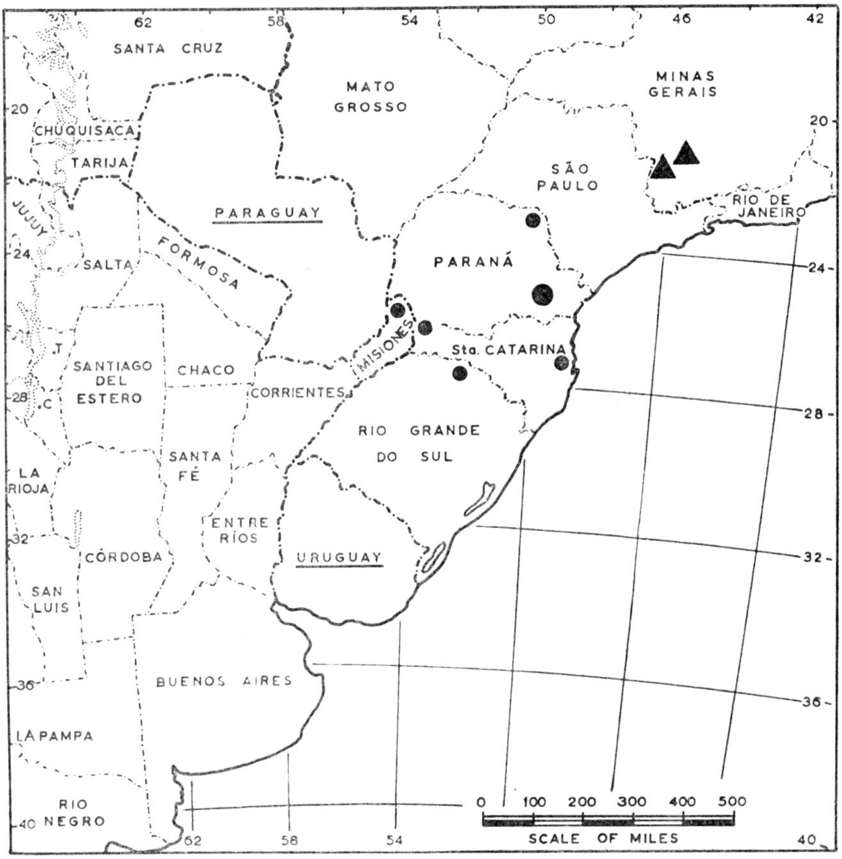

▲ *Solanum calvescens* Bitt.

● *Solanum chacoense* Bitt. subsp. *muelleri* (Bitt.) Hawkes & Hjerting

MAP 4. Distribution of *S. calvescens* Bitt. and *S. chacoense* Bitt. subsp. *muelleri* (Bitt.) Hawkes & Hjerting.

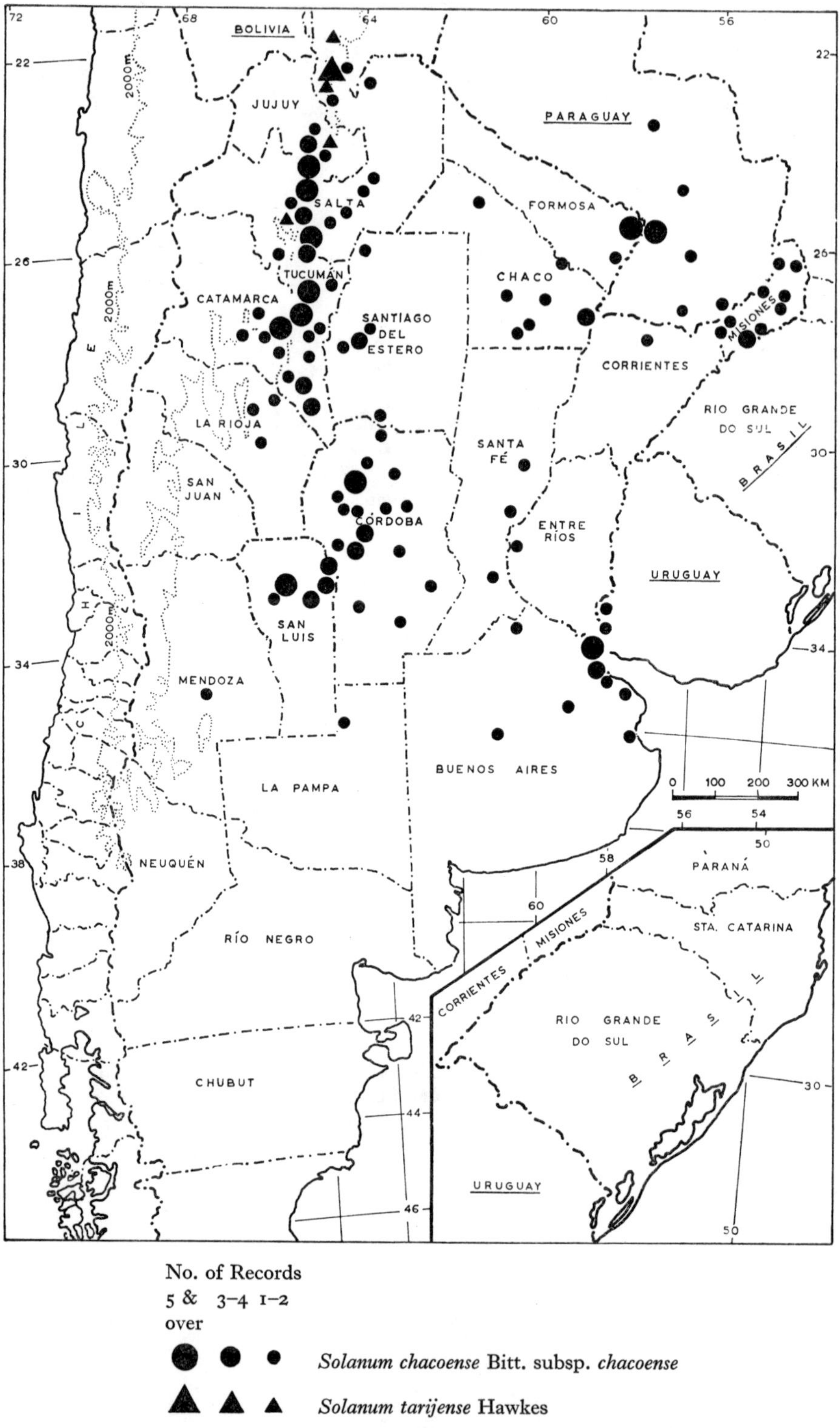

MAP 5. Distribution of *S. chacoense* Bitt. subsp. *chacoense* and *S. tarijense* Hawkes.

Flowering times in subsp. *chacoense* vary enormously. In the plains it has generally been collected in flower in February to October, with a peak in April, and another small peak in October. In the mountains, collections of flowering material have been made in November to April, with a peak in February. For the country as a whole, collections have been made in every month, but very few in July to September.

Subsp. *muelleri* has been obtained in February to May and again in October and November.

1. Subsp. **chacoense** is the most widespread of the two subspecies and occurs in a very large range of habitats, as the following list, composed from actual field records, makes clear:

Woods, forests, and forest borders, shade of *Alnus*, *Podocarpus*, *Acacia*, *Lithraea*, *Celtis*, *Bougainvillea*; rough herbage; natural herbage; grassy fields; abandoned cultivations; as a common field weed in cotton, maize, wheat, and sugar-cane fields; by river banks, marshes, drainage ditches; on dry hillsides, pathsides, roadsides, railway embankments, stone walls. Found both in shade and full sun. Apparently has no soil preference, since it has been recorded for damp black humiferous soil and for clayey, sandy, stony, and calcareous soils.

From a phytogeographical point of view, the range of subsp. *chacoense* is also remarkably wide, and it occurs quite commonly in at least four of the main phytogeographical provinces recognized by Cabrera (1953). The following list gives some details of its distribution.

(*a*) *Western subtropical province*. Here subsp. *chacoense* ascends from the Chaco phytogeographical province into the western subtropical woodlands of Tucumán, Salta, and Jujuy. It is sparsely distributed in the lower Tucumán and Orán districts, (as a rule in open places or in cultivations), but is much more commonly found in the Montano district ((1200–)1500 m and over). Probably most of these mountain forms possess genes derived by introgression from other species (e.g. *S. microdontum*) and are not typical of the species as a whole (see later, pp. 212–19). They are rarely found in dense forest, generally occurring in open places by streams, paths, forest clearings, and cultivated fields.

(*b*) *Eastern subtropical province*. Occurs chiefly in the 'Mixed Forest' (Selvas Mixtas) district of Misiones and south-east Paraguay; possibly grows also in the gallery woods (galería paranense y uruguayense) of the Paraná and Uruguay rivers, in the Field or Savanna district ('Distrito de los Campos') and in the Araucaria Forest district (Distrito de los Pinares) of eastern Misiones. In this province, although it is chiefly found in forest clearings, roadsides, cultivated and grass fields, etc., in a few localities it occurs as a component of natural forest vegetation.

(*c*) *Chaco province*. As the Latin name implies this species was first found in the Chaco and was at one time thought to be confined to it. Subsp. *chacoense* is chiefly found on the eastern and western edges of the Chaco where the rainfall is higher. In the drier central region it is confined probably entirely to irrigated fields, ditches, waysides, etc. In the Eastern Chaco district it is found in south-east Paraguay, and the Argentine provinces of Formosa and Chaco, grading into the eastern subtropical woodlands. In the west and south-west part of the Chaco province it is found fairly abundantly in the low foothills of the Chaco Serrano district in Salta, Jujuy, Tucumán, Catamarca, Córdoba, La Rioja, and San Luis. From the northern of these provinces it spreads upwards in places into the western subtropical woodlands, as we have already indicated.

(*d*) *Espinal province*. In this province subsp. *chacoense* is found predominantly at the base of the Córdoba sierra and in San Luis. It does not seem to extend into the Ñandubay district of northern Entre Ríos and Corrientes, but may perhaps have been previously overlooked. It is, however, quite frequent in the Tala district of the La

Plata coast of Buenos Aires province. In the rest of the Espinal province it is very rare and is possibly confined chiefly to river banks, irrigated fields, and a few natural habitats. Unfortunately, few of the collections from this province bear adequate habitat data.

(*e*) *Prepuna province*. As with the preceding, it would seem that subsp. *chacoense* is not really native to this province, but occurs only as a weed of irrigated soils and river banks, chiefly in prov. Catamarca. However, it is possible that some introgressed forms in this area may be able to survive in the natural vegetation, especially in the wetter transition areas between this and the western subtropical forest zone.

(*f*) *Monte province*. We have only a few gatherings of subsp. *chacoense* from this dry, rather warm province and it is doubtful whether the species is native here. These few records are from provs. Mendoza, La Rioja, and Catamarca, probably of adventive plants growing in irrigated fields or ditches. Nevertheless, some of the records for prov. Catamarca seem to indicate that the species may form a component of the natural vegetation in parts of that province.

(*g*) *Pampa province*. Subsp. *chacoense* apparently occurs only very sparsely in this province; it has been seen by us only from a few places in Buenos Aires, one in Córdoba and one in La Pampa. Here again it would seem to be an adventive or ruderal plant which does not form part of the natural vegetation. (See also Ratera 1942.)

To summarize the above information on the habitat range of subsp. *chacoense* we can say that it occurs abundantly in and appears to be native to the eastern and western borders of the Chaco, in the lower reaches of the western subtropical forests and the western portions of the eastern subtropical forests. It is also quite frequent on the coastal Tala belt of the Espinal province. Elsewhere, however, where the rainfall is lower and the normal vegetation either xerophytic or composed of dense grasslands, subsp. *chacoense* can survive only in subclimax vegetation on river banks, waysides, drainage ditches, irrigated fields, and the like. Its distribution must therefore have once been much more limited than it is today.

2. Subsp. ***muelleri*** is much more restricted in its range, both geographically and ecologically, than subsp. *chacoense*.

It has been recorded as growing among grass and other herbage, on river banks and railway tracks, in fields, and on forest margins and clearings, from 0 to 350 m altitude. It seems to prefer damp places with a high rainfall and occurs only in Cabrera's eastern subtropical forest province. Here it is found both in the 'Mixed Forest' district of Misiones and Sta Catarina state, and in the Araucaria Forest district of Rio Grande and Paraná (possibly also Misiones). There is no direct evidence to suggest that it grows in the tropical forest coast belt of Sta Catarina and Paraná, but nevertheless the recorded occurrences cannot be far from the boundary of this vegetational zone. Further information by collectors in southern Brazil would be of the greatest value in tracing the exact boundaries and ecological preferences of this little-known subspecies.

TAXONOMIC NOTES

1. Subspecies *chacoense*

S. chacoense was described in 1911 by Hassler under the name *S. guaraniticum*. This was unfortunately a homonym, since the epithet *guaraniticum* had already been used by St. Hilaire in 1825. The mistake was pointed out by G. Bitter in 1912 who put forward the name *S. chacoense* as a replacement. A further publication by Hassler later in the year proposed that the name *S. bitteri* should be used instead of *S. chacoense*, since he himself had used the latter name some years back for a subspecies of *S. nigrum*. Subsp. *chacoense* of *S. nigrum* was raised to specific rank by Bitter in another publication, also in 1912, but given the name *S. pulchrilobum*; Hassler concluded that the epithet *chacoense* should be kept for this latter taxon when raised to specific rank. This is of course not necessary under the present International Code of Nomenclature (1961), and Bitter was quite correct in retaining the name *S. chacoense* for Hassler's original wild potatoes described first as *S. guaraniticum*.

Despite this superficially correct solution of the nomenclatural problem it remains an unfortunate fact that the material was collected and the descriptions made of this species by Hassler, though the finally correct name was applied by Bitter who had apparently never seen Hassler's type material. The type specimens for Hassler's varieties and forms are in Herb. Geneva, still with his original labels '*S. guaraniticum* Hassl.' and with no comment or name alteration by Bitter. The only type specimen that we know that Bitter had seen at this period was a Berlin sheet of *Hassler 2805* (Field Museum negative 2749), now destroyed; in no other European or American herbarium are any of Hassler's specimens of *S. chacoense* to be found with Bitter's annotations, and this lack of knowledge of the variation within *S. chacoense* no doubt induced Bitter to describe another form of it from Tucumán as a distinct species, *S. subtilius*. We shall return to this later.

Hassler later (1917) accepted Bitter's use of the name *S. chacoense*, publishing some notes on the presence of calcium oxalate in the calyx of that species, in which the varieties *angustisectum* and *latisectum* and the forms *glabrescens* and *pilosulum* occur, all linked with the name *S. chacoense* Bitt. Hassler also mentions a third form of var. *latisectum*, f. *plurijugum*, but no description is given. This form was based on *Hassler 11849*, and must in our opinion also be assigned to *S. chacoense*, though some specimens in the collection do not seem to be very typical.

Hassler's description of *S. chacoense* (*S. guaraniticum*) in 1911 was based on four specimens from the Chaco region of south Paraguay which he placed as the following varieties and forms:

var. *angustisectum* (*Hassler Nos. 470* and *2805*). See Fig. 15 and Plate 35.
var. *latisectum* f. *pilosulum* (*No. 3674*). See Plate 34.
var. *latisectum* f. *glabrescens* (*No. 326*). See Plate 36.

Correll (1962) suggested *Hassler 2805* as the lectotype for *S. chacoense* and we should here like to designate one of the *Geneva* sheets of this number as the LECTOTYPE for *S. chacoense*. Many other gatherings of the species were also made by Hassler, but apparently subsequent to the publication of his descriptions. Although var. *angustisectum* agrees well with the description and with the generally accepted connotation of this species, var. *latisectum* f. *pilosulum*, on the other hand, does not. It is, in fact, a form of *S. commersonii* subsp. *malmeanum*, as we have already seen, agreeing with it in the short semi-rosette habit, shortly petiolulate sub-decurrent leaflets and rather broad terminal, as well as the reduced inflorescence and long, broadly triangular calyx acumens ($\pm$1 mm) which are not to be found on *S. chacoense* except in the introgressed mountain forms in the north-western part of Argentina and in subsp. *muelleri*. The long thick frequent hairs on the sepals are also not typical of *S. chacoense* in this region, though completely normal for *S. commersonii* subsp. *malmeanum* (compare Plates 34 and 35). We have therefore removed var. *latisectum* f. *pilosulum* from *S. chacoense* and placed it with *S. commersonii* subsp. *malmeanum*.

The other form of var. *latisectum*, namely, f. *glabrescens*, is based on a rather small portion of a plant with cordate leaf bases and well-developed petiolules. The terminal leaflet is rather wider than normal for *S. chacoense*, but the calyx is so typical for this species (glabrescent, with acumens not longer than 0·3 mm) that we have no hesitation in retaining f. *glabrescens* within *S. chacoense*.

Whilst dealing with *S. chacoense* from Paraguay we should also mention the species *S. renggeri* of Schlechtendal. The only descriptive notes given are 'Small insipid

FIG. 15. *S. chacoense* Bitt. One of Hassler's specimens of '*S. guaraniticum* Hassl. var. *angustisectum*' Hassl. (see Plate 35 for the other), collected near Villeta, Paraguay, by Hassler (*No. 470*—Herb. Geneva). Note the close similarity in leaf and leaflet shape between this and Bitter's '*S. subtilius*' (Fig. 16).

tubers, extremely watery, from Rengger's observations not altered by cultivation. In the kingdom of Paraguay.' These phrases might be taken as descriptive of every species of wild potato so far known, apart from the locality. From this, however, we are bound to assume that Rengger (1835) observed the species now known to us as *S. chacoense*, though of course it is possible that he may have seen *S. commersonii* subsp. *malmeanum*. In view of the completely inadequate description given by Schlechtendal the name should certainly be discarded, as also the name *S. paraguariense*, applied to it quite without justification, by Dunal (in de Candolle 1852).

Before the description of *S. chacoense* was published by Hassler the species had been collected from many parts of its range, notably in the surroundings of Buenos Aires and the regions of Tucumán, and had been incorrectly referred to a number of very unrelated species.

A collection made by Bâcle, for instance (*117* in Hb. Geneva), from Buenos Aires, was described by Dunal as *S. tuberosum* var. *glabriusculum* Dun. (1852), whilst Weddell (1857) considered this 'variety' to spread into Chile as a wild plant and to be cultivated throughout the Andes! There is no doubt that *Bâcle No. 117* is referable to *S. chacoense*, though the facts are not so easily verifiable when we consider the collections made by Tweedie in Buenos Aires and by Gillies in Mendoza. If we refer to J. D. Hooker's *S. commersonii* var. *glabriusculum* we shall see that a collection of Tweedie from Buenos Aires is quoted. In fact there are two Tweedie collections from Buenos Aires at Kew, one being of *S. commersonii* and the other of *S. chacoense*, so that we cannot tell which of these (? perhaps both) Hooker had in mind when describing his var. *glabriusculum*. It seems very probable, however, that it was the specimen of *S. commersonii*, since he mentions the leaf form and the pubescent calyx, and quotes also the undoubted *S. commersonii* figured by Sabine (1824). The other collection quoted by Hooker for this variety is *Bridges 401* from Chile, referable not to *S. chacoense* or to *S. commersonii* but to *S. maglia*.

A number of the older collectors and even some of the modern ones have referred to gatherings of *S. chacoense* under the name of *S. caldasii* var. *glabrescens* [*Kuntze* (s.n. Dec. 1890), *Stuckert 3505* from Córdoba; *Lorentz and Hieronymus 1050*, from Tucumán; *Schickendantz 56*, from Catamarca; *Knappe* (s.n.)]. Although the so-called *S. caldasii* var. *glabrescens* is to be found only in southern Chile, the previous lack of authoritative taxonomic works on the tuber-bearing *Solanum* species in Argentina is sufficient reason for this confusion. Unfortunately, even at the present day *S. chacoense* is known to some plant-breeders by the name of *S. caldasii* or *S. caldasii* var. *glabrescens* and we consider it therefore of both theoretical and practical interest to attempt to unravel the nomenclatural skein which surrounds it.

S. caldasii was first described by Dunal (1816) from Quito, Ecuador, on the basis of plants collected by Humboldt and Bonpland. It is in fact a synonym for *S. ochranthum* (series Juglandifolia), which Dunal described at the same time.[1]

Dunal mentioned *S. caldasii* again (1852) in his treatise on *Solanum* in de Candolle's *Prodromus*, ascribing it then to Kunth, although the Solanums in Humboldt, Bonpland, and Kunth's *Nov. Gen. et Spec.* did not appear in print until 1818, two years after Dunal's own publication. In addition, Kunth specifically states that he has seen no specimens and relies for his descriptions on information sent by Dunal! The same error, incidentally, occurred with *S. ochranthum* Dun. (*not* H. B. K.).

Not content with this confusion, Dunal incorrectly placed a completely distinct

[1] There is some confusion about the collections of *S. ochranthum* and *S. caldasii* in the Paris and Montpellier herbaria. On the whole, the evidence suggests that both species were described by Dunal from one and the same collection (*Bonpland 3054*). Since they are undoubtedly synonymous, and since Dunal later confused *S. caldasii* with the Chilean *S. palustre*, it seems better to retain the name *S. ochranthum* and to allow the name *S. caldasii* to sink into well-deserved oblivion.

Chilean species, *S. palustre* Poepp. (series Etuberosa), as a synonym of *S. caldasii*. He also renamed as *S. caldasii* var. *glabrescens* a glabrescent collection which had been previously described by Schlechtendal as *S. palustre* var. *glabrescens*. We now know that this glabrescent variety has nothing to do with *S. caldasii*, *S. ochranthum*, *S. palustre*, or *S. chacoense*, but should really be placed with *S. brevidens* (series Etuberosa).

Even after Bitter's publication of *S. chacoense* in 1912, many collections were erroneously identified as *S. caldasii* or *S. caldasii* var. *glabrescens*. Heckel (1915) persisted in this use, and seed of *S. chacoense* was still offered under the name of *S. caldasii* in Botanic Gardens seed-exchange lists as recently as 1959.

This unfortunate species received yet another name (though without a Latin diagnosis) from Louis Planchon (1909*c*), viz. 'pseudomaglia', and this was later changed by potato-breeders into *S. pseudomaglia*. Planchon described it in this way because he thought it was a mutation from *S. maglia*, which had been produced by Heckel. From the description, and also from a living collection kindly given by the late Dr. R. N. Salaman in 1939, it is quite certain that the so-called *S. pseudomaglia* is none other than *S. chacoense*.

Very shortly after the publication of *S. chacoense*, based on Paraguayan plants, Bitter (1913) described what we now consider to be a form of the same species from north-west Argentina (prov. Tucumán), as *S. subtilius* (see Fig. 16). A number of botanists have in fact agreed with Bitter in separating *S. subtilius* from *S. chacoense* (see Brücher 1954, etc.). Whilst not going so far as to consider these two taxa as separate species we began this work by thinking that the north-west Argentine mountain and valley forms might perhaps be distinguished as a separate subspecies based on *S. subtilius*. It has become increasingly clear, however, in examining some hundreds of collections of *S. chacoense* and so-called '*S. subtilius*' from different regions, that there are no constant characters that may be used to separate them. Many mountain forms are of delicate habit and do not form the 'sub-rosette' in the lower part of the stem which seems to be very frequently seen in *S. chacoense* from the plains; but the basal sub-rosette may be seen also in mountain forms whilst the delicate tall-growing habit is to be found occasionally in collections from near the La Plata estuary. One has the impression that non-rosette forms of *S. chacoense* are commoner in the north-western provinces but this character is almost impossible to tell from most herbarium specimens, and is certainly not very geographically constant. Observations in the field are urgently needed in different provinces, altitudes, and phytogeographical regions to provide more information on this problem.

Meanwhile, if one compares the descriptions and type specimen of Hassler's *S. chacoense* var. *angustisectum* with that of Bitter's *S. subtilius* (see Figs. 15 and 16, and Plate 35) one is forced to the conclusion that there are practically no differences at all between them, and certainly fewer than those to be seen between different individuals of a single population, which, as every collector of this difficult group will know, can be very varied indeed. The distinction between them can certainly not be maintained, therefore.

We must now mention nine species described by the Russian taxonomists and potato-breeders S. M. Bukasov and V. Juzepczuk, and two species and one variety proposed by Hawkes in 1944.

Of these, the following eight were from the mountain provinces of north-west Argentina:

S. jujuyense Hawkes (Jujuy). See Plate 45.
S. horovitzii Buk. var. *multijugum* Hawkes[1] (Jujuy).

[1] 'Var. *glabristylum*' *nom. nud.* was published in error; it is a nomenclatural synonym of var. *multijugum*.

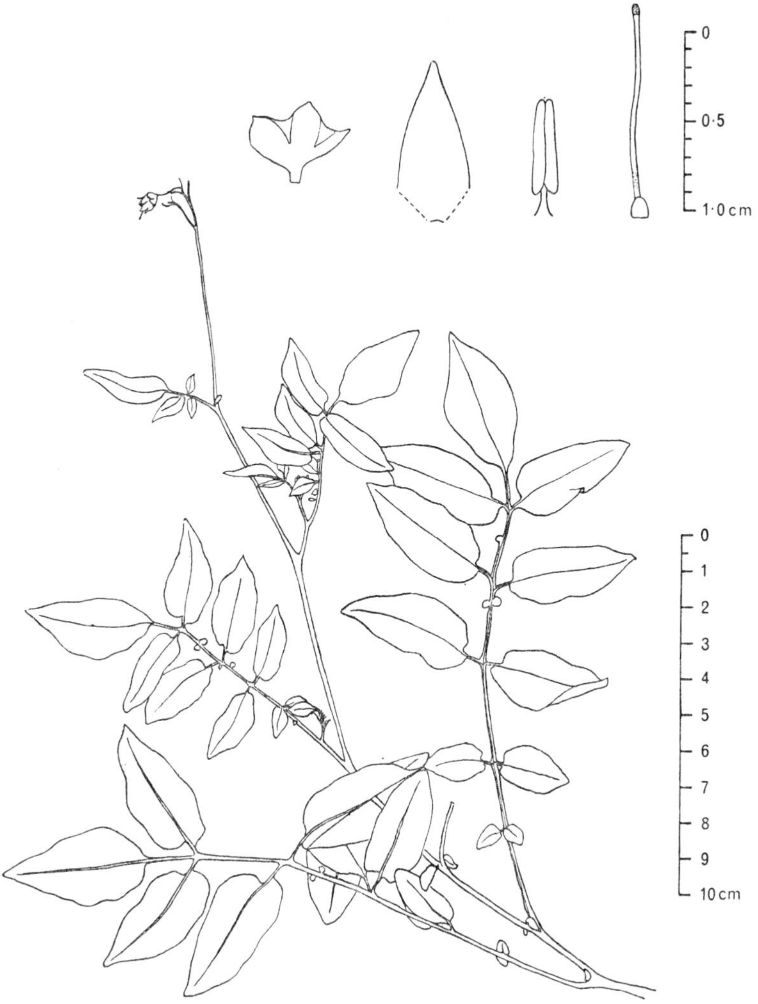

FIG. 16. *S. chacoense* Bitt. Type specimen of '*S. subtilius*' Bitt., collected by Lillo in the vicinity of Tucumán, Argentina (*No. 319*—Herb. Paris). This compares very closely with the Hassler specimen (*No. 470*) from Paraguay, shown in Fig. 15.

S. knappei Juz. et Buk. (Salta). See Plate 44.
S. horovitzii Buk. (Salta). See Fig. 17.
S. saltense Hawkes (Salta). See Plate 46.
S. parodii Juz. et Buk. (Tucumán). See Plate 42.
S. schickii Juz. et Buk. (Tucumán). See Plate 43.
S. emmeae Juz. et Buk. (Unknown, possibly Tucumán). See Plate 47.

Two species were from the lower-altitude regions of west-central Argentina:

S. garciae Juz. et Buk. (Córdoba). See Plate 41.
S. gibberulosum Juz. et Buk. (Córdoba). See Plate 40.

Two were from the low-lying coastal plains of the La Plata region:

S. boergeri Buk. (Uruguay). See Plate 38.
S. laplaticum Buk. (Buenos Aires). See Plate 39.

In a previous work the species *S. laplaticum* was placed by Hawkes (1956*b*) as a synonym of *S. commersonii* on the basis of the description and the statement by Bukasov (1937*b*) that: 'The species *S. commersonii* sensu stricto, *S. henryi*, *S. ohrondii*, and *S. laplaticum* form *S. commersonii sp. coll.*, with its area the La Plata region of Argentina and Uruguay.' On seeing the type specimen in the course of the present work, however, we have been convinced that *S. laplaticum* must be placed within *S. chacoense*. In its long petiolules, many-forked peduncles, and small glabrous calyx this species really comes within the confines of *S. chacoense*, and not of *S. commersonii*, as Bukasov at first supposed.

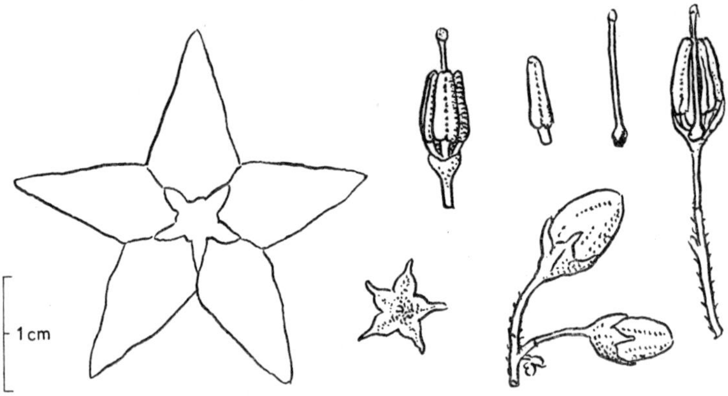

FIG. 17. *S. chacoense* Bitt. Dissected flowers and buds of '*S. horovitzii*' Buk. (from Bukasov 1940*a*). Note the 1 mm long calyx acumen. This species, of which no specimen has been seen by us, was collected by S. Horovitz in 1935 at Potrero de Diaz, prov. Salta, Argentina.

In the present treatment we have grouped these eleven microspecies and one variety as synonyms of *S. chacoense*, not because we consider that they are similar in every detail to the type specimen of that species, but because we are convinced that the concept of *S. chacoense* must be widened to include the variation shown by them and by other collections which have not received species names.

A comparison of the most important diagnostic features of these eleven microspecies (including also *S. chacoense* var. *angustisectum* and *S. subtilius*) is presented in Table 8. The data are taken from published descriptions and verified where possible by an examination of the actual type specimen or an authenticated collection. The collections are arranged approximately in order of altitude above sea level so far as this is known; where the altitudes are presented in square brackets they have been calculated on the basis of the locality given in the description. The dimensions in square brackets are those from type or authenticated plant material, where they differ from the published descriptions.

The table does not, of course, represent the total range of variability within the species *S. chacoense*, or even a completely random sample of it, but only of some biotypes from certain areas which were at one time described as new species.

If, in looking at this Table, we compare the expression of the same character in different species we shall see that *S. schickii* seems much more vigorous than the others, with longer stem and leaf, and larger leaflets, though the flower dimensions do not exceed those of other species.

The character of stem winging is of much greater interest, however, since it is particularly evident in those species from the north-western provinces of Tucumán (*S. parodii*, *S. schickii*), Salta (*S. knappei*, *S. horovitzii*, *S. saltense*), and Jujuy (*S. jujuyense*), though a species with the wings confined to narrow lines or ribs is also found in the same region (*S. subtilius*). We have verified by a careful comparison of many collections from all parts of Argentina, Paraguay, and Uruguay that all the forms from the plains and mountain outliers (e.g. Córdoba) possess very narrow wings or none at all. On the contrary, a rather high proportion of collections (though not all) from the foothills and lower slopes of the Andes in Tucumán, Catamarca, Salta, and Jujuy possess quite broad wings either straight or marginally crinkled.

Pubescence is difficult to compare on the basis of the descriptions provided. Generally the hairs are sparse in *S. chacoense*, or more or less completely absent. A peculiar feature of *S. gibberulosum* from Córdoba is that there are sparse hairs with swollen bases or cushions of tissue, which are quite stiff and easily discernible when running the tips of the fingers along the stem or leaf rhachis. These gibbous hairs are not mentioned for the other described species but are nevertheless more common in *S. chacoense* than one would suppose from the descriptions. Several collections grown by us from the Córdoba region (CPC: 1157, 2134, 2141) and one from Salta (1761) have shown gibbous hairs, though these collections do *not* correspond in other features with *S. gibberulosum*. There is thus probably a localization of the gene or genes which control the growth of gibbous hairs in the Córdoba region, but since these are not in any way correlated with other characters we cannot regard *S. gibberulosum* as a distinct species. It would be of value here for collectors in the field to record with some precision the type of hair growth in different parts of the range of *S. chacoense*.

A character that we considered might be of some value in separating the mountain from the plain forms, was the growth habit and basal rosette formation. However, it is sometimes quite difficult to decide on the presence or absence of a basal rosette, even with plants grown all together under comparable conditions, whereas those taken in the wild vary in habit enormously according to the surrounding vegetation. It seems possible that some selective advantage might accrue to plants in mountain woods that were capable of vigorous and unrestricted stem growth, the reverse being true for plants in short grass which would probably do better with a rosette of leaves at the base. Nevertheless, a constant genotype that could give either a long-stemmed or rosette-forming phenotype in response to the appropriate conditions might be just as effective. To solve this problem more observations, transplant experiments, and progeny tests are obviously needed.

The leaflet number may vary from 3 to 7 pairs of main laterals with a varying number of interjecteds. The general dissection of the leaf is undoubtedly genotypically controlled, but is susceptible to some variation due to environmental differences. On the whole, the number of interjected leaflets is rather low, with the exception of *S. horovitzii* and possibly *S. boergeri* (although the description mentions 3–9 pairs, the type specimen of *S. boergeri* possesses only 0(–1½) pairs). Furthermore, some other collections from the mountains and valleys of north-west Argentina certainly show a larger number of interjected leaflets than those from the plains; thus, *S. horovitzii* var. *multijugum* (not shown in the table) possesses a 7–8-jugate leaf with 18–21 interjected leaflets. Nevertheless, many mountain forms exhibit a very poorly dissected leaf (*S. saltense*) and we can only say that if we were given a specimen with four or more pairs of interjected leaflets we should suspect that it had been collected from some fairly high mountain habitat.

The length/breadth ratio of *S. chacoense* subsp. *chacoense* leaflets seems to vary from about 1·7 to about 3 (contrast subsp. *muelleri*, however, p. 220) but follows no very regular pattern. Thus Hassler, himself, pointed out this variability when he created vars. *angustisectum* and *latisectum*. The only exception to the L/B ratios of 1·7–3 lies in the description of *S. garciae* which gives a ratio of 1–1·6. Nevertheless the only available authenticated specimen (the type description having been made from living plants), CPC: 1158, had a ratio of about 2·3, which falls within the range of the other microspecies.

The petiolule length is in general rather great in collections taken at altitudes of under 1000 m (2 to 15 mm) but becomes shorter in collections taken at over 1000 m (1–4 mm) from the provinces of Tucumán, Salta, and Jujuy. On examining specimens of species which are said to have a petiolule range of 2–8 mm or 2–11 mm (*S. knappei*, *S. parodii*) we find that the smaller dimension rarely occurs (in lowest leaflets or in basal leaves only) and that in fact the long petiolule is very typical and easily visible. An exception to this rule is (again) *S. garciae* from Córdoba which, although from low altitudes, is said to have a short petiolule, but that of the authenticated specimen was always 3–4 mm in well-developed leaves.

The number of flowers in the inflorescence and the length of peduncle and pedicel differ greatly and apparently without any well-defined pattern. The position of the pedicel articulation has been used in distinguishing the microspecies within *S. chacoense* since it varies from near the base to some way above the centre and is generally to be found in the middle third. There is a very rough positive correlation between the position of the articulation in the middle third of the pedicel and the altitude of the plant above sea level; thus the collections from over 1000 m show a supra-median articulation, that of *S. jujuyense* being the highest (two-thirds above the base).[1]

Whilst the shape of the calyx lobes varies erratically from short ovato-triangular (cf. *S. subtilius*) to quite long ovate the constancy of the acumen or 'mucron' is of much greater importance. This is short or rudimentary, rarely more than 0·3 mm long, in most microspecies from the plains, including *S. subtilius*, but is considerably longer (0·5–1·0 mm) in *S. parodii*, *S. schickii*, *S. horovitzii*, *S. saltense*, *S. jujuyense*, and *S. boergeri*; we can see from the table that most of these are from fairly high altitudes in Tucumán, Salta, and Jujuy, though *S. boergeri* is from the low, coastal plains, and proves something of an exception.

The lobes of the invariably white corolla vary in dimensions so that the length/breadth ratio may be as much as 1·75 (*S. subtilius*) or as little as 1·1 (*S. schickii*, which has remarkably broad lobes).

Anther dimensions seem to vary to some extent, and are perhaps a little larger (6–7 mm) in the mountain forms. The length of style and length of the section covered with papillae are quite variable and show no particular correlations so far as we have been able to see. The variation in stigma shape is extraordinarily great. In some forms the stigma is long (up to 1·5 mm in *S. emmeae*) obovate, with apical and basal grooves. In others it is quite ovoid or even capitate. So far as we can see, this great variability seems to occur at random, though perhaps more observations might indicate a certain correlation with geographical distribution, altitude, or with purely morphological characters.

The foregoing survey of variation in these microspecies points towards some sort of correlation of certain characters with higher altitude, as follows:

1. Broad stem wings.
2. No leaf rosetting at stem base.
3. Short petiolule (1–3 mm).
4. Pedicel articulation well above the centre.
5. Calyx acumens 0·5–1 mm long or longer.
6. Some pubescence on calyx.

[1] But some possible hybrids with *S. kurtzianum* possess a low articulation (see p. 236).

:hacoense *Bitt.*

ɔrovitzii	*S. schickii*	*S. jujuyense*	*S. saltense*	*S. emmeae*
ɔo m)	(±1300 m)	1300 m	1500 m	—
	Tucumán	Jujuy	Salta	—
'e Diaz	Siambón	San Antonio	Q. San Lorenzo	—
ds	—	Grass fields	Pathsides	—
	Schick	*Balls 5921* (p.p.)	*Balls 5936* (p.p.)	*Knappe*
1047)				
	75–90 cm	20–55 cm	40–60 cm	35 cm
base	? None	Few at base	Many at base	Some
traight	**Broad straight**	**Narrow crisped**	**Broad crisped**	Very narrow
	Glabrescent	Sparse	Sparse	Glabrous
	Green and dark purple	—	Green and purple	Green, purple near nodes
	None	**None**	**None**	Present
	15–28 cm	7–11 cm	5·5–11 cm	6–18 cm
	5	(3–)4–5	3(–4)	3–4[–5]
	0–3	0–2	0(–1)	0–3
	2–4 mm	**1–3 mm**	**1–3 mm**	1–10 mm
	2·6–2·7	2–2·3	2·2–3·0	2·1–2·5
	Sparse	Sparse	**Frequent**	Glabrescent
	7–22	7–16	7–19	4–11
	—	2	—	2
	4–9 cm	7–9 cm	4–6 cm	9[–13] cm
n	13–22 mm	8–20 mm	10–25 mm	15–25 mm
entre	**Above centre**	**⅔ Above base**	**Above centre**	**Above centre**
:	Glabrous	Sparse	Frequent	Glabrescent
	2·5 mm	1–2 mm	2 mm	2·5 mm
mm	**0·75 mm**	**0·5–1 mm**	**1·0 mm**	0·3 mm
	Glabrous	**Frequent**	**Frequent**	Glabrescent
	3·5 cm	2·4–2·8 cm	2·5–2·8 cm	±3 cm
	1·1–1·2	1·2–1·4	1·2–1·4	1·2–1·3
2·2 mm	7×2 mm	6–7×2 mm	6–7×2 mm	7·5×1·75 mm
	2 mm	1 mm	2 mm	1·5–2 mm
n	11 mm	8–10 mm	10–11 mm	13 mm
	½ way	½ way	Glabrous	½ way
tate	Ovoid	Long	Capitate	Clavate
	0·75 mm	—	—	1·5 mm
	Emarginate	—	Emarginate	Emarginate
imen d (very	No type specimen found. Only poor authenticated material seen	Type specimen examined	Type specimen examined	Type specimen examined

ies no appropriate information available.
l hairs and glands on type specimens.

TABLE 8. *A comparison of the more important diagnostic characters of the microsp*

Data presented	*S. chacoense* v. *angustisectum*	*S. boergeri*	*S. laplaticum*	*S. gibberulosum*	*S. garciae*	*S. subtilius*	*S. parodii*	*S. knappei*	
Altitude	(±70 m)	(±sea level)	(±sea level)	(±290 m)	(±400 m)	(±450 m)	(? 370–600 m)	(? 1000–1200 m)	(±1
Country, province, etc.	Paraguay	Uruguay	Argentina	Argentina	Argentina	Argentina	Tucumán	Salta	Salta
Locality	nr. Asunción	nr. Mercedes	nr. Buenos Aires	nr. Córdoba	nr. Córdoba	nr. Tucumán	Río Chico	Salta	Potrero
Habitat	Wood margins	—	—	—	—	Fields	—	—	Cultd. fi
Collector and number	*Hassler 470, 2805*	*T. Henry*	*Millán G–972*	*Millán G–1183*	*G. García (Millán 1201)*	*Lillo 319*	*Millán G–980*	*Knappe*	*Horovitz (Millá*
Stem									
Length	32 cm	30–35 cm	20 cm	22 cm	12–23 cm	30 cm	25–30 cm	±35 cm	±45 cm
Branching	Some	Some	None	Some near base	—	None	None	Many at base	Many at
Wings	None	Very narrow	Very narrow	None	None	None	**Broad crisped**	**Narrow crisped**	**Broad**
Hairs	Few	Glabrescent	Few	**Gibbous**	Sparse	Sparse	Sparse	Glabrous	—
Coloration	—	? Green	? Green	Green and violet	Green	—	Green and violet	Green and violet	—
Leaf									
Sub-rosetting	—	Sometimes present	Apparently none	?Present	Present	**Absent**	? Present	Partial	—
Length	12–22 cm	17–27 cm	11·3 cm	8–15 cm	5·5–11 cm	16–17 cm	10–23 cm	9–18 cm	—
Pairs of laterals	5–6	5–6	4	(2–)4–5	3[–4]	5	5–7	4–5	5–7
Pairs of interjecteds	1–3	0–9	None	0–2	0–2	1–2	0–3	0(–4)	4–8
Petiolule length	5–8 mm	10–15 mm	7–9 mm	2–12 mm	**1–4 mm**	[2–]5–6 mm	2–11 mm	2–8 mm	**2–3 m**
L/B of laterals	3	1·8–2·4	1·7	1·7–1·9	1·0–1·6[–2·3]	2·3–2·4	2–2·8	2[–2·3]	2·75–3
Hairs	Glabrescent	Very sparse	Sparse	**Gibbous**	Sparse	Sparse	Sparse	Glabrescent	—
Inflorescence									
No. of flowers	15–30	3–9	4–8[–12]	3–7	5–8	±11	6[–13]	[2–]12–24	20
No. of forks	3–4	1–4	1–2	2	[2]	1–2	2	1	—
Peduncle length	7–24 cm	6·5–15 cm	5[–10] cm	2–5 cm	6[–8] cm	5–7·2 cm	5 cm	5–9 cm	8 cm
Pedicel length	8–21 mm	14–28 mm	10–18 mm	16–19 mm	28–32 mm	8–9 mm	14–15 mm	17–25 mm	16–22 n
Pedicel articulation	At or below centre	Below centre	[near base]	Below centre	Below centre	Slightly below centre	Slightly below centre	Below centre	**Above**
Hairs	—	Pubescent	—	**Gibbous**	Sparse	Subglabrous	**Bristles and glands***	Glabrous	Pubesce
Calyx									
Lobes, length of	2 mm	2·5 mm	1·5 mm	3·5 mm	2 mm	—	2·3 mm	1·5–2 mm	—
Acumens, length of	0·3 mm	0·5[–0·75] mm	0·3 mm	Rudimentary	0·5 mm	Rudimentary	±0·5 mm	0·3 mm	0·5[1–1
Hairs, density of	Glabrescent	Pubescent	Glabrescent	Sparse	Sparse	Glabrescent	Glabrous	Glabrous	—
Corolla									
Diameter	—	2·0–2·5 cm	1·7–2·1 cm	±3·0 cm	2·7–3·0 cm	1·6 cm	±2 cm	3–4 cm	2·8 cm
L/B of lobes	—	0·8–1·4	—	—	—	1·75	1·3	1·25	1·3–1·4
Anthers	4·5×1·75 mm	4·5–6×1·5 mm	5×1·5 mm	6×2 mm	4·5×2 mm	5×1·3 mm	5×1·5 mm	7×1·75 mm	5–7×1·
Length of filaments	1·5 mm	Slender	—	1·5 mm	0·75 mm	1·5 mm	2 mm	1 mm	1 mm
Style									
Length	—	10–12 mm	±7–9 mm	'Long'	9 mm	10 mm	9 mm	[8–]13 mm	9·5–11
Papillae	—	? None	None	½ way	⅜ way	None	⅓ way	[½ way]	—
Stigma									
Shape	—	Capitate	Long obovate	Oblong ovoid	Obovate	Long ellipsoid	Obovate	Broad obovate	Small c
Length	—	0·5 mm	1·0 mm	0·75 mm	0·75 mm	—	0·5 mm	[0·75–]1·5 mm	—
Apex	—	Entire or bifid	Emarginate	Entire	—	—	Emarginate	[Subentire]	Entire
Notes	Type specimens examined	Type specimen K. 1070 examined	Type specimen (very poor) K. 597 examined	Authenticated specimen P.G.S. 0016 examined	Authenticated specimen CPC: 1158 examined	Type specimen examined	Type specimen examined	Type specimen examined	Type s exan poor

Calculated altitudes given in brackets, thus: ( ).
When type specimens differ from description the information is placed in square brackets.
Important diagnostic characters shown in bold type.

— Si
* No

After making this detailed analysis of the type specimens of the microspecies within *S. chacoense*, therefore, we were able to say that on the whole a characteristic regional facies was evident in the north-west mountain forms growing at altitudes of over 1000 m. Nevertheless, we did not find, when analysing variation in other supplementary material, that all the 'high-altitude' characters were invariably correlated or linked together in every collection or biotype that we examined.

Bearing in mind the fact (to be discussed in more detail below) that *S. chacoense* hybridizes frequently with mountain species such as *S. microdontum* and others in the mountain valleys of north-west Argentina, it seemed probable to us that a certain amount of introgression was taking place (see Hawkes 1962*b*). In other words the germ plasm of *S. chacoense* in these regions was becoming contaminated with that of another species and in this process the species itself was becoming considerably altered. Probably *S. chacoense* was at first predominantly a plains species, not able to ascend to altitudes of over 1000 m. But once in the possession of germ plasm from *S. microdontum* it was able to increase its range of biotypes and its ecological adaptation to an enormous extent. This may very possibly have been a comparatively recent event, perhaps having been initiated only in the seventeenth century after the Spanish colonists had provided a range of intermediate habitats of cultivated fields and waste places for the two species *S. microdontum* and *S. chacoense* to come into contact and hybridize. Evidently the hybrids were vigorous and fully capable of back-crossing to the *S. chacoense* parent, forming various series of *S. chacoense*-like biotypes which, by natural selection, became able to colonize areas where certainly the *S. chacoense* parent had not previously been able to penetrate. This process of introgression is, as pointed out by Edgar Anderson, a most important factor in evolution, comparable to, but perhaps capable of an even greater degree of development than that involved in allopolyploidy.

It is clear that we have here a fluid situation, where a distinct series of forms, differing greatly from the typical forms of *S. chacoense*, is nevertheless linked to it by all sorts of intermediates. From a nomenclatural point of view we have a situation of some difficulty. Obviously, one cannot consider that the mountain forms are a new and distinct species, though perhaps one day they may evolve into one. Neither can they be considered as a regional or geographical subspecies, in the conventional sense in which that term is normally used. They vary too much in the combinations of characters to be found in them, so that no constant combination of correlated characters is presented, to which we can apply a subspecific epithet. We shall therefore not classify them into a separate subspecies for the time being, until a clearer view can be obtained as a result of further work on living collections. If at a later date a subspecific epithet is used for these introgressed forms we consider that the name *saltense* should be used.

We have deliberately suggested the subspecific epithet *saltense*, derived from the *S. saltense* which was described by Hawkes in 1944, since the type specimen possesses all the characters of the mountain forms combined in the single specimen. Although the species name '*saltense*' is a homonym, due to the publication by Morton of a *S. saltense* in 1944, this does not render the name *saltense* invalid at the subspecific level.

We shall no doubt be criticized for not retaining the name '*subtilius*' which was suggested in 1956 (Hawkes 1956*b*, p. 61) as the type for these mountain forms; and we began this present study by believing that *S. subtilius* might indeed be used as the basis for the mountain subspecies. Further studies indicated that this solution was impossible. Apart from the fact that *S. subtilius* comes from low altitudes (fields near Tucumán at about 450 m only) it possesses none of the distinctive characters of the introgressed mountain forms, with the possible exception of the absence of a basal rosette. The material of *Lillo 319* (in Herb. Paris) on which this species is based does not include the lower part of the stem, and apparently no isotype specimens exist. However, a later collection by Lillo (*No. 12910*) from possibly the same region shows a certain amount of stem rosetting, though it is evident that most stems are elongated through growing amongst tall grasses and other rank vegetation. Again, if we refer to the characters of *S. subtilius* given in Table 8 we can see, in the absence of stem wings, 5–6 mm petiolule, slightly sub-median articulation, rudimentary calyx acumens and glabrous calyx, a very close approximation to the type specimen of *S. chacoense* var. *angustisectum* from Paraguay. It seems, therefore, impossible to consider *S. subtilius* as anything but very typical *S. chacoense*.

At this point it would perhaps be well to pause for a moment to consider the views expressed by Brücher (1953, 1954, 1956*a*) on *S. chacoense* and its relatives. In his review of the Commersoniana Brücher (1956*a*) draws attention to the wide variability within this series, and rightly concludes that a number of the species names are no more than synonyms, as we ourselves have pointed out (see also Hawkes 1956*b*). Brücher has not made a critical study of the type material of *S. chacoense* or of the various microspecies mentioned above; neither has he assessed the range of variability of *S. chacoense* in parts of Argentina other than the north-western provinces. Consequently, he follows Bukasov in erroneously placing *S. laplaticum* under *S. commersonii* instead of under *S. chacoense*.

By an adherence to rigid phytogeographical concepts he claims that *S. chacoense* should be considered as confined to the 'dry forests' of the Chaco zone, together with *S. garciae*, *S. cordobense* (= *S. garciae*), *S. gibberulosum*, etc. The mountain or near-mountain biotypes of *S. chacoense*, including *S. parodii*, *S. schickii*, *S. emmeae*, *S. knappei*, *S. horovitzii*, *S. setulosistylum*, *S. saltense*, and *S. jujuyense* he thinks should be classified as *S. subtilius*. His distribution map is incorrect in a number of respects, and by no means gives a true picture of species distribution, since he has presumably not constructed it from the localities of material actually examined, but from the literature, which at times (as in the cases of *S. rionegrinum* and *S. pampasense*) he has misinterpreted.

Brücher correctly points out the considerable variation amongst the mountain forms (classed by him under *S. subtilius*), such as the more strongly developed calyx acumens of biotypes from Jujuy and the greater leaf dissection in the Jujuy–Salta border region (cf. *S. horovitzii*). On the other hand, he gives no key for the identification of *S. subtilius*, and his description of it would equally well fit *S. chacoense*.

In the section on distribution (p. 196) we mentioned that *S. chacoense* extended for some 300 km into Bolivia. Three of its forms in that country were described by Professor M. Cárdenas (1956) as new species, as follows:

S. arnezii Cárd., from dept. Chuquisaca, prov. Tomina. Alt. 2200 m. *Cárdenas 5076*.

S. caipipendense Cárd., from dept. Santa Cruz, prov. Cordillera, near Caipipendi. Alt. 1100 m. *Cárdenas 5072*, and also from Lagunillas in the same province. Alt. 1100 m. *Cárdenas 5072a*.

S. cuevoanum Cárd., from dept. Chuquisaca, prov. Luis Calvo, Cuevo. Alt. 800 m. *Cárdenas 5073*.

All three microspecies differ sharply from each other, and if one considers them by themselves, as Cárdenas probably did, without reference to the wide range of material available from Argentina and elsewhere, then one might well think of them as distinct species. However, in the light of the variation seen in the material from Argentina, and having in mind the rather wide definition of *Solanum* species that we are using in this work, it becomes clear that we must group all three species as synonyms of *S. chacoense*.

Both *S. caipipendense* and *S. cuevoanum* are typical of the plains forms of *S. chacoense*, with very narrow stem wings, twice-forked peduncle, glabrous leaf, stem, inflorescence and calyx, and below-centre articulation. The calyx acumens of

S. cuevoanum are 0·5 mm or less and quite glabrous, as also are those of the type specimen of *S. caipipendense* (*Cárdenas 5072*). The acumens of the other specimen of *S. caipipendense* (*Cárdenas 5072a*), however, are 1 mm long and rather pubescent; furthermore, the flowers are said to be blue-lilac, though the specimen seen by us had what appeared to be white flowers. Evidently some introgression is taking place in the populations from which this specimen was gathered.

Finally, *S. arnezii* is a highly introgressed typical mountain form, very similar to some of the Salta and Jujuy plants (index figure 65, see below, p. 216) we have already mentioned. The type specimen of *S. arnezii* possesses a narrow delicate stem, wide stem wings, puberulent leaf, stem, inflorescence and calyx, and long calyx acumens, to 1 mm long in flower and even longer in fruit. The petiolules are much shorter than those of *S. cuevoanum* and *S. caipipendense*, which again is a feature of mountain introgressed forms of *S. chacoense*.

It would seem that introgression of genes into the higher altitude forms of *S. chacoense* is taking place in Bolivia also, just as in Argentina. Whether these genes are derived from *S. microdontum* in Bolivia is not quite certain, though there is nothing in the morphology of *S. arnezii* that would induce us to think that some other, quite distinct species was involved. Furthermore, as we shall show later (pp. 335–6), *S. microdontum* extends northwards into Bolivia, where it has been described by Cárdenas under the names of *S. cevallos-tovari* and *S. higueranum*. It thus seems certain that *S. chacoense* spreads into Bolivia and that its pattern of variation in that country is probably roughly similar to that in Argentina.

Correll's treatment (1961, 1962) of *S. chacoense* follows largely that proposed by Hawkes (1956*b*, 1963), which itself was based on the studies we had initiated jointly on this species. Correll groups most of the lower-altitude microspecies under typical *S. chacoense*, including *S. subtilius* from Tucumán and *S. cuevoanum* from Bolivia. He proposes that *Hassler 2805* from Paraguay should be the lectotype for *S. chacoense*, and since this was one of the two specimens cited for Hassler's first-described variety *angustisectum*, we are in complete agreement with this choice. We furthermore consider that one of the Geneva specimens of *Hassler 2805* should be considered as the lectotype for this species. This specimen of Hassler is most typical of the plains form of *S. chacoense* and seems to be in every way suitable for a lectotype. (See Plate 35.)

In addition to the typical *S. chacoense*, Correll describes three distinct forms, f. *caipipendense*, f. *gibberulosum*, and f. *pilosulum*. The first-mentioned form is based on Cárdenas's *S. caipipendense*, which we have discussed above, and is said by Correll to be 'merely a tall robust form of typical *S. chacoense*'. We consider, for the reasons already given, that the form or species *caipipendense* has no taxonomic value and should be treated as a synonym of *S. chacoense*.

The second form, *gibberulosum*, is based on *S. gibberulosum*, of Juzepczuk and Bukasov, and includes material chiefly from Córdoba, with gibberulous hairs and shiny coreaceous leaves, etc. As we have already pointed out, although typical gibberulous pubescence is found mainly on plants from Córdoba, specimens with hairs intermediate between gibberulous and normal are known from other provinces, whilst in the Córdoba material there seems to be very little evidence of correlation between gibberulous hairs and other taxonomic characters. It would seem to be much better to disregard this form, *gibberulosum*, in view of the immense amount of variation in *S. chacoense* which has not received taxonomic recognition, and which in any case is far better left out of any formal taxonomic scheme.

For his third form Correll uses Hassler's name *pilosulum*, taking it to include not only the pubescent-calyxed forms of var. *latisectum* from Paraguay, but also a number of other elements from the north-western provinces of Argentina. We have already shown that the Paraguayan material grouped by Hassler under *S. chacoense* var. *latisectum* forma *pilosulum*, as well as Chodat's *S. commersonii* f. *pubescens*, should really be grouped under *S. commersonii* subsp. *malmeanum*. Correll also includes, quite erroneously in our opinion, the introgressed mountain forms of *S. chacoense*, namely, *S. horovitzii*, *S. saltense*, and *S. jujuyense*. He also places in this forma the non-introgressed microspecies *S. garciae* and *S. boergeri*, the former from Córdoba and the latter from Uruguay. The highly introgressed Bolivian microspecies *S. arnezii* is kept as a distinct species and even included in a distinct series, Yungasensa, together with *S. muelleri* (see below, p. 222). We disagree most strongly with this taxonomic scheme, since Correll unites under one forma, *pilosulum*, two completely discordant elements, *S. commersonii* subsp. *malmeanum* from the region of Paraguay, on the one hand, and, on the other, the mountain introgressed forms of *S. chacoense*, chiefly on the presence of some pubescence and the rather more developed calyx acumens. It is clear, however, that Correll has some doubts as to the validity of this grouping, since he says, 'Some of the plants possess the laxly branched inflorescence of typical *S. chacoense*, while others, as represented by Hassler's collection *No. 11849*, have the characteristically compact inflorescence of *S. commersonii*.' Furthermore, he says, 'These rather heterogeneous pubescent plants, which perhaps represent a mixing of *S. chacoense* and *S. commersonii* (as we interpret these species), *S. kurtzianum*, and possibly other species, are for the time being relegated to the appropriate epithet, *pilosulum*.' The form *pilosulum* as interpreted by Correll is obviously highly artificial and should be discontinued.

Correll finally describes another species, *S. limense*, which he places in series Conicibaccata. The collection on which this species was based was made by Dr. J. Soukup at Lima, Peru (*No. 3555*) in 1948. It seemed to us when we examined it in about 1956 to be so unlike anything that had ever been seen in Lima, on the west coast of Peru, and so like *S. chacoense* from the lowlands of Argentina or Paraguay, that we assumed that the label had been affixed to the plant in error or that the plant had been grown in Dr. Soukup's experimental garden from tubers or seeds received from elsewhere. We had therefore decided to relegate it to the 'dustbin' of collections for which there is strong doubt as to the validity of the label or the provenance of the material. This so-called *S. limense*, we are convinced, has nothing to do with series Conicibaccata, and in any case no berry was preserved on the specimen. It is quite a typical lowland form of *S. chacoense* with twice-forked peduncle, very narrow stem wings, glabrous calyx with 0·3 mm acumens and stellate white corolla whose lobes are 8–9 mm long ×5–6 mm wide, not 10×9 mm as shown in Correll's figure (Correll 1962, p. 175, Fig. 61). Ochoa (1962) agrees with us here, since he states (p. 297): '*S. limense* is in my opinion a synonym of *S. chacoense*; the type was collected from plants grown in Lima derived from wild material imported from Argentina.'

This unfortunate species, *S. chacoense*, has now received in various ways some twenty-two species names, not including all the varieties and forms; this is in part a penalty it has had to pay for its wide geographical range and high degree of morphological variability. But it has also received perhaps more than its fair share of nomenclatural synonyms and errors of other nature.

We return now to a more detailed consideration of the source of introgressed genes in *S. chacoense*. We previously mentioned that *S. microdontum* probably constituted the fount of such genes, enabling *S. chacoense* to spread into areas that the original species would not have been able to colonize. We arrived at this conclusion by correlating and extrapolating the characters expressed in subsp. *chacoense* in their most extreme form (Anderson 1949). From this study we were able to postulate that the unknown donor of the 'contaminating' genes would possess broad crisped stem wings, no basal rosette, shortly petiolulate or subsessile leaflets, frequent hairs on leaves, peduncles and calyces, high pedicel articulation, long calyx acumens, white corolla, and fairly large anthers. It would also need to be a mountain species whose altitude range overlapped that of *S. chacoense* at its lowest point.

The answer to this problem was not too difficult; indeed only one species fulfils all these conditions, despite the fact that the leaf shape is very dissimilar from that of *S. chacoense*. We have little hesitation, therefore, in postulating that *S. microdontum* is the species that has 'contaminated' *S. chacoense* in the way we have described. Although the leaf of *S. microdontum* is not well dissected and may often be completely simple, with an invariably well-developed broad terminal leaflet, this character does not appear to have entered into *S. chacoense* and seems always to be correlated with the broader corolla type of *S. microdontum*. Perhaps this is due to the fact that we automatically identify as *S. microdontum* any collection with this particular leaf and corolla type. We shall return to this point, however, in discussing *S. microdontum* itself (see p. 339). There is much evidence from collected specimens and records (see hybrids, p. 229) that these two species frequently hybridize in certain regions. Brücher (1953), for example, compared artificial hybrids of *S. chacoense* and *S. microdontum* with naturally occurring hybrid populations, which apparently were well adapted, even spreading into habitats which the parents themselves could not colonize. He noted that the hybrids were not so fertile as the parents, though the observations by other workers that F_1 hybrids between South American diploid species are as fertile as the parents themselves are directly opposed to this statement (see also 'Fertility', p. 226).

In order to verify the hypothesis mentioned above, namely that *S. chacoense* varies in the direction of *S. microdontum* in the mountains due to introgressive hybridization, we scored all the herbarium specimens of *S. chacoense* available to us at the time (170 gatherings) for the following characters: (1) width and degree of crisping of stem wing, (2) greatest petiolule length, (3) stem and rhachis hairs, whether normal or gibbous, (4) position of pedicel articulation, (5) length of calyx acumen, (6) calyx pubescence, (7) stigma length, (8) style length, (9) greatest number of interjected leaflets.

Some of the more interesting results of this survey are to be seen in Figs. 18 and 19, Map 6, and Tables 9 and 10. (See also Hawkes 1962*b*.) Fig. 18 is a scatter diagram in which length of calyx acumens is plotted against petiolule length; data on width of stem wings, pedicel articulation, calyx pubescence, and altitude are included in the metroglyphs (Anderson 1960). We see here a strong tendency to short acumen (0·25–0·5 mm) in the low-altitude forms of *S. chacoense* from the plains (0–450 m), whilst a tendency to a lengthening of the acumen takes place in the collections from medium altitudes (500–1450 m), and this becomes even more marked at high altitudes (1500–2500 m). A general correlation of short petiolule, long calyx acumen, wide stem wing, above-centre pedicel articulation, and high altitude is seen at the top left-hand corner of the diagram and the reverse situation at the bottom right.

The variation in acumen length shows more clearly in the frequency histograms (Fig. 19). At 2000 m and over, only 8 per cent of specimens (actually, one specimen) has an acumen of 0·6 mm or less. In the 1500–1950 m and 1000–1450 m belts 50 per cent have a short acumen, 60 per cent in the 500–950 m belt, and 83 per cent in the 0–450 m belt.

Stem wing width, petiolule lengths, and calyx pubescence are shown in Table 9. From this one can observe that stem wings are absent or very narrow (to 0·5 mm) in 92 per cent of the plains forms, in 86 per cent of the 500–950 m forms, in 52 per cent of the 1000–1450 m forms, in 41 per cent of the 1500–1950 m forms, and only 25 per cent of the 2000–2500 m forms. In other words, stem wings are very frequent at high altitudes, but their frequency falls considerably as we pass below the 1000 m contour. The typical long petiolule (8 mm or more) of *S. chacoense* is found in a decreasingly small proportion of specimens examined as we ascend from plains to sierra, until, as we pass over the 2000 m contour, no plants at all have a petiolule of over 8 mm in length.

Calyx pubescence is perhaps even more striking. We scored the calyx pubescence into three categories: dense (pubescence abundant), medium (sparse hairs), absent (glabrous or almost so) according to previously set-up standards. Nearly all the plains specimens from 0 to 450 m have a glabrous calyx (92 per cent). The percentage falls to 68 in the 500–1450 m belt, and 48 in the 1500–1950 m range. It falls even lower, to 17 per cent, in the belt above 2000 m. No specimens from the plains possess dense calyx pubescence, but this appears in material from 500 to 950 m (only 4 per cent) and rises to 50 per cent in the 2000–2500 m belt. In this higher region, especially in prov. Jujuy, the whole plant is quite pubescent (see *H. P. & R. 277*, *P. & H. P–349, 958*, for example, mentioned on p. 184, and other high-altitude collections).

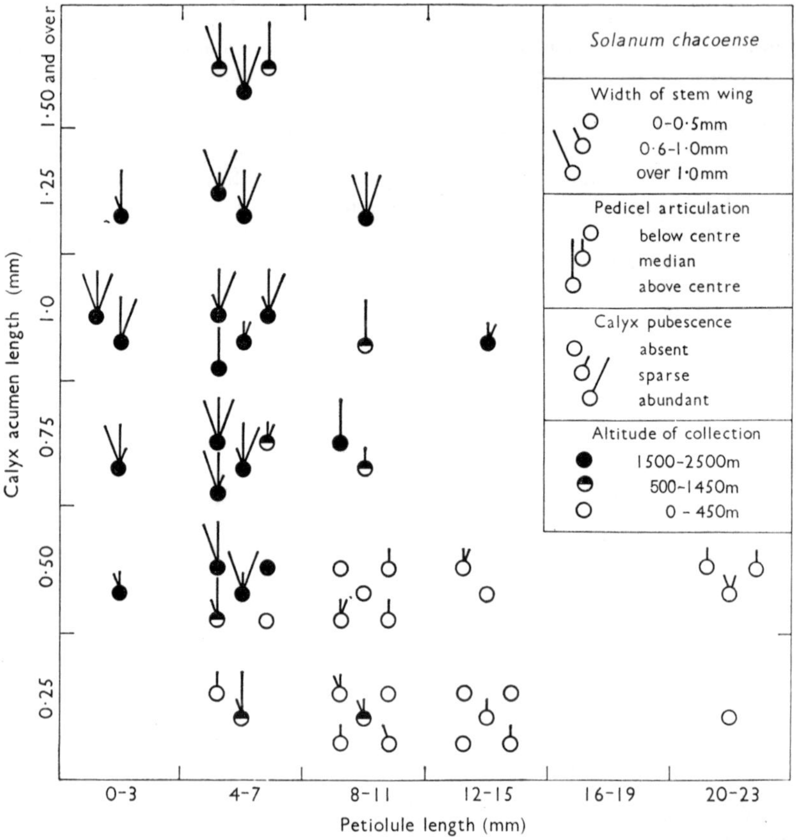

FIG. 18. Pictorialized scatter diagram for *S. chacoense* of fifty-one plants taken from different parts of its geographical range (see Maps 5 and 6) and at different altitudes. (For further explanation see p. 213.) (After Hawkes 1962*b*.)

The characters of stigma length, style length, and interjected leaflet number showed no geographical pattern in their expression, as we have already seen from the study of the microspecies type specimens. Pedicel articulation tended to be low in the lowland and above median in the highland forms, again as in the type specimens.

It also seemed to us to be of interest to inquire into the extent of introgression of *S. microdontum* genes into *S. chacoense* in a geographical or spatial sense. *S. microdontum* is not normally found lower than 1500 m, and never lower than 800 m. Yet we can see from Table 9 that its effect can be felt in the *S. chacoense* specimens from 500 to 1500 m also.

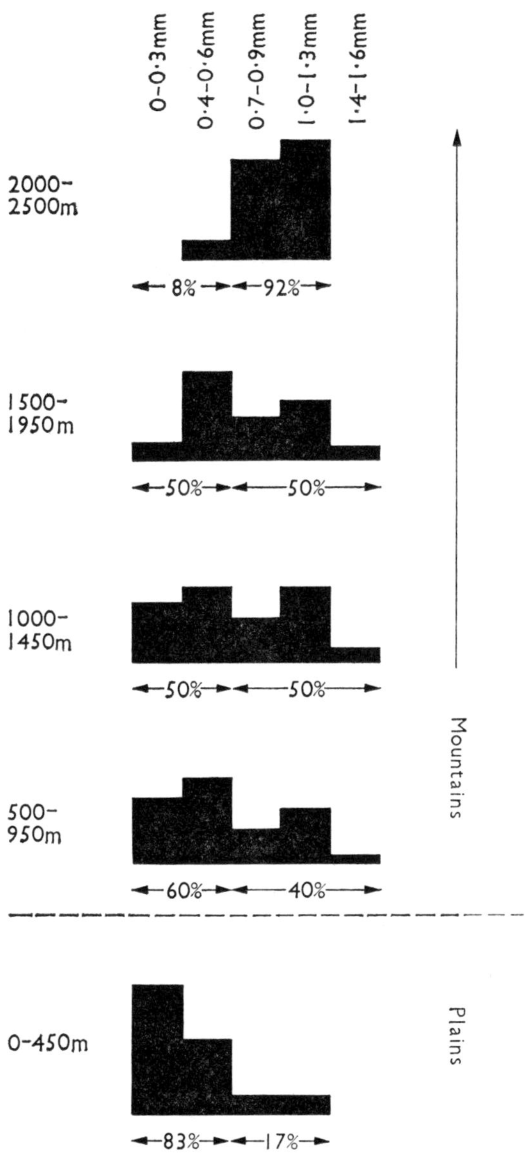

FIG. 19. Frequency histograms of *S. chacoense* calyx acumen lengths at different altitudes. (After Hawkes 1962*b*.)

TABLE 9

Percentage frequencies of S. chacoense *plants at different altitudes with characters as shown below* (from Hawkes 1962*b*)

Altitude Range	Stem wing width		Petiolule length		Calyx pubescence		
	0–0·5 mm	0·6 mm and over	0–7 mm	8 mm and over	Dense	Medium	Absent
2000–2500 m	25	75	100	0	50	33	17
1500–1950 m	41	59	79	21	29	23	48
1000–1450 m	52	48	81	19	11	21	68
500–950 m	86	14	64	36	4	28	68
0–450 m	92	8	43	57	0	8	92

The lowest altitude category, 0–450 m, includes most of the geographical range of *S. chacoense*, from the base of the Andes to Paraguay and Brazil. Can one see any changes in these forms, resulting from *S. microdontum* introgressions, as one passes from the mountains eastwards across the plains? Apparently, one can.

To illustrate this we have used yet another of the methods proposed by Dr. Edgar Anderson (1949), namely the 'hybrid index'. To obtain this we have given a figure of from 0 to 5 for each character, according to its expression, arranged so that the higher number consistently applies to one species (*S. microdontum*) and the lower to the other (*S. chacoense*).

Scoring four characters (stem wing, pedicel articulation, calyx acumen, and calyx pubescence) for five plants[1] taken at random from every province, the total maximum index would be 100 (4 × 5 × 5). The figures for the various provinces may be seen on Map 6. Those provinces with no hybrid index figure but which occur within the range of the species had fewer than five collections made in them; for this reason no data have been included.

The provinces with the highest index figures are Jujuy (77) and Salta (68). These are followed by two provinces in which the average index figure is reduced because of the forms of *S. chacoense* from low-lying areas in the eastern parts of the provinces: Tucumán (33), Catamarca (27).

The two provinces of Santiago del Estero and Córdoba lie well away from the main mountain chain, yet still give quite high index figures (31 in each case). The index figure for Mendoza and San Luis (37) is also high. Although the Córdoba and San Luis forms may be showing some variation which has nothing to do with introgression, since in any case *S. microdontum* does not extend so far to the south, it seems more than likely that the high index figure for Santiago del Estero is due to a flow of *S. microdontum* genes on to the plains in this region.

To clarify the situation a little better we have attempted a breakdown of the hybrid index figures (see Table 10), to make it possible to compare similar altitude bands in the different provinces. The slight discrepancies in some cases between the data given on the map and those shown in the table are due to the fact that some collections had no altitude data, or that there were insufficient collections from certain altitudes to show in the table. The higher figure (31) for the 0–450 m range of Santiago del Estero

[1] Or multiples of five in some cases; the score was then reduced to a possible total of 100 by dividing it by the appropriate figure.

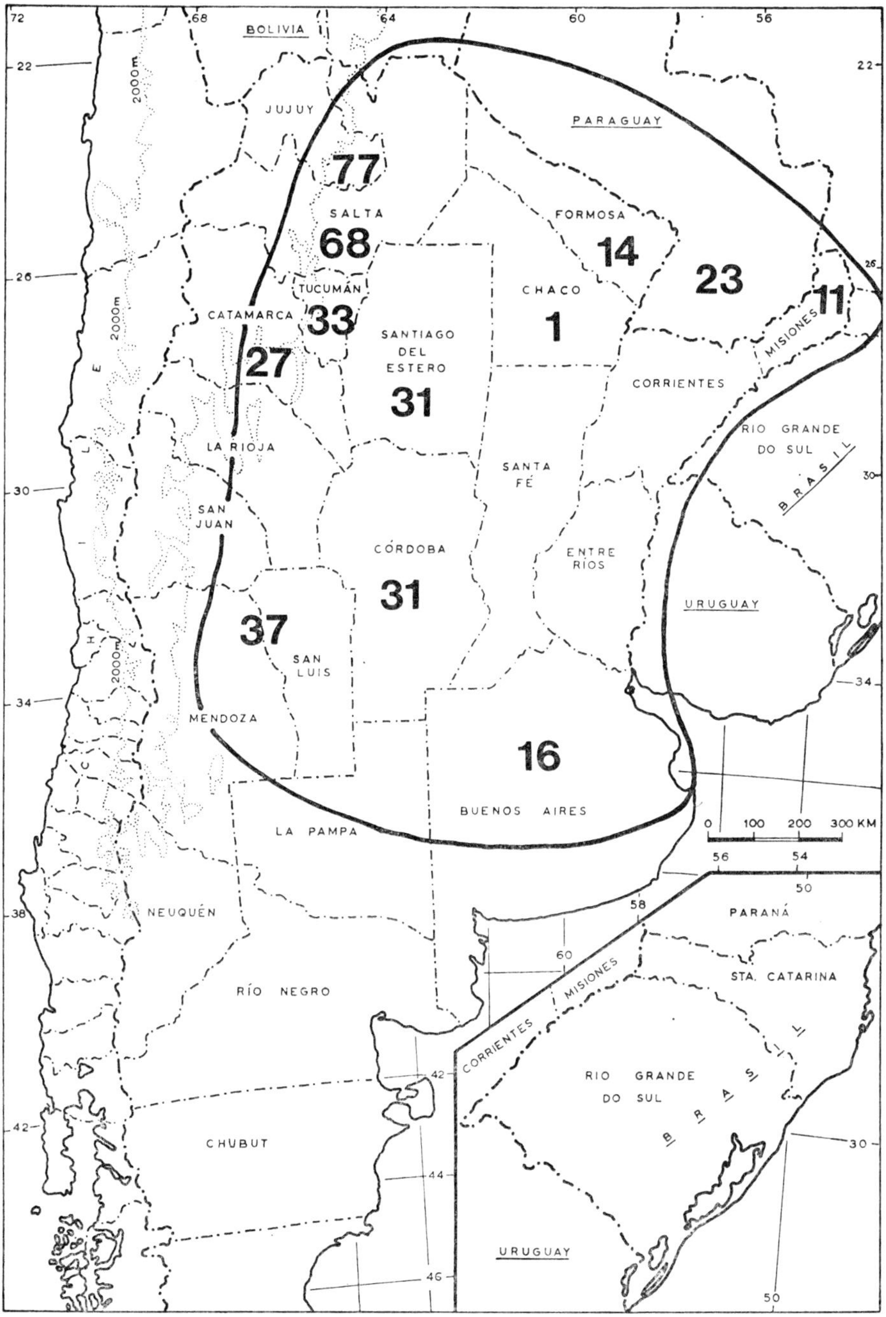

MAP. 6. *S. chacoense* Bitt. Hybrid indices (see text p. 216 for explanation). (Modified from Hawes, 1962*b*.)

stands out very clearly, as does also the figure of 23 for Tucumán. The table also shows the very much greater extent of introgression in Salta, on the one hand, compared with Tucumán and Catamarca, on the other, at comparable altitudes. We shall discuss the significance of the Córdoba results later on in this section.

TABLE 10

Hybrid indices of S. chacoense *in various Argentine provinces* (*and in Paraguay*) *at different altitudes* (modified from Hawkes 1962*b*)

Altitude	Province										
	Jujuy	Salta	Tucumán	Catamarca	Córdoba	Santiago del Estero	Buenos Aires	Misiones	Formosa	Chaco	Paraguay
2000–2500 m	74										
1500–1950 m	—	76	34	37							
1000–1450 m	—	60	36	32							
500–950 m	—	—	29	—	34						
0–450 m	—	—	23	(16)*	22	31	16	11	14	1	23

* Some altitudes uncertain.

Paraguay, which lies at an even greater distance from the foot of the Andes, shows an inexplicably high hybrid index figure (23). Very little is known of the wild potatoes in that country, even though the type collections of *S. chacoense* were made there. Later work might show, for instance, that some hybridization with *S. commersonii* subsp. *malmeanum* is taking place, though we cannot be sure that this is so until adequate living collections or mass samplings are made.

S. chacoense seems to exist in a very pure, and non-introgressed state in prov. Chaco, though it seems probable that an index figure of up to about 16 may not be indicative of hybridization, or the presence of *S. microdontum* genes, but may merely be an expression of the natural variation within the species. Unfortunately, we have insufficient data from provinces Entre Ríos, Santa Fé, and Corrientes or from the neighbouring republic of Uruguay. The single collections of *S. chacoense* from each of these areas were not sufficient for a reasonably reliable hybrid index figure to be derived.

The results from scoring hair type were disappointing. On the whole, pubescence is scanty in *S. chacoense*, and hairs when present are generally of the normal sub-appressed multicellular type. Six of our specimens showed unmistakably gibberulous hairs, and these specimens were mainly from Córdoba, though one gathering from San Luis with hairs of this type was also seen. Hairs that were roughly intermediate between gibberulous and normal were noted on two specimens from Tucumán, two from Catamarca, and one from Santiago del Estero. In addition, a very dense coarse pubescence was observed on three specimens from San Luis. The characteristic gibberulous pubescence was thus mainly confined to Córdoba, and we were at first inclined to think that all these specimens might be placed in Juzepczuk and Bukasov's *S. gibberulosum*. Yet in other characters our specimens did not agree with the description of that species, and it seems more probable that the gene or genes which control this type of hair in *S. chacoense* are chiefly localized in Córdoba, but are not correlated with genes controlling the expression of other characters in any way.

We have seen that the pattern of variability of *S. chacoense* in north-west Argentina, and also probably in Bolivia, can be adequately explained only by postulating introgression of genes from another species, probably *S. microdontum*. Since *S. chacoense* also forms hybrids with species such as *S. spegazzinii* and *S. kurtzianum* in certain regions it will be of interest to see whether these species play a role similar to *S. microdontum* in those areas.

The evidence at first sight appeared to indicate that this could well be so. We knew that *S. chacoense* formed hybrids with *S. spegazzinii* and *S. kurtzianum* in some abundance in specific areas and we were therefore inclined to suppose that the distinct forms of *S. chacoense* from provs. San Luis and Córdoba were due to introgression from *S. kurtzianum*. Our 1965–6 expedition showed us that this was not so, however. We found no *S. kurtzianum* in those provinces, and the pattern of variation seemed rather to indicate that the particular coarse pubescence and a number of other characters were indicative of a natural development of the species, not of genes flowing into it from some other species. Thus, the high 'hybrid index' figures of 22 and 34 for Córdoba in Table 10 and the high figures for Córdoba (31) and Mendoza/San Luis (37) in Map 6 may be no more than an expression of the natural variation pattern. We shall discuss this in more detail in Appendix III.

Forms of *S. chacoense* collected in prov. Jujuy, dept. Tumbaya, Volcán (*Castillón 354*; *Garolera & Romero* (s.n.); *H.P. & R. 277*; *P. & H. 958*; *Sleumer 3566*) with strong leaf pubescence may possibly be introgressants with *S. spegazzinii*, which has been recorded in that locality. It must be noted, however, that *S. microdontum* is known from a little higher up the valley slopes, and could possibly have been involved in these rather distinct Volcán biotypes. Equally distinct, perhaps, are the variants with highly dissected leaves from Jujuy, dept. San Antonio, El Potrerillo (*Balls 5921* p.p.) that were described by Hawkes as *S. horovitzii* var. *multijugum*; they formed part of a population in which less dissected leaves were seen on many plants (described by Hawkes as *S. jujuyense*), and since there is no other species with extreme leaf dissection in the vicinity it is probable that the high degree of leaf dissection in this area is due to gene mutations within *S. chacoense* itself.

We have been considering in the work mentioned above only the morphological evidences of introgression. Interesting corroborative evidence was recently obtained by Cockerham (in Anon. 1962, 1965) who showed that a gene controlling reaction to virus Y in *S. chacoense* was either identical to or an allele of a gene controlling necrotic reaction to virus Y in *S. microdontum* subsp. *gigantophyllum*. The two forms in question were from Salta province, San Lorenzo, and were CPC: 51*b* ('*S. saltense*') and 51*a* ('*S. simplicifolium*') respectively. From this result and further data obtained by Cockerham (personal communication) it seems highly probable that the distribution of this virus Y resistance gene may also be taken as evidence of introgression from one species into the other. At the moment, however, we are not certain as to which of the two species is the original source of this gene, though the indications are that it may well be *S. chacoense*.

The variation problem of this subspecies of *S. chacoense* is therefore one of great interest; we can discern a well-marked pattern in it, of which part at least seems to agree with the hypothesis that introgression of genes has taken place from the species *S. microdontum*, enabling it to colonize areas and extend into altitudes and habitats where the pure *S. chacoense* had previously been unable to penetrate.

The next phase of investigation would be to follow up this survey based on herbarium material with one based on living plants in the field. Populations should be scored, as was done by Anderson (1954, pp. 26–27) for *Tradescantia virginiana*. It would be particularly interesting, for example, to score populations of *S. chacoense* along the roads and railways in the various provinces, since it is probable that this species has chiefly spread along the communication routes and by means of rivers. In 1965–6 we made some advances in this direction, as reference to Appendix III will show. More information is needed, also, of the variability of *S. chacoense* in the provinces of Mendoza, Santa Fé, Corrientes, and Entre Ríos. Work of such nature would not be costly to carry out (beyond the necessity of having the use of a serviceable vehicle) and would be of immense interest in demonstrating the dynamic nature of the variability in our wild potato species.

Note: Occasional teratological phenomena were observed by Roig (1956) in the flowers of *S. chacoense* from prov. Córdoba (*Roig 1154*, La Cumbrecita, dept. Calamuchita. 18–27 Dec. 1955. Alt. 1450 m). Amongst other abnormalities an increase or decrease of floral parts, fusions of various types, and petaloidy of stamens were noted. Some 25 per cent out of a total of 554 flowers were deformed in this way.

2. Subspecies ***muelleri***

Having dealt at some length with subsp. *chacoense* we shall by contrast be able to dispose of subsp. *muelleri* in a short space of time, since comparatively little is known about it.

The species *S. muelleri* was described by Bitter in 1913 from material cultivated in the Vienna Botanic Garden in 1869 (see Fig. 20). It was collected by F. Müller in Rio Grande do Sul, Brazil, in the form of seeds and sent to Vienna by Professor Hildebrand (Freiburg im Breisgau). After a careful study of this material, together with other collections of the same taxon made by various collectors in Rio Grande do Sul, Paraná, and Santa Catarina in Brazil, and Misiones in Argentina, we have come to the conclusion that *S. muelleri* does not differ specifically from *S. chacoense* but should be given subspecific rank only. It represents the most easterly facies of the very polymorphic species *S. chacoense*, and, although fairly distinct, it is not so far removed from that species as one might have supposed from the over luxuriant type specimen from the Vienna Botanic Garden.

In the same year, 1913, Bitter described a variety *grandifrons* of the Mexican species, *S. jamesii*. The material was cultivated in the botanic gardens of Kew under the name of *S. suaveolens*, and a herbarium specimen was taken in 1874.[1]

No locality data for this material are known and Bitter linked it with *S. jamesii* instead of with *S. muelleri*, remarking: 'This variety, whose exact home I unfortunately do not know, remarkably exceeds, as does the above described variety [var. *ripicola* —now placed under *S. brachistotrichum*], the dimensions found up to now in the forms of *S. jamesii*. Particularly striking is the much greater number of leaflet pairs (5–7) compared with the rest of the forms of *S. jamesii*, which are only 2–4-jugate.'

It is clear that Bitter's *S. jamesii* var. *grandifrons* is none other than *S. muelleri*. The 5–7-paired leaflets of var. *grandifrons* compare quite well with the 4–6-paired leaflets of *S. muelleri*, and the highly characteristic very oblique bases of the lateral leaflets are also noted in both descriptions. In short, allowing for the fact that the specimens were made from material grown in botanic gardens under culture conditions not perhaps by any means identical, the match between them is quite good. It seems likely that seeds were distributed by F. Müller or by Professor Hildebrand to several botanic gardens, some of which retained the original locality data whilst others did not.

A most characteristic feature of subsp. *muelleri* is the long, narrow leaflet which is frequently $3\frac{1}{3}$–4 times as long as broad in the type specimen. This attribute is

[1] This specimen has certainly nothing to do with *S. oxycarpum* as Correll (1962, p. 274) thinks.

FIG. 20. *S. chacoense* Bitt. subsp. *muelleri* (Bitt.) Hawkes & Hjerting. Type specimen (in Herb. Vienna) grown at Vienna Botanic Gardens from seeds collected by F. Müller in Río Grande do Sul State, Brazil. This specimen possesses much larger leaves and smaller flowers than material growing naturally in South America (see Plate 48) but demonstrates well the long, narrow leaflets (length/breadth = 3·3–4·3) characteristic of this subspecies.

shown clearly in *Petersen & Hjerting P–66* from Misiones (see Plate 48), where the leaflets range from 3 to 4·3 times as long as broad, and are 5-paired with generally no interjected leaflets, though occasionally 1–3 are produced. The type material also possesses few (1–3) or no interjected leaflets whilst the Kew sheet ('var. *grandifrons*') has none. A collection of Müller (*No. 154*) taken in 1868 from Santa Catarina also matches the type specimens very well.

Schenck's collection from Blumenau, Santa Catarina (*No. 947*), which was in the Berlin herbarium and is known to us from a photograph made by the Chicago Natural History Museum, seems to be less typical of subsp. *muelleri*. Nevertheless, it was identified by Bitter as *S. muelleri*, though Schenck considered it to belong to *S. chacoense*. We place it with subsp. *muelleri* provisionally, but in view of the fact that the leaflets are only three times (or slightly less) as long as broad it may be an intermediate or hybrid form between the two subspecies. The collection made by Rambo (*No. 28190*) from Sarandí in Rio Grande do Sul State is also of this type. On the other hand, *Ule 696*, also from Blumenau, is a good fit with subsp. *muelleri*.

Evidently some range of variability, as yet imperfectly known because of the scarcity of collections available, must be allowed for in this subspecies.

Correll (1962) has retained *S. muelleri* as a distinct species and has linked it with the Bolivian species *S. arnezii* (classed by us as a mountain biotype of *S. chacoense*), *S. trigalense* (= *S. tarijense*, see p. 242) and *S. yungasense*, grouping the four species into a new series, Yungasensa. He also describes a form of *S. muelleri* under the name *densipilosum*, based on a collection of Smith, Reitz, and Caldato (*No. 9516*) from Chapecó, Santa Catarina, Brazil. (See Plate 49.)

We have been able to examine the collection of Smith *et al.* and can confirm that it compares fairly well with subsp. *muelleri* apart from the fact that the whole plant is covered with long spreading hairs. The calyx acumens are short (0·5 mm) and the corolla more pubescent than in the typical form. Glandular hairs are frequent on the peduncles and pedicels. This is an extremely interesting variant of subsp. *muelleri* that would repay study in the living state.

We shall be dealing with the hybridization of '*S. muelleri*' and *S. chacoense* in the section on Fertility and Crossability (p. 227), and shall only remark here that both F_1 and F_2 hybrids of *S. chacoense* and *S. muelleri* are fully fertile and show no loss of vigour or thriftiness. Whilst this fact by itself is not enough to make it essential to group them as one and the same species, it would surely indicate that they should not be separated into two distinct series, as Correll has done. In view of the basic morphological similarity between them and the fact that they replace each other geographically, coupled with the crossability data to which we have just referred, we consider that *S. chacoense* and *S. muelleri* can most satisfactorily be considered as two subspecies within the same species. Even if they are maintained as distinct species they are certainly too closely related to be separated into two distinct series. We cannot agree, either, with Correll's grouping of Bitter's *S. jamesii* var. *grandifrons* as a form of the Mexican species *S. oxycarpum*.

NOTES ON LIVING PLANTS

1. Subspecies *chacoense*

We have grown over fifty separate collections of this subspecies, obtained from many different parts of its distribution area. There is no doubt whatsoever that subsp. *chacoense* is highly variable, especially in the north-west mountain zone where introgression from other species seems to be taking place. Morphological variation observed in herbarium specimens proved to be very largely genetically determined, when we raised plants from seeds or tubers of the same collection and grew them under uniform conditions.

One of the most difficult points to determine from herbarium specimens is the presence or absence of a basal rosette of leaves due to a shortening of the basal internodes. Many collectors have taken only the top of the plant for pressing, so that one cannot know if a rosette was present or not. In other cases the plant grew amongst tall grass, herbs, or bushes and has been drawn up by lack of light or competition with other plants, so that any genetic tendency to rosette growth would have been masked by the habitat factor.

When growing collections in Cambridge, Birmingham, and Vandel taken from the plains of Argentina and Paraguay at altitudes of 600 m or less we have been able to observe that, whatever the habit of the original plants, the European-grown material showed a marked rosetting in the early stages. When the plant becomes older the stem elongates and sometimes branches widely once or twice. The leaves on this upper stem are often smaller and more delicate than those at the base of the plant, which form the rosette. In very old plants the basal leaves may even wither away, so that the rosette form is lost. Nevertheless, it is always present in young plants, and is a most characteristic feature.

Material collected in the higher regions from about 1000 m upwards rarely shows rosetting in our experimental plots, though the stem may not always be very tall. Extreme forms, with no rosette at any stage, and thin highly branched stems, were the collections of *Balls 5936* (p.p.) (CPC: 51*b*), *5921* (CPC: 84. Type for *S. jujuyense*), *Petersen & Hjerting P–349*, and *Sleumer 3566*.

The amount of stem branching is also rather characteristic in *S. chacoense*. In the typical forms it is sparse, and the branches are short and diverge widely from the main stem. In the introgressed mountain forms the branches are long and frequent.

Stem colour in mountain forms is often intense, and laid down in a mottled or reticulate pattern. In plains forms it is normally absent or confined to touches of pigment round the leaf axils. Stem wings are developed very characteristically in many of the mountain forms, and this feature is constant under cultivation. Crisped wings with purple pigmentation are also seen in some mountain forms of *S. chacoense*.

Differences of leaf dissection are to a large extent genotypically controlled and observable in plants cultivated under uniform conditions, as also is the pubescence of leaf, stem, flower stalks, and calyx, and the form of this latter organ. On the whole, the leaflets of plains forms seem to be broader, more oblong, and less acuminate than those of the mountain forms.

Corolla differences may also be seen, both in respect of the width of the lobes and the colour. In some collections, e.g. *Sleumer 3566*, the colour is a creamy white, though generally a fairly pure white corolla is developed.

Berries are often abundant and in many cases are spotted with white dots and blotches, sometimes slightly raised above the surrounding areas. They may be 2 cm in diameter, or more.

The tubers are round to oval, often sharp-pointed or 'beaked' at the stolon end, and are developed generally on pale brown stolons, though in some mountain forms the stolons and roots are purple. The tubers themselves seem to be larger in the plains forms, attaining dimensions of 3×5(–6) cm and weights of 5–15 grammes each. The mountain forms generally bear smaller tubers of about 1×2 cm, though occasionally they attain dimensions of 3×4 or 3×5 cm. The weight is 1–6 grammes per tuber (up to 30 grammes according to Brücher 1956*a*). In colour they vary from brownish yellow to purple, and are either smooth or verrucose with enlarged lenticels.

To arrive at a better understanding of the differences between mountain and plains forms of *S. chacoense* we grew two small families of seedlings in the field at Birmingham, one from seeds collected in prov. Jujuy at 2300 m (*Petersen & Hjerting P–349*. H. 16.

dept. Tilcara, Yaquis Pampa) and the other from prov. Chaco at ±50 m (*Schultz 9027*. H. 87. dept. Resistencia, Colonia Benítez). The seedlings were scored for stem winging, leaf rosetting, leaf dissection, petiolule length, leaf and calyx pubescence, and calyx acumen length.

The mountain plants showed no rosetting in six out of eight seedlings, and only very slight rosetting in the other two; low rosette growth was well marked in all the plains seedlings. Stem wings were seen in the mountain and not in the plains seedlings, though leaf dissection was rather similar in both. Petiolule length was greater in the plains seedlings (±11 mm) than in the mountain ones (2–5 mm). Pubescence of leaf and calyx was variable in the mountain seedlings, ranging from dense to very sparse; pubescence was completely absent in all the plains seedlings. The acumen length was also more variable (0·25–0·5 mm) in the mountain than in the plains (constant at 0·25 mm) seedlings.

On the whole, then, there seemed to be much less variation in the plains seedlings than in the mountain ones, even though these latter were much more similar to each other than one would have expected in a simple F_2 or backcross progeny of *S. chacoense* × *S. microdontum*. One may therefore tentatively conclude from this rather small sample that at least some of the introgressed forms of *S. chacoense* in the north-west mountainous regions of Argentina at relatively high altitudes have become reasonably stabilized by selection. Further work on a larger scale is clearly needed to verify this conclusion, however.

The frequency with which subsp. *chacoense* invades cultivations as a weed is well known, and we have found it frequently in cotton, maize, and bean fields. It thus brings itself to the notice of the peasants and farmers to an extent much greater than most wild potatoes in Argentina; consequently it has received a number of native names, indicative of its wild or 'untamed' nature:

Sacha papa	Wild, forest, or monte potato
Papa del monte	Mountain or forest potato
Papa del diablo	Devil potato
Papa yuta	Partridge potato
Papa silvestre	Wild potato
Papa del zorro	Fox potato

2. Subspecies ***muelleri***

We have grown two collections of this subspecies (*Petersen & Hjerting P–66* and *Olavo J. Boock* (s.n.)), and have observed that it retains its characteristic leaf form quite well under cultivation. The dimensions do not attain to those of the type specimen, which was also cultivated, but this may possibly have been grown under more favourable conditions. There seems to be little evidence of a basal rosette in this subspecies, despite the fact that it grows in the plains. We must not forget, however, that it is to be found normally in subtropical forests where there would be no particular selective advantage in a rosette habit, whereas the ability to produce a long stem among bushes and other rank vegetation would have probably been of selective advantage.

The 6–7-jugate leaf with very oblique leaflet bases and lax inflorescence in plants grown at Birmingham is similar to the development in Misiones province. The flowers are normal in size, as contrasted with their very small and probably atypical size on the type specimen grown in the Vienna Botanic Garden.

S. chacoense is of great value in breeding and has been shown to carry resistance to the following diseases: wart, scab, powdery scab, early blight, charcoal rot, *Verticillium* wilt, skin spot, blackleg, bacterial ring rot, brown rot, horsenettle cyst and Osborne's cyst nematodes, Colorado beetle, *Epilachna*, leaf hopper, flea beetle, aphids, and viruses X, A, Y, F, and leaf-roll. It shows resistance to drought but not to frost and some clones have exhibited high starch, dry matter, and protein. Altogether, *S. chacoense* combines more qualities of interest to plant breeders than perhaps any other species of wild potato, with the possible exception only of *S. demissum* and *S. stoloniferum*. (See Chapter 5 for further details.)

CYTOLOGY

The chromosome number of *S. chacoense* was determined as $2n = 24$ by Smith (1927) and as $n = 12$ by Vilmorin and Simonet (1927). These latter workers used material named '*S. caldasii*' and '*S. pseudo-maglia*', which, as we have already seen, are names formerly applied incorrectly to *S. chacoense*. Since that time many cytologists have verified these counts (e.g. Ratera, 1938*b*, 1940, 1954) and have never found triploid forms, in contrast to *S. commersonii* where they are not uncommon.

We ourselves have determined the somatic chromosome number of over 50 collections of *S. chacoense* from various parts of its distribution range; in only *one* of these (CPC: 1720, from Argentina, Río de la Plata) was a somatic number of 36 recorded. From morphological and phytogeographical observations we believe this collection to be a true autotriploid.

Meiosis is quite regular in this species, with generally 12 bivalents at first metaphase (Longley and Clark 1930, Propach 1937, Choudhuri 1943, Okuno 1951, Matsubayashi 1955, Magoon *et al.* 1958, Masutani 1962*a*). Meiosis is also remarkably regular in hybrids of *S. chacoense* with other diploid species such as *S. verrucosum*, *S. phureja*, *S. goniocalyx*, and *S. catarthrum* (Emme 1937, Propach 1940, Howard and Swaminathan 1952, Magoon *et al.* 1958).

References to cytological investigations of triploid hybrids between *S. chacoense* and tetraploid species are frequent in the literature. With species such as *S. acaule* and *S. tuberosum* the triploids behave cytologically as though there were no differentiation between the genomes of these species and *S. chacoense* or of the constituent genomes of the tetraploids themselves (see Propach 1937, Prakken and Swaminathan 1952). Triploid crosses of *S. chacoense* with the Mexican species *S. stoloniferum* show a tendency to 12 bivalents + 12 univalents at meiotic metaphase, thus indicating that the two species may have one genome in common whilst the other of *S. stoloniferum* may be rather distinct (Matsubayashi 1955, Kawakami and Takayama 1955, Kawakami and Matsubayashi 1956, 1958).

Tetraploid hybrids result often from the cross *S. tuberosum* × *S. chacoense* due to the functioning of a diploid pollen grain from *S. chacoense*. These are similar to hybrids of *S. tuberosum* × (colchicine-tetraploid) *S. chacoense*, and both types of hybrid apparently give mostly 24 bivalents at meiosis. This would indicate autosyndetic pairing, but Oppenheimer (1933) and Propach (1938*a*) consider that this is probably not the correct interpretation since quadrivalents are also seen at first metaphase. Even though all four chromosome sets in the hybrid do not differ genomically to any great extent they can be assumed to exhibit preferential pairing up to a certain point; nevertheless, the evidence from quadrivalent formation and even more so from the results of breeding experiments indicates that allosyndetic pairing takes place, and useful selection for progeny with characters from both parents can be made. In this connexion the work of Kawakami and Matsubayashi (1957) on hybrids of *S. tuberosum* × *S. chacoense*, compared with tetraploid *S. chacoense*, is worth noting. Tetraploid forms of *S. chacoense* were made by Lebedeva (1965) and are of interest in their resistance to Colorado beetle and *Epilachna*. The comparison of amphiploid plants of the hybrid *S. chacoense* × *S. stenotomum*, by Masutani (1962*b*), showed mean configurations of quadrivalents and trivalents of from 1·93 to 2·26, compared with 2·39 in the autotetraploids. The differences are perhaps not very significant, indicating little genome differentiation between these two species.

More complex hybrids with hexaploids such as *S. demissum* have also been analysed cytologically (see Matsubayashi 1962), but since this latter species does not come within the area under discussion we shall not describe these in detail (see also Propach 1937, Beamish 1955).

FERTILITY AND CROSSABILITY

Like most of the other diploid potato species *S. chacoense* is normally self-incompatible, but fertile enough when biotypes with different *S*-gene combinations are crossed together. Some authors, however (Stout and Clark 1924, Pal and Pushkarnath 1951, Torka 1951), have found a complete lack of self-incompatibility in certain of the clones they investigated.

Ratera (1943*b*) states that at Villa Ortúzar *S. chacoense* was visited by the bumble bee (*Xylocopa ciliata*) as well as the common bee; it possessed abundant pollen, being 96 per cent fertile.

The ability of *S. chacoense* to form artificial or natural hybrids with other potato species is considerable. This fact, together with its important characters of resistance to viruses Y and A, and to Colorado beetle, have caused it to be one of the most popular and widely grown species, both in plant-breeding stations and botanic gardens.

Most attempts at hybridizing *S. chacoense* with diploid species in series Commersoniana, Conicibaccata, and Cuneoalata have been successful, and the F_1 hybrids have shown good fertility. It has also been hybridized with all the diploid cultivated potato species and many of the wild ones in series Tuberosa, and with *S. verrucosum* in the Mexican series Demissa. Here again, the F_1 hybrids were fertile. Choudhuri (1944), however, records reciprocal differences in the fertility of hybrids of *S. chacoense* with certain diploid species.

A reduction in fertility and sometimes a greater crossing difficulty ensues in hybrids with tetraploid species, such as *S. tuberosum*, *S. acaule*, *S. fendleri*, and *S. stoloniferum*, since the F_1 is generally triploid (sometimes tetraploid in *S. tuberosum* crosses, however). Hybrids with *S. demissum* (6*x*) are also possible, though these also are not very fertile. Their chromosome number is 48, or 60, due to an 'unreduced' *S. chacoense* gamete functioning. (See Beamish 1955.)

We shall now consider the artificial hybrids of *S. chacoense* made by ourselves and others with species in the same general geographical region so that some comparison with naturally occurring hybrids can be made.

We have not attempted a large crossing programme *within* the confines of *S. chacoense* itself, though this would have been interesting as it would have served to confirm or contradict our views on the specific boundaries of this very widespread species. However, some results in this sphere have been obtained by previous workers. Thus Longley and Clark (1930) crossed *S. chacoense* with the form known as '*S. caldasii* var. *glabrescens*'. Both F_1 and F_2 generations were fertile and vigorous. Emme (1938) obtained F_2 families from crossing '*S. gibberulosum*' (from Córdoba) with '*S. knappei*' (from Salta, mountain form) and '*S. schickii*' (from Tucumán) in different combinations. All these former species have been placed by us under *S. chacoense*, and they all gave healthy and vigorous F_2 families, as we should have expected.[1]

To elucidate the position of the former species '*S. muelleri*', which in the present treatise we have placed as a subspecies of *S. chacoense*, we used the same method of

[1] Contrast this with the 'breakdown' in F_2 generation seen in the same author's crosses of the good species, *S. phureja* × *S. chacoense*, *S. chacoense* × *S. goniocalyx* and reciprocal, and in Propach's (1940) crosses of *S. verrucosum* × *S. chacoense* and *S. chacoense* × *S. spegazzinii*.

raising F_2 families. Subsp. *muelleri* shows a constant moderate degree of distinctness from *S. chacoense* even when grown under identical conditions, but the differences are small, and seem to indicate subspecific, rather than specific rank.

The collection of subsp. *muelleri* (*P. & H. P–66*) from prov. Misiones was hybridized with a typical plains form of *S. chacoense* (CPC: 2062 from Buenos Aires province). The F_1 (family 120) showed good vigorous uniform plants with an intermediate leaflet form and the short basal internodes and semi-rosette habit of the subsp. *chacoense* parent. Two F_2 families (218, 219) were raised from separate intersib crosses and these showed good seed germination and vigour. There was some lack of uniformity in plant height and leaflet width, as was to be expected, but all plants were equally vigorous. We therefore concluded that '*S. muelleri*' could with justification be ranked as a subspecies of *S. chacoense*, geographically localized to the eastern forest zone, but not differing sufficiently in its genetic composition to warrant its retention as a species.

Attempts at crosses of *S. chacoense* with *S. commersonii* (using both subspecies) have indicated a good specific distinction between them (see p. 178).

Two small F_1 families (Nos. 244 and 273) were raised from the cross *S. chacoense* (*H.P.R. 356*, from Salta) × *S. tarijense* (*H. & R. 151*, from the same province but a different locality). These families were intermediate between the two parents, showing some of the glandular pubescence of *S. tarijense*. The inflorescence was once-branched and the calyx acumens were rather shorter than in *S. tarijense* (1–1·5 mm) though longer than in *S. chacoense*. Two F_2 families (294, 295) were raised from intersib crosses within family 273 in 1963 and showed the usual signs of genetic breakdown between species that are genetically distinct. Of the 59 seedlings planted out 13 were vigorous, 20 more or less medium-sized, 4 small, and 22 small to dwarfed and chlorotic which died rather early. A completely graded range of vigour and longevity could be seen, of which the numbers quoted above are given for the sake of convenience and are not intended to show clear-cut divisions. A few berries were produced on some of the larger plants only.

Crosses with species outside series Commersoniana were also effected. Thus an F_1 of *S. chacoense* (CPC: 2135, from Buenos Aires) × *S. infundibuliforme* (*Balls 6126*, from Bolivia, Villazón—only a few km from the Argentine border) was raised (Family 10), and cytological examination showed an unduly high proportion of diads and triads in the maturing pollen grains. The hybrids themselves were intermediate and showed some leaflet decurrency.

Reciprocal hybrids of *S. chacoense* (*P. & H. 1889*, from Tucumán) × *S. microdontum* (intermediate form) (*P. & H. P–364*, from Salta) showed a nicely intermediate and highly fertile F_1. The leaves were 2- to 3-jugate (occasionally 4-jugate) and compared closely with naturally occurring hybrids (false '*S. bijugum*') between the two species. The semi-stellate white corollas, long calyx acumens, and fairly dense pubescence are also distinctive features. (See Plates 50 and 51.)

A form collected in prov. Salta (*H.P.R. 287*), itself a probable natural hybrid of *S. chacoense* × *S. microdontum* subsp. *gigantophyllum*, gave a most heterogeneous family when crossed with *S. chacoense* (*P. & H. 1889*). This is to be expected, since the family could be regarded as a first backcross generation to *S. chacoense*. On the whole the seedlings approached the *S. chacoense* parent, though some differed little from a simple F_1.

Brücher (1953) also crossed *S. microdontum* subsp. *gigantophyllum* and *S. chacoense* together. He records that the characters of the former parent, such as the pubescence, broad leaflets, and distinctive calyx teeth were prominent in the early growth stages, while those of *S. chacoense*, such as the strong leaf dissection, became evident later.

Reciprocal crosses of *S. vernei* subsp. *ballsii* (*P. & H. P–351*, from Jujuy) and *S. chacoense* (*P. & H. 1889*, from Tucumán) gave vigorous F_1 families, as also did the combination *S. chacoense* (*H.P.R. 356*, from Salta) × *S. vernei* subsp. *ballsii* (*P. & H.*

P–351). In all F_1 families the plants were fairly uniform, lacking the dense white felting on the lower leaf surface of subsp. *ballsii*, and developing 5–6 primary and up to 9 pairs of interjected leaflets. The stem was winged and the flowers were pale mauve, varying in intensity a little from seedling to seedling. There was also some variation in the length/breadth ratio of the corolla lobes. The berries possessed slightly raised white spots.

The cross *S. gourlayi* (*P. & H. P–365*) × *S. chacoense* (CPC: 2435/1—*P. & H. P–76*) was successful, giving a limited number of seeds. Family 289 raised from this cross in two successive seasons gave only 52 seedlings from over 100 sown, of which 40 died rather soon. The 12 remaining were fairly uniform and fertile. From these, two F_2 families (337 and 338) were raised. Of a total of 270 seeds sown, only 25 germinated; some died early and none was very thrifty. Again, an F_2 breakdown showed here, indicating genome incompatibility between the two species, *S. chacoense* and *S. gourlayi*.

Crosses were also made between *S. spegazzinii* (*P. & H. 1900a*, from prov. La Rioja) and *S. chacoense* (*H.P.R. 356*, from Salta, and also *P. & H. 1889*, from Tucumán). The F_1 hybrids exhibited a leaf very similar to that of *S. chacoense*, though perhaps shorter, and pale mauve flowers intermediate in shape and colour between those of the two parents.

We also crossed *S. chacoense* (*H.P.R. 356*, from prov. Salta) as male parent, with a curious form of *S. spegazzinii* (*H.P.R. 334*, also from Salta province). The latter parent differs from the usual form of *S. spegazzinii* in the very coarse hairs and the mauve and white corolla. The F_1 was fairly uniform, and showed dominance from the *S. spegazzinii* parent in respect of leaf type, long spreading hairs, and corolla colour and form. The flowers were larger, however, with a longer style and obovate stigma. (See Plate 54.)

A vigorous and healthy F_1 family of the cross *S. chacoense* (*P. & H. 1889*, from Tucumán) × *S. kurtzianum* (*Ruiz Leal 11906*, from Mendoza) was raised (see Plate 56), and was seen to be intermediate between the two parents, although the seedlings varied a little in corolla colour; the upper surface was white, whilst the lower varied from white to pale lilac. There was also some fluctuation in corolla shape, which varied from the true rotate shape of *S. kurtzianum* to the stellate form of *S. chacoense*. The articulation of the pedicel was approximately median, showing dominance of *S. chacoense* in the F_1 for this character. In the reciprocal cross six fruits were formed but contained no seeds.

An F_1 family was also raised from the cross *S. maglia* (CPC: 2057, from Mendoza) × *S. chacoense* (*H.P.R. 56*, from Yala, Jujuy). The plant possessed 2–3-jugate leaves, white substellate flowers and anthers and filaments quite distinct from each other. The peculiar *S. maglia* stamens seem rather recessive in this cross, therefore.

NATURAL HYBRIDS

(*a*) With ***S. commersonii***

The frequency of natural hybrids between *S. chacoense* and *S. commersonii* must be very low, judging from the difficulty with which they can be made artificially (see p. 177). We have already discussed on p. 179 under *S. commersonii* the specimens that may, with some doubt, be classified as hybrids of this type.

(*b*) With ***S. tarijense***

Three natural hybrids of *S. chacoense* × *S. tarijense* have been observed by us during the present study.

The first two are from prov. Salta, dept. Orán, Urundel (*A. T. Hunziker 2012* and *2046*), on the eastern slopes of the Argentine Andes, not far from the Bolivian frontier.

Although we have no records for the occurrence of the parents in the immediate vicinity, they have been collected in the same general region (see records for *S. chacoense* and *S. tarijense* from Jujuy, depts. Ledesma and Valle Grande; Salta, depts. Santa Victoria, San Martín, and Iruya, pp. 184, 186 and 241), and under the same ecological conditions.

The leaf type of these hybrids comes, on the whole, closest to that of *S. chacoense*, but the long petiolules, glandular pubescence, and long calyx acumens of *S. tarijense* demonstrate clearly that the specimens are true hybrids.

The third collection (*H.P.R. 302*) is from much further south (Salta, Chicoana, Escoipe). *S. chacoense* has been collected from the same locality and virtually from the same altitude (*H.P.R. 307*), but *S. tarijense* had never previously been seen so far south. On the evidence of this hybrid we made a successful search in 1966 for the other parent, *S. tarijense*, and found it growing quite close to the locality of the hybrids (see Appendix III).

All three hybrids are closely similar to the artificial hybrids between *S. chacoense* and *S. tarijense* referred to in the previous section (p. 227), and there seems little doubt that further search in the regions where the first two were found would reveal the presence of both parent species there as well.

(*c*) With ***S. microdontum*** subsp. ***microdontum***

Hybrids between *S. chacoense* and the typical subspecies of *S. microdontum* are hardly known, perhaps because the latter is restricted to rather inaccessible regions not generally collected in. Yet Hjerting and Rahn collected subsp. *microdontum* (*220*) and *S. chacoense* (*221*) in the same area (Salta, Santa Victoria, Toldos, Q. de Astillero, 1600–1700 m) without finding hybrids. A hybrid collection was made in the Orán district of the same province (*P. & H. P–5*); the *S. microdontum* parent is known from about 40 km to the west (*Fabris* et al. *5860*) whilst *S. chacoense* has been found not too far away. Possibly further search in these regions would reveal the presence of both parents and hybrids in the same locality, since there is no evidence of a crossing barrier between the two species.

(*d*) With ***S. microdontum*** subsp. ***gigantophyllum***

S. chacoense hybridizes with this subspecies quite frequently, as the data on Table 11 show, where the two species overlap in the subtropical woodlands, of the 'selva tucumano-boliviana', i.e. the Montano, Orán, and Tucumán districts. Normally *S. chacoense* is found only near the roads and round cultivated places in the open or in half shade, and especially where the soil is disturbed. It spreads up, probably with man, into the forest paths and clearings where it is able to hybridize with subsp. *gigantophyllum*. We have already discussed the strong evidence for introgression of the latter species into the former, and have included all such introgressed forms with *S. chacoense* itself. Only obvious hybrids are noted here.

Many collections of Hjerting, Petersen, and Rahn were made in zones of hybridization, where all intermediates between the two parental types were observed. For convenience, these collectors made three or four collections from each area, though this is not meant to indicate that only one or two hybrid variants, together with the two parents, were seen. Such samplings from the parental and hybrid populations are as follows: *H.P.R. 51, 52, 53*; *H.P.R. 285, 286, 287*; *H.P.R. 308, 309, 310*; *H.P.R 370, 371, 377a, 377b*; *P. & H. 24, 26, 31, 32*. (See Table 11 and Plate 50.)

The normal altitude range of the hybrids is from 1300 to 2350 m, though two collections come from altitudes of 600 and 900 m, well outside the range of subsp. *gigantophyllum*.

Tubers of the hybrid collection *H.P.R. 287* from Salta, Quebrada de Yacones,

TABLE 11

Natural hybrids between S. chacoense *and other species. (Only includes those where one or both parents have been collected from the same area. See text (pp. 237–9) for the remainder of hybrid records)*

Locality	Collection number and altitude		
	S. chacoense	Hybrid	*S. tarijense*
Salta, Chicoana, Escoipe	*H.P.R. 307.* 2200 m [*H.H.R. 3703.* 2075 m]*	*H.P.R. 302.* 2100 m	[*H.H.R. 3702.* 2075 m]*
	S. chacoense	Hybrid	*S. microdontum* subsp. *gigantophyllum*
Jujuy, Capital, Lagunas de Yala	*H.P.R. 52.* 2100 m *O'Donell 4870.*	*Garolera & Romero* (s.n.). *H.P.R. 53.* 2100 m *T. Meyer 16965.*	*Abbiatti & Claps 885.* *H.P.R. 51.* 2100 m *O'Donell 2970.*
Salta, Caldera, Quebrada de Yacones	*H.P.R. 286.* 1650–1700 m	*H.P.R. 287.* 1650–1700 m	*H.P.R. 285.* 1650–1700 m
Salta, Chicoana, Pie de Cuesta del Obispo	*H.P.R. 309* (p.p.). 2350 m	*H.P.R. 309* (p.p.). 2350 m *H.P.R. 310.* 2350 m	*H.P.R. 308.* 2350 m
Salta, Guachipas, Pampa Grande	*H.P.R. 370.* 1700 m *H.P.R. 377a.* 1700 m	*H.P.R. 377b.* 1700 m	*C. Spegazzini* (various collections). *H.P.R. 371.* 1700 m
Tucumán, Tafí, El Siambón	*Lillo 165.* *Lorentz & Hieronymus 1050.* *Olea 302.* *P. & H. 940. c.* 1300 m *Rothe* (LIL 136859). *Schick* ('*S. schickii*'). *Sleumer 72.* 1300 m	*Sleumer 71.* 1300 m	*Lillo 1031.* 1100 m *Sleumer 96.* 1250 m *Lorentz & Hieronymous 802.* ('*S. gigantophyllum*')
Tucumán, Tafí, Yerba Buena	*Venturi 221* (p.p.). 600 m	*Venturi 221* (p.p). 600 m	—
Catamarca, Andalgalá 'Yunca Suma'	*P. & H. P–24.* 1450 m *P. & H. P–31.* 1400 m	*P. & H. P–32.* 1450–1500 m	*P. & H. P–26.* 1450–2350 m
Catamarca, Andalgalá, Quebrada frente del Alto de Las Juntas	*Sleumer 2273* (p.p.). *c.* 1600 m	*Sleumer 2273* (p.p.). *c.* 1600 m	—

	S. chacoense	Hybrid	*S. microdontum* (unknown subspecies)
Salta, San Martín, Río Tartagal	*Schulz & Varela 5226.*	*Eyerdam & Beetle 22640.* 900 m	—
	S. chacoense	Hybrid	*S. spegazzinii*
Salta, Guachipas, Pampa Grande	*H.P.R. 364, 370, 377a, 379, 380.* 1700–1900 m	*C. Spegazzini* (s.n.) ('*S.×setulosistylum*')	[*H.H.R. 3646, 3930.* 2650–90 m]*
Catamarca, Belén, Belén to Pozo de Piedra	*Sleumer & Vervoorst 2326.* 1300 m *Sleumer & Vervoorst 2644.* 1650–1700 m	*Sleumer & Vervoorst 2428, 2429, 2430, 2431, 2433. c.* 1900 m *2435.* 1900–2000 m *2436.* 1700–1800 m	*Sleumer & Vervoorst 2434.* 1900–2000 m
Beyond Pozo de Piedra, above Granadillas			*Sleumer & Vervoorst 2489. c.* 2650 m *2490.* 2700 m *2534.* 2800 m
Catamarca, Pomán, Qu. del Río Saujil	*Vervoorst 3525. c.* 1600 m	*Vervoorst 3526. c.* 1600 m	
	S. chacoense	Hybrid	*S. kurtzianum*
Catamarca, Andalgalá, Las Estancias	*Rodríguez V. 908.* 1500 m	*Sleumer 1642.* 1520 m (road between Alamito and Las Estancias)	
Catamarca, Pomán, between El Simbolar and Los Ciénegos	*Vervoorst 3427* (p.p.). *c.* 2000 m	*Vervoorst 3427* (p.p.). *c.* 2000 m	

* Hawkes, Hjerting, and Rahn 1966 collections. See Appendix III for details.

were grown by us and exhibited the very characteristic intermediate form of *S. chacoense* × subsp. *gigantophyllum* hybrids, with 2-jugate leaf, crisped narrow stem wings and semi-stellate white corolla. This and other 2-jugate forms caused us at first to confuse such hybrids with the species '*S. bijugum*', described by Bitter. Later, however, we were able to show that this species represented only a genetic variant of *S. microdontum* and had nothing to do with hybrids of the sort we have been describing, many of which, on the other hand, coincided well with our artificially produced F_1 families between *S. chacoense* and *S. microdontum*, as we have already shown (p. 227).

Such hybrid populations were also remarked on by Brücher (1953), who observed them in the Río Potrero (Catamarca) area and concluded that he was dealing with a natural hybrid population of F_1 and F_2 biotypes. He considered that these forms maintained themselves by vegetative propagation and were sexually sterile, with unviable pollen. All plants sampled cytologically possessed twenty-four somatic chromosomes, as was to be expected, but we consider that he was incorrect in assuming that all plants were sterile, since the evidence for fertility in F_1 and even F_2 and backcross hybrids between these species is quite strong. Brücher further concluded that the hybrid segregates were able to penetrate into drier areas of hot sandy river banks as well as fire-scarred mountain slopes where neither of the two parents was able to spread. Besides a wide range of intergrading morphological variation he also observed differential resistance to insect attack, dependent on the gene combinations inherited from the two parents. If such hybrid derivatives were better adapted to certain habitat conditions and could be 'fixed' in some way other than tuber propagation (i.e. by amphiploidy) we should be tempted to look on them as new taxa. However, there is insufficient genome differentiation between the two species to make this possible, even though we suspect that by introgression from subsp. *gigantophyllum*, *S. chacoense* has extended its range and variability to a very great extent. The hybrids show a complete range of variation between the two parents and, in our opinion, should not be given taxonomic or nomenclatural rank.

(*e*) Probable hybrid of ***S. chacoense*** with ***S. microdontum*** of unknown subspecies

A collection made by Eyerdam and Beetle (*22640*) from the very low altitude of 900 m probably represents a hybrid formed in the mountains much higher up and washed down by the Río Tartagal to the position in which it was found. Possibly subsp. *microdontum* was involved in this cross, though little is known of the distribution of the two subspecies of *S. microdontum* in that area.

(*f*) Possible hybrid of ***S. chacoense*** with ***S. vernei*** or ***S. microdontum***

Only one possible hybrid between *S. chacoense* and *S. vernei* has been observed. This is a gathering by Garolera and Romero from Jujuy, Termas de Reyes.

Both *S. chacoense* and *S. microdontum* are known to occur in that area, whilst *S. vernei* is not. The specimen might therefore be a *S. chacoense* × *S. microdontum* hybrid, probably an F_2 segregate or a backcross to *S. chacoense*. Yet it possesses some curious features, such as a somewhat velvety pubescence on the lower leaf surface and marked stem pigmentation. These characteristics could have been derived from *S. vernei*, though our specimen certainly does not match any of the artificial F_1 hybrids raised by us from *S. chacoense* × *S. vernei*, or indeed from *S. chacoense* × *S. microdontum*. A search was made in this locality in 1966 for *S. vernei* or hybrids of it with *S. chacoense* but without success. It is possible, however, that *S. vernei* could have been washed down the valley from its usual habitat higher up, and before it died out might have formed a short-lived hybrid with *S. chacoense*. (See also Appendix III.)

The curious absence of undoubted *S. chacoense* × *S. vernei* hybrids may probably be explained most satisfactorily by reference to the altitude ranges of these two species. *S. chacoense* extends up to 2000 m, and occasionally to 2350 m or over. The range of *S. vernei* extends down to 2200 m and occasionally to 1850 m. There is thus a gap of 200 m between their usual ranges, though exceptional overlaps occur. Evidently these are so rare as to make the chances of natural hybridization extremely unlikely. We can contrast this situation with *S. microdontum* in relation to *S. chacoense*, which overlap in their *usual* ranges for about 600 m, so allowing natural hybrids to be produced with some frequency.

(*g*) With ***S. spegazzinii***

(*S.*×*setulosistylum* Bitt. (in *Reprium nov. Spec. Regni veg.* **12**, 450–2, 1913))
(*S.*×*puberulofructum* Corr. (*Wrightia* **2**, 193, 1961))

In the drier valleys of north-west Argentina *S. chacoense* comes into contact with the more drought-loving species, *S. spegazzinii*, forming hybrids with it in several localities.

The 'species' *S.*×*setulosistylum*, described by Bitter in 1913 on the basis of a collection made by Carlos Spegazzini at Pampa Grande, prov. Salta (see Fig. 21), is undoubtedly a hybrid of this kind. The highly dissected leaf, with characteristically long, narrow interjected leaflets, is very similar to that of *S. spegazzinii*, though the lateral leaflets seem to show some influence from *S. chacoense*. The corolla is more or less stellate, white or pale mauve in colour, and strongly recalls the corolla of the Sleumer and Vervoorst hybrid collections from Pozo de Piedra, to be described below.

Correll (1962) treats *S.*×*setulosistylum* as a good species, though he concedes that Hawkes's suggestion, made in 1956, that it was a hybrid derived from *S. chacoense*×*S. leptophyes*,[1] is not unlikely. He further states that the assumption first needs verification through experimentation before adoption, and with that we would certainly agree. However, we consider that the experimental hybrids we have been able to make between *S. chacoense* and other species, together with the populations of seedlings that we have raised from natural hybrids, give a strong indication of the hybrid nature of *S.*×*setulosistylum*. Apart from the type specimen,[2] Correll examined a number of collections from north-west Argentina, some of which we would undoubtedly place as hybrids, on the basis of the intermediate morphological features between the two parental species, *S. chacoense* and *S. spegazzinii*. Some of the other specimens cited would seem to be pure *S. spegazzinii*, however.

Although part of the Sleumer and Vervoorst collections from Pozo de Piedra was classified by Correll as *S.*×*setulosistylum*, another part was named *S. puberulofructum* Corr. chiefly on the basis of the curious puberulent apiculate fruits; it must be remembered, however, that only immature fruits were seen, and it is doubtful whether this character would persist until maturity. (See Appendix III.) Another collection, (*Sleumer & Vervoorst 2434*), is cited as the type of yet another new species, *S. sleumeri*.

It seems to us now, though admittedly for several years we were not too sure, that these curious collections from Pozo de Piedra are hybrids (see further, p. 235), occurring together with their parents (see Table 11), and that the segregation of characters has led Correll, most understandably, to think of them as distinct species. Brücher (1964*a*) considers *S.*×*setulosistylum* to be no more than *S. chacoense*, though he notes that occasional examples with long (1·5 mm) calyx acumens and low articulation are found.

[1] At that time it was thought that all forms of *S. spegazzinii* should be included under *S. leptophyes*. We now think that they may best be considered as two distinct species.

[2] Correll reported that he was unable to see setae on this specimen. They are, however, present.

We shall now return to a more detailed discussion of the type material of *S.* × *setulosistylum* from Pampa Grande. Style setae are to be seen in the single type specimen but in no other gatherings of this hybrid or of *S. spegazzinii* itself from any part of its range. Nor are such setae known in *S. chacoense* outside the Pampa Grande area.

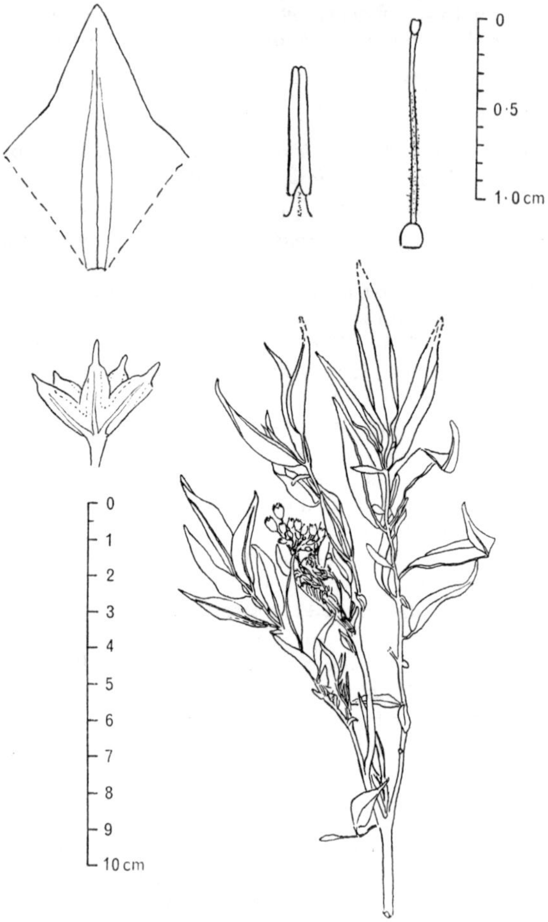

FIG. 21. Natural hybrid of *S. chacoense* Bitt. × *S. spegazzinii* Bitt. Type specimen of '*S.* × *setulosistylum*' Bitt., collected at Pampa Grande, prov. Salta, Argentina by Carlos Spegazzini. This shows the highly dissected leaf and narrow leaflets of *S. spegazzinii* and the white stellate corolla of *S. chacoense* (cf. Sleumer and Vervoorst's collections from Pozo de Piedra, Plates 52, 53). Note also the well-marked style setae, which are occasionally found in material of other species from Pampa Grande.

Bearing in mind the significance of the Pampa Grande region for *Solanum* taxonomy, Hjerting, Petersen, and Rahn sampled in great detail the wild potatoes growing there during their visit in 1956. The styles of all these collections were examined subsequently, but in only one collection of *S. chacoense* (*H.P.R. 380*) were style setae discovered; even then they occurred only in two out of the four specimens of the collection. Two further visits were made to Pampa Grande in 1966. On that occasion we found style setae on one collection of *S. chacoense* and two of *S. microdontum* from the general region. (See Appendix III.) We accordingly decided to examine the styles of every other collection available to us from the area to see whether setae could be found in other species, too.

Setae were seen in one collection of *S. microdontum* subsp. *gigantophyllum*, gathered by Carlos Spegazzini in Pampa Grande, and used by Bitter as the type for *S. simplicifolium* var. *mollifrons*. Bitter mentions the presence of 2–4-celled acute hairs on the style of this specimen, though in this case he does not call them setae. We were not able to find setae on the styles of any other collections from this area, however.

Hjerting, Petersen, and Rahn did not find *S. spegazzinii* or *S.* × *setulosistylum* at Pampa Grande during their visit in 1956; we were fortunate enough to discover *S. spegazzinii* there in 1966 though we did not see *S.* × *setulosistylum* (see Appendix III). *S. chacoense* and *S. microdontum* are extremely common, though the hybrids between them are not frequent at Pampa Grande. Possibly the climate and vegetation may have changed since Spegazzini's visit in 1897. Or, what is more likely, Spegazzini's collections may have been made in some other part of this rather extensive Estancia not visited by Hjerting, Petersen, and Rahn in 1955 or by ourselves in 1966. A further search is still needed to clear up this point.

From the data available it would seem most likely that a mutation controlling the development of style setae took place in this area, probably in *S. chacoense*, and has spread by means of natural hybridization both into *S. microdontum* and into the *S. chacoense* × *S. spegazzinii* hybrid (= *S.* × *setulosistylum*) collected by Carlos Spegazzini in 1897. The fact that only six out of a very large number of collections made at Pampa Grande possess style setae would rather indicate that selection pressure in favour of this mutant cannot be very high. Perhaps its frequency might be greater in the little-explored region to the south or west of Pampa Grande, and there is no doubt that the whole area would repay further study.

The facts allow us to conclude in any case that style setae cannot be considered as a species-diagnostic character in this group, and we therefore recommend that the use of the name *S.* × *setulosistylum* be discontinued.

Apart from this single specimen from Salta (Pampa Grande) and one other from Tucumán (Valle de Tafí) all the other recorded hybrids of *S. chacoense* × *S. spegazzinii* are from Catamarca. Here, thanks to the remarkably detailed and interesting collections made by Sleumer and Vervoorst in depts. Belén and Pomán, two important series of hybrid derivatives have been discovered.

In the vicinity of the town of Belén (1300 m) and again at 11 km (1650–1700 m) along the road from Belén which ascends to Pozo de Piedra *S. chacoense* was collected by Sleumer and Vervoorst in the dry prepuna vegetation (*2326* and *2644*). At about 1700 m a zone of hybridization began, extending along the same path for a distance, from 11 to 16 km from Belén (*Sleumer & Vervoorst 2436*, *2428*, *2429*, *2430*, *2431*, and *2433*). Here were gathered hybrids of great exuberance and vigour, with the leaf form of *S. spegazzinii*, although the actual dimensions and degree of dissection were greater than that of either of the parents. The flowering, also, seemed to be very prolific in these hybrids, and the presence of some young green berries is evidence that they were not all sterile. In most of the collections from approximately 1900 m (*2428* (see Plate 52), *2429*, *2430*, *2431*; but **not** *2436*) the flowers are large, white, and stellate, showing the overwhelming influence here of *S. chacoense*. A slight purple flush develops in the older flowers of *2429*, however. In the last and first collections of this group, *2433* possesses white flowers that are constantly violet externally, whilst *2436* has violet flowers.

It should be noted that *Sleumer & Vervoorst 2645*, which was also collected in the same area (but at 1700 m), has been identified as *S. kurtzianum*. It seems very possible that *S. kurtzianum* may have had some influence in the formation of this complex series of hybrids at about 1900 m altitude, even though they appear to be composed mainly of *S. chacoense* × *S. spegazzinii* crosses. According to Sleumer and Vervoorst the whole area was very dry, with an annual precipitation of probably about 230 mm and a dry loess type of soil.

Further away from Belén, at 15–18 km from the town, on the same path and in the same soil conditions, but at the slightly higher altitude range 1900–2000 m, a much smaller influence of *S. chacoense* and correspondingly a greater one of *S. spegazzinii* is seen. Thus, collection *2435* has pale violet flowers and greenish-purple puberulent berries, though the leaf dissection is similar to the 11–16 km group. It was this collection that was named by Correll '*S. puberulofructum*', and although the puberulent fruits are curious, all the other characters fit well into the series of hybrids we have been discussing (see Plate 53). Collection *2434* is, in our opinion, fairly normal *S. spegazzinii* though Correll makes it the type specimen of his new species, *S. sleumeri*. Unfortunately, no collections were made from just higher up in this region, but it seems fairly safe to presume that more *S. spegazzinii* would have been found there, as it was seen not far away, at altitudes of 2600 m and over.

Such collections as these, made with care and very fully documented, are a magnificent aid to the taxonomist, and the collectors cannot be praised too highly. (See Appendix III for a further discussion of this interesting region.)

Vervoorst's collections in Pomán, Río Saujil, are unfortunately not so complete, but the data are of great utility. With collection *3525* (*S. chacoense*), made at 1600 m, Vervoorst records that *S. chacoense* forms almost pure stands in damp places in the bottom of the quebrada. The hybrid (note with *3526*) is said not to be as abundant as *S. chacoense*, alternating with it and growing in damp places where there are still primitive cultivation terraces. In its leaf it recalls *S. chacoense*, but the purple flowers reveal its hybrid nature. Although *S. spegazzinii* was not collected in the vicinity it is very likely to occur there.

(*h*) With ***S. kurtzianum***

S. chacoense has undoubtedly formed hybrids here and there with *S. kurtzianum*, a species that, like *S. spegazzinii*, seems to be drought resistant to some extent, but occurs on the whole at a rather lower altitude range.

The hybrids retain the fairly moderate dissection of *S. chacoense* leaves (in contrast to the highly dissected leaf of the *S. chacoense* × *S. spegazzinii* hybrids); the pedicel articulation seems to be low, as in *S. kurtzianum*, but may be up to median or slightly above in some plants (see note on median pedicel articulation in artificial hybrid section, p. 228). The calyx acumens are well developed but not clearly delimited from the lobes; the corolla is white or with a faint streak of mauve on the outside centre of each petal. These natural hybrids match the artificial ones fairly closely.

Vervoorst's collection *3427*, which we have split into *S. chacoense* and its hybrid with *S. kurtzianum* (see Plate 55), was described as follows by Sleumer: 'The material of *3427* was collected in a place where even today ancient Indian cultivation terraces may be seen. One part of the material is certainly identical with *S. chacoense* . . . whilst the other part possesses flowers tinted outside with pale violet: influence of another species with violet flowers?'

Seeds from berries taken from the probable hybrid plants were germinated at Birmingham (Accession H. 61) and gave rise to a mixed population of hybrid seedlings, ranging from ones much closer to *S. chacoense* to others very much closer to *S. kurtzianum*. They therefore behaved in the way expected from our assumption, and that of Sleumer, that the original collection from which the seeds were taken was some sort of complex hybrid, or perhaps even an F_1 between *S. chacoense* and *S. kurtzianum*.

Another hybrid of *S. chacoense* with *S. kurtzianum* from prov. Catamarca was collected by Sleumer (*1642*) in dept. Andalgalá. *S. chacoense* was obtained by Rodríguez V. (*908*) in the same locality and at roughly the same altitude. *S. kurtzianum* has not been recorded for that area, though it is known from not far away (*P. & H. P–40*. Rio Pantanillos, 1800 m).

Hybrids of the same nature have also been collected in prov. La Rioja, dept. Capital, by R. Alaníz (*60*) and by A. P. Rodrigo (*2069*). The parents are not yet known for these exact localities, though they have been recorded in the same general region. However, we have some doubt about these two collections; they may be nothing more than *S. chacoense*.

Three collections of Ruiz Leal (*4281*, *4327*, *22078*) from prov. Mendoza are also hybrids of this kind. They possess a median to below-centre pedicel articulation, semi-stellate corolla, long calyx acumens and rather coarse hairs on stems, leaves, peduncles, pedicels, and calyx. For most collections the corolla colour is not noted. The third collection was actually gathered by Brücher and determined by him as *S. ruiz-lealii* (see below). *S. kurtzianum* is known from many collections made in this province, whilst *S. chacoense* was obtained at Rama Caída (CPC: 2140) in the same province, though at a probably lower altitude. Further collections are needed to verify the extent to which *S. chacoense* has spread into Mendoza province and how much it has been able to hybridize there with *S. kurtzianum*.

Brücher (1962) described a new species, *S. ruiz-lealii*, which seems to be no more than a hybrid of *S. chacoense* × *S. kurtzianum*. He indicates that *S. ruiz-lealii* differs from *S. chacoense* in the calyx, with 2 mm long lobes and scattered hairs and in the large rotate or subrotate corolla. From *S. kurtzianum* it is said to differ in the white flowers and fruit without white spots. We have seen a specimen kindly sent to us by Professor Ruiz Leal (*22078*) which seems to be identical with material we ourselves have placed as hybrids of *S. chacoense* × *S. kurtzianum*. We also made collections of this 'species' in our 1965–6 expedition from an authenticated locality, and were able to examine further living and dried material in Professor Ruiz-Leal's collections. All evidence strengthened our view as to the hybrid nature of this 'species'. (See also Appendix III.)

LOCALITIES OF NATURAL HYBRIDS

(*a*) With ***S. commersonii***. See p. 180.

(*b*) With ***S. tarijense***

Prov. Salta, dept. Chicoana, 2 km arriba de Escoipe. Alt. 2100 m. 22 Feb. 1956. *Hjerting, Petersen & Rahn 302* (J.G.H., J.P.H.).

Prov. Salta, dept. Orán, Urundel, Arroyo Nipán. 12 May 1942. *A. T. Hunziker 2012* (LIL, US). (Classified by Correll, 1962, p. 224, as *S. trigalense* [≡ *S. tarijense*].)

Prov. Salta, dept. Orán, Urundel, 14 May 1942. 'En terreno recién desboscado; frecuente; flores blancas.' *A. T. Hunziker 2046* (CORD, LIL, SI).

(*c*) With ***S. microdontum*** subsp. ***microdontum***

Prov. Salta, dept. Orán, Quebrada del Diablo. Alt. 400–500 m. 25 Oct. 1948. 'En la selva subtropical.' *Petersen & Hjerting P–5* (J.G.H.). (Possibly washed down the river from a higher altitude.)

(*d*) With ***S. microdontum*** subsp. ***gigantophyllum***

Prov. JUJUY:

Dept. Capital, Lagunas de Yala. 13 Jan. 1947. *Garolera & Romero* (s.n.) (LIL 207809).

Dept. Capital, Lagunas de Yala. Alt. 2100 m. 1 Feb. 1956. 'En el cerco de piedras, cerca de la casa.' *Hjerting, Petersen & Rahn 53* (J.G.H., J.P.H.).

Dept. Capital, Lagunas de Yala. 13 Feb. 1951. *T. Meyer 16965* (LIL).

Dept. Capital, Lagunas de Yala. Alt. 2000 m. 17 Feb. 1959. *Ross & Vervoorst 76* (Max.).

Prov. SALTA:

Dept. Caldera, Quebrada de Yacones. Alt. 1650–1700 m. 20 Feb. 1956. *Hjerting, Petersen & Rahn 287* (J.G.H., J.P.H.).

Dept. Chicoana, un poco arriba de la Hostería, Pie de Cuesta del Obispo. Alt. 2350 m. 22 Feb. 1956. *Hjerting, Petersen & Rahn 309* (p.p.) (J.G.H.).

Locality data as for 309 above. *Hjerting, Petersen & Rahn 310* (J.G.H.).

Dept. Guachipas, Estancia Pampa Grande, cerca de la represa. Alt. 1700 m. 25 Feb. 1956. *Hjerting, Petersen & Rahn 377b* (J.G.H., J.P.H.).

Prov. TUCUMÁN:

Dept. Tafí, El Siambón, en la chacra del Sr. Dionisio Olea. Alt. 1300 m. 15 Jan. 1950. 'Spontanea.' *Sleumer 71* (LIL).

Dept. Tafí, Yerba Buena. Alt. 600 m. Feb. 1919. 'Faldas de los cerros. Flor blanca. Pl. 30–60 cm.' *Venturi 221* (p.p.) (LIL, SI, US).

Prov. CATAMARCA:

Dept. Andalgalá, Estancia 'Yunka Suma', cerca del edificio principal. Alt. 1450–1500 m. 18 Jan. 1949 and 21 Feb. 1950. *Petersen & Hjerting P–32* (LIL, J.G.H., J.P.H.). See Plate 50.

Dept. Andalgalá, quebrada en frente del Alto de las Juntas. Alt. *c.* 1600 m. 19 Jan. 1952. 'Fl. alb.' *Sleumer 2273* (p.p.) (LIL).

(*e*) Probable hybrid of ***S. chacoense*** with ***S. microdontum,*** subspecies unknown.

Prov. Salta, dept. San Martín, Río Tartagal, 25 km west of Manuela Pedraza. Alt. 900 m. 26 Oct. 1938. 'In sandy mud along shady river bank. Perennial herb, 30–50 cm; petals white, stamens yellow.' *Eyerdam & Beetle 22640* (GH, K, UC).

(*f*) Possible hybrid with ***S. vernei*** (or with ***S. microdontum***)

Prov. Jujuy, dept. Capital, Termas de Reyes. 12 Jan. 1947. *Garolera & Romero* (s.n.) (LIL 184221, 184234).

(*g*) With ***S. spegazzinii***

Prov. SALTA:

Dept. Guachipas, Pampa Grande. Jan. 1897. *C. Spegazzini* (s.n.) (LPS 14215). (Type for '*S. setulosistylum*' Bitt.) See Fig. 21.

Prov. TUCUMÁN:

Dept. Tafí, Tafí del Valle, playa del Río Churqui. Alt. 2080 m. 16 Dec. 1912. *Castillón 2284* (LIL).

Prov. CATAMARCA:

Dept. Belén, Pozo de Piedra, km 15–16. Alt. *c.* 1900 m. 2 Feb. 1952. 'Fl. alb. magni, fr. virid.' *Sleumer & Vervoorst 2428* (LIL, J.G.H.). See Plate 52.

Dept. Belén, Pozo de Piedra, km 15–16. Alt. *c.* 1900 m. 1 Feb. 1952. 'Fl. alb. magni, postea violascentes.' *Sleumer & Vervoorst 2429* (LIL).

Dept. Belén, Pozo de Piedra, km 15–16. Alt. *c.* 1900 m. 1 Feb. 1952. 'Fl. alb. magni.' *Sleumer & Vervoorst 2430* (LIL).

Dept. Belén, Pozo de Piedra, km 15–16. Alt. *c.* 1900 m. 1 Feb. 1952. 'Fl. alb. magni, fr. virid.' *Sleumer & Vervoorst 2431* (LIL, J.G.H.).

Dept. Belén, Pozo de Piedra, km 15–16. Alt. 1900 m. 1 Feb. 1952. 'Fl. alb., extrinsecus violascentes.' *Sleumer & Vervoorst 2433* (LIL).

Dept. Belén, Pozo de Piedra, lugares secos (Loess) a lo largo del camino, km 15–18. Alt. 1900–2000 m. 25 Jan.–2 Feb. 1952. 'Fl. viol. clar.; fr. viridi-violasc. Hojas angostas.' *Sleumer & Vervoorst 2435* (L, LIL, US, J.G.H.). (This collection forms the type of *S. puberulofructum* Correll.) See Plate 53. (Close to *S. spegazzinii* Bitt.)

Dept. Belén, en la orilla del camino de Yacutula a Pozo de Piedra. Km 11–13. Alt. *c.* 1700–1800 m. 24 Jan. 1952. 'Fl. violasc. Folia valde angusta. Terreno muy seco (Loess). Precip. anual más o menos 230 mm.' *Sleumer & Vervoorst 2436* (LIL, US, J.G.H.). (Close to *S. spegazzinii* Bitt.)

Dept. Pomán, Saujil, El Potrero, quebrada Río Saujil. Alt. *c.* 1600 m. 18 Feb. 1952. 'Flor morada. No es tan abundante como la de flor blanca; alterna con ella; crece en lugares húmedos, donde hay andenes primitivos.' *Vervoorst 3526* (J.G.H.).

Exact locality unknown: Material sent to Birmingham by Dr. H. Ross, under number 51/151. EBS 212, and named *S. setulosistylum*. Grown from seed. Accession H. 41 (J.G.H.).

(*h*) With ***S. kurtzianum*** (≡ *S. ruiz-lealii*)

Prov. CATAMARCA:

Dept. Andalgalá, camino entre Alamito y Las Estancias. Alt. *c.* 1520 m. 20 Feb. 1951. 'Fl. palide viol.' *Sleumer 1642* (LIL, US).

Dept. Pomán, entre El Simbolar y Los Ciénegos, donde edificio vieja escuela. Alt. *c.* 2000 m. 27 Feb. 1952. 'Faldos secos, cerca campos cultivados, donde se observan andenes primitivos. Fl. blanco-crema.' *Vervoorst 3427* (p.p.) (J.G.H.). See Plate 55.

Prov. LA RIOJA:

Dept. Capital, Los Duraznillos, al km 32. Alt. 1200 m. 10 March 1942. 'Herbácea, flor blanca.' *R. Alaniz 60* (LIL, LP, SI).

Dept. Capital, Quebrada Los Sauces, Sierra de Velasco. Feb. 1939. 'Flor blanca.' *A. P. Rodrigo 2069* (LP). *Note*: These two collections may be *S. chacoense* and not the hybrid with *S. kurtzianum*.

Prov. MENDOZA:

Dept. Junín, Rodríguez Peña. Cultivada en dept. Capital, La Ciudad. Feb. 1937. *Ruiz Leal 4327* (Ruiz Leal).

Dept. San Carlos, Arroyo Aguanda cerca de Chilecito. 10 Feb. 1962. Leg. & det. H. Brücher. 'Fl. blanca.' *Ruiz Leal 22078* (Ruiz Leal).

Dept. San Rafael. Cultivada en Rodríguez Peña (dept. Junín) de tubérculos silvestres. 5 Jan. 1937. *Ruiz Leal 4281* (LIL, Ruiz Leal, J.G.H.).

5. SOLANUM TARIJENSE Hawkes

(*Bull. imp. Bur. Pl. Breed. Genet., Camb.* 114–15, 1944).
S. zudaniense Cárdenas (*Boln Soc. peru. Bot.* **5**, 31–32, 1956).
S. trigalense Cárdenas (ibid. **5**, 41–42, 1956).
S. berthaultii Hawkes f. *zudaniense* (Càrd.) Correll (*Wrightia* **2**, 184, 1961).

Herbaceous, tuberiferous, with a strong but not unpleasant characteristic odour. Tubers round to oval, up to 4 cm long, with pale yellowish-brown skin and white flesh, borne on long stolons.

Stem ascending to a height of 20–40 cm, branched or simple, angled, with short internodes 2–5 cm long, often with slightly crisped or straight wings decurrent from the leaf bases, very sparsely endowed with weak spreading hairs and *very frequent stalked glands*.

Leaf to 20 cm long × 9 cm broad, with (3–)4–5 pairs of primary leaflets which are rather remotely placed from each other, giving the leaf an 'open' appearance; interjected leaflets 0–12-jugate, 0–3 pairs between each pair of primary leaflets, orbicular, cordate, or ovate, sessile or very shortly petiolulate, 2–10 mm long; lateral primary leaf segments *long petiolulate*, the petiolules 3–15 mm long, the leaflets themselves ovate to ovate-lanceolate or elliptic with *obtuse apex* which is often slightly acuminate and a rather cordate not very oblique base; uppermost laterals 32–55 × 15–27 mm; terminal similar to laterals but sometimes slightly smaller; margins of leaflets often undulate, markedly denticulate, the teeth often with coarse hairs at their apices; leaves rather pale green, often folded longitudinally, endowed above and below *over the whole surface with small stalked glands* and weak spreading hairs; both hairs and glands are also particularly visible on rhachis and petiolules; semilunar pseudostipulate leaves well marked, up to 1 cm or more long.

Inflorescence 5–20 flowered; peduncle furcate, to 8 cm long below and 2–3 cm above the fork, thickly covered with glands and sparse to frequent weak spreading hairs; pedicels 10–25 mm long, *curved and thickened towards the apex*, articulated at or above the centre, 8–12 mm long above the articulation, with glandular and non-glandular pubescence similar to that of the peduncles.

Calyx 6–12 mm long, the free part of the lobes 3–9 mm, including the *well-marked sublinear 1·5–7 mm long acumens*; external pubescence as for peduncle and pedicels.

Corolla *white or cream, stellate*, very reflexed, about 25–35 mm diam. when spread out flat, with lobes, 7–13 mm wide at the base and 10–14 mm long.

Filaments glabrous, about 2 mm long in the free part; anthers 5–7 mm long × 2 mm wide near the base.

Style 10–14 mm long, exserted for 3–5 mm from the anther column, covered with frequent minute papillae from slightly above the base to about $\frac{1}{2}$–$\frac{3}{5}$ its length; stigma slightly thicker than the style apex, globular to clavate.

Berries globular or slightly ovate, green with slightly raised white spots and streaks. Chromosome number: $2n = 24$.

AFFINITIES

S. tarijense is placed in the series Commersoniana by virtue of its stellate corolla. It is distinguished from *S. commersonii* by the well-differentiated leaflets with long petiolules, the white corolla (except in the case of introgressed forms), and the dense covering of short stalked glands. From *S. chacoense* it is distinguished by the long and often irregularly lengthed calyx acumens, and also by the characteristic glandular pubescence. This latter feature also serves to distinguish it easily from the other two species in series Commersoniana, *S. calvescens*, and *S. yungasense*.

Its possible relationship to *S. berthaultii*, a blue-flowered glandular species with semi-stellate to pentagonal corolla from Bolivia, remains a matter of doubt. We consider that the similarities between *S. berthaultii* and *S. tarijense* may probably be explained by postulating that *S. berthaultii* represents the product of natural hybridizations between *S. tarijense* and some blue-flowered species from higher altitudes in series Tuberosa. Detailed collections of dried and seed material from Bolivia are needed to verify this hypothesis.

SPECIMENS EXAMINED

SOUTH BOLIVIA

Dept. CHUQUISACA:

Prov. Azero, between Monteagudo and Lagunillas. Petrol district. Alt. 800–900 m. July 1941. *Cárdenas 3515* (CPC: 1727) (P.G.S., J.G.H.).

Dept. TARIJA:

Tarija, above the left bank of the river a little way above Tarija. Alt. 2050 m. 26 Feb. 1939. 'Growing in light, sandy loam under a spiny leguminous shrub amongst other herbage. Seemed to be confined entirely to this one position.' *Balls 6093* (CPC: 144) (K, P.G.S.—Lectotype, J.G.H.).

Tarija. *Cárdenas* (s.n.) (J.G.H. No. H. 17). See Plate 57.

ARGENTINA

Prov. JUJUY:

Dept. Valle Grande, Calilegua. 10 May 1942. *A. T. Hunziker 1970* (LIL, SI).

Prov. SALTA:

Dept. Iruya, San Pedro, arriba de la escuela. Alt. 2300–2800 m. 5 Apr. 1952. *Petersen & Hjerting P–363* (LIL, J.G.H., J.P.H.).

Dept. Santa Victoria, Santa Victoria, en el mismo pueblo. Alt. 2500 m. 8 Feb. 1956. 'Aroma. También común en los campos de maíz.' *Hjerting & Rahn 151* (C, J.G.H., J.P.H.).

Dept. Santa Victoria, 1–4 km de Pie de Cuesta de Mesón hacia Santa Victoria. Alt. 2450 m. 15 Feb. 1956. *Hjerting & Rahn 237* (C, J.G.H., J.P.H.).

Dept. Santa Victoria, Santa Victoria. Alt. 2385 m. 20 Jan. 1943. *T. Meyer 4624* (LIL).

Dept. Santa Victoria, Santa Victoria. Alt. 2385 m. Jan. 1943. *T. Meyer 4891* (LIL).

Dept. Santa Victoria, Santa Victoria, Quebrada de San Felipe. Alt. *c.* 2400 m. 5 Feb. 1953. 'Fl. alb., Papa silvestre.' *Sleumer 3785* (LIL, J.G.H.).

Dept. Santa Victoria, Cuesta de Mesón. Alt. 2600 m. 12 Feb. 1953. 'Fl. alb.' *Sleumer 3999* (LIL, J.G.H.).

Dept. Santa Victoria, Arroyo San Felipe Antiguo. Alt. *c.* 2450 m. 15 Feb. 1953. 'Fl. blanca. Un poco abajo del rancho de Don Rafael Aparicio, en la orilla del arroyo.' *Sleumer 4023* (LIL, J.G.H.). See Plate 58.

Probably ***S. tarijense***:

Prov. SALTA:

Dept. Santa Victoria, La Huerta (cerca de Santa Victoria). Alt. 2700 m. 16 Feb. 1956. 'Sin flor.' *Hjerting & Rahn 242* (J.G.H.).

DISTRIBUTION

Eastern Andes of Bolivia and north-west Argentina. (See Map 5.)

It has been recorded for Bolivia in depts. Tarija, Chuquisaca, Cochabamba, and Santa Cruz.

In Argentina *S. tarijense* is apparently confined to prov. Jujuy, dept. Valle Grande, and to prov. Salta, depts. Chicoana (see Appendix III), Iruya, and Santa Victoria. It may also occur in prov. Salta, depts. Orán and La Viña, on the evidence of certain probable hybrids of *S. tarijense* with other species in those departments (see pp. 228 and 245).

HABITAT

Grows at medium high altitudes from (800–)2050 to 2800 m, but is probably most frequent in the 2400–2600 m range.

It seems to prefer open places amongst grass, not or only to a limited extent overshadowed by other vegetation, and apparently never in very humid regions. It has also been recorded from dry stony soil amongst cacti and thorny xerophilous shrubs, in maize fields, and in walls.

The vegetational zones which it inhabits in Argentina would apparently correspond approximately with the drier part of the Montano district ('pajonales') of the Western Subtropical province and parts of the Prepuna province, according to Cabrera's phytogeographical scheme. In Bolivia it would possibly correspond with the less dry parts of the Prepuna province, and may even extend into the higher parts of the Chaco province, though confirmatory evidence is required for this.

TAXONOMIC NOTES

S. tarijense was described first by Hawkes (1944) from plants collected by Balls, Gourlay, and Hawkes in Bolivia, dept. Tarija, near Tarija (*Balls* 6093; CPC: 144). No holotype was then cited, and we now designate one of the sheets in herb. P.G.S. as LECTOTYPE for *S. tarijense*. This species was included in series Commersoniana because of the stellate corolla that is so typical of this series. Bukasov (1955*b*) separated a series Glabrescentia from series Commersoniana, and included in it not only the glabrescent species *S. chacoense* and its related microspecies (all grouped by us within *S. chacoense* itself), but also the very glandular species *S. tarijense*. We have already mentioned the inadvisability of separating series Commersoniana into a number of smaller series, and even if Glabrescentia were retained one could hardly group the densely glandular-pubescent *S. tarijense* in it if the character of glabrescence were the basis for separating this new series. Indeed, Bukasov specifically states, in the Latin description given in a later publication (Bukasov & Kameraz 1959), that the series Glabrescentia is distinguished by the glabrescent leaves (*folia glabrescentia*). He then says that *S. tarijense* should be included in it, stating, however, that '*S. tarijense* is somewhat differentiated from the species previously listed because of its thick glandular pubescence'.

Cárdenas (1956) considers *S. tarijense* as a species intermediate between series Commersoniana and Tuberosa, and therefore presumably referable to neither series.

Correll (1962) separates *S. tarijense* into a new series, Tarijensa; he also places *S. berthaultii* in the same series, including as forms of *S. tarijense* Cárdenas's species *Solanum vallegrandense* and its variety *pojoense*. On the other hand, Cárdenas's *S. trigalense*, which we have placed as a synonym of *S. tarijense*, has by Correll been relegated to yet another new series, Yungasensa.

Since we are concerned primarily with the potatoes of Argentina, Paraguay, and other countries to the south and east of that region, we should not perhaps dwell too long on problems concerned with Bolivian species. However, the fact that *S. tarijense* just enters the northern tip of Argentina makes it appropriate to say a few more words about this rather difficult group. We are convinced that no useful purpose is served by creating two distinct series for part of the Commersoniana, and we cannot agree with Bukasov or Correll in this respect. It seems more and more clear that *S. tarijense* has formed hybrids with certain blue-flowered species in the valleys of north-west Argentina and east Bolivia, and that there has been some gene flow from these hybrids into *S. tarijense* itself. We shall describe on p. 245 the hybrids examined by us in Argentina, together with the introgressed forms of *S. tarijense* in the same area.

It seems very likely that the Bolivian species, *S. berthaultii*, represents the product of hybridization between *S. tarijense* and some high mountain species with blue flowers, and that Cárdenas's *S. zudaniense* (from Bolivia, dept. Cochabamba, Zudañez) and *S. trigalense* (from Bolivia, dept. Santa Cruz, Trigal) are biotypes of *S. tarijense* which show some influence from this primary hybridization. Indeed, *S. trigalense* seems to us to be practically pure *S. tarijense*, whilst *S. zudaniense* is not too far removed from the typical form, even though Correll places it as a forma of *S. berthaultii*. On the other hand, *S. vallegrandense* and its variety *pojoense* seem to us to be much closer to *S. berthaultii*, that is to say, they show a much greater influence from the non-glandular blue-flowered species from higher altitudes; and we cannot agree with Correll that they should be classed as a variety of *S. tarijense*, unless, indeed, the whole group of variants is so placed. Until we are in a position to know more about the detailed variation of *S. tarijense* and *S. berthaultii* in Bolivia, we shall place *S. zudaniense* and *S. trigalense* as synonyms of *S. tarijense*; and *S. vallegrandense* with its variety *pojoense* as synonyms of *S. berthaultii*. We do not consider it advisable to speak of the introgressed blue-flowered forms of *S. tarijense* from prov. Salta as *S. berthaultii*, since it is rather doubtful whether the same specific source of the blue-flowered character would be involved in Argentina.

NOTES ON LIVING PLANTS

When grown at Vandel under short days plants of *S. tarijense* gave an average tuber weight of 8 grammes. They were not frost-resistant in the field.

Field and glasshouse plants both in Denmark and England show a typically glistening appearance, especially when seen in sunshine, due to their dense glandular pubescence. The aroma when the leaves are rubbed between finger and thumb is strong but quite pleasant, a little like that of the tomato.

A variation in leaf dissection has been observed in various lines of *S. tarijense* when grown from seed. Thus, the type collection (*Balls 6093*, CPC: 144) gave seedlings with primary leaflets from 4–5-jugate, and interjecteds from 1–9 pairs per leaf. The collection (H. 17) sent by Professor Cárdenas several years later from approximately the same locality possessed 4–5 pairs of primary lateral leaflets and 3–5 pairs of interjected ones (see Plate 57). In the collection from the Bolivian department of Chuquisaca (*Cárdenas 3515*, CPC: 1727) the plants were taller, more branched, and with longer leaves; 3–5-jugate primary laterals, and only 0–3-jugate interjected leaflets were observed in various seedlings. In these biotypes also there was a larger amount of spreading non-glandular pubescence as compared with that of the type collection.

Although the flowers in *S. tarijense* are typically white there seems to be a slight tendency to a very pale blue, mauve, or lilac tinge in some seedlings of certain collections from Bolivia (H. 17) and Argentina (*Hjerting & Rahn 151*). Segregation for coloured and white flowers in seedling populations in several collections from north Argentina indicates introgression of genes from a blue or purple-flowered species, possibly *S. vernei*, *S. vidaurrei*, or *S. gourlayi*. These collections will be dealt with in detail below when the problem of natural hybridization is being discussed. The calyx of all seedlings shows the well-marked linear acumens, though there is a great variation in their length (1·5–4 mm), sometimes even in the same flower (cf. *S. microdontum*). The stigma varies in dimensions slightly from collection to collection but is generally rather longer than broad, with a well-marked basal sinus.

Fruit production is not excessive in most lines that we have grown; the berries are globular, with typical raised white spots or blotches which often become confluent, forming raised lines or streaks.

Resistance to wart, early blight, *Verticillium* wilt, common eelworm, horsenettle cyst and Osborne's cyst nematodes, *Epilachna*, leaf hopper, flea beetle, and virus X (see Chapter 5 for further details) has been noted in *S. tarijense*.

CYTOLOGY

A chromosome number of $2n = 24$ was first established by Hawkes (1944) for this species. Since that time we have made counts of other collections (CPC: 1727; H. 17; *Hjerting & Rahn 151, 234, 238*) of both pure and introgressed *S. tarijense*; in every case a count of $2n = 24$ was recorded.

Magoon, Hougas, and Cooper (1958*a*) studied pollen mother cell meiosis in *S. tarijense* and its hybrid with *S. phureja*. Although they found small percentages of anomalies, meiosis was, on the whole, quite regular, both in parents and hybrids; they could find little significant difference between the species and hybrids in this respect, and even the percentage stainable pollen was about the same. Beamish and Shejkal (1955) and Beamish *et al.* (1957) made colchicine-tetraploids of *S. tarijense* and the cross *S. tarijense* × *S. phureja* in an attempt to cross them with the cultivar Katahdin of *S. tuberosum*. Meiosis in the c-tetraploid of *S. tarijense* is not described, but that in the c-amphiploid referred to above is irregular, with an average of 18·08 bivalents and 2·8 trivalents+quadrivalents per cell. The seedlings from the cross with *S. tuberosum* showed 1·8–3·5 trivalents+quadrivalents per cell.

FERTILITY AND CROSSABILITY

Self-incompatibility is to be found in *S. tarijense*, though Marks (personal communication) succeeded in obtaining selfed seed in one instance. Choudhuri (1944) also mentions that this species is self-sterile.

Hybrids with other diploid species are quite easy to obtain, but according to Choudhuri (1944) some reciprocal crosses are sterile. Thus, *S. chacoense* × *S. tarijense* is fertile: the reciprocal is sterile. *S. phureja* × *S. tarijense* is sterile: the reciprocal is fertile.

Fertile hybrids of *S. tarijense* by *S. infundibuliforme*, *S. microdontum*, *S. goniocalyx* and *S. stenotomum* have also been recorded by Choudhuri. Grun (1961) made reciprocal hybrids between *S. tarijense* and *S. kurtzianum*.

In crosses of *S. demissum* × *S. tarijense* (Hougas 1955; Hougas *et al.* 1955) tetraploid hybrids were obtained. Walker (1955) reports few seeds and many aborted embryos for this type of cross, similar to the results obtained from *S. demissum* × *S. chacoense* crosses.

Families of seedlings from the cross *S. chacoense* (*H.P.R. 356*) × *S. tarijense* (*H. & R. 151*; see also p. 227) were raised by us and showed good vigour. In most characters they were intermediate between the two parents; a noticeable feature was the much sparser glandular pubescence on the leaves and stems of the F_1 hybrids, though the peduncles and pedicels were still fairly densely glandular, and the calyx acumens were quite well marked.

We have also been able to hybridize *S. vernei*, subsp. *ballsii* (*P. & H. P–351*) with S. tarijense (*H.P.R. 151*). From seven berries only two seeds giving one F_1 plant were produced. This was not very thrifty and did not flower, possibly due to the dry weather in the year that it was grown (1959). The pubescence was a mixture of glandular and non-glandular hairs, typical of each parent. A cross of *S. kurtzianum* (*Ruiz Leal 11906*) × *S. tarijense* (CPC: 1727/3) was not very successful, since it gave poor seeds, only one of which germinated (Family 122). This plant was unthrifty and died young. This result contrasts with the success recorded by Grun (see above) for *S. tarijense* × *S. kurtzianum* crosses.

NATURAL HYBRIDS

(*a*) With ***S. chacoense.*** See p. 228.

(*b*) With ***S. microdontum.***

The only natural hybrid of *S. tarijense* × *S. microdontum* recorded by us (*Hjerting & Rahn 167*; see Plate 59) possesses white flowers but no fruits, leaves with 2–3 pairs of broad lateral leaflets, decreasing rapidly in size towards the leaf base, a large broad terminal leaflet, hardly any interjected leaflets, and a mixture of glandular and acute spreading hairs on the pedicels. The tubers are quite large. Both *S. tarijense* (*H. & R. 237*) and *S. microdontum* were found growing in the area of this hybrid. Possibly the typical subspecies (subsp. *microdontum*) was involved here, for it has been recorded from the same area, though 100 m lower (*H. & R. 236*). A collection of *S. microdontum* from the same locality and altitude as the hybrid (*H. & R. 168*) has been placed by us in a group of intermediate forms which show characteristics of both subspecies (see p. 338). It would therefore be unwise to be too definite about the subspecies of *S. microdontum* involved in this cross, since it seems to have taken place in a zone of intergradation between subsp. *microdontum* and subsp. *gigantophyllum.*

(*c*) With hybridization and introgression from one or more blue-flowered species (***S. vernei, S. vidaurrei,*** or ***S. gourlayi***)

The hybrids to be discussed in this section were found in prov. Salta at altitudes of 2400–2600 m in the region of Santa Victoria (Cuesta de Mesón and San Felipe) where the distribution areas of *S. tarijense*, *S. vernei*, *S. vidaurrei*, and *S. gourlayi* come into contact.

Most plants possess pale blue to pale lilac-coloured flowers, but are on the whole vegetatively similar to *S. tarijense*; in some instances, however, the glandular pubescence is less evident. The corolla lobes are wider than those of pure *S. tarijense*, with some demarcation between the lobe and the acumen; this feature is especially well marked in *Sleumer 3786*. (See Plate 60.) The following collections approach *S. tarijense* very closely, differing for the most part only in the pale mauve or lilac-coloured flowers and slightly sparser glandular pubescence: *Sleumer 4002, 4024*; *H. & R. 234, 238*. More intermediate collections are those of *Sleumer 3786, 4000*; and *Hjerting & Rahn 169*, where the leaflets are narrower and the petiolules shorter.

We were fortunate in being able to grow seedlings from a number of these collections (*Sleumer 3786, 4024*; *H. & R. 234, 238*). In these families some segregation for flower colour and leaf type was observed. All the plants were, in general, similar to *S. tarijense*; a few were a little more like *S. vidaurrei* or possibly *S. gourlayi*, though the resemblance was not well marked. The white-verrucose fruit character of *S. tarijense* also segregated in these families.

Whilst it has not been possible, with the material available to us, to solve the problem of the source of the blue-coloured flowers and other characters in these hybrids, it would seem most probable that *S. gourlayi* or *S. vidaurrei* might be the species involved. The terrain in the Santa Victoria region where the hybrids occur is very complex, and the ecological barriers between most species that grow in this region are, to some extent, broken down. It seems certain that little (if any) of the *S. tarijense* in that region is uncontaminated with genes from these blue-flowered species, though further collections would be needed to verify the point.

(*d*) With ***S. spegazzinii***

A probable hybrid of *S. tarijense* × *S. spegazzinii* was collected by Ross and Vervoorst near La Viña in prov. Salta (No. 60). It combines the dense coarse pubescence of *S. spegazzinii* with the glandular pubescence of *S. tarijense*, and exhibits a leaf dissection and leaflet dimensions roughly intermediate between the two species. The calyx acumens are fairly long (cf. *S. tarijense*) and the corolla white and sub-stellate.

Although the *S. spegazzinii* parent is known from the vicinity, *S. tarijense* has not been recorded from this locality. However, *S. tarijense* was collected by us in 1966 (see Appendix III) from Escoipe in dept. Chicoana, only 30–40 km to the north. There seems no reason, therefore, why it should not occur in the La Viña region also.

LOCALITIES OF NATURAL HYBRIDS

ARGENTINA

Prov. SALTA:

(*a*) With ***S. chacoense.*** See p. 237.

(*b*) With ***S. microdontum.***

Dept. Santa Victoria, Pie de Cuesta de Mesón. Alt. 2500 m. 9 Feb. 1956. 'Estéril? Flor blanca.' *Hjerting & Rahn 167* (J.G.H., J.P.H.). See Plate 59.

(*c*) With hybridization and introgression from one or more blue-flowered species (***S. vernei, S. vidaurrei,*** or ***S. gourlayi***)

Dept. Santa Victoria, Cuesta de Mesón. Alt. 2600 m. 9 Feb. 1956. 'Flor ligeramente lila.' *Hjerting & Rahn 169* (J.G.H., J.P.H.).

Dept. Santa Victoria, Pie de Cuesta de Mesón. Alt. 2450 m. 15 Feb. 1956. 'Frutos con puntos verdes. Flor blanca.' *Hjerting & Rahn 234* (C, J.G.H., J.P.H.).

Dept. Santa Victoria, 4 km de Pie de Cuesta de Mesón hacia Santa Victoria. Alt. 2500 m. 15 Feb. 1956. 'Flor lila clara; con frutos.' *Hjerting & Rahn 238* (J.G.H., J.P.H.).

Dept. Santa Victoria, Quebrada de San Felipe. Alt. 2400 m. 5 Feb. 1953. 'Fl. viol. pálid.' *Sleumer 3786* (H. 57) (LIL, J.G.H.). See Plate 60.

Dept. Santa Victoria, Cuesta de Mesón. Alt. *c.* 2600 m. 12 Feb. 1953. 'Fl. viol. pálid.' *Sleumer 4000* (LIL, J.G.H.).

Dept. Santa Victoria, Cuesta de Mesón. Alt. *c.* 2500 m. 12 Feb. 1953. 'Fl. viol. pálid.' *Sleumer 4002* (LIL, J.G.H.).

Dept. Santa Victoria, Arroyo San Felipe Antiguo. Alt. *c.* 2450 m. 15 Feb. 1953. 'Fl. viol. En la orilla del arroyo debajo del rancho de Don Rafael Aparicio.' *Sleumer 4024* (H. 58) (LIL, J.G.H.).

(*d*) With ***S. spegazzinii***

Dept. La Viña, near La Viña, Quebrada de Molina. Alt. 1300 m. 15 Feb. 1959. 'Bergquebrachowald, an feuchter Stelle, im Schatten.' *Ross & Vervoorst 60* (Max.).

SERIES ACAULIA JUZ.

(*Izv. Akad. Nauk SSSR* **2**, 316, 1937, *nomen nudum*;
ex Buk. & Kameraz, *Bases of Potato Breeding*, 21, 1959)

PLANTS herbaceous, low, generally more or less stemless but occasionally forming long stems under appropriate growth conditions, bearing stolons and tubers; leaves interruptedly imparipinnate, with typically obtuse leaflets auricled at the base on the acroscopic side. Peduncles very short or absent. Pedicel articulation absent or shown only by a ring of pigment, very rarely well marked, when present situated very high up, just below the calyx. Corolla small, blue-purple to white, rotate, with very short lobes. Berries persistent, not falling off at maturity because of the lack of well-developed abscission layer, more or less globose, partly penetrating into the soil because of the positively geotropic curvature of the pedicels.

General distribution: extending from northern Peru (dept. Cajamarca) through Bolivia to north-west Argentina, occurring as a sequence of well-marked geographical races or subspecies, at very high altitudes (2600 to 4650 m).

Chromosome numbers: $2n = 48, 72$.[1]

The series was not validated by a Latin description until 1959, though a short English description was published by Hawkes in 1956 and one in German by Brücher in 1959.

Although it has been customary to place only *S. acaule* in this series, Bukasov (in Bukasov and Kameraz 1959) considers that the two hybridogenic cultivated species, *S. juzepczukii* and *S. curtilobum*, as well as the wild species, *S. bruecheri*, should also be included in series Acaulia. Since these species are natural hybrids of *S. acaule* with species in other series (Tuberosa or Megistacroloba) it is difficult from a logical point of view to decide how to place them, since there is equal justification for putting them in the series where the other parents belong, or indeed in creating new series for them. On the whole, we are in favour of an empirical approach here, and have therefore left the two cultivated species in series Tuberosa where they were originally placed and in which they can be most easily identified. There is some doubt about the parentage of *S. bruecheri* as described by Correll (since some specimens seem to be hybrids of *S. infundibuliforme* × *S. megistacrolobum*, rather than *S. acaule* × *S. megistacrolobum*, which was what Brücher originally thought) but we have for the time being placed this species in series Acaulia and discussed it in the section dealing with *S. acaule* hybrids (p. 266).

Brücher (1959*a*) considered that series Acaulia contained two species, the tetraploid *S. acaule* and the diploid *S. aemulans*. Bukasov viewed the series as containing a number of species, one of which we treat as a subspecies (*S. punae*) and the others (*S. depexum*, *S. schreiteri*) we consider to be no more than minor variants of the single polymorphic species, *S. acaule*.

Species of this series included in the present treatment:

S. ACAULE Bitt.

[1] Brücher (1959*a*) claims $2n = 24$ for '*S. aemulans*'.

6. SOLANUM ACAULE Bitter

(in *Reprium nov. Spec. Regni veg.* **11**, 391–3, 1912)

S. acaule var. *subexinterruptum* Bitt. (in *Reprium nov. Spec. Regni veg.* **11**, 393–4, 1912).
S. acaule var. *caulescens* Bitt. (ibid. **12**, 453–4, 1913).
S. depexum Juz. (*Izv. Akad. Nauk SSSR* **2**, 317–18, 1937).
S. schreiteri Buk. *nomen nudum* (*Vest. sots. Rasteniev.* No. 4, 13, 1940).
S. acaule var. *checcae* Hawkes (*Bull. imp. Bur. Pl. Breed. Genet., Camb.* 115, 1944).
S. uyunense Cárd. (*Boln Soc. peru. Bot.* **5**, 33–35, 1956).

A low-growing rosette-forming species with generally very short stems, though these may be 10–15 cm or more under shade or amongst rocks and other vegetation. Underground stolons often long and thick, bearing round or elongated tubers up to at least 2 cm diam. or growing up through the soil to form small daughter rosettes.

Leaf (3–)6–20(–30) cm long, rather narrow, with petiole almost sheathing stem at base; lateral leaflets smaller than terminal, (2–)4–5(–7)-paired according to biotype and soil conditions, 12–30(–40)×(5–)10–20(–25) mm, narrow to broad ovate-elliptic to oblong, obtuse and rounded or somewhat acute at the apex, the acroscopic side lobed at the base and the basiscopic side lobed or markedly decurrent at the base, according to the subspecies; interjected leaflets 0–2(–4)-paired; terminal leaflet up to 70×50 mm, wider and larger than laterals, orbicular to obovate-rhomboid; leaf surface somewhat rugose; leaf and stem with sparse to frequent short crisped sub-adpressed hairs or with long weak spreading ones, according to subspecies.

Inflorescence 1–10-flowered; *peduncle very short or absent*; pedicels 13–35 mm long; *articulation absent* or marked by a ring of pigment or a change in colour or thickness of the pedicel, sometimes quite well marked, always (when visible) *placed high up*, some 2–4 mm below calyx; pedicel glabrous or provided with moderately sparse to frequent hairs and occasional stalked glands.

Calyx small, 4–6 mm, campanulate, with triangular lobes and short acumen, 0·5–2·5 mm; pubescence as for pedicel.

Corolla *very rotate*, 8–20(–25) mm diam., generally blue or violet, occasionally pale mauve to white; *lobes very short*, giving corolla a 10-lobed appearance, with small broad-triangular acumen 1·5–2·5 mm long.

Filaments slender, 1·2–2·0 mm long; anthers 2–3 mm long.

Style slender, 4–6·5 mm long, with minute papillae on lower $\frac{1}{3}$–$\frac{1}{2}$; stigma large, globose to ovoid, 0·5–1 mm diam.

Berries orbicular to ovoid or ovoid-cordate, sometimes acutish at the apex, 1–2 cm diam., not separating at the articulation at maturity. Fruiting pedicels bent downwards, apparently pushing fruits into the soil.

Chromosome number: $2n = 48$ (72).

AFFINITIES

S. acaule is apparently not very closely related to any other species of wild potato, although there are certain points of similarity between subsp. *aemulans* and species in series Megistacroloba. It is easily distinguished by its low rosette habit, leaflets auriculate basally on the acroscopic side and often slightly decurrent on the basiscopic side, very short peduncle, pedicels with the articulation generally absent or only weakly defined and then placed very high up, small rotate corolla with very short lobes, short style, and large capitate stigma.

KEY TO SUBSPECIES

1. Upper lateral leaflets broadly decurrent; leaflets decreasing rapidly in size from apex to base of leaf; pedicel articulation always clearly visible subsp. ***aemulans***

Upper lateral leaflets not or only slightly decurrent; leaflets not decreasing rapidly in size from apex to base of leaf (except in very high mountains); pedicel articulation generally invisible 2

2. Pubescence of crisped shortish subadpressed hairs; articulation sometimes visible subsp. ***acaule***

Pubescence of long weak spreading hairs; articulation hardly visible (Central and Northern Peru) { subsp. ***punae*** / subsp. ***albicans*** }

1. Subspecies ***acaule***

(For synonyms see p. 248)

Lateral leaflets not much shorter than terminal, not decreasing in size very rapidly from apex to base of leaf (except in juvenile leaves or on plants growing at very high altitudes—4000 m or over); uppermost one or two pairs of leaflets not or only slightly decurrent; articulation invisible or sometimes weakly marked, rarely distinct; fruits remaining on plant when mature. Pubescence of fairly short crisped hairs.

2. Subspecies ***aemulans*** (Bitter & Wittmack) Hawkes & Hjerting, *comb. nov.*

(*Scott. Pl. Breed. Stn. Rec.* 116, 1963)

S. aemulans Bitt. & Wittm. (in Engler, *Bot. Jb.* **50,** Suppl. 553–5, 1914).
S. depexum Juz. var. *chorruense* Hawkes (*Bull. imp. Bur. Pl. Br. Genet., Camb.* 115, 1944).
S. acaule var. *aemulans* (Bitt. & Wittm.) Corr. (*Wrightia* **2,** 169, 1961).

Differs from subsp. *acaule* in the short sublyrate leaf with 0–4 pairs of lateral leaflets decreasing in size rapidly from apex to base of leaf, and of which the upper 1–2 pairs are broadly decurrent; terminal leaflet much larger than laterals, long elliptic or long obovate; pedicel articulation well marked, though fruits apparently not abscissing when mature. Flower colour varies from blue to white. Pubescence as for subsp. *acaule.*

3. Subspecies ***punae*** (Juz.) Hawkes & Hjerting, *comb. nov.*

(*Scott. Pl. Breed. Stn. Rec.* 117, 1963)

S. punae Juz. (*Izv. Akad. Nauk SSSR* **2,** 316–17, 1937).
S. acaule var. *punae* (Juz.) Hawkes (*Bull. imp. Bur. Pl. Breed. Genet., Camb.* 23, 1944). See p. 261.

4. Subspecies ***albicans*** (Ochoa) Hawkes, *comb. nov.*

(*Scott. Pl. Breed. Stn. Rec.* 117, 1963)

S. acaule Bitt. var. *albicans* Ochoa (*Agronomía, Lima,* **27,** 363–4, 1960). See p. 262.

SPECIMENS EXAMINED

1. Subspecies *acaule*

BOLIVIA

Dept. Tarija, Puna Patanca. 8 Jan. 1904. *Fiebrig 3429* (S—Lectotype, F, G, HBG, JE, M, NY, SI, W, Z). See Fig. 22 and Plate 61.

ARGENTINA

Prov. JUJUY:

Dept. Cochinoca, Sierra de Cochinoca, entre Abra Pampa y Rinconada. 29 Jan. 1940. 'Flor azul.' *Cabrera 7727* (GH, LP).

Dept. Cochinoca, Abra Pampa, Cerro Huancar. Alt. 3500 m. 23 Jan. 1948. 'Arenales. Flores azules.' *Cabrera 9420* (BAB, LP, P.G.S.).

Dept. Cochinoca, Abra Pampa, Azul Pampa. Alt. *c.* 3600 m. 31 Dec. 1900. Leg. Fritz Claren, *F. Kurtz 11303* (CORD, S).

Dept. Cochinoca, Cochinoca. Alt. *c.* 3500 m. 1 Jan. 1901. Leg. Fritz Claren, *F. Kurtz 11327* (p.p.) (BAF, CORD, S).

Dept. Humahuaca, Sierra de Zenta. Alt. 4500 m. Mar. 1931. 'Flor azul morada.' *Budín 1501* (SI).

Dept. Humahuaca. Sierra de Zenta. Alt. 4500 m. Feb. 1931. 'Color de la flor azul violáceo.' *Budín 7439* (LIL).

Dept. Humahuaca, Tres Cruces. Alt. 3800 m. 24 Nov. 1916. 'Corola azul.' *Lillo 18154* (LIL).

Dept. Humahuaca, El Aguilar, cerca del Molino de Mina Aguilar. Alt. 4100 m. 28 Mar. 1952. *Petersen & Hjerting P–358* (p.p.) (LIL, J.G.H.).

Dept. Humahuaca, Sierra del Aguilar, arriba de la Mina Aguilar. Alt. 4650 m. 29 Mar. 1952. *Petersen & Hjerting P–361* (LIL, J.G.H., J.P.H.). See Plate 64.

Dept. Humahuaca, Chaupi Rodero, entre Iturbe y Cóndor. Alt. 3600 m. 9 Apr. 1952. *Petersen & Hjerting P–369* (LIL, J.G.H., J.P.H.).

Dept. Humahuaca, Mina Aguilar. Alt. 4500 m. 18 Jan. 1953. 'Flor violácea.' *Sleumer 3462* (LIL, J.G.H.).

Dept. Rinconada, Pirquitas, Quebrada Carabajal. Alt. 4200 m. 21 Jan. 1948. *Cabrera 9391* (BAB, P.G.S.).

Dept. Rinconada, Rinconada. Alt. 3900 m. 24 Jan. 1948. 'Flor azul.' *Cabrera 9447* (LP).

Dept. Santa Catalina, La Ciénaga, Río de San Juan Mayo. 19 Jan. 1960. *Cigliano* (s.n.) (LP).

Dept. Santa Catalina, Santa Catalina. Alt. 3400–4300 m (3650 m). 9 Jan. 1901. 'In arenos ripariis.' Leg. Fritz Claren, *F. Kurtz 11396* (CORD, S).

Dept. Tilcara, above Tilcara, San Gregorio. Alt. 3900 m. 11 Feb. 1939. *Balls 5987* (p.p.) (K, P.G.S.).

Dept. Tilcara, route from Tilcara to San Antonio at Laguna Colorada. Alt. 4100 m. 13 Feb. 1939. 'Growing in patches of rich deep soil at camping site.' *Balls 6025* (P.G.S.).

Dept. Tilcara, near Tilcara, Chorro Valley. Alt. 3650 m. 13 Feb. 1939. 'Deep rich humus, very moist, among herbage.' *Balls 6028* (CPC: 115) (J.G.H., P.G.S.).

Dept. Tilcara, Laguna Colorada. Alt. 4000 m. 19 Jan. 1906. 'Arenoso.' *Budín 23* (LIL).

Dept. Tilcara, Huacalera, Pampa Corral. Alt. 4000 m. 1 Mar. 1955. 'Flores azules.' *Cabrera 12112* (LP).

Dept. Tilcara, Abra de Punta Corral. Alt. 3700 m. 20 Mar. 1952. 'Sin flor.' *Petersen & Hjerting P–338* (J.G.H.).

Dept. Tilcara, Estancia Vieja, cerca de la casa. Alt. 3700 m. 21 Mar. 1952. 'En un corral.' *Petersen & Hjerting P–344* (LIL, J.G.H., J.P.H.).

Dept. Tilcara, Abra de Mal Paso, Abra Mayo. Alt. 3300 m. 21 Mar. 1952. *Petersen & Hjerting P–346* (J.G.H.).

Dept. Tilcara, entre Quebr. del Chorro y Laguna Colorada. Alt. 3500 m. 24 Mar. 1952. *Petersen & Hjerting P–353* (p.p.) (as seedlings raised at Vandel in 1955—J.G.H., J.P.H.).

Dept. Tilcara, Quebrada de San Gregorio. Alt. 4000 m. 24 Mar. 1952. *Petersen & Hjerting P–357* (LIL, J.G.H., J.P.H.).

Dept. Tilcara, Tilcara, arriba de San Gregorio. Alt. *c.* 3900–4050 m. 27 Jan. 1953. 'Fl. violácea.' *Sleumer 3589* (LIL, J.G.H.).

Dept. Tilcara, Abra de Remate, subida. Alt. *c.* 3950 m. 23 Feb. 1953. 'Fruit.' *Sleumer 4114* (p.p.) (LIL, J.G.H.).

Dept. Tumbaya, Quebrada de Punta Corral. Alt. 3800 m. 20 Mar. 1952. *Petersen & Hjerting P–341* (J.G.H.).

Dept. Tumbaya, San Antonio (Abra) de los Pibes. Alt. 3850 m. 5 Jan. 1953. 'Flor violácea.' *Sleumer 3252* (LIL, J.G.H.).

Dept. Tumbaya, Volcán. Alt. 2600 m. 5 Mar. 1929. 'Fl. blanca. En los prados.' *Venturi 9526* (GH, US).

Dept. Yavi, Cajas. Alt. 3900 m. 6 Feb. 1956. 'Growing together with *S. infundibuliforme*, *S. megistacrolobum* and their hybrids.' *Hjerting, Petersen & Rahn 109* (J.G.H., J.P.H.).

Dept. Yavi, Yavi. Alt. 3450 m. 16 Feb. 1953. *Petersen & Hjerting 970* (seeds only. Herb. made at Vandel in 1956. J.G.H., J.P.H.).

Dept. Yavi, Cajas, 23 km este de Yavi. Alt. 3950 m. 16 Feb. 1953. *Petersen & Hjerting 974* (J.G.H.).

Dept. Yavi, Cangrejillos, prope Yavi. 31 Dec. 1901. 'Loco uliginoso graminoso.' Leg. R. E. Fries, *Regnell 979* (GOET, P, S).

Dept. Yavi, subida a la Abra de Lizoite. Alt. 4400 m. 1 Feb. 1953. 'Flor violácea.' *Sleumer 3674* (LIL, J.G.H.).

[Dept. unknown], Purrahuas. Apr. 1922. Leg. F. Chausette, *Hauman 13* (BRLU).

Prov. Jujuy or Salta, La Peña, inter Jujuy et Salta.

Rosenstiel (s.n.) (LE, WIR) (type of *S. depexum* Juz.). See Plate 62.

Prov. Jujuy or Salta, San José, inter Jujuy et Salta, 1932.

Schick (s.n.) (WIR, No. K. 3.) (type of *S. schreiteri* nom. nud.). See Plate 63.

Prov. SALTA:

Dept. Cachi, Cuesta del Obispo. Alt. 3720 m. 21 Jan. 1947. *Garolera & Romero* (s.n.) (LIL 184338).

Dept. Cachi, Cerro de Cachi. Alt. 3500 m. 10 Mar. 1927. 'Entre los pastizales del cerro.' *Venturi 6993* (US).

Dept. Cafayate, Sierra de los Quilmes. Alt. 3780 m. 27 Jan. 1943. *Castellanos* (s.n.) (BA 46991*a*).

Dept. Chicoana, Cuesta del Obispo. Alt. 3400 m. 22 Feb. 1956. *Hjerting, Petersen & Rahn 312* (C, J.G.H., J.P.H.).

Dept. Los Andes, Chorillos. Mar. 1930. *Budín 15* (BA).

Dept. Los Andes, San Antonio de los Cobres, Quebradita. 29 Jan. 1944. 'Flores azules. Tuberígeno.' *Cabrera 8264* (LP, US).

Dept. Los Andes, Región de la Puna, Cerros. 21 Jan. 1949. *Cabrera & Schwabe 11* (LP).

Dept. Los Andes, San Antonio de los Cobres. 21 Feb. 1927. *Castellanos* (s.n.) (BA 27/902).

Dept. Rosario de Lerma, quebrada entre Muñaño y Tastil. Alt. 3900 m. 9 Feb. 1946. 'Flor azul.' *Cabrera 8998* (LP).

Dept. Rosario de Lerma, El Alisal. Alt. 2800 m. 10 Jan. 1917. 'Flor violácea azulada. Laderas.' *Lillo 16672* (LIL).

Dept. Rosario de Lerma, El Alisal. Alt. 2800 m. 10 Jan. 1914. 'Flores violetas azuladas; habita en laderas.' *Rodríguez 1258* (BAF, LIL).

Dept. Santa Victoria, Mesón, entre Santa Victoria y Toldos. Alt. 3150 m. 10 Feb. 1956. *Hjerting & Rahn 173* (C, J.G.H., J.P.H.). See Plate 65.

Prov. TUCUMÁN:

Dept. Tafí, Sierras Calchaquíes, La Queñoa. Alt. 3000 m. 28 Jan. 1933. 'En cerro muy destrozado, flor azul.' *Burkart 5274* (GH, SI).

Dept. Tafí. Verano 1906–7. *Castillón 468* (LIL).

Dept. Tafí, Infiernillo, La Quesería. 22 Dec. 1913. *Castillón 3195* (LIL).

Dept. Tafí, Cerro Muñoz, Ciénaga Grande. Jan. 1915. *Castillón* (s.n.) (LIL 62592).

Dept. Tafí, La Queñoa (Cerro San José). Alt. 2800 m. Mar. 1933. 'Flor azul.' *R. Díaz* (s.n.) (LIL 73623).

Dept. Tafí, km 83 entre Infiernillo y Tafí, cerca de Quebrada Honda. Alt. 3000 m. 19 Jan. 1956. *Hjerting, Petersen & Rahn 2* (J.P.H.).

Dept. Tafí, La Ciénaga. Alt. 2400 m. 13 Dec. 1888. 'Campos secos'. *Lillo 1231* (LIL).

Dept. Tafí, Cerro Muñoz. Alt. 4000 m. 25 Feb. 1905. 'Prados (raro).' *Lillo 4212* (LIL).

Dept. Tafí, Cerro Muñoz. Alt. 3900 m. 27 Jan. 1908. 'Prados entre rocas.' *Lillo 7403* (LIL).

Dept. Tafí, La Ciénaga, Sierra de Tucumán. 10–14 Jan. 1874. *Lorentz & Hieronymus 662* (CORD, GOET).

Dept. Tafí, Infiernillo. Alt. *c.* 3000 m. 1 Nov. 1947. *O'Donell 4596* (LIL).

Dept. Tafí, Cumbre de Chasquivil. 11 Jan. 1945. 'Flor azul.' *Olea 221* (LIL).

Dept. Tafí, Cumbre de Chasquivil. 12 Jan. 1945. 'Flor azul.' *Olea 253* (LIL).

Dept. Tafí, La Queñoa. Alt. 3000 m. 27 Jan. 1933. 'Común en los suelos arenosos de las cumbres.' *L.R. Parodi 10775* (WIR).

Dept. Tafí, Infiernillo, Quebrada Honda. Alt. 3000–3200 m. 16 and 20 Feb. 1952. *Petersen & Hjerting P–329* (C, J.G.H., J.P.H.).

Dept. Tafí, Cumbres de Calchaquíes, Laguna Verde. Alt. 4200 m. 17 Feb. 1952. *Petersen & Hjerting P–330* (≡ *Sparre 9617*) (LIL, J.G.H.).

Dept. Tafí, Infiernillo, Quebrada de Alazán. Alt. 3900 m. 17 Feb. 1952. *Petersen & Hjerting P–331* (≡ *Sparre 9650*) (LIL, J.G.H.).

Dept. Tafí, Cumbres de Calchaquíes, cerca de Cerro Negrito. Alt. 4000 m. 12 Feb. 1952. *Petersen & Hjerting P–335* (≡ *Sparre 9546*) (LIL, J.G.H.).

Dept. Tafí, Lara. Alt. 3200 m. 20 Feb. 1912. 'Valles; flor azul.' *Rodríguez 577* (LIL, SI).

Dept. Tafí, Cumbre de Cajón (Distrito de Colalao del Valle). Alt. 3900 m. 18 Feb. 1927. 'Flor azul.' *Schreiter 4841* (LIL).

Dept. Tafí, San José, La Queñoa. Alt. 2800 m. Papas mandadas a Dr. Baur. Dec. 1931. *Schreiter 6965* (LIL).

Dept. Tafí, Los Cardones (Sta. Rosa). Alt. 2800 m. Dec. 1931. 'Flor azul.' *Schreiter 7097* (LIL).

Dept. Tafí, Cuesta de los Cardones (Las Lagunas a Amaicha). Alt. 3500 m. Dec. 1931. 'Flor azul.' *Schreiter 7103* (LIL).

Dept. Tafí, San José, La Queñoa. Alt. 2800 m. Jan. 1933. 'Flor violácea.' *Olea* (*Schreiter 8757*) (LIL).

Dept. Tafí, La Ciénaga. Alt. 2800 m. 29 Jan. 1950. 'Flor violácea.' *Sleumer 228* (LIL).

Dept. Tafí, Infiernillo. Alt. 3100 m. 23 Feb. 1949. *Sparre 5989* (LIL).

Dept. Tafí, Cerro Negrito, Las Lagunas. Alt. 4200 m. 26 Feb. 1949. *Sparre 6102* (LIL).

Dept. Tafí, Sierra del Cajón. Alt. 3000 m. 28 Apr. 1926. 'En una quebrada. Color de la flor amarillo (??).' *Venturi 6672* (US).

Dept. Trancas, San Pedro de Colalao. Jan. 1916. *Castillón* (s.n.) (LIL 62590).

Dept. Trancas, El Chorro, Cumbres de Calchaquíes. Alt. 2800 m. 1–5 Jan. 1914. 'Papa de yuta', 'Laderas y mesadas. Rastrera. Flor azul.' *Rodríguez 1258* (LP, SI).

Dept. Trancas, Quebrada de Gualinchay, Casa de Piedras. Alt. 2000(?)–2500 m. Dec. 1917. *Schreiter 352* (LIL, MVM—Osten 12397).

Dept. Trancas, Cumbre Alta del Chorro. Alt. 4000 m. Dec. 1917. 'Flor violácea obscura.' *Schreiter 353* (LIL, MVM—Osten 12396).

Dept. Trancas, Cumbre del Chorro. Alt. 3800 m. 21 Apr. 1926. 'Terreno pedregoso. Flor violácea.' *Venturi 7778* (US).

Dept. unknown, Cumbres de Calchaquíes, Alt. 4200 m. 30 Jan. 1907. 'Lugares arenosos. Corola azul obscura.' *Lillo 5521* (G, LIL).

Prov. CATAMARCA:

Dept. Andalgalá, Río Potrero sup., arriba de la Choyana. Alt. 3600 m. 27 Feb. 1951. 'Fr., cerca del arroyo, raro.' *Sleumer 1838* (LIL).

Dept. Andalgalá, Cuesta de la Negrilla, cerca de Mina de Capillitas. Alt. 3000–3100 m. 2 Feb. 1952. 'Flor violácea.' *Sleumer 2689* (LIL, J.G.H.).

Dept. Andalgalá, subida al Cerro Yutuyaco desde Capillitas. Alt. 3600–3800 m. 3 Mar. 1952. 'Flor violácea, fruto verde.' *Sleumer 2750* (LIL, J.G.H.).

Dept. Belén, Quebrada del Río Blanco, Las Mesaditas, arriba de Granadillas. Alt. 2800 m. 28 Jan. 1952. 'Flor violácea.' *Sleumer & Vervoorst 2532* (LIL, US, J.G.H.).

Dept. Tinogasta, Reales Blancos a La Tranca. Alt. 3500 m. 4 Feb. 1930. 'Flor blanca.' *Schreiter 6198* (LIL).

Dept. Tinogasta, Reales Blancos a La Tranca. Alt. 3500 m. 4 Feb. 1930. *Schreiter 6203* (LIL).

Prov. LA RIOJA:

Dept. Famatina, entre Cuevas de Noroño y Encrucijada. Alt. 2900 m. 8 Mar. 1956. *Petersen & Hjerting 1897* (C, J.G.H., J.P.H.). See Plates 66 and 67.

2. Subspecies *aemulans*

ARGENTINA

Prov. JUJUY:

Dept. Tilcara, near Tilcara, Chorro Valley. Alt. 3800 m. 13 Feb. 1939. *Balls 6026* (CPC: 82) (type of *S. depexum* Juz. var. *chorruense* Hawkes) (K, P.G.S.).

Dept. Tilcara, near Tilcara, Chorro valley. Alt. 3700 m. 13 Feb. 1939. *Balls 6037* (P.G.S.).

Dept. Tilcara, entre Quebr. del Chorro y Laguna Colorada. Alt. 3300 m. 24 Mar. 1952. *Petersen & Hjerting P–353* (p.p.) (LIL, J.G.H.).

Dept. Tilcara, Casa Colorada, subida a la Abra de Remate. Alt. 4000 m. 23 Feb. 1953. 'Flor violácea.' *Sleumer 4113* (LIL, J.G.H.).

Prov. LA RIOJA:

Dept. Chilecito, Sierra Famatina, cercanías de la mina 'El Oro'. 23–25 Jan. 1879. *Hieronymus & Niederlein 415* (p.p.) (CORD).

Dept. Famatina, Sierra Famatina, La Encrucijada. 29 Jan.–2 Feb. 1879. 'Flor coerulea.' *Hieronymus & Niederlein 474* (CORD, F—part of Berlin type, rest destroyed, G—Lectotype (upper plant)). See Fig. 23.

Dept. Famatina, Sierra Famatina, Mina San Juan. Alt. 3050 m. 3 Mar. 1906. 'Fruct. virid. Sub arbusculis raro.' *F. Kurtz 13698* (CORD).

Dept. Famatina, Sierra Famatina, La Mesada (Río Amarillo superior). Alt. 3500 m. 15–16 Mar. 1906. 'Ob. sub arbustis passim. Fr. ex viridis brunnei.' *F. Kurtz 13800* (CORD).

Dept. Famatina, La Encrucijada, Alt. 3150 m. 8 Mar. 1956. 'Varias plantas de sombra.' *Petersen & Hjerting 1898a.* (Accession H. 146) (C, J.G.H., J.P.H.). (Topotype.) See Plates 68 and 69.

Dept. Famatina, abajo de La Encrucijada. Alt. 3000 m. 8 Mar. 1956. *Petersen & Hjerting 1899b* (J.G.H.).

Dept. Famatina, abajo de La Encrucijada. Alt. 2950 m. 8 Mar. 1956. *Petersen and Hjerting 1900b* (J.G.H., J.P.H., herbarium specimen from seed grown as H. 147).

Dept. unknown, Sierra Famatina, Real Viejo. Alt. 3000–3500 m. 1 Mar. 1907. 'In declivibus graminosis areolas quasi cultas formans. Fl. sordide albi, tubera (nuex Juglandis forma) esculenta.' *F. Kurtz 14719* (CORD).

DOUBTFUL SUBSP. ***aemulans***:

Prov. Tucumán, dept. Tafí, Cerro Muñoz, Ciénaga Grande. Alt. 3400 m. 28 Jan. 1903. 'Prados.' *Lillo 2957* (LIL).

DISTRIBUTION

S. acaule extends through the Andes of South America from north Peru, through Bolivia, to north-west Argentina (Sierra Famatina, prov. La Rioja) (see Map 7). Its north–south distribution is thus about 3000 km, which is far greater than that of *S. chacoense*; the east–west spread is very narrow, however, since it is confined to the higher altitudes of the Andes. Further botanical exploration may possibly show that *S. acaule* extends also into northern Chile; it seems quite certain, however, that it does not occur in Ecuador or Colombia.

In Argentina subsp. *acaule* is to be found in provinces Jujuy, Salta, Tucumán, Catamarca, and La Rioja.

Subsp. *aemulans* is to be found chiefly in prov. La Rioja, but it has also been collected in prov. Jujuy, dept. Tilcara.

HABITAT

This species spreads into higher altitudes than any other wild potato, almost to the lower limit of the permanent snow. The range of subsp. *acaule* is (2600–)2800–4400 (–4650) m; subsp. *aemulans* is known only from 2950 to 4000 m at present.

The frost-resistant properties of *S. acaule* are well known, and it may be considered as the ecological vicariad in South America of the Mexican species *S. demissum*, though it extends to far higher altitudes than *S. demissum*.

Subsp. *acaule* is often found growing with species in series Megistacroloba (*S. megistacrolobum* and *S. sanctae-rosae*) and with *S. gourlayi* in series Tuberosa. It grows less commonly with *S. infundibuliforme* (series Cuneoalata), which although overlapping in its altitude range, normally prefers much drier habitats. Subsp. *acaule* is found only in the wetter regions, whilst Megistacroloba species are to be found also in the rather

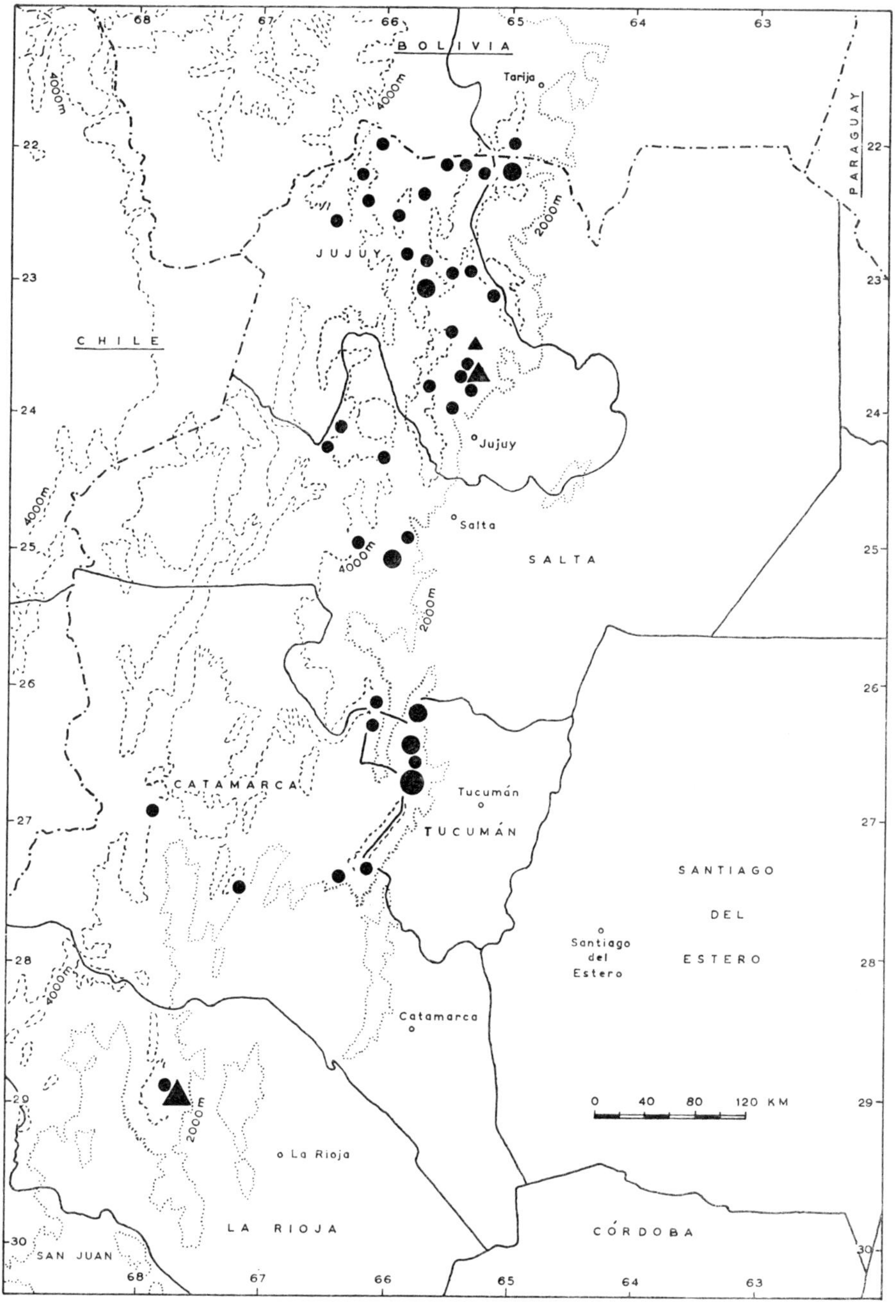

No. of Records

5 & over | 3–4 | 1–2

● ● • *Solanum acaule* Bitt. subsp. *acaule*

▲ ▲ ▴ *Solanum acaule* Bitt. subsp. *aemulans* (Bitt. & Wittm.) Hawkes & Hjerting

MAP 7. Distribution of *S. acaule* Bitt. subsp. *acaule* and subsp. *aemulans* (Bitt. & Wittm.) Hawkes & Hjerting.

drier parts, together with *S. infundibuliforme*, and these drier regions are not normally colonized by *S. acaule*.

S. acaule occurs in the humid part of the Puna province and in the Montano district ('pajonales') of the Eastern Subtropical province of Cabrera (1953). It also grows, though apparently very rarely, in the lower and wetter parts of the High Andean province. It is a very common plant of springs and flushes in high Andean grassland, in irrigation ditches (Brücher 1959*a*) and amongst stones and in cultivated fields where it often may become a troublesome weed. It is frequent in the proximity of dwellings, where the soil is disturbed; it is often seen also in corrals and cattle enclosures and in the walls around them. Here the soil is generally very rich, which induces *S. acaule* to develop very large leaves and stems, to such an extent that it may at first sight be confused with its cultivated hybrid derivative, *S. juzepczukii*. In its natural habitats it is generally found with species of *Cajophora*, *Oxalis*, *Plantago*, *Ranunculus*, *Calceolaria*, and other high-Andean species. (See Plate 13.)

The above paragraph refers chiefly to subsp. *acaule*, since it has been much better collected and observed than subsp. *aemulans*. This latter subspecies is found growing with *S. megistacrolobum* in the northern part of its range and with *S. spegazzinii* in the southern part. Subsp. *acaule* is also occasionally found with *S. spegazzinii*, and both subspecies apparently form hybrids with it. Subsp. *aemulans* occurs very rarely with the typical subspecies; it seems to be adapted to similar ecological conditions so far as we know at present, though perhaps not to the very wide altitudinal range of subsp. *acaule*.

TAXONOMIC NOTES

1. Subspecies *acaule*

Bitter's type description of *S. acaule* was based on the Berlin sheet (now destroyed) of *Fiebrig 3429*, collected at Puna Patanca, in southern Bolivia, very close to the Argentine border. Many isotype specimens are to be found in the European herbaria, and we designate the Stockholm sheet as LECTOTYPE for *S. acaule*. (See Plate 61 and Fig. 22.) Fiebrig's collection was made very close to those quoted above from prov. Jujuy, dept. Yavi (*P. & H. 970, 974*; *H. P. & R. 109*; *Sleumer 3674*). Fiebrig's plant had from one to three pairs of interjected leaflets per leaf, though the number is not specified in Bitter's description. The species was also noted by Bitter (1913) for localities in northern Argentina, provs. Jujuy, Tucumán, and Catamarca. In the 1912 publication Bitter described a variety, *subexinterruptum*, from La Paz, north Bolivia, collected by Buchtien (s.n.), differing from the type specimen in the almost complete absence of interjected leaflets. He also described a variety *caulescens* in 1913 from plants taken originally at La Viacha, near La Paz, and grown at Grenoble in France by Professor Claude Verne. It is a well-known fact that *S. acaule* when removed from its high Andean habitat and grown in the richer, moister soil of experimental plots or under glass, loses its rosette habit and produces sometimes quite a long stem. The present writers have observed plants in South America even, which, happening to grow in a favoured habitat amongst grass or other herbage or in the shelter of stones and rocks, produced stems of some 10 cm (*Balls 6028*, *Petersen & Hjerting P–341*, *Hjerting & Rahn 173*. (See Plate 65.)) Plants grown in Europe have attained a height of 60 cm. Bitter's var. *caulescens* is nothing more than a growth form, therefore, and the name should be discarded.

Bitter's other variety, *subexinterruptum*, is also of doubtful utility. It has now been possible to examine a large range of material from Bolivia and northern Argentina, including the type specimens of *S. acaule* and var. *subexinterruptum* and to grow plants under uniform environmental conditions both in England and Denmark. The presence of interjected leaflets seems to be probably controlled by one or a few genes and indeed varies also with slight differences of habitat or degree of development of the individual plant. By itself it is of insufficient importance for distinguishing a variety, especially since the character occurs erratically in all parts of the range of this species and even in plants from a single population. Thus, although it is quite certain that genotypic variation takes place in respect of this character, it is not sufficiently constant in any one area, and is not linked with other variations for it to be worthy of taxonomic rank. We therefore advocate that the use of the name *subexinterruptum* be discontinued.

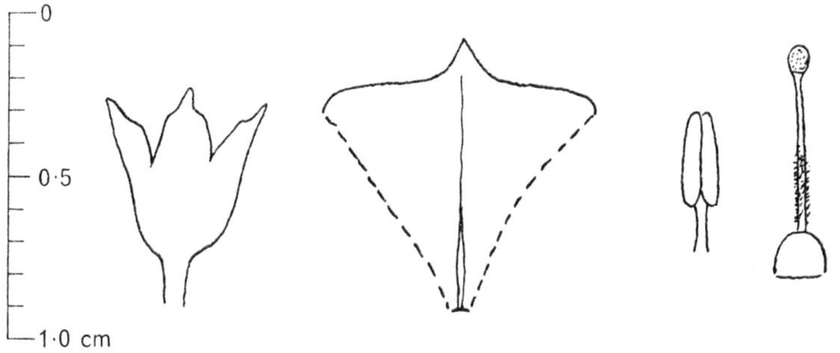

FIG. 22. *S. acaule* Bitt. Flower dissection of Isotype (*Fiebrig 3429*—Herb. Hamburg) collected at Puna Patanca, south Bolivia, very close to the Argentine border. (See Map 7 and Plate 61.) Note the very short corolla lobes, short style, and globular stigma.

The Russian taxonomist S. W. Juzepczuk described two more species closely related to *S. acaule*, namely *S. depexum* and *S. punae*, the former from north Argentina, collected by Rosenstiel (s.n.) at La Peña between Jujuy and Salta (see Plate 62), the latter from Central Peru. He considered both these species to be no more than geographical races of *S. acaule*, but goes on to say 'being myself, on principle, an opponent of triple nomenclature as applied to geographical races, I have given both species special binomials'.

S. depexum was retained as a species by Hawkes (1944) who followed Juzepczuk in considering that the short scattered hairs, slightly waved leaf margins, absence of interjected leaflets, and certain other minor characters were sufficient to distinguish it specifically from *S. acaule*. The above-mentioned work was written during war-time when type and other material was not available from European and American herbaria. Later on, a comparison of *S. depexum* material from Russia with Bitter's type of *S. acaule* showed, however, that the two are conspecific. The differences come well within the range of *S. acaule* as shown by the herbarium material we have examined. The presence of a slightly visible articulation in *S. depexum* is of no value as a specific character since this feature is to be found in several collections of *S. acaule* from the southern part of its range.

The following specimens show a partially visible articulation: *P. & H. P–329, P–330, P–331, P–335, 1897*; *Cabrera 9391*; *Balls 6028*; *Lillo 5521*; *Regnell 979*; *Rodríguez 1258*; *Schreiter 6203*; *Sleumer 3674*; *Sparre 5989, 6102*; *Venturi 6993*; in the following it is quite well marked: *O'Donell 4596*, *P. & H. P–369*, *Rosenstiel* (s.n.) (type for *S. depexum*). By referring to the detailed list of localities given above it will be seen that these collections have been taken from all parts of the range of *S. acaule* in north Argentina. It is of interest to note that in no material so far examined from Peru or Bolivia is the articulation visible on dried herbarium material.

The absence of interjected leaflets in *S. depexum* is also of little value, as we have seen above when discussing var. *subexinterruptum*; furthermore, even authenticated specimens of *S. depexum* possess a few interjected leaflets here and there, though the type description denies their presence categorically. In our opinion, therefore, *S. depexum* is identical with *S. acaule* and the use of this name should be discontinued. (See Hawkes 1956*b*.)

A further taxonomic and nomenclatural problem that needs discussing at the present time is the position of *S. schreiteri*, mentioned by S. M. Bukasov and other plant-breeders as a *nomen nudum*. Plants of this name grown in Cambridge from 1938 onwards from tubers kindly provided by Professor Bukasov did not differ significantly enough from *S. acaule* to warrant the retention of *S. schreiteri* as a separate species. (See Hawkes, 1944, p. 23; 1956*b*, p. 69.) Bukasov's herbarium specimen of *S. schreiteri* has also been examined by us and found not to possess an articulation. It comes from San José, between Jujuy and Salta, probably not far from the type locality of *S. depexum*. (See Plate 63.)

One other kind of variability ought also to be mentioned here, namely, corolla coloration. The normal colour in *S. acaule* is a rather clear blue or bluish-purple, but occasional collections possess a very pale-blue or mauve corolla and in some the pigment is almost absent. White-flowered forms from Peru were previously described by one of us (Hawkes 1944, p. 22) as var. *checcae*, differing also by the larger leaves and other characters. It is doubtful whether this should be maintained as distinct either, in view of the wide variability now known to us from the whole range of the species. White or pale mauve corollas are to be seen in the following Argentine collections: *Schreiter 6198*, *Venturi 9526*. Brücher (1959*a*) also, mentions white-flowered biotypes from the Aconquija mountains. *Solanum uyunense*, described by Cárdenas in 1956, seems to us, and to Correll also, to be nothing more than a large growth form of *S. acaule*, though it seems doubtful whether it can have anything to do with subsp. *aemulans*. The enormously long leaves (*c.* 30 cm) are no doubt a response to the shady damp conditions in which it was growing. The leaflet dimensions, pedicels, flowers, and fruits are all typical of *S. acaule*, subsp. *acaule*, and there seems no reason for retaining *S. uyunense* as a distinct species.

2. Subsp. *aemulans*

It seems curious that Bitter and Wittmack, when describing *S. aemulans*, did not consider that it emulated *S. acaule*, to which it is in fact closely related, rather than *S. demissum* from Mexico, which they thought it emulated and to which they thought it might be related. There is no doubt that *S. aemulans* is closely related to *S. acaule* and bears only a superficial similarity to *S. demissum*.

Bitter and Wittmack's type material was a specimen of Hieronymus and Niederlein (*No. 474*), collected at La Encrucijada in 1879. According to Bitter, two gatherings of Kurtz should also be identified as *S. aemulans*; these were *No. 14719* from Real Viejo, 3000–3500 m, 1 March 1907; and *No. 13698* from Mina San Juan, 3050 m, 'Below small trees, rare, with submature globose green berries, 15–16 mm diam.' These specimens were deposited in the Berlin herbarium and were destroyed during the Second World War, but duplicates exist in Córdoba, as well as another collection *No. 13800*.

The type specimen of Hieronymus and Niederlein was also destroyed, except for a single leaf donated to the Chicago Field Museum. An excellent photograph of the Berlin type is in existence, and isotype material can be seen at Geneva and Córdoba.

From a careful comparison of the isotype material and the type photograph of *S. aemulans* with *S. acaule*, and a study of living material of the two taxa collected at or close to their respective type localities by Petersen and Hjerting, it becomes clear that *S. aemulans* cannot be regarded as a distinct species. In flower form, habit, inflorescence type, and pubescence the two taxa are identical; only in the leaf dimensions and dissection do they differ materially. We therefore suggest that *S. aemulans* should rank as a subspecies of *S. acaule*, and we designate the upper plant of the Geneva sheet of *Hieronymus and Niederlein 474* as LECTOTYPE for *S. acaule* subsp. *aemulans* (Bitt. et Wittm.) Hawkes et Hjerting. (See Fig. 23.)

By referring to the drawings and photographs of subsp. *aemulans* (Fig. 23 and Plates 68, 69) it will be seen that the terminal leaflet is much larger than the laterals, which tend to decrease very rapidly in size towards the leaf base. In some plants this is very well marked indeed, so that the laterals are reduced to one or two pairs or are even absent altogether. In others, on the other hand, the lateral pairs are three to four in the best-developed plants, though even so, they are much smaller than the terminal leaflet.

When first examining the isotype material of *S. aemulans* (*Hieronymus & Niederlein 474*, from Geneva and Córdoba, see Fig. 23) we were struck by its general resemblance to species in series Megistacroloba, especially in the shape and dissection of the leaf. An examination of the flower, however, showed the typical very short corolla lobes of *S. acaule*, and left us in no doubt of the true affinities of this taxon.

Living topotype material obtained by Petersen and Hjerting in 1956 (see Plate 69) demonstrated beyond any doubt that this taxon had a concrete reality, but was sufficiently close to *S. acaule* to be considered as a distinct subspecies. This view was strengthened by cytological examinations, which showed that subsp. *aemulans* had 48 somatic chromosomes,[1] just as the typical subspecies. As we shall see later, material of subsp. *aemulans*, grown in Birmingham and Vandel, came closer in its leaf dissection to subsp. *acaule* than did the same genetic line growing in its native land. It should be noted that material of subsp. *acaule* growing in very exposed conditions at high altitudes shows a tendency to reduction, both in size and number, of the lateral leaflets as compared with the terminal; it is then very difficult to distinguish from subsp. *aemulans*. If ripe berries or mature tubers are collected and grown in a more favourable environment the typical leaf form of subsp. *acaule* is restored.

Certain collections from prov. Jujuy (previously classified by Hawkes as *S. depexum* var. *chorruense*) have been placed by us with some reserve under subsp. *aemulans*, since there are some intermediates between these and typical subsp. *acaule*. In leaf form and dissection they agree very well with subsp. *aemulans* (see Hawkes 1944, pp. 23, 115); yet they are separated from it geographically by a distance of over 600 km. (*c*. 400 miles) and no collections of subsp. *aemulans* have been seen by us from the intervening areas. Yet Brücher (1959*a*) states: 'In a northerly direction *S. aemulans* may spread to the pre-andean sierras of Catamarca province, where I have found related forms in the Carreta Mountains [dept. Andalgalá] at 3400 m.' We have not seen Brücher's specimens, but if the material referred to is indeed subsp. *aemulans* it would provide an interesting link in the range of this subspecies. It is possible also that subsp. *aemulans* might have been otherwise overlooked in provs. Catamarca, Tucumán, and Salta, and that what we have assumed to be stunted forms of subsp. *acaule* from high altitudes are not in every instance this subspecies but might sometimes in fact be subsp. *aemulans*. In cases where living material has been available we have verified the presence of subsp. *acaule* by culture experiments; but we have seen much herbarium material where seeds or tubers have not been available, and it is possible that some of this might have been subsp. *aemulans*. An alternative explanation for the absence of subsp. *aemulans* collections from the provinces of Catamarca, Tucumán, and Salta, is that it might conceivably have died out from these areas, where it was once widespread. Since the presence of a well-marked pedicel articulation leads us to the conclusion that subsp. *aemulans* is the oldest or most ancestral of the three subspecies of *S. acaule*, a disjunct distribution of a number of relic populations would not be an unreasonable state of affairs to find in this subspecies.

[1] Said by Brücher (1959*a*) to possess $2n = 24$ chromosomes, however.

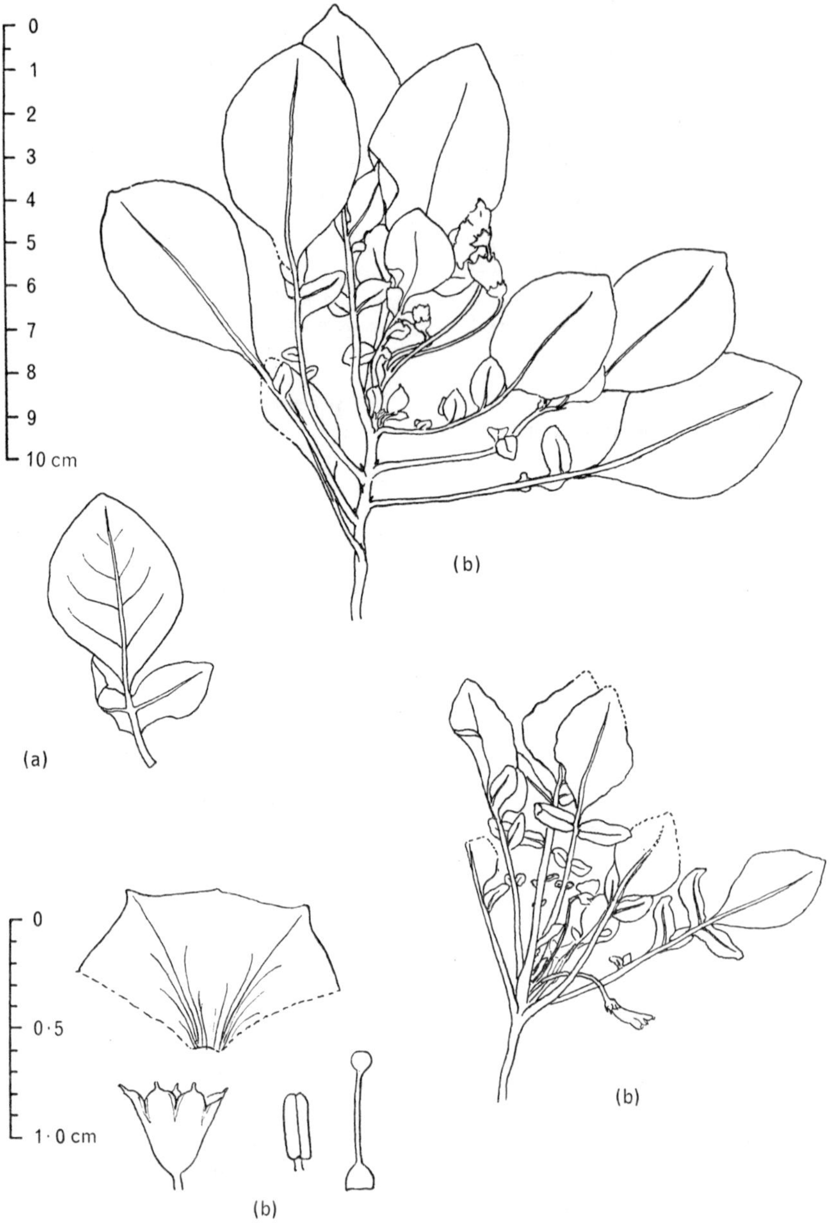

FIG. 23. *S. acaule* Bitt. subsp. *aemulans* (Bitt. & Wittm.) Hawkes & Hjerting.

(*a*) Part of type specimen, formerly in Berlin, now in Chicago.

(*b*) Lectotype from herb. Geneva (upper plant). Collected by Hieronymus and Niederlein (*No. 474*) from La Encrucijada, Sierra Famatina, prov. La Rioja, Argentina. Note the floral parts which are extremely similar to those of subsp. *acaule*, and the difference of leaf, with enlarged terminal and fewer smaller lateral leaflets that are somewhat decurrent and not basally auriculate as are those of subsp. *acaule*. (See also Plates 68, 69.)

In his recent publication on the taxonomy of series Acaulia, Brücher (1959*a*) deals with the systematic relationships of *S. acaule* and *S. aemulans*. He rightly emphasizes the effect of habitat differences on leaf dissection and stem length, and speaks of racial differentiation within the species leading to differences of flower colour and other features. We do not agree with him that *S. punae* is a complete synonym of subsp. *acaule*, since in our opinion the combination of characters seen in it, together with its geographical location, make it worthy of subspecific rank. Brücher's treatment corresponds with ours in the relegation of *S. depexum*, *S. schreiteri*, var. *subexinterruptum* and var. *caulescens* as complete synonyms of *S. acaule*. We cannot concur with him that var. *chorruense* (now placed by us with subsp. *aemulans*) and *S. punae* (= subsp. *punae*) are 'no more than habitat modifications of *S. acaule*' ('sind nichts weiter als Standorts-Modifikationen von *S. acaule* Bitter'). We have grown both of these taxa in company with typical *S. acaule*, both in glasshouse and field, and they have completely kept their distinctive features.

Brücher retains *S. aemulans* as a distinct species; this seems to us to be incorrect, for the reasons we have already given. One of the points stressed by him is the different chromosome number of *S. aemulans* ($2n = 24$) as contrasted with *S. acaule* ($2n = 48$). Here again, we do not agree. We have made several counts on our material from the type locality (*P. & H. 1898a*) and these have been kindly checked also by Dr. P. Jacobsen. All counts gave $2n = 48$.

Subsp. *aemulans* grows abundantly with *S. spegazzinii* in Sierra Famatina and the two species apparently form hybrids. One of the leaf pictures shown by Brücher (1959*a*, Fig. 7, right-hand leaf) would certainly seem to be *S. spegazzinii* and not *S. aemulans*, as he claims. Possibly, counts were made on the *S. spegazzinii* material by mistake; and this species is, of course, a diploid. It is interesting, and perhaps significant, that Correll (1962, p. 344) states: 'A collection by Brücher, without number, from Sierra de Famatina, Argentina, in January 1958, is labelled *S. aemulans*. Brücher also states on the label that the plants, which would normally have 48 chromosomes, had only 24. A duplicate of this collection provided me by Dr. Brücher consists not only of plants of *S. acaule* var. *aemulans*, but also plants of what I consider to be *S. venturii*, a species which Hawkes and Hjerting thought might have 24 chromosomes. It appears that the chromosome count might have been based on plants of this latter species.'

It is of course possible that subsp. *aemulans* may occur in two distinct cytotypes and that Brücher collected the diploid form and Petersen and Hjerting collected the tetraploid form. Only careful mass collections of herbarium specimens and tubers from the type and other localities will be able to help us solve this problem.

It is interesting to note that the typical subspecies also spreads south into the Sierra Famatina (see *P. & H. 1897*, Plate 66); Brücher also records it from there (*No. 1022*), though in his map (1959*a*, p. 152, Fig. 2) he separates the distribution areas of the two subspecies and places the southern limit of subsp. *acaule* at about the latitude of Tucumán.

Correll's treatment of *S. acaule* agrees with ours in most respects, and he rightly includes *S. depexum*, *S. schreiteri*, and *S. uyunense* as synonyms. He does not separate *S. punae* as a subspecies or other variant, as we have done, though he includes *S. aemulans* as a variety of *S. acaule*. Since Correll's use of the taxonomic category 'variety' is roughly equivalent to our use of the subspecific category, we find ourselves in very good agreement with him on this latter point.

Although not strictly within the area under consideration the species described by Juzepczuk as *S. punae* deserves further mention. It differs from subsp. *acaule* in the smaller habit, shorter stem, smaller flower, narrower leaf, and the presence of *long, weak, spreading hairs* over the whole plant. The articulation is never visible, except as a change in colour of the pedicel which can be seen in fresh material only.

This entity, *S. punae*, which we have here placed as a subspecies, was designated by Hawkes (1944, p. 23) as a variety of *S. acaule*. Contrary to Juzepczuk's and Hawkes's earlier surmise, it has nothing to do with *S. acaule* var. *subexinterruptum* Bitt.

Another extremely interesting variant of *S. acaule* was described by Ochoa (1960) from north Peru, dept. Cajamarca, as *S. acaule* var. *albicans*. This is characterized by the possession of white flowers and dense long white hairs over most of the green parts. The articulation is said to be hardly visible. According to Correll (1962) var. *albicans* possesses a chromosome number of $2n = 72$. One of us has collected and examined this extremely interesting variant of *S. acaule* and can verify its distinctive morphology as well as the hexaploid chromosome number; it also should be more appropriately classed as a subspecies, rather than a variety (see Hawkes 1963, p. 117).

Returning again to the pedicel articulation in *S. acaule*, it will be remembered that this feature is often completely invisible on dried material. Bitter (1912), when first describing this species, considered quite correctly that it should be included in section Tuberarium, even though the pedicel articulation was lacking in the specimen he examined. The uniqueness of *S. acaule* was further emphasized by Juzepczuk (1937), who created the series Acaulia for it and added the two microspecies *S. depexum* and *S. punae*, to which we have already referred. In subsp. *aemulans* and some forms belonging to the typical subspecies from northern Argentina, the articulation is emphasized to a greater or lesser degree. Nevertheless, even in these forms with a well-articulated pedicel the fruits never drop off (Bukasov 1955*b*) at maturity, as they do in other species of subsection Hyperbasarthrum. Bukasov also found the same feature in *S. demissum* from Mexico, and considers that there is some causal connexion between the cold climate and the retention of the fruit in the high alpine regions where these two species grow.

Juzepczuk considered that series Acaulia, and hence *S. acaule* itself, was of very recent origin, and that the absence of a pedicel articulation was a recent, derived character. If this hypothesis is correct, then subsp. *aemulans* would be the most primitive of the four subspecies, followed by subsp. *acaule*, subsp. *punae*, and subsp. *albicans*, since the pedicel articulation becomes more and more obscure as one passes northwards from Argentina, and the subspecies themselves become more distinctive and less like any other wild potato species known to us. The supposed 'ancestral' subspecies, *aemulans*, on the other hand, is vegetatively quite similar to certain species in series Megistacroloba, in particular to *S. megistacrolobum* and *S. sanctae-rosae*.[1] Although species in series Megistacroloba possess a semi-stellate corolla, the effect of chromosome doubling might well be to widen and shorten the corolla lobes (cf. very rotate corollas in hexaploid Demissa species). We do not of course know whether *S. acaule* arose by simple chromosome doubling of a diploid species or whether it arose by amphiploidy; the fact that it behaves as an allopolyploid now may merely be the result of selection pressure for fertility on a previously relatively infertile autotetraploid.

We may conclude then, by tentatively suggesting that *S. acaule* may have arisen from one or more diploid ancestors in the region of north-west Argentina, and that its most primitive subspecies is subsp. *aemulans*, whilst the most advanced one is the hexaploid subsp. *albicans* from northern Peru.

NOTES ON LIVING PLANTS

Resistance to low temperatures has been remarked by many authors for *S. acaule* (see Chapter 5). Firbas and Ross (1961) place it in the most frost-resistant group, together with the Peruvian species *S. chomatophilum* and *S. multidissectum* and the cultivated *S. juzepczukii* and *S. curtilobum*.

[1] Correll also points out the close resemblance of some plants of subsp. *aemulans* to *S. megistacrolobum* or to a hybrid of *S. acaule* × *S. megistacrolobum*.

Brücher (1954, 1959*a*) stated that *S. acaule* could be seen growing in the Aconquija mountains west of Tucumán at 4400 m, at times covered with hoar frost. He considered also that it could withstand black frosts of −10° C during the growing period, and that the tubers also were frost-resistant. Firbas and Ross showed, however, that there was no correlation between the frost resistance of leaves and tubers in this species.

Another valuable quality in *S. acaule* is immunity to virus X. Besides this, resistance has also been shown to occur to the following: wart, scab, powdery scab, early blight, *Verticillium* wilt, bacterial ring rot, brown rot, common eelworm, Horsenettle cyst and Osborne's cyst nematodes, Colorado beetle, *Epilachna*, and viruses A and leaf-roll. (See Chapter 5.)

Genetic differences in stem length were observed in seedlings of different accessions when grown at Birmingham under comparable conditions, and these were also verified in similar plantings at Vandel. Some of the longest-stemmed forms of subsp. *acaule* were from north-western Argentina, the tallest being *Petersen & Hjerting 1897* from prov. La Rioja, Sierra Famatina. Interestingly enough, this is one of the most southerly collections of subsp. *acaule* yet made. (See Plate 67.)

The stem of subsp. *aemulans* was as high as that of the above-mentioned gathering of subsp. *acaule* (*P. & H. 1897*), and the typical widely decurrent leaflet bases were well marked.

The other extreme, with very short stems, was subsp. *punae* from central Peru. A general trend towards longer stems is evident, therefore, in a north–south direction but this is only approximate, and one cannot point to a very exact correlation in *S. acaule* between stem length and latitude.

Material of subsp. *aemulans* (*Petersen & Hjerting 1898a*) from Sierra Famatina when grown in Birmingham and Vandel showed a better development of the lateral leaflets than did most plants of the original herbarium specimens (see Plate 69). The lateral leaflets were 3–4-paired with 0–2 pairs of interjecteds (originals had 1–3-paired laterals with 0–1 pairs of interjecteds); the upper laterals were widely decurrent and all laterals were considerably smaller than the terminal leaflet. The articulation of the pedicel was well marked and the berries had a few small white spots.

In the collections of subsp. *aemulans* from prov. Jujuy, dept. Tilcara (formerly *S. depexum* var. *chorruense*), the leaflet bases were not so widely decurrent, although the terminal leaflet was very enlarged, compared with the laterals.

The collection of *Petersen & Hjerting P–353* from Jujuy, dept. Tilcara, Chorro valley, was made at two different altitudes (3300 and 3500 m), and the seeds from each locality were unfortunately mixed together. Plants raised from these were of two different types; most were white or very pale lilac-flowered forms of subsp. *aemulans*, though a few blue-flowered seedlings referable to subsp. *acaule* occurred also. From the original herbarium sheets it is apparent that the 3300 m gathering was subsp. *aemulans*, and the white-flowered seedlings must have come from this altitude; the intensity of pigmentation in the white- to lilac-flowered forms varied from year to year, probably due to different weather conditions.

Under artificial short days at Vandel the tubers of subsp. *acaule* varied in weight from 1 to 3 grammes approximately; those of subsp. *aemulans* averaged 2–5 grammes approximately. Frost resistance was observed in most accessions.

CYTOLOGY

A somatic number of $2n = 48$ was first established by Rybin (1930, 1933) and has since been verified by numerous workers (Propach 1937, 1938*a*; Lamm 1943, 1945, 1953; Hawkes 1944; Brücher 1959*a*, etc.). All three of the southern subspecies possess the same chromosome number, $2n = 48$, according to our own investigations (but see discussion of Brücher's account of $2n = 24$ for subsp. *aemulans* on p. 261). A triploid collection ($2n = 36$), collected by Brücher from near Mina Aguilar at 4000 m was stated by him to be a hybrid between *S. acaule* and *S. megistacrolobum*. This was later named by Bukasov (in Bukasov and Kameraz 1959) as *S. bruecheri* (*nom. nud.*) and was finally described by Correll (1961), though without any reference to the chromosome number or probable parentage.

Meiotic studies of *S. acaule* by many workers (notably Lamm 1943, 1945, Swaminathan 1954) have shown regular bivalent pairing, indicating that this species is a good allotetraploid. Propach (1937), from a study of *S. acaule* × *S. chacoense* hybrids, however, considered that all three parent genomes were similar and that the allopolyploidy of *S. acaule* was only apparent, due to some 'safety catch' mechanism interpolated in the formation of allopolyploid species of this type. It would perhaps be more reasonable to explain Propach's results on the basis of the differential affinity of the chromosomes. Thus, each chromosome will pair with its exact homologue in tetraploid *S. acaule*. In the triploid hybrid, however, when the exact homologue is absent, each chromosome shows no more affinity with its homeologous partners from *S. acaule* itself than it does with those of *S. chacoense*; and hence trivalents will often be formed.

Colchicine-induced octoploids of *S. acaule* have been produced by several cytologists (Lamm 1943, 1945, Swaminathan 1954, Lebedeva 1955, 1959 *a* and *b*). None of these authors observed higher multivalent configurations than quadrivalents; and this again points to the functional diploidy (allopolyploidy) of tetraploid *S. acaule*. The fact that Swaminathan's hybrids of 6*x* (*S. acaule* × *S. microdontum*) also showed thirty-six bivalents at meiosis, bears out the hypothesis that *S. acaule* chromosomes will form only bivalents when their exact homologue is present. Swaminathan concludes that the two genomes of *S. acaule* are well differentiated, and he suggests that *S. acaule* probably arose as a segmental allotetraploid from diploid species belonging to other groups in *Solanum*. This suggestion seems very probably correct, and we have already pointed out that a species in series Megistacroloba might have been one of the ancestral forms in the evolution of the series Acaulia.

FERTILITY AND CROSSABILITY

S. acaule is one of the few species of tuber-bearing *Solanum* that is not only fully self-fertile, but normally pollinates itself without any need for insect pollen vectors. Self-fertility is found in all polyploid potato species and a few diploid ones also, though generally some agent is required to transfer pollen from anthers to stigma. In *S. acaule*, however, not even that is needed, and it would seem that *S. acaule* has developed this self-fertility as a direct response to the lack of pollinizing insects in the cold windswept plateaux and hillsides of the High Andes. Additional evidence for this hypothesis is provided by the fact that the same mechanism is found in the ecological vicariad of *S. acaule* in Mexico, *S. demissum*.

In spite of its chromosome number *S. acaule* forms hybrids only with some difficulty with other tetraploid species such as *S. tuberosum* and *S. stoloniferum*, though diploid species may be used freely as male parents. Less success is usual when the *S. acaule* is used as male parent.

The following species are recorded in the plant-breeding literature as having been crossed with *S. acaule*: *S. bukasovii*, *S. bulbocastanum*, *S. cardiophyllum*, *S. chacoense*, *S. demissum*, *S. goniocalyx*, *S. microdontum* (see p. 343), *S. phureja*, *S. pinnatisectum*, *S. sparsipilum*, *S. stenotomum*, *S. stoloniferum*, *S. tuberosum*. A few of these, it will be noted, are sympatric with *S. acaule*.

We made the cross of *S. acaule* (H. 129/1) × subsp. *aemulans* (CPC: 82/1) which gave vigorous uniform F_1 and F_2 families, thus indicating that there are no genetic barriers between the two subspecies, which are genomically very similar.

We have also been able to cross *S. acaule* (*Petersen & Hjerting P–329*) with the sympatric species *S. sanctae-rosae* (*P. & H. P–328*) and have obtained hybrids (Family 234) which are apparently sterile and certainly triploid ($2n = 36$).[1] *S. acaule* was used as the female parent in this cross. Both parents were from the same locality, namely, Tucumán, dept. Tafí, Infiernillo, Quebrada Honda. Although they were found growing together in a mixed community no natural hybrids between them were discovered.

The family of seedlings of this cross were of two very distinct morphological types. The 'A' type seedlings were large and vigorous, and showed many intermediate features between the two parents. The 'B' type were small and stunted, looking more similar to small specimens of *S. sanctae-rosae*, yet with wider terminal leaflets than are normally exhibited in that species. Further investigations are needed to elucidate the nature of these two very distinct morphological types of F_1 hybrid.

The artificial hybrid *S. megistacrolobum* × *S. acaule* was also obtained by us, using the following material: *S. megistacrolobum*: (*Petersen & Hjerting P–366*) from prov. Jujuy, Humahuaca, between Cóndor and Abra del Cóndor; *S. acaule*: (*Petersen & Hjerting P–329*) from prov. Tucumán, dept. Tafí, Infiernillo, Quebrada Honda. The resulting somatic chromosome number of $2n = 36$ showed that the F_1 was a true hybrid; this was confirmed by the form of the plant, intermediate in most characters between the two parents. It compares very well, also, with the naturally occurring hybrid between these species reported by Brücher (1959*a*) and named by Bukasov and by Correll as *S. bruecheri*.

We were also able to cross subsp. *aemulans* (*Petersen & Hjerting 1898a*) with *S. kurtzianum* (*Ruiz Leal 11906*) in both directions. *S. kurtzianum* × subsp. *aemulans* gave only one seedling, fully intermediate between the parents and with the expected number of somatic chromosomes, namely $2n = 36$.[1] The reciprocal cross gave several seedlings, none of them fertile, and also triploid.[2] They were fully intermediate between the two parents and apparently identical to the *S. kurtzianum* × subsp. *aemulans* seedling.

Most breeders have found that the cross *S. acaule* × diploid species gives highly sterile triploid hybrids. When *S. acaule* is crossed with *S. tuberosum* the resulting tetraploids, if produced at all, are also very sterile, owing to irregular chromosome pairing. Often hexaploid hybrids are produced from this cross, due to the functioning of non-reduced *S. acaule* egg cells. These hybrids are reasonably fertile, as also are the hybrids of octoploid *S. acaule* × *S. tuberosum*. Despite its frost resistance and immunity to virus X (Ross & Baerecke 1951), therefore, *S. acaule* is not easy to use as initial material in plant breeding. Furthermore, it should preferably be used as female parent in crosses[3] since the reciprocal is rarely successful, even when octoploid *S. acaule* is employed. Von Wangenheim (1954) found that this was due to a complete suppression of pollen-tube growth of *S. acaule* in the style of *S. tuberosum*. When tetraploid *S. acaule* was used as the female parent fertilization took place but all embryos aborted at an early stage owing apparently to disturbances in endosperm growth. This was not so marked when octoploid *S. acaule* was used as female parent.

To sum up, then, we may say that artificial hybrids with species in series Commersoniana, Demissa, Longipedicellata, Cuneoalata, Megistacroloba, and Tuberosa may be quite possible in most cases. It is particularly interesting to note, also, that Dionne (1963) has been able to obtain crosses between *S. acaule*, on the one hand, and series Bulbocastana, Cardiophylla, and Pinnatisecta, on the other. Some of these triploid hybrids were treated with colchicine to form fully fertile hexaploids showing regular bivalent pairing. Furthermore, the hexaploids appear to be quite capable of forming

[1] Counts kindly made by P. Jacobsen of the Botanical Institute, University of Copenhagen.

[2] Counts kindly made by O. Mattsson of the Botanical Institute, University of Copenhagen.

[3] See, however, our hybrids of *S. kurtzianum* × *S. acaule* subsp. *aemulans*.

hybrids with both polyhaploid and tetraploid *S. tuberosum*. It may therefore be possible to transfer valuable qualities of *Phytophthora* resistance from these Mexican series to *S. tuberosum*, using *S. acaule* as a bridge, and to add at the same time the quality of frost resistance from this latter species.

NATURAL HYBRIDS

S. acaule forms spontaneous hybrids in Peru and Bolivia with the cultivated diploid species *S. stenotomum* to give the triploid frost-resistant cultigen *S. juzepczukii* (Hawkes 1962*a*). This later species has been collected in Argentina, where it is cultivated in one or two localities (see p. 430). It certainly did not originate in Argentina, since the cultivated parent, *S. stenotomum*, is not grown in that country so far as we know.

Hybrids between *S. acaule* and other wild species are probably most infrequent; this is not so surprising when we remember that *S. acaule* is habitually self-pollinated and normally occurs at very high altitudes where the insect pollen vectors would probably not be very active owing to the cold and windy conditions. In addition, of course, *S. acaule* is taxonomically rather remotely related to many of the other wild Argentine species, so that the hybrids, when formed, are generally sterile, due partly of course to their unbalanced chromosome complement.

(*a*) With ***S. megistacrolobum***

Although *S. megistacrolobum* and *S. acaule* are frequently seen growing together (Brücher 1957*a*, Fig. 3, for example), we ourselves have only seen three possible hybrid collections. The first is part of *Balls 5986* from above Tilcara, where *S. acaule* and *S. megistacrolobum* were growing with the probable hybrids in mixed populations. The second is a part of *Sleumer 4111* from above Tilcara, Abra de Remate. (See Plate 70.) It is improbable that either of these represents the cross *S. acaule* × *S. gourlayi*, and we have seen no satisfactory sheets of that combination (see Table 12). The third collection of probable *S. acaule* × *S. megistacrolobum* is *Brücher 557*, determined by Brücher as *S. bruecheri*, and collected in prov. Jujuy, dept. Humahuaca, Rancho bei Iturbe, at 3600 m.

There are several confusing points about *S. bruecheri* which should be mentioned here. The first concerns the locality of the type collection: Correll's type description is based on a sheet of *Brücher 557* in Herb. Lundell, and the locality quoted is: 'Puesto I, Tilcara, 3650 m', with the date: 'March 1955'. The specimens of *Brücher 557* seen by us in four different Argentine herbaria and with the labels written by Brücher himself bear the following details of the locality, altitude and date: 'Rancho bei Iturbe, Jan. 1960. Alt: 3600 m.' All points differ except the number, therefore. Brücher also notes on the label: 'locus classicus', which seems to indicate that he thinks that this is not the type collection, but material taken at some other time from the type locality. However, the Iturbe material is clearly not taken from the type locality even. We must therefore consider the type of *S. bruecheri* to be the Lundell sheet of *Brücher 557*, whilst the rest of this number *cannot* be thought of as isotypes.

The next point of ambiguity concerns the name, *S. bruecheri*, which was proposed by Bukasov (in Bukasov and Kameraz 1959) for some hybrids of *S. acaule* × *S. megistacrolobum* described by Brücher (1954, 1959*a*) from yet a third locality (Jujuy, dept. Humahuaca, between Tres Cruces and Mina Aguilar at an altitude of over 4000 m). With this material, which is triploid, Brücher stated that neither of the assumed parents grew on the dunes which it inhabited, though they were apparently found in the general vicinity. Brücher therefore claimed that the hybrid was vigorous enough to colonize areas on which the parents were unable to survive.

TABLE 12

Natural hybrids between S. acaule *and other species in Argentina*

Locality	Collection number and altitude		
	S. acaule subsp. *acaule*	Hybrid	*S. megistacrolobum**
Jujuy, Tilcara, San Gregorio	*Balls 5987* (p.p.). 3900 m	*Balls 5986* (p.p.). 3700 m	*Balls 5981*. 3700 m *Balls 5986* (p.p.). 3700 m *Balls 5987* (p.p.). 3900 m *Balls 5989*, *5991*. 3900 m
Jujuy, Tilcara, Casa Colorada, path to Abra de Remate	*Sleumer 4114* (p.p.). 3950 m	*Sleumer 4111* (p.p.). 3950 m	*Sleumer 4111* (p.p.). 3950 m *Sleumer 4114* (p.p.). 3950 m
	S. acaule probably subsp. *acaule*	Hybrid	*S. spegazzinii*
Salta, Cafayate, Sierra de los Quilmes	*Castellanos* (s.n.) (BA 46991a). 3780 m	*Castellanos* (s.n.) (BA 46991b). 3780 m	*Castellanos* (s.n.) (BA 46985, 46986).
	S. acaule probably subsp. *aemulans*	Probable hybrid	*S. spegazzinii*
La Rioja, dept. Famatina, between Cuevas de Noroño and La Encrucijada	— — *P. & H. 1900b*. 2950 m *P. & H. 1899b*. 3000 m	*P. & H. 1896b*. 2900 m *P. & H. 1900c*. 2950 m *P. & H. 1901*. 2950 m *P. & H. 1899a*. 3000 m	*P. & H. 1895*. 2900 m *P. & H. 1896a*. 2900 m *P. & H. 1900a*. 2950 m — —
La Rioja, dept. Famatina, La Encrucijada	*P. & H. 1898a*. 3150 m *Hieronymus & Niederlein 474*.	*P. & H. 1898b*. 3150 m	—

* The following specimens of *S. gourlayi* occurred in the same areas: 1. San Gregorio: *Balls 5979* (3700 m); *5987* (p.p.), *5990* (both at 3900 m). 2. Casa Colorada: *Sleumer 4106* (3800 m).

From an examination of Correll's description and the figure in his 1962 publication, as well as the 'rest' of the *Brücher 557* collection to which we have already referred we think that *S. bruecheri* is probably a hybrid of *S. acaule* × *S. megistacrolobum*. If this is so, it follows that the original hybrids collected by Brücher at a locality between Tres Cruces and Mina Aguilar must also be given the name *S. bruecheri*, even though they form no part of the type. We have also been able to see other material (paratypes) cited by Correll (1961, 1962), namely, a photograph of *Correll 667* and *682*. The first of these is probably a hybrid of *S. acaule* with *S. infundibuliforme* or *S. gourlayi*, whilst the second may be a *S. infundibuliforme* × *S. megistacrolobum* hybrid. We should need to see a population sample from which these collections were made or a sample from the same area. Unfortunately, conditions in 1966 were too dry for us to be able to accomplish this.

Correll is evidently uncertain as to the affinities of *S. bruecheri*. In the original publication (1961) he places it in series Cuneoalata, but transfers it to Megistacroloba in 1962, noting that: 'This species shows some affinity with several species, but mainly with *S. acaule*, *S. sanctae-rosae* and *S. infundibuliforme*' (Correll 1962, p. 311). He makes no mention of Brücher's or Bukasov's publications and does not anywhere mention its possible hybrid nature. In view of the element of doubt and uncertainty about this species we shall leave it as one of 'probable hybrid origin' for the time being.

We have seen collections of *S. acaule* (*P. & H. P–358* (p.p.), *P–361*; *Sleumer 3462*) and *S. megistacrolobum* (*P. & H. P–358* (p.p.), *P–359*) from the region where Brücher's hybrids were found. Seeds of the mixed collection *Petersen & Hjerting P–358*, when sown in Birmingham, gave rise to plants which were exclusively *S. acaule*, however. Apparently none of the *S. megistacrolobum* plants was fruiting. No hybrids were observed, though we should not expect to obtain hybrids from seeds collected in this way, unless they are being frequently formed by natural cross-pollinations. This is certainly not the case, as evidence their rarity in nature. Tuber collections might have revealed some hybrid clones, but unfortunately the tubers collected did not survive the journey to Europe.

No natural hybrids between *S. acaule* and *S. sanctae-rosae* have been recorded, even though we have been able to make an artificial hybrid between them. Perhaps further search in the regions where these two grow together will reveal such hybrids.

(*b*) Hybrid between ***S. acaule*** subsp. ***acaule*** and ***S. spegazzinii***

Natural hybridization of *S. acaule* and *S. spegazzinii* apparently takes place occasionally, but it is not always possible to be certain as to which subspecies of *S. acaule* is involved in each case.

A hybrid of this type was collected by Castellanos (s.n.) in the Sierra de los Quilmes (Salta, dept. Cafayate) at 3780 m, together with *S. acaule* itself (see Plate 71). *S. spegazzinii* was also gathered by Castellanos in the same Sierra, though on a different occasion and without any altitude data (see Table 12).

It is difficult to be certain of the subspecies of *S. acaule* involved in this cross, though subsp. *acaule* is the more likely, since it was found growing with the hybrids, whereas subsp. *aemulans* has not been recorded for that region. Nine plants were included in the hybrid collection, and these show a reasonable uniformity and no berry formation. One may therefore conclude that they are sterile triploid F_1 hybrids. The plants are very pubescent dwarf semi-rosettes with 3–5-jugate leaves, rather narrow sessile lateral leaflets and 0–2 interjected leaflets. Leaflet shape is roughly intermediate between the two parents, whilst the leaflet position and number closely follow *S. acaule*.

The general appearance of the inflorescence is remarkably similar to that of *S.* × *juzepczukii* (triploid hybrid of *S. acaule* × *S. stenotomum*). The articulation is well marked and high, and the flowers drop off easily (contrast *S. acaule*, with no flower drop at the articulation). In form and colour the corolla closely resembles that of *S. acaule*, with a very rotate form and blue colour; it is *c.* 1·5 cm diam., with flat lobes. The style is only 6 mm long, showing dominance from the *S. acaule* parent.

(*c*) Probable hybrids between ***S. acaule*** subsp. ***aemulans*** and ***S. spegazzinii***

In the higher altitudes of Sierra Famatina, in La Rioja province, Petersen and Hjerting made an interesting series of collections which indicate that *S. acaule* subsp. *aemulans* grows in mixed communities with *S. spegazzinii*.

Thus *S. spegazzinii* spreads up from 1900 m to nearly 3000 m, whilst subsp. *aemulans* is known from about 2900 to 3150 m. Here, in the zone of overlap between the two species, a number of probable hybrids between them have been identified by us (see Table 12). These plants possess rather short stems and large leaves, with broadly decurrent sessile leaflets. The influence of subsp. *aemulans* shows much more clearly in the young leaves, whilst the older leaves show a stronger influence of *S. spegazzinii*, since the leaflets begin to be petiolulate, and certain oblong interjected leaflets occur here and there. (See also Plate 72.)

None of the probable hybrids was producing berries, and we should expect them to be sterile in any case. All the seeds collected from this area grew either into subsp. *aemulans* (*P. & H. 1898a*—H. 146) or into *S. spegazzinii* (*P. & H. 1895*—H. 143; *P. & H. 1896a*—H. 144; *P. & H. 1900a*—H. 147). Chromosome counts revealed only tetraploids (subsp. *aemulans*) and diploids (*S. spegazzinii*). Unfortunately none of the tubers collected by Petersen and Hjerting survived the journey, so there is no means of verifying the presence of triploid hybrids from the material available to us at present. Random tuber collections from this high-altitude zone of 2600–3150 m in Sierra Famatina are needed, to verify our hypothesis that hybrids are formed. The herbarium material is intermediate in nature between *S. spegazzinii* and subsp. *aemulans*, leaving one in little doubt about its hybridity. It should be noted that subsp. *acaule* was also found in this region.

Hybrids between *S. acaule* and *S. infundibuliforme* have not been recorded, whilst there is some doubt about the formation of natural hybrids between *S. acaule* and *S. gourlayi* (see p. 266).

LOCALITIES OF NATURAL HYBRIDS

(*a*) With ***S. megistacrolobum***

Prov. JUJUY:

Dept. Humahuaca, Rancho bei Iturbe. Alt. 3600 m. Jan. 1960. 'Fl. viol. Locus classicus. *S. bruecheri*, det. Brücher.' *Brücher 557* (BAA, CORD, SI, Ruiz Leal).

Dept. Tilcara, San Gregorio. Alt. 3700 m. 15 Feb. 1939. *Balls 5986* (p.p.) (P.G.S., J.G.H.).

Dept. Tilcara, Casa Colorada, camino a la Abra de Remate. Alt. *c.* 3950 m. 23 Feb. 1953. *Sleumer 4111* (p.p.) (J.G.H.). See Plate 70A.

(*b*) Hybrid between ***S. acaule*** subsp. ***acaule*** and ***S. spegazzinii***

Prov. SALTA:

Dept. Cafayate, Sierra de los Quilmes. Alt. 3780 m. 27 Jan. 1943. *Castellanos* (s.n.) (BA 46991b). See Plate 71.

(*c*) Probable hybrids between ***S. acaule*** subsp. ***aemulans*** and ***S. spegazzinii***

Prov. LA RIOJA, dept. Famatina:

Sierra Famatina, between Cuevas de Noroño and La Encrucijada. Alt. 2900 m. 8 Mar. 1956. *Petersen & Hjerting 1896b* (J.G.H., J.P.H.). See Plate 72.

Sierra Famatina, La Encrucijada. Alt. 3150 m. 8 Mar. 1956. *Petersen & Hjerting 1898b* (J.G.H., J.P.H.).

Sierra Famatina, below La Encrucijada. Alt. 3000 m. 8 Mar. 1956. *Petersen & Hjerting 1899a* (J.G.H., J.P.H.).

Sierra Famatina, below La Encrucijada. Alt. 2950 m. *Petersen & Hjerting 1900c* (J.G.H.).

Sierra Famatina, below La Encrucijada. Alt. 2950 m. *Petersen & Hjerting 1901* (J.G.H., J.P.H.).

Series CUNEOALATA Hawkes

(*Bull. imp. Bur. Pl. Breed. Genet., Camb.* 118, 1944)

SMALL straggling tuberiferous herbs with pinnatifid leaf, the rachis bearing narrow, wedge-shaped, decurrent wings between each pair of lobes or segments; interjected leaflets absent. Corolla white to purplish, sub-stellate to rotate with well-developed petal acumens. Fruits globular.

General distribution: south Bolivia, northern Chile, and north-west Argentina at altitudes of 2450 to 4100 m.

Chromosome number: $2n = 24$.

In Hawkes's original publication several other species were proposed, in addition to the type species, *S. infundibuliforme* Phil. Further study (Hawkes 1956*b*) showed that these other species, with the exception of *S. xerophyllum*, should be considered as phenotypic variants only. Brücher (1957*a*) agreed that the previously described micro-species were no more than biotypes, or phenotypic variants, even including amongst them the species, *S. xerophyllum*. Having now seen more material, we agree with Brücher that this latter species is also no more than an extreme variant, growing in poor soil or under conditions of great exposure.

Correll (1962) has retained *S. xerophyllum* as a distinct species. He considers that Cuneoalata is closely allied to series Megistacroloba, and that some hybridization may take place between the two series.

Species of this series included in the present treatment:

S. INFUNDIBULIFORME Phil.

7. SOLANUM INFUNDIBULIFORME Philippi

(*An. Mus. nac. Chile*, 2nd ed. Bot. 65, 1891)

S. infundibuliforme Phil. var. *angustepinnatum* Bitt. (*in Reprium nov. Spec. Regni veg.* **11**, 388–9, 1912).
S. platypterum Hawkes (*Bull. imp. Bur. Pl. Breed. Genet., Camb.* 118, 1944).
S. microphyllum Hawkes (ibid.; *non S. microphyllum* Dun. *Hist. Solan.* 187, 1813).
S. glanduliferum Hawkes (*Bull. imp. Bur. Pl. Breed. Genet., Camb.* 118–19, 1944).
S. pinnatifidum Cárd. (*Revta Agric. Univ. Cochabamba* **2**, 33–34, 1944; *non S. pinnatifidum* Ruiz et Pavon, *Flora Peruv.* **2**, 37, 1798–1802).
S. xerophyllum Hawkes (*J. Linn. Soc. Lond.* **53**, 108, 1945).

Plant small, graceful, tuber-bearing.

Stem slender, often flexuose, branched, erect or decumbent, and sparsely to densely pubescent, 3–30 cm long; vertical subterranean rootstock bearing long creeping stolons with small tubers ½–1 cm diam.

Leaf 2–10 cm long × 1–6 cm broad, lyrate to *imparipinnatisect* or almost imparipinnate, with (1–)2–3(–5) pairs of lateral lobes and *no interjected lobes*; lateral lobes linear, linear-falcate, linear-lanceolate or lanceolate, up to 20 × 7 mm but often only 1·5–4 mm wide, with obtusish apex and *base continued into a decurrent wing*, which is often broad (up to 5 mm above) but may be very narrow; terminal leaflet *generally larger than laterals* (in lower juvenile leaves and always on some plants 3–4 times the size of the upper laterals), linear-lanceolate to elliptic-lanceolate or rhomboid-lanceolate, to 50×20 mm but generally smaller; leaves and stem provided with sparse to frequent short thick acute hairs (occasionally longer and sometimes intermixed with minute glands, visible easily only on living plants). Small semi-lunate pseudostipular leaves sometimes present, sometimes absent or replaced by minute leafy shoots.

Inflorescence (1–)3–7(–10)-flowered; peduncle simple or forked above, 0·5–6·5 cm long, clothed as the pedicels with generally upward-pointing short acute hairs of differing size and sometimes with numerous stalked glands barely visible on dried specimens; pedicels 5–35 mm long, with the articulation situated more or less at the centre or in the upper third, sometimes only 2·5 mm below the calyx.

Calyx 5–8 mm long, with 3–6 mm long lobes of which the acumens are 1–4 mm long and often reflexed and linear; pubescence similar to that of peduncles and pedicels.

Corolla diam. 12–25 mm, varying in colour from white, white with mauve interpetalar membranes, white with mauve on median part of petals, to entirely pale mauve or even deep bluish-violet; corolla shape varying from *semi-stellate to completely rotate* (circular with well-marked acumens as in the Mexican series Longipedicellata) according to the degree of development of the interpetalar membranes; lobes (including acumens) about 4–10 mm long×7–10 mm broad, the acumens themselves 3–5 mm long.

Anthers 4–6 mm long; filaments 1–2 mm long.

Style slender, 5–10 mm long, with minute papillae towards the base, sometimes apparently glabrous; stigma globular to ovoid, much thicker than style apex.

Berries globular, about 1 cm diam., green or green splashed with white.

Chromosome number: $2n = 24$.

AFFINITIES

Possibly most closely related to series Megistacroloba but easily distinguished by its delicate habit, thin branched stem, small leaves with no interjected leaflets, pinnatisect leaf blade with lanceolate to linear lobes, bearing a wedge-shaped decurrent wing from their bases, and well-developed peduncle. Corolla variable in shape and of little diagnostic value.

SPECIMENS EXAMINED

ARGENTINA

Prov. JUJUY:

Dept. Cochinoca, Abra Pampa, Cerro Huancar. 23 Jan. 1948. 'Rocas.' *Cabrera 9426* (LP).

Dept. Cochinoca, Cochinoca. 3 Feb. 1937. *Castellanos* (s.n.) (BA 20171).

Dept. Cochinoca, Cochinoca. Alt. *c.* 3500 m. 1 Jan. 1901. 'Fl. saturate coerul.' Leg. Fritz Claren, *F. Kurtz 11327* (p.p.) (S).

Dept. Cochinoca, Abra Pampa, playa del Río Grande. Alt. 3400 m. 16 Feb. 1929. 'Flower white.' *Venturi 9394* (US).

Dept. Humahuaca, entre Iturbe y Tres Cruces. 10 Feb. 1927. *Castellanos* (s.n.) (p.p.) (BA 4078).

Dept. Humahuaca, along stream among low shrubs on altiplano, about 7 km south of Tres Cruces. 25 Feb. 1960. 'Fls. blue-purple.' *Correll, Dodds, Brücher & Paxman A660* (p.p.) (K).

Dept. Humahuaca, cerca de Iturbe, *c.* 25 km de Humahuaca hacia Tres Cruces. Alt. 3360 m. 3 Feb. 1956. *Hjerting, Petersen & Rahn 64* (J.G.H., J.P.H.).

Dept. Humahuaca, entre Humahuaca y Tres Cruces, 30 km antes de Tres Cruces. Alt. 3550 m. 3 Feb. 1956. 'Flor casi blanca (el centro lila, los puntos blanquitos.)' *Hjerting, Petersen & Rahn 75* (C, J.G.H., J.P.H.).

Dept. Humahuaca, entre Humahuaca y Tres Cruces, 30 km antes de Tres Cruces. Alt. 3550 m. 3 Feb. 1956. 'Same locality as 75, but lilac-flowered types.' *Hjerting, Petersen & Rahn 76* (p.p.) (C, J.G.H., J.P.H.). See Plate 77A.

Dept. Humahuaca, entre Humahuaca y Tres Cruces, 28 km antes de Tres Cruces. Alt. 3600 m. 3 Feb. 1956. 'Flor blanca; 15–20 cm alto.' *Hjerting, Petersen & Rahn 79* (J.G.H., J.P.H.).

Dept. Humahuaca, entre Humahuaca y Tres Cruces, 28 km antes de Tres Cruces. 3600 m. 3 Feb. 1956. 'Flor lila hasta casi blanca.' *Hjerting, Petersen & Rahn 80* (C, J.G.H., J.P.H.).

Dept. Humahuaca, El Aguilar, cerca del Molino de Mina Aguilar. Alt. 4100 m. 28 Mar. 1952. *Petersen & Hjerting P–360* (CPC: 2477) (LIL, J.G.H., J.P.H.).

Dept. Humahuaca, Iturbe, cerca de la estación. Alt. 3350 m. 31 Mar. 1952. 'En tierra muy seca.' *Petersen & Hjerting P–362* (CPC: 2478) (LIL, J.G.H.).

Dept. Humahuaca, entre Cóndor y Abra del Cóndor. Alt. 3800 m. 9 April 1952. 'Entre piedras; muy seco.' *Petersen & Hjerting P–367* (CPC: 2479) (LIL, J.G.H., J.P.H.).

Dept. Humahuaca, km. 1313, entre Iturbe y Tres Cruces. Alt. 3500 m. 9 Feb. 1953. *Petersen & Hjerting 963* (J.G.H., J.P.H.).

Dept. Humahuaca, Tres Cruces. Alt. 3800 m. 20 Feb. 1959. 'Trockenpuna mit *Pseudobaccharis*.' *Ross & Vervoorst 84* (Max.).

Dept. Rinconada, 38 km oeste de Abra Pampa hacia Mina Pirquitas. Alt. 3800 m. 10 Feb. 1953. *Petersen & Hjerting 965* (LIL, J.G.H., J.P.H.). See Plates 75 and 76.

Dept. Susques, Susques. 5 Mar. 1927. *Castellanos* (s.n.) (BA 27/909).

Dept. Tilcara, Quebrada del Huasamayo. Alt. 2450 m. 10 Feb. 1939. *Balls 5956* (CPC: 52 and 85) (K, UC, P.G.S., J.G.H.) (Type collection for '*S. glanduliferum*' Hawkes).

Dept. Tilcara, above Tilcara, San Gregorio. Alt. 3300 m. 11 Feb. 1939. *Balls 5978* (P.G.S., J.G.H.).

Dept. Tilcara, above Tilcara, San Gregorio. Alt. 3650 m. 15 Feb. 1939. 'In sandy soil amongst stones, often sheltered by other vegetation.' *Balls 5980* (CPC: 75) (UC, US, P.G.S.) (part of this collection formed the type for '*S. xerophyllum*' Hawkes).

Dept. Tilcara, Abra de Punta Corral. Alt. 3700 m. 20 Mar. 1952. *Petersen & Hjerting P–339* (LIL, J.G.H., J.P.H.).

Dept. Tilcara, subida a la Abra de Remate. Alt. *c.* 3700–3800 m. 23 Feb. 1953. 'Fl. viol. clar.' *Sleumer 4105* (LIL, J.G.H.).

Dept. Tumbaya, Abra de los Pibes a Purmamarca. 18 Feb. 1937. *Castellanos* (s.n.) (BA 20174).

Dept. Tumbaya, Nevado de Chañi. Alt. 3500 m. 11 Dec. 1901. 'Inter saxa loco umbroso.' Leg. R. E. Fries, *Regnell 893* (S).

Dept. Tumbaya, San Antonio (Abra) de los Pibes. Alt. 3850 m. 5 Jan. 1953. 'Fl. prob. violasc.' *Sleumer 3254* (LIL, J.G.H.).

Dept. Yavi, Cangrejillos. 28 Jan. 1948. 'Rocas. Flor azul obscura.' *Cabrera 9481* (LP).

Dept. Yavi, La Quiaca. Alt. 3600 m. Feb. 1916. *Hauman* (s.n.) (BRLU).

Dept. Yavi, Cajas. Alt. 3900 m. 6 Feb. 1956. *Hjerting, Petersen & Rahn 110* (C, J.G.H., J.P.H.). (Possibly some introgression from *S. megistacrolobum*, see No. 111, section on hybrids, p. 282.)

Dept. Yavi, La Quiaca. Alt. 3450 m. 12 Feb. 1940. 'Planta herbácea de 20–30 cm de altura; flor morada.' *T. Meyer* (s.n.) (LIL 33409, SI).

Dept. Yavi, Cajas, 25 km este de Yavi. Alt. 3950 m. 16 Feb. 1953. *Petersen & Hjerting 975* (LIL, J.G.H., J.P.H.).

Dept. Yavi, Quebr. de Toro Ara. Alt. 3450 m. 21 Feb. 1959. 'Im Monte und Nardengebüsch, bevorzugt feuchtere seitenwände.' *Ross & Alandia 86* (Max.).

Dept. Yavi, La Quiaca-Villazón. Alt. 3442 m. 24 Jan. 1940. 'Flores azul obscuro.' *Schreiter 11198* (CORD, LIL, SI).

Dept. Yavi, alrededores de Yavi. Alt. 3500–3600 m. 30 Jan. 1953. 'Lugares secos.' *Sleumer 3608* (J.G.H.).

Dept. Yavi, Rincón de Cajas. Alt. 4000 m. 31 Jan. 1953. 'Fl. viol. pálido. Entre rocas. Rara!' *Sleumer 3626* (LIL, J.G.H.).

Dept. probably Tumbaya. In mountains just below Tascal. 26 Feb. 1960. 'Fls. purplish, rotate-pentagonal; fruits orbicular, white-spotted, 2 cm diam.' *Correll, Dodds, Brücher & Paxman A678* (K).

Prov. SALTA:

Dept. Los Andes, San Antonio de los Cobres. 21 Jan. 1949. 'Cerros. Flor azul.' *Cabrera & Schwabe 10* (LP).

DISTRIBUTION

In Argentina *S. infundibuliforme* has been collected in the dry inter-Andean valleys of provs. Jujuy and north-west Salta. (See Map 8.) Although not recorded for Catamarca and La Rioja there is reason to believe that it may also occur in the unexplored western regions of these latter provinces. It has also been recorded for Chile, provs. Antofagasta, Atacama, and Tarapacá[1] (where the type specimen was collected) and for Bolivia in depts. Tarija, Potosí, Chuquisaca, and Cochabamba.

HABITAT

The altitude range of this species is from (2450–)3400–3900(–4100) m.

It grows in very dry situations, generally in sandy soil or amongst stones and is usually found in semi-desert conditions growing in 'open' communities with *Opuntia* and *Trichocereus* species. It is thus found chiefly in the drier parts of the Puna province, and *very* occasionally in the wetter parts of the Prepuna province of Cabrera's (1953) phytogeographical scheme. In some localities it may be found together with *S. megistacrolobum* Bitt., *S. oplocense* Hawkes, or *S. gourlayi* Hawkes, but very rarely with *S. acaule* Bitt., which is to be found in much wetter regions. Its distribution and habitat preferences have also been noted for Chile and Bolivia by Montaldo and Sanz (1962).

Its preference for very dry places indicates that *S. infundibuliforme* might be a xerophyte and therefore of interest in breeding commercial potatoes for resistance to drought. Apart from certain hybrids made for cytological and taxonomic work no attempt has been made to see whether this species could be of value in breeding potatoes for dry regions.

TAXONOMIC NOTES

The type collection of *S. infundibuliforme* was made by Philippi in north Chile, prov. Tarapacá, near Calcalhuay (now in Bolivia, dept. Potosí), probably in 1885, the description being published in 1891. No date of collection was given in the original

[1] Now in Bolivia, dept. Potosí, prov. Nor Lipez.

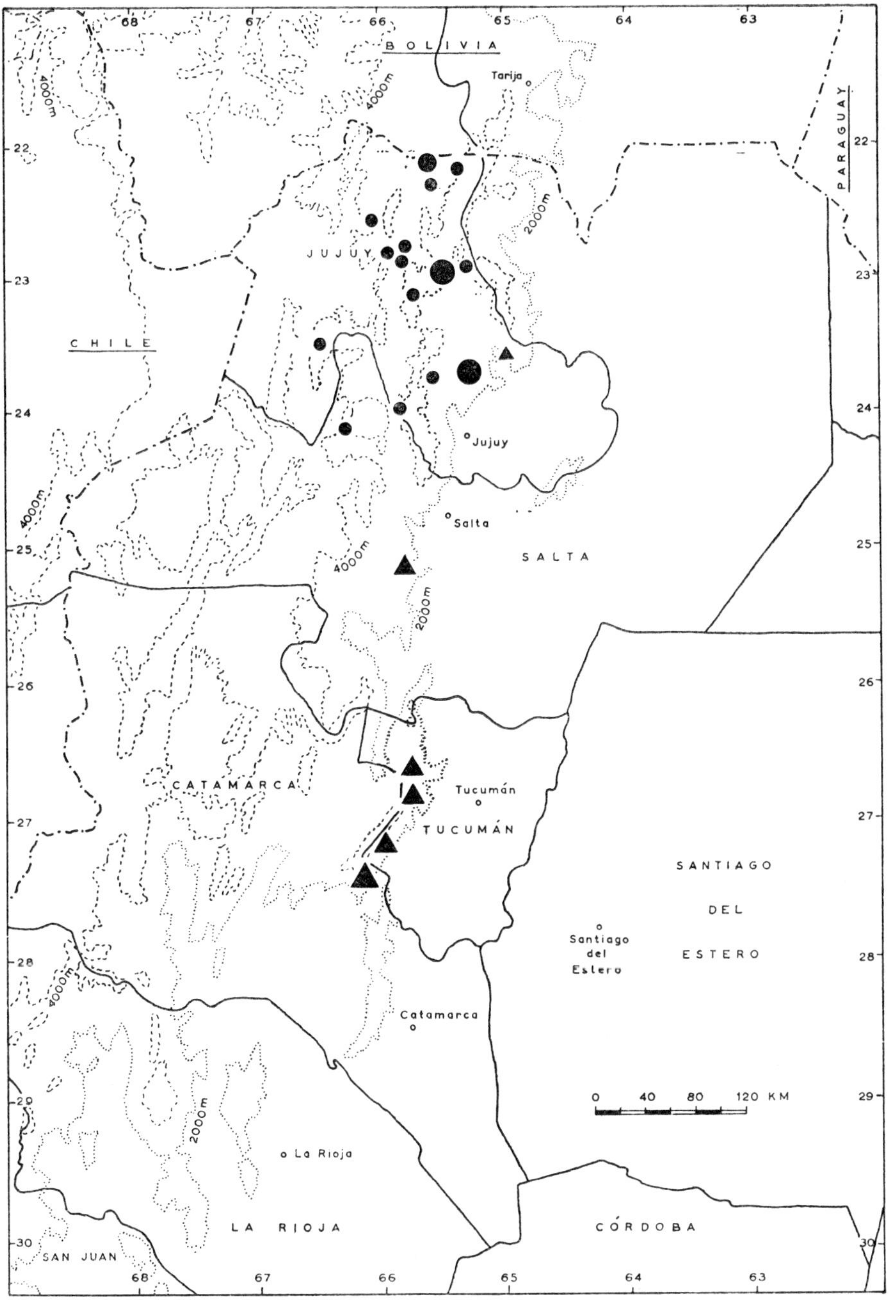

No. of Records

5 & over | 3–4 | 1–2

● ● • *Solanum infundibuliforme* Phil.

▲ ▲ ▲ *Solanum venturii* Hawkes & Hjerting

MAP. 8. Distribution of *S. infundibuliforme* Phil. and *S. venturii* Hawkes & Hjerting.

diagnosis but Bitter labelled as *S. infundibuliforme* one of the two plants on the Philippi sheet in Berlin, collected in March 1885; the other he named var. *angustepinnatum*. This Berlin sheet is now destroyed, but good photographs exist at Chicago and elsewhere. (See Plate 74.) According to Muñoz Pizarro (1960) the type for *S. infundibuliforme* still exists and is to be found in the Museo Nacional de Historia Natural at Santiago, Chile, under the number 55467. One of us was able to visit this herbarium in December 1965, thanks to the kindness of the Director, Carlos Muñoz Pizarro. The specimen 55467 referred to above bore the label 'Calcalhuay Jan. 1886', though the publication of R. A. Philippi where the description is to be found mentions that the expedition took place from December 1884 to March 1885. Those taking part were R. A. Philippi's son, Federico, and his grandson, Oton, as well as Carlos Rahmer and a Sr. Pablo Ortega. Federico Philippi described this trip in an article 'Reise nach der Provinz Tarapacá' (in *Verh. dt. wiss. Ver. Santiago Chile* **1**, Valparaiso) in November 1885, stating that he and his son left Tarapacá for Santiago on 12 March of that year because of ill health. There is no mention in this or any other publication known to us of a further expedition in 1886. The 1885 collection of Berlin (now destroyed) does not bear the name Calcalhuay, whereas the 1886 one of Santiago does. We are inclined to think that both collections were made in 1885 and that the date 1886 was an error. Clearly, more documentary evidence is needed, however.

Turning now to the Santiago specimen, 55467, a comparison of it with the type description reveals an extraordinarily close similarity. The stem is *exactly* 22 cm long, with 10 mm long internodes; the largest terminal leaflet is *exactly* 38 mm long and 13 mm wide; the peduncle is *exactly* 6·5 cm long, pedicels *exactly* 3·5 cm long, and so forth. All these measurements are given in the original description, and fit the specimen so precisely that we are forced to conclude that this was the actual specimen used by R. A. Philippi despite the discrepancy in date. Since Philippi did not actually cite a specimen with his description we designate this one, No. 55467 in Herb. Mus. Hist. Nat. Santiago, as LECTOTYPE of *S. infundibuliforme* Phil., since it agrees extremely well with the original description and does not differ in any essential detail. (See Plate 73.)

The great variability in leaf and flower of *S. infundibuliforme* induced one of us (J.G.H., 1944) to describe three extreme variants of it as distinct species, namely *S. platypterum*, *S. glanduliferum*, and *S. microphyllum* (= *S. xerophyllum*). Subsequent examination of a wider range of material has shown us, however, that leaf and flower characters in this species show continuous variation from one extreme to the other, and cannot be used as a basis for separating distinct species, subspecies, or even varieties. Furthermore, with the material now at our disposal we cannot distinguish any geographical clines or subspecies. For a similar reason, also, Bitter's variety *angustepinnatum* cannot be retained. Collections made by Cárdenas from Bolivia (dept. Cochabamba) which were classified under *S. xerophyllum* (= *S. pinnatifidum* Cárdenas) may possibly be distinct but the scanty material at present available to us from this region makes us unwilling to pronounce on their taxonomic position.

As regards leaf form there seems to be continuous variation from those with very wide decurrent leaflet wings (previously classed as *S. platypterum*) to ones with small leaves and extremely narrow wings (*S. xerophyllum*). Furthermore, some leaves exhibit a terminal leaflet much longer and broader than the laterals (Berlin specimen of *S. infundibuliforme*), whilst in others the difference is less well marked (var. *angustepinnatum* and *S. platypterum*) and in others again is not manifest at all (*S. xerophyllum*). The whole range of leaf variation is often to be seen within a single population, whilst a considerable variation often occurs within the same plant. It seems fairly certain that most of this variability is determined by the environmental factors such as soil fertility and moisture, degree of shading, etc., and also by the degree of development of the individual plant. Lower leaves tend to possess broader wings and more enlarged terminals whilst in the upper leaves the terminals are reduced in size relative to the laterals, and the wings are narrowed. This is to be seen on plants grown under experimental conditions in Birmingham and Vandel. No doubt some of the variation in leaf form is genetically determined, especially the relative size of the terminal leaflet, but the environmental plasticity of this species seems to be greater than in many others from Argentina and generally masks the genetic differences.

The type of leaf pubescence does not vary so greatly as the leaf form, though the hairs are longer in some collections than in others and are often clearly visible, standing out from the edge of the leaflets. Glandular pubescence is generally uncommon, though occasionally developed on the upper portions of the plant. It was abundant on the upper portions of the type specimen for *S. glanduliferum* (*Balls 5956*) though this is not very evident on herbarium material. This collection was taken at the extreme lower altitude limit for *S. infundibuliforme* (2450 m) and was probably brought down from higher regions by the river Huasamayo, near whose banks it was growing. Here again, the presence of glandular hairs is not linked with other characters or found in any one region, so that we are unable to use the name *glanduliferum* to designate any well marked or constant form.

No less variable is the form of the corolla in this species. The very name '*infundibuliforme*', which was applied by Philippi to indicate the funnel-shaped corolla, is inappropriate. In all specimens with well-developed flowers the corolla is spreading or even reflexed; it only becomes funnel-shaped in wilted, old, or semi-withered flowers, or in specimens taken late at night or early in the morning. This character, however, applies to all the tuber-bearing *Solanum* species, whose flowers exhibit 'sleep movements', closing at night and opening in the morning. Thus the character of funnel-shaped flowers is of no diagnostic value whatsoever.

The shape of the corolla is of great taxonomic importance in most species of *Solanum*, but in *S. infundibuliforme* it shows a remarkably wide range, due partly to genetical factors and partly to its stage of development. In some collections the corolla is semi-stellate (see Plate 75), notably in *Balls 6126* from Villazón, Bolivia (type collection of *S. platypterum*), *Petersen & Hjerting P–362* and *H.P.R. 75* and *110*. In others it is predominantly rotate (see Plate 76) and very close to that of *S. stoloniferum*, a Mexican species of quite distinct affinities. Collections exhibiting this corolla form are *Balls 5956* (type for *S. glanduliferum*), *Meyer* (s.n.) (LIL 33409), and *Petersen & Hjerting 975*. In other collections the corolla varies in form from one extreme to the other (*Petersen & Hjerting P–360, 965; H.P.R. 76* (p.p.)). Under conditions at Cambridge the semi-stellate corolla of *Balls 6126* and the rotate corolla of *Balls 5956* never varied very much, so that in some cases at least the shape seems to be genetically determined. For taxonomic purposes the corolla form is of little use in this species, and we have not been able to determine any geographical correlations with it.

In colour, also, there is a wide range of variation, since in some collections the corolla is white, in others it varies from white to blue-violet and in others it is entirely blue-violet. In the following collections only violet flowers are noted: *Balls 6126*, Philippi's type from Calcalhuay, *Balls 5980, Petersen & Hjerting P–360, P–362, Cabrera 9481, Cabrera & Schwabe 10, Kurtz 11327* (p.p.), *Meyer* (s.n.) (LIL 33409), and *Schreiter 11198*. Pale violet flowers are noted for Bitter's var. *angustepinnatum, H.P.R. 76* (p.p.), *Sleumer 3626* and *4105*. Many collections show a tendency to a paling of the colour on the centre of the lobes and petal apices so that in some the only purple coloration is to be seen on the interpetalar membranes: *Petersen & Hjerting P–367* and *963, H.P.R. 75* (plants with parti-coloured flowers taken from amongst the self-coloured plants of *H.P.R. 76* from the same population), *H.P.R. 80* (parti-coloured flowers from a population of which completely white-flowered plants were given the number *79*; see below). The reverse patterning, e.g. white interpetalar membranes and pale-purple coloration on petal centres and acumens, was seen in *Balls 5956* (type of *S. glanduliferum*) but in no other collections so far seen. Pure white flowers are to be seen in *Venturi 9394* and *H.P.R. 79* (see above).

From the amount of variability described above one might imagine it to be impossible to identify *S. infundibuliforme* in the field or in the herbarium. This is not so, however, since the species is a very distinctive one and may be easily recognized by the pinnatisect leaf, rather narrow lobes, absence of interjected leaflets, decurrent leaflet wings and short stiff acute hairs. The plants are always small and delicate and cannot easily be confused with any other Argentine tuber-bearing *Solanum*.

Since we began this revision of the Argentine potato species a paper was published on the series Cuneoalata by Brücher (1957*a*) in which he independently arrived at many of our present conclusions. After stressing the xeromorphic nature of the species in this series and mentioning their apparent frost resistance he points out that they have not been utilized by plant-breeders to any extent. Furthermore, Brücher rightly concludes that the variants previously described by Hawkes as separate species are due largely to differences of habitat, acting on very similar genotypes, as we have already pointed out and as was indicated by Hawkes in 1956.

We cannot, however, agree with Brücher in attributing the glandular pubescence of some forms of *S. infundibuliforme* ('*S. glanduliferum*') and their more rotate corolla to hybridity, the other parent being, according to him, *S. megistacrolobum* ('*S. tilcarense*'). On the contrary, the corolla of this latter species is even more stellate than that of the most stellate (and, incidentally, non-glandular) form of *S. infundibuliforme*; in contrast to this supposition of Brücher, we can only reiterate that the plant of '*S. glanduliferum*' (*Balls 5956*. Alt. 2450 m) with a very rotate corolla was found at a very low altitude, where *S. megistacrolobum* most certainly would not occur, and even if it did do so, this latter species has no glandular hairs and could not therefore have contributed them to its hybrid offspring. We shall see later (p. 282) that hybrids between *S. infundibuliforme* and *S. megistacrolobum* occur at higher altitudes (3550–3900 m), but they are morphologically completely dissimilar to the low altitude (2450 m) glandular forms of *S. infundibuliforme*. Correll (1962) mentions one of his collections (*A660*) from Jujuy, near Tres Cruces, which he thinks is a mixture of *S. infundibuliforme* and var. *angustepinnatum*, though one plant has leaves of both variants. This condition is quite normal, as we have already pointed out, but the wider leaf is normally the juvenile and the narrower the adult form, rather than the reverse, as suggested by Correll.

We have examined the Kew sheet of this collection, and can confirm that it is all good *S. infundibuliforme*; a photograph of another sheet, sent us by Correll, is a mixture of *S. infundibuliforme* and hybrids of this species with *S. megistacrolobum*. These hybrids, shown also as the left-hand plants in Correll's Fig. 129, possess very broad terminal leaflets and broad decurrent wings, with rather few (*c.* 2) pairs of lateral lobes, showing the strong influence of *S. megistacrolobum*. Hjerting, Petersen, and Rahn collected hybrids of this nature in the same region (*No. 76* (p.p.)), together with *S. infundibuliforme* (*Nos. 75, 76* (p.p.)) and *S. megistacrolobum* (*Nos. 77, 78*); and Correll also obtained *S. megistacrolobum* (*No. A661*) in the vicinity of his *A660* collection. There is therefore little doubt that the two species form hybrids in this area and that some of the special attributes described by Correll for his collection *A660* are due to hybridization with *S. megistacrolobum*.

NOTES ON LIVING PLANTS

The following collections of *S. infundibuliforme* from Argentina and Bolivia have been grown in Europe by the present authors: *Balls 5956, 6126*; *Petersen & Hjerting P–360, P–362, P–367, 963, 965* (see Plate 76). All gave a rather delicate moderately branched plant with decumbent stem. The lower leaves possessed a moderately wide rhachis wing which became much narrower in the upper ones. The flower colour varied according to the collection, as has already been mentioned.

It is of some interest to note the great differences in stem length and leaf-lobe development in genetically similar material when growing in its natural habitat under poor dry soil conditions, on the one hand, and when cultivated in Europe in rich soil with adequate moisture, on the other.

Thus for instance, the collection of *Petersen & Hjerting P–360* in Argentina grew only 7·5 cm high and the lobes of the 3–3·8 cm long leaves were never more than 3·5 mm wide, with wings only up to 1·5 mm wide. In contrast, material of this line grown from original seed in Denmark was 30 cm tall, the leaves were up to 9·5 cm long, the leaflet lobes to 12 mm wide and the wings to 4 mm wide.

The other collections mentioned above as having been cultivated in Europe showed the same tendencies, though it should be pointed out that not all the Argentine plants were as small and under-developed as the example we have taken. In some of these original collections (*Petersen & Hjerting P–362*, *965* (see Plate 75) for example) there is a mixture of plants with poorly developed and with well-developed leaves; and these latter plants are then very similar in leaf type, though much smaller in plant height, to the luxuriant European-grown seedlings of the same line.

Under artificial short days at Vandel the average tuber weight for different plants varied from 1·4 to 1·7 grammes per tuber approximately. All seedlings showed a considerable resistance to frost in the field, but were much attacked by *Phytophthora*.

S. infundibuliforme has also been shown to be resistant to common eelworm and to virus A (see Chapter 5).

CYTOLOGY

The somatic chromosome number of *S. infundibuliforme* was determined as $2n = 24$ by Hawkes (1944) and this was confirmed by Choudhuri (1943, 1944) and Brücher (1957*a*). Meiotic studies were made by Choudhuri (1943, 1944) who observed small percentages of irregularities such as incomplete bivalent pairing, restitution nuclei with unreduced chromosome number, and occasional anaphase bridges. A hybrid with *S. chacoense* (as *S. parodii*) showed pollen mother cells with almost complete bivalent pairing, however.

FERTILITY AND CROSSABILITY

S. infundibuliforme is self incompatible but sets seed freely when clones with different incompatibility genes are crossed together. Hybrids of this species with *S. chacoense* (as *S. parodii*), *S. phureja* (and also as *S. rybinii*, *S. kesselbrenneri*), *S. goniocalyx*, *S. tarijense*, and *S. microdontum* were made by Choudhuri (1944). The present authors have also obtained hybrids with *S. goniocalyx*, *S. chacoense*, *S. megistacrolobum*, *S. gourlayi*, *S. spegazzinii*, *S. vernei*, and *S. demissum*.

In all these the decurrent wings on the leaf rhachis are developed to a greater or lesser degree though they may be suppressed on all but a few leaves in each plant, as in the hybrid with *S. chacoense*, which has already been described (p. 227). The hybrid, family 253, of *S. infundibuliforme* (*P. & H. 965*, from Jujuy, Rinconada) with *S. megistacrolobum* (*P. & H. P–366*, from Jujuy, Humahuaca) is vigorous and approximately intermediate between the two parental species. (See Plate 79.) It is extremely close in leaf and floral features, when grown at Vandel, to the natural hybrids (*H.P.R. 76* (p.p.); *Sleumer 4112* (p.p.). See Plates 77 and 78) noted below, and its leaves match almost exactly those of the hybrid plants in Correll's Fig. 129 to which we have already referred. Family 297, raised from a cross between *S. infundibuliforme* (*P. & H. P–360* from Humahuaca) and *S. vernei* (CPC: 2413/1 from Tucumán) was, on the whole, vigorous though not very uniform. Most plants exhibited characters from each parent quite clearly and many flowered and formed berries freely. One F_2 family was raised from this cross, family 339. Fifty seeds were sown, 20 germinated, and 17 survived to be transplanted into the field. The seedlings were disparate in growth and vigour, and 10 of them died later. At the end of the season only 7 plants remained, and 1 of these had flowered and produced a single berry. Genetic breakdown, due to incompatibility of genomes, seems to have taken place here, indicating that *S. infundibuliforme* and *S. vernei* are rather distantly related.

Hybrids of *S. gourlayi* (*P. & H. P–365*, from Salta, Iruya, Rodero Colorado) with *S. infundibuliforme* (*P. & H. P–360*, from Jujuy, Humahuaca, El Aguilar) were not very vigorous. Of twenty seeds sown only three plants could be raised; when grown in the field at Birmingham they showed some leaf decurrency, but not as much as was exhibited by the hybrids between *S. infundibuliforme* and *S. megistacrolobum*. Finally, the hybrids of *S. infundibuliforme* (*P. & H. P–362*, from Jujuy, Humahuaca, Iturbe) × *S. spegazzinii* (*H.P.R. 334*, from Salta, Piedra Molino to Cachi) showed clearly the finely dissected leaf and coarse long pubescence of the latter species. Characters of *S. infundibuliforme* were not so obvious here.

Triple hybrids of *S. demissum* × (*S. chacoense* × *S. infundibuliforme*) were also successful and fertile when crossed with *S. tuberosum*. It is possible therefore to introduce *S. infundibuliforme* into crosses with *S. tuberosum* without any difficulty.

NATURAL HYBRIDS

(*a*) With *S. megistacrolobum*

Where the ranges of *S. infundibuliforme* and *S. megistacrolobum* overlap, the two species occasionally form hybrids. This takes place in the drier part of the range of *S. megistacrolobum* (sometimes at lower altitudes) and the wetter part of the distribution area of *S. infundibuliforme*. Natural hybrids of this type were collected by Hjerting, Petersen, and Rahn, by Correll, by Sleumer, and by Castellanos; they agree very well with the artificial hybrids grown by us in 1957 (Family 253), in the 1–3-jugate leaf with widely triangular decurrent wings and in the corolla form. Compare Plates 77, 78, and 79. Furthermore, in three cases both parents and hybrids have been collected from the same areas, as shown in Table 13.

With some collections it will be seen that *S. infundibuliforme* was gathered at lower altitudes, the hybrid higher up where the two species were intermingling, and *S. megistacrolobum* higher still. This arrangement of species is of course dependent on the fact that the climate gradually becomes wetter in ascending from the dry rain-shadow valleys up on to the more open mountain slopes, where the drought-loving species *S. infundibuliforme* becomes replaced by the more moisture-requiring *S. megistacrolobum*.

No record exists for *S. megistacrolobum* in the San Antonio de los Cobres area, but the hybrid is very similar to the others just mentioned. It seems very probable that *S. megistacrolobum* may exist in that region too, since the altitude would be about 3800 m and the climate is cold. It has been recorded a little to the south, also, in dept. Cachi, as well as in many places to the north-east of San Antonio de los Cobres.

(*b*) With *S. gourlayi*

Hybridization between *S. infundibuliforme* and *S. gourlayi* appears to be rare, or possibly such hybrids have been hitherto overlooked. Sleumer collected one hybrid of this nature (*3253* p.p.), growing with *S. infundibuliforme* (*3254*) and probable *S. gourlayi* (*3253* p.p.). (See Table 13.) A good blending of the characters of both

TABLE 13

Natural hybrids of S. infundibuliforme *with* S. megistacrolobum *and* S. gourlayi *in Argentina*

Locality	Collection, number, and altitude		
	S. infundibuliforme	Hybrid	*S. megistacrolobum*
Jujuy, Humahuaca, Tres Cruces	*H.P.R. 75, 76* (p.p.). 3550 m *H.P.R. 79, 80*. 3600 m *Correll, etc. A. 660* (p.p.).	*H.P.R. 76* (p.p.). 3550 m *Correll, etc. A. 660* (p.p.)	*H.P.R. 77, 78*. 3600 m *Correll, etc. A. 661, A. 665*.
Jujuy, Yavi, Cajas	*Sleumer 3626*. 4000 m *H.P.R. 110*. 3900 m *P. & H. 975*. 3950 m	 *H.P.R. 111*. 3900 m	*P. & H. 973*. 3950 m *H.P.R. 113, 114*. 3900 m *Sleumer 3628*. 3900–4100 m
Jujuy, Tilcara, Abra de Remate, subida.	*Sleumer 4105*. 3700–3800 m	*Sleumer 4112* (p.p.). 3850–3900 m	*Sleumer 4112* (p.p.). 3850–3900 m *Sleumer 4111* (p.p.). 3950 m *Sleumer 4114* (p.p.). 3950 m
Salta, Los Andes, San Antonio de los Cobres	*Cabrera and Schwabe 10*.	*Castellanos* (s.n.). (BA 27/903)	Not recorded
	S. infundibuliforme	Hybrid	*S. gourlayi* (?)
Jujuy, Tumbaya, San Antonio (Abra) de los Pibes.	*Sleumer 3254*. 3850 m *Castellanos* (s.n.). (BA 20174)	*Sleumer 3253* (p.p.). 3850 m	*Sleumer 3253* (p.p.). 3850 m

species is seen in the hybrid, though the upper leaves seem particularly close to those of *S. infundibuliforme*. (See Plate 80.)

Although *S. infundibuliforme* and *S. spegazzinii* grow in dry places and might be expected to occur together and form hybrids, these have not yet been recorded. *S. infundibuliforme* has been very occasionally found growing with *S. acaule*, but again, no natural hybrids between these species have been seen.

LOCALITIES OF NATURAL HYBRIDS

ARGENTINA

(*a*) With *S. megistacrolobum*

Prov. JUJUY:

Dept. Humahuaca, along stream among low shrubs on altiplano, about 7 km south of Tres Cruces. 25 Feb. 1960. 'Fls. blue-purple.' *Correll, Dodds, Brücher & Paxman A660* (p.p.) (photograph of LL sheet).

Dept. Humahuaca, entre Humahuaca y Tres Cruces, 30 km antes de Tres Cruces. Alt. 3550 m. 3 Feb. 1956. 'Same locality as 75 but lilac-flowered types.' *Hjerting, Petersen & Rahn 76* (p.p.) (J.G.H.). See Plate 77B.

Dept. Tilcara, Casa Colorada, camino al Abra de Remate. Alt. 3850–3900 m. 23 Feb. 1953. *Sleumer 4112* (p.p.) (J.G.H.). See Plate 78A.

Dept. Yavi, Cajas. Alt. 3900 m. 6 Feb. 1956. *Hjerting, Petersen & Rahn 111* (J.G.H., J.P.H.).

Prov. SALTA:

Dept. Los Andes, San Antonio de los Cobres. 22 Feb. 1927. *Castellanos* (s.n.) (BA 27/903).

(*b*) With *S. gourlayi*

Prov. JUJUY:

Dept. Tumbaya, San Antonio (Abra) de los Pibes. Alt. 3850 m. 5 Jan. 1953. 'Fl. viol.' *Sleumer 3253* (p.p.) (J.G.H.). See Plate 80.

SERIES MEGISTACROLOBA CÁRDENAS & HAWKES

(*J. Linn. Soc. Lond.* **53**, 93, 1945)

RATHER short-stemmed or straggling tuber-bearing herbs, occasionally growing as rosettes. The lateral lobes or segments are always much smaller than the terminal and are sometimes completely absent, but when present are broadly decurrent on to the rhachis on the basiscopic side, much more so than in series Cuneoalata; interjected leaflets normally not developed; peduncle very short; pedicel long, normally with very high articulation which is always well marked (contrast Acaulia). Corolla sub-stellate to rotate, purple. Fruits globular.

General distribution: north Peru, southwards through Bolivia to north-west Argentina, at altitudes of 2000 to 4500 m.

Chromosome number: $2n = 24$.

Series Megistacroloba was described in 1945 by Cárdenas and Hawkes, based on the type species, *S. megistacrolobum* Bitt., and including certain other allied species from Peru and Bolivia. Bukasov (1955*b*) illegitimately separated the Bolivian and north Argentine species of this series from the Peruvian ones then known (*S. raphanifolium* and *S. hawkesii*) under the new name Alticolae. This was incorrect, since the type species for the series Megistacroloba, namely, *S. megistacrolobum*, was included in the new series Alticolae, whilst only the two Peruvian species mentioned above were retained in series Megistacroloba. Bukasov considered the group to be artificial, but even if this were so the nomenclatural solution he proposed was incorrect, and in 1959 (in Bukasov and Kameraz 1959) he reversed his decision and returned to the original plan of including all these species in the one series, Megistacroloba.

Brücher (1959*b*) used a slight variant of Bukasov's name, evidently in error, calling the series Alticola instead of Alticolae. He made no study of the Peruvian species but considered that the Bolivian and Argentine species grouped by Cárdenas and Hawkes (1945) and by Hawkes (1956*b*) under Megistacroloba, should now be placed under Alticola. This is merely a question of name changing that is quite unjustified and is against the laws of priority. The name Alticola should be discontinued, therefore.

Correll (1962) has retained almost completely the original concept of series Megistacroloba, merely adding certain new species, such as the Peruvian *S. hastiforme*, *S. chavinense*, and *S. sogarandinum*, together with the new hybridogenic species, *S. bruecheri*.

We agree with Correll that series Megistacroloba is probably closely related to Cuneoalata. The characteristic broadly ovate sessile decurrent leaf segments or lobes and wide rhachis wings serve to distinguish Megistacroloba from Cuneoalata, whose leaves possess very narrow lobes. The habit, form of inflorescence, and corolla also serve to distinguish these two series, and we are quite convinced that Megistacroloba is a unified natural group of species which are all closely related to each other.

We consider, furthermore, that series Megistacroloba bears some relationship to series Acaulia and is probably in some way ancestral to it. It is possible that certain diploid ancestral forms of Megistacroloba may have taken part in the formation of the Acaulia tetraploids, though we are not as yet certain which is the other ancestor. It may perhaps be a species in series Tuberosa, and research is at present in progress to elucidate this problem.

Species of this series included in the present treatment:

S. MEGISTACROLOBUM Bitt.
S. BOLIVIENSE Dun.
S. SANCTAE-ROSAE Hawkes.

8. SOLANUM MEGISTACROLOBUM Bitter

(in *Reprium nov. Spec. Regni veg.* **10,** 536–7, 1912)

S. alticola Bitt. (in *Reprium nov. Spec. Regni veg.* **12,** 5–6, 1913).
S. tilcarense Hawkes (*Bull. imp. Bur. Pl. Breed. Genet., Camb.* 119–20, 1944).
S. alticola var. *xanthotrichum* Hawkes (ibid. 120).

A dwarf, rosette-forming species or with weak decumbent stems when growing amongst stones or other vegetation; vertical underground stem long, producing several long stolons which bear small tubers up to 1·5–2(–3) cm diam.

Stem 0·5–30 cm high, often weakly angled from node to node and with marked decurrent lines, sparsely to densely pilose.

Leaf in a rosette or more widely spaced in longer-stemmed plants; lamina simple or *pinnatilobed* to *sublyrate*, gradually decurrent into the petiole or in rosette plants sometimes sharply truncate; terminal leaflet or lamina of simple leaf varying from long oblong, long ovate, lanceolate or oblanceolate to elliptic, rhomboid, rhomboid-cordate or subtriangular (the latter chiefly in small rosette plants—cf. *S. tilcarense*); margin entire or slightly undulate; dimensions from 2–9(–13) cm long × 1–4(–7·5) cm broad; apex acute to sub-obtuse but *never acuminate*; lateral lobes much smaller than the terminal, 1–3(–5)-paired or sometimes completely absent (cf. *S. alticola*), ovate to oblong, decreasing in size rapidly from apex to base of leaf, generally *broadly decurrent below on the basiscopic side*, sometimes rounding into an auricle on the acroscopic side, apically rounded, obtuse, or rarely acute, *never acuminate*; largest lateral lobe about 2–4 × 0·7–2·75 cm but lobes often much smaller; *interjected leaflets completely absent* (except in some hybrid forms); leaf surface clothed with sparse to frequent white or yellowish rather thick adpressed hairs, generally sparser beneath, and also on the margins. Pseudostipular leaves completely absent.

Inflorescence lax, 1–4(–8)-flowered; *peduncle very short*, or absent in rosette forms, up to 5·5 cm long in decumbent plants, *generally unforked*; pedicels short to very long (2–)4–6(–9) cm, articulated in upper third, 5–10(–15) mm below calyx base; peduncles and pedicels sparsely to densely covered with erecto-patent thickish hairs of unequal lengths and occasional glands.

Calyx about 4–9 mm long, with ovate-triangular lobes and generally well-marked acumens; pubescence as for pedicels.

Corolla *semi-stellate* generally large (1·7–)2·5–3·5 cm diam., varying in colour from deep or medium purple to pale lilac, the paler colour more common; lobes (including acumens) 7–15 mm broad × 5–10 mm long, thus always slightly broader than long; *acumens not well delimited from lobes*, 3–5 mm long and broad.

Anthers 4–5 mm long × 1·5–2 mm broad; filaments up to 1 mm long × 1 mm broad at base, narrowing above.

Style thick, 7–11 mm long, glabrous or with minute papillae below; stigma slightly thicker than style apex, capitate to clavate, varying from entire to apically and basally bifid or grooved.

Berries green, globular, up to about 2 cm diam. (to 2·5 cm; *fide* Brücher).
Chromosome number: $2n = 24$.

AFFINITIES

S. megistacrolobum may be distinguished from the closely related species *S. sanctae-rosae* by the rounded obtuse widely decurrent lateral leaf lobes and obtuse or acute (never acuminate) terminal lobe. The flowers are generally of a lighter blue colour, also. It differs from *S. boliviense* by the short stem and more typically semi-stellate corolla and from *S. acaule* by the markedly semi-stellate corolla, and well-marked pedicel articulation.

SPECIMENS EXAMINED

BOLIVIA

Dept. Tarija, Puna Patanca. Alt. 3700 m. 8 Jan. 1904. *Fiebrig 2618* (p.p.) (S—Lectotype, BG, BM, F, G, GOET, HBG, JE, LD, LIL, M, NY, SI, W). See Fig. 24.

ARGENTINA

Prov. JUJUY:

Dept. Cochinoca, al N. de Tres Cruces. Alt. 3725 m. 6 Feb. 1921. *Castellanos 49* (BA).

Dept. Cochinoca, Abra Pampa. Alt. 3400 m. 17 Feb. 1929. 'Fl. violáceo-obscura. Falda del cerro.' *Venturi 9393* (US).

Dept. Humahuaca, Mina Aguilar. Alt. 4000 m. 12 Jan. 1948. 'Flor azul.' *Cabrera 9171* (BAB, LP, J.G.H.). Rosette plants.

Dept. Humahuaca, Mina Aguilar. Alt. 4300 m. 15 Jan. 1948. 'Flor azul.' *Cabrera 9272* (LP).

Dept. Humahuaca, along stream among low shrubs on altiplano, about 7 km south of Tres Cruces. 25 Feb. 1960. *Correll, Dodds, Brücher & Paxman A. 661* (K.).

Dept. Humahuaca, along stream among low shrubs, about 8 km south of Tres Cruces. 25 Feb. 1960. 'Leaves large; fls. pale lavender.' *Correll, Dodds, Brücher & Paxman A. 665* (K).

Dept. Humahuaca, entre Humahuaca y Tres Cruces, 28 km antes de Tres Cruces. Alt. 3600 m. 3 Feb. 1956. *Hjerting, Petersen & Rahn 77 and* 78 (C, J.G.H., J.P.H.). Stem 28 cm and 3 cm, respectively, the former being shade plants. See Plates 83, 84, and 85.

Dept. Humahuaca, 3 km norte de Tres Cruces hacia Abra Pampa. Alt. 3700 m. 3 Feb. 1956. *Hjerting, Petersen & Rahn 85* (C, J.G.H., J.P.H.). Rosette plants.

Dept. Humahuaca, El Aguilar, cerca del Molino de Mina Aguilar. Alt. 4100 m. 28 Mar. 1952. *Petersen & Hjerting P–358* (p.p.) (LIL, J.G.H.). Stem 10 cm.

Dept. Humahuaca, El Aguilar, cerca del Molino de Mina Aguilar. Alt. 4100–4300 m. 28 Mar. 1952. *Petersen & Hjerting P–359* (LIL, J.G.H., J.P.H.). Stem from 0·5 to 15 cm.

Dept. Humahuaca, entre Cóndor y Abra del Cóndor. Alt. 3600–4100 m. 9 Apr. 1952. *Petersen & Hjerting P–366* (LIL, J.G.H., J.P.H.). Stem 5–25 cm. (See also p. 295, natural hybrids.)

Dept. Humahuaca, entre Cóndor y Abra del Cóndor. Alt. 3700 m. 9 Apr. 1952. 'Crece bajo arbustos.' *Petersen & Hjerting P–368* (p.p.) (J.G.H.). Atypical leaf; stem 3 cm.

Dept. Humahuaca, Tres Cruces. Alt. 3800 m. 20 Feb. 1959. 'Trockenpuna mit *Pseudobaccharis*.' *Ross & Vervoorst 85* (Max.).

Dept. Rinconada, 38 km oeste de Abra Pampa hacia Mina Pirquitas. Alt. 3800 m. 10 Feb. 1953. *Petersen & Hjerting 964* (J.G.H., J.P.H.). Rosette plants.

Dept. Tilcara, above Tilcara, San Gregorio. Alt. 3700 m. 15 Feb. 1939. *Balls 5981* (CPC: 76) (P.G.S.). Stem 1–15 cm.

Dept. Tilcara, above Tilcara, San Gregorio. Alt. 3700 m. 15 Feb. 1939. *Balls 5986* (p.p.) (K, UC, US, P.G.S., J.G.H.) (Part of this collection formed the types for '*S. alticola* Bitt. var *xanthotrichum*' Hawkes and '*S. tilcarense*' Hawkes.) Stem 1–6 cm. See Figs. 26 and 27.

Dept. Tilcara, above Tilcara, San Gregorio. Alt. 3900 m. 15 Feb. 1939. *Balls 5987* (p.p.) (P.G.S.). Rosette plants.

Dept. Tilcara, above Tilcara, San Gregorio, Alt. 3900 m. 15 Feb. 1939. *Balls 5989* (P.G.S.). Rosette plants.

Dept. Tilcara, above Tilcara, San Gregorio. Alt. 3900 m. 15 Feb. 1939. *Balls 5991* (P.G.S.). Stem 1–2·5 cm.

Dept. Tilcara, above Tilcara, San Gregorio. Alt. 3950 m. 15 Feb. 1939. *Balls 6012* (P.G.S., J.G.H.). Rosette plants.

Dept. Tilcara, above Tilcara, San Gregorio. Alt. 3950 m. 15 Feb. 1939. *Balls 6023* (P.G.S.). Rosette plants.

Dept. Tilcara, above Tilcara, San Gregorio. Alt. 4000 m. 15 Feb. 1939. *Balls 6024* (P.G.S.). Rosette plants.

Dept. Tilcara, entre Quebrada del Chorro y Laguna Colorada. Alt. 3300 m. 24 Mar. 1952. *Petersen & Hjerting P–353* (p.p.) (J.G.H.). Short stems.

Dept. Tilcara, Quebrada de San Gregorio. Alt. 4000 m. 24 Mar. 1952. *Petersen & Hjerting P–355* (LIL, J.G.H.). Rosette plants.

Dept. Tilcara, Tilcara, arriba de San Gregorio. Alt. 3850–3900 m. 27 Jan. 1953. 'Fl. viol. pálida.' *Sleumer 3590, 3591* (LIL, J.G.H.). Stem 1–5 cm.

Dept. Tilcara, Casa Colorada, camino a la Abra de Remate. Alt. *c.* 3950 m. 23 Feb. 1953. 'Fl. violác. pálida.' *Sleumer 4111* (p.p.) (LIL, J.G.H.). Stem 1–10 cm. See Plate 70B.

Dept. Tilcara, Casa Colorada, camino a la Abra de Remate. Alt. 3850–3900 m. 23 Feb. 1953. 'Fl. violác. pálida.' *Sleumer 4112* (p.p.) (LIL, J.G.H.). Stem *c.* 2 cm. See Plate 78B.

Dept. Tilcara, Abra de Remate, subida. Alt. *c.* 3950 m. 23 Feb. 1953. 'Fruto.' *Sleumer 4114* (p.p.) (J.G.H.). Rosette plants.

Dept. Tumbaya, Volcán. Alt. 2700 m. 3 May 1929. 'Fl. blanca.' *Venturi 9521* (US).

Dept. Yavi, Cangrejillos. 28 Jan. 1948. *Cabrera 9459* (LP).

Dept. Yavi, Cajas. Alt. 3900 m. 6 Feb. 1956. 'Una planta de sombra.' *Hjerting, Petersen & Rahn 113 and 114* (C, J.G.H., J.P.H.). Sun plants with stems to 2 cm long. Shade plants with stems 3·5–23·5 cm.

Dept. Yavi, Cerro Negro. Alt. 4000 m. 27 Feb. 1940. 'Flor morada.' 'N. V. Aparoma.' *T. Meyer* (s.n.) (LIL 33398). Rosette plants.

Dept. Yavi, Cajas, 23 km este de Yavi. Alt. 3950 m. 16 Feb. 1953. *Petersen & Hjerting 973* (LIL, J.G.H., J.P.H.). Stem 2–6 cm. See Plates 81 and 82.

Dept. Yavi, Rincón de Cajas. Alt. 3900–4100 m. 31 Jan. 1953. 'Fl. violác. pálida.' *Sleumer 3628* (LIL, J.G.H.). Rosette plants.

Prov. SALTA:

Dept. Cachi, Cuesta del Obispo. Alt. 3720 m. 21 Jan. 1947. *Garolera & Romero* (s.n.) (LIL 184321 and 184325). Rosette plants.

Dept. Cachi, Piedra Molino. Alt. 3700 m. 22 Feb. 1956. 'Frutos.' *Hjerting, Petersen & Rahn 316* (C, J.G.H., J.P.H.). Stem 2–10 cm.

Dept. Caldera, Quebrada del Río Potrero (del) Castillo. Alt. *c.* 3800 m. 16 Mar. 1952. 'Rara. Fl. viol.' *Sleumer & Vervoorst 2975* (LIL). Stem 10 cm.

Dept. Chicoana, Cuesta del Obispo. Alt. 3400 m. 22 Feb. 1956. *Hjerting, Petersen & Rahn 314* (J.G.H., J.P.H.). Stem 5–12 cm.

Dept. Santa Victoria, entre Rodeopampa y Patahuasi. Alt. 3900 m. 16 Feb. 1956. 'Con frutos.' *Hjerting & Rahn 251* (C, J.G.H., J.P.H.). Stem 5–13 cm.

Dept. Santa Victoria, entre Lizoite y Abra de Lizoite. Alt. 3450–3700 m. 17 Feb. 1956. 'Sigue por lo menos hasta 3900 m o más.' *Hjerting & Rahn 255* (C, J.G.H., J.P.H.). Stem 5–15 cm.

Dept. Santa Victoria, Mal Paso (Patahuasi) a Rodeopampa. Alt. *c.* 3600 m. 3 Feb. 1953. 'Fl. violác. (pálida).' *Sleumer 3745* (LIL, J.G.H.). Stem 2–5 cm. (Introgression from *S. boliviense*?)

Dept. Santa Victoria, entre Abra Delgada y Abra Peña. Alt. *c.* 3500–3600 m. 7 Feb. 1953. 'Fl. violác. intensa.' *Sleumer 3839* (LIL, J.G.H.). Stem 4–25 cm. (Introgression from *S. boliviense*?)

Dept. Santa Victoria, Cuesta de Mesón. Alt. *c.* 2600–3000 m. 12 Feb. 1953. 'Fl. violác.' *Sleumer 4003* (LIL, J.G.H.). Stem 6–10 cm. (Introgression from *S. boliviense*?)

DOUBTFUL RECORD:

Prov. JUJUY:

Dept. Tilcara, Estancia Vieja, cerca de la casa. Alt. 3700 m. 21 Mar. 1952. *Petersen & Hjerting P–343* (LIL, J.G.H., J.P.H.).

DISTRIBUTION

S. megistacrolobum is a species of high mountain slopes and plateaux, spreading from the Lake Titicaca region of south Perú and north Bolivia southwards into north-west Argentina, where it is confined to the provinces of Jujuy (depts. Cochinoca, Humahuaca, Rinconada, Tilcara, Tumbaya, and Yavi) and Salta (depts. Cachi, Caldera, Chicoana, and Santa Victoria)[1] (see Map 9). (It has also been collected in Salta, depts. Guachipas and San Carlos. See Appendix III.)

Further south, in provs. Tucumán and Catamarca, it is replaced at roughly similar (but on the average rather lower) altitudes and habitats by the closely related species *S. sanctae-rosae*.

HABITAT

This species occurs in situations with a fairly high summer rainfall, at altitudes of (2600–)3000–4100(–4300) m. It is normally found in the wetter Puna vegetational province and occurs only rarely in the drier Puna and the upper Prepuna province. On the eastern slopes of the Andes it occurs in the upper Montano district ('pajonales') of the Western Subtropical vegetational province, as defined by Cabrera (1953).

In the wetter part of its range *S. megistacrolobum* is to be found in company with *S. acaule* and occasionally with *S. gourlayi*, *S. microdontum*, and *S. boliviense*. In the drier part it comes into contact with *S. infundibuliforme*.

It is found generally in high Andean meadows, in sandy and stony places, by the sides of paths and walls, and around dwellings or old field enclosures. Correll (1962) states that it also occurs amongst bushes and by streams.

[1] According to Brücher (1959*b*) this species is said to extend as far south as latitude 28°. It is thus implied, though no collection data are given to support this, that *S. megistacrolobum* extends into provs. Catamarca and Tucumán. Our own results indicate that this species does not spread further south than about 25° 50′ (Pampa Grande, dept. Guachipas, and Cuesta del Obispo, depts. Chicoana and Cachi, prov. Salta) and is entirely absent from Tucumán and Catamarca. Although we have not been able to examine Brücher's specimens, we have studied a very large number of collections of wild potatoes from Catamarca and Tucumán, and have never seen anything from these provinces identifiable with *S. megistacrolobum*. Possibly he confused it with *S. sanctae-rosae*.

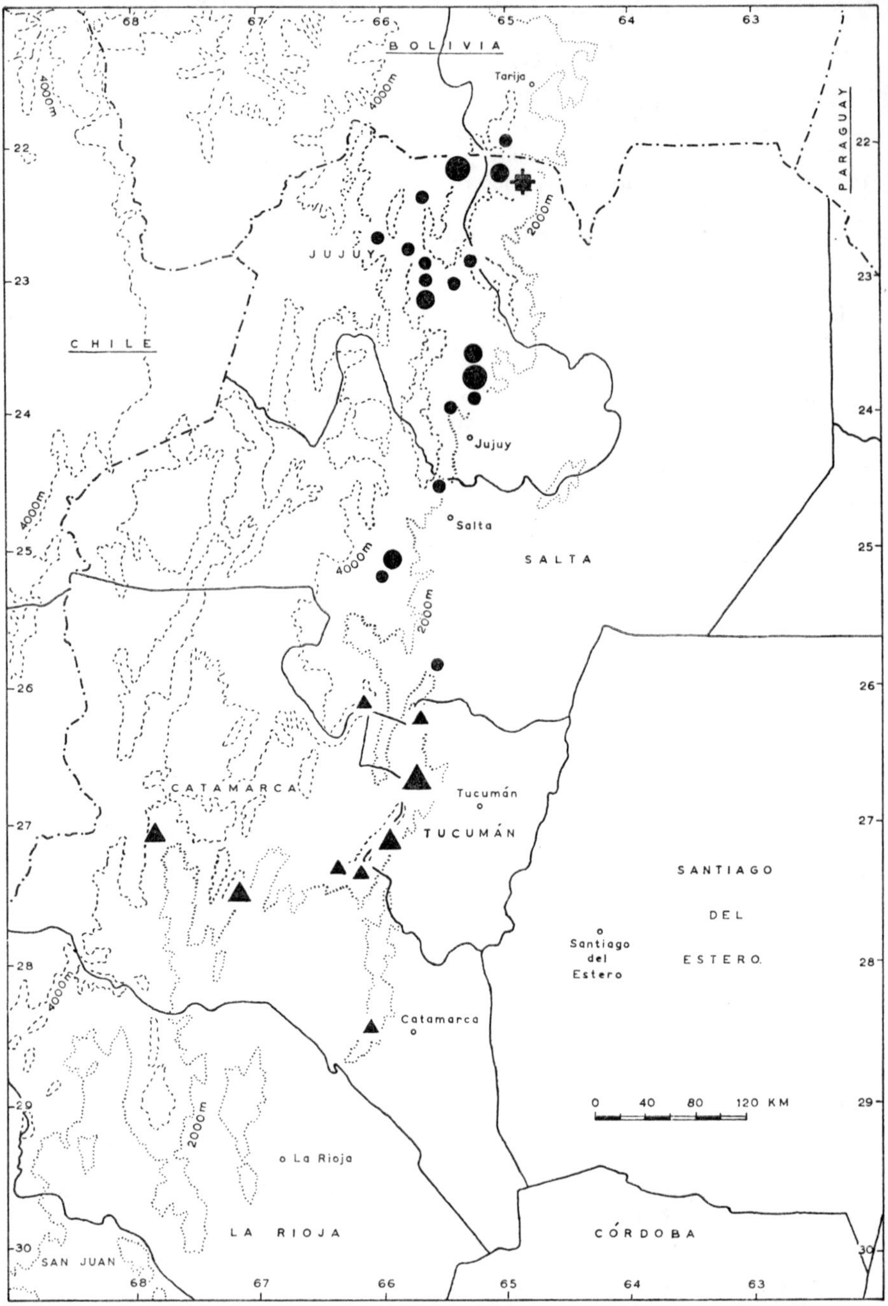

No. of Records
5 & over 3–4 1–2

Solanum megistacrolobum Bitt.

Solanum sanctae-rosae Hawkes

Solanum boliviense Dun.

MAP. 9. Distribution of *S. megistacrolobum* Bitt., *S. sanctae-rosae* Hawkes, and *S. boliviense* Dun.

TAXONOMIC NOTES

S. megistacrolobum was described by Bitter in 1912 from material collected by Fiebrig (*No. 2618* p.p.) at Puna Patanca in Bolivia (see Fig. 24) just across the northern Argentine border from the eastern part of prov. Jujuy, dept. Yavi, where many of our own collections were made. This was typical of the shade form of the species with lax decumbent stems, long oblong or oblanceolate leaves, and a few pairs of lateral leaf

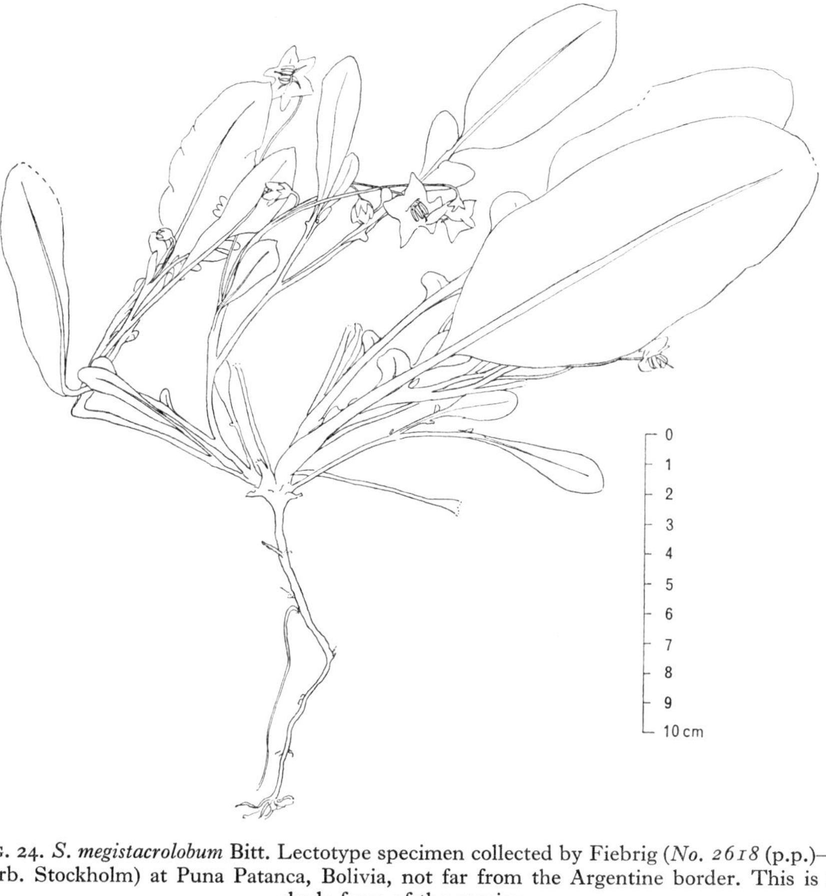

FIG. 24. *S. megistacrolobum* Bitt. Lectotype specimen collected by Fiebrig (*No. 2618* (p.p.)—Herb. Stockholm) at Puna Patanca, Bolivia, not far from the Argentine border. This is a shade form of the species.

lobes. The locality given: 'In damp irrigated soil. By settlements' reinforces this conclusion. Bitter's type specimen from Herb. Berlin has now been destroyed, except for a fragment in Chicago. We therefore designate the Stockholm sheet of *Fiebrig 2618* (p.p.) as LECTOTYPE of *S. megistacrolobum*. (See Fig. 24.)

In 1913 Bitter described a further species, *S. alticola*,[1] from specimens obtained by G. Mandon in north Bolivia, prov. Larecaja, near Mt. Sorata (*No. 398*). (See Fig. 25.) These plants possessed short stems or grew as stemless rosettes with simple lanceolate leaves, and no lateral leaf lobes whatsoever. They are typical of the growth

[1] Bitter's name was *S. alticolum*, but we have followed *Index Kewensis* (Suppl. 5, 1921) in correcting this orthographic error.

habit of *S. megistacrolobum* when fully exposed to the sun in poor soil and with very little vegetation in its immediate vicinity. The total absence of lateral leaf lobes is, however, largely genetically controlled, as we shall see later.

Brücher (1959*b*) retained Bitter's species, *S. alticola*, asserting that it was quite distinct from *S. megistacrolobum*. With this we cannot agree, for the reasons already given.

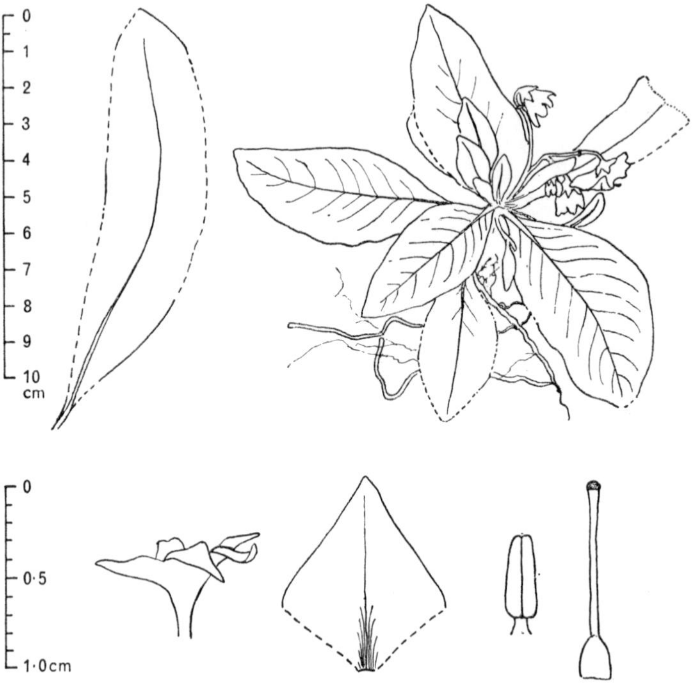

FIG. 25. *S. megistacrolobum* Bitt. One of the specimens from the type-sheet of '*S. alticola*' Bitt., collected by G. Mandon (*No. 398*—Herb. Paris) near Mt. Sorata, prov. Larecaja, north Bolivia. This is a form of *S. megistacrolobum* with entire leaves and very short stem, which cannot be considered specifically distinct because it is connected to the type by numerous intermediates (see also Fig. 27). Note the similarity of corolla between this and Fig. 24.

Plants collected by Balls and Hawkes in 1939 from Tilcara, prov. Jujuy were described by Hawkes (1944) as a new species, *S. tilcarense*, since although the plants were mostly growing as rosettes the terminal leaflet was rhomboid to sub-triangular and a number of lateral lobes were formed. (See Fig. 26.) Variants in the same population with 0–1(–2) very small lateral lobes and dense yellowish pubescence were further distinguished as *S. alticola* var. *xanthotrichum*. (See Fig. 27.) At this time the intergrading forms between *S. megistacrolobum*, *S. alticola*, and *S. tilcarense* were considered as due to hybridization between these three species.

After revising the much more extensive collections made by Sleumer, Petersen and Hjerting, Hjerting, Petersen and Rahn, and Hjerting and Rahn in Jujuy and Salta we have come to the conclusion that the above-mentioned taxa should all be grouped under *S. megistacrolobum*. We are convinced that the high range of variability, due partly to environmental and partly to genetical causes, does not transgress the limits of the one species, *S. megistacrolobum*. The variation in leaf dissection, for instance, may be seen in Table 14, where the collections are placed according to the number of lateral leaf lobes. It will be noted that quite a few collections range from

FIG. 26. *S. megistacrolobum* Bitt. Type specimen of '*S. tilcarense*' Hawkes, collected by Balls, Gourlay, and Hawkes (*Balls 5986* (p.p.)—Herb. P.G.S.) at San Gregorio, Tilcara, prov. Jujuy, Argentina. This is merely a shorter-stemmed form, growing in full sun, of *S. megistacrolobum* with fairly good leaf dissection. See also Fig. 25.

FIG. 27. *S. megistacrolobum* Bitt. Type specimen of '*S. alticola* Bitt. var. *xanthotrichum*' Hawkes, collected by Balls, Gourlay, and Hawkes (*Balls 5986* (p.p.)—Herb. P.G.S.) at San Gregorio, Tilcara, prov. Jujuy, Argentina. This is a form from full sun with stronger pubescence and less leaf dissection than the type (cf. '*S. alticola*', Fig. 25).

Table 14

Variation in leaf dissection in S. megistacrolobum

Leaf simple or with 1 minute pair of lateral lobes		Leaf with 1–2 pairs of lateral lobes		Leaf with 2–3 or more pairs of lateral lobes
				Balls 5981
Balls 5986 (p.p.)	↔	Balls 5986 (p.p.)	↔	Balls 5986 (p.p.)
Balls 5989				Balls 5987 (p.p.)
				Balls 5991
		Balls 6012	↔	Balls 6012
				Balls 6023
		Balls 6024	↔	Balls 6024
Cabrera 9459		Cabrera 9171		Cabrera 9272
Garolera & Romero (s.n.)		Castellanos 49		Fiebrig 2618 (p.p.) (Type. Bolivia)
Hjerting & Rahn 251	↔	Hjerting & Rahn 251	↔	(Hjerting & Rahn 251. Plants grown at Vandel)
,, ,, 255	↔	,, ,, 255	↔	,, ,, 255
Hjerting, Petersen & Rahn 77	↔	Hjerting, Petersen & Rahn 77	↔	Hjerting, Petersen & Rahn 77
,, ,, ,, 78	↔	,, ,, ,, 78	↔	,, ,, ,, 78
		,, ,, ,, 85	↔	,, ,, ,, 85
,, ,, ,, 113	↔	,, ,, ,, 113		,, ,, ,, 114
,, ,, ,, 314				
,, ,, ,, 316	↔	,, ,, ,, 316		
		T. Meyer (s.n.) (LIL 33398)		
		Petersen & Hjerting P–353 (p.p.)		
		,, ,, P–355	↔	Petersen & Hjerting P–355
		,, ,, P–358 (p.p.)		
Petersen & Hjerting P–359	↔	,, ,, P–359		
,, ,, P–366	↔	,, ,, P–366	↔	,, ,, P–366 (and Vandel plants)
				,, ,, P–368 (p.p.)
,, ,, 964	↔	,, ,, 964		
,, ,, 973	↔	,, ,, 973	↔	,, ,, 973 (and Vandel plants)
Sleumer 3590	↔	Sleumer 3590	↔	Sleumer 3590
,, 3628		,, 4111 (p.p.)	↔	,, 4111 (p.p.)
,, 3745		,, 4114 (p.p.)		,, 4112 (p.p.)
,, 3839				
,, 4003				
Venturi 9393		Sleumer & Vervoorst 2975		

simple-leaved to 1-, 2-, or 3-jugate ones, and from this we have been forced to conclude that it is impossible to separate out the simple-leaved plants as a species distinct from *S. megistacrolobum*, since the intergrading is so gradual.

Differences in stem length and leaf size may be largely due to environmental causes. The stem lengths of most specimens are given in the locality list, but where the stems are very short we have merely noted 'rosette plants'. Although most collectors do not give habitat notes sufficient for us to be able to relate these measurements to the size of surrounding plants or amount of shading (see, however, *Hjerting, Petersen & Rahn 77 & 78; 113 & 114*) our personal observations in the field have shown us that there is a good correlation between stem length and degree of shading. Differences in leaf dissection, however, are controlled mainly by genetical differences, and are maintained when plants are grown in the field in Europe (see below).

The situation in *S. megistacrolobum*, therefore, is very similar to that of *S. sanctae-rosae*, where leaf dissection is influenced partly by environment but largely by the genetical constitution of the population under observation.

There is no doubt that the phenotype or 'growth form' most frequently seen in nature is the short rosette, similar to *S. tilcarense* and *S. alticola* (see Figs. 25–27) whilst the type specimen of *S. megistacrolobum* (see Fig. 24) collected by Fiebrig is a shade form growing in very rich soil. Plants cultivated in Europe, therefore, tend to show the more 'typical' *S. megistacrolobum* growth form, whilst those in nature are generally closer in habit to the other two 'species' mentioned above. Thus the solution to the problem of *S. alticola*, *S. tilcarense*, and *S. megistacrolobum* was only made possible by cultivation of seed and tuber collections under uniform environmental conditions, and the comparison of such long-stemmed plants with the rosette-forming original collections made in Bolivia and Argentina.

Although the flower colour of *S. megistacrolobum* tends to be a light blue-purple, it varies even within populations to some extent. Very much darker-flowered biotypes occur in prov. Salta, dept. Santa Victoria, possibly due to introgression with *S. boliviense*. In general, however, light blue-purple flowers are typical of *S. megistacrolobum* and serve as an additional character for separating it from the closely related *S. sanctae-rosae*, which has dark blue-purple flowers.

NOTES ON LIVING PLANTS

Plants raised at Birmingham and at Vandel are rather long-stemmed, but have retained the leaflet type seen in the original herbarium specimens, except that there is a general trend for the addition of one pair of lateral leaf lobes and the occasional production of ill-defined interjected lobes, as indicated in Table 15.

Glasshouse-grown material often showed more leaf lobes than field-grown plants and in Table 15 the number of leaf lobes found on these glasshouse plants is placed in parenthesis. Some slight variations between seedlings of the same collection were also noted.

In habit the plants were never as rosette-forming as plants collected in Argentina, except in the very youngest stages of growth. Compare also Plates 81 and 82; 84 and 85.

The leaves mostly possessed an odour when rubbed lightly between the fingers, due to the presence of rather sparse glandular hairs. The smell is quite distinctive, is apparently confined to series Megistacroloba, and is considered by some to be unpleasant and by others to be not unlike the smell of parsley (*Petroselinum crispum*). The leaf colour is bright green in most collections, but in *Hjerting, Petersen & Rahn 316* a dark reddish-purple coloration is visible on the mesophyll of the under surface of the leaves. In this collection, also, the stem is longer than is usual for this species.

The high pedicel articulation, semi-stellate corolla, and non-acuminate leaflet apices are all completely constant under cultivation.

TABLE 15

Comparison of degree of leaf lobing in Argentine and European material of S. megistacrolobum

Collector	No.	Pairs of lateral lobes		Pairs of interjected lobes	
		In Argentina	In Europe	In Argentina	In Europe
Petersen & Hjerting	*P–366*	2–3	3–4	0	0 (–2)
,, ,,	*973*	1–2	2–3	0	0
Hjerting & Rahn	*255*	1–2	2–3	0	0
Hjerting, Petersen & Rahn	*316*	0–2	0 (–2)	0	0
Sleumer	*4114* (p.p.)	1–2	2–3 (–4)	0	0 (–2)

For further explanation see text p. 293.

Berry production under European conditions in the field seems to be negligible. In *Hjerting, Petersen & Rahn 316*, however, a few green slightly axially compressed berries covered with small white raised spots were produced. On some seedlings the berries were darkly flushed with red-purple pigment.

Under artificial short days at Vandel the mean tuber weight varied from 1 to 3 grammes. The plants were quite frost-resistant in the field but some were very susceptible to *Phytophthora infestans* (notably *Hjerting & Rahn 255* and *Hjerting, Petersen & Rahn 316*). Frost resistance was noted also by Brücher (1959*b*) in plants growing at high altitudes (4300 m) in north-west Argentina. Resistance to common eelworm and to virus leaf-roll have also been demonstrated for this species (see Chapter 5).

CYTOLOGY

The somatic chromosome number of *S. megistacrolobum* was reported by Hawkes (1956*b*) to be $2n = 24$ on the basis of counts from a number of different seedlings. We have confirmed this figure by further counts on other collections and have found no deviation from the simple diploid number of $2n = 24$.

No meiotic studies have been made on this species or on hybrids between it and others, so far as we are aware.

FERTILITY AND CROSSABILITY

S. megistacrolobum is self-incompatible but quite fertile when different clones are crossed. It forms hybrids easily with species in series Megistacroloba and Cuneoalata, and would probably do so also with other series, such as Commersoniana and Tuberosa.

Pandey (1960, 1962) studied the genetic basis of self-incompatibility in *S. megistacrolobum*, and concluded that one series of S-alleles could be postulated to explain his results; these acted gametophytically in the pollen and independently in the style. Colchicine-tetraploids were mostly self-compatible.

We have succeeded in crossing *S. megistacrolobum* with *S. acaule* (see p. 265), with *S. infundibuliforme* (see p. 279) and with *S. raphanifolium*. The leaf type in the F_1 *S. infundibuliforme* × *S. megistacrolobum* seedlings was very close to that of the probable hybrids between these two species collected in Argentina (*Sleumer 4112* (p.p.), *Hjerting, Petersen & Rahn 111*). In the artificial hybrids the wings running down from the leaf lobes were quite broad, about intermediate between the two parents in width (see

Plate 79), and all seedlings were fairly similar in this respect. The flowers were still semi-stellate, but the corolla lobes were rather broader than in *S. megistacrolobum* itself. The high articulation of *S. megistacrolobum* was dominant in all seedlings. So far, an F_2 generation has not been raised from these seedlings. Several attempts were made to cross *S. megistacrolobum* with *S. microdontum* (both subsp.). A few fruits were formed, containing very few seeds and giving rise to one poor plant only. No success was obtained in reciprocal attempts at crossing *S. megistacrolobum* (*H. P. & R. 316*, *P. & H. P–366*) with *S. gourlayi* (*P. & H. P–365*).

NATURAL HYBRIDS

(*a*) Natural hybrids between ***S. megistacrolobum*** and ***S. acaule*** (= *S.* × *bruecheri* Corr.) and between ***S. megistacrolobum*** and ***S. infundibuliforme*** have already been described (see pp. 266 and 280).

(*b*) With ***S. boliviense***

Three collections from prov. Salta, dept. Santa Victoria (*Hjerting & Rahn 170* (see Plate 86) and *246*; *Sleumer 4040*) were obtained from the general distribution area of *S. boliviense* and would seem to be hybrids of *S. megistacrolobum* with that species (see Table 16). These possess rather long, narrowly oblong or oblanceolate *simple* leaves and longer stems (up to 40 cm long) than is usual in *S. megistacrolobum*. Other collections from the same area with simple leaves, which we have placed under *S. megistacrolobum* itself (see p. 287) may nevertheless owe this reduction in leaf dissection to an introgression of genes from *S. boliviense*. These collections are *Sleumer 3745*, *3839*, and *4003*. Other simple-leaved collections are known from regions where *S. boliviense* has not been recorded, and we cannot therefore be certain that in all cases the simple-leaved condition is due to the influence of *S. boliviense*. More living collections are needed before this problem can be solved.

(*c*) With ***S. microdontum*** subsp. intermediate

Hybrids of *S. megistacrolobum* with *S. microdontum* subsp. intermediate have been found in prov. Jujuy, dept. Tumbaya in the region of Volcán at altitudes of 2500–3000 m. (*Schreiter 2895* (see Plate 87) and *Venturi 9536*) (see Table 16). In these specimens the terminal leaflets are much wider and the laterals more distinct and less decurrent than in *S. megistacrolobum* itself. Furthermore, in some specimens the flowers are white, though in others they are apparently blue. There is no reason why these two species should not hybridize, since their altitude ranges overlap, even though *S. microdontum* is generally to be found in regions of higher rainfall, but we had no success in raising hybrids of this kind. Possibly hybrids between these two species might occur in other parts of north-west Argentina also.

(*d*) With ***S. gourlayi*** or ***S. oplocense***

Some of the plants raised at Vandel from original seed of *P. & H. P–366* (Jujuy, Humahuaca, between Cóndor and Abra del Cóndor. Alt. 3900 m) were possibly hybrids of *S. megistacrolobum* × *S. gourlayi*, or *S. megistacrolobum* × *S. oplocense*. The original herbarium specimens of this collection were good *S. megistacrolobum*, but it is possible that the seeds on these plants were of a hybrid nature. It is known that *S. gourlayi* and *S. oplocense* grow in the same general region (cf. *P. & H. P–368* (p.p.) probable *S. gourlayi*, p. 373) and the morphology of the Vandel plants points strongly to the fact that they may be hybrids of this type. (See Table 16.)

Unfortunately, we were not able to make artificial hybrids for comparison with the naturally occurring ones, though the marginal toothing of the leaves gives some

TABLE 16

Natural hybrids between S. megistacrolobum *and other species in Argentina*

Locality	Collection number and altitude		
	S. megistacrolobum	Hybrid	*S. boliviense*
Salta, Santa Victoria, Cuesta de Mesón	*Sleumer 4003*. 2600–3000 m	*H. & R. 170*. 2800–3100 m	(Known only from Cuesta de San José, in Santa Victoria dept., not far away)
Salta, Santa Victoria, entre Rodeopampa y Patahuasi	*H. & R. 251*. 3900 m	*H. & R. 246*. 3100–3400 m	
Salta, Santa Victoria, Rodeopampa (Cementerio)		*Sleumer 4040*. *c*. 3000 m	
	S. megistacrolobum	Hybrid	*S. microdontum* subsp. intermediate
Jujuy, Tumbaya, Volcán	*Venturi 9521*. 2700 m	*Venturi 9536*. 2700 m	*Venturi 9529*. 2700 m
Jujuy Volcán, Loma del Tambo		*Schreiter 2895*. 2500–3000 m	
	S. megistacrolobum	Hybrid	*S. gourlayi* or *S. oplocense*
Jujuy, Humahuaca, entre Cóndor y Abra del Cóndor	*P. & H. P–366*. 3600–4100 m (original herbarium specimens)	*P. & H. P–366*. 3600–4100 m (plants grown at Vandel from original seeds)	*P. & H. P–368* (p.p.). 3700 m ('probable *S. gourlayi*')

indication that the other parent may have been *S. oplocense*. On the other hand, the collection of *S. gourlayi* from not far away (*P. & H. P–368* (p.p.)) shows that that species was growing in the vicinity and could have been the other parent.

LOCALITIES OF NATURAL HYBRIDS

(*a*) With ***S. acaule*** and ***S. infundibuliforme***. (See pp. 269 and 282.)

(*b*) With ***S. boliviense***

Prov. SALTA:

Dept. Santa Victoria, Cuesta de Mesón. Alt. 2800–3100 m. 9 Feb. 1956. *Hjerting & Rahn 170* (C, J.G.H., J.P.H.). See Plate 86.

Dept. Santa Victoria, entre Rodeopampa y Patahuasi. Alt. 3100–3400 m. 16 Feb. 1956. *Hjerting & Rahn 246* (J.G.H., J.P.H.).

Dept. Santa Victoria, Rodeopampa (cementerio). Alt. *c.* 3000 m. 16 Feb. 1953. 'Fl. violac. (viol. clara).' *Sleumer 4040* (LIL, J.G.H.). Stem 10–15 cm.

(*c*) With ***S. microdontum*** subsp. intermediate

Prov. JUJUY:

Dept. Tumbaya, Volcán, Loma del Tambo. Alt. 2500–3000 m. 22 Feb. 1924. 'Flor de color azul obscuro.' *Schreiter 2895* (LIL). See Plate 87.

Dept. Tumbaya, Volcán. Alt. 2700 m. 4 May 1929. 'En los prados. Flor amarillenta.' *Venturi 9536* (US).

(*d*) With ***S. gourlayi*** or ***S. oplocense***

Prov. JUJUY:

Dept. Humahuaca, entre Cóndor y Abra del Cóndor. Alt. 3900 m. 9 Apr. 1952. *Petersen & Hjerting P–366*. Plants grown from original seed at Vandel in 1956. The original herbarium collection was of good *S. megistacrolobum*.

(Possible hybrids of *S. megistacrolobum* × *S. microdontum* subsp. *microdontum*—see Appendix III.)

9. SOLANUM BOLIVIENSE Dunal

(In *DC. Prodr.* **13**, 1, 43, 1852)

Plants sometimes growing as rosettes but generally with decumbent or weakly ascending stem, with short underground rootstock and tuber-bearing stolons.

Stem very short in rosette plants, otherwise up to 45 cm long, *bent from node to node*, slightly *woody*, angled but unwinged.

Leaf ovate, obovate, elliptic-lanceolate, elliptic, oblanceolate, etc., *generally simple* or with occasionally one or very rarely two pairs of small lateral lobes or leaflets which are up to 2 × 0·8 cm but generally much smaller; lamina (or terminal leaflet in leaves with lateral lobes) 1·5–11·5 cm long × 1–4 cm wide, obtuse or acutish at apex, often very narrowly cuneate at base; petiole very short or absent, never more than 2 cm long; leaf and stem clothed with sparse to frequent short rather thick adpressed hairs.

Inflorescence rather few flowered (2–6); peduncle generally unforked, 2–4·5 cm long; pedicels long, straight, 1·5–4 cm long, *articulated in upper third*, the articulation about 5–8 mm below calyx base; peduncles, pedicels, and calyx provided with sparse to frequent rather spreading multicellular and unicellular hairs.

Calyx about 4·5–7 mm long; lobes ovate-lanceolate with short acumens (1–2 mm) not well delimited from lobes, the lobe and acumen about 2–4·5 mm long.

Corolla *deep purple, pentagonal to rotate*, generally large, (1·7–)2·0–3·0(–3·5) cm diam.; *lobes rather wide*, (9–)13–20 mm and only 5–10 mm long including the 2–3 mm acumen, the shoulders of the lobes smoothly rounded or even almost obsolete.

Filaments 0·5–2 mm long × 1 mm broad; anthers yellow, 4–6 mm long.

Style 8–11 mm long, with minute papillae in the lower third; stigma clavate to capitate, not much thicker than the style apex.

Berries round, to 2 cm diam., dark green, with numerous slightly elevated white blotches, not visible in the dried state.

Chromosome number: $2n = 24$ (determined from Bolivian plants).

AFFINITIES

S. boliviense is placed in series Megistacroloba, together with *S. megistacrolobum* and *S. sanctae-rosae*, by virtue of the very enlarged terminal leaflet or completely simple leaf blade, the low straggling growth and tendency to rosette formation under appropriate habitat conditions. It also agrees with those species in the high pedicel articulation. It is distinguished from *S. megistacrolobum* and *S. sanctae-rosae* by the rather zigzag stem, when this is developed, the generally simple leaf, and the larger more rotate deep-purple corolla in which the lobes are well defined and clearly delimited from the acumens.

SPECIMENS EXAMINED

ARGENTINA

Prov. SALTA:

Dept. Santa Victoria, Cuesta de San José (entre Sta. Victoria y Toldos). Alt. 2900 m. 11 Feb. 1956. *Hjerting & Rahn 186* (J.G.H.).

Dept. Santa Victoria, Cuesta de San José. Alt. 2800 m. 11 Feb. 1956. *Hjerting & Rahn 187* (J.G.H., J.P.H.).

Dept. Santa Victoria, Cuesta de San José. Alt. 2650 m. 11 Feb. 1956. *Hjerting & Rahn 193* (C, K, J.G.H., J.P.H.). See Plate 89.

DISTRIBUTION

S. boliviense was collected first in Bolivia, dept. Chuquisaca, and has now been recorded by us for the first time in Argentina (see Map 9). It is probably to be found continuously at the upper limits of the sub-tropical mountain forests from Chuquisaca southwards through Tarija into north-west Argentina (prov. Salta, dept. Santa Victoria).

HABITAT

In Argentina *S. boliviense* grows on the very humid east-facing slopes of the Andes at altitudes of 2650–2900 m. It is to be found in the upper part of the Montano district ('pajonales') of the Western Subtropical province of Cabrera (1953). It mostly occurs amongst long grass and scrub vegetation, and is to be found occasionally with *S. megistacrolobum*, *S. microdontum*, and possibly also with other tuber-bearing *Solanum* species.

TAXONOMIC NOTES

The species under discussion was first described by Dunal in 1852 from plants collected by d'Orbigny (*No. 1170*) at Chiquisaca (Sucre) (see Fig. 28) and deposited in the Paris herbarium. Isotype specimens are also to be found at Montpellier and Vienna. Subsequently, Bitter (1912) made a new description from the Vienna isotype material. The species was previously collected by Weddell in 1845–6, but this gathering seems to have been overlooked by Dunal and Bitter. No more collections of

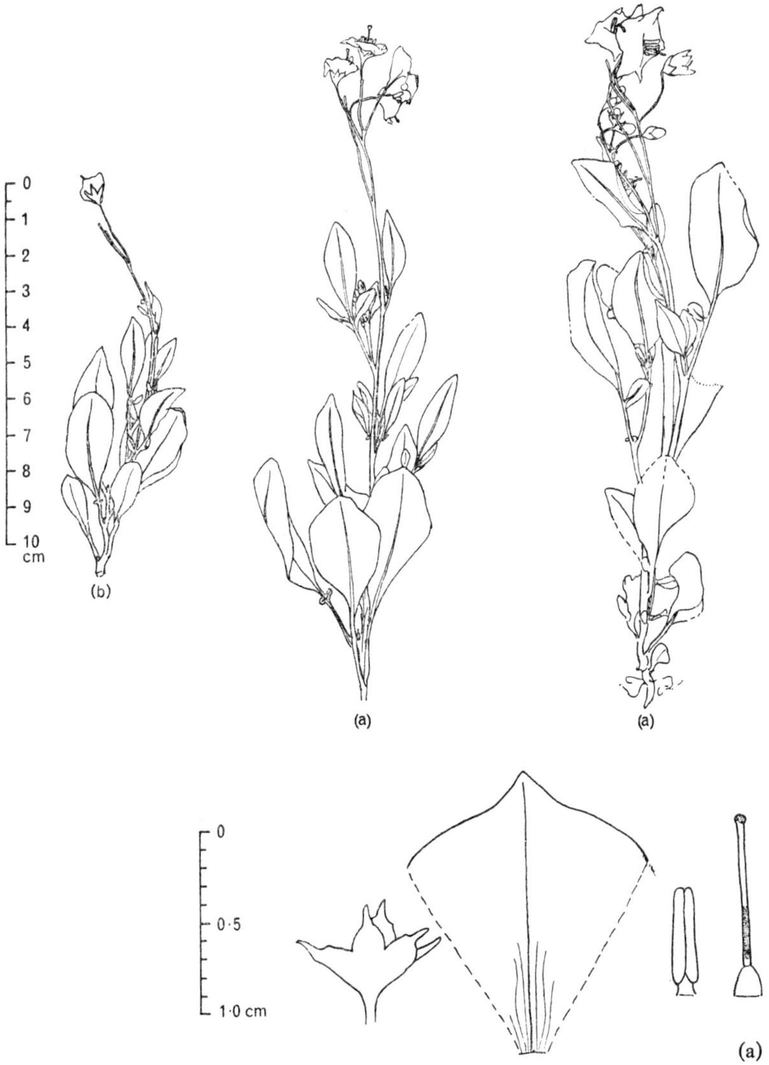

FIG. 28. *S. boliviense* Dun. (*a*) Type specimen (Herb. Paris) and (*b*) Isotype specimen (Herb. Montpellier) of *d'Orbigny 1170*, collected at Chuquisaca (= Sucre), Bolivia. Stem erect; leaf simple or with 1–2 pairs of minute lateral lobes. Corolla lobes much broader and flatter than in *S. megistacrolobum*.

S. boliviense were made, so far as is known, until those of Balls, Gourlay, and Hawkes in 1939 (see Plate 88 and Hawkes 1944) and Cárdenas in 1949, all from the region of Sucre. The above description is based both on the type and isotype material of d'Orbigny 1170, the collections of Balls, Hawkes, and Gourlay and of Cárdenas, and the Argentine specimens of Hjerting and Rahn. Although the present records for this species are very widely separated it is probable that further search on the eastern slopes of the Andes of southern Bolivia would reveal a continuous geographical range.

The Argentine plants of *S. boliviense* (see Plate 89) seem to possess longer, narrower leaves than the Bolivian ones though in floral details they agree very well with them. Possibly some introgression with *S. megistacrolobum* or *S. microdontum* has taken place in this former region. Strongly marked violet-purple veins may be observed on the leaves, especially on the under surface, on certain plants (*H. & R. 193*); this, again, may indicate introgression from *S. megistacrolobum*.

In the present work we have included *S. boliviense* in the series Megistacroloba although it was placed in Tuberosa in previous works (Hawkes 1944, 1956*b*) owing to the rotate to pentagonal corolla. However, the points of resemblance between *S. boliviense* and *S. megistacrolobum* in leaf characters, inflorescence type, and high articulation of pedicels and calyx are quite strong. The transition from semi-stellate to rotate corolla which occurs in *S. boliviense* has also been made in *S. raphanifolium* and it now seems to us that *S. boliviense* should be grouped in series Megistacroloba.

NOTES ON LIVING PLANTS AND CYTOLOGY

Unfortunately, none of the tubers collected by Hjerting and Rahn could be induced to sprout at Birmingham. The original plants bore no berries, and we were therefore unable to grow Argentine material here.

Bolivian material (*Balls 6188*; CPC: 93) was grown for several years at Cambridge. It produced straggling plants with weak stems and generally simple leaves; the flower was deep purple, with a rather strong acrid odour. (See Plate 88.) Frost resistance and some evidence of eelworm resistance and of Colorado beetle resistance have been noted in *S. boliviense* (see Chapter 5).

Cytological examination of the Bolivian collection showed the somatic chromosome number to be $2n = 24$.

FERTILITY AND CROSSABILITY

Little is known about the fertility of *S. boliviense* since only one clone was grown at Cambridge. This clone showed self-incompatibility, and there is reason to suppose that *S. boliviense* is similar to most of the other diploid *Solanum* species in this respect. The few attempts at crossing it with other species were not successful, though it is most probable that hybrids could be made between *S. boliviense* and a number of other diploid species in Megistacroloba, Tuberosa, and related series if more material were available.

NATURAL HYBRIDS

Three probable hybrids between *S. boliviense* and *S. megistacrolobum* have already been mentioned (p. 295). They were discovered in the Santa Victoria region of Salta province by Hjerting and Rahn in 1956, and also by Sleumer in 1953. Some other collections from this area with simple leaves which we have placed under *S. megistacrolobum* (p. 287), may possibly owe this character to an introgression of genes from *S. boliviense*, and the whole region needs careful re-examination to solve the problems that have arisen.

Hybridization is also probably taking place between *S. boliviense* and *S. microdontum* somewhere in the area of La Huerta–Rodeopampa at 2700–3000 m, west of Santa Victoria in prov. Salta. Even though we have no collections of simple hybrids between these two species, certain collections of *S. microdontum* (*H. & R. 240, 241*; *Sleumer 3752, 4037, 4038*) show strong variation in the direction of *S. boliviense* (see p. 338). Part of the Sleumer collection *No. 3752* of *S. microdontum* is particularly close in habit and leaf type to *S. boliviense*, even though the corolla of all specimens is uniformly white. Since hybrids of *S. boliviense* with *S. megistacrolobum* (*Sleumer 4040*) are known from the same locality it seems highly likely that a fresh search there would reveal the presence of *S. boliviense* also. (See also Plate 102.)

There is no doubt that the Rodeopampa region would well repay further study. It seems that the three species, *S. microdontum*, *S. megistacrolobum*, and *S. boliviense* are here forming a natural complex of hybrids which has as yet been imperfectly sampled. A detailed study might reveal that new adaptive norms were being established here among the hybrid populations, combining favourable gene combinations from all three species.

10. SOLANUM SANCTAE-ROSAE Hawkes

(*Ann. Mag. nat. Hist.* Ser. 12, **7**, 702–3, 1954)

Plant herbaceous, tuberiferous, small, rosette-forming. Tubers generally small.

Stem short, (2–)8–16 cm long (or occasionally up to 50 cm), erect, slender, somewhat flexuose, not winged, provided with sparse, short, adpressed hairs.

Leaf densely following, 6–10(–16) cm long, rather stiff and erect, imparipinnatisect; lateral leaf lobes sessile (0–)2–3(–4)-jugate, with no interjected leaflets, considerably smaller than the terminal, very broadly decurrent basiscopically in most leaves, ovate-lanceolate, *apically acuminate*, (0·4–)1·5–2·5(–5) × (0·2–)0·5–1·0(–3·5) cm; terminal lobe much larger than laterals, rhomboid to ovate-rhomboid or even elliptic or oblong, gradually attenuated at base into the winged rhachis, apically acute or subacuminate, (2·5–)4–5(–12) × (1–)1·5–2·5(–5) cm; leaf provided on both surfaces with sparse to fairly dense short hairs, shorter below and sometimes more pronounced on the margins.

Inflorescence terminal, few—3–4(–7)-flowered. Peduncle 3–4(–10) cm long, sometimes apically bifurcate. Pedicels 1·7–2·5(–3·5) cm long, *articulated in the upper third*, the articulation generally only 2–3 mm (sometimes, however, to 6 mm) below calyx base; peduncles and pedicels provided with sparse to moderately frequent short adpressed hairs.

Calyx campanulate, to 7 mm diam., with triangular lobes narrowed gradually into an acumen, the lobes and acumen totalling about 4–5 mm; calyx covered externally with dense unequal-lengthed hairs.

Corolla *sub-stellate, dark blue-purple*,[1] about 2–3(–4) cm diam., the broadly triangular lobes 8–10(–15) mm broad × 8(–13) mm long including the not very sharply delimited acumen.

Anthers about 4·5–6 mm long; filaments 1 mm long.

[1] In his most recent publication on '*S. catamarcae*', Brücher (1959*b*) asserts that the flower colour may vary from dark blue to pale whitish-blue. Yet in a previous work (1954) he states that the flowers are always dark blue or violet, and that only hybrids with '*S. microdontum*' (really *S. venturii*) show light blue to white flowers. We agree with this former statement. (See also Hawkes 1954, p. 706.) Pure *S. sanctae-rosae* always maintains dark purple-blue flowers, and a dilution of this colour is seen only in hybrids with *S. venturii* (see also p. 309).

Style about 8–9 mm long, apparently glabrous; stigma much larger than style apex, globose, about 1 mm diam.

Berries orbicular, up to 2 cm diam., becoming yellow at maturity.

Chromosome number: $2n = 24$.

AFFINITIES

S. sanctae-rosae is placed in series Megistacroloba Cárd. et Hawkes by virtue of the semi-stellate corolla with broad triangular lobes, the high pedicel articulation and enlarged terminal leaf lobe and lateral lobes. It differs from *S. megistacrolobum* in the often rhomboid acute terminal lobe and the invariably *acuminate* laterals. In all other species of this series the laterals are obtuse or rounded. It is the most southerly representative of Megistacroloba, a series of species stretching from north Peru (*S. sogarandinum* Ochoa, *S. hastiforme* Corr.) through Bolivia (*S. boliviense* Dun., *S. toralapanum* Cárd. et Hawkes, *S. ellipsifolium* Cárd. et Hawkes, and *S. megistacrolobum* Bitt.) and into north-west Argentina. Here, *S. megistacrolobum* Bitt. is to be found in the northernmost provinces (Jujuy and Salta) and is replaced to the south at rather lower altitudes in southern Salta, Tucumán, and Catamarca by *S. sanctae-rosae*.

SPECIMENS EXAMINED

ARGENTINA

Prov. SALTA:

Dept. Cafayate, Sierra de los Quilmes. 10 Jan. 1943. *Castellanos* (s.n.) (BA 47012).

Prov. TUCUMÁN:

Dept. Chicligasta, Estancia Las Pavas, La Cascada a Las Cuevas. Alt. 3000 m. 10 Mar. 1949. *T. Meyer 15075* (p.p.) (LIL).

Dept. Chicligasta, Quebrada de Las Pavas, entre Puesto La Cascada y Las Cuevas. Alt. 3200–3400 m. 8 Mar. 1949. 'Generalmente crece en tierra arenosa y en lugares secos a pleno sol.' *Petersen & Hjerting P–56* (p.p.) (J.G.H., J.P.H.).

Dept. Chicligasta, Estancia Santa Rosa. Alt. 3600 m. 14 Jan. 1927. 'Flor morada. Entre los pastos.' *Venturi 4737* (S-Holotype, BA, GH, LIL, LP, SI, US). See Fig. 29(*a*) and Plate 90.

Dept. Tafí, Cerro Muñoz. Alt. 3800 m. 2 Jan. 1914. *Castillón 3264* (LIL).

Dept. Tafí, 84 km from Tucumán on way to Cafayate. 28 Feb. 1960. 'In open rocky grassy soil. Fls. large, bluish-purple.' *Correll, Dodds, Brücher & Paxman A702* (p.p.) (K).

Dept. Tafí, La Ciénaga. *Descole* (s.n.) (LIL 101446).

Dept. Tafí, km 83, entre Infiernillo y Tafí, cerca de Quebrada Honda. Alt. 3000 m. 19 Jan. 1956. *Hjerting, Petersen & Rahn 3 and 4* (J.G.H., J.P.H.).

Dept. Tafí, Cerro Muñoz, Ciénaga Grande. Alt. 3400 m. 28 Jan. 1903. *Lillo 2998* (LIL).

Dept. Tafí, Tafí del Valle, Río Blanco. Alt. 2600 m. 7 Feb. 1908. 'Arenales. Flor azul.' *Lillo 7562* (LIL).

Dept. Tafí, Sierra de Tucumán, La Ciénaga. 10–17 Jan. 1874. *Lorentz & Hieronymus (664)* (CORD).

Dept. Tafí, Infiernillo, Quebrada Honda. Alt. 3200 m. 16 Feb. 1952. *Petersen & Hjerting P–328* (CPC: 2483) (C, J.G.H., J.P.H.). See Plates 91 and 92.

Dept. Tafí, Infiernillo, Río Tafí y alrededores. Alt. 2750–3100 m. 1 Jan. 1953. *Petersen & Hjerting 858a* (C, J.G.H., J.P.H.).

Dept. Tafí, Quebrada Honda, Calchaquíes. Alt. 3700 m. 1 Jan. 1953. *Petersen & Hjerting 858b* (J.G.H.).

Dept. Tafí, Weg Santa María–Tafí. Alt. 2950–3300 m. 13 Feb. 1959. 'Feuchte alpine Matte; im Gras; Wegrand; im *Satureja parviflora*; u.s.w.' *Ross & Vervoorst 38, 39, 40, 42, 50* and *51b* (Max.).

Dept. Tafí, La Ciénaga, Morro de las Aguadas. Alt. 3200 m. 28 Jan. 1950. 'Fl. viol.' *Sleumer 176* (LIL).

Dept. Tafí, Tafí del Valle, Quebrada de la Bolsa. Alt. 2900 m. 25 Jan. 1950. 'Fl. viol.' *Sleumer 256* (LIL).

Dept. Tafí, Infiernillo. Alt. 3100 m. 23 Feb. 1949. *Sparre 5976* (LIL).

Dept. Tafí, Infiernillo. Alt. 3200 m. 23 Feb. 1949. *Sparre 6021* (LIL).

Dept. Tafí, Infiernillo, Quebrada Honda. Alt. 3300 m. 3 Feb. 1952. *Sparre 9840* (LIL, J.P.H.).

Dept. Trancas, Rodeo de la Pirca (Quebrada del Chorro). Alt. 2500 m. 23 Feb. 1920. 'Flor violeta obscura.' *Schreiter 1210* (LIL).

Prov. CATAMARCA:

Dept. Andalgalá, Río Potrero sup., Queñoa Pelada. Alt. 2750 m. 27 Feb. 1951. 'Fl. viol.-caerulesc.' *Sleumer 1831* (LIL).

Dept. Andalgalá, entre Río Lampacillo y Río Potrero sup. Alt. 2700–2900 m. 27 Feb. 1951. 'Fl. caerul.' *Sleumer 1840* (LIL, J.G.H.).

Dept. Andalgalá, Cuesta de la Negrilla cerca de la Mina de Capillitas. Alt. 3000–3100 m. 2 Mar. 1952. 'Fl. viol.' *Sleumer 2686* (LIL). See Fig. 29(*b*).

Dept. Andalgalá, arroyo en frente de la Mina de Capillitas. Alt. *c.* 2900 m. 3 Mar. 1952. 'Fl. viol., en lugar húmedo asombroso.' *Sleumer 2749* (LIL, US, J.G.H.).

Dept. Belén, El Rodeo a Quebrada de los Potrerillos arriba de Granadillas. Alt. *c.* 2550 m. 26 Jan. 1952. 'Fl. viol.' *Sleumer & Vervoorst 2488* (LIL, US, J.G.H.). (Atypical growth form.)

Dept. Belén, Quebrada del Río Blanco, arriba de Granadillas. Alt. 2500 m. 28 Jan. 1952. 'Fl. violasc.' *Sleumer & Vervoorst 2517* (LIL).

Dept. Belén, 'Las Bayas', en cerros al oeste de Granadillas. Alt. 2700–2800 m. 28 Jan. 1952. 'Fl. viol.' *Sleumer & Vervoorst 2550* (LIL, US, J.G.H.).

Dept. Capayán, cuesta a la cumbre del Alto de los Chiqueritos, próximo a Las Angosturas. Alt. *c.* 3000 m. 29 Feb. 1952. 'Entre rocas, en el dominio de la estepa graminosa puneña; fl. azul morada; planta algo crasa.' *Vervoorst 3484* (LIL, J.G.H.).

Dept. Pomán, entre Las Ciénagas y Las Angosturas, por el Alto de la Cruz, 25 km al S.E. de Pomán. Alt. *c.* 2500 m. 29 Feb. 1952. 'En un faldeo seco con exposición al NE, en el dominio de los pastizales, entre arbustos; fl. azul.' *Vervoorst 3482* (p.p.) (J.G.H.). See Plate 96B.

Dept. Tinogasta, Reales Blancos a La Tranca. Alt. 3500 m. 4 Feb. 1930. 'Fl. violácea.' *Schreiter 6199, 6200*, and *6201* (LIL).

(Dept. Santa María?) Cerro de Suasayalo. *Castillón 11976* (LIL).

Locality unknown, possibly Tucumán, dept. Tafí, Sierra de Tucumán, La Ciénaga. 1874. 'Bltn. violett.' *Lorentz & Hieronymus 664* (GOET).

Locality unknown. Grown at Max-Planck Institut, Voldagsen über Elze, Germany. Specimens kindly sent by Dr. H. Ross, under number E.B.S. 493 (J.G.H.).

ABERRANT BIOTYPES, POSSIBLY CONTAMINATED WITH GENES FROM OTHER SPECIES:

Prov. CATAMARCA:

Dept. Andalgalá, Capillitas. Mar. 1877 (?) *Schickendantz 226* (p.p.) (B-photo, CORD). The photograph in Chicago Field Museum of the destroyed Berlin specimen is labelled: '*S. catamarcae* Bitt.' '*S. utile* Klotzsch var. *meridionale* Wittm.' *nomina nuda*. The Córdoba sheet of this collection contains two specimens: (*a*) Labelled

'*S. verrucosum* forma *genuina*', which is referable to *S. spegazzinii*. (*b*) Labelled '*collinum* (syn. *S. collinum* Dun.)', which is referable to *S. sanctae-rosae* though is not typical of that species. A label by F. Kurtz states that (*b*) was *Schickendantz No. 225*, whilst (*a*) was *Hieronymus & Niederlein 474*. See Plate 95.

Dept. Andalgalá, Capillitas. 1881? *Schickendantz 280* (LP). Leaflet apices obtuse. Flower as *S. sanctae-rosae*.

DISTRIBUTION

S. sanctae-rosae is one of the few endemic potato species in Argentina, and occurs mostly in the high mountains of prov. Tucumán, Depts. Chicligasta, Tafí, and Trancas, and in prov. Catamarca in depts. Andalgalá, Belén, Capayán, Pomán, and Tinogasta. There is also one record for it in prov. Salta, dept. Cafayate, just over the border from prov. Tucumán. (See Map 9.)

HABITAT

S. sanctae-rosae occurs in regions of moderate to high rainfall at altitudes of (2500–) 2800–3500(–3800) m above sea level, and thus at rather lower altitudes than its nearest relative, *S. megistacrolobum*, to the north, whose geographical range it does not overlap so far as we know. It grows in the more humid part of the Puna province and the upper part of the Montano district ('Pajonales') of the Western Subtropical province of Cabrera (1953), in high Andean grassland, in open ground, in sandy and stony places, amongst rocks, in the walls of old corrals, and occasionally in the shade of low shrubs. It occurs together with *S. acaule* in the lower part of the range of that species and is also occasionally associated with *S. spegazzinii*. In the lower parts of its range (2500–2800 m) it is sometimes found with *S. venturii*, with which it crosses freely, forming a large number of hybrids and hybrid segregates of varying leaf shape and flower colour (see section on hybrids, p. 309).

TAXONOMIC NOTES

The species was described by Hawkes (1954) on the basis of a collection by Venturi (*4737*) made at Estancia Santa Rosa, dept. Chicligasta, prov. Tucumán (see Fig. 29(*a*) and Plate 90). Some forms of *S. sanctae-rosae* apparently approach very nearly a specimen collected by Schickendantz (*226*) from Catamarca, dept. Andalgalá, Capillitas, formerly in Berlin but now destroyed. This specimen, of which photographs exist in the Chicago Field Museum and elsewhere (see Plate 95), was labelled '*S. catamarcae*' Bitt., but apparently the species was never described and the name must rank as a *nomen nudum*. A Córdoba sheet of possibly the same collection contains two specimens, one of which is referable to *S. sanctae-rosae*. In view of the fact that the photograph of the Berlin specimen does not agree very well with *S. sanctae-rosae* and no knowledge exists of the flower colour, hairiness, etc., of this collection, it was thought better in 1954 not to use the specimen or name for the species we are dealing with.

'*S. catamarcae*' has unfortunately been taken up by Brücher (1954, 1959*b*). From the description and figure given by him it seems fairly probable that *S. sanctae-rosae* is in fact being discussed, though no collections are cited. He states, furthermore, that the species '*S. catamarcae*' was named by Hawkes *S. venturii*. This is, of course, quite incorrect since the latter name was not published until 1960 (see p. 349). He is also incorrect in stating that the growth period of '*S. catamarcae*' is confined to the months of December to February. Reference to the collections examined by us (pp. 302–3) will show that these have been made between January and March, but the vegetative season of the species is undoubtedly much longer, certainly at least four months.

The variation in *S. sanctae-rosae*, as shown in the collections that we have cited above, is considerable. What we have considered to be typical for this species is to be found in a wide range of altitudes (2500 to 3800 m, but mostly above 2800 m), growing in exposed situations, or on very poor soil. This form has the typical narrow rhomboid terminal leaf lobes and 2–3 pairs of acuminate decurrent laterals; it generally grows as a rosette or semi-rosette and corresponds with the smaller dimensions given in the description. The following collections conform to this growth habit: *Venturi*

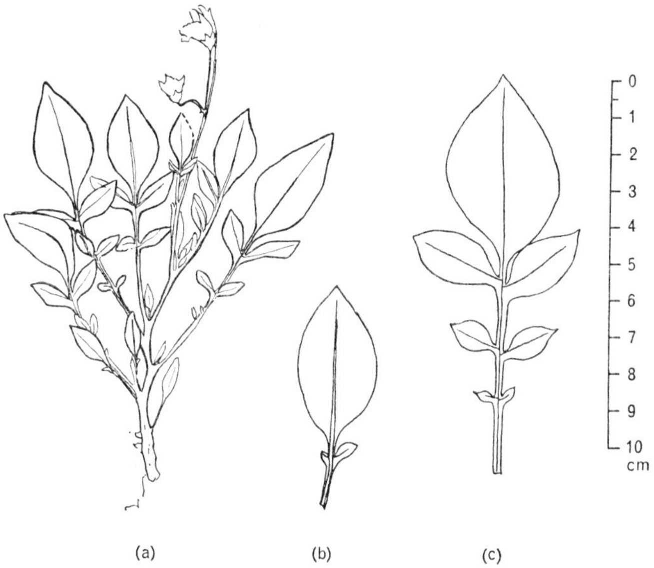

FIG. 29. *S. sanctae-rosae Hawkes.*

(*a*) One plant from type sheet (*Venturi 4737*—Herb. Stockholm), from Estancia Sta. Rosa, dept. Chicligasta, prov. Tucumán, Argentina (see also Plate 90).

(*b*) Sub-simple leaf from *Sleumer 2686* (Herb. Lillo), from near Mina Capillitas, dept. Andalgalá, prov. Catamarca.

(*c*) Leaf from plant grown at Birmingham, 1953 of *P. & H. P–328*, from Infiernillo, dept. Tafí, prov. Tucumán.

(From Hawkes 1954.)

4737 (Type, see Plate 90 and Fig. 29(*a*)), *Castillón 3264*, *Descole* (s.n.), *Hjerting, Petersen & Rahn 3*, *Lillo 7562*, *T. Meyer 15075* (p.p.), *Petersen & Hjerting P–56* (p.p.) (some plants only), *Petersen & Hjerting P–328* (plants not growing in shade. See Plate 91), *Petersen & Hjerting 858a and b*, *Schreiter 1210*, *Sleumer 176*, *1840*, *2686* (some plants only of all three collections), *Sparre 9840*, *Vervoorst 3482* (p.p.).

Another set of plants is characterized by a marked reduction of the lateral leaf lobes (see Fig. 29(*b*) and some of the plants in Plate 91), so that the leaf may be simple or only unijugate, the laterals when they do occur being much smaller than in the type and the terminals rather longer and certainly broader, about $2–5 \times 1–3{\cdot}5$ cm and not so definitely rhomboid, sometimes elliptic or even oblong. The laterals show decurrency and acumination, which is perhaps the most constant and characteristic feature of the species. A deep purple corolla, high articulation, and large globular stigma is also always met with. The average altitude of this sub-simple-leaved growth type is roughly similar to the first one though its range is a little less, namely 2500–3500 m. The following specimens are to be included in the sub-simple type of *S. sanctae-rosae*: *Castellanos* (*47012*), *Lorentz & Hieronymus 664* (grading to typical), *Lillo 2998*, *Schreiter 6199*, *6200*, *6201* (one plant), *Sleumer 176* (some plants), *Sleumer 256*, *Sleumer 1840* (two plants), *Sleumer 2686* (one plant, see Fig. 29(*b*)), *Sleumer & Vervoorst 2517*.

The fact that some collections contain plants grading from one growth form to the other indicates that it is not possible to distinguish two separate botanical varieties or forms. Nevertheless, it seems fairly certain that the differences between them are not due to environment only, since both are to be found in roughly the same altitude range and in apparently similar habitats. There is evidently a number of genes tending to cause a reduction in lateral lobe development, and possibly occurring at random throughout the population.

It is possible that this form with sub-simple leaves might be due to some introgression of genes into *S. sanctae-rosae* from the sub-simple leaved species *S. venturii*, which overlaps the range of *S. sanctae-rosae* between 2500 and 2800 m or even higher. Obvious hybrids between these two species are to be found in approximately the same altitude zone as *S. sanctae-rosae* itself, though very infrequently above 3000 m. (See Table 17.)

The third type of variation seems to be due solely to the effect of a richer, damper, or more shady environment. Under these conditions the plants of either of the two above-mentioned types may form a thick vegetative mass, perhaps to 50 cm tall, the lobes become broader, and the terminal attains dimensions of 6×3, $7{\cdot}5 \times 4$, $8 \times 4{\cdot}5$, or even up to 12×5 cm, losing its characteristic rhomboid shape and becoming ovate, elliptic, etc.; the laterals are always decurrent and apically acuminate, the articulation of the pedicel high and the corolla dark purple with broad triangular lobes. The amount of leaf dissection varies according to which of the basic genotypes mentioned previously has been subjected to the richer environment. The altitude range of the plants with more or less normal leaf dissection but more luxuriant habit is 2700–3500 m (the normal up to 3800 m) which is in agreement with our concept that this is a growth form confined to more sheltered or richer habitats. We should therefore not expect it to occur at the very highest limit of the altitude range where the climatic conditions are most severe.

The range of plants with luxuriant growth and sub-simple leaf is from 2550 to 2900 m and is thus much lower than the general range for the species as a whole. Some of these specimens are quite difficult to separate from the hybrids of *S. sanctae-rosae* and *S. venturii* which itself is only found up to 2800 m (see list, p. 350). The following collections of the luxuriant growth type possess more or less normally developed lateral leaflets: *Hjerting, Petersen & Rahn 4*, *Petersen & Hjerting P–56* (p.p.) (some plants only), *Petersen & Hjerting P–328* (plants mentioned by collectors as growing in shade. See also below, CPC: 2483, grown from seed of *P–328* and conforming to this growth pattern), *Schreiter 6201* (one plant), *Sleumer 1831*, *Sleumer 1840* (one plant), *Sparre 5976*, *Sparre 6021*, *Vervoorst 3484*.

Plants with sub-simple leaves (0–1 lateral pairs) growing apparently under moister conditions and also conforming to the larger measurements given above are: *Castillón 11976*, *Sleumer 2749*, *Sleumer & Vervoorst 2488*, *Sleumer & Vervoorst 2550*.

It must be emphasized that we have here picked out the extreme types in what are often populations showing all gradations from one extreme to the other. Many collections show at least two of the phenotypes mentioned whilst *Sleumer 1840* shows three. Although it may be argued that such collections represent mixtures, this is sufficient to illustrate our point: the populations in many cases are segregating for the genetic differences involved, or are found in both dry and moist habitats, and every plant is likely to differ from its neighbour, except when they belong to the same vegetative clone and are growing under similar conditions, in the number of pairs of lobes and their dimensions.

TABLE 17

Natural hybrids of S. sanctae-rosae *with* S. venturii *and* S. spegazzinii

Locality	Collection number and altitude		
	S. sanctae-rosae	Hybrid	*S. venturii*
Tucumán, Chicligasta, Estancia Las Pavas, La Cascada a Las Cuevas	*T. Meyer 15075* (p.p.). 3000 m *P. & H. P–56* (p.p.). 3200–3400 m	*T. Meyer 15075* (p.p.). 3000 m *T. Meyer 14948*. 2800 m *P. & H. P–56* (p.p.). 3200 m	*Venturi 3271*. 2600 m (And two collections at lower altitudes: 2100–2550 m)
Tucumán, Chicligasta, Estancia Santa Rosa	*Venturi 4737*. 3600 m	*Venturi 4727* (p.p.). 3600 m	*Venturi 4727* (p.p.). 3600 m
Tucumán, Tafí, Infiernillo, Quebrada Honda to Río Tafí	*P. & H. P–328*. 3200 m *P. & H. 858a*. 2750–3100 m *Sparre 5976*. 3100 m *Sparre 6021*. 3200 m *Sparre 9840*. 3300 m	*P. & H. P–326*. 2900 m *P. & H. P–327*. 2900 m *P. & H. P–333*. 2900 m *P. & H. P–334*. 2700 m *P. & H. P–336*. 2700 m	
Tucumán, Tafí, San José		*P. & H. 899*. 2400 m	*P. & H. 898a*. 2150–2200 m *P. & H. 898b*. 2650 m *R. Diaz 9583*. 2800 m *Sparre 5835*. 2100 m *Venturi 3565*. 2700 m
Catamarca, Andalgalá, Río Lampacillo a Río Potrero superior	*Sleumer 1840*. 2700–2900 m	*Sleumer 1839*. 2700–2900 m	
	S. sanctae-rosae	Hybrid	*S. spegazzinii*
Tucumán, Tafí, road from Santa María to Tafí del Valle, near Quebrada Honda	*H. P. & R. 3, 4*. 3000 m *P. & H. P–328*. 3200 m *P. & H. 858a*. 2750–3100 m *Ross & Vervoorst 38, 39, 40, 42, 50, 51b*. 2950–3300 m *Sparre 5976, 6021, 9840*. 3100–3300 m	*Ross & Vervoorst 43* (p.p.). 3100 m	*Ross & Vervoorst* 43 (p.p.). 3100 m *P. & H. 856b*. 2800 m *P. & H. P–332b*. 2900 m
Catamarca, Andalgalá, region of Mina Capillitas	*Sleumer 2686*. 3000–3100 m *Sleumer 2749*. *c*. 2900 m [*Schickendantz 226* (p.p.), *280*; doubtful *S. sanctae-rosae*]	*Sleumer 2687* (p.p.). 3000–3100 m *Sleumer 2688* (p.p.). 3000–3100 m	*Ross & Vervoorst 25*. 2700 m *Sleumer 2687* (p.p.), *2688* (p.p.). 3000–3100 m *Schickendantz 226* (p.p.).
Catamarca, Belén, El Rodeo a Quebrada de los Potrerillos	*Sleumer & Vervoorst 2488*. 2550 m	*Sleumer & Vervoorst 2485*. 2600 m	*Sleumer & Vervoorst 2489*. 2650 m; and *2490*. 2700 m
Catamarca, Belén, Quebrada del Río Blanco, Las Mesaditas		*Sleumer & Vervoorst 2533 and 2535*. 2800 m	*Sleumer & Vervoorst 2534*. 2800 m
Catamarca, Pomán, entre Las Ciénagas y Las Angosturas	*Vervoorst 3482* (p.p.). 2500 m	*Vervoorst 3482* (p.p.). 2500 m	

It thus becomes evident from a study of the very complete collections of Petersen, Hjerting, Sleumer, and Vervoorst, that *S. sanctae-rosae* is a species of considerable plasticity and that the phenotypic variability is due both to environmental and genetical differences. It is well known that most species of wild potato are extremely plastic, and *S. sanctae-rosae* proves to be no exception to this rule.

NOTES ON LIVING PLANTS

Seeds of *Petersen & Hjerting P–328* were grown at Cambridge and Birmingham under the number CPC: 2483. The seedling progeny was uniform, with leaf development as for the herbarium specimen of this collection from a shade locality. (See Plate 92.) Leaves were 2–3-jugate and all plants possessed a curious but characteristic odour, faintly suggestive of parsley (*Petroselinum crispum*). The corolla was typically semi-stellate as in other members of series Megistacroloba, and of a blue-purple colour; the diameter was 3 cm, the broadly triangular lobes being about 15 mm broad at the base, 9–10 mm long, including the not very clearly delimited acumen. The dimensions of stamens and pistil do not differ materially from those given in the general description.

Under artificial short days at Vandel the average tuber weight for different plants varied from 0·5 to 1 grammes per tuber in most cases, and occasionally tubers of up to 4 grammes were found. The plants were quite frost-resistant in the field. *S. sanctae-rosae* is reported in the literature as showing resistance to eelworm, flea beetles, aphids, virus leaf-roll, and frost. (See Chapter 5.)

CYTOLOGY

The somatic chromosome number of *S. sanctae-rosae* was shown by Hawkes (1954) to be $2n = 24$. Brücher (1954), also, mentions $2n = 24$ for his '*S. catamarcae*', which as we have mentioned above, is probably identical with *S. sanctae-rosae*. So far as the writers are aware, no other work on the cytology of this species has been reported in the literature.

FERTILITY AND CROSSABILITY

In common with other diploid species *S. sanctae-rosae* is self-incompatible but fully fertile otherwise. It would probably hybridize with a large range of species in series Megistacroloba, Cuneoalata, and Commersoniana, though pollinations have not yet been carried out.

The cross between *S. acaule* and *S. sanctae-rosae* has already been mentioned on p. 265. The F_1 seedlings were apparently all sterile triploids.

We also raised a hybrid progeny from the cross *S. sanctae-rosae* (*P. & H. P–328*) × *S. toralapanum* (CPC: 1773—from Bolivia) which is of interest in view of the apparent close relationship between *S. toralapanum* and *S. megistacrolobum*. The F_1 plants (Family 130) were fairly uniform and quite fertile. Two F_2 families (185, 221) were grown, but, curiously enough, differed considerably from each other. In family 185 two sowings, totalling 65 seeds, were made; 31 seedlings survived to be planted into the field and were very disparate, finally giving only 6 good plants. On the other hand, 30 seeds of family 221 gave 27 seedlings of which 21 were planted in the field and exhibited good uniformity and thriftiness. Although there seems to be some evidence of genetic differentiation between *S. sanctae-rosae* and *S. toralapanum* the results are inconclusive. The work should be repeated, using different accessions.

The F_1 hybrids raised from the cross *S. microdontum* (intermediate) (*P. & H. P–364*) × *S. sanctae-rosae* (*P. & H. P–328*) were vigorous (Family 235), with low spreading habit and 1–2-jugate acuminate leaves; narrow wings were present on petiolules and leaf rhachis. The flowers were large, purple to mauve, with paler petal streaks. No F_2 was raised. Crosses of *S. sanctae-rosae* with *S. microdontum* subsp. *micodontum* (H. & R. 226) and subsp. *gigantophyllum* (H.P.R. 293) were also made and small F_1 families raised at Vandel.

An F_1 progeny from the cross *S. vernei* subsp. *ballsii* (*P. & H. P–351*) × *S. sanctae-rosae* (*P. & H. P–328*) was also grown and the seedlings were uniform, with deep blue-violet corollas. Their vigour was poor, but this may have been due to the bad season (1959) in which they were grown.

NATURAL HYBRIDS

It is evident that *S. sanctae-rosae* readily forms hybrids in certain regions where its distribution area overlaps those of other species (see Table 17).

No hybrids with *S. acaule* have been seen by us and it is doubtful whether the two species normally cross in nature, even though hybrids have been produced artificially (see p. 265). Brücher (1954), also, does not record hybrids of this type, though he states that the two species may be found growing together.

(*a*) With ***S. venturii***

The distribution areas and altitude ranges of *S. sanctae-rosae* and *S. venturii* overlap for a considerable distance, and hybrids between them are consequently not uncommon. We have recognized as hybrids those plants which show a dilution of the flower colour, especially those with white central petal streaks and white apices, indicating hybridization with a white-flowered species (cf. artificial *S. microdontum* × *S. sanctae-rosae* crosses). A lowering of the articulation, a rounding of the apex of the leaf lobes, and an enlargement of the terminal leaflet, or a completely imparripinnate leaf instead of an imparripinnatisect one are also good indications of hybridity. Unfortunately, we have not been able to make artificial hybrids since we did not possess living material of *S. venturii*. However, the artificial *S. microdontum* × *S. sanctae-rosae* hybrids are very similar to the natural ones which we assume to be *S. sanctae-rosae* × *S. venturii*. That *S. microdontum* is the other parent, and not *S. venturii*, is out of the question because the habitat and altitude ranges of the two species do not overlap. We have also been able to cultivate clonal material of a putative *S. sanctae-rosae* × *S. venturii* hybrid (*P. & H. P–336*) which had twenty-four somatic chromosomes and morphological characters nicely intermediate between the two probable parents (see notes on p. 311). Our experience in general of hybrids between wild species with blue flowers and those with white ones is that the blue shows incomplete dominance in the F_1 (mauve, white, or pale streaks, etc.) and gives plants with a smoothly graded range of colours, from white to blue in the F_2 generation. Leaf characters are similarly graded. On the whole, then, even when artificial hybrids cannot be made, one can make a fairly accurate prognostication as to the form they would take from experience with other artificial hybrids and the examination of herbarium specimens.

The plants which we have classified as hybrids of this nature occur not only from 2500–2800 m, where the two parent species overlap in their altitudinal ranges, but also up to 3200 m (even to 3600 m?). It is possible that the altitude data are at fault in one case (*Venturi 4727*); apparently, however, the hybrids have migrated into higher altitudes than can be colonized by the pure species, *S. venturii*. Detailed collections from these higher altitudes would enable us to see this more clearly, and to decide whether *S. venturii* hybrids are following the type of evolutionary development which we have already seen in *S. chacoense*.

Brücher (1954) mentions hybrids with '*S. microdontum*', by which he presumably means *S. venturii* (see p. 352); the hybrids are stated to be found in the contact zone between the two species '*S. catamarcae*' (= *S. sanctae-rosae*) and '*S. microdontum*' (= *S. venturii*). However, when describing his work on *S. venturii* he mentions certain collections (e.g. *T. Meyer 14948*, *Sparre 5753*) as being hybrids with *S. vernei*. This is certainly incorrect, the other parent being not *S. vernei* but *S. sanctae-rosae*.

The occurrence of *S. venturii* in the same locality as *S. sanctae-rosae* has not been verified in every region where the putative hybrids between them were collected. Both parents, together with the hybrids, have been seen in some areas, which leaves us in very little doubt as to the nature of the hybrids. (See Table 17.) This is especially evident in prov. Tucumán, dept. Chicligasta, Estancias Santa Rosa, and Las Pavas (see collections of *Venturi 4727* (p.p.), *T. Meyer 14948*, *15075* (p.p.), and *P. & H. P–56* (p.p.)).

In other districts, notably prov. Tucumán, dept. Tafí, Infiernillo, Quebrada Honda, the *S. venturii* parent has not been collected, though it has certainly been obtained from the neighbouring La Ciénaga and San José regions. The position is rather reversed at San José, since most of the collections here are of *S. venturii*, only one hybrid has been seen, and *S. sanctae-rosae* has not been obtained at all.

One could postulate three reasons for this curious situation: (1) The 'missing parent' may actually occur there but has been missed or overlooked. (2) The 'missing parent' once occurred there, but has become absorbed into the dominant species and no longer occurs in the pure state in that particular area. (3) Rather long-range pollination occurs between the two species. This is certainly, on the face of it, rather unlikely, but we cannot prove or disprove this suggestion until more is known about the range of flight and general biology of the insect pollinators. Only careful observations in the field by both botanists and entomologists could resolve this problem. (See Plates 93 and 94.)

(*b*) With ***S. spegazzinii***

Natural hybrids between *S. sanctae-rosae* and *S. spegazzinii* occur occasionally in the altitude range 2500–3100 m in provs. Catamarca and Tucumán, in probably rather drier situations than the ones with *S. venturii*.

One of the areas where *S. sanctae-rosae*, *S. spegazzinii*, and the hybrids between them seem to occur with some frequency is in prov. Catamarca, dept. Andalgalá, Mina Capillitas. Good hybrids were collected by Sleumer (*2687* (p.p.), *2688* (p.p.)); typical *S. sanctae-rosae* was also obtained by the same collector (*2686*, *2749*), and also good *S. spegazzinii* (*2687* (p.p.), *2688* (p.p.)).

Thanks to the kindness of the collector we were able to grow a seedling progeny of *Sleumer 2688*, the seeds presumably having been obtained from one of the hybrid plants. There was some variation in leaf dissection and leaflet width, though on the whole the seedlings resembled *S. spegazzinii* much more than *S. sanctae-rosae*. The corolla colour and form were quite variable, and a curious feature of these hybrids was the strong tendency to anther abortion. Indeed, in some flowers the stamens were completely absent.

The collections of Ross and Vervoorst (*25*) and of Schickendantz (*226* (p.p.)) also belong to *S. spegazzinii*. The rest of *Schickendantz 226* and also *Schickendantz 280* are rather aberrant *S. sanctae-rosae*. We have classified them provisionally with that species, but they might possibly represent an F_2 segregate or some type of backcross with *S. sanctae-rosae*. We have already noted that *Schickendantz 226* which was labelled by Bitter '*S. catamarcae*', but not described, and which Brücher wanted to use as a type for this species (which would have included *S. sanctae-rosae* also) is a mixed collection, part of which is *S. spegazzinii*.

Another interesting series of parents and hybrids are the collections made by Sleumer and Vervoorst (*2485*, *2488*, *2489*, *2490*) in prov. Catamarca, Belén, El Rodeo, to Los Potrerillos. Here *S. sanctae-rosae* seems to occur lower down than the hybrids and

S. spegazzinii. It would be interesting to know whether differences of rainfall were operative here. Other interesting collections were those of Sleumer and Vervoorst from Belén, Las Mesaditas; and of Vervoorst from Pomán, between Las Ciénagas and Las Angosturas. (See Plate 96.)

The collections of these two botanists in Catamarca province have been of the greatest value in elucidating problems of distribution and species hybridization in areas to which very few other collectors have penetrated.

LOCALITIES OF NATURAL HYBRIDS

(*a*) With ***S. venturii***

ARGENTINA

Prov. TUCUMÁN:

Dept. Chicligasta, Estancia Las Pavas, La Cascada a Las Cuevas. Alt. 2800 m. 9 Mar. 1949. 'Flor morada.' *T. Meyer 14948* (LIL). (Leaf intermediate between the two species. Purple flower and high articulation of *S. sanctae-rosae*.)

Dept. Chicligasta, Estancia Las Pavas, La Cascada a Las Cuevas. Alt. 3000 m. 10 Mar. 1949. *T. Meyer 15075* (p.p.) (LIL). (Terminal leaflets very broad.)

Dept. Chicligasta, Quebrada de Las Pavas, entre Puesto La Cascada y Las Cuevas. Alt. 3200 m. 8 Mar. 1949. *Petersen & Hjerting P–56* (p.p.) (J.G.H.). (Enlarged terminal leaflet, ovate to oblong.) See Plate 93.

Dept. Chicligasta, Estancia Santa Rosa. Alt. 3600 m. 6 Jan. 1927. 'Fl. blanca. Falda del cerro.' *Venturi 4727* (p.p.) (BA, GH, LIL, LP, SI, UC, US). (Altitude probably not correct. Some of these plants are very close to *S. venturii*, but others are definite hybrids.)

Dept. Tafí, 84 km from Tucumán on way to Cafayate. 28 Feb. 1960. 'In open rocky grassy soil. Fls. large, bluish-purple.' *Correll, Dodds, Brücher & Paxman A702* (p.p.) (Photograph of LL?)

Dept. Tafí, Río Tafí, cerca de la Quebrada Honda. Alt. 2900 m. 14 Feb. 1952. 'Flor lila clara.' *Petersen & Hjerting P–326* (J.G.H., J.P.H.). (Leaf 2-jugate. Flower pale lilac; articulation almost median.)

As above, but 'Flor azul y blanca'. *Petersen & Hjerting P–327* (LIL, J.G.H., J.P.H.). (Leaf like that of *S. megistacrolobum*, but flower large, rotate, white and blue, typical of hybrids.)

Dept. Tafí, Infiernillo, entre km 83 y Río Tafí. Alt. 2900 m. 4 and 20 Feb. 1952. 'Flor azul y blanca.' *Petersen & Hjerting P–333* (LIL, J.G.H., J.P.H.). (Probably a hybrid segregating population with leaf type varying from one similar to *S. venturii* to typical *S. sanctae-rosae*.)

Dept. Tafí, Infiernillo, entre km 83 y Río Tafí. Alt. 2700 m and 2900 m. 20 and 27 Feb. 1952. 'Flor azul, en general con un poco de blanca.' *Petersen & Hjerting P–334* (J.G.H.). (Poorly acuminate lateral leaflets, long oval terminal leaflet, larger paler flowers with white acumens.)

Dept. Tafí, Río Tafí, cerca de Quebrada Honda. Alt. 2700 m. 27 Feb. 1952. 'Flor blanca con un poco de violeta.' *Petersen & Hjerting P–336* (LIL, J.G.H.). (This has a leaf with 0–2 pairs of lateral lobes and very large terminal. Specimens grown from tubers at Cambridge (CPC: 2463) possessed a sublyrate leaf, with 2–3 pairs of lateral lobes and very large terminal. The flowers were pale blue, paling to white central petal streaks, typical of experimental hybrids between white- and blue-flowered species. Chromosome number: $2n = 24$.) See Plate 94.

Dept. Tafí, Estancia San José (de Chasquivil). Alt. 2400 m. 4 Jan. 1953. *Petersen & Hjerting 899* (J.G.H.). (A hybrid segregate with large ovate terminals, rounded very small laterals and large pale mauve flowers.)

Dept. Tafí, Ñuñorco. Alt. 3000 m. 25 Jan. 1943. *Rohmeder* (s.n.) (LIL 88939). (Leaf good *S. sanctae-rosae*, but flower pale lilac; articulation median.)

Dept. Tafí, Peña de la Flora (Cerro del Cajón). Distrito de Colalao del Valle. Alt. 3700 m. 18 Feb. 1927. 'Flor azul.' *Schreiter 4842* (LIL). (Terminal leaflet long-elliptic. Altitude probably not correct.)

Dept. Tafí, Tafí del Valle. Alt. 2400 m. 13 Feb. 1949. *Sparre 5753* (LIL). (Leaves ovate-oblong, simple or unijugate. One leaf with widely decurrent laterals. Flower and pedicel intermediate between the two species.)

Prov. CATAMARCA:

Dept. Andalgalá, entre Río Lampacillo y Río Potrero superior. Alt. 2700–2900 m. 27 Feb. 1951. 'Fl. alb.' *Sleumer 1839* (LIL, J.G.H.). (Flower and inflorescence of all specimens fairly typical of *S. venturii*. Leaf and habit of one specimen close to *S. venturii*, of the other five specimens closer to *S. sanctae-rosae*.)

(*b*) With ***S. spegazzinii***

ARGENTINA

Prov. TUCUMÁN:

Dept. Tafí, road from Santa María to Tafí del Valle, km 84, near Quebrada Honda. Alt. 3100 m. 13 Feb. 1959. 'Hochgrasflor unter Felsen.' *Ross & Vervoorst 43* (p.p.) (Max.).

Prov. CATAMARCA:

Dept. Andalgalá, Cuesta de la Negrilla, cerca de la Mina de Capillitas. Alt. 3000–3100 m. 2 Feb. 1952. 'Fl. pallide viol.' *Sleumer 2687* (p.p.) (J.G.H.). (*Note*: The LIL sheet seems to be good *S. spegazzinii*.)

Dept. Andalgalá, Cuesta de la Negrilla, cerca de la Mina de Capillitas. Alt. 3000–3100 m. 2 Feb. 1952. 'Fl. viol.' *Sleumer 2688* (p.p.) (LIL, US, J.G.H.).

Dept. Belén, Quebrada de los Potrerillos to El Rodeo (Puesto 2600 m), arriba de Granadillas. Alt. 2600 m. 26 Jan. 1952. 'Fl. viol.' *Sleumer & Vervoorst 2485* (LIL, J.G.H.).

Dept. Belén, Quebrada del Río Blanco, Las Mesaditas, arriba de Granadillas. Alt. 2800 m. 28 Jan. 1952. 'Fl. violasc.' *Sleumer & Vervoorst 2533* (LIL, J.G.H.); *2535* (LIL, US, J.G.H.).

Dept. Pomán, entre Las Ciénagas y Las Angosturas, por el Alto de la Cruz, 25 km al S.E. de Pomán. Alt. *c.* 2500 m. 29 Feb. 1952. 'En un faldeo seco con exposición al N.E., en el dominio de los pastizales, entre arbustos; fl. azul.' *Vervoorst 3482* (p.p.) (LIL, J.G.H.). See Plate 96A.

SERIES TUBEROSA RYDB.

(*Bull. Torrey bot. Club* **51**, 146–8, 1924, p.p., *nomen nudum*)

sensu BUKASOV (*Izv. Akad. Nauk SSSR.* 711–32, 1938) *et*
sensu HAWKES (*Rep. Scott. Pl. Breed. Stn.* 78–79, 1956).
TUBEROSA (Rydb.) Buk., *sensu stricto* (*ex* Buk. & Kameraz. *Bases of Potato Breeding* 18, 1959).
ANDIGENA Buk. (ibid. 24, 1959).
TRANSAEQUATORIALIA Buk. (ibid. 21, 1959).
VAVILOVIANA Buk. (ibid. 18, 1959).
ANDREANA Hawkes (*Bull. imp. Bur. Pl. Breed. Genet., Camb.* 50, 1944, *nomen nudum*).
MINUTIFOLIOLA Corr. (*Tex. Res. Found. Contrib.* **4**, 216–18, 1962).

TUBER-BEARING herbs with elongated stem and imparipinnate to simple leaves; peduncle bifurcate, well developed; pedicels normally not articulated near the base or the apex; corolla rotate to semi-stellate. Berry more or less globular.

General distribution: Venezuela, Colombia, Ecuador, Peru, Bolivia, Argentina, and Chile. Especially concentrated in north-west Argentina, Bolivia, and Peru. Altitudes from sea level to *c.* 3500 m.

Chromosome numbers: $2n = 24, 36, 48, 60$.

The series Tuberosa was designated by Rydberg in 1924, but unfortunately no description of any sort was given. Furthermore, Rydberg included, besides the type species *S. tuberosum*, species that we now consider to belong to series Longipedicellata, Demissa, and Conicibaccata, all of these being from Mexico and Central America.

In 1938 (and also in 1939*a*) Bukasov removed the species of series Demissa and Longipedicellata from Tuberosa, though Latin descriptions of these series were not published before 1959. The one species from Rydberg's Tuberosa that belongs to series Conicibaccata, namely *S. agrimonifolium*, was transferred to that series by Hawkes in 1956.

In the 1938 publication of Bukasov to which we have referred, various sub-series or groups are mentioned:

A. EUTUBEROSA, which includes the species of the coasts of Chile (*S. maglia*, *S. tuberosum* s. str., etc.).
B. ANDIGENA, comprising nearly all the cultivated species, though not *S. tuberosum* from Chile.
C. ARACCIANA, with the wild species of Peru and Argentina then known to Bukasov in the living state.
D. AJANHUIRI, represented by the single cultivated species, *S. ajanhuiri*.
E. TRIGONOHYPSA, with the two hybrid cultivated species, *S. juzepczukii* and *S. curtilobum*.

Hawkes's 'Revision' of 1956 included 39 wild and 8 cultivated species in Tuberosa, without any subdivisions. The second edition of this work (1963) had 57 wild and 7 cultivated species. Both editions included short descriptions in English, and the 1956 edition therefore contained the first actual description of this series to be published, though unfortunately not in Latin.

In 1955 Bukasov again (1955*b*) attempted a classification of series Tuberosa and its allies, though no descriptions were given. The old series Tuberosa was then divided into the following series:

1. TUBEROSA s. str. (equivalent to the old subgroup EUTUBEROSA, but with the addition of the Peruvian coastal species).
2. ANDIGENA (equivalent to the other four subgroups given above).
3. ANDREANA (including *S. andreanum* from Colombia and Ecuador, together with one or two rather discordant elements from those countries that are now grouped in series Conicibaccata and Piurana).

Finally, in 1959 (Bukasov & Kameraz 1959) we see Bukasov's more considered views in which yet another system is proposed, and the various series are given formal Latin descriptions, as follows:

I. TRANSAEQUATORIALIA. This, contrary to the strict meaning of the name, includes all the species of wild potatoes from the Andes of South America, both *north* and *south* of the equator, that were not included in any of the other series.
II. ANDIGENA. This series now includes all the Andean cultivated species.
III. VAVILOVIANA. This small series includes in reality only two Peruvian coastal species, *S. wittmackii* and *S. weberbaueri*.
IV. TUBEROSA. Contains all the other wild species from the coasts of Peru and Chile, and the Chilean forms of *S. tuberosum*.

In his monograph of 1962 Correll followed Bukasov's classification approximately, but placed two species in a separate series, Minutifoliola. Furthermore, he did not deal with the cultivated species at all, and therefore did not mention the series Andigena, though he placed the Peruvian and Chilean coastal species (other than Vaviloviana) in series Tuberosa. The cultivated species were discussed by Dodds in the same work, but were given a completely different treatment. These were all 'lumped' under the species *S. tuberosum*, except for the hybrid species *S. juzepczukii* and *S. curtilobum*. Curiously enough, Dodds's groups ANDIGENA, etc., are minor categories below the rank of species, whilst those of Bukasov are above the rank of species.

We have said enough to show the state of confusion that is caused by this proliferation of series, which are changed so radically in each succeeding publication. Although a complete standardization and a rigid treatment would be inadvisable, since it would make no allowance for the incorporation of fresh knowledge, we feel that nothing of value can come from these continual and perhaps rather arbitrary changes. In the present state of knowledge it is surely better to adhere to a fairly conservative and broad system of classification without very many subdivisions, since this would mean that changes could be made in our concept of species relationships without necessitating a corresponding change in the classification at every step. We therefore consider that the concept of series Tuberosa put forward by Bukasov in 1938 (though without the five subdivisions) is probably correct for the present state of our knowledge. This broad treatment was followed by Hawkes (1956*b*, 1963), by Brücher (1957*b*), and by Ochoa (1962). Any further subdivisions, such as those of Bukasov and Correll, seem to be based chiefly on geographical and ecological considerations. Geo-ecological classifications could be regarded as valid so long as it was clearly understood that they were 'special purpose' classifications with no attempt to trace phylogenetic relationships, but evolutionary concepts appear in almost every paragraph of Bukasov's latest work (1959 with Kameraz), and in our opinion are unjustifiable, since they are based largely, though not entirely, on geographical distribution data. Thus, Bukasov's geographical/taxonomic separation of coastal from mountain species in Chile and in Peru is clearly wrong. Detailed morphological studies have shown that the coastal species are not related so much to each other as to the mountain species near to each corresponding coastal area. The only possible migratory routes for coastal species in Peru and northern Chile is from mountains to coast along the valleys, whilst wide stretches of desert absolutely prevent longitudinal migrations at sea level. Thus Ochoa (1962) has shown that coastal species are frequently found in the high mountains behind the coastal belt (e.g. *S. medians*, *S. chancayense*, etc.). In the present book we show the same for *S. maglia*, which occurs on the coast of Central Chile. It has also been discovered in the mountains of Argentina (Mendoza) and is clearly morphologically related to the group of Argentine species which includes *S. kurtzianum*. Just as wrong are the attempts made by Bukasov to classify the various forms of *S. chacoense* into groups of species adapted to the pampas, the chaco, the pampa-sierras, etc. Geographical separation can be the basis for species evolution, but until it can be shown to have taken place we should assume that it has not. Similarly, the cultivated and wild forms of *S. tuberosum* in southern Chile and in Patagonia are morphologically closely similar to the tetraploid cultivated forms from the Andes. They have been shown by one of us (1956*a*) to be subspecifically though not specifically distinct from each other, and the Chilean *S. tuberosum* cannot be claimed to bear a special relationship with *S. maglia* just because it occurs in the same lowland belt, or with *S. medians*, for the same reason, since these species are separated by several thousands of kilometers of desert.

Only detailed experimental studies on series Tuberosa, in the wide sense, will afford the right kind of evidence for us to piece together, little by little, the true affinities of the species in this group. As yet, we cannot see our way clearly enough to be able to do this, and in any case the criterion of expediency and comparative ease of identification must carry weight in a group of plants which are notoriously difficult to identify. For this reason, too, we maintain that a broad classification is necessary. Any more detailed classification is at the present state of knowledge, in our view, not only almost meaningless, but will tend to retard progress in a study of the affinities of the species concerned.

Species of this series included in the present treatment:

(*a*) WILD:

- *S. MICRODONTUM* Bitt.
- *S. VENTURII* Hawkes & Hjerting.
- *S. VERNEI* Bitt. & Wittm.
- *S. GOURLAYI* Hawkes.
- *S. OPLOCENSE* Hawkes.
- *S. VIDAURREI* Cárdenas.
- *S. SPEGAZZINII* Bitt.
- *S. KURTZIANUM* Bitt. & Wittm.
- *S. MAGLIA* Schlechtd.

(*b*) CULTIVATED:

- *S.× JUZEPCZUKII* Buk.
- *S. TUBEROSUM* L.
- *S.× CURTILOBUM* Juz. & Buk.

11. SOLANUM MICRODONTUM Bitter

(In *Reprium nov. Spec. Regni veg.* **10**, 535–36, 1912)

emend. Hawkes & Hjerting (*Phyton*, Graz **9**, 144–6, 1960).
S. bijugum Bitt. (In *Reprium nov. Spec. Regni veg.* **10**, 533, 1912).
S. cevallos-tovari Cárd. (*Boln Soc. peru. Bot.* **5**, 13–15, 1956).
S. higueranum Cárd. (ibid. **5**, 20–21, 1956).

Herbaceous, bearing stolons up to 2 m long, and small to medium-sized spherical verrucose white to brown tubers, 0·5–4 × 0·5–5 cm.

Stem delicate to thick, according to subspecies, upright or decumbent, 0·1–1(–2) m long, (1·5–)2–10(–20) mm diam., simple to (especially in old decumbent specimens) very branched, green or more or less densely sprinkled with purple anthocyanin; decurrent wings from leaf bases 0–5 mm wide, according to subspecies, green or pigmented, straight, undulate or crisped, with the margin entire or irregularly crenulate, glabrescent, or furnished with a few large coarse hairs, chiefly on the teeth.

Leaf *generally large*, sometimes simple but normally with 1–2(–3) pairs of lateral leaflets or lobes which are smaller than the terminal leaflet and *decrease in size rapidly towards the leaf base*; interjected leaflets absent or very rare. Terminal leaflet (or leaf lamina in simple-leaved forms) (3–)5–18(–37) × (1·5–)3–9(–16) cm, varying from lanceolate to ovate-lanceolate, *elliptic*, broad ovate or rotundo-cordate, with acute to acuminate apex and the base varying from cordate to rounded or cuneate, generally long decurrent. Lateral leaflets sessile or shortly petiolulate (to 8 mm), up to 15 × 7 cm, though often much less, obliquely elliptic to ovate or ovate-lanceolate, apically obtuse to acuminate, basally auricled on the basiscopic side or decurrent on to the rhachis; occasionally the lateral leaflets take the form of irregularly ovate obtuse lobes projecting from a wide rhachis wing. *Margins of all leaflets typically minutely denticulate*, the teeth ending in hairs, but sometimes the margins are more or less entire, though still pubescent; lamina typically densely pubescent above and more sparsely below; hairs *shining, silky, closely adpressed*, multicellular, *directed acroscopically and towards the leaf margins, often pale straw-coloured when young*. Pseudostipular leaves semilunate, clasping the stem, to 2 × 1 cm but often much smaller.

Inflorescence 5–20(–35)-flowered. Peduncles once-forked, 3–8 cm long, frequently provided with a reduced simple leaf and a narrow crenulate wing; clothed with sparse to dense large and small sub-adpressed multicellular hairs. Pedicels 10–25 mm long, gradually thickening above the articulation, articulated at or slightly below the centre, or sometimes up to $\frac{2}{3}$ or $\frac{3}{4}$ the length above the base, provided with hairs as the peduncles and a few inconspicuous glands.

Calyx campanulate, with short lobes and *long acumens of unequal length and width*; lobes (including acumens) 3–8 mm long × 1·5–3 mm broad; *acumens linear to subspathulate*, 0·5–6 mm long; calyx generally covered with frequent multicellular hairs of unequal lengths which are adpressed to spreading.

Corolla *white* (pale mauve in some introgressed forms only), *pentagonal to rotate*, (15–)20–35(–40) mm diam.; lobes 5–18 mm broad × 6–15 mm long, of which the *large acumen* is 4–6 mm long.

Anthers 4·5–8 mm × 1·5–2 mm, slightly tapering towards the apex; filaments 1–3 mm long × 0·5–1·5 mm broad.

Style 8–12 mm long, glabrous or provided with sparse to dense papillae in the lower third, occasionally mixed with a very few 1–4-celled acute hairs; stigma small, rotund to clavate, barely thicker than style apex, entire or with apical and basal grooves or sinuses.

Berries round or apically slightly conical when young, smooth, pale green and almost glaucous.

Chromosome number: $2n = 24$.

AFFINITIES

S. microdontum is closely related to *S. venturii* but can be distinguished from it by the strong pubescence, generally winged stem and more robust habit, as well as the denser inflorescence and larger flowers. Subsp. *gigantophyllum* is much larger and more robust in every way than *S. venturii*; subsp. *microdontum* may be sometimes of about the same size and general dimensions as *S. venturii* but can easily be distinguished by the coarse pubescence on all green parts—a feature entirely lacking in *S. venturii*, which possesses characteristic short triangular 3-celled hairs only. The habit of *S. venturii* is much more decumbent than that of subsp. *microdontum*, and furthermore the two taxa are not found in the same area.

S. microdontum differs from the white-flowered *S. maglia* by the dense pubescence and absence of gibberulous hairs; it also possesses a normal, distinct demarcation between anthers and filaments, whilst in *S. maglia* (and also in *S. calvescens*) the anthers and filaments are not sharply differentiated.

S. microdontum differs from the Bolivian *S. virgultorum*, which has a similar leaf and habit, in flower colour, and a few other features. It is probably related also to the Bolivian species *S. alandiae* and *S. gandarillasii*, though it differs from them considerably in several characters.

From the Argentine species *S. vernei* it seems to be morphologically very distinct, differing in leaf form and flower colour most markedly; yet, as we shall see later, the two species seem to be closely related genetically (see p. 344).

KEY TO SUBSPECIES

Stem slender, 1·5–3(–5) mm diam., with wings 0–1(–2·5) mm wide, which are generally straight and non-denticulate marginally; terminal leaflets normally 8 × 5 cm or less subsp. ***microdontum***

Stem robust, 3–10(–20) mm diam., with wings 2–5 mm wide, which are generally undulate to crisped and marginally crenulate to denticulate; terminal leaflets normally 8–18 × 4–9 cm subsp. ***gigantophyllum***

(*Note*: Forms intermediate between the above subspecies also occur in the zone of overlap between them.)

1. Subspecies *microdontum*

(For synonyms see p. 316)

Plant delicate, spreading, with tubers 0·5–1 cm diam. *Stem slender, to 50 cm tall*, 1·5–3(–5) mm diam.; *wings often absent, 0–1(–2·5) mm wide*, generally *straight* or sometimes undulate, marginally non-denticulate, glabrescent. Leaf with the uppermost pair of lateral leaflets often not much smaller than the terminal leaflet; *terminal leaflet (3–)5–8(–13) × (1·5–)3–5(–7) cm*; laterals up to 8 × 5 cm. Inflorescence fewer-flowered, 1–8(–15), than subsp. *gigantophyllum*.

2. Subspecies *gigantophyllum* (Bitter) Hawkes & Hjerting

(*Phyton*, Graz **9**, 144–6, 1960)

S. gigantophyllum Bitter (in *Reprium nov. Spec. Regni veg.* **11**, 368–9, 1912).
S. simplicifolium Bitter (ibid. **11**, 369–70, 1912).
S. simplicifolium subsp. *gigantophyllum* Bitter (ibid. **12**, 445, 1913).
S. simplicifolium var. *metriophyllum* Bitter (ibid. **12**, 445, 1913).
S. simplicifolium var. *mollifrons* Bitter (ibid. **12**, 445–6, 1913).
S. simplicifolium var. *trimerophyllum* Bitter (ibid. **12**, 446, 1913).
S. simplicifolium var. *variabile*[1] Brücher & Ross (*Lilloa* **26**, 465–6, 1953).
S. tafiense Lillo, *nomen nudum*.

Plant robust, upright or decumbent and much branched, with tubers from 1–4 × 1–5 cm. *Stem thick, 0·3–1(–2) m tall, robust,* (2–)*3–10*(–20) *mm diam.*; *wings always present and generally well developed,* (1–)*2–5 mm wide, generally undulate to crisped,* marginally denticulate to *irregularly crenulate*, and with a few large coarse hairs chiefly on the teeth. Leaf with the uppermost lateral leaflets normally much smaller than the terminal, this latter (7–)8–18(–37) × (3–)4–9(–16) cm; laterals up to 15 × 7 cm. Inflorescence 5–20(–35)-flowered.

When growing under exceptionally poor and dry soil conditions the leaf size of subsp. *gigantophyllum* may be reduced to that of subsp. *microdontum*. Under these conditions, however, the internodes become very short (0·5–1 cm) and no lateral leaflets are produced, whereas the internodes of subsp. *microdontum* of comparable plant size are of normal length (2–5 cm).

In general there is no difficulty in placing a specimen into one or other of the two subspecies unless it comes from the zone of intergradation between them. The forms from this zone are roughly intermediate in all characters between the two subspecies.

SPECIMENS EXAMINED

1. Subspecies *microdontum*

ARGENTINA

Prov. JUJUY:

Dept. Tilcara, near Tilcara, Chorro Valley. Alt. 2450–3350[2] m (down to San Antonio). 13–14 Feb. 1939. *Balls 6040* (CPC: 107) (K, UC, P.G.S., J.G.H.).

Dept. Tilcara, near Tilcara, Chorro Valley. Alt. 2450–3350[2] m. 14 Feb. 1939. *Balls 6042* (p.p.) (P.G.S.).

Dept. Tilcara, near Tilcara, Chorro Valley. Alt. 3500[2] m. 14 Feb. 1939. 'Around the edges of disused fields.' *Balls 6053* (CPC: 109) and *6053 a* (p.p.) (CPC: 113, p.p.) (K, P.G.S.).

Dept. Tilcara, Quebrada Estancia Vieja, Abra Mayo. Alt. 3000 m. 21 Mar. 1952. 'Flor blanca.' *Petersen & Hjerting P–347* (J.G.H.).

Dept. Valle Grande, Serranía de Calilegua, Mesías. 18 Feb. 1965. 'Flor blanca.' *Fabris, Crisci & Petriella 5860* (LIL).

[1] This was incorrectly named var. *variabilis* by the original authors and has been corrected in the present account.

[2] Altitude probably incorrect; should be ±3200 m.

Prov. SALTA:

Dept. Santa Victoria, Toldos, near Bermejo [formerly in Bolivia]. Alt. 1900 m. 26 Nov. 1903. 'Humid forest.' *Fiebrig 2253* (B—photo of now destroyed type for '*S. bijugum*' Bitt.). See Plate 98.

Dept. Santa Victoria, Toldos, near Bermejo [formerly in Bolivia]. Alt. 1800 m. Dec. 1903. *Fiebrig 2498* (F, GOET, M, SI, W (Lectotype of *S. microdontum*)). See Fig. 30 and Plate 97.

Dept. Santa Victoria, Mesón (entre Santa Victoria y Toldos). Alt. 3150 m. 10 Feb. 1956. 'Todas hojas simples.' *Hjerting & Rahn 174* (J.G.H., J.P.H.).

Dept. Santa Victoria, Cuesta de San José. Alt. 2700 m. 11 Feb. 1956. 'Generalmente 1 par de hojitas.' *Hjerting & Rahn 189* (C, J.G.H., J.P.H.).

Dept. Santa Victoria, Cuesta de San José. Alt. 2650 m. 11 Feb. 1956. *Hjerting & Rahn 191* (C, J.G.H., J.P.H.), *192* (J.G.H., J.P.H.).

Dept. Santa Victoria, cerca de Toldos, El Nogalar. Alt. 1850 m. 11 Feb. 1956. *Hjerting & Rahn 194* (C, J.G.H., J.P.H.).

Dept. Santa Victoria, Toldos, Quebrada del Astillero. Alt. 1600–1700 m. 13 Feb. 1956. 'En bosque de *Alnus* y *Podocarpus*.' *Hjerting & Rahn 220* (C, J.G.H., J.P.H.).

Dept. Santa Victoria, cerca de Toldos, El Nogalar. Alt. 1700–1850 m. 14 Feb. 1956. 'Se encuentra entre 1600–2500 m, en el bosque húmedo.' *Hjerting & Rahn 226* (C, J.G.H., J.P.H.), *227* (J.G.H.). See Plates 99 and 100.

Dept. Santa Victoria, cerca de Toldos, El Nogalar. Alt. 1850–1900 m. 14 Feb. 1956. '30 plantas de 200 m de camino.' *Hjerting & Rahn 228* (J.G.H.).

Dept. Santa Victoria, cerca de Toldos, entre El Nogalar y San José. Alt. 1950 m. 14 Feb. 1956. 'En bosque puro de *Podocarpus*.' *Hjerting & Rahn 230* (C, J.G.H., J.P.H.), *231* (J.G.H., J.P.H.). See Plate 101.

Dept. Santa Victoria, cerca de Toldos, entre Nogalar y Estancia San José. Alt. 2500 m. 14 Feb. 1956. 'En bosque húmedo de *Alnus* y *Podocarpus*.' *Hjerting & Rahn 232* (J.G.H., J.P.H.).

Dept. Santa Victoria, cerca de Toldos, entre Río Huaico Grande y Río San José. Alt. 2450–2500 m. 14 Feb. 1956. 'En bosque húmedo de *Alnus* y *Podocarpus*.' *Hjerting & Rahn 233* (C, J.G.H., J.P.H.).

Dept. Santa Victoria, Pie de Cuesta de Mesón. Alt. 2400 m. 15 Feb. 1956. *Hjerting & Rahn 236* (J.G.H., J.P.H.).

Dept. Santa Victoria, La Huerta (cerca de Santa Victoria). Alt. 2700 m. 16 Feb. 1956. *Hjerting & Rahn 240* and *241*. (J.G.H., J.P.H.). (These two collections exhibit some characteristics of subsp. *gigantophyllum*.)

Dept. Santa Victoria, Rodeopampa. Alt. 3000 m. 3 Feb. 1953. 'Fl. alb., calyx obscure castan. Frequens.' *Sleumer 3752* (LIL, J.G.H.). (Curious plants with erect growth, long narrow leaves and brown calyx.) See Plate 102.

Dept. Santa Victoria, Toldos, pie del 'Astillero.' Alt. *c.* 1600 m. 8 Feb. 1953. 'Dentro de la selva de pino y aliso.' *Sleumer 3882* (LIL, J.G.H.).

Dept. Santa Victoria, Toldos, pie del 'Vallecito.' Alt. *c.* 1650 m. 9 Feb. 1953. 'Fl. blanc. En la selva de pino y aliso, cerca de la junta del Río Astillero y Río Toldos.' *Sleumer 3883* (LIL, J.G.H.).

Dept. Santa Victoria, El Nogalar, cerca de Toldos. Alt. *c.* 1850 m. 10 Feb. 1953. 'Fl. blanc. Selva subtropical de pino, cedro salteño, nogal; frecuente.' *Sleumer 3937* and *3938* (LIL, J.G.H.).

Dept. Santa Victoria, Toldos, pie de la Cuesta de San José. Alt. *c.* 2450 m. 11 Feb. 1953. 'Fl. blanc. Frecuente en la selva de alisos cerca del "rancho".' *Sleumer 3962* (LIL, J.G.H.).

Dept. Santa Victoria, La Huerta-Rodeopampa. Alt. *c.* 2900 m. 16 Feb. 1953. 'Fl.

alb., cáliz obscuro.' *Sleumer 4037* (LIL, J.G.H.). (Leaf rather larger than normal; calyx dark brown; cf. *Sleumer 3752*.)

Dept. Santa Victoria, La Huerta-Rodeopampa. Alt. *c.* 2800 m. 16 Feb. 1953. 'Fl. alb., cáliz más o menos obscuro (marrón).' *Sleumer 4038* (LIL, J.G.H.). (Habit and leaf normal; calyx dark brown; cf. *Sleumer 3752, 4037*.)

Dept. Santa Victoria, La Huerta. Alt. 2700 m. 16 Feb. 1953. 'Fl. blanc., cáliz clar. verd.' *Sleumer 4039* (LIL, J.G.H.).

2. Subspecies ***gigantophyllum***

BOLIVIA

Dept. TARIJA:

La Aguada, 33 km from Tarija on the road to Villazón. Alt. 3350 m.[1] 27 Feb. 1939. *Balls 6125* (K, UC, US, P.G.S., J.G.H.).

Cuesta de Sama, 32 km north-west of Tarija, on road to Villazón. Alt. *c.* 2800 m. 12 Feb. 1937. 'Gulch in steep boulder-strewn slope, among bushes and tall grass. Bush to 50 cm, fl. white.' *James West 8338* (GH).

ARGENTINA

Prov. JUJUY:

Dept. Capital, Termas de Reyes. 28 Feb. 1945. 'Flor blanca. La raíz lleva tubérculos.' *Abbiatti & Claps 786* (LP).

Dept. Capital, Laguna de Yala. 28 Feb. 1945. *Abbiatti & Claps 885* (LP).

Dept. Capital, Termas de Reyes. 17 Feb. 1940. 'Fl. blanca.' *Burkart & Troncoso 11127* (SI).

Dept. Capital, Yala. Alt. 1445 m. 7 Jan. 1947. *Garolera & Romero* (s.n.) (LIL 187915).

Dept. Capital, Termas de Reyes. 12 Jan. 1947. *Garolera & Romero* (s.n.) (LIL 197183).

Dept. Capital, Lagunas de Yala. Alt. 2100 m. 1 Feb. 1956. *Hjerting, Petersen & Rahn 51* (C, J.G.H., J.P.H.).

Dept. Capital, de Yala a Lagunas de Yala. 8 Apr. 1945. *O'Donell 2970* (LIL).

Dept. Capital, Tal von Yala. Alt. 2000 m. 18 Feb. 1959. 'Nogalwald durchsetzt m. subtrop. Elementen.' *Ross & Vervoorst 71* (Max.).

Dept. Santa Bárbara, Cerro Centinela. Alt. 2650 m. 11 Feb. 1964. 'Hierba 40 cm.' *H.A. Fabris 5143* (LP).

Dept. Santa Bárbara, senda del Fuerte al Cerro El Centinela. 16 Feb. 1965. 'Flores blancas, alt. 0·40 m.' *Legname & Cuezzo 5105c* (LIL, LP).

Prov. SALTA:

Dept. Caldera, Quebrada de Yacones. Alt. 1650–1700 m. 20 Feb. 1956. *Hjerting, Petersen & Rahn 285* (C, J.G.H., J.P.H.).

Dept. Caldera, Cuesta del Carancho, arriba del Potrero del Castillo. Alt. *c.* 3200 m. 13 Mar. 1952. 'Fl. alb.' *Sleumer & Vervoorst 2892* (p.p.) (LIL).

Dept. Capital, Quebrada de San Lorenzo. Alt. 1530 m. 7 Feb. 1939. *Balls 5936* (p.p.) (CPC: 51*a*) (K, UC, P.G.S., J.G.H.). See Plate 103.

Dept. Capital, Quebrada de San Lorenzo. Alt. 1500 m. 20 Feb. 1956. 'Frutos.' *Hjerting, Petersen & Rahn 293* (C, J.G.H., J.P.H.). See Plate 103.

Dept. Capital, Quebrada de San Lorenzo, near Salta. 9 Mar. 1873. *Lorentz & Hieronymus* (s.n.) (Berlin type specimen of '*S. simplicifolium*' Bitt. now destroyed. Only photograph seen.)

[1] This altitude may be exaggerated and should possibly be only about 2900 m.

Dept. Capital, San Lorenzo. 19 Feb. 1949. *Montenegro 437* (LIL).

Dept. Capital, Quebrada de San Lorenzo. Alt. 1400–1500 m. 19 Mar. 1952. 'Topotypus.' *Sleumer & Vervoorst 3030* (LIL, J.G.H.).

Dept. Capital, San Lorenzo. Alt. 1200 m. 21 Jan. 1941. *Villalón* (s.n.) (LIL 36743).

Dept. Chicoana, entre Chicoana y Escoipe, cerca de Mal Paso. Alt. 1650 m. 22 Feb. 1956. 'Frutos.' *Hjerting, Petersen, & Rahn 298* (C, J.G.H., J.P.H.). See Plates 104 and 105.

Dept. Chicoana, un poco arriba de la Hostería, Pie de Cuesta del Obispo. Alt. 2350 m. 22 Feb. 1956. *Hjerting, Petersen, & Rahn 308* (J.G.H., J.P.H.).

Dept. General M. M. de Güemes, entre Caldera y Carmen, km 1650, camino de La Cornisa. Alt. 1450 m. 27 Jan. 1956. *Hjerting, Petersen, & Rahn 15* (J.G.H., J.P.H.).

Dept. Guachipas, 3 km de Ojo de Agua hacia Guachipas. Alt. 1900 m. 24 Feb. 1956. *Hjerting, Petersen, & Rahn 363* (C, J.G.H., J.P.H.).

Dept. Guachipas, Pampa Grande, cerca de la represa. Alt. 1700 m. 24 Feb. 1956. *Hjerting, Petersen, & Rahn 371* (J.G.H., J.P.H.).

Dept. Guachipas, Ojo de Agua, entre Guachipas y Pampa Grande. Alt. 1900 m. 25 Feb. 1956. *Hjerting, Petersen, & Rahn 378* (C, J.G.H., J.P.H.).

Dept. Guachipas, Cuesta de Lajar. 15 Feb. 1959. 'Subalpine Matten in Umgebung von Wasserlauf.' *Ross & Vervoorst 61* (Max.).

Dept. Guachipas, Pampa Grande. Jan. 1897. *C. Spegazzini* (s.n.) (LPS 14209, Type for '*S. simplicifolium* Bitt. var. *mollifrons*' Bitt.). See Fig. 33.

Dept. Guachipas, Pampa Grande. Jan. 1897. *C. Spegazzini* (s.n.) (LPS 14213).

Dept. Guachipas, Pampa Grande. Jan. 1897. *C. Spegazzini* (s.n.) (LPS 14218, Type for '*S. simplicifolium* Bitt. var. *trimerophyllum*' Bitt.). See Fig. 34.

Dept. Guachipas, Pampa Grande. Jan. 1897. *C. Spegazzini* (s.n.) (LPS 14219, Syntype for '*S. simplicifolium* Bitt. var. *metriophyllum*' Bitt.). See Fig. 32.

Dept. Rosario de Lerma, Quijano. 4 Feb. 1941. *A.T. Hunziker 1278* (CORD, US).

Dept. Rosario de Lerma, El Manzano. Alt. 2000 m. 12 Feb. 1941. 'Flor blanca.' *T. Meyer 3771* (LIL).

Dept. Santa Victoria, Santa Victoria, Arroyo San Felipe Antiguo. Alt. *c.* 2450 m. 15 Feb. 1953. 'Frecuente en la chacra de Don Rafael Aparicio.' *Sleumer 4022* (LIL, J.G.H.).

Dept. La Viña, Coronel Moldes, La Hoyada. 14 Nov. 1942. *Burkart 13218* (SI).

Dept. La Viña, Coronel Moldes, en La Hoyada a 20 km de Cnl. Moldes. Alt. *c.* 1200 m. 2 Feb. 1941. *A. T. Hunziker 1078* (CORD).

Dept. La Viña, Coronel Moldes. Alt. 1104 m. Cultivado en la Est. Exp. de Coronel Moldes. 'Tiene pequeñas papitas.' *A. T. Hunziker 2519* (CORD).

Dept. La Viña, Quebrada del Churqui, camino La Viña-Amblayo. 18 March 1943. 'Frecuente. En lugares húmedos y de sombra. Estolones muy largos, hasta de 2 m. Aún no había formado tubérculos. Corola blanca.' *A. T. Hunziker 2590* (CORD, J.G.H.).

Dept. La Viña, Coronel Moldes. 15 Dec. 1945. 'Papa de la Ollada', 'Crece silvestre en las quebradas de las montañas'. *Mintzer* (s.n.) (BAL, J.G.H.).

Probably prov. Tucumán (but Salta when collected): Arcas-Trancas. Feb. 1897. *C. Spegazzini* (s.n.) (LPS No. 14217, Syntype for '*S. simplicifolium* Bitt. var. *metriophyllum*' Bitt.).

Probably prov. Tucumán (but Salta when collected): Tacanas. *C. Spegazzini*? (s.n.) (LPS No. 14220. Syntype for '*S. simplicifolium* Bitt. var. *metriophyllum*' Bitt.).

Prov. TUCUMÁN:

Dept. Burruyacú, Cerro del Campo. Alt. 1900 m. Mar. 1918. 'Color de la flor blanco.' *Bailetti 185* (LIL).

Dept. Burruyacú, Burruyacú. Jan. 1911. *Stuckert 22009* (LPS).

Dept. Chicligasta (?), La Banderita, on long rock slide of mountain slope. Alt. 2500 m. 29 Feb. 1960. 'Fls. creamy white; fruits light green, ovoid and cordate at base, 2 cm long.' *Correll, Dodds, Brücher, & Paxman A705* (K).

Dept. Chicligasta, Cuesta del Clavillo. Alt. 1500 m. 12 Feb. 1949. 'Corola blanca.' *T. Meyer 14731* (LIL).

Dept. Chicligasta, Puente Arroyo Calao. 15 Feb. 1949. *T. Meyer 14774* (LIL).

Dept. Chicligasta, El Bolsón a La Cascada. Alt. 2500 m. *T. Meyer 14892* (LIL).

Dept. Chicligasta, Río Cochuna. 4 Mar. 1941. *O'Donell 117* (LIL).

Dept. Chicligasta, Cuesta del Clavillo. Alt. 1400–1800 m. 9 June 1949. *Petersen & Hjerting P–22*. (Only living material seen.)

Dept. Chicligasta, Quebrada Las Pavas, entre Puesto Santa Rosa y La Cascada. Alt. 1050–2250 m. 6 Mar. 1949. 'Prefiere lugares húmedos. Tallo alado.' *Petersen & Hjerting P–53* (J.G.H., J.P.H.).

Dept. Chicligasta, Río Cochuna. Alt. 1070 m. 11 Feb. 1959. 'Tucumaner Regenwald.' *Ross & Vervoorst 5* (Max.).

Dept. Chicligasta, Estancia Las Pavas. Alt. 1200 m. 9 Dec. 1925. 'Flor blanca. Altura 70 cm. Entre los pastos.' *Venturi 4015* (GH, LIL, SI, US).

Dept. Chicligasta, Estación de Piscicultura. Alt. 1000 m. 26 Nov. 1926. 'Fl. blanca. En los prados.' *Venturi 6064* (GH, LIL, US).

Dept. Monteros, Quebrada de Las Sosas, km 36. Alt. 1300 m. 27 Feb. 1956. 'Frutos.' *Hjerting, Petersen, & Rahn 399* (J.G.H., J.P.H.).

Dept. Monteros y Tafí, Santa Lucía, El Nogalar. 10 Jan. 1951. *T. Meyer 16480* (LIL).

Dept. Monteros y Tafí, el camino entre Acheral y Tafí del Valle, entre km 30 y 46. Alt. 1000–1700 m. 7 Nov. 1948 to 26 Feb. 1949. 'En la selva, en lugares húmedos y sombreados. Hojas muy variables.' *Petersen & Hjerting P–9* (J.G.H., J.P.H.).

Dept. Río Chico, Escaba, Cumbre de Marbay. Alt. 2350 m. 24 Dec. 1913. 'Matorral; flor crema clara. 120 cm de altura.' *Monetti 1830* (LIL).

Dept. Tafí, Valle de Tafí. 1908. *Bruch* (s.n.) (p.p.) (LP).

Dept. Tafí, Las Juntas. Jan. 1943. *Descole 1581* (LIL).

Dept. Tafí, Cuesta de Garabatal. *Descole 1586* (LIL).

Dept. Tafí, Mollar. Alt. 1900 m. 27 Feb. 1956. *Hjerting, Petersen, & Rahn 395* (J.G.H., J.P.H.).

Dept. Tafí, El Siambón, Río Matadero, margen del arroyo. Alt. 1100 m. 21 Nov. 1888. *Lillo 1031* (LIL).

Dept. Tafí, Las Piedras, márgenes del arroyo. 8 Jan. 1899. *Lillo 2201* (LIL).

Dept. Tafí, Cuesta de Anfama. Alt. 2500 m. 12 Feb. 1905. 'Entre las rocas, lugares sombríos; corola blanca.' *Lillo 3990* (LIL).

Dept. Tafí, Tafí del Valle, quebrada al norte. Alt. 2300 m. 19 Feb. 1905. 'Parajes arenosos, húmedos y sombríos. Corola blanca.' *Lillo 4148* (LIL).

Dept. Tafí, Tafí del Valle, quebrada al norte. Alt. 2300 m. 5 Mar. 1905. 'Bosque, lugares arenosos. Corola blanca.' *Lillo 4320* (LIL).

Dept. Tafí, Anfama. Alt. 1800 m. 23 Jan. 1907. 'Campos cultivados.' *Lillo 5445* (GOET, LIL).

Dept. Tafí, quebrada al norte de Tafí del Valle. Alt. 2200 m. 11 Feb. 1908. 'Playa arenosa; flor blanca.' *Lillo 7603* (LIL).

Dept. Tafí, en los pendientes y las quebradas de la Cuesta del Garabatal, cerca de Siambón, Sierra de Tucumán. 27 Jan. 1874. *Lorentz & Hieronymus 802* (CORD, Lectotype for *S. microdontum* Bitt. subsp. *gigantophyllum* (Bitt.) Hawkes & Hjerting; GOET). See Fig. 31.

Dept. Tafí, Mollar, cerca del río. Alt. 2100 m. 25 Feb. 1949. 'En pasto.' *Petersen & Hjerting P–49* (J.G.H.).

Dept. Tafí, Tafí del Valle, El Mollar. Alt. 2000 m. Jan. 1892. *Schickendantz* (s.n.) (LIL 62597).

Dept. Tafí, Morteritos (Quebrada de Tafí). Alt. 2000 m. Dec. 1917. 'Flor blanca.' *Schreiter 350* (LIL).

Dept. Tafí, Casita a Morteritos. Alneta (Alisales). Alt. 2000 m. Dec. 1917. *Schreiter* (s.n.) (MVM—Corn. Osten 12399).

Dept. Tafí, Cuesta oriental de Lara. Alt. 3500 m. (?) 15 Feb. 1920. 'Flor blanca.' *Schreiter 1211* (LIL).

Dept. Tafí, El Siambón. Alt. 1250 m. 16 Jan. 1950. 'Fl. crem.' *Sleumer 96* (LIL).

Dept. Tafí, Tafí del Valle, La Angostura. Alt. 1900 m. 24 Jan. 1950. *Sleumer 184* (LIL).

Dept. Tafí, Tafí del Valle. Alt. 2000 m. 11 Feb. 1949. *Sparre 5630 and 5631* (LIL).

Dept. Tafí, La Ciénaga. Alt. 2500 m. 18 Feb. 1949. 'Arroyo.' *Sparre 5930* (LIL).

Dept. Trancas, San Pedro de Colalao, Cuesta de Juchayo. Alt. 2000 m. 19 Jan. 1915. *Castillón* (*4*) (LIL 62604).

Dept. Trancas, San Pedro de Colalao, Cuesta de Juchayo. Alt. 2000 m. Summer 1916. *Castillón* (s.n.) (LIL 62605).

Dept. Trancas, Cumbres de Calchaquíes, Filo del Medio. Alt. 2500 m. 17 June 1949. 'En pasto.' Herb. specimen made at Est. Exp. 'El Suncho', Catamarca, 4 Mar. 1950. *Petersen & Hjerting P–89* (J.G.H., J.P.H.).

Dept. Trancas, Cuesta Grande del Chorro. Alt. 3000 m. 1 Feb. 1927. 'Flor blanca.' *Schreiter 4843* (LIL).

Prov. CATAMARCA:

Dept. Ambato, Las Juntas, Quebrada Río La Salvia. 15 Mar. 1959. 'Flor blanca.' *Villa Carenzo & Legname 1136 and 1187* (LIL).

Dept. Andalgalá, El Candado. Alt. 2700 m. 5 Feb. 1916. 'Planta alta, robusta, de flores azules, blancas o azuladas, papas grandes venenosas. Abundantes.' *P. Jörgensen 1311* (p.p.) (GH, LIL, MVM, UC).

Dept. Andalgalá, Estancia 'Yunka Suma', en la selva. Alt. 1450–2350 m. Jan. 1949. 'En lugares húmedos y sombreados. Hojas muy variables, tallos con alas.' *Petersen & Hjerting P–26* (J.G.H., J.P.H.). See Plate 3.

Dept. Andalgalá, Estancia 'Yunka Suma', Puesto de Vallecito. Alt. 2000 m. 16 Jan. 1949. 'Hojas lucientes pubescentes; tallo con alas violetas.' *Petersen & Hjerting P–27* (J.G.H., J.P.H.).

Dept. Andalgalá, Valle del Río Pisavil. Alt. 1950 m. 9 Feb. 1949. 'En pasto; tallo con alas.' *Petersen & Hjerting P–37* (J.G.H., J.P.H.).

Dept. Andalgalá, en la barranca del Río Chacra, cerca de la Estancia 'Yunka Suma'. Alt. 1350 m. 11 Feb. 1950. 'Fl. alb.-crem., 30–200 cm.' *Sleumer 14* (LIL).

Dept. Andalgalá, Los Queñoales, arriba de la Mesada Las Rosas. Alt. 2250–2400 m. 15 Jan. 1952. 'Fl. alb.' *Sleumer 2251* (LIL, US, J.G.H.).

Dept. Belén, El Rodeo, arriba de Granadillas. Alt. 2500 m. 26 Jan. 1952. 'Fl. blanc. 160 cm alta!' *Sleumer & Vervoorst 2482* (LIL, US, J.G.H.).

No exact locality given. *Brücher* (s.n.) (sent by Professor H. Ross, as EBS 184. CPC: 2417. No herbarium specimen).

Prov. LA RIOJA:

(See Appendix III.)

3. Probably subspecies ***gigantophyllum***

Prov. TUCUMÁN:

Dept. Tafí, Cumbre de Anfama. Alt. 2700 m. 9 Mar. 1919. 'Color de la flor violáceo.' *Schreiter 2208* (LIL). (Apparently good subsp. *gigantophyllum* in stem and leaf characters, but flower said to be violet; possibly due to introgression from *S. vernei*.)

4. ***S. microdontum*** (Intermediate forms between subspecies ***microdontum*** and subspecies ***gigantophyllum***.)

(See notes on living plants, p. 341)

Prov. JUJUY:

Dept. Tilcara, Quebrada Estancia Vieja, Cuevas. Alt. 2300 m. 22 Mar. 1952. 'Flor blanca.' *Petersen & Hjerting P–348* (LIL, J.G.H., J.P.H.). (Fairly close to subsp. *microdontum*.)

Dept. Tilcara, Quebrada del Chorro. Alt. 2700–3200 m. 27 Mar. 1952. 'Flor blanca.' *Petersen & Hjerting P–350* (CPC: 2481) (LIL, J.G.H., J.P.H.).

Dept. Tumbaya, Volcán, falda del cerro. Alt. 2700 m. 6 May 1929. *Venturi 9529* (US).

Prov. SALTA:

Dept. Iruya, entre San Pedro y Rodero Colorado. Alt. 2800 m. 6 Apr. 1952. *Petersen & Hjerting P–364* (LIL, J.G.H., J.P.H.).

Dept. Orán, San Andrés. Alt. 1800 m. 8 Feb. 1945. *Pierotti 245* (LIL).

Dept. Santa Victoria, Pie de Cuesta de Mesón. Alt. 2500 m. 9 Feb. 1956. 'Todas las hojas simples.' *Hjerting & Rahn 168* (J.G.H., J.P.H.).

DISTRIBUTION

1. Subspecies ***microdontum*** is very restricted in its range in the area dealt with here, though it extends northwards into Bolivia for at least 500 km, and possibly even more. It occurs sparingly on the east Andean slopes of Argentina in Jujuy province, depts. Tilcara and Valle Grande, and has been collected many times in Salta province, dept. Santa Victoria, very close to the Bolivian border. The type collection was made in this area, at Toldos, which was once in Bolivia but now lies just inside Argentina. (See Map 10.)

Probably further search would establish an extension of this subspecies into the departments of Iruya and Orán (prov. Salta), and possibly even into the department of Ledesma (prov. Jujuy).

2. Subspecies ***gigantophyllum*** is much more common and widespread in Argentina than the type subspecies; it has been recorded from the following provinces and departments:

Prov. Jujuy, depts. Capital and Santa Bárbara;
Prov. Salta, depts. Caldera, Capital, Chicoana, General M. M. de Güemes, Guachipas, Rosario de Lerma, Santa Victoria, La Viña;
Prov. Tucumán, depts. Burruyacú, Chicligasta, Monteros, Río Chico, Tafí (very abundant), Trancas;
Prov. Catamarca, depts. Ambato, Andalgalá, Belén;
Prov. La Rioja, dept. Chilecito (see Appendix III).

It has also been found in south Bolivia, dept. Tarija (see Map 10).

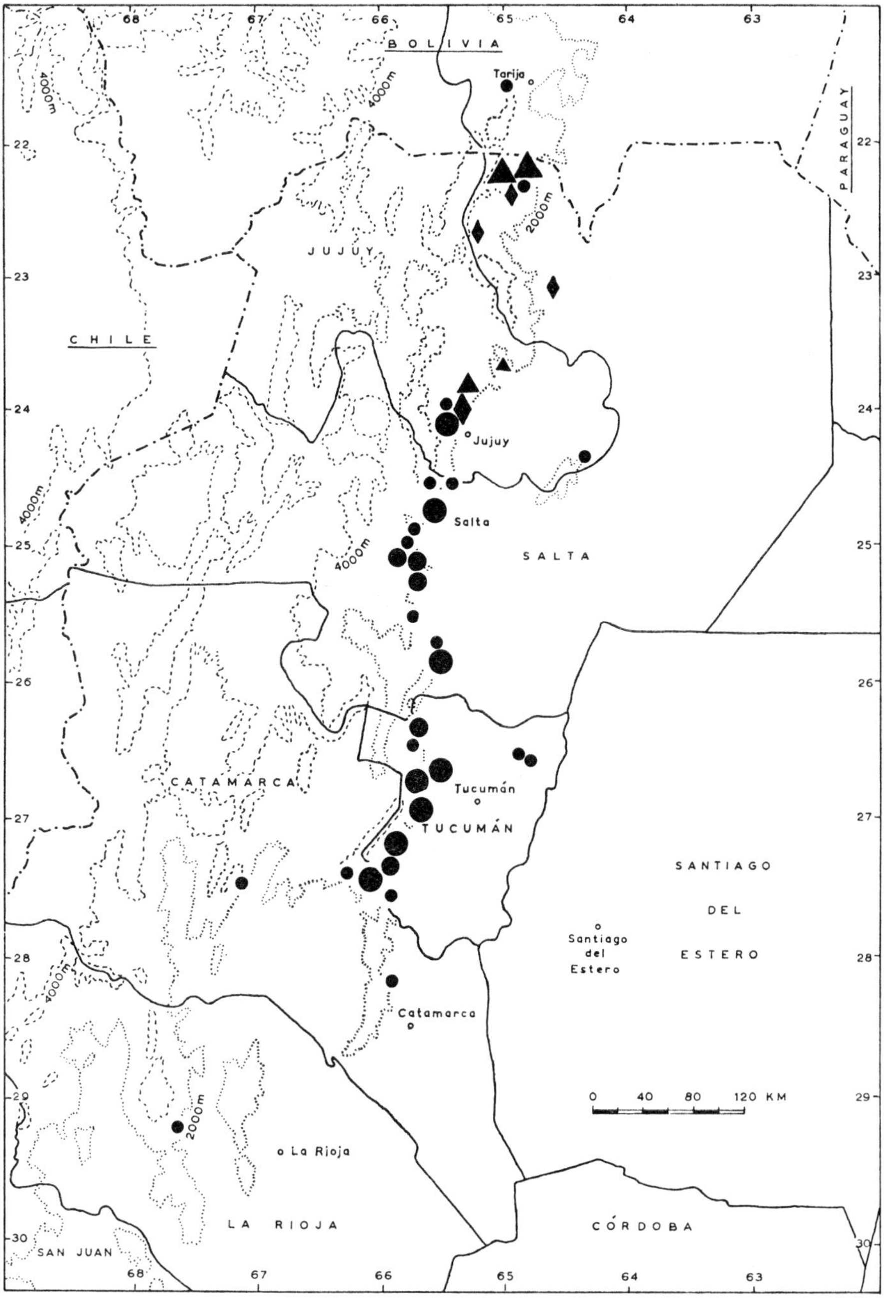

No. of Records
5 & over 3–4 1–2

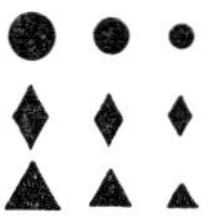

Solanum microdontum Bitt. subsp. *gigantophyllum* (Bitt.) Hawkes & Hjerting

Solanum microdontum Bitt. Intermediates

Solanum microdontum Bitt. subsp. *microdontum*

MAP 10. Distribution of *S. microdontum* Bitt. subsp. *microdontum* and subsp. *gigantophyllum* (Bitt.) Hawkes & Hjerting.

The two subspecies are not complete geographical vicariads, though by and large this is so. Subsp. *gigantophyllum* is very rare indeed north of the tropic of Capricorn ($23\frac{1}{2}$° S.) whilst subsp. *microdontum* has hardly been seen to the south of that line.

HABITAT

1. Subspecies **microdontum** has been recorded for a wide altitudinal range, as follows: (1600–)1800–2700(–3200) m above sea level.

It is found typically in the rain and cloud forest, known as the Tucumán–Bolivian selva. On Cabrera's (1953) phytogeographical scheme it would occur chiefly in the Montano district and possibly sometimes also in the upper Orán district of the Western Subtropical province. It also extends upwards into the alpine meadow formation as defined by Lorentz (1876), Holmberg (1898), and Lillo (1919). This would presumably correspond with the upper, more open part of the Montano district, which we have classified as 'pajonales'.

From altitudes of 1600 to about 2000 m subsp. *microdontum* grows normally under the dense shade of *Podocarpus*, *Cedrela*, and *Juglans australis* trees. Above this level, from about 2000 to 2700 m, the dominant trees are *Alnus jorullensis* ('Aliso') and *Podocarpus parlatorei*; subsp. *microdontum* is found quite abundantly in this zone in Santa Victoria department. From 2600 to about 2850 m the *Alnus* thins out into isolated clumps of trees with grass and bushes between, but the potato species continues up through this zone and beyond it, despite the increase of light, though it usually keeps to the less-open parts. Thus, from 2600 m right up to its highest limit at 3200 m it is found increasingly in more open communities (though always in valleys and on slopes with abundant rainfall) in full light amongst tall grass and with shrubby Compositae (*Baccharis*, etc.), Ericaceae, and Myrtaceae. At these high altitudes it is associated often with *S. vernei* subsp. *ballsii* (prov. Jujuy, dept. Tilcara) and occasionally with *S. boliviense*, *S. gourlayi*, and possibly even *S. megistacrolobum*. At medium altitudes here and there it may possibly come into contact with *S. tarijense*.

In the lower zone from 1600 to 2600 m (and perhaps even lower, though so far no one has collected it well down into the Orán forests) it grows in dense shade, forming small spreading plants with few small flowers, in contrast to the many-flowered specimens from higher up which are found in somewhat more open habitats.

Where the natural vegetation is disturbed and the forest cover partially removed subsp. *microdontum* may come into contact with the ruderal species *S. chacoense*, though so far only one possible hybrid between them (*P. & H. P–5*) has been discovered.

2. Subspecies **gigantophyllum** has been recorded for slightly lower levels than the typical subspecies, as follows: (1000–)1200–2500(–3200) m. Confirmatory collections from the lower and higher ranges of both subspecies would be needed before we could be certain that this apparent slight altitude difference was a real one.

It is to be found in the Tucumán–Bolivian selva or rain and cloud forest chiefly on the east-facing slopes of the Andes. It extends through the upper part of the Tucumán and Montano districts in the south but probably does not occur (or only rarely) in the Orán district to the north, of the Western Subtropical province of Cabrera (1953). It is most abundant in Salta and Tucumán, and both there and in Catamarca it appears in the myrtaceous forests in not very dense shade from 1000 m to about 1500 m. Thereafter it extends into the *Alnus jorullensis* zone and upwards to the *Polylepis* region at 2500 m. In many parts of Catamarca and in La Rioja further south and west than the limits of the rain forest zone, subsp. *gigantophyllum* is found at altitudes of about 1800 to 2500 m in 'pajonales' and scrub.

In Tucumán and Catamarca it hardly extends above 2500 m, but in Salta, Jujuy, and south Bolivia it seems to extend upwards to 3200 or 3350 m, in more open grassy areas ('pajonales'—cf. subsp. *microdontum*), though only in isolated instances. In the southern part of its range, subsp. *gigantophyllum* is not generally present at the higher altitudes but is replaced there in the more open grassland by *S. venturii*.

Brücher (1954) records subsp. *gigantophyllum* from four different localities with a wide pluviometric range, e.g. (i) Laguna del Tesoro, 2144 mm per year, (ii) Río Potrero, 1175 mm per year, (iii) Mesada de los Hoyos–Suncho, 413 mm per year, (iv) Agua de Palomas, Campo de Pucará, 264 mm per year. All these points are situated in the Sierra de Aconquija, south-west of Tucumán, and are only 30 km apart as the crow flies. This demonstrates the plasticity of *S. microdontum* subsp. *gigantophyllum* which is able to grow under such a very wide range of rainfall conditions.

Although subsp. *gigantophyllum* is primarily a plant of damp shady places in rain and cloud forests it sometimes colonizes river banks and terraces, and may also be seen amongst stones and grass, around walls, and occasionally as a weed in cultivated fields. It is not so common in open, grassy and bushy places, and both subspecies require abundant shade and moisture for optimum growth. When they occur in very dry places they become stunted and show atypical growth and development.

TAXONOMIC NOTES

S. microdontum was first described by Bitter in 1912, on the basis of a collection made by Fiebrig (*No. 2498*) near the village of Toldos (see Fig. 30 and Plate 97). At the time the collection was made (1903) Toldos was situated in Bolivia, but it was subsequently transferred to Argentina.

This one collection was all we knew of *S. microdontum* (in the original sense) until the 1939 expedition of Balls, Gourlay, and Hawkes, and the subsequent ones of Petersen and Hjerting in 1952, Sleumer in 1953, Hjerting and Rahn in 1956, and the 1966 expedition of Hawkes, Hjerting, and Rahn.

Bitter also described another species, *S. bijugum*, in the same publication. This was based on another collection of Fiebrig from Toldos ('in a damp wood, flowering', 26 November 1903, *No. 2253*, Herb. Berlin), but there was apparently only a single specimen of this collection, which was destroyed in the Second World War. Nevertheless, a good photograph of it was made by the Field Museum, Chicago. (See Plate 98.)

From a study of this single photograph we were at first inclined to the view that it typified the bijugate natural hybrids between *S. chacoense* and *S. microdontum*. This was later proved to be incorrect, but the true affinities of the two species, *S. microdontum* and *S. bijugum*, were not made clear until Hjerting and Rahn were able to bring back extensive collections from the Toldos and Santa Victoria region in 1956. Even though some of the isotype specimens of *S. microdontum* exhibited an occasional bijugate leaf we had previously considered this species to be quite distinct specifically from *S. bijugum*, and to possess, in the main, simple or subsimple leaves. Hjerting and Rahn's collections proved beyond any doubt that *S. microdontum* and *S. bijugum* were in fact two biotypes of the same species, since a whole range of intermediate forms in the type locality linked them together. Furthermore, Hjerting and Rahn took random leaf-sample collections from populations at several different points (see p. 337) and were able to demonstrate quite clearly that a range of biotypes with simple to bijugate leaves could be seen *within one population*. The two species are obviously synonymous.

Although the description of *S. microdontum* is preceded by that of *S. bijugum* in the same publication by the space of a few pages we have considered it best to treat *S. bijugum* as the synonym to be discarded,[1] since there is no extant type material available for it. On the other hand, although the original type sheet of *S. microdontum*

[1] Brücher (1966) wrongly contests this.

FIG. 30. *S. microdontum* Bitt. Lectotype specimen of *Fiebrig 2498* (Herb. Vienna), collected at Toldos, near Bermejo, prov. Salta, Argentina (formerly Bolivia). Leaves simple to 1-jugate. Stem slender.

at Berlin was destroyed there is isotype material available for study at Vienna, Munich, Göttingen, Chicago (one leaf from Berlin), and San Isidro (as well as Utrecht according to Correll 1962). We therefore designate the Vienna sheet of *Fiebrig 2498* as a LECTOTYPE (previously designated incorrectly as neotype in Hawkes & Hjerting 1960) (see Fig. 30), since it agrees very well both with the original description and the now destroyed Berlin sheet, of which a photograph still exists. Correll (1962) is in agreement with this treatment of Bitter's two species, *S. microdontum* and *S. bijugum*.

We must now consider the taxonomy and nomenclature of subsp. *gigantophyllum*. Up to 1960 this was known as a distinct species, under the name *S. simplicifolium*, or, occasionally, *S. gigantophyllum*. We pointed out in 1960 that neither of these species could be maintained as distinct from *S. microdontum* on the specific level, and we shall give our evidence for this decision shortly. Meanwhile, we shall deal with the infra-specific variants that have been described within this complex.

Bitter described *S. gigantophyllum* in 1912, with *S. simplicifolium* directly following it in the same publication. The former was based on a collection of Lorentz and Hieronymus, made at Siambón in prov. Tucumán, in 1874. (See Fig. 31.) The gigantic terminal leaflets of this specimen, reaching dimensions of 23 × 10 cm, makes the name chosen particularly apt. The description of *S. simplicifolium* was based on material obtained by the same collectors in the San Lorenzo valley near Salta, in 1873. The type sheet was in Berlin and is now destroyed; no other specimens of this collection seem to exist, even though Correll (1962) thought that a collection of Spegazzini (La Plata No. 14209) might represent part of the type collection. This is not so, since Spegazzini's collection was from Pampa Grande, some way from San Lorenzo, and was collected in 1897, not 1873. It hardly seems worth designating a neotype for this species, however, since in any case it is completely synonymous with the previously described *S. gigantophyllum*, for which a lectotype can be found (see below, p. 335).

In the following year, 1913, Bitter again returned to the problem of these two 'species', and under the heading of *S. simplicifolium* wrote: 'In the province of Salta an abundance of forms of this species occurs, as is evident from the material which Professor Spegazzini has sent me for investigation. These newly added types force me to lay more stress on the close relationship between *S. simplicifolium* . . . and *S. gigantophyllum*, by grouping *S. gigantophyllum* as a subspecies under *S. simplicifolium*. The name for a species representing the fusion of the two types which I published in juxtaposition must be chosen by me, as the future monographer of this genus,[1] according to its utility, so that in the future the name *S. gigantophyllum* shall not be put in front according to its priority. I consider *S. gigantophyllum* to be an extremely developed type of the polymorphic *S. simplicifolium* and suggest, therefore, naming the plant formerly called *S. gigantophyllum* from now on as *S. simplicifolium* Bitt. subsp. *gigantophyllum* (Bitt. pro sp.) nov. subsp.'

We agree with Bitter in uniting these two taxa into one species, but we would go further, and state that there is no real subspecific difference between them either. It is true that whereas *S. gigantophyllum* possessed one pair of lateral leaflets, *S. simplicifolium* was completely devoid of these; but when one examines the abundant material collected in Tucumán, Salta, Jujuy, Catamarca, and La Rioja it is quite evident that there is absolutely no geographical or morphological distinction between these two taxa described by Bitter. They are not only linked by innumerable intermediates, but the variation now known far surpasses that known to Bitter.

For a similar reason we consider that Bitter's varieties *metriophyllum*, *mollifrons*, and *trimerophyllum* are either habitat variants or at most slightly different biotypes, not even worthy of varietal rank, despite the fact that Bitter distinguishes them from *S. simplicifolium* and subsp. *gigantophyllum* by their thinner styles, which are clothed

[1] Unfortunately, a goal that Bitter never attained.

FIG. 31. *S. microdontum* Bitt. subsp. *gigantophyllum* (Bitt.) Hawkes & Hjerting. Lectotype specimen (*Lorentz & Hieronymus 802*—Herb. Córdoba) collected at Cuesta del Garabatal, near Siambón, prov. Tucumán, Argentina.

with papillae in the lower third. The same argument, that there are all types of intermediates linking these variants with the type specimen, would apply equally forcibly to Brücher and Ross's variety *variabile*, which also should not be retained as a distinct taxon.

Var. *metriophyllum*, which was originally collected in three different places in Salta and Tucumán provinces, seems almost identical with the type specimen of *S. simplicifolium*, which also comes from Salta province. The simple leaves are broadly elliptic, up to 9·5 × 5·5 cm, as compared with the elliptic to lanceolate leaf, to 13·5 × 5 cm, of *S. simplicifolium*. This size difference is negligible, whilst the shape difference is quite unimportant when one examines the inter-plant variation within a single population of this species. (See Fig. 32.)

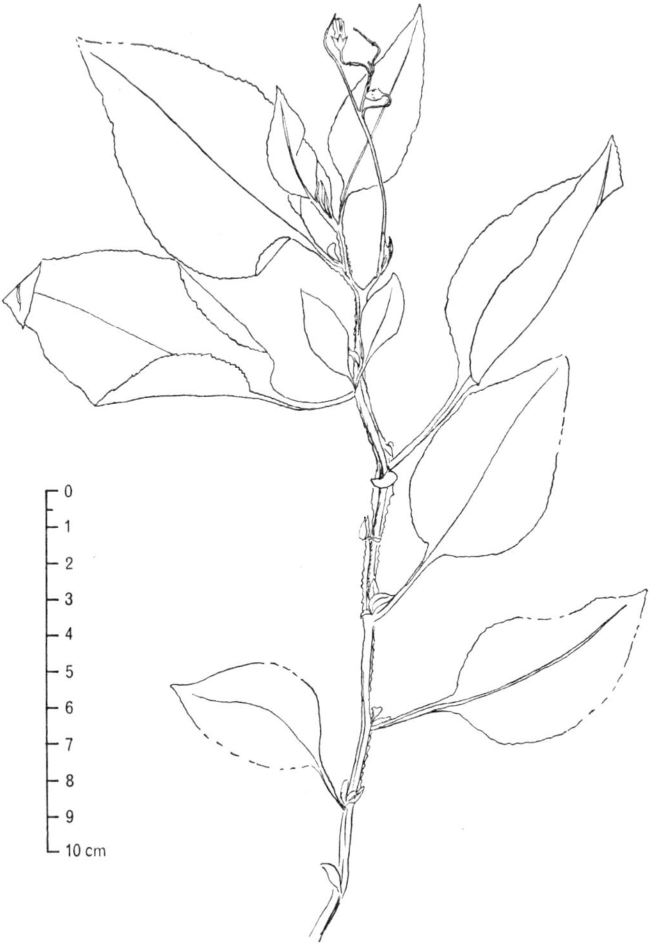

FIG. 32. *S. microdontum* Bitt. subsp. *gigantophyllum* (Bitt.) Hawkes & Hjerting. Syntype of '*S. simplicifolium* Bitt. var. *metriophyllum*' Bitt. from Pampa Grande, prov. Salta, Argentina, collected by C. Spegazzini (s.n.—Herb. Inst. Spegazzini 14219). Note the typical stem wings for this subspecies.

Var. *mollifrons* is also from Salta province, at Pampa Grande, where one of the var. *metriophyllum* collections was made. It also possesses simple leaves (or sometimes minute laterals), but the leaves are small (only to 6 × 4 cm), densely following, and covered with a yellowish-white silky pubescence. We have seen this combination of characters both in the field and the herbarium, on subsp. *gigantophyllum* growing in full sun, and we do not doubt that habitat differences can account for it quite adequately. The smallness of the anthers (only 4·5 × 1·5 mm) is due also, in our opinion to the same cause. (See Fig. 33.)

A curious feature, probably genetically controlled, is the presence of 1–4-celled acute hairs on the lowest third of the style, mixed with round-ended single-celled

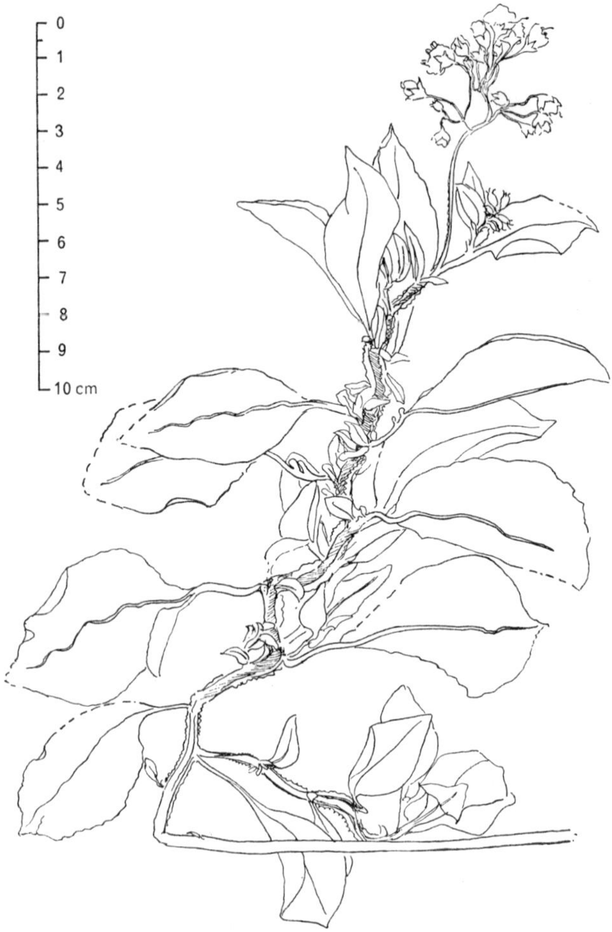

FIG. 33. *S. microdontum* Bitt. subsp. *gigantophyllum* (Bitt.) Hawkes & Hjerting. Type specimen of '*S. simplicifolium* Bitt. var. *mollifrons*' Bitt. from Pampa Grande, prov. Salta, Argentina, collected by C. Spegazzini (s.n.—Herb. Inst. Spegazzini 14209). The small leaves are covered with a dense yellow-white silky pubescence, undoubtedly due to the fact that this plant grew in full sun.

papillae of the usual type. Such acute hairs may also be seen on the style of *S.* × *setulosistylum* (*S. chacoense* × *S. spegazzinii*) from the Pampa Grande area. We have already described (p. 234) the survey made by Hjerting, Petersen, and Rahn, and by Hawkes, Hjerting, and Rahn, at Pampa Grande (see Appendix III) to see how common the style setae character was. Style setae were seen in two collections of *S. chacoense* and two of *S. microdontum* subsp. *gigantophyllum*, and so have been seen in a total of six collections, though not in every plant of each collection.

Of the two collections of subsp. *gigantophyllum* collected by Hjerting, Petersen, and Rahn, at Pampa Grande, one (*No. 378*) had a completely glabrous style, whilst the other (*No. 371*) possessed dense long *obtuse* papillae on the basal third. Style setae were seen on some plants of *Hawkes, Hjerting, & Rahn 3680*. No other collections of subsp. *gigantophyllum* or of the hybrids between this and *S. chacoense* showed style papillae that were in any way different from the normal type, apart from the collections that we have already mentioned.

It would therefore seem that there is no value in the presence or absence of stylar papillae or in their shape for differentiating the species in Pampa Grande, or indeed in any other part of Argentina and adjacent regions under discussion in the present work.

The third of Bitter's varieties, *trimerophyllum*, also from Pampa Grande, possesses one pair of lateral leaflets which are much better developed than those of the type specimen of subsp. *gigantophyllum*, attaining a size (according to Bitter) of 8·5 × 3·5 cm. Compared with the lateral leaflets on subsp. *gigantophyllum* (3·5 × 1·5 cm) the differences would seem at first sight to be rather significant. These dimensions for subsp. *gigantophyllum* are taken from Bitter's description made from a specimen, which was presumably destroyed during the Second World War at the Berlin Herbarium. Another specimen of the same collection is available at Córdoba (*Lorentz & Hieronymus 802*, now designated by us as lectotype; see below), and it is interesting to note that this plant possesses lateral leaflets as large as 9 × 3·7 cm, and thus even a little larger than those of var. *trimerophyllum*. Of course the terminal leaflets of var. *trimerophyllum* are not as large as the largest mentioned in the type description of subsp. *gigantophyllum*, though they are quite as big as the smaller ones. There is thus a complete overlap of the type specimens of these two taxa, subsp. *gigantophyllum* and var. *trimerophyllum*, which renders it inadvisable to maintain them as separate entities. (See Fig. 34.)

Var. *variabile*, described by Brücher and Ross from plants collected by Brücher in Catamarca (exact locality not specified) might well include most collections of this species, since the authors show that the lower leaves are simple, the middle ones 1–2-jugate, and the upper ones 2-jugate. Only a very few biotypes of subsp. *gigantophyllum* possess genotypically controlled simple leaves (those from San Lorenzo, for example), whilst most develop simple leaves in the juvenile state only, or when growing under poor conditions of soil or humidity.

We hope that we have been able to make it clear that the variability of this species, *S. simplicifolium*, together with Bitter's subsp. *gigantophyllum*, is so continuous, both in leaf dissection and pubescence, that none of the above-mentioned varieties can be maintained. Neither can any part of it be separated off as a distinct, giant-leaved subspecies.

Until 1956, then, we considered that *S. simplicifolium* was a good species, possessing a wide range of infra-specific variability, and well removed from the then little-known *S. microdontum*. The relationships of these two species to each other were not clear owing to our very incomplete knowledge of *S. microdontum*; and we were also not happy about the relationships of *S. microdontum* with the newly recognized but as yet undescribed species, *S. venturii*. Both *S. microdontum* and *S. venturii* seemed roughly similar in habit, but whereas the latter species exhibited a southern high-altitude distribution (Tucumán, Catamarca; 2000–2800 m), the former was known only from further north (Salta, Jujuy) and had a much wider altitude range (1600–3200 m).

Brücher (1954) was convinced that the southern high-mountain forms, which we have here grouped as *S. venturii*, were part of *S. microdontum*, which he maintained as a distinct species. We shall show later that this view is incorrect, and that *S. venturii* is a good species, related to *S. microdontum* but certainly not synonymous with it (see p. 352).

The living collections made by Hjerting and Rahn in 1956 in the Toldos region of Salta province enabled us to solve the problem of *S. microdontum*, *S. simplicifolium*, and *S. venturii*. Material was grown at Birmingham and Vandel from 1956 onwards, and an examination was made of the detailed herbarium specimens collected at the same time. On the basis of these collections we were able to show quite clearly that:

(i) *S. microdontum* was completely synonymous with *S. bijugum*, since they were connected by a whole range of morphological intermediates.

FIG. 34. *S. microdontum* Bitt. subsp. *gigantophyllum* (Bitt.) Hawkes & Hjerting. Type specimen of '*S. simplicifolium* Bitt. var. *trimerophyllum*' Bitt. from Pampa Grande, prov. Salta, Argentina, collected by C. Spegazzini (s.n.—Herb. Inst. Spegazzini 14218). Some development of lateral leaflets can be seen in this specimen.

(ii) *S. microdontum* was not clearly distinct specifically from *S. simplicifolium*; in floral characters it was almost identical with *S. simplicifolium*, though it differed in leaf size and in the size of the stem and width of stem wings; pubescence was identical in both taxa. These similarities and rather small differences were of an order to warrant subspecific distinction, but no more.

(iii) *S. microdontum* was completely distinct specifically from *S. venturii*, though the two species were obviously closely related.

Since *S. microdontum* was described earlier than *S. simplicifolium*, it therefore takes priority over the latter taxon. This is unfortunate in many ways, since *S. simplicifolium* has become well known amongst plant-breeders. Another difficulty arises when we attempt to use the word '*simplicifolium*' to designate the southern subspecies of *S. microdontum*; there is an already existing subspecies name—'*gigantophyllum*', which, as we have already seen, can also be applied to this group of plants. According to the International Rules of Nomenclature then, the epithet *gigantophyllum* must be used for this southern subspecies of *S. microdontum*, and the name *S. simplicifolium* should be discontinued.

In view of the fact that Bitter did not designate a type specimen for subsp. *gigantophyllum*, we are left in some doubt as to which sheet of *Lorentz & Hieronymus 802* he used for his description, though we have assumed that he used one in the Berlin Herbarium, which would now have been destroyed. However, there is another example of this collection in Göttingen, and a better one at Córdoba, Argentina. *We therefore designate the Córdoba sheet of* Lorentz & Hieronymus 802 *as* LECTOTYPE *for S. microdontum* subsp. *gigantophyllum* (Bitt.) Hawkes & Hjerting. (See Fig. 31.)

Although it is to be regretted that such a well-known name as *S. simplicifolium* should fall into disuse, in many ways the name change is to be welcomed. Few biotypes of this subspecies are constantly simple-leaved, whereas all possess very large leaves (even up to 37 cm long!) and hence the epithet '*gigantophyllum*' is much more apt than '*simplicifolium*'. It also indicates the contrast between this subspecies and the smaller-leaved subspecies *microdontum* in a way which the word '*simplicifolium*' would certainly not do.

Finally, we should perhaps mention the *nomen nudum*, *S. tafiense*, written by Lillo on the labels of certain of his collections of subsp. *gigantophyllum* from prov. Tucumán, dept. Tafí. So far as we are aware the name was written on herbarium labels only and was never published; its use should therefore be discontinued. The name *S. pteropodum*, which appears on the label of the sheet of *Lorentz & Hieronymus 802* seen by Bitter should also be disregarded. Bitter (1912) states at the end of his description of this taxon: 'Hieronymus remarks on the herbarium label that Grisebach considered this plant to be *S. pteropodum* Kth. As far as I know, this absolutely erroneous determination has not been taken up in the literature.' The species *S. pteropodum* was described by Dunal (not by Kunth), is non-tuberiferous, and is native to Quito, Ecuador. It can have nothing to do with subsp. *gigantophyllum*, and the application of the name *S. pteropodum* is evidently merely due to an erroneous determination on Grisebach's part.

Before leaving purely nomenclatural matters we should mention the species *S. cevallos-tovari* and *S. higueranum*, which were described by Professor M. Cárdenas from specimens collected by him in Bolivia, dept. Santa Cruz, prov. Valle Grande. The syntype material of *S. cevallos-tovari* (*Cárdenas 5087*) possesses stems with very narrow wings (0·5–1 mm wide), 2-jugate leaves with the terminal leaflet to about 10×6 cm, largest laterals up to $8 \times 5{\cdot}5$ cm, few-flowered inflorescence, densely pubescent pedicels articulated $\pm$ medianally, long linear calyx acumens which are of unequal lengths in the same flower, and white corolla. In all its characters *S. cevallos-tovari* matches extremely well the typical subspecies of *S. microdontum*, and we have no hesitation in declaring it a synonym of that species. *S. higueranum* is said to be a taller, more upright species than the preceding (up to 150 cm), with basally decurrent lateral leaflets which seem to be a little narrower than those of *S. cevallos-tovari*. The flower characters seem to be the same, however, and we can therefore confidently state, from a careful examination of isotype material (*Cárdenas 5067*), that *S. higueranum* is also synonymous with *S. microdontum* subsp. *microdontum*.

S. cevallos-tovari was said to have been collected 'among bushes in wet forest edges', and at 2700 m altitude; it presumably grew in the 'selva Tucumano-Boliviana' in similar phytogeographical conditions to *S. microdontum* in Argentina. The habitat of *S. higueranum* is not stated, but the author mentions that it reminds him of *S. violaceimarmoratum* var. *papillosum* 'from similar habitat'. Since we know that this latter species also grows in the selva Tucumano-Boliviana, it can be assumed that *S. higueranum* is adapted to growth in wet mountain forests and is ecologically similar to *S. microdontum*. Thus, in both morphological and physiological characters Cárdenas's two Bolivian species are closely similar to *S. microdontum* subsp. *microdontum*, and may be considered as synonymous with it.

Although considering it to be closely related to *S. microdontum*, Correll (1962) has retained *S. simplicifolium* as a distinct species, under which he groups as synonyms vars. *metriophyllum* and *mollifrons*. Bitter's var. *trimerophyllum* is retained as a distinct variety by Correll, in which he includes all those specimens with one or two pairs of lateral leaflets; under it are grouped var. *variabile* of Brücher and Ross, together with Bitter's subsp. *gigantophyllum*. This separation of var. *trimerophyllum* would seem to be for convenience only, since the suppression of lateral leaflets may be conditioned both by genotypic and environmental factors, and can show considerable variation in one and the same plant. Correll, furthermore, thought that 'some of the plants placed in var. *trimerophyllum* may represent hybrids between this species and *S. microdontum*'. We cannot agree with this, since the two taxa, whether regarded as subspecies or species, show intergrading or 'hybrid' forms only in the northern provinces of Jujuy and Salta, whereas most of the '*trimerophyllum*' forms have been collected in the southern part of the range of *S. simplicifolium* (= *S. microdontum* subsp. *gigantophyllum*) in Tucumán and Catamarca provinces, where the typical subspecies of *S. microdontum* does not occur.

Correll retains *S. microdontum* in the narrower sense, to include what we regard as the typical subspecies only. He also groups with it as synonyms, as we have done, *S. bijugum*, *S. cevallos-tovari*, and *S. higueranum*. In the list of specimens examined he mentions several collections from Catamarca, Tucumán, and south Salta provinces. Apart from the Brücher material, which we have not seen, we regard the other collections as identifiable with subsp. *gigantophyllum*, and not with subsp. *microdontum*.

Having disposed of the nomenclatural details we shall now consider in a more general way the range of variation to be seen in both subspecies of *S. microdontum* when growing in their natural habitat in north-west Argentina.

1. Subspecies *microdontum*

Most collections of this subspecies have been made from rather deep shade (see habitat notes, p. 326) where the plants do not develop a very dense pubescence and where flower production is rather sparse. The sun plants, however, show the typical rather shining white to yellowish pubescence much more clearly; the flowers are more abundant and the internodes are rather shortened. Both sun and shade plants develop a slender, often branched stem with wings that are normally no wider than about 1 mm, and which may be generally straight or a little undulate. The wings in subsp. *microdontum* are never so conspicuous or so striking as in subsp. *gigantophyllum*. The characteristic calyx acumens of different lengths and the white corolla with somewhat broadly triangular lobes, so characteristic of the whole species, are well seen both in sun and shade plants, too. One can always clearly see the small teeth on the leaf margins which give the species its name.

A continuous range of leaf dissection may be seen in this subspecies, from completely simple leaves to ones with two or even three pairs of lateral leaflets; the uppermost pairs are frequently smaller than the terminals but may occasionally be more or less the same size. The commonest plant type is that having a range of leaf dissection, from simple entire lower leaves to 1-jugate upper ones.

Basing our observations on herbarium specimens we have concluded that completely simple-leaved examples are not very common (*H. & R. 174, 192, 240*; *Sleumer*

3882, 4037, 4038); plants with a range of 0–1-jugate or 0–2-jugate leaves are commonest (15 samples), whilst those with a 1–2-jugate leaf range are less common (7 samples). Only two collections (*Balls 6040*; *Sleumer 3752*) showed a range of 0–3-jugate leaves.

Hjerting and Rahn's collections *227* and *230* (see Plate 101) from Toldos were constantly uni-bi-jugate and very similar to the type photograph of Bitter's '*S. bijugum*'; it should be explained, however, that these were deliberately selected from a more variable population in each case, the plants from the same population with 0–1-jugate leaves being gathered separately as numbers *226* (see Plate 99) and *231*, respectively. A similar selection was deliberately made by Hjerting and Rahn in respect of their collections *240* and *241*. These were gathered from an area in which small groups of plants with 0–1-jugate leaves (*241*) were scattered through a population of simple-leaved ones (*240*).

Other collections were made so as to show the range of variation within a population, as, for example, *H. & R. 233*, where the eighteen plants gathered under this number showed the following leaflet pair range: 0 (1 plant), 0–1 (3 plants), 0–2 (1 plant), 1 (5 plants), 1–2 (4 plants), 2 (4 plants). A similar range may be seen amongst the five plants in *H. & R. 220*, and other examples could be cited to illustrate the same trend. Even Bitter's type description, based on *Fiebrig 2498*, illustrates the fact that a wide variation within a single plant is a common occurrence: 'folia partim fere simplicia pinnis lateralibus $\pm$ve deficientibus, nonnumquam (*in eodem specimine*) *tamen etiam bijuga, foliolis lateralibus bene evolutis*' (our italics).

Although, as we have pointed out, the degree of leaf dissection is due partly to the ontogenetic stage reached by a plant at any given time, this cannot wholly explain the differences observed. We have already shown that, in *S. sanctae-rosae*, for example, the differences of leaf dissection depend partly on genetic differences between biotypes. Evidence to support this in the case of subsp. *microdontum* comes from a collection of Hjerting & Rahn (*No. 228* from Toldos, El Nogalar) in which samples from thirty plants from a distance of 200 m along the side of a path were taken at standard intervals. All plants were at an approximately equivalent stage of development (initiation of flowering), and a short piece of stem with flowers and the top 3–5 leaves was taken in each case. Table 18 shows the results of scoring leaf dissection range in this sample.

TABLE 18

Variation in leaf dissection within a population of S. microdontum *subsp.* microdontum *from the type locality* (*Hjerting & Rahn No. 228*)

Range of lateral leaflet pairs per sample	Number of plants	Percentage value
0	8	27
0–1	17	56
1	3	10
1–2	2	7

The sample is admittedly rather small but the results indicate that within a small and uniform area plants at an equivalent stage of development still clearly vary in leaf dissection. Although final proof is needed in the form of transplants into a uniform introduction garden we can assume with some degree of confidence that most of this variation is due to genotypic differences.

Enough has been said to illustrate the wide and continuous range of biotypes in this subspecies with regard to leaf dissection to show that it is impossible to maintain two distinct specific entities here in the way that Bitter attempted to do.

A curious type of variation may be seen in some of Sleumer's collections from Salta, Santa Victoria, La Huerta to Rodeopampa (see p. 301). In certain gatherings from this region (*Sleumer 3752, 4037, 4038*) the calyx is of a curious dark-brown colour. This feature is also seen, though not to such a marked degree, in Hjerting and Rahn's collections *240* and *241* from the same region. (See also Appendix III.) Sleumer's collection *3752* is highly variable, some plants possessing a green and others a brown calyx, and with the leaf varying from simple to bijugate in different plants. Two of the four plants (see Plate 102) are particularly striking, with a stiff erect habit and long, narrow, erect completely simple leaves which remind one very much of *S. boliviense*. The latter species has not been collected just here though the hybrid of *S. boliviense* × *S. megistacrolobum* was found in this same area (*H. & R. 170, 246*; *Sleumer 4040*). The whole region of Santa Victoria is very rich in potato material and has certainly not yet been adequately explored from this point of view. The plants under consideration at the moment cannot be considered as simple hybrids with the purple-flowered *S. boliviense*, since their corollas are pure white. From the leaf and habit it is possible that they represent *S. microdontum* × *S. boliviense* back-crossed several times with *S. microdontum*, or are introgressed forms. A search in the area to discover whether the variation in subsp. *microdontum* there is due to contamination with *S. boliviense* genes would certainly be most rewarding.

Before leaving subsp. *microdontum* it will be convenient to mention certain gatherings from provs. Jujuy (especially in the Chorro valley and its vicinity) and Salta (see p. 324), which seem to be intermediate in some respects between this and subsp. *gigantophyllum*. The two subspecies have been found in more or less the same region, and the forms under discussion might represent hybrids or transitional forms. They sometimes show a rather intermediate leaf size between the two subspecies or possess rather wider and more crisped wings than is normal for subsp. *microdontum*. Since seedling progenies were raised from some of the collections we shall return to this topic in more detail later (p. 341) when discussing living plants grown in Europe.

2. Subspecies ***gigantophyllum***

Nearly all the collections of this subspecies that we have examined show the characteristic strong vigorous growth, thick stem (1–2 cm), wide crisped wings extending right up to the base of the peduncle, and large leaves that are so typically associated with this taxon.

The rather coarse shining white to yellowish hairs that are particularly visible in sun forms (cf. 'var. *mollifrons*') are also most characteristic.

As in subsp. *microdontum*, differences in leaf dissection both within and between plants are noticeable. In addition, there are indications of a sorting out of different leaf types in different parts of the geographical range of this subspecies, as was pointed out by Brücher (1954). To test our general impressions and Brücher's statement we scored all the collections then available to us (seventy-four separate gatherings) for the maximum number of leaflet pairs and sorted the data into groups, according to latitude (see Table 19).

Although rather more collections would certainly give a more reliable figure, especially if they were obtained on a more random basis, a strongly marked trend is discernible. Simple-leaved collections are very frequent in the north (about 70 per cent), decreasing to about 50 per cent at Pampa Grande, and further diminishing to about 30 per cent in the region of Tucumán. It is interesting to note that the type collection of the simple-leaved '*S. simplicifolium*' was obtained from near Salta (*c.* 25° S.), whilst the 1-jugate '*S. gigantophyllum*' was gathered near Tucumán (*c.* 27° S.). It is a little doubtful whether any of the Tucumán plants are genetically incapable of producing a few lateral leaflets under the appropriate conditions, and it seems probable that most or all could do so, since none of the collections we have grown from there were completely simple-leaved. Many of the Salta plants, on the other hand, probably do not possess genes for lateral leaflet production. Certainly *Balls 5936* (p.p.) (CPC: 51*a*) from the type locality of '*S. simplicifolium*' has never produced lateral leaflets under any conditions over a period of twenty years' cultivation.

TABLE 19

Change of leaf type in S. microdontum *subsp.* gigantophyllum *at different latitudes in north-west Argentina*

Latitude south of equator and approximate locality	Leaf type			
	Simple		With lateral leaflets	
	Number of samples	%	Number of samples	%
24° City of Jujuy	6	67	3	33
25° City of Salta	10	71	4	29
26° Pampa Grande	8	53	7	47
27° City of Tucumán	10	28	26	72

It was pointed out by Choudhuri (1944, 1948) that compound leaf is dominant to simple in crosses of '*S. simplicifolium*' × *S. phureja*. We have obtained confirmatory evidence of this also in crosses between clones of subsp. *gigantophyllum* with simple and compound leaves (Family 140).

It therefore seems fairly reasonable to suppose from the data available that the differences in leaf dissection of subsp. *gigantophyllum* between the northern and southern parts of its range are real, though the change is a gradual one. Compound leaves can certainly still be found in collections from the north, and vice versa in the south. The phenomenon may best be described as a cline, though why it should have arisen we cannot yet tell. Perhaps an introgression of genes from *S. chacoense* in the south could account for it, since hybrids between the two species are well known in prov. Tucumán (see p. 213). Yet hybrids of the same type are also seen in Pampa Grande and elsewhere further north. Alternatively, this trend might be due to the selective action of gradually changing environmental conditions as one passes from north to south; this might either produce an adaptive response directly, by the genes controlling leaf dissection, or by some other genes closely linked with them; or of course it might be a pleiotropic effect. Further detailed studies are needed here to throw more light on this interesting problem.

In a study of variation within '*S. simplicifolium*' (*S. microdontum* subsp. *gigantophyllum*), Brücher and Ross (1953) distinguished three groups of biotypes, based mainly on the degree of leaf dissection, as follows:

Group 1. Leaf simple, sometimes auriculate or 1-jugate; style glabrous or papillose towards the base. Here are included '*S. simplicifolium*' in the narrow sense and its varieties '*mollifrons*' and '*metriophyllum*'.

Group 2. Leaf simple below, unijugate in the upper part of the stem; lateral lobes petiolulate; style papillose for about one-third length above the base or glabrous. They include here subsp. *gigantophyllum* and var. *trimerophyllum*.

Group 3. Leaf simple below, unijugate half-way up the stem and bijugate towards the stem apex.

Although they stated earlier in the same work that a systematic grouping of the various gatherings studied by them would be extremely difficult because of lack of correlation between the leaf and other characters, and that the biotypes all represent the products of natural hybridization, Brücher and Ross designate the more highly dissected forms of group 3 as a distinct variety, *S. simplicifolium* var. *variabile*. As we have already pointed out, there is little value in designating varieties of this kind in a subspecies with a very continuous range of variation in many features, few of which seem to show any close correlation; it is surely better, therefore, to discontinue the use of this name.

Apart from the leaf variation in subsp. *gigantophyllum* we can also see enormous differences in plant size, probably chiefly due to habitat differences. The wide stem wing and thick succulent stem is very characteristic for shade plants growing in moist soil, though genetically controlled variation in wing width and shape can be seen. Pigmentation of stem and wings is seen in different collections, ranging from none at all through a light marbling or veining of purple to completely dark purple. This may be due to slight introgressive hybridization with *S. vernei*, even though the flowers themselves remain white (see p. 345 in section on Natural Hybrids).

Similarly, the tubers are normally pale to medium brown; but in one collection (*P. & H. P–53*, from Quebrada Las Pavas, Chicligasta, Tucumán) the tubers were rosy purple. This, also, may be due to introgression. In nature the tuber size varies from 1 × 1 cm to 4 × 4·5 cm, though slightly smaller or slightly larger dimensions are sometimes seen. The weight varies from 5 to 20, or even occasionally 30 grammes each. Brücher (1953) records that the tubers possess a bitter taste but are occasionally eaten by shepherds.

Variation may also be noted in the pubescence of the style, which may be glabrous, papillose, or even very occasionally setose. Style length and stigma dimensions also vary slightly.

NOTES ON LIVING PLANTS

1. Subspecies *microdontum* (and intermediates)

We were very fortunate in being able to grow living material of this subspecies, collected probably in the exact locality and altitude of the type specimen (*H. & R. 226*, from prov. Salta, Santa Victoria, Toldos, 1800 m). The original herbarium specimens matched very well the types of *S. microdontum* and '*S. bijugum*', the leaves varying from sub-simple to 2-jugate. (See Plates 99 and 100.)

When grown at Vandel and at Birmingham this collection showed very clearly the typical pubescence, unequal-lengthed calyx acumens, white corolla, and general leaf form of what we had known up to then as *S. simplicifolium*, and we were at once left in no doubt that they were conspecific. Yet our plants of *S. microdontum* differed from *S. simplicifolium* in the more delicate habit, narrower and more branched stems, narrow straight (or only slightly crisped) wings, and small leaves. In contrast to the poorly flowering material from the Toldos forests its blooming was very luxurious in Europe, and differed in no fundamental way in this respect from *S. simplicifolium*.

The short leaves were rather arched, with the terminal leaflets pointing obliquely downwards; and they varied in dissection from simple to 2-jugate. The median leaflet veins were dark purple and the stem also was purple-mottled.

A number of other collections was grown, which, from the evidence of the original herbarium sheets and the variable seedling progenies, indicated that the original gatherings were probably hybrids or intermediates between the two subspecies (see collection notes, p. 324).

Thus, *P. & H. P–348* from Jujuy, Tilcara, Quebrada de Estancia Vieja possessed too wide and crisped a stem wing for subsp. *microdontum*, though the leaf dimensions fitted fairly well. In a seedling progeny of ten plants the leaf dimensions remained fairly constant but the stem wings varied from dark straight to green crisped on different plants; the width varied also.

Plants from the next valley to the north, Quebrada del Chorro (*P. & H. P–350*), gave several highly variable progenies, segregating both to typical subsp. *microdontum* and to subsp. *gigantophyllum*. The leaves varied in size, and the dissection fluctuated from 0–2-jugate. Stem wings varied from absent to wide and crinkled. The original Chorro valley plants also showed some range, though not as much as was revealed in these progeny tests. Slight purple flushes in the corolla of certain seedlings indicated possible gene flow from *S. vernei* subsp. *ballsii* into this highly heterogeneous population. Purple-flowered or semi-purple-flowered variants were seen by Balls, Gourlay, and Hawkes (*Balls 6041*), and designated by Hawkes (1944, p. 41) *S. microdontum* f. *purpureum* (*nom. nud.*). These, also, are due to gene flow from *S. vernei* subsp. *ballsii*. (See also, Natural Hybrids, p. 345.)

The original specimens of *P. & H. P–364* from further north in prov. Salta, dept. Iruya, between San Pedro and Rodero Colorado were, on the whole, rather closer to subsp. *gigantophyllum* in leaf dimensions, though the delicate stem and wings were similar to those of *S. microdontum*. Seedlings grown at Birmingham and Vandel were rather more robust in stem and wing dimensions, whilst the leaf type was intermediate between the two subspecies.

A tuber weight of 2–10 grammes was noted in Vandel for different plants of these intermediate forms.

These three collections were from an altitude range of 2300–2800 m in the valleys of the eastern Andes of north-west Argentina. Since subsp. *gigantophyllum* has been collected just beyond the Argentine–Bolivian border at higher altitudes (*Balls 6125*; altitude said to be 3350 m, but probably only about 2900 m) and in Santa Victoria at similar ones (*Sleumer 4022*; Alt. 2450 m) it would seem that it may spread thinly over this general area, hybridizing or intergrading freely with subsp. *microdontum* in several places. Other intermediates were collected by Hawkes, Hjerting, and Rahn in 1966 and are described in Appendix III.

2. Subspecies *gigantophyllum*

Variation in plants grown at Vandel and Birmingham was similar to that already noted from those in Argentina. Stem and wing pigmentation varied from absent to dark purple; and when the pigmentation of the stems was intense the purple coloration could also sometimes be seen on the leaflet veins and even on the calyx acumens. The variation of leaf form and dissection and the stem diameter mirrored closely the variation seen in South America. (Compare leaflet number in Plates 104 and 105.) The white pentagonal corolla and the calyx acumens of different lengths were most characteristic for all the lines cultivated.

Tuber production under long days in the open was very poor, and under short days it was moderate. In Vandel a tuber weight of 3·5–5·5 grammes was obtained.

All plants were very susceptible to frost in the experimental plots, despite Brücher's

(1954) claim that a resistance to $-5°$ C could be observed in Argentina. *Phytophthora* attack was moderate in Vandel, though finally all plants succumbed.

S. microdontum has been recorded as exhibiting resistance to the following diseases and pests: wart, scab, powdery scab, powdery mildew, *Verticillium* wilt, blackleg, aphids, and viruses X, A, Y, and leaf-roll. Possible resistance to late blight and a lower degree of resistance to leaf hoppers and flea beetles have also been noted. (For further details, see Chapter 5.)

CYTOLOGY

A somatic chromosome number of $2n = 24$ was established by Hawkes (1944) and by Choudhuri (1944) for subsp. *gigantophyllum*; this was confirmed later by Brücher (1954) with several different gatherings; and other workers have also obtained counts of twenty-four somatic chromosomes. No counts on pure subsp. *microdontum* have been made, though $2n = 24$ has been established by us for the intermediate form (*P. & H. P–350*; CPC: 2481), and it is unlikely that there will be any difference between the two subspecies in this respect.

According to von Wangenheim (1957) and Magoon, Hougas, and Cooper (1958*a*), meiosis is regular in pollen mother cells, though the latter workers found a certain very small percentage of univalents, quadrivalents, bridges, and fragments in subsp. *gigantophyllum* and its hybrid with *S. vernei*. Occasional univalents, bridges, and fragments were also observed by Choudhuri (1944) in the cross *S. microdontum* (subsp. *gigantophyllum*) × *S. phureja*. Frequent diploid cells were seen at M II, due to fusion of plates.

Gottschalk (1954) and Gottschalk and Peters (1955) concluded from their pachytene studies that *S. microdontum* was particularly primitive, in that it possessed only eight distinct chromosome types and was thus considered to have diverged less than any other potato species from the hypothetical ancestor with a basic number of six chromosomes. Their work was repeated by von Wangenheim (1957), who was not able to agree with their results or their conclusions. He was not convinced that only eight distinct chromosome types could be distinguished; furthermore, he found very regular pachytene pairing, both in *S. microdontum* and in its hybrids with *S. vernei* and with *S. goniocalyx*. Such regularity would be impossible if *S. microdontum* differed from all other species so fundamentally in its chromosome architecture. Furthermore, there seem to be close genetic links between *S. microdontum* and *S. vernei*, in their ability to form fertile F_1 and F_2 hybrids when crossed together; such a situation could not be envisaged if the two species were as distinct as Gottschalk claims. Finally, there appear to be no morphological or other features pointing to an ancestral condition for *S. microdontum*, apart from the one mentioned by Gottschalk.

FERTILITY AND CROSSABILITY

Choudhuri (1944) pointed out that *S. microdontum*, in common with most other diploid species of potato, was self-incompatible; this has been generally confirmed by other workers, though Malheiros-Gardé (1959*a*) obtained one berry with nineteen seeds after repeated selfings. Pollen fertility is high according to Pushkarnath (1953)—up to 92 per cent, and Ratera (1947*b*)—up to 90 per cent at Villa Ortúzar, Buenos Aires.

Successful crosses of *S. microdontum* (always using subsp. *gigantophyllum*) have been made by various workers with *S. chacoense*, *S. tarijense*, *S. acaule*, *S. infundibuliforme*, *S. vernei*, *S. maglia*, and *S. kurtzianum* (as well as with the diploid cultivated species *S. stenotomum*, *S. goniocalyx*, and *S. phureja*). Choudhuri (1944) found one-way incompatibility between *S. microdontum* and certain other species, and certain differences were observed by Malheiros-Gardé (1959*a*) in backcrosses of F_1 hybrids to the original parents, using *S. vernei* and *S. kurtzianum*. Magoon *et al.* (1958) were able to make successful triple hybrids of the type (*S. vernei* × *S. microdontum*)4x × *S. tuberosum*, using artificial tetraploid plants of the initial cross when hybridizing it with the cultivated potato. Swaminathan's hybrid of *S. acaule* × *S. microdontum* (Swaminathan 1950, 1951, 1954) was sterile as a triploid, with a high frequency of trivalents. It thus conformed to the usual pattern of *S. acaule* × diploid crosses in exhibiting the cytological characteristics of an autotriploid. The colchicine-hexaploid of this cross showed very regular bivalent pairing and 61 per cent fertile pollen, thus suggesting that the two genomes of *S. acaule* and that of *S. microdontum* are basically similar, and will form multivalents when their exact homologues are not present, but will pair differentially when the exact homologue is present, as is the case in the amphiploid. This phenomenon of differential pairing is not uncommon in the genus *Solanum*. The c-tetraploid of the cross *S. kurtzianum* × *S. microdontum* subsp. *gigantophyllum* (Swaminathan 1956) showed that tetrasomic inheritance of certain characters and disomic inheritance of others took place. It was concluded that these results lent support to the idea that these species have evolved by means of cryptic structural differentiation.

Crosses of *S. microdontum* with diploid species from South America other than the area we are considering have generally been successful, as also with the cultivated *S. tuberosum* and the wild Mexican species, *S. demissum*. With other Mexican species (Malheiros-Gardé 1959*b*) little success has been obtained, owing to the death of the hybrid embryos consequent on abnormal development and degeneration of endosperm.

Our own hybrids of *S. microdontum* with *S. chacoense* and with *S. sanctae-rosae* have already been mentioned (see pp. 227 and 308). F_2 hybrids of *S. toralapanum* (a Bolivian species allied to *S. megistacrolobum*) × *S. microdontum* subsp. *gigantophyllum* (families 186, 222) showed either breakdown or poor seed germination. The F_1 (Family 131) was quite vigorous.

We made various hybrids of collections within *S. microdontum*, crossing three of the intermediate forms (*P. & H. P–348*, *P–350*, *P–364*, from provs. Jujuy and Salta) with subsp. *microdontum* (*H. & R. 226* from prov. Salta). All the F_1 seedlings were typically subsp. *microdontum*, though some variation in stem wings was observed. Even the F_1 progeny from subsp. *gigantophyllum* (*H.P.R. 293*) crossed with an intermediate form (*P. & H. P–350*) seemed predominantly similar to subsp. *microdontum*. Further work is needed to confirm this apparent dominance of subsp. *microdontum* over subsp. *gigantophyllum*. No F_2 progenies were raised.

A cross of the two intermediates *P–348* and *P–350* gave a very uniform F_1, quite similar to the parents but with no segregation of any characters typical of subsp. *gigantophyllum*. This rather indicates that the intermediates are balanced stable forms which cannot be regarded merely as hybrids between the two subspecies.

An F_1 family (140) between two accessions of subsp. *gigantophyllum* (CPC: 2417/3 from prov. Catamarca; and CPC: 51*a* from prov. Salta, San Lorenzo) was uniform and vigorous. The F_2 family (188) was very vigorous and uniform, apart from four chlorotic seedlings, which were evidently due to the presence of recessive semi-lethal genes in the homozygous state, unmasked by the intersibbing of the F_1 seedlings. There was no genetic breakdown in this cross.

A number of crosses of the intermediate forms was made, both with *S. vernei* subsp. *vernei* and with subsp. *ballsii*. The hybrid of subsp. *vernei* (49/180, a collection of Brücher, via Professor H. Ross) × *S. microdontum* (*P. & H. P–364*—intermed. subsp.) showed pale lilac large flowers and a semi-felted under-surface to the leaflets; the petiolules were quite long, up to 8 mm, and the leaves also were long. (See Plate 107.)

A cross of subsp. *vernei* (CPC: 2414—EBS: 197, from Brücher, via Professor H. Ross) × *S. microdontum* subsp. *gigantophyllum* (*Balls 5936* (p.p.)—CPC: 51*a*) gave a uniform F_1 (Family 13) in which the leaves were 3–4-jugate with large terminal leaflet and laterals decreasing rapidly towards the leaf base. The flowers were white or barely tinged with mauve externally, or with a mauve flush except for the acumens. Intersib crosses were made and two F_2 families raised (159 and 178). A wide segregation for leaf type was obtained, from simple to 4-jugate; and segregation for flower colour also took place, giving a range from almost pure white to quite dark purple. All plants made good growth and showed no indication of genetic breakdown. One must therefore conclude that, although the two species are distinct in nature and show few intermediates, their genetic architecture is so similar that any combination of genes from them in the F_2 is viable (see pp. 56 and 116–18 for a general discussion on this theme).

A backcross from the F_1 to *S. vernei* (Family 124) was not so uniform and thrifty as the F_1 and F_2 generations, and many seedlings did not develop properly. This may be due to inbreeding depression, however. Although the leaf in this backcross generation was much closer to that of *S. vernei*, the flowers were mixtures of mauve and white, clearly showing its hybrid nature.

Several crosses were also made between the intermediate forms of *S. microdontum* and *S. vernei* subsp. *ballsii* (*P. & H. P–364* × *P. & H. P–351*, and reciprocal; *P. & H. P–350* × *P. & H. P–351*, and reciprocal). In these the thick velvety pubescence of subsp. *ballsii* and the short to absent petiolules were clearly dominant in the F_1. There was a number of differences between these last-mentioned crosses in respect of flower colour, possibly due to the fact that part of *P. & H. P–350* was not pure *S. microdontum* but had some contamination of genes already from *S. vernei* subsp. *ballsii* (see p. 341). The crosses which did not involve *P. & H. P–350*, on the other hand, were identical in flower colour, showing only a slight variation from lilac to pale lilac. By contrast, the seedlings of the cross *350* × *351* were all mostly white-flowered, whilst those of the reciprocal were mostly mauve; this may indicate that pigment suppressor genes are being inherited cytoplasmically in this case.

In most seedlings the auricled leaflet base of the subsp. *ballsii* parent was most noticeable; generally also the leaf was shorter than in the *S. microdontum* × subsp. *vernei* crosses; and this again demonstrates that there are such clear subspecific differences between subsp. *vernei* and subsp. *ballsii* that they can be seen even in their hybrids with *S. microdontum*.

A cross of *S. maglia* (CPC: 2057 from Mendoza) × *S. microdontum* (*P. & H. P–350* from Jujuy) gave an F_1 family with a rather wide variation range. Some seedlings had sub-simple and others 3-jugate leaves with several pairs of interjected leaflets. Purple leaf pigmentation was observed in some plants. The loose anther column of *S. maglia* was dominant, and also the typical *S. maglia* filaments.

NATURAL HYBRIDS

(*a*) Natural hybrids between *S. microdontum* and the following species have already been discussed:

S. chacoense (pp. 229–32).
S. tarijense (p. 245).
S. megistacrolobum (p. 295).

We have mentioned, also, the possible introgression of genes from *S. boliviense* or some other pigmented species into certain collections (*Sleumer 3752, 4037, 4038*) in Prov. Salta, dept. Santa Victoria, which cannot be classed as simple hybrids, since in most characters they conform with *S. microdontum* (see pp. 301, 338).

Intermediate forms between the two subspecies of *S. microdontum*, which might possibly be considered as hybrids, have also been referred to (p. 341).

Natural hybrids of *S. microdontum* with *S. vernei* are not very common when one considers that the two species overlap in their altitude ranges from approximately 1850 to 3200 m. Their ecological requirements may perhaps be distinct enough to keep them apart. Thus *S. microdontum* is generally confined to the forests, whilst *S. vernei* is a species of less dense shade, growing typically on the damp river terraces or 'mesaditas', amongst bushes and long grass, and under the partial shade of trees. Our 1966 observations indicate that a difference of flowering time might also constitute an effective barrier in nature. (See Appendix III.) Since the two species are easily crossed experimentally (pp. 343–4), forming vigorous and fertile F_1 and F_2 hybrids, their distinctness in nature must be maintained by other than purely genetical barriers; presumably their different ecological requirements or differences in flowering time act as sufficiently strong isolating mechanisms. Possibly any hybrids that are formed may be poorly adapted to the parental habitats and therefore not able to survive in competition with the parents or other plants in those habitats. Where the habitats themselves are 'hybridized', then hybrids may be expected to survive.

If we refer back to the altitude and habitat range of *S. microdontum* (p. 326) we shall see that from 2600 to 3200 m this species is found increasingly in more open communities, since the tree cover is thin from 2600 m and disappears at altitudes of 2850 m and over. Thus in this part of its altitude range the ecological barrier between *S. microdontum* and *S. vernei* would be expected to have disappeared, and hybrids would be more frequent.

This hypothesis seems to be reasonably well substantiated. Of the nine natural hybrid collections with altitude data between these two species identified by us, as many as six were from the 2700 to 3200 m range, where *S. microdontum* grows in the open and has been frequently found in close proximity to *S. vernei* (chiefly subsp. *ballsii*). Only three collections of hybrids are from much lower down, namely, *H.P.R. 306* (2200 m), *P. & H. P–33* (2050–2200 m) and *P. & H. P–42* (2200 m). The first was collected in grassy, shrubby places where the natural forest cover had been removed; the second was from an intermediate natural habitat in light shade; and the last was from a river bank, where the natural vegetation had been disturbed.

It is interesting to note, also, that a number of our collections of *S. microdontum* and *S. vernei* from high altitudes (*P. & H. P–348*; *P–350*; *P–351*) were difficult to classify and not genetically 'pure' species, even allowing for the natural heterozygosity normally found in potato species. It would seem possible that the lack of natural barriers between the two species in the high-altitude zones of Jujuy and Salta might be the reason for this, and that introgression of genes from each into the other is taking place in that region.

We shall now discuss the hybrids of *S. microdontum* and *S. vernei* by reference to the subspecies.

(*b*) Hybrids of subsp. ***microdontum*** × ***S. vernei*** subsp. ***ballsii***

These have been found in prov. Jujuy, dept. Tilcara and in prov. Salta, dept. Santa Victoria, mostly in company with the parents (see Table 20). The hybrids may be recognized by the more dissected leaf and range of flower colour, from violet through lilac to white; some specimens possess a rather dark corolla and leaves similar to *S. microdontum* (see *Balls 6041*, Plate 106), and were previously named *S. microdontum* f. *purpureum*, *nom. nud.*, whilst in others the typical leaf of subsp. *ballsii* is apparent. Probably not only F_1 but also F_2 and backcross combinations may be encountered.

TABLE 20

Natural hybrids between S. microdontum *and other species in Argentina*

Locality	Collection number and altitude		
	S. microdontum subsp. *microdontum*	Hybrid	*S. vernei* subsp. *ballsii*
Jujuy, Tilcara, Chorro Valley	*Balls 6040.* ,, *5042* (p.p.). 2450–3200 m	*Balls 6041.* 2450–3200 m *P. & H. P–352.* 2850 m	*Balls 6042* (p.p.) 2450–3200 m *Balls 6043.* 3200 m *P. & H. P–351.* 2850 m
Salta, Santa Victoria, La Huerta	*H. & R. 240, 241.* 2700 m *Sleumer 4039.* 2700 m	*H. & R. 243.* 2700 m	(Not observed, but *H. & R. 253* was from Lizoite, not far away)
	S. microdontum subsp. *gigantophyllum*	Hybrid	*S. vernei* subsp. *vernei* and intermediates
Salta, Chicoana, above Escoipe	*H.P.R. 308.* 2350 m	*H.P.R. 306.* 2200 m	(Known only from above 2750 m. Intermediate form)
Tucumán, Trancas, Cuesta Grande del Chorro	*Schreiter 4843.* 3000 m	*Schreiter 4840* (p.p.). 3000 m	*Schreiter 4840* (p.p.). *Schreiter 5749.* 3000 m.
Catamarca, Andalgalá, Estancia 'Yunka Suma'	*P. & H. P–26.* 1450–2350 m *P. & H. P–27.* 2000 m	*P. & H. P–33.* 2050–2200 m	*P. & H. P–25.* 1950–2450 m *P. & H. P–34.* 2800 m
	S. microdontum subsp. *intermediate*	Hybrid	*S. vernei* subsp. *ballsii*
Jujuy, Tumbaya, Volcán	*Venturi 9529.* 2700 m	*Sleumer 3567.* *c.* 2850 m	*Castillón 247.* 3000 m *Sleumer 3568.* *c.* 2850 m
	S. microdontum subsp. *gigantophyllum*	Hybrid	*S. spegazzinii* or *S. vernei*
Catamarca, Andalgalá, El Candado	*Jörgensen 1311* (p.p.). 2700 m	*Jörgensen 1311* (p.p.). 2700 m	*S. spegazzinii*: *Jörgensen 1311* (p.p.). 2700 m (*S. vernei* was collected at Río Pisavil, not far away. *P. & H. P–39.* 2250–2500 m)

(*c*) Hybrids of subsp. ***gigantophyllum*** × ***S. vernei*** subsp. ***vernei*** (and intermediates)

These hybrids have been found in the provinces of Salta (dept. Chicoana), Tucumán (depts. Tafí, Trancas), and Catamarca (dept. Andalgalá), where the distribution areas of the two taxa under consideration overlap. In many cases here also the two parental forms have been collected with the hybrids (see Table 20). On the whole, these hybrids tend to be rather more robust than those in category (*b*), often possessing thick stems, crisped stem wings, longer leaves, and well-marked petiolules, which are indicative of the more robust subsp. *gigantophyllum*, involved in the cross. All types of flower colour from purple to white, and of leaf dissection from simple to compound, can be seen in the hybrids, in the same way as has already been described for the hybrids of subsp. *microdontum* with subsp. *ballsii*. (See Plate 108.) It should be noted that the Salta material of *S. vernei* is rather intermediate between the two subspecies.

(*d*) Hybrid of subsp. intermediate × ***S. vernei*** subsp. ***ballsii***

This combination is rare, and we have seen only one apparently authentic collection (*Sleumer 3567*), from Jujuy, dept. Tumbaya (see Table 20). Here again, both parents are known from the same region. The stem is very robust in this specimen and the leaflets show the typical basiscopic lobe of subsp. *ballsii*. Although the flowers are white it is probable that a larger collection of plants from the same area would reveal a range of colour types similar to those established in (*b*) and (*c*) above. (See Appendix III.)

(*e*) Hybrid of subsp. ***gigantophyllum*** × ***S. vernei*** or ***S. spegazzinii***

Amongst the material of subsp. *gigantophyllum* collected by P. Jörgensen at El Candado, Andalgalá, prov. Catamarca, were three sheets of a hybrid with a blue-flowered species. The collection notes refer to a mixture of blue- and white-flowered forms with 'large tubers'. We cannot be certain whether *S. vernei* or *S. spegazzinii* was the other parent of this hybrid, though from the blue flowers and large tubers mentioned in the collection note it is perhaps more likely to have been *S. vernei*.

S. spegazzinii is normally found in very much drier regions than *S. microdontum* and *S. vernei*. The locality of Jörgensen's hybrid, El Candado, lies on the southern flanks of the Aconquija range, half-way between the humid eastern slopes and the arid western ones. Since part of Jörgensen's collection from the same locality has been identified by us as *S. spegazzinii* (see Table 20) it seems probable that this species, and not *S. vernei*, is the other parent. It is to be hoped that further collections and observations will clarify this problem.

(*f*) Hybrids of subsp. ***gigantophyllum*** × ***S. kurtzianum*** (= *S.* × *rechei* Hawkes & Hjerting. *Scott. Pl. Breed. Stn. Rec.* 146, 1963)

This hybrid combination is very localized, since the two parent species overlap only in a very restricted area, as can be seen by comparing their distribution ranges in Maps 10 and 13. We have seen that *S. microdontum* is to be found in areas of high rainfall, whilst *S. kurtzianum* is a plant of dry Monte, Prepuna, and other rather low rainfall phytogeographical provinces. The hybrid between these species has been collected in the area of Chilecito and Guanchín Viejo at the southern end of Sierra Famatina in La Rioja province. (See Plate 109.) Although the two parent species occur also in fairly close proximity in the Pozo de Piedra area, Catamarca province (see Appendix III), hybrids between them have not yet been seen there.

Several collections of this hybrid, *S.* × *rechei*, were made by Castellanos and by Reche and were known to us prior to the 1966 expedition (see Plates 109 and 110). Since we had no idea that this hybrid combination was possible in nature or that subsp.

gigantophyllum occurred so far south or could come into contact with *S. kurtzianum* because of their very differing growth requirements, we assumed that the Castellanos and Reche collections represented a new species. This we named in honour of the latter collector, since he had made various collections of it, including living ones. Curiously, even the living material seemed to be rather uniform, possibly because two collections may have been simple F_1 hybrids. Another collection varied rather more, however. Our 1966 collections (noted and discussed in Appendix III) made it clear, however, that *S.* × *rechei* was a hybrid of *S. microdontum* subsp. *gigantophyllum* × *S. kurtzianum* since we found it in company with the two parents in most cases, and it seemed to show quite a degree of hybrid sterility. The hybrid always occurred in the zone of contact between the dry prepuna zone (where *S. kurtzianum* grew) and the wetter grasslands ('pajonales') of slightly higher altitude, occupied by subsp. *gigantophyllum*.

LOCALITIES OF NATURAL HYBRIDS

(*a*) With ***S. chacoense, S. tarijense,*** and ***S. megistacrolobum.*** (See pp. 237, 246, and 297.)

(*b*) Hybrids of subsp. ***microdontum*** × ***S. vernei*** subsp. ***ballsii***

Prov. JUJUY:

Dept. Tilcara, near Tilcara, Chorro Valley. Alt. 2450–3350[1] m. 14 Feb. 1939. *Balls 6041* (CPC: 108) (K, P.G.S.). (*S. microdontum* Bitt. f. '*purpureum*' Hawkes *nom. nud.*) See Plate 106.

Dept. Tilcara, Quebrada del Chorro. Alt. 2850 m. 24 Mar. 1952. *Petersen & Hjerting P–352* (J.G.H.).

Prov. SALTA:

Dept. Santa Victoria, entre Abra Delgada y Abra de San José. Alt. 3200 m. 11 Feb. 1956. *Hjerting & Rahn 178* (J.G.H., J.P.H.).

Dept. Santa Victoria, La Huerta (cerca de Santa Victoria). Alt. 2700 m. 16 Feb. 1956. 'Flor blanca.' *Hjerting & Rahn 243* (J.G.H., J.P.H.). (Long, oblong leaflets.)

(*c*) Hybrids of subsp. ***gigantophyllum*** × ***S. vernei*** subsp. ***vernei*** and intermediates

Prov. SALTA:

Dept. Chicoana, 3 km arriba de Escoipe, cerca de la Hostería. Alt. 2200 m. 22 Feb. 1956. 'Flor suavemente lila y blanca.' *Hjerting, Petersen, & Rahn 306* (J.G.H., J.P.H.). (Intermediate form of *S. vernei* probably involved.)

Prov. TUCUMÁN:

Dept. Tafí, Valle de Tafí. 1908. *Bruch* (s.n.) (p.p.) (LP).

Dept. Trancas, Cuesta Grande del Chorro. Alt. 3000 m. 1 Mar. 1927. 'Flor azul.' *Schreiter 4840* (p.p.) (LIL).

Prov. CATAMARCA:

Dept. Andalgalá, Estancia 'Yunka Suma', entre Casa de Piedras y Los Queñoales. Alt. 2050–2200 m. 21 Jan. 1949. 'Tal vez una mutación; la flor lila clara, tallo alado.' *Petersen & Hjerting P–33* (J.G.H., J.P.H.). See Plate 108.

Dept. Andalgalá, Arroyo Charquiadero, al borde del río. Alt. 2200 m. 12 Feb. 1949. 'Flor lila, tubérculos pequeños.' *Petersen & Hjerting P–42* (J.P.H.).

[1] Altitude should probably be less than the figure stated, not more than 3200 m.

(*d*) Hybrid of subsp. intermediate × ***S. vernei*** subsp. ***ballsii***

Prov. JUJUY:

Dept. Tumbaya, subida a la Abra Grande de Volcán. Alt. *c.* 2850 m. 23 Jan. 1953. 'Fl. alb.' *Sleumer 3567* (LIL, J.G.H.).

(*e*) Hybrid of subsp. ***gigantophyllum*** × ***S. vernei*** or ***S. spegazzinii***

Prov. CATAMARCA:

Dept. Andalgalá, El Candado. Alt. 2700 m. 2 Feb. 1916. 'Planta alta robusta, de flores azules o blancas; papas grandes, venenosas, abundantes.' *Jörgensen 1311* (p.p.) (MVM).

(*f*) Hybrids of subsp. ***gigantophyllum*** × ***S. kurtzianum***

Prov. LA RIOJA:

Dept. Chilecito, Sierra Famatina, Guanchín Viejo. 25 Jan. 1928. *Castellanos* (s.n.) (BA 28/345 two sheets. Formerly designated as Cotype, now redesignated as HOLOTYPE and ISOTYPE for *S. rechei* Hawkes & Hjerting). See Plate 109.

Dept. Chilecito [road from Chilecito to Mina de Oro], Las Talas. 21 Apr. 1960. *Reche* (s.n.) (E.B.S. 2082—accession number in German collection. Max.). See Plate 110.

Dept. Chilecito, La Puntilla. 26 Apr. 1960. *Reche* (s.n.) (E.B.S. 2083—accession number in German collection. Max.).

[Dept. Chilecito?] Famatina. Apr. 1960. *Reche* (s.n.) (Accession H. 614 in Birmingham as seed, formerly designated as Cotype. Some segregation occurs when seedlings of this line are compared.)

Note: Further collections noted in Appendix III.

12. SOLANUM VENTURII Hawkes et Hjerting

(*Phyton, Graz.* **9**, 140–4, 1960)

Plant herbaceous, small and delicate, bearing underground stolons with small globular tubers about 5–10 mm diam.

Stem *decumbent*, *weak*, *branched*, generally 10–30(–60) cm long, occasionally shorter, *slender*, 1·5–3·0 mm diam., *unwinged*, sparsely pubescent.

Leaf thin, frequently eaten by insects, often with purple veins; lateral leaflets (0–)1–3-jugate, much smaller than the terminal, up to 20(–70) mm long, *decreasing in size rapidly towards the base of the leaf*, ovate, cordate or orbicular, sessile or shortly (0–3(–8)mm) petiolulate, apically obtuse or acute; terminal leaflet orbicular, broad-ovate, elliptic or elliptic-oblong, apically acute to acuminate, basally cordate, truncate or broadly cuneate (3·5–)6–10(–13) cm long × (2–)4–6(–10) cm broad; interjected leaflets often absent or 1–2-jugate, ovate to orbicular, up to 10 mm long, sessile; pubescence on upper and lower surfaces of *very sparse short 3(–4)-celled triangular white to purple hairs*; margins strongly denticulate; pseudostipular leaflets generally well developed, semi-lunate, 0·5–1 cm long.

Inflorescence lax (3–)6–18-flowered; peduncle once or twice forked, *short below the fork* (0·5–5 cm), 3–5 cm long above the fork; pedicels about (1–)2–2·5 cm long in the flowering state, articulated more or less in the centre though sometimes slightly below or up to two-thirds the length from the base; peduncles and pedicels *glabrous*, or with very sparse, short, few-celled hairs similar to those on the leaf.

Calyx broadly campanulate, 5–7 mm long, with short lobes and well-defined acumens which vary a little in length in the same flower; lobes and acumens 3–6 mm long, the acumens themselves 2–4 mm long; pubescence of very sparse to moderately frequent *short few-celled transparent triangular hairs.*

Corolla white, pentagonal to rotate, 1·75–3(–3·5) cm diam.; lobes rather narrow, about 8–10(–15) mm broad and 8–10(–12) mm long, including the sometimes not very well-defined acumens.

Anthers 5–7 × 1·75–2 mm; filaments 1–2 mm long.

Style 8–11 mm long, generally *gradually swollen above,* with *quite long dense papillae on the lower third*; stigma capitate to conical, slightly thicker than the swollen style apex, entire or barely emarginate.

Berries more or less globose or slightly longer than broad, small, about 1 cm diam. Chromosome number $2n = 24$ (36). (See also Appendix III.)

AFFINITIES

S. venturii is apparently closely related to *S. microdontum,* agreeing with it in the enlarged terminal and reduced lateral leaflets, the general form of the calyx and the form and colour of the corolla. It differs from it in the different ecological adaptation, weak decumbent habit, small sparse tubers, thin branched unwinged stem, sparse pubescence of triangular 3(–4)-celled hairs on all green parts, shorter peduncle below the fork, smaller corolla, and dense quite long papillae on the lowest third of the style. It possibly represents a recently derived endemic separation from the ancestral stock of *S. microdontum.*

SPECIMENS EXAMINED

ARGENTINA

Prov. JUJUY:

Dept. Valle Grande, Serranía de Calilegua, Tolditos. Alt. 2600 m. 15–17 Feb. 1965. 'Flor blanca.' *H. A. Fabris 5931* (LIL, LP).

Prov. SALTA:

(See Appendix III.)

Prov. TUCUMÁN:

Dept. Chicligasta, El Bolsón a La Cascada. Alt. 2100 m. 9 Mar. 1949. *T. Meyer 14868* (LIL).

Dept. Chicligasta, Quebrada Las Pavas, de Puesto Bolsón hasta Puesto La Cascada. Alt. 2200–2550 m. 6–7 Mar. 1949. 'En lugares húmedos. Hojas a veces con nervaduras rojizas. Tallo no alado, tubérculos muy chicos.' *Petersen & Hjerting P–54* (J.G.H., J.P.H.). (Perhaps some introgression from *S. microdontum* subsp. *gigantophyllum.*)

Dept. Chicligasta, Estancia Las Pavas, Puesto La Cascada. Alt. 2600 m. 17 Mar. 1924. Fl. blanca; entre los pastos.' *Venturi 3271* (GH, LIL, US).

Dept. Chicligasta, Estancia Santa Rosa. Alt. 3600 m. 6 Jan. 1927. 'Fl. blanca. Falda del cerro.' *Venturi 4727* (p.p.) (BA, GH, LIL, LP, SI, UC, US). (Altitude probably incorrect. Probably all specimens show introgression with *S. sanctae-rosae.*)

Dept. Tafí, Cumbres de San José, La Queñoa. Alt. 2800 m. Mar. 1933. 'Flor blanca.' *R. Diaz 9583* (LIL).

Dept. Tafí, La Ciénaga. 14 Feb. 1905. 'Lugares húmedos. Corola blanca.' *Lillo 4028* (LIL).

Dept. Tafí, Sala Chasquivil. 14 Jan. 1945. 'Flor blanca.' *Olea 270* (LIL).

Dept. Tafí, Estancia San José (de Chasquivil). Alt. 2150–2200 m. 4 Jan. 1953. *Petersen & Hjerting 898a* (K-Holotype, C, LIL, J.G.H., J.P.H.). See Plate 111.

Same collection data but altitude 2650 m. *Petersen & Hjerting 898b* (J.G.H., J.P.H.).

Dept. Tafí, Quebrada de La Junta. Alt. 1500 m. 21 Nov. 1921. 'Flor blanca.' *Schreiter 1917* (LIL). (Altitude probably incorrect.)

Dept. Tafí, Tafí del Valle, Cerro Pelado. Alt. 2200 m. 24 Jan. 1950. 'Fl. alb.' *Sleumer 151* (LIL).

Dept. Tafí, La Ciénaga, cerca de la casa grande. Alt. 2500 m. 27 Jan. 1950. 'Fl. alb.' *Sleumer 300* (LIL).

Dept. Tafí, San José. Alt. 2100 m. 17 Feb. 1949. *Sparre 5835* (LIL).

Dept. Tafí, Cerro San José. Alt. 2700 m. 10 Feb. 1925. 'En los prados.' *Venturi 3565* (BA, LIL, SI, US). (Probably some introgression from *S. sanctae-rosae*.)

Prov. CATAMARCA:

Dept. Andalgalá, Estancia 'Yunka Suma', entre Puesto Vallecito y Puesto Chapina. Alt. 2150–2250 m. 16–24 Jan. 1949. 'En un lugar húmedo, en pasto alto. Flor blanca, hojas glabras, tallo glabro, sin alas.' *Petersen & Hjerting P–28* (C, J.G.H., J.P.H.).

Dept. Andalgalá, Estancia 'Yunka Suma', entre Los Queñoales y Mesadas de Vallecito. Alt. 2350–2800 m. 16–23 Jan. 1949. 'Prefiere lugares húmedos. Tallo y hojas glabras. Sin alas.' *Petersen & Hjerting P–30* (LIL, J.G.H., J.P.H.). See Plate 112.

Dept. Andalgalá, Río Pisavil, cerca del río. Alt. 2250–2500 m. 10 Feb. 1949. 'En pasto.' *Petersen & Hjerting P–38* (J.G.H., J.P.H.).

EXACT LOCALITY UNKNOWN:

Surroundings of Tucumán. *Brücher* (s.n.) E.B.S. 457. Herbarium specimen made by Professor H. Ross in Germany (J.G.H.).

DISTRIBUTION

S. venturii is one of the few species of tuber-bearing Solanums endemic to the Argentine republic, having been found in provinces Jujuy (dept. Valle Grande), Salta[1] (dept. Chicoana), Tucumán (depts. Chicligasta and Tafí), and Catamarca (dept. Andalgalá). (See Map 8.)

HABITAT

S. venturii is a typical species of very humid grasslands at altitudes of (1500[2]–)2000–2800 m, perhaps sometimes higher. It is never to be found in dense shade, always amongst high grass and occasionally in the partial shade of bushes or trees. The restricted altitudinal and ecological range of this species and its superficial similarity to *S. microdontum* has caused it to be overlooked until recently. It occupies approximately the upper part of the Montano district ('pajonales') of the Western Subtropical vegetational province according to Cabrera's phytogeographical scheme (1953) or the region of Aliso, Queñoa, and alpine meadows of Lorentz (1876), Holmberg (1898), and Lillo (1919). The growth period is from November to March, perhaps beginning even earlier, in October.

It is often found growing near to *S. vernei* and *S. microdontum* subsp. *gigantophyllum* but does not normally seem to form hybrids with them; in the upper part of its range it comes into contact with *S. sanctae-rosae* with which it appears to hybridize freely (see pp. 309–10).

[1] See Appendix III for collection details of Salta material.

[2] The single record for 1500 m may be an error (see collection of Schreiter, *No. 1917*).

TAXONOMIC NOTES

In 1952, when the material so generously sent on loan by the Instituto Miguel Lillo of Tucumán was being revised for the first time, two collections by Venturi (*Nos. 3271* and *3565*) seemed to be specifically distinct from all other forms hitherto described in the Aconquija mountains of Tucumán and Catamarca. The name *S. venturii* was provisionally applied to them though no description was published at that time. These were low-growing semi-rosette forms with flowers that varied from white to blue. Later, however, when the more numerous collections of Petersen and Hjerting and of Sleumer became available to us it was realized that these two collections of Venturi were unfortunately not typical of the species but were extreme forms with the low habit and blue flowers due to introgression from *S. sanctae-rosae*. It was therefore decided not to designate *Venturi 3271* as the type, although this had been the original intention, but to select a specimen that showed the normal growth of this species and had no obvious introgression of genes from any other species. Accordingly, we have designated the Kew specimen of *Petersen & Hjerting 898a* as the Holotype for *S. venturii*. (See Plate 111—Isotype.)

For some time the differences between *S. venturii* and *S. microdontum* subsp. *microdontum* were thought perhaps to be of little significance, and it was not until we were able to obtain abundant material of the latter species from Santa Victoria that we were quite certain of their specific distinctness. We began, indeed, by considering *S. venturii* as a high-altitude southern subspecies of *S. microdontum* (= *S. simplicifolium*); we later thought it might be equivalent to subsp. *microdontum* from Santa Victoria, and ended by being convinced that it was indeed a good species in its own right. The very characteristic habit, loose inflorescence, and hair type served finally to distinguish *S. venturii* from *S. microdontum*, though the two species are most certainly closely related.

Brücher (1954) makes the understandable mistake of equating *S. venturii* with *S. microdontum* subsp. *microdontum*, having only seen an isotype specimen of Fiebrig's original collection of *S. microdontum* from Salta, Sta. Victoria (Toldos). Nevertheless, one is left in doubt as to the material Brücher had in mind, since he mentions the 'easily visible loose pubescence, above all on the leaf surface, with relatively large (to 1 mm), thick and brightly shining hairs'.[1] This feature is definitely *not* characteristic of *S. venturii*, but of *S. microdontum*, and unless we can see Brücher's original specimens from Aconquija we cannot say with certainty whether they should be placed with *S. venturii* or should be considered as small stunted forms of *S. microdontum* subsp. *gigantophyllum*. It is quite certain that they cannot belong to *S. microdontum* subsp. *microdontum*, since this subspecies is never to be found so far south. In spite of Brücher's published statements, however, it is certain that he did make collections of the genuine *S. venturii* since we have been able to see dried material of a living collection (EBS 457) sent by him to Professor H. Ross at the Max Planck Institute at Voldagsen, West Germany (now transferred to Cologne). This gathering, of which Professor Ross kindly sent us dried leaves and flowers and photographs of living plants, is undoubtedly *S. venturii*, though unfortunately Brücher gave no exact locality information to Ross beyond the general phrase 'surroundings of Tucumán'.

Spontaneous hybrids of '*S. microdontum*' × '*S. simplicifolium*' (= *S. microdontum* subsp. *gigantophyllum*) are mentioned by Brücher as occurring in the upper *Polylepis* region, and marked by a sturdier growth than what he considers to be the 'typical *S. microdontum*'. We ourselves have not observed hybrids between *S. venturii* and

[1] 'Ein charakteristisches Merkmal, das selbst bei noch so wechselndem Habitus in meinen Sammlungen stets auftritt, ist eine deutliche lockere Behaarung der Pflanze, vor allem der Blattspreiten, mit relativ grossen (bis 1 mm) dicken und hellglänzenden Haaren.'

S. microdontum subsp. *gigantophyllum*, and we consider it likely that his so-called hybrids may in fact be the more typical *S. venturii*, or hybrids with *S. sanctae-rosae*.

Another set of hybrids mentioned by Brücher[1] is considered by him to be derived from crosses between *S. microdontum* and *S. vernei*. One of these (*Meyer 14948*—Herb. Lillo 2445731) is a hybrid of *S. venturii* × *S. sanctae-rosae*; another (*Sparre 5895*—Herb. Lillo 244683) is a hybrid of *S. venturii* × *S. vernei*; a further collection (*Sparre 5976*—Herb. Lillo 244503) we consider to be a luxuriant growth form of *S. sanctae-rosae*; we have been unable to trace the collector or number of the fourth collection mentioned—Herb. Lillo 382491.

A further error made by Brücher (1954) is that he states that Hawkes used the name *S. venturii* for Brücher's '*S. catamarcae*' (probably = *S. sanctae-rosae*). We are unable to say on what grounds he bases this statement since *S. venturii* was not published until 1960, some six years later.

Until 1966 we thought that *S. venturii* was wholly confined to the Aconquija mountains of Catamarca and Tucumán provinces. During our recent expedition, however, we discovered it also, growing quite abundantly, in the area of Cuesta del Obispo in Salta province. We also were surprised to see a collection from Jujuy, made by Fabris (*5931*), during our recent visit to Argentina. It seems probable, therefore, that *S. venturii* might be found in other regions of the north-west and perhaps spreads further north than we know it at present.

As we have said before, it is a species that has been overlooked because of its close resemblance to *S. microdontum*. Living collections are scarce because of its very small tubers, its tendency to grow in remote and inaccessible places, and the confusion with *S. microdontum* that we have just mentioned. Its potentialities as a parent for breeding for disease resistance have thus not been very much investigated.

NOTES ON LIVING PLANTS

We ourselves have not yet cultivated *S. venturii* in Europe though we now have living material available from the 1966 expedition. The living collections made by Petersen and Hjerting previously were found to be either *S. sanctae-rosae* or hybrids of *S. venturii* × *S. sanctae-rosae*. When cultivated by Hjerting at the experimental station 'El Suncho' of the University of Tucumán it retained its characteristic habit and remained quite distinct from *S. microdontum*. The collection sent by Brücher to Ross (EBS 457), to which we have already referred, was kept under observation for a number of years, and we are indebted to Professor Ross for the following information (see also Goffart & Ross 1954, under '*S. microdontum*'): 'The plants were not particularly vigorous and the tuber production was weak; the terminal leaflet was always large, sometimes without side leaflets and sometimes with one or two pairs. Pedicels were about 2 cm long, articulation median and calyx sparsely hairy.'

The dimensions of leaf and flower and the characteristic pubescence on the specimens sent us by Ross come well within the range mentioned in our description given above, which is made from wild material from Argentina. In the form of the corolla this species is seen to be very similar to *S. microdontum*, the lobes being steeply sloping though fairly distinguishable from the acumens.

Although Ross mentions in his correspondence to us that *S. venturii* is susceptible to *Phytophthora*, Brücher (1954) notes that it is not attacked in the field under natural conditions in Argentina. Brücher also mentions that the growing season is short and that the tubers are small, 1 cm diam., 1–3 grammes in weight, and with a tender skin.

[1] Unfortunately no collector or number is given, only the Herb. Lillo registration number—a practice to be deprecated, since it makes it extremely difficult for the information to be checked by other workers.

We do not include here other observations made by Brücher, since, as we have already stated, there is some doubt as to whether all his observations were made on genuine *S. venturii*. Resistance to eelworm has been recorded for this species (see Chapter 5).

CYTOLOGY

Goffart and Ross (1954), Brücher (1954), and Ross (in correspondence) all give a chromosome number of $2n = 24$ for *S. venturii*. Although some of Brücher's counts may have been made on material that did not belong to *S. venturii*, there is no doubt that the count mentioned by Ross for EBS 457 was made on the genuine species. It is fairly safe to assume, therefore, that *S. venturii* is a diploid species, as are so many of the other wild Argentine potatoes. Furthermore, a hybrid of *S. sanctae-rosae* ($2n = 24$) × *S. venturii* (*Petersen & Hjerting P–336*, CPC: 2463) which was grown in Cambridge for several years also had twenty-four chromosomes in root-tip mitoses.[1]

FERTILITY AND CROSSABILITY

Very few data are available on the self- or cross-fertility of this species, but from a knowledge of the frequency with which it forms hybrids in nature we consider that it is likely to cross quite readily with a number of other diploid species.

Ross (in correspondence) states that natural hybrids of *S. venturii* and *S. sanctae-rosae* were found in the experimental field in Germany and that (see Goffart & Ross 1954) he had obtained a hybrid between *S. venturii* and *S. stoloniferum*. Beyond this, however, we have no other information on the crossability of this species or the fertility of hybrids between it and other wild or cultivated potatoes.

NATURAL HYBRIDS

(*a*) Hybrids between ***S. venturii*** and ***S. sanctae-rosae*** have already been described (see p. 309) occurring in the zone of overlap between the two species (2500–2800 m) and also extending 400 m higher than the pure *S. venturii* (–3200 m), and 100 m lower (to 2400 m) than pure *S. sanctae-rosae*. We must, therefore, assume that the hybrids collected from above the upper altitude limit of *S. venturii* (2800 m) are not simple F_1 or F_2 segregates but are possible backcrosses with *S. sanctae-rosae* or represent *S. sanctae-rosae* with introgression of *S. venturii* genes. In fact, these high-altitude collections show a much greater similarity to *S. sanctae-rosae* than the lower-altitude hybrids, which tend more towards the *S. venturii* habit and leaf type. The only one we have had an opportunity of growing was *Petersen and Hjerting P–336* (see Plate 94) from 2700 m altitude; this was from the zone of overlap of the two species and showed about equal influence from each. Hybrid collections in the living state from altitudes of 2800–3200 m would be very valuable in demonstrating the amount of influence of *S. venturii* in this region. Further collections might also show that *S. venturii* extended into the 2800–3200 m zone in a pure state or, on the other hand, was only to be found introgressed with *S. sanctae-rosae* in that particular zone.

So far as we are able to ascertain, despite Brücher's statement referred to above (p. 352), *S. venturii* does not seem to form natural hybrids with *S. microdontum*, although *S. venturii* (alt. range 2000–2800 m) completely coincides in its altitude range with part of the range of *S. microdontum* (1000–3350 m). The probable reason for this is that the two species, although occurring in the same region of the high rainfall eastern-facing Aconquija slopes, are to be found in different habitats. *S. venturii* is a plant of open grassland or partial shade of bushes, whereas *S. microdontum* is adapted to the denser shade of forests. Nevertheless, in the upper part of its range *S. microdontum* grows in the open, since trees are here very sparse (*Polylepis* zone)

[1] Triploid forms are also known for this species (see Appendix III).

and do not afford dense shade. Another possible explanation of the lack of natural hybrids might be that there was some genetical barrier to the hybridization of the two species or that F_1 hybrids between them were unthrifty, weak, and infertile. It is to be hoped that Argentine botanists may be able to make observations and collections of *S. venturii* and *S. microdontum* in the 2000–2800 m region where the former species occurs, in order to solve this problem.

(*b*) Probable hybrids with ***S. vernei.***

S. vernei, in contradistinction to *S. microdontum*, is a species of rather less shaded places and it therefore comes into contact with *S. venturii* over a certain part of its range (2000–2800 m). Because of this altitudinal and habitat coincidence one might expect that hybrids would occur quite frequently. In point of fact, however, they do not, or at least we have not been able to discover them. This may possibly be due to the earlier flowering of *S. vernei* to which we have already called attention in the section on *S. microdontum*.

Only three probable hybrids of *S. venturii* × *S. vernei* have been seen by us, namely *Petersen & Hjerting P–55b*, *Meyer 14881*, and *Sparre 5895*.

Table 21 shows that one or both of the parental species has been found in the same region.

TABLE 21

Probable natural hybrids between S. venturii *and* S. vernei

Locality	Collection number and altitude		
	S. venturii	*Hybrid*	*S. vernei*
Tucumán, Chicligasta, Quebrada Las Pavas	*P. & H. P–54.* 2200–2550 m *Venturi 3271.* 2600 m	*P. & H. P–55b.* 2800 m *Meyer 14881.* 2800 m	*P. & H. P–55a.* 3100 m *Venturi 3082, 4685.* 3200 m
Tucumán, Tafí, San José	*Venturi 3565.* 2700 m *R. Díaz 9583.* 2800 m	*Sparre 5895.* 2700 m	

Petersen and Hjerting's collection *55b* possesses a short leaf and rapidly diminishing laterals but no flowers. Although it might possibly be a hybrid of *S. vernei* (or *S. sanctae-rosae*) × *S. microdontum* we consider, on the whole, that *S. venturii* is more likely to be the other parent. It will be remembered that *S. sanctae-rosae* and the hybrid of *S. sanctae-rosae* × *S. venturii* have also been collected in this area (see pp. 309–10). One cannot exclude the possibility therefore that more than two species have been involved in some of the hybrids found in the area of Quebrada Las Pavas and similar regions.

Meyers' collection *14881* is from virtually the same place; it possesses a 3-jugate leaf with rapidly diminishing laterals and especially large broad obovate terminal leaflets. There are up to five pairs of small interjected leaflets and the flowers are large and purple in colour. This collection is much closer to *S. vernei* and might be an F_2 segregate or backcross. It resembles *S. venturii* in the small delicate leaf hairs and large broad terminal leaflet.

Sparre's collection is much more like a simple F_1 hybrid; it resembles *S. venturi* in leaf shape but possesses the large purple flowers of *S. vernei*. The pubescence is fairly intermediate between the two possible parents.

LOCALITIES OF NATURAL HYBRIDS

(*a*) With ***S. sanctae-rosae*** (see pp. 311–12)

(*b*) Probable hybrids with ***S. vernei***

Prov. TUCUMAN:

Dept. Chicligasta, La Cascada a Las Cuevas. Alt. 2800 m. 9 Mar. 1949. 'Flor morada.' *T. Meyer 14881* (LIL).

Dept. Chicligasta, Quebrada Las Pavas, entre La Cascada y Las Cuevas. Alt. 2800 m. 8 Mar. 1949. *Petersen & Hjerting P–55b* (J.G.H.).

Dept. Tafí, San José, La Banderita. Alt. 2700 m. 18 Feb. 1949. *Sparre 5895* (LIL).

13. SOLANUM VERNEI Bitter et Wittmack

(in Engler, *Bot. Jb.* **50**, Suppl., 550–1, 1914)

Herbaceous, tuber-bearing, with long rather shallow stolons which are often purple in colour. Tubers 1–10 (or more) per plant, *large*, up to 8 × 10 cm, from 8–70(–168) grammes in weight, round or elongated, somewhat flattened, yellow to brown or dark purple and rather verrucose.

Stem (15–)50–100(–200) cm tall, 5–10(–20) mm diam., *purple mottled, generally winged*, moderately to densely pubescent; wings 1–5 mm, straight or crisped, often deeply pigmented with purple anthocyanin.

Leaf large, 7–40 cm long × 5–18·5 cm broad; lateral leaflets (3–)4–5(–7)-jugate, with (0–)2–10(–15) pairs of rotund, ovate or cordate sessile interjected leaflets up to 12 × 10 mm; laterals broad ovate-lanceolate to oblong-lanceolate, obovate or elliptic, oblique at the base, the apex obtuse to acute or acuminate, sessile or with a short 5(–15) mm long petiolule, the upper one or two pairs often narrowly decurrent; dimensions 3–10·5 × 1·8–4·8 cm; terminal often slightly larger than upper laterals but similar in shape to them, 3·8–11·5 × 2·2–6 cm; upper leaf surface dark to medium green, clothed with short adpressed sparse to frequent hairs not visible except with a lens; lower surface more or less *densely whitish-grey tomentose*, especially well marked in subsp. *ballsii*; pseudo-stipular leaflets generally well marked, semilunate, up to 25 mm long.

Inflorescences frequent, 2–15(–25)-flowered; peduncle once or twice-forked, 1·5–13 cm long; pedicels 1·5–3(–5) cm long, the articulation varying in position from $\frac{1}{3}$ to $\frac{2}{3}$ above the base but generally placed quite near the centre; peduncle and pedicels clothed with short often crisped hairs, rather sparsely in subsp. *vernei* and more densely in subsp. *ballsii.*

Calyx campanulate, 6–10 mm diam.; lobes lanceolate, 1–2·5 × 3–5 mm, including the 1–3 mm long acumens, the *outer surface dark purple*, with pubescence similar to that of peduncles; the *acumens often vary in length* in the same flower.

Corolla *large, showy, deep violet purple*, 2·5–4·5(–5) cm diam.; lobes 8–12 mm long including acumen × 8–18 mm broad at the base; *acumen large*, to 5 mm long × 10 mm broad, not well demarcated from the generally steeply sloping lobes, thus giving the corolla a *semi-stellate* appearance.

Anthers 5·5–7 × 2–2·5 mm; filaments 1–3 mm long.

Style 10–13 mm long, papillose in the lower $\frac{1}{3}$–$\frac{1}{2}$; stigma subglobose to conical, up to 1 mm long, apically entire or notched.

Berries abundant, more or less globular or slightly cordate, to 2 cm diam., green or green with white slightly raised spots which are not visible in the dried state.

Chromosome number: $2n = 24$.

AFFINITIES

S. vernei is a large, handsome, easily recognizable species with large, dark purple stems and flowers. In its typical growth form it cannot very well be confused with any other of the north-west Argentine tuber-bearing Solanums.

It is perhaps fairly closely related to *S. microdontum*, from which it may be distinguished by the larger tubers, pubescence, greater degree of leaf dissection, and large deep violet-purple corolla. It agrees with that species, however, in the tall growth, winged stem, and unequal-lengthed calyx acumens. From *S. spegazzinii* it may be distinguished by the larger tubers, wider leaflets, pubescence type, and the shape, colour, and size of the corolla. To that species also it may in some way be related.

KEY TO SUBSPECIES

Lateral leaflets (except upper pair) clearly petiolulate; leaflet apex markedly acuminate, base gently rounded (provs. south Salta, Tucumán, Catamarca) Subsp. ***vernei***

Lateral leaflets sessile or very shortly petiolulate; leaflet apex obtuse to acute or barely acuminate, the base markedly auriculate on the basiscopic side (provs. Jujuy and north Salta) Subsp. ***ballsii***

Note: Intermediate forms occur sometimes in southern Salta province.

1. Subspecies *vernei*

Distinguished from subsp. *ballsii* by the (3–)4–5(–6)-jugate *petiolulate* lateral leaflets (except upper pair) which are ovate-lanceolate and *apically acuminate* with a cuneate to *gently rounded base on the basiscopic side* and often decurrent; the interjected leaflets are *fewer but longer, generally ovate to cordate and acute*; pubescence on under surface of leaf not so dense, though this varies with habitat; the flowers are larger (up to 4·5 cm) than in subsp. *ballsii* and on longer pedicels.

2. Subspecies *ballsii* (Hawkes) Hawkes et Hjerting, *comb. nov.*

(*Scott. Pl. Breed. Stn. Rec.* 149, 1963)

S. ballsii Hawkes (*Bull. imp. Bur. Pl. Breed. Genet., Camb.* 121, 1944).

Distinguished from subsp. *vernei* by the 3–4(–6)-jugate *generally sessile* broad ovate to elliptic-oblong lateral leaflets with *acute to obtuse apex* and *markedly auricled base on the basiscopic side*, never decurrent; leaflet margins typically *repand*; the interjected leaflets are *more frequent and orbicular to broad ovate or even reniform, and apically obtuse*; pubescence on under surface of leaf always very dense, greyish white, and felted; the pedicels not more than 2 cm long and the corolla not larger than 3 cm in diameter.

SPECIMENS EXAMINED

1. Subspecies *vernei*

ARGENTINA

Prov. TUCUMÁN:

Dept. Chicligasta, Quebrada Las Pavas, entre Las Cuevas y La Cascada. Alt. 3100 m. 8 Mar. 1949. 'Comúnmente bajo *Polylepis australis*; prefiere tierra humífera y sombra. Flor morada, tallo alado; tub. grandes.' *Petersen & Hjerting P–55a* (J.G.H.).

Dept. Chicligasta, Estancia Las Pavas, Puesto del Rayo. Alt. 3200 m. 12 Mar. 1924. 'Flor morada. Laderas del cerro.' *Venturi 3082* (GH, LIL, SI, US).

Dept. Chicligasta, Estancia Las Pavas. Alt. 3200 m. 10 Dec. 1926. 'Flor rosada. En los pastizales.' *Venturi 4685* (BA, GH, LIL, LP, S, SI, UC, US).

Dept. Río Chico, Escaba. Alt. 2100 m. 26 Dec. 1913. '50 cm alto.' *Monetti 1803* (LIL).

Dept. Tafí, Tafí del Valle, La Casita. Jan. 1912. *Castillón 2147* (LIL).

Dept. Tafí, Tafí del Valle, Cerro Muñoz, Los Corrales. Alt. 3000 m. 29 Jan. 1912. *Castillón 2564* (LIL).

Dept. Tafí, 81 km from Tucumán on way to Cafayate. 28 Feb. 1960. 'Plants up to 4 ft high, robust; fls. purplish. In open rocky grassy soil.' *Correll, Dodds, Brücher & Paxman A. 703* (K).

Dept. Tafí, La Ciénaga. Jan. 1943. *Descole 1519* (LIL).

Dept. Tafí, La Ciénaga. *R. Díaz* (s.n.) (LIL 101501).

Dept. Tafí, km 83, entre Infiernillo y Tafí del Valle, cerca de Quebrada Honda. Alt. 3000 m. 19 Jan. 1956. *Hjerting, Petersen & Rahn 1* (J.G.H., J.P.H.).

Dept. Tafí, Tafí del Valle, Río Infiernillo. Alt. 2400 m. 17 Dec. 1888. 'Barrancas.' *Lillo 1265* (LIL).

Dept. Tafí, entre Tafí del Valle y La Ciénaga. Alt. 2600 m. 18 Dec. 1888. 'Barrancas del arroyo; quebrada al E. del camino.' *Lillo* (s.n.) (LIL 62608).

Dept. Tafí, Tafí del Valle, La Quebradita. Alt. 2300 m. 14 Dec. 1908. 'Bajo alisos, lugares húmedos. Flor rojo-violácea. Yerba de 1 m. Tiene tubérculos.' *Lillo 8744* (LIL).

Dept. Tafí, La Ciénaga. Alt. 2500 m. 'Bosques de alisos.' *Lillo 8791* (LIL).

Dept. Tafí, Sierra de Tucumán. 1 Jan. 1872. *Lorentz* (s.n.) (CORD).

Dept. Tafí, pr. La Ciénaga. 10–17 Jan. 1874. *Lorentz & Hieronymus 708* (B—photo of now destroyed type, CORD—Lectotype, GOET). See Plate 113 and Fig. 35.

Dept. Tafí, Infiernillo, Río Tafí. Alt. 2800–3000 m. 29 Jan.–27 Feb. 1952. 'Flor morada.' *Petersen & Hjerting P–324* (C, J.G.H., J.P.H.).

Dept. Tafí, Río Tafí y afluentes (Infiernillo). Alt. 2800 m. 1 Jan. 1953. *Petersen & Hjerting 857* (J.G.H., J.P.H.).

Dept. Tafí, laderas, camino a Colalao del Valle. Alt. 2800 m. Dec. 1913. 'Flor azul.' *Rodríguez 1204* (BA, BAF, LIL, LP, SI).

Dept. Tafí, La Ciénaga, en el arroyo del Pabellón grande. Alt. 2720 m. 29 Jan. 1950. 'Fl. viol.' *Sleumer 198* (LIL).

Dept. Trancas, Casa de Piedras (Gualinchay). Alt. 2000 m. Dec. 1917. 'Fl. violeta obscura.' *Schreiter 632* (LIL).

Dept. Trancas, Cuesta Grande del Chorro. Alt. 3000 m. 1 Feb. 1927. 'Fl. azul.' *Schreiter 4840* (p.p.) (LIL).

Dept. Trancas, Cuesta occidental del Chorro. Alt. 3000 m. 1 Mar. 1927. 'Fl. azul obscura.' *Schreiter 5749* (LIL).

Prov. Tucumán, locality unknown. *Schreiter 9957* (LIL).

Prov. Tucumán, dept. unknown. Surroundings of Tucumán. *Brücher 50299* (CPC: 2414) and *Brücher 50305* (CPC: 2413) (both P.G.S.).

Prov. CATAMARCA:

Dept. Andalgalá, Río Pisavil. 10 Feb. 1949. *T. Meyer 14763* (LIL).

Dept. Andalgalá, Estancia 'Yunka Suma', entre Casa de Piedras y Los Queñoales. Alt. 1950–2450 m. 16–21 Jan. 1949. 'Generalmente bajo *Polylepis australis*, en lugares muy húmedos. Flor morada. Tubérculos hasta 5×6 cm, de color marrón.' *Petersen & Hjerting P–25* (C, LIL, J.G.H., J.P.H.). See Plates 114 and 115.

Dept. Andalgalá, Estancia 'Yunka Suma', Mesadas del Vallecito. Alt. 2800 m. 23 Jan. 1949. 'Hojas gruesas (el sol?). Flor morada.' *Petersen & Hjerting P–34* (J.G.H.).

Dept. Andalgalá, Río Pisavil, cerca del río. Alt. 2250–2500 m. 10 Feb. 1949. 'Bajo *Polylepis australis*, en tierra húmeda y muy buena. Flor morada.' *Petersen & Hjerting P–39* (C, J.G.H., J.P.H.). (Some plants with paler lilac flowers, showing introgression from *S. microdontum* subsp. *gigantophyllum*.)

Dept. Andalgalá, Rio Potrero. Alt. 2600 m. 13 Feb. 1942. *Rohmeder* (s.n.) (LIL 29310).

Dept. Andalgalá, Las Mesadas, Est. Exp. 'El Suncho'. (Alt. 1620 m). 16 Feb. 1950. 'Cult., proced. del Valle de Río Chacra, cerca de "Yunka Suma", *c.* 2150 m. Fl. viol.' *Sleumer 15* (LIL).

Dept. Andalgalá, Río Potrero sup., arriba de 'Queñoa Pelada'. Alt. 2750 m. 27 Feb. 1951. 'Solo sicco. Fl. viol.' *Sleumer 1830* (LIL, US).

Dept. Capayán, Las Angosturas, quebrada del arroyo. Alt. *c.* 2400 m. 29 Feb. 1952. 'En un faldeo de escombros, donde se cría en poblaciones puras; fl. azul aterciopelada; con tubérculos.' *Vervoorst 3478* (LIL, J.G.H.). (Leaflets rather narrow.)

2. Subspecies *ballsii*

ARGENTINA

Prov. JUJUY:

Dept. Tilcara, Chorro Valley, above San Antonio. Alt. 3500[1] m. 13 Feb. 1939. 'Steep shaded bank, soft loam among shrubs. No flowers; tubers to 4 cm long, light yellowish-brown.' *Balls 6038* (CPC: 104) (P.G.S., J.G.H.).

Dept. Tilcara, Chorro Valley, above San Antonio. Alt. 2450–3350[1] m. 14 Feb. 1939. 'In rich soil generally amongst other taller plants and often in shade. Often found with *Balls 6040* (*S. microdontum* subsp. *microdontum*). Plants to 30–50 cm. Stem stout, winged. No flowers. Tubers large, to 5 cm, light brown.' *Balls 6042* (p.p.) (CPC: 105) (P.G.S.—Lectotype, J.G.H.). See Fig. 36 and Plate 116.

Dept. Tilcara, Chorro Valley, above San Antonio. Alt. 3300[1] m. 14 Feb. 1939. 'Growing in small stony stream. Plants to 30 cm. Stem stout, flexuose, dark purple, very hairy. Not flowering. Tubers to 5 cm long, pale yellowish-brown.' *Balls 6043* (CPC: 112) (P.G.S., J.G.H.).

Dept. Tilcara, Chorro Valley. Alt. 3500[1] m. 13 Feb. 1939. 'In old fields, walls, etc., amongst herbage. Stem up to 45 cm. Flowers purple, to 3 cm. Berries globular, green, spotted white. Tubers to 5 cm long.' *Balls 6050* (CPC: 103) (P.G.S., J.G.H.).

Dept. Tilcara, Chorro Valley. Alt. 3500[1] m. 14 Feb. 1939. 'Tubers to 7 cm long.' *Balls 6053a* (p.p.) (CPC: 113 p.p.) (P.G.S., J.G.H.).

Dept. Tilcara, Quebrada del Chorro. Alt. 2850–3200 m. 24 Mar. 1952. *Petersen & Hjerting P–351* (LIL, J.G.H.). (Original herbarium collection was good subsp. *ballsii*, but some seedlings were *S. microdontum*.)

Dept. Tumbaya, Estación Volcán, Quebrada del Volcán. Alt. 3000 m. 19 Jan. 1927. 'Flor morada.' *Castillón 247* (LIL).

Dept. Tumbaya, Tumbaya, subida a la Abra de Tiraxi. Alt. 2780 m. 31 Dec. 1952. 'Fl. violac.' *Sleumer 3204* (LIL, J.G.H.).

Dept. Tumbaya, Abra Grande de Volcán. Alt. c. 2850 m. 23 Jan. 1953. 'Fl. violác.' *Sleumer 3568* (LIL, J.G.H.).

Prov. SALTA:

Dept. Santa Victoria, Lizoite, cerca del río. Alt. 3450 m. 17 Feb. 1956. *Hjerting & Rahn 253* (C, J.G.H., J.P.H.).

[1] Altitudes probably too high; should not exceed 3200 m.

3. *S. vernei.* Intermediate forms between subspecies *vernei* and subspecies *ballsii*

ARGENTINA

Prov. SALTA:

Dept. Cachi, Cuesta del Obispo. Alt. 3720 m.(?) 21 Jan. 1947. *Garolera & Romero* (s.n.) (LIL 194388, 202462).

Dept. Caldera, Cuesta del Carancho, arriba del Potrero del Castillo. Alt. *c.* 3200 m. 13 Mar. 1952. *Sleumer & Vervoorst 2892* (p.p.) (LIL, US, J.G.H.).

Dept. Caldera, Quebrada del Río Potrero Castillo. Alt. *ca.* 3200 m. 16 Mar. 1952. 'Sine flores.' *Sleumer & Vervoorst* 2974 (LIL).

Dept. Chicoana, Cuesta del Obispo. Alt. 2750–3050 m. 22 Feb. 1956. *Hjerting, Petersen & Rahn 311* (C, J.G.H., J.P.H.).

DISTRIBUTION

S. vernei is to be found in Argentina, provs. Jujuy, Salta, Tucumán, and Catamarca. Subsp. *vernei* is confined to provs. Tucumán (depts. Chicligasta, Río Chico, Tafí, and Trancas) and east Catamarca (depts. Andalgalá and Capayán).[1] Subsp. *ballsii* occupies the more northern part of the range, occurring chiefly in prov. Jujuy (depts. Tilcara and Tumbaya), though it has been recorded once for prov. Salta, dept. Sta. Victoria. Intermediate forms occur in prov. Salta, depts. Cachi, Caldera, and Chicoana. (See Map 11.)

Although *S. vernei* has not yet been recorded for Bolivia it is very likely that subsp. *ballsii* may spread northwards into that country.

HABITAT

Subsp. *vernei* has been recorded at altitudes of (1950–)2200–2800(–3200) m, whilst subsp. *ballsii* is found rather higher, from (2450–)2700–3450 m.

Both subspecies normally grow in the Montano district of the Western Subtropical province of Cabrera's phytogeographical scheme (1953). They are to be found in the cloud forest region on the eastern slopes of the Andean cordillera and pre-cordillera, and prefer the very damp humus-rich soil of valley bottoms, stream sides, and river terraces. They are frequently seen amongst bushes, stones, and field borders or in the partial shade of trees such as *Polylepis australis*, *Sambucus peruviana*, and *Alnus jorullensis* in the vegetational zones that are characterized by these species. Neither subspecies of *S. vernei* is to be found in really dense shade (contrast *S. microdontum*) and both may be seen in full sun amongst bushes and long grass; under these conditions the dimensions of all parts of the plant are smaller and the pubescence more dense. On the whole, subsp. *ballsii* is found more frequently in the open, perhaps because of its slightly higher altitude range, whereas subsp. *vernei* seems to prefer the partial shade of *Polylepis*, *Alnus*, *Sambucus peruviana*, and other trees.

TAXONOMIC NOTES

S. vernei was first described by Bitter and Wittmack (Wittmack 1914) on the basis of plants collected by Lorentz and Hieronymus in 1874 (*No. 708*) in the mountains of Tucumán. It was named after Professor Claude Verne of Grenoble who made many introductions of wild potatoes into European botanic gardens.

The type specimen of *Lorentz & Hieronymus 708* was deposited in Berlin and is now destroyed. Good photographs of this exist, however (see Plate 113). There are

[1] We saw it also in prov. Salta, dept. Guachipas on our 1966 expedition (see Appendix III).

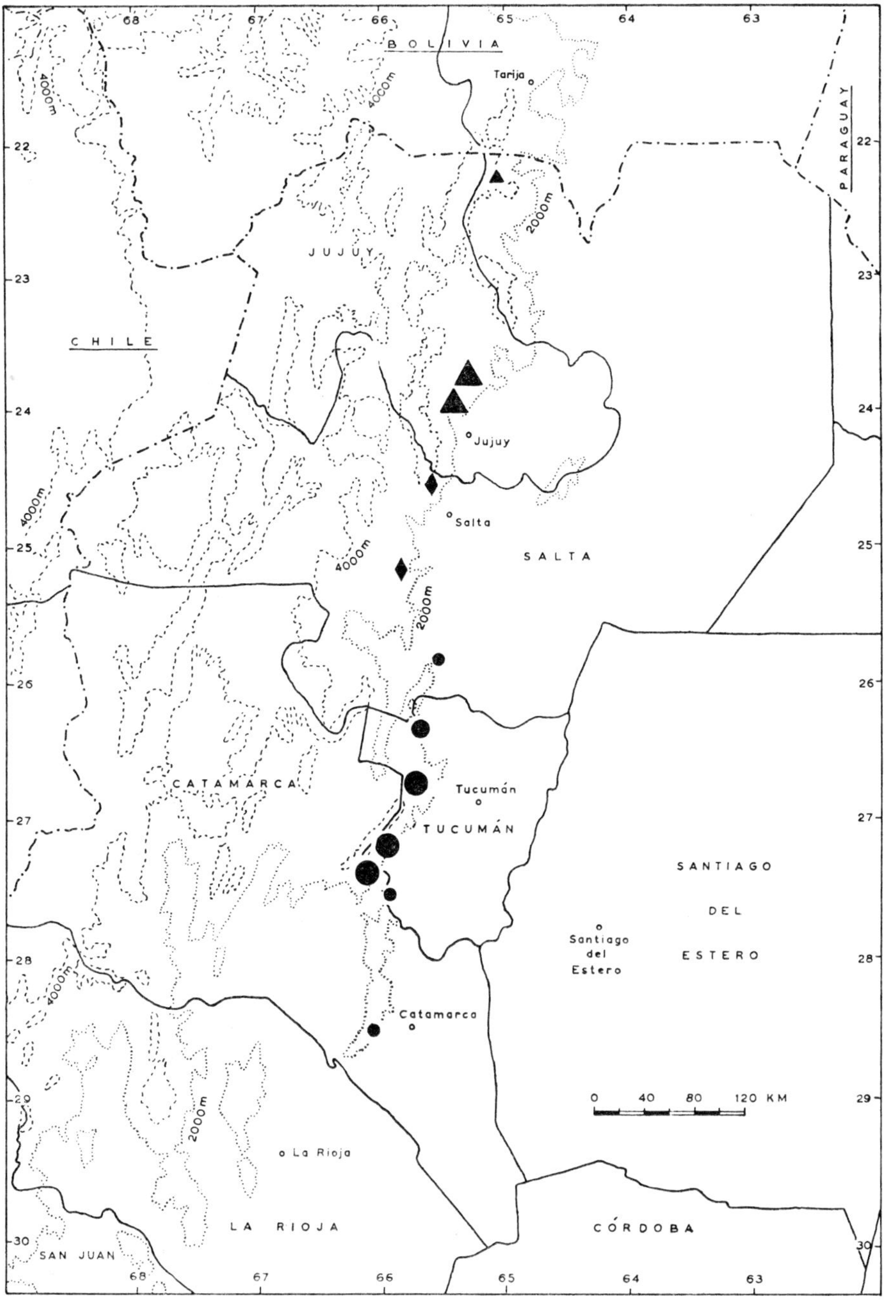

No. of Records
5 & over 3–4 1–2

Solanum vernei Bitt. et Wittm. subsp. *vernei*

Solanum vernei Bitt. et Wittm. Intermediates

Solanum vernei Bitt. et Wittm. subsp. *ballsii* (Hawkes) Hawkes & Hjerting

MAP 11. Distribution of *S. vernei* Bitt. & Wittm. subsp. *vernei* and subsp. *ballsii* (Hawkes) Hawkes & Hjerting.

two isotype specimens known to us from Göttingen and Córdoba, respectively. Since the Göttingen sheet is not in very good condition we designate the Córdoba sheet of *Lorentz & Hieronymus 708* as LECTOTYPE for *S. vernei* Bitt. et Wittm. (See Fig. 35.)

Subsp. *ballsii* was described from material collected by Balls, Gourlay, and Hawkes in 1939 in the province of Jujuy, northern Argentina. It was then considered to be a distinct species, since at that time *S. vernei* was not known in the living state and no type material was available for study. No type specimen was designated by Hawkes in 1944 and we have therefore considered it advisable to do so in the present publication. Material of *Balls 6042* (p.p.) was figured on p. 44 of the publication where '*S. ballsii*' was first described and we hereby designate the specimen in Herb. P.G.S. on which the figure was based as the LECTOTYPE for *S. vernei* Bitt. & Wittm., subsp. *ballsii* (Hawkes) Hawkes & Hjerting. (See Fig. 36 and Plate 116.)

From Bitter and Wittmack's description of the substellate corolla of *S. vernei*, Hawkes (1944) considered it possible that this species might belong to series Commersoniana. However, when isotype material was discovered in the Córdoba and Göttingen herbaria in 1948 it was at once realized that this species could belong to no other series but Tuberosa. Furthermore, with the large range of collections made available from sources already mentioned in an earlier section of this publication, it was no longer possible to maintain *S. ballsii* as a separate species, and we have accordingly grouped it here as a geographical subspecies of *S. vernei*.

In 1953 Brücher, who made numerous field observations on *S. vernei* and *S. ballsii*, considered them to be two distinct species, differing chiefly in their height, leaf size and pubescence, and the size of the tubers (see Brücher and Ross 1953); later, however, Brücher (1954, 1956*b*, 1957*b*) came to consider *S. ballsii* as a synonym of *S. vernei*, or at best a local variety.

In 1951, and again in 1954 and 1957*b*, Brücher put forward the hypothesis that *S. vernei* was a morphological and cytological 'missing link' in the evolution of the cultivated tetraploid potato *S. tuberosum*;[1] in other words, he considered *S. vernei* to be the ancestral diploid species, or at least to have played a great part in the phylogeny of our cultivated tetraploid potatoes. He based his argument chiefly on the very large tuber size, the alleged close morphological similarity between the two species, and the fact that *S. vernei* is apparently eaten by peasants at certain times of the year when no other source of food is available. He states, moreover, that *S. vernei* is even at times confused by the peasants with the true cultivated potato. He also quotes the older botanists such as Lillo, Grisebach, and Lorentz, who thought that *S. vernei* was probably none other than a form of *S. tuberosum*. *S. vernei* is, indeed, found frequently in field borders, where at first glance it appears to have escaped from former cultivations. Finally, he reports the discovery of a 'tetraploid biotype' of *S. vernei* in prov. Jujuy at 4000 m altitude, similar in chromosome number to the tetraploid *S. tuberosum* and (therefore?) in some way related to it.

We find ourselves unable to agree with this hypothesis for the origin of *S. tuberosum* for the following reasons. In the first place, *S. vernei* is readily distinguishable from *S. tuberosum*, not only by reason of the tall stem, very large dark purple flowers, and velvety pubescence on the under surface of the leaf, but also by the distinct tuber form and pigmentation. Secondly, *S. vernei* occupies a very distinct ecological zone of high rainfall 'cloud forest' in which none of the primitive cultures of *S. tuberosum* is to be found. Only one diploid species of cultivated potato, *S. phureja*, is commonly grown in this zone, but only occurs from north Bolivia northwards to Colombia and Venezuela, where *S. vernei* is never to be found. On the other hand, all the other cultivated potato species are to be found primitively in the high windswept valleys and plateaux

[1] He referred always to *S. andigenum*, which is now generally considered to be a geographical subspecies of *S. tuberosum*.

FIG. 35. *S. vernei* Bitt. & Wittm. Lectotype specimen of *Lorentz & Hieronymus 708* (Herb. Córdoba), collected near La Ciénaga, dept. Tafí, prov. Tucumán. Compare with Plate 113 of the former type in herb. Berlin, now destroyed.

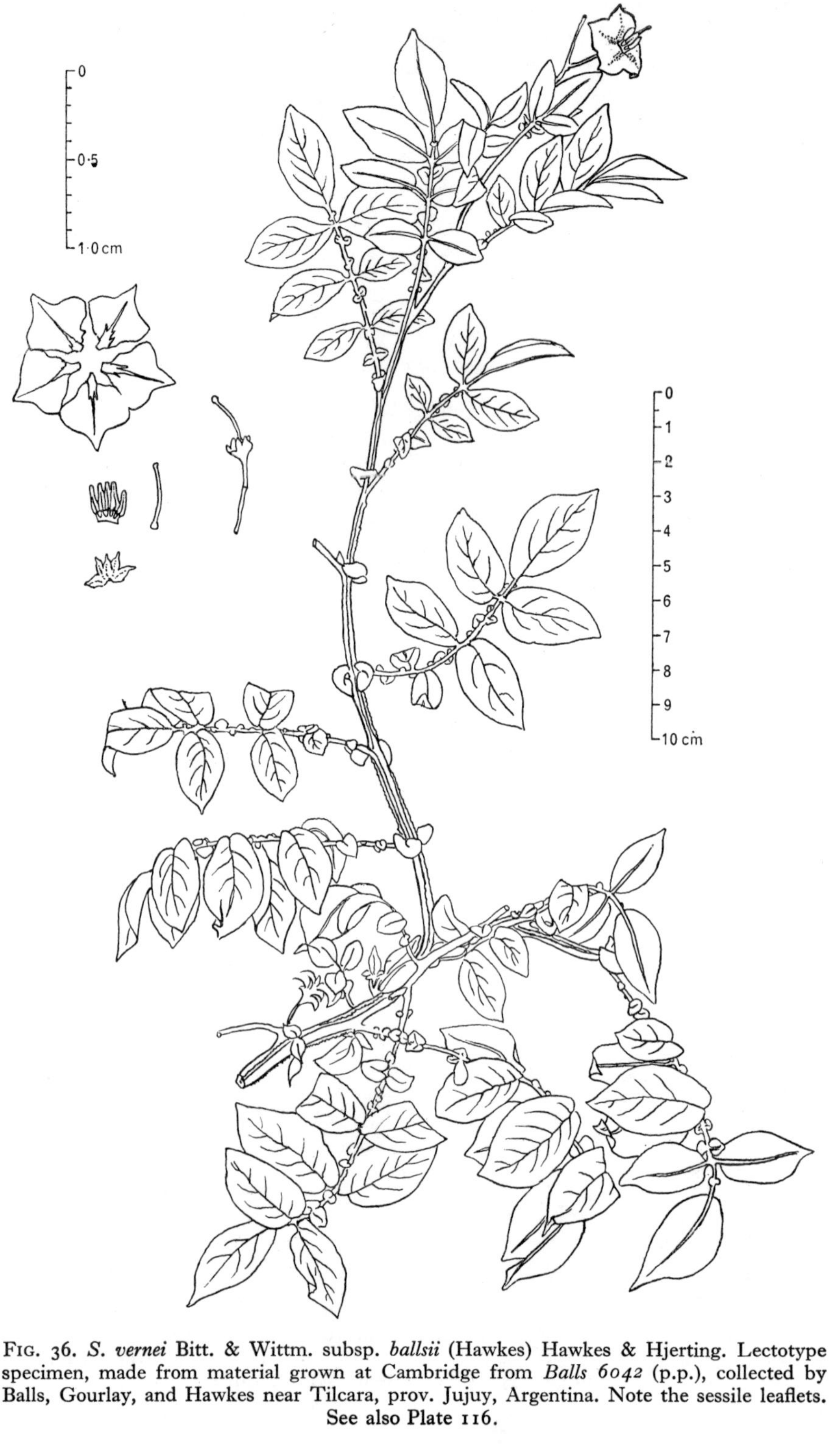

FIG. 36. *S. vernei* Bitt. & Wittm. subsp. *ballsii* (Hawkes) Hawkes & Hjerting. Lectotype specimen, made from material grown at Cambridge from *Balls 6042* (p.p.), collected by Balls, Gourlay, and Hawkes near Tilcara, prov. Jujuy, Argentina. Note the sessile leaflets. See also Plate 116.

of the puna and páramo regions. No doubt mutations for large tuber size have occurred from time to time in wild potatoes (cf. *S. polyadenium* from Mexico); and under certain circumstances the biotypes possessing them were taken into cultivation; we cannot argue, however, in the reverse direction, that *because* a wild potato has large tubers therefore it *must* once have been cultivated. Again, our own observations from Mexico southwards to Argentina have shown that peasants will eat tubers of many different kinds of wild potato if they lack other food; no one would suggest, on the basis of these observations, however, that all these potatoes are closely related to *Solanum tuberosum*. Wild potatoes are very apt to grow in field borders, amongst piles of stones, and in walls; generally this is because the group as a whole cannot compete well with natural vegetation and will therefore colonize 'open' habitats where the pressure of competition is somewhat relaxed. Sometimes longer stoloned and less well-yielding segregants within the sexually reproducing cultivated species also become established as weeds of the fields themselves (see Hawkes 1944, p. 103). Thus, since potatoes of fields and field borders may be truly wild, occasional escapes, or reversions from earlier cultivations, we cannot draw any conclusions from the behaviour of *S. vernei* in this respect without other confirmatory evidence.

Again, we should expect the region where potato cultivation originated to be particularly rich both in cultivated species and in the amount of their genetic variability. In north-west Argentina, however, there is little genetic variability amongst cultivated potatoes compared with the very rich store of diversity in south Peru to north Bolivia where potato cultivation is generally considered to have originated. Thus the distribution area of *S. vernei* does not coincide with the area of highest genetic variability of *S. tuberosum* subsp. *andigena* or with that of any of the more primitive diploid cultivated species such as *S. stenotomum* and *S. phureja*. In north-west Argentina a rather limited number of clones of the tetraploid *S. tuberosum* subsp. *andigena* is to be found, together with one or two clones of the specially derived triploid *S. juzepczukii* and the pentaploid *S. curtilobum*.

Although Brücher identified a collection from south Peru (*Marín 1892*—Herb. Lillo 301351) as *S. vernei*, this belongs in fact not to this species but to the widespread weed species *S. sparsipilum*. The evidence on hand shows, therefore, that *S. vernei* does not coincide in its distribution area with that of the primitive diploid cultivated potatoes or with that of the greatest diversity of the tetraploid ones.

Finally, we should like to return to the tetraploid form from Jujuy (Tilcara, Corral, 4000 m), collected by Brücher and identified by him (1951, 1954) as *S. vernei*, which we have unfortunately not been able to examine. We have seen no collection of *S. vernei* from altitudes of over 3450 m, and we consider it more likely that the plant in question may have been a stunted form of *S. oplocense*, which, as we shall show later (p. 382), sometimes occurs as a tetraploid and in fact has been collected from the Humahuaca region not very far away from Tilcara, at an altitude of 3500 m. Alternatively, since Brücher considers that *S. vernei* and *S. tuberosum* subsp. *andigena* are so similar morphologically, his tetraploid form might perhaps have been a variety of subsp. *andigena*, growing from a tuber discarded by a muleteer or escaped from cultivation in some way. Even if it is a true autotetraploid form of *S. vernei*—and this hypothesis must not be completely discarded, since a tetraploid form of *S. oplocense* is known—there is still insufficient evidence for regarding this alleged tetraploid biotype of *S. vernei* as in some way ancestral to the tetraploid cultivated potato.

We should also like to dispose here of the erroneous suggestions (see Brücher and Ross 1953) that the Bolivian species *S. berthaultii* extends into northern Argentina and that it might have been formed from *S. vernei* by spontaneous chromosome doubling. *S. berthaultii* is a diploid, not a tetraploid species, from the region of Cochabamba, Bolivia, connected with the stellate-flowered and highly glandular *S. tarijense*; its distribution area does not extend into Argentina, and its ecological requirements are quite distinct from those of *S. vernei*. Even if *S. berthaultii* were later shown to be derived by hybridizations between *S. tarijense* and some blue-flowered Bolivian species it still could have no connexion with *S. vernei*, and one can only conclude that Brücher's tetraploid material of '*S. berthaultii*' was wrongly identified by him.

To sum up, then, we are firmly of the opinion that *S. vernei* has played no part in the evolution of cultivated potatoes and is in no way related to the Bolivian species *S. berthaultii*.

S. vernei has so far been recorded only in north-west Argentina, though one would expect it to spread into southern Bolivia. It may perhaps be related to the little-known Bolivian species *S. liriunianum* Cárd. et Hawkes, from Cochabamba, but it certainly does not spread as far north as southern Peru. Its not infrequent hybridizations in nature with *S. microdontum* have undoubtedly caused some rather limited introgression of genes from each species into the other. Thus we occasionally find *S. microdontum* with darkly pigmented stems, rhachides, and leaflet veins on the one hand, whilst certain forms of *S. vernei* with wide crisped stem wings and rather paler flowers probably owe these characters to germ-plasm of *S. microdontum*.

NOTES ON LIVING PLANTS

According to Brücher and Ross (1953) *S. vernei* may yield up to 556 grammes of tubers per plant. In Vandel, subsp. *vernei* gave tubers weighing up to 35 or even 55 grammes each and measuring 3×4 to 7×8 cm. Subsp. *ballsii* produced somewhat smaller tubers, weighing up to 20–45 grammes each. The tubers of *S. vernei*, although large, are of poor and bitter flavour after cooking. (See Plate 115.)

Some of the seedlings grown by us possessed rather more dissected leaves than in plants grown in the wild, with up to 7 or 8 pairs of lateral leaflets and 18 pairs of interjecteds, especially in subsp. *ballsii, Petersen & Hjerting P–351*. In this line also the flowers were more rotate and smaller than is general for *S. vernei*.

When grown at Birmingham the articulation in some lines was as high as $\frac{2}{3}$–$\frac{4}{5}$ above the base, though in others much lower. This agrees with Brücher's (1954) field observations in Argentina. In subsp. *ballsii* (CPC: 105) the berries were definitely pointed, almost cordate, and white spotted. A waviness or crisping of the leaflet margins is also noticeable on living plants of this subspecies; Brücher and Ross (1953) also claim a smaller tuber size for subsp. *ballsii*, and this agrees with our own observations. Most clones possess deep anthocyanin pigmentation on stems, wings, peduncles, pedicels, calyx, leaf rhachis, and leaflet nerves. Jones, Plaisted, and Peterson (1963) record a chlorophyll mutant, designated *gg*, which reduces the quality of the chlorophyll in the leaf tissue.

Brücher (1954) states that *S. vernei* is field resistant to *Phytophthora* and is able to withstand 5° C of frost. Viirsoo (1954*a*) records a resistance to −5° C in Humahuaca.

S. vernei has been shown to be of very great value in breeding for eelworm resistance. It has also been shown to possess resistance to the following pests and diseases: late blight, wart, scab, *Verticillium* wilt, skin spot, flea beetles, and viruses A and Y. (For further details see Chapter 5.)

CYTOLOGY

Chromosome counts of subsp. *ballsii* by Hawkes (1944) revealed a somatic number of $2n = 24$. This was verified by Brücher (see Brücher & Ross 1953) in the typical subsp. *vernei*, which gave $2n = 24$ in eight samples analysed. Hybrids of *S. vernei* × *S. tuberosum*, according to Goffart and Ross (1954), gave three plants with $2n = 48$, owing presumably to the functioning of an unreduced egg cell from *S. vernei*. Triploid seedlings were not obtained in this cross. Gottschalk and Peters (1955), on the basis of pachytene chromosome studies, claim a strong similarity between *S. vernei* and *S. chacoense*, with eight chromosomes of similar type. On the other hand, *S. microdontum* (as *S. simplicifolium*) is thought to be very dissimilar to *S. vernei* in this respect. They strongly disagree with Brücher that *S. vernei* had anything to do with the origin of the cultivated potato, since they can see few similarities between the chromosomes of the two species.

Their work was repeated by von Wangenheim (1957) and by Fiedler and Schreiter (1959). None of these workers obtained results in agreement with those of Gottschalk and Peters on the morphology of the pachytene chromosomes.

Von Wangenheim effected pachytene analyses of *S. vernei* crossed with *S. spegazzinii* and with *S. microdontum*, respectively. In all this material meiosis was extremely regular, and at diakinesis only a slight statistically insignificant rise in frequency of univalents could be observed in the hybrids as compared with *S. vernei* itself. Some diads and monads were observed in *S. microdontum* × *S. vernei* hybrids, due to degeneration of the spindle mechanism at second anaphase, but the proportion of these to normal tetrads was influenced by differing weather conditions.

Magoon, Hougas, and Cooper (1958*a*) obtained results consistent with those of von Wangenheim, in their meiotic analysis of *S. vernei* × *S. microdontum* hybrids. Meiosis was mostly regular but small percentages of univalents, quadrivalents, and other irregularities were observed. These, apart from quadrivalents, were also seen in *S. vernei* itself.

It seems fairly clear from the foregoing that Gottschalk and Peters were wrong in claiming large chromosome differences between *S. vernei* and *S. microdontum*, since pairing at pachytene and later stages in the hybrids was shown by von Wangenheim and others to be very regular—a condition that could not be expected if the chromosomes were largely non-homologous. Since the hybrids of *S. vernei* with *S. microdontum* and with *S. spegazzinii* do not differ significantly from *S. vernei* itself in their meiotic behaviour we must conclude that there is a close genomic similarity between these species.

Lebedeva (1959*a*, *b*, 1960) compared diploid and colchicine-tetraploid *S. vernei* in morphological features. The leaflets of the tetraploid were wider than in the diploid but were similar in number; the corolla of the tetraploid was larger and the lobes flatter. Furthermore, the tetraploid was more vigorous, yielded better than the diploid under long day, and crossed readily with *S. tuberosum*.

FERTILITY AND CROSSABILITY

In common with most other diploid potato species, *S. vernei* is self-incompatible. However, when clones with different *S* genes are crossed the seed set is good. Brücher and Ross (1953) record good pollen with 95–98 per cent germination. Viirsoo (1954*b*) mentions that the pollen is abundant and 100 per cent fertile in Argentina. Hybrids of *S. vernei* with diploid species, in series Commersoniana, Demissa, Megistacroloba, and Tuberosa are easy to make and quite fertile.

We have crossed *S. vernei* subsp. *ballsii* with *S. chacoense*, *S. tarijense*, *S. sanctae-rosae*, *S. toralapanum*, *S. microdontum*, *S. gourlayi*, *S. spegazzinii*, *S. verrucosum*, *S. phureja*, and *S. stenotomum*. We hybridized subsp. *vernei* with *S. infundibuliforme*, *S. microdontum*, *S. spegazzinii*, and *S. kurtzianum*. No success was obtained in reciprocal crosses with the tetraploid weed species *S. sucrense* and only limited success with *S. tuberosum* (Goffart & Ross 1954), when tetraploid seedlings were produced. The use of colchicine-induced tetraploids of *S. vernei* in crosses with *S. tuberosum* is recommended by Goffart and Ross (1954), Jones (1954), Magoon *et al.* (1958), and Rothacker (1958). Several authors reported satisfactory F_1 plants but poor seed setting in the F_2 and backcross generations. Beamish *et al.* (1957) made tetraploids from a *S. vernei* × *S. microdontum* cross in an attempt to hybridize it with *S. tuberosum*. The hybrids were described later by Magoon *et al.* (1958). (See also our own hybrids, pp. 343–4.)

Our own crosses between the hexaploid species *S. demissum* and *S. vernei* gave three seedlings, only one of which survived long enough for a chromosome count to be made. This was a tetraploid, but its fertility and viability were low.

The hybrids with *S. chacoense*, *S. tarijense*, *S. infundibuliforme*, *S. sanctae-rosae*, and *S. microdontum* have already been reported on (see under these species). A hybrid of *S. gourlayi* (*P. & H. P–365*) × *S. vernei* subsp. *ballsii* (*P. & H. P–351*) gave a reasonably vigorous F_1 family of plants, roughly intermediate between the two parents. The leaves were similar in size and shape to *S. gourlayi* with about the same number of lateral leaflets; subsp. *ballsii* characters were the whitish pubescence on the under surface, larger number of interjected leaflets (5–7 pairs), and enlarged basiscopic lobe at the base of the lateral leaflets. The flowers approximated to *S. gourlayi* in size and colour. The reciprocal hybrid was also made but not sown out.

S. vernei (CPC: 2414/3 from Tucumán) crossed with *S. oplocense* (CPC: 2116 from Bolivia) gave only three seeds and these proved to be inviable.

We also hybridized *S. vernei* subsp. *vernei* (CPC: 2414/3 from prov. Tucumán) with *S. spegazzinii* (*Vervoorst 3159*, H. 68, from prov. Catamarca). The F_1 seedlings (family 143) were of intermediate size, with short internodes, narrow stem wings, and a smaller (2·5 cm diam.) paler purple corolla than that of *S. vernei*. The leaf showed strongly the influence of the *S. spegazzinii* parent, since the petiolules were extremely long, up to 2 cm, and pointed forwards towards the leaf apex at an angle of about 45°. Numerous rather narrow interjected leaflets, very reminiscent of *S. spegazzinii*, were present, both on rhachis and petiolules. The leaflet shape was approximately that of subsp. *vernei*, and some whitish pubescence was visible on the under surfaces. Intersib crosses between F_1 plants gave an unthrifty F_2 family (190) showing that genomically the plants were not very similar.

Hybrids of *S. spegazzinii* (aberrant form, *H.P. & R. 334* from prov. Salta) × *S. vernei* subsp. *ballsii* (*P. & H. P–351* from prov. Jujuy) gave a more sturdy plant than the above-mentioned F_1 hybrids, with much shorter petiolules and some sub-sessile primary and interjected leaflets. This difference could partly be accounted for by the different subspecies of *S. vernei* used and the larger size of *H.P. & R. 334*. The rather coarse spreading hairs on most parts of the plant were obviously inherited from *S. spegazzinii*. The rich purple corolla had lighter regions on the 'interpetalar membranes' and was about 3·5 cm diam.; the calyx acumens were linear and 3 mm long.

Hybrids of *S. kurtzianum* (*Ruiz Leal 11906* from prov. Mendoza) × *S. vernei* subsp. *vernei* (CPC: 2413, from prov. Tucumán) were rather similar in general habit and appearance to the hybrids of subsp. *vernei* with *S. spegazzinii*. In the *S. kurtzianum* hybrids the petiolules were long, but reached a maximum of only 13 mm, and the interjected leaflets were much less frequent. The plant was taller, with longer internodes and well-marked stem wings. The flowers were dark purple, the colour of *S. vernei* flowers, but intermediate in size between those of each parent. The articulation was median or supra-median, and it seems that the low articulation of *S. kurtzianum* is recessive in hybrids between that species and others with a high articulation. Various attempts to make intersib hybrids of this cross failed.

NATURAL HYBRIDS

Natural hybrids of *S. vernei* with certain other diploid species have been seen from time to time in Argentina, depending for their frequency largely on the amount of coincidence in altitudinal range and moisture requirements of the species involved.

(*a*) With ***S. chacoense, S. tarijense, S. microdontum,*** and ***S. venturii.*** Hybrids with *S. chacoense* are rare, or perhaps do not occur at all, since the altitude ranges of *S. vernei* and *S. chacoense* hardly overlap. Only one possible hybrid (*Garolera and Romero* (s.n.) from Jujuy, Termas de Reyes) has been seen (see p. 232).

The occurrence of hybrids with *S. tarijense* is also rather doubtful. Certain forms of *S. tarijense* in prov. Salta, dept. Santa Victoria, show introgression with a blue-flowered species which might be *S. vernei*, *S. vidaurrei*, or *S. gourlayi*. These are mentioned in the natural hybrid section of *S. tarijense* (see p. 245). If *S. vernei* is the blue-flowered species involved, then subsp. *ballsii* must have been the one concerned.

Hybrids of *S. vernei* with *S. microdontum* have also been dealt with (pp. 345–7). We find crosses of subsp. *vernei* with *S. microdontum* subsp. *gigantophyllum*, of subsp. *ballsii* with subsp. *microdontum*, and finally, of subsp. *ballsii* with *S. microdontum* subsp. intermediate. The distribution areas of subsp. *vernei* and subsp. *microdontum*, as well as of subsp. *ballsii* and subsp. *gigantophyllum*, do not overlap, and in consequence no hybrids are to be found.

Three collections of probable hybrids between *S. vernei* and *S. venturii* have been recorded in prov. Tucumán, depts. Chicligasta and Tafí, at 2800 and 2700 m respectively. These are dealt with more fully on p. 355.

(*b*) Hybrids between ***S. vernei*** and ***S. spegazzinii.*** These have been seen in the transitional zone between the two species on passing from wetter to drier regions. The hybrids are distinguished by the larger number of smaller and narrower leaflets and the smaller paler flowers when compared with *S. vernei* itself.

Six collections of hybrids have been recognized, three of them from prov. Tucumán, dept. Tafí, Río Infiernillo, at altitudes of 2800 to about 3000 m (*P. & H. P–332a, 856a*; *Castillón 3192*). In this area, as one travels north-westwards, through the Infiernillo pass, one comes to the more arid prepuna vegetation of the western slopes of the Aconquija range. (See Table 22.)

Table 22

Natural hybrids between S. vernei *and* S. spegazzinii

Locality	Collection number and altitude		
	S. vernei	Hybrid	*S. spegazzinii*
Tucumán, Tafí, Infiernillo	*H.P. & R. 1.* 3000 m	*Castillón 3192.* [*c.* 3000 m.]	
	P. & H. P–324. 2800–3000 m	*P. & H. P–332a.* 2900 m	*P. & H. P–332b.* 2900 m
	P. & H. 857. 2800 m	*P. & H. 856a.* 2800 m	*P. & H. 856b.* 2800 m
			Ross & Vervoorst 43 (p.p.). 3100 m (Occurs frequently on the western side of the Infiernillo pass)
Tucumán, Trancas, Gualinchay, and Cuesta Grande del Chorro	*Schreiter 632.* 2000 m	*Schreiter 4837.* 1500 m	
	Schreiter 4840 (p.p.). 3000 m	*Spegazzini* (s.n.).	*Schreiter 4839.* 3300 m
Catamarca, Andalgalá, El Candado, and La Playa		*P. Jörgensen 1311* (p.p.).	*P. Jörgensen 1311* (p.p.).

The collection *P. & H. P–332* was gathered from a hybrid population and the herbarium material was divided into hybrid forms (*P–332a*, see Plates 117 and 118) and *S. spegazzinii* (*P–332b*). The collection was grown in England both from tubers (CPC: 2452) and seed (CPC: 2485). The tuber collection (2452) proved to be quite good *S. spegazzinii*, thus agreeing with one of the original herbarium specimens, separated as *P–332b*. Since only one clone within this collection was grown we were not able to ascertain the masked variability within it, if such existed, by raising a family of seedlings from it. The leaflets were typically narrow, with 5–6 pairs of laterals and a large number of interjecteds; the flowers were small, purple, and of the *S. spegazzinii* type. Plants raised from original seed (CPC: 2485), all with twenty-four somatic chromosomes, showed a wide range of variation and appeared to be typical F_2 segregants. Some seedlings were quite similar to *S. vernei*, others to *S. spegazzinii*, and others again showed combinations of characters from the two species. (See Plate 118.) Some of the intermediates were not unlike our artificial F_1 hybrids of *S. vernei* × *S. spegazzinii*, and the family as a whole showed the range of variation seen in the F_2 hybrids derived from this cross. There seems to be little doubt, therefore, as to the hybrid nature of *P. & H. P–332*.

Another collection from the same area, but taken in the following year, *P. & H. 856a*, was essentially similar to *P. & H.P–332a*. It was part of a mixed collection, consisting of 'good' *S. vernei* (857), 'good' *S. spegazzinii* (*856b*), and obvious hybrids between them (*856a*) all of which were found growing together in the same area. It would obviously be of the greatest interest to continue work here on modern biosystematic lines, so as to assess the nature of the hybrid populations and their effect (by introgression) on the parent species.

In addition to the Infiernillo hybrids we have also recognized two collections from prov. Tucumán, dept. Trancas (*Schreiter 4837* and *Spegazzini* (s.n.)) as probably *S. vernei* × *S. spegazzinii* hybrids (see Table 22). The former occurs at a lower altitude (only 1500 m, if this is correct) but seems quite genuine. The leaf of both collections is essentially similar to that of our artificial hybrids (Family 143) between *S. vernei* and *S. spegazzinii* and the flower also is of intermediate type. *S. vernei* has been obtained near by (*Schreiter 632*), whilst the same species (*Schreiter 4840* (p.p.)), as well as *S. spegazzinii* (*Schreiter 4839*), have been collected further west and higher up, from 3000 to 3300 m. Unfortunately no altitude data exist for the Spegazzini collection.

A further hybrid between *S. vernei* and *S. spegazzinii*, together with *S. spegazzinii* itself, was collected by P. Jörgensen (*No. 1311* (p.p.)), in prov. Catamarca, dept. Andalgalá. So far, the *S. vernei* parent has not been seen in that locality (see Table 22).

Apparently *S. vernei* and *S. kurtzianum* do not normally occur sufficiently close together for hybrids to be formed. However, a collection made by Vervoorst (*3478*) in Catamarca, Capayán, Las Angosturas, which we have placed as *S. vernei*, may be influenced by *S. kurtzianum* to a slight degree, since the leaflets seem to be rather too narrow for typical *S. vernei*. New collections from about 2400 m in the quebrada of Las Angosturas are needed to throw light on this problem.

LOCALITIES OF NATURAL HYBRIDS

(*a*) With ***S. chacoense, S. tarijense, S. microdontum***, and ***S. venturii*** (see pp. 238, 246, 348–9, and 356)

(*b*) With ***S. spegazzinii***

Prov. SALTA: (or Tucumán, dept. Trancas?)

N. Camino Cuesta del Arca-Trancas. Dec. 1896. *C. Spegazzini* (s.n.) (GOET). (Although Salta is noted, this collection was probably obtained in Tucumán at the

Cuesta Grande del Chorro, where Schreiter obtained both *S. spegazzinii* and *S. vernei* (*Nos. 4839, 4840* (p.p.)), as well as the hybrid (*4837*); see below.)

Prov. TUCUMÁN:

Dept. Tafí, Tafí del Valle, Infiernillo. 22 Dec. 1913. *Castillón 3192* (LIL).

Dept. Tafí, Infiernillo, Río Tafí. Alt. 2900 m. 27 Feb. 1952. *Petersen & Hjerting P–332a* (CPC: 2485, seed collection) (J.G.H.). See Plates 117 and 118.

Dept. Tafí, Río Tafí y afluentes (Infiernillo). Alt. 2800 m. 1 Jan. 1953. *Petersen & Hjerting 856a* (J.G.H., J.P.H.).

Dept. Trancas, Gualinchay (Distrito San Pedro de Colalao). Alt. 1500 m. 1 Feb. 1927. 'Flor azul.' *Schreiter 4837* (LIL).

Prov. CATAMARCA:

Dept. Andalgalá, La Playa. 10 Feb. 1917. 'Abundante.' *P. Jörgensen 1311* (p.p.) (US).

14. SOLANUM GOURLAYI Hawkes

(*Bull. imp. Bur. Pl. Breed. Genet., Camb.* 120–1, 1944)

S. leptophyes Bitt., f. *gourlayi* (Hawkes) Corr. (*Wrightia* **2**, 188, 1961).

Herbaceous, tuber-bearing. Tubers up to 2·5 cm long, oval, pale brown, with smooth skin.

Stem short, (5–)10–30 cm long in wild plants, profusely branched from the base or sometimes simple, slightly flexuose, very narrowly winged, provided with sparse slender hairs; internodes 0·5–4 cm long.

Leaves densely following, often in a rosette or semi-rosette in plants with short stems, (5–)7–10(–14) cm long × (2·5–)5–9 cm broad. Leaflets (2–)3–4(–5)-jugate, with 0–2(–4) pairs of interjected leaflets which are small, sessile, and orbicular (larger and more numerous in cultivation); lateral leaflets ovate-lanceolate to rhomboid or pentagonal, rather broad, (1·5–)2–3·5(–4·5) cm long × (0·5–)1·0–1·75(–2·5) cm broad, *the two distal pairs typically subequal*, diminishing in size rapidly after this towards the leaf base, with short (0–)2–3(–5) mm petiolules, basally truncate or cordate, sometimes very narrowly decurrent on to the rhachis in wild plants (but not when cultivated); terminal leaflet frequently longer and broader than laterals, broadly lanceolate, rhomboid, pentagonal, or even (in young plants) triangular-cordate; all leaflets clothed rather sparsely above and slightly more densely below with rather adpressed multicellular hairs.

Inflorescence 1–6(–12)-flowered, barely exceeding the leaves. Peduncle 2·5–6 cm long, once-forked above. Pedicels 1–2·5 cm long, articulated ½ to ¾ length above the base, 3–9 mm long above the articulation, clothed (as is also the peduncle) with rather sparse adpressed multicellular hairs.

Calyx 5–8 mm long, the lobes broadly ovate, with linear acumens 2·3 × 0·5–1 mm; pubescence similar to peduncles and pedicels.

Corolla medium to dark purple in the wild plants, 2–3 cm diam., *pentagonal to semi-stellate*, the lobes 9–15 mm wide at the base and 6–8(–10) mm long including the 4–6 mm long × 4–5 mm wide acumens.

Anthers 4–6 mm long × 1·5–2 mm wide at the base; filaments about 0·5–1·5 mm long.

Style 8–10 mm long, clothed with small papillae for two-thirds of its length above the glabrous base; stigma short, capitate, slightly thicker than the style apex, entire or emarginate.

Berries oval, 1·5–2·5 cm long × 1–2 cm diam., grey, marbled with dark green.

Chromosome number: $2n = 24$.

AFFINITIES

S. gourlayi may be distinguished from *S. vidaurrei* by the shorter and relatively broader leaf, with wider generally petiolulate and normally not decurrent leaflets with no well-marked marginal teeth or hairs. In floral characters the two species are not very dissimilar. *S. gourlayi* is also rather similar to the little-known species, *S. oplocense*. This latter species differs from *S. gourlayi* in the leaf with thick hairs, often cuneate leaflet base decurrent on to the rhachis, denticulate leaf margin provided with hairs on the teeth, very short filaments and minute capitate stigma. *S. gourlayi* differs from *S. spegazzinii* in the relatively broader leaf, fewer lateral and interjected leaflets, and truncate to cordate and not cuneate or decurrent leaflet bases. The species are rather similar in floral characters.

SPECIMENS EXAMINED

ARGENTINA

Prov. JUJUY:

Dept. Tilcara, above Tilcara, San Gregorio. Alt. 3700 m. 11 and 15 Feb. 1939. 'In sandy soil amongst stones.' *Balls 5979* (CPC: 74) (K, P.G.S.—Lectotype, J.G.H.). See Fig. 37 and Plate 119.

Dept. Tilcara, above Tilcara, San Gregorio. Alt. 3900 m. 15 Feb. 1939. *Balls 5987* (p.p.) (P.G.S., J.G.H.).

Dept. Tilcara, above Tilcara, San Gregorio. Alt. 3900 m. 11 Feb. 1939. 'Rough stony ground, sheltered to some extent by other plants. Stem 6 cm. long; small flowers. Tubers 1·2–1·6 cm, oval, smooth skinned.' *Balls 5990* (CPC: 80) (P.G.S.).

Dept. Tilcara, Quebrada de San Gregorio. Alt. 4000 m. 24 Mar. 1952. *Petersen & Hjerting P–356* (LIL, J.G.H., J.P.H.).

Dept. Tilcara, above Tilcara, Garganta del Diablo. Alt. 2600 m. 19 Feb. 1959. 'Geröllhang zwischen Kakteen u. *Abromeitiella*.' *Ross & Vervoorst 79* (Max.).

Dept. Tilcara, Tilcara, arriba de San Gregorio. Alt. *c.* 3850 m. 27 Jan. 1953. 'Fl. violác. Rarísima.' *Sleumer 3592* (LIL).

Dept. Tilcara, Casa Colorada, subida a la Abra de Remate. Alt. 3800 m. 23 Feb. 1953. 'Fl. viol.' *Sleumer 4106* (LIL, J.G.H.).

Dept. Tilcara, Casa Colorada, subida a la Abra de Remate. Alt. 3500 m. 23 Feb. 1953. 'Fl. viol.' *Sleumer 4107* (LIL, J.G.H.).

Dept. Tumbaya, Quebrada de Punta Corral. Alt. 3800 m. 20 Mar. 1952. *Petersen & Hjerting P–340* (J.G.H.).

Dept. Tumbaya, Quebrada de Punta Corral. Alt. 3800 m. 20 Mar. 1952. 'Flor lila clara.' *Petersen & Hjerting P–342* (LIL, J.G.H., J.P.H.).

Prov. SALTA:

Dept. Iruya, Rodero Colorado, arriba del Valle Delgado. Alt. 3300 m. 7 Apr. 1952. 'El fruto en forma de huevo, gris, marmorado en verde obscuro, 2 × 2·5 cm.' *Petersen & Hjerting P–365* (LIL, J.G.H., J.P.H.). (CPC: 2480. $2n = 24$.) See Plates 120 and 121.

Dept. Santa Victoria, Cuesta de San José, (entre Santa Victoria y Toldos). Alt. 2650 m. 11 Feb. 1956. *Hjerting & Rahn 190* (J.G.H., J.P.H.). See Plate 122.

Probably *S. gourlayi*

Prov. JUJUY:

Dept. Humahuaca, entre Cóndor y Abra del Cóndor. Alt. 3700 m. 9 Apr. 1952. 'Crece bajo arbustos.' *Petersen & Hjerting P–368* (p.p.) (J.G.H.).

Dept. Tilcara, cerca del Garganta del Diablo. Alt. 2850 m. 19 Mar. 1952. 'Entre piedras y cactus.' *Petersen & Hjerting P–337* (J.G.H.).

Dept. Tilcara, Tilcara, Puesto Amarillo. Alt. *c.* 3000 m. 25 Feb. 1953. 'Fl.? Sobre terreno muy seco.' *Sleumer 4119* (LIL, J.G.H.).

Dept. Tumbaya, 'Encrucijada', arriba de Huachichocana. Alt. *c.* 3600 m. 4 Jan. 1953. 'Fl. viol.' *Sleumer 3217* (LIL, J.G.H.).

Dept. Tumbaya, San Antonio (Abra) de Los Pibes. Alt. 3850 m. 5 Jan. 1953. 'Fl. violác.' *Sleumer 3253* (p.p.) (LIL).

DISTRIBUTION

S. gourlayi is known up to now only from the far north of Argentina, in the high Andes of provs. Jujuy (depts. Tilcara and Tumbaya and also possibly Humahuaca) and Salta (depts. Iruya and Santa Victoria) (see Map 12). It probably spreads northwards into south Bolivia, though so far only one authentic record for it from that country is known (*Correll* et al. *B645*).

HABITAT

S. gourlayi is found at altitudes of (2600–)3300–4000 m in moderately humid parts of the upper Prepuna, in the Puna province and in the upper part of the Montano district ('pajonales') of the western Subtropical province (Cabrera 1953). It grows in stony or sandy soil, often sheltered to some extent by grass and small shrubs or other vegetation. In the drier part of its range it occurs with cacti and bromeliads and is occasionally found with *S. infundibuliforme*, whilst in the wetter part it may come into contact with *S. acaule*. In most regions it grows with *S. megistacrolobum*, and has once (*H. & R. 190*) been recorded growing with *S. microdontum* and *S. boliviense* under conditions of greater humidity and lower altitude (2650 m) than appear to be usual for *S. gourlayi*. There is also some evidence to show that *S. gourlayi* occurs with *S. vidaurrei* and with *S. oplocense* in certain regions and that natural hybrids between these species may be formed.

TAXONOMIC NOTES

This comparatively little-known species was collected by Balls, Gourlay, and Hawkes in 1939 at San Gregorio in the mountains east of Tilcara, prov. Jujuy (*Balls 5979*, *5987* (p.p.), *5990*). It was described by Hawkes in 1944, who considered *Balls 5979* to be the type. In order to avoid any future confusion we now designate one of the P.G.S. specimens of *Balls 5979* as the LECTOTYPE for this species. (See Fig. 37 and Plate 119.) Tuber material collected in 1939 unfortunately failed to grow, and living collections were not made again until the expedition of Petersen and Hjerting in 1952. Unfortunately, even now, only one living collection has been available for study, and most of our ideas on the distribution and altitude range of *S. gourlayi* are based on rather poor stunted collections, often without flowers. It is likely that *S. gourlayi*, in common with many other species growing in rather dry regions, does not grow or flower well in many years owing to lack of moisture, and in consequence many of the herbarium specimens are very stunted. A wider range of living material is needed before we can make any very valid statements on its growth and distribution, and the account which follows must be regarded as provisional.

Apparently environmental conditions can cause some modification in the leaf of *S. gourlayi*. The type collection (*Balls 5979*) possesses well-developed leaves, 3–5-jugate, with relatively short broad leaflets, 3–5 mm long petiolules, truncate to rounded leaf bases, and 0–2 pairs of interjected leaflets. The other collections quoted above probably represent different biotypes, or perhaps ecads, with less well-developed leaves, or with shorter petiolules and fewer interjected leaflets. Such is the growth

FIG. 37. *S. gourlayi* Hawkes. Lectotype specimen of *Balls 5979* (Herb. P. G. S.) collected by Balls, Gourlay, and Hawkes at San Gregorio, Tilcara, prov. Jujuy, Argentina. This plant shows sessile to subsessile leaflets with some slight decurrency.

form of *P. & H. P–365* whose original specimens possess short petiolules and very few or absent interjected leaflets with rather cordate leaf bases. (See Plate 120.) These specimens would seem to differ quite considerably from the type, yet, as we shall see in the section dealing with living plants, this collection when grown in Europe (see Plate 121) showed the typical features of *S. gourlayi* as expressed in the type specimen. By inference, then, we have placed other collections from Argentina under *S. gourlayi* when they agreed with the original wild collection of *P. & H. P–365*, concluding that they were habitat modifications only (*Balls 5987* (p.p.), *P. & H. P–340*, *P–342*, *Sleumer 4106*, *4107*). Original wild collections which matched the type specimen well were *P. & H. P–356*, *H. & R. 190* (see Plate 122) and *Ross & Vervoorst 79*. Presumably these were collected in a wetter region or during a year when the rainfall was higher than average.

It is sometimes impossible to make an identification of potato specimens that is 100 per cent certain, and several of the very stunted forms might be referable to some other species, such as *S. oplocense* or *S. vidaurrei*. For that reason, and because it seems probable that hybrids might occur amongst these three species, we have designated certain forms as 'Probably *S. gourlayi*'. The better-developed flowering specimens of *Sleumer 3217* induce us to consider that this collection might be a hybrid of *S. gourlayi* with *S. oplocense*, since the leaflets are more decurrent than is normal for *S. gourlayi*. However, *S. oplocense* has not been recorded so far south,[1] and until more is known about the distribution and variation of these two species it will probably be better to leave the determination of Sleumer's specimen undecided.

A collection of Petersen and Hjerting from prov. Jujuy, dept. Humahuaca, between Tres Cruces and Iturbe (*No. 962*), which we at first thought should be placed as *S. gourlayi*, seems now to fit better with *S. oplocense*. It is of particular interest in being tetraploid, and may on further investigation perhaps be placed as an amphiploid hybrid between the two species. This fact further underlines the urgent necessity for more living collections from prov. Jujuy to help solve such problems.

Correll (1961, 1962) groups *S. gourlayi* as a form of *S. leptophyes*, stating that it 'graduates from typical *S. leptophyes* in being usually less pubescent and in having somewhat broader leaflets and few or usually no interstitial leaflets. The lateral leaflets are also usually, but not always, more or less decurrent on the leaf rachis. In this latter characteristic it shows some resemblance to *S. infundibuliforme* of series Cuneoalata.'

Undoubtedly *S. gourlayi* is related in some way to *S. leptophyes*, and it may finally be necessary to place it as a southern subspecies of that species. The situation is at present obscure because of a general lack of knowledge of material of this group of plants from Bolivia, and indeed an incomplete knowledge of *S. gourlayi* itself. It is not surprising, therefore, that Correll has classified this species in a somewhat different way from ourselves, and much more needs to be known about the pattern of variability of *S. gourlayi*, *S. oplocense*, *S. vidaurrei*, *S. spegazzinii*, and *S. leptophyes* before a generally acceptable scheme of classification of this group can be advanced.

One final point remains. We have stated above that no or only very slight leaflet decurrency is to be seen in *S. gourlayi*, whilst Correll in the passage quoted above considers that decurrency is usual, though not always present. This discrepancy may be explained by the fact that we have excluded from this species certain forms with decurrent leaflets that Correll has associated with it. These are *Correll* et al. *A666* (considered by us to be a form of *S. oplocense*), *Correll* et al. *A677*[2] (placed by us as a form intermediate between *S. spegazzinii* and *S. vidaurrei*, cf. *Hjerting*, *Petersen & Rahn 264*, p. 392) and *Sleumer 4119* (placed by us as 'probably *S. gourlayi*'). More collections of living material from provs. Jujuy and Salta are needed to solve the problems of these aberrant forms; until that is done we can only say that we are not certain as to how they should best be classified.

NOTES ON LIVING PLANTS

Only one living collection has been available for study, *Petersen & Hjerting P–365* (CPC: 2480). See Plates 120, 121. This attains a height of 70 cm in the field at Vandel and in Birmingham, and shows vigorous growth with dense foliage. The leaves are (3–)4(–5)-jugate, with 0–2 (occasionally to 6) pairs of interjected leaflets and long petiolules (to 5 mm or occasionally even to 13 mm). The leaflets are normally broad ovate-lanceolate with a truncate to rounded base and no decurrency. Leaf and stem pubescence is normally sparse and there are generally no marginal hairs to the leaflets;

[1] See Appendix III, p. 486, however.

[2] Only photograph of this collection seen.

if marginal hairs are present they are small and inconspicuous. It is worth noting, however, that plants of *P. & H. P–365* grown at Vandel were quite hairy on the lower leaf surface.

The flowers are borne on pedicels articulated about three-quarters length above the base. The corolla varies a little in colour from seedling to seedling but shows a constant pentagonal to semi-stellate shape with large 4–5 mm long acumens which are not well delimited from the lobes.

The tuber weight averaged about 5·5 grammes in one season at Vandel. Frost resistance of this clone at Vandel was not very well marked. *S. gourlayi* has been noted as resistant to eel-worm and flea beetles and to bear some resistance to Colorado beetle. (See Chapter 5.)

CYTOLOGY

A chromosome number of $2n = 24$ was established for *P. & H. P–365* (CPC: 2480). The count was made in 1957 and was verified from further fixations in the same year.

No meiotic investigations have yet been carried out on this species.

FERTILITY AND CROSSABILITY

Very little information on the fertility of *S. gourlayi* is available at present, and no natural seed was set in the field at Birmingham or Vandel.

Hybrids of *S. gourlayi* × *S. chacoense*, *S. gourlayi* × *S. infundibuliforme*, and *S. gourlayi* × *S. vernei* subsp. *ballsii* have already been mentioned in other sections (pp. 228, 280, and 368).

The F_1 hybrid (Family 290) between *S. gourlayi* and *S. sparsipilum* (a diploid weed species from Bolivia and Peru) was on the whole rather uneven in the first sowing but fairly uniform in another sowing in the following year. Two F_2 families (320, 321) were grown. Of a total of 100 seeds sown, 87 germinated, 44 survived to be planted out in the field, and 33 finally survived to the end of the season. In the wide range of size and thriftiness the seedlings exhibited a clear genetic breakdown, showing that there is genomic incompatibility between the two species involved.

Attempts to cross *S. gourlayi* with different lines of *S. megistacrolobum* (P. & H. P–316, P–366) in both directions were unsuccessful.

NATURAL HYBRIDS

(*a*) *S. gourlayi* seems to form only a limited number of natural hybrids. Thus, we have seen no hybrids with *S. acaule* and it is possible that there might be genetical or other barriers preventing natural hybridization between these two species. Doubtful hybrids of ***S. tarijense*** × ***S. gourlayi*** and ***S. megistacrolobum*** × ***S. gourlayi*** have already been mentioned (pp. 245 and 295).

Only a single hybrid collection of ***S. infundibuliforme*** × ***S. gourlayi*** (*Sleumer 3253*) (p.p.) has been recorded. This was growing in company with *S. infundibuliforme* itself, and probable *S. gourlayi*. (See p. 280 and Plate 80.)

(*b*) Probable hybrids between ***S. gourlayi*** and ***S. vidaurrei*** were collected by Hjerting and Rahn in Santa Victoria (*Nos. 176, 254*), prov. Salta.

H. & R. 176 from between Mesón and Abra Delgada is roughly intermediate in leaf type between the two species; the leaflets are narrower than those of *S. gourlayi* though not so narrow as those of *S. vidaurrei*; the pubescence is denser than in *S. gourlayi*, well-marked marginal hairs are present, and interjected leaflets are absent or very rare.

H. & R. 254 from between Lizoite and Abra de Lizoite is obviously a range of hybrid segregates, with parental forms also (see Plate 123). The twenty-three specimens may be roughly sorted as follows:

(i) Fairly typical *S. gourlayi*: 2 specimens
(ii) Fairly typical *S. vidaurrei*: 7 specimens
(iii) Intermediates of all types: 14 specimens

In fact, the variation is probably continuous, and represents no doubt a segregating population of hybrid recombinants between the two species. *S. vidaurrei* has been collected by Sleumer from Lizoite (*No. 4059* from 3450 m—c.f. altitude of *H. & R. 254*: 3450–3650 m), but *S. gourlayi* is known only from Cuesta de San José, 40–45 km away (*H. & R. 190*: 2650 m). There seems to be some evidence, then, that *S. gourlayi* and *S. vidaurrei* are the parents of the crosses we have been discussing.

(*c*) Evidence for hybridization between ***S. gourlayi*** and ***S. oplocense*** is meagre but there are at least some signs that it takes place, as we have already stated.

Thus *Sleumer 3217* ('Probable *S. gourlayi*'—see p. 375) may be one example of this type of hybrid, whilst the tetraploid form of *S. oplocense* to be discussed shortly (p. 382) may be another. Insufficient evidence is available to make it profitable to discuss these possible hybrids any further, however.

LOCALITIES OF PROBABLE NATURAL HYBRIDS

(*a*) With ***S. tarijense, S. infundibuliforme***, and ***S. megistacrolobum*** (see pp. 246, 282, and 297)

(*b*) With ***S. vidaurrei***

Prov. SALTA:

Dept. Santa Victoria, entre Mesón y Abra Delgada. Alt. 3400 m. 11 Feb. 1956. *Hjerting & Rahn 176* (J.G.H., J.P.H.).

Dept. Santa Victoria, entre Lizoite y Abra de Lizoite. Alt. 3450–3650 m. 17 Feb. 1956. *Hjerting & Rahn 254* (J.G.H., J.P.H.). (Plants range from good *S. vidaurrei* through intermediates to good *S. gourlayi*.) See Plate 123.

(*c*) With ***S. oplocense***

No authentic records.

15. SOLANUM OPLOCENSE Hawkes, *emend* Hawkes et Hjerting

(*Bull. imp. Bur. Pl. Breed. Genet., Camb.* 119, 1944)

S. subandigenum Hawkes, var. *camarguense* Cárdenas (*Boln. Soc. peru. Bot.* **5**, 25–26, 1956).

Herbaceous, tuberiferous, often with a long rootstock and long creeping stolons which bear small pale brown tubers up to 2·5 cm diam.

Stem often short, (5–)30–70 cm tall, 2–4 mm diam., frequently flexuose and with short internodes, with decurrent lines or narrow wings from the leaf bases, provided with sparse to quite frequent hairs of varying lengths and sometimes a few stalked glands.

Leaves generally closely following each other on the stem, *often aggregated into a rosette in the shorter-stemmed plants*, rather broad in relation to their length, (6–)10–15

(–18) cm long × (4–)5·5–8(–11) cm broad, (1–)3–4(–5)-jugate; lateral leaflets ovate to elliptic, occasionally oblong, apically acute or slightly acuminate, *basally gradually attenuated and decurrent into a generally well marked petiolule* (1–)3–7(–10) mm long; largest laterals 20–70 × 10–30 mm; interjected leaflets 0–2(–3)-paired, never more than one pair between each two pairs of primary laterals, ovate to orbicular, to 7 mm long, sessile or subsessile; terminal leaflet about the same size as or a little larger than the uppermost laterals, narrowly rhomboid, elliptic, or obovate, much wider in the more juvenile leaves and then *sometimes subcordate to hederiform*, when fully developed 35–40(–60) mm × 12–18(–20) mm, in juvenile leaves to 45 × 37 or 33 × 33 mm, etc.; *rhachis frequently with narrow wings* running down from the leaflet bases on the basiscopic sides of the petiolules; upper leaf surface clothed with a few glands and with sparse to frequent thick shortish hairs which are thinner and more frequent on the lower surface; margins denticulate, with a thick short acroscopic hair on the apex of each minute tooth.

Inflorescence few, 2–6(–12)-flowered. Peduncle short, 4–6 cm long, once-forked, occasionally bearing a reduced leaf on its lower half. Pedicels short, (7·5–)15–25 mm long, articulated generally above the centre but *occasionally with much lower, almost basal articulation*, clothed like stem and pedicels with dense uni–multicellular hairs and a few minute glands.

Calyx broadly campanulate, the lobes slightly reflexed in the flowering state; calyx itself 4–8(–10) mm long, with lobes 2·5–5 mm long, ± unequal-lengthed, including the 2–3 mm long narrow triangular to linear acumens; pubescence as for pedicels.

Corolla violet to purple with a pale star, *pentagonal to semi-stellate*, 15–25(–30) mm diam.; free parts of petals 6–7(–9) mm long × 8–9(–13) mm wide at the base; acumen 4–6 mm long, *often not well delimited from the base of the lobe.*

Filaments *very short*, 0–0·5(–1) mm long; anthers subsessile *strongly tapered* from base to apex, 4–6(–8) mm long × 2–2·5 mm at the base, narrowing to ± 1 mm at the apex; *base of anther often 4-lobed.*

Style slender, 10–11 mm long, exserted 5–6 mm, with small papillae in the lower half; *stigma minute* (except in tetraploid forms), generally capitate to oval, the same thickness as the style apex or slightly thicker.

Berries globular to slightly oval, 7–9 mm diam. (perhaps larger?), pale green.
Chromosome number: $2n = 24, 48$.

AFFINITIES

S. oplocense is probably rather closely related to *S. gourlayi*, *S. vidaurrei*, and *S. spegazzinii*, agreeing with them in possessing a sub-stellate to pentagonal corolla and in its ability to grow in the rather drier parts of the mountains of north-west Argentina and south Bolivia.

The chief distinguishing features of *S. oplocense* are its not very well dissected leaves, petiolulate leaflets with attenuated base running down the basiscopic side of the petiolule on to the narrowly winged rhachis, the frequently denticulate leaf margin, the very short filaments, strongly tapered anthers from base to apex, frequently with two extra basal lobes, and the minute stigma.

From *S. gourlayi* it differs in the coarser pubescence, marginal leaflet denticulation, basally attenuated and decurrent leaflets, and form of stamens and stigma. *S. oplocense* differs from *S. vidaurrei* in the petiolulate broader leaflets and the form of stamens and stigma. From *S. spegazzinii* it differs in stamen and stigma form, the less well-dissected leaf, and denticulate leaflet margins.

SPECIMENS EXAMINED

BOLIVIA

Dept. CHUQUISACA:

Prov. Nor Cinti, Purón near Tacaquira. Alt. 2800 m. 28 Feb. 1959. *Alandia, Rimpau & Ross 140* (Max.) ($2n = 24$, *fide* Ross).

Prov. Nor Cinti, near Camargo. Alt. 2200 m (or higher?). Feb. 1949. 'On clayish soil. 30–70 cm. Flower violaceous, white star.' *Cárdenas 5078* (K, J.G.H., Isotypes of *S. subandigenum* Hawkes, var. *camarguense* Cárdenas). See Plate 127.

Dept. POTOSÍ:

Prov. Sud Chichas, Villazón, Quebrada de La Quiaca. Alt. 3450 m. 23 Feb. 1959. *Alandia & Ross 106* (Max.). ($2n = 48$, *fide* Ross.)

Prov. Sud Chichas, between Oploca and Oro Ingenio. Alt. 3800 m. 28 Feb. 1939. 'Growing amongst loose shaley stones on the railway embankment.' *Balls 6127* (K, P.G.S.—lectotype, J.G.H.). See Plate 124.

Prov. Sud Chichas, Salo district near Tupiza. Alt. 3400 m. 'Sandy dry soil.' *Braun* (s.n.) (CPC: 2116—herb. specimen from field, Cambridge—P.G.S., J.G.H.). ($2n = 24$.) See Plate 126.

Prov. Sud Chichas, near Oro Ingenio. 'Under thickets of *Schinus molle*.' *Cárdenas* (s.n.) (CPC: 2503—herb. specimen from field, Birmingham—J.G.H.). ($2n = 24$.) See Plate 125.

Prov. Sud Chichas, near Mojo. 23 Feb. 1960. 'In *Acacia* thicket; flowers blue.' *Correll, Dodds, Brücher & Paxman B659* (K).

Dept. TARIJA:

Prov. Méndez, Cienaguillas. Alt. 3200 m. 26 Feb. 1959. *Alandia, Rimpau & Ross 133* (Max.).

Prov. Méndez, Agua de Torre, near Sarcara. Alt. 3000 m. 26 Feb. 1959. *Alandia, Rimpau & Ross 134* (Max.). ($2n = 24$, *fide* Ross.)

ARGENTINA

Prov. JUJUY:

Dept. Humahuaca, near Azul Pampa. 25 Feb. 1960. 'On rocky, bushy slope. Flowers purple, rotate—pentagonal.' *Correll, Dodds, Brücher & Paxman A666* (K, upper plant).

Dept. Humahuaca, entre Iturbe y Tres Cruces, km 1313. Alt. 3500 m. 9 Feb. 1953. *Petersen & Hjerting 962* (J.G.H., J.P.H.) (H. 35, $2n = 48$). See Plates 129 and 130.

Dept. Yavi, Yavi, in den Marmorklippen des Río Sococha. Alt. 3500 m. 24 Feb. 1959. 'Mattenformation.' *Alandia & Ross 123* (Max.) ($2n = 48$, *fide* Ross). See Plate 128.

DISTRIBUTION

S. oplocense is known in Argentina from five collections only, all made in prov. Jujuy, depts. Humahuaca, Tumbaya (see Appendix III), and Yavi (see Map 12). It has been collected in south Bolivia more frequently, namely, in dept. Potosí, prov. Sud Chichas; dept. Tarija, prov. Méndez; and dept. Chuquisaca, prov. Nor Cinti.

It is reasonable to suppose that *S. oplocense* is not quite as rare in northern Argentina as these five records would seem to indicate, but that it has previously been overlooked.

HABITAT

S. oplocense is to be found at altitudes of 2800–3500 m[1] and may perhaps extend up to 3850 m or even to 4000 m. It seems to inhabit rather dry regions and often grows in sheltered ravines in the puna. It is often found amongst loose earth and stones, amongst thorn bushes, under the shade of *Prosopis ferox*, and with other drought and cold resistant plants. On Cabrera's phytogeographical system it would be placed in the upper part of the Prepuna and lower part of the Puna provinces. Apparently it grows in drier regions than those normally inhabited by *S. gourlayi* but we do not yet know the extent to which these two species overlap ecologically. It has been recorded in Bolivia quite close to *S. vidaurrei* and it could possibly be found in the same regions as *S. infundibuliforme*. On the whole, the data to hand for this species are very meagre, and the above account must be regarded as provisional.

TAXONOMIC NOTES

This little-known species was described by Hawkes in 1944 on the basis of a collection made in 1939 by Balls, Gourlay, and Hawkes (*Balls 6127*, see Plate 124) on the railway embankment between Oploca and Oro Ingenio in south Bolivia, dept. Potosí [*not* Tarija, as indicated on the original description]. The plants were obviously stunted by the dry conditions and poor soil, and represented a rather special habitat form of this species. As no holotype was indicated in the original publication we think it advisable to designate as LECTOTYPE the specimen in P.G.S. of *Balls 6127* that was figured in 1944 by Hawkes, though not then designated as holotype.

A topotype collection was made by Cárdenas in 1954 (CPC: 2503, see Plate 125), and another collection from the vicinity of Salo, Tupiza was made by Sr. Otto Braun (CPC: 2116, see Plate 126) at about the same time. Fortunately, both these latter collections have been studied in Birmingham in the living state.

Later, in 1956, Cárdenas described what he considered to be a variety of the tetraploid weed species *S. subandigenum*,[2] namely var. *camarguense*, from Camargo, about 100 km to the north-east of the type locality of *S. oplocense*. Cárdenas recently kindly sent isotype material of this taxon (see Plate 127) to Birmingham and we have been able to compare it with *S. subandigenum* and *S. oplocense*. In 1956 Hawkes considered that the tetraploid weed species *S. subandigenum* was no more than escaped forms of the cultivated tetraploid *S. tuberosum* subsp. *andigena*, and could not be considered as a separate taxon. However, var. *camarguense* of Cárdenas is, in our opinion, quite distinct specifically from *S. tuberosum* subsp. *andigena*, and we consider that it should more correctly be placed as a synonym of *S. oplocense*.

Although the type specimens of *S. oplocense* and var. *camarguense* are very dissimilar, the two collections of *S. oplocense* made by Cárdenas and Braun link the two extremes rather conclusively. Furthermore, we have recently been fortunate in being able to examine the collections of Ross and his collaborators (see Plate 128) which were identified by Cárdenas as *S. camarguense* (having later considered that it should be raised to specific rank). These match very well indeed with the Cárdenas and the Braun collections (CPC: 2503 and 2116) in leaf, pubescence, and flower type. The latter collections were grown in Birmingham and agree well with Ross's collections, both the originals and those grown from the same material in Cologne. Undoubtedly the original collections of *S. oplocense* were stunted and atypical for the species as a whole, though they possess enough features in common with the Cárdenas topotype

[1] Bolivian data also used here. Cárdenas's type collection of var. *camarguense* was recorded for 2200 m, but this is probably wrong, since Camargo is situated at an altitude of 2770 m, according to our maps.

[2] Should be *S. subandigena* according to *Index Kewensis*.

collection (CPC: 2503) to leave us in no doubt about them. This situation shows clearly the misleading nature of statements made by Correll and other authors that all comparisons must be made on material collected in the original localities. If this had remained so for *S. oplocense* and var. *camarguense* we should never have seen the connexion between them. Only when we compared the living collections cultivated under similar or roughly similar conditions in Europe was it possible to solve the problem of the relationship between *S. oplocense* and var. *camarguense*. With the other species from similar rather dry or cold regions, *S. gourlayi*, the same difficulties were experienced in placing original South American specimens into the correct species. Living collections grown in Europe (or in any experimental field where conditions for potato growth are reasonable and there is not too much virus infection) have time and again provided the key to taxonomic position and relationships.

As a consequence we have found it necessary to emend the original description of *S. oplocense* by adding details from material collected by Ross and others in South America.

As we shall see later, in the section on living plants, most of the plants collected by Ross in South America match our Birmingham-grown plants of *S. oplocense* from the type locality remarkably well.

Correll (1961) classified certain collections made by him in southern Bolivia (*Correll, Dodds, Brücher & Paxman 632, 649, 657, 659*) as '*S. tacnaense* var. *sandemanii* (Hawkes) Correll'. We have seen the Kew sheet of one of these and photographs of two others and are convinced that these also belong to *S. oplocense* and have nothing to do with the species under which Correll classified them.

S. oplocense is of further interest in that it occurs in two chromosome races, one with 24, the other with 48 somatic chromosomes. We thought at first that the tetraploid form ought to be classified as *S. gourlayi* (Hawkes 1958), since it differs in some features from the type of *S. oplocense* and even from the material of Cárdenas and of Braun grown at Birmingham. However, as our concept of the ranges of variability in the two species was made clearer it became obvious that the tetraploid fitted best into *S. oplocense*. Ross reports, also (personal communication), that tetraploid forms of the Bolivian var. *camarguense* are known.

Only three living collections have been studied by us, two of them diploid (from Bolivia) and the third tetraploid. Professor Ross kindly informs us that there were two diploid and two tetraploid forms in his collections of *S. oplocense*; one of the tetraploids (*No. 123*) was from Argentina and the other three collections (two diploids and a tetraploid) from Bolivia. Apart from this information we have no idea of the frequency of the tetraploid form, nor can we be certain that it can always be recognized on morphological features alone.

We have already mentioned the tetraploid collection of so-called *S. vernei* (*No. 200*) described by Brücher in various publications (1951, 1954, etc. See p. 365), and we have suggested that this form might in reality belong to *S. oplocense*, or even to *S. tuberosum* subsp. *andigena*. Unfortunately Brücher has not published any description or figure of his tetraploid collection and we have not been able to examine his material. Our own tetraploid collection, although gathered in prov. Jujuy, was obtained some distance away from that of Brücher, and we have as yet no adequate information about the distribution or frequency of *S. oplocense* in Argentina.

NOTES ON LIVING PLANTS

(*a*) DIPLOIDS

The only two diploids so far studied by us are the collections of Cárdenas (CPC: 2503, topotype, see Plate 125) and Otto Braun (CPC: 2116 from Salo, Tupiza, see Plate 126), both from south Bolivia. The following notes refer to these, therefore, since no living diploid Argentine material of *S. oplocense* has yet been collected.

In both collections the short 2–4(–5)-jugate leaf with 0–2 pairs of interjected leaflets was very characteristic, as also were the winged petiolules and rhachis. The attenuation of the leaflet base and the winged petiolule and rhachis were not apparent in the type specimen of Balls, Gourlay, and Hawkes, which was rather stunted; these features show up fairly well on Birmingham-grown topotype material and are even better developed on the Tupiza material (CPC: 2116) whose leaves match perfectly with Cárdenas's and Ross's collections of var. *camarguense*.

The stiff white hairs on the leaflet margins set on the apices of small teeth are most characteristic in the plants grown by us, and agree well with the marginal hairs on plants collected in Bolivia. In the living topotype material an interesting feature is that some of the terminal leaflets of the lower leaves are lobed in the same way as the type specimen. This tendency to terminal leaf lobing seems to be confined, so far as we know at present, to the type locality.

On the living plants grown in Europe the glistening glandular pubescence on peduncle and pedicels is most noticeable. Corolla form and colour agree well with collections from South America and are reasonably constant. The calyx varies from regular to a little irregular in form, whilst the stamen morphology is constant and characteristic. The very conical anther column in the living plant is also most distinctive.

A curious feature in these two collections is the presence of one or two extra lobes at the anther base. We do not yet know whether this is characteristic of the species as a whole since it is not mentioned by Cárdenas for his var. *camarguense*. The lobes are present, though not quite so well marked, in the tetraploid form of *S. oplocense* also. Unfortunately the isotype specimen of var. *camarguense* sent us by Cárdenas had no flowers. In one seedling of CPC: 2503 the lobes at the base of the anthers were not well marked and the style was pigmented a deep purple, the same colour as the corolla, except just above the base. The anthers and filaments remained completely devoid of pigment.

The minute stigma, no thicker than the style apex, was very characteristic for both collections and would appear to be a useful taxonomic feature, though Cárdenas makes no special mention of stigma size in his description of var. *camarguense*. Only one of the south Bolivian collections of *S. oplocense* made by Ross has a flower with an intact stigma, and this also is very small.

(*b*) TETRAPLOID

One collection of tetraploid *S. oplocense* (*P. & H. 962* (H. 35), from Argentina, prov. Jujuy, dept. Humahuaca) has been under observation at Vandel and Birmingham (see Plate 130). Whilst attaining about the same height as the diploid it is more delicate,[1] with branched, lax semi-decumbent stems and more delicate leaves. The leaf dissection was about the same as that of the diploid but the leaflets were, on the whole, narrower, being ovate-lanceolate rather than ovate or elliptic. The lateral leaflets are not so decurrent on to the rhachis, as in the diploid. In floral characters the diploid and tetraploid forms were very similar, though the filaments of the tetraploid were a little longer (to 1·5 mm) than those of the diploid. The form of the anther, including the curious basal lobing, was also similar. The stigma of the tetraploid was a little larger on the original material, though on living collections it was as small as on typical diploid collections.

Tuber yields in the tetraploid were poor, the tubers themselves averaging barely 1 gramme in weight. The frost resistance of this tetraploid line was good.

It seems curious that the tetraploid is more delicate and possesses narrower leaflets

[1] The tetraploid collections made by Ross, however, seem to be quite robust.

than the diploid forms, since autotetraploids commonly possess wider leaflets and thicker stems than the corresponding diploids. In view of this we should like to put forward the tentative suggestion that the tetraploid form might be an amphiploid hybrid of *S. oplocense* and *S. vidaurrei*, or possibly *S. oplocense* × *S. gourlayi*. However, the data and collections at present available are insufficient for us to be more definite, and it is to be hoped that new samples of living material of this interesting species will be collected.

S. oplocense (unspecified whether 2*x* or 4*x* cytotypes) has been shown to exhibit resistance to eelworm and to frost (see Chapter 5).

CYTOLOGY

A somatic chromosome number of $2n = 24$ was established for CPC: 2116 and for CPC: 2503 at the Potato Genetics Station in Cambridge. The collection of *P. & H. 962* (H. 35) from Argentina was found by us to possess a somatic chromosome number of $2n = 48$; the original counts were made in 1954 and were verified from new fixations in 1957. Tetraploid forms were also collected by Alandia and Ross (written communication), both in Argentina (*No. 123*) and Bolivia (Villazón, *No. 106*). They also made diploid collections (*Nos. 134, 140*) from Bolivia, but so far no authentic diploids have been obtained in Argentina.

No meiotic investigations have yet been made on this species.

FERTILITY AND CROSSABILITY

The diploid forms of *S. oplocense* are apparently self-incompatible. Not many crosses have been attempted with them, nor have we had much success with the crosses that we have attempted. No success was obtained in pollinations with *S. microdontum*, whilst in the cross *S. maglia* (2*x*) × *S. oplocense* (CPC: 2116) only one berry with two seeds was produced, and these did not germinate. *S. vernei* × *S. oplocense* (CPC: 2116), also, only yielded three empty seeds. A good seed set was obtained in the combination *S. sucrense* × *S. oplocense* (CPC: 2116), but again, no germination took place. There seems to be an indication of crossability barriers of some type here, though more work is needed to supplement these few results.

With the tetraploid, which does not set natural seed in field or glasshouse, very few crosses have been attempted. The cross *S. tuberosum* (Beltsville 2159) × *S. oplocense* 4*x* (*P. & H. 962*) was successful, but the seed germination of the hybrids was poor. Only five seedlings could be raised and these seemed intermediate between the two parents. No cytological analysis of these seedlings has yet been made.

NATURAL HYBRIDS

We have no authentic records of natural hybrids between *S. oplocense* and other species, though it is possible that hybrids might be formed with *S. megistacrolobum* (see p. 295), *S. gourlayi*, *S. vidaurrei*, or even with *S. infundibuliforme*. Certain specimens mentioned on p. 373 as 'probably *S. gourlayi*', especially *Sleumer 3217*, may perhaps be *S. gourlayi* × *S. oplocense* hybrids, but at the moment there is insufficient evidence for us to be sure of this. We mentioned, also, that the tetraploid form of *S. oplocense* may be an amphiploid hybrid between *S. oplocense* and *S. vidaurrei*, or *S. gourlayi*, but more collections are needed before this problem can be solved.

16. SOLANUM VIDAURREI Cárdenas[1]

(*Boln Soc. peru. Bot.* **5**, 26–30, 1956)

Herbaceous, tuberiferous, 10–50 cm tall, with very thin creeping stolons bearing small, 5–10 mm diam., light-brown tubers.

Stem simple to much branched, straight or flexuose, *slender*, only 1·5 mm diam., rather glabrous below, with decurrent lines, and scattered short white hairs above.

Leaves often aggregated into a semi-rosette in plants with very short stems, 5–15 cm long, and 3–9 cm broad, 3(–5)-jugate; *interjected leaflets absent* or occasionally 1–2-jugate, ovate, sessile to subsessile, up to 13×6 mm; lateral leaflets *lanceolate to narrow lanceolate*, the uppermost from 20–50 mm long×3–18 mm broad, *generally sessile* or occasionally with petiolules to 5 mm long, the upper pair sometimes broadly decurrent on to the rhachis, but normally only with a very narrow wing running down the petiolules on to the very narrowly winged rhachis; terminal leaflet often a little larger than upper laterals, *lanceolate to oblanceolate*, 20–50×5–18 mm; *all leaflets rather pale yellowish green*, provided with short thick quite frequent white hairs (also often on rhachis), the *margins generally denticulate* and each tooth provided with one or two short acroscopic hairs (cf. *S. oplocense*).

Inflorescence 2–7-flowered, not or only just surpassing the leaves. Peduncle 3–4·5 cm long, generally not forked, provided with numerous short thick white hairs. Pedicels 10–15 mm long, articulated in the upper third or occasionally at the centre, provided with pubescence of same type as the peduncle.

Calyx *c.* 5 mm long, dark green, with dense white hairs of same type as on peduncles and pedicels; lobes short, but provided with narrowly triangular to linear acumen 1–2 mm long.

Corolla *pentagonal* (? rotate), 10–15(–25?) mm diam., blue-violet; lobes 4–7 (or more ?) mm long, including the well-marked 2–5 mm acumen, puberulent on both surfaces.

Filaments 1 mm long; anthers 5 mm long, gradually tapering from base to apex.

Style *c.* 8 mm long, exserted about 2–3 mm beyond anther column, minutely papillose below; stigma dark green, globose, manifestly larger than style apex.

Berries not seen.
Chromosome number: $2n = 24$.[2]

AFFINITIES

S. vidaurrei seems to be part of the complex of species (or possibly subspecies) represented in north Bolivia by *S. leptophyes* and *S. pachytrichum* and in south Bolivia and north-west Argentina by *S. oplocense* and *S. gourlayi*. Further south still (provs. Salta, Tucumán, Catamarca, La Rioja, and San Juan), it is replaced by *S. spegazzinii*. We have followed Cárdenas in retaining it as a separate species, for the time being.

It differs from *S. spegazzinii* in the smaller number of generally sessile lateral leaflets, the interjected leaflets few or absent, the well-marked marginal hairs, and the low delicate habit. From *S. oplocense*, to which it is also similar in many respects, it

[1] Description modified from that of Cárdenas.
[2] We are indebted to Professor H. Ross for this information, based on Bolivian plants.

differs in the much narrower leaflets, very narrowly winged rhachis, distinct form of anthers, and the larger globose stigma. It shares with *S. oplocense*, however, the feature of a generally denticulate leaflet margin with hairs set apically on the teeth. From *S. gourlayi* it differs in the relatively narrower leaf, narrow, generally sessile leaflets, and distinctive leaf margin. From *S. infundibuliforme*, which also occurs in the same area, it can be distinguished by the very narrowly winged rhachis, generally denticulate leaf margins, and lanceolate leaflets which are only rarely broadly decurrent on to the rhachis, and are apically acuminate (obtuse in *S. infundibuliforme*).

The most distinctive features of *S. vidaurrei* are the delicate habit, generally 3(–5)-jugate leaf with narrowly lanceolate sessile to subsessile leaflets, general absence of interjected leaflets, normally denticulate leaflet margin, and frequent short thick white hairs, as well as the yellowish-green leaves and stems.

SPECIMENS EXAMINED

BOLIVIA

Dept. CHUQUISACA:

Prov. Nor Cinti, 2 km before Chichamayo. Alt. 3300 m. 28 Feb. 1959. '*Polylepis* slope with *Gentiana* species.' *Alandia, Rimpau & Ross 145* (Max.).

Prov. Nor Cinti, near Camargo. Alt. 2400 m. Feb. 1949. *Cárdenas 4514* (J.G.H.).

Prov. Nor Cinti, at about 150 km on road from Potosí to Camargo (near Muyuquira). Alt. 2200 m (or higher, probably *c.* 3250 m). 'Loose sandy soil under bushes, by a stone bridge.' *Cárdenas 5075* (J.G.H.—Isotype). See Plate 131.

Dept. TARIJA:

Prov. Méndez, between Obispo and Las Carreras. Alt. 3150 m. 23 Feb. 1953. *Petersen & Hjerting 1008* (J.G.H., J.P.H.).

Same locality, but at 2500 m. *Petersen & Hjerting 1010* (J.G.H., J.P.H.).

Note: Other collections have been recorded from further north.

ARGENTINA

Prov. SALTA:

Dept. Santa Victoria, Cuesta de Mesón. Alt. 2700 m. 12 Feb. 1953. 'Fl. violác. pálida.' *Sleumer 3998* (LIL, J.G.H.).

Dept. Santa Victoria, Lizoite. Alt. 3450 m. 17 Feb. 1953. 'Fl. violác. clar.' *Sleumer 4059* (LIL, J.G.H.). See Plate 132.

Probably ***S. vidaurrei***

Prov. JUJUY:

Dept. Humahuaca, Tres Cruces. 2 Feb. 1944. *Soriano 647* (SI).

DISTRIBUTION

S. vidaurrei is really a south Bolivian species which apparently hardly penetrates into Argentina (see Map 12). It is known in Argentina so far by two collections from prov. Salta, dept. Sta. Victoria, not far from the Bolivian border, and from a doubtful collection in prov. Jujuy, dept. Humahuaca. From Bolivia it is known from dept. Tarija, prov. Méndez and dept. Chuquisaca, prov. Nor Cinti. Collections of probable *S. vidaurrei* have been seen from dept. Chuquisaca, prov. Yamparaez and dept. Potosí, provs. Frías and Linares.

Forms intermediate between *S. vidaurrei* and *S. spegazzinii* have been collected in Argentina, prov. Jujuy, depts. Tilcara and Tumbaya (see p. 402). It is therefore possible, though by no means certain, that *S. vidaurrei* may occur also in prov. Jujuy, depts. Cochinoca, Humahuaca, and Yavi, and prov. Salta, dept. Iruya. Although Correll (1962) mentions a collection of this species from Catamarca, much further south (*Sleumer & Vervoorst 2535* p.p.), we consider this collection to be a hybrid of *S. sanctae-rosae* × *S. spegazzinii*, and to have no affinity whatsoever with *S. vidaurrei*.

HABITAT[1]

In Argentina, *S. vidaurrei* has been recorded for the Montano district ('pajonales', 'prados alpinos') of the Western Subtropical province, according to the phytogeographical scheme of Cabrera (1953). In Bolivia, it was recorded by Ross (1960*a*) from mesophile bush with mountain meadow elements such as *Polylepis* and *Gentiana* species; so here again, it was probably growing in fairly damp grassland or alpine meadows ('prados alpinos') at fairly high altitudes. Our information is insufficient to say more exactly what plants it is normally associated with, or whether it prefers wetter or drier habitats.

The altitude range of *S. vidaurrei* appears to be approximately 2700–3500 m[2] but these figures may need to be adjusted when more material is available.

S. vidaurrei has been collected in Bolivia not far from *S. oplocense* but the two species have not been recorded growing together. In prov. Salta, Santa Victoria, it grows with *S. gourlayi* and seems to hybridize with it (see p. 376) in one or two places.

TAXONOMIC NOTES

S. vidaurrei was described by Cárdenas in 1956 on the basis of a collection made by him (*5075*) in south Bolivia, dept. Chuquisaca, prov. Nor Cinti, about 150 km from Potosí on the road to Camargo (see Plate 131). It seems to grade into *S. pachytrichum* and *S. leptophyes* further north, but we have not been able to obtain sufficient material from Bolivia, especially in the living state, to be certain about this. To the south, it may also intergrade with *S. spegazzinii*, or be worthy of not more than subspecific rank. For the time being, since we are not certain of the connexions of this species in Bolivia, and since, according to Ross (verbal communication), it remains distinct from *S. spegazzinii* when grown under identical conditions in Cologne, we have considered that it should best be retained as a distinct species. No doubt it comes within the general affinities of *S. leptophyes*, *S. pachytrichum*, *S. oplocense*, *S. gourlayi*, and *S. spegazzinii*, but seems to be fairly uniform in a small area of southern Bolivia, just spreading southwards into north Argentina.

A slight resemblance between *S. vidaurrei* and *S. infundibuliforme* can be seen, but this might perhaps be attributable to convergent evolution rather than relationship by descent from a recent common ancestor. The possibility of some introgression between these two species should not be overlooked, however. Correll (1962) considered that *S. vidaurrei* might be allied both to *S. infundibuliforme* and to *S. spegazzinii*.

NOTES ON LIVING PLANTS AND CYTOLOGY

We have not grown *S. vidaurrei*, either from Argentine or Bolivian material, but Ross states (personal communication) that the Bolivian plants are highly distinctive and of a characteristic yellow-green colour. The calyx acumens were rather short on

[1] Notes based partly on Bolivian collections.

[2] Cárdenas gives 2200 m for the altitude of the type collection but this is probably wrong, since the maps give 3250 m for Muyuquira, which lies about 150 km from Potosí (see Cárdenas, loc. cit., p. 30) on the road to Camargo.

flowers taken from living plants at Cologne, and the articulation was only about 2–3 mm below the base of the calyx. The corolla diameter varied from 20–25 mm with lobes about 7–8 mm broad at the base, and 7 mm long, of which 4–5 mm was the well-marked acumen. The filaments were only about 1 mm and the anthers 5 mm long, tapering from base to apex. The style was about 7·5 mm long, papillose on the lower half, and with a globose stigma, thicker than the style apex.

A chromosome number of $2n = 24$ was reported by Ross for Bolivian material collected near Chichamayo (personal communication). No counts of Argentine material have yet been made.

FERTILITY AND ARTIFICIAL AND NATURAL HYBRIDS

We have no information on the fertility of *S. vidaurrei*, but it is probably self-incompatible, since this is the usual condition for diploid *Solanum* species.

No hybridizations have been attempted by us and none are reported in the literature.

Doubtful hybrids between *S. tarijense* and *S. vidaurrei* have already been mentioned on p. 245.

Natural hybridization between *S. vidaurrei* and *S. oplocense* may take place and we have already tentatively suggested (p. 383) that the tetraploid *S. oplocense* might be an amphiploid hybrid of *S. oplocense* and *S. vidaurrei* or possibly *S. oplocense* and *S. gourlayi*.

Hybrids between *S. vidaurrei* and ***S. gourlayi*** have been discussed on p. 376, and occur in Santa Victoria, where the geographical distribution and ecological range of the two species overlap.

17. SOLANUM SPEGAZZINII Bitter, *emend.* Hawkes et Hjerting

(In *Reprium nov. Spec. Regni veg.* **12**, 449–50, 1913)

S. famatinae Bitt. et Wittm. (in Engler, *Bot. Jb.* **50**, Suppl., 552–3, 1914).
S. sleumeri Corr. (*Wrightia* **2**, 195–6, 1961).

Herbaceous, tuberiferous, with thin long stolons and pale brown tubers, which are round to oval, 0·5–2 cm long.[1]

Stem (3–)10–40(–100) cm long, slender, (1–)2–4(–5) mm diam., unbranched and erect when small, or occasionally very branched, especially in the larger plants, which are often weak and decumbent; stem generally somewhat flexuose from node to node, with decurrent lines from the leaf bases or narrow wings up to 1 mm wide; stem pubescence of sparse to fairly dense coarse spreading hairs which are generally rather short (longer in some forms from prov. Salta, however).

Leaf varying in size from (3·5–)6–20(–30) cm long × (2–)4–10(–16) cm broad, often obovate in outline and widest towards the middle or nearer the apex. Lateral leaflets (3–)4–5(–7)[2]-jugate, *frequently only sub-opposite* or even alternating with each other on the rhachis, commonly *three times as long as broad* but ranging from 2–4 times as long as broad; largest laterals on each leaf varying in size from (1·0–)2·0–5·5(–8·5) cm long × (0·4–)0·6–1·6(–2·5) cm broad, and in shape varying from ovate, elliptic, *elliptic-lanceolate, narrowly lanceolate, or narrowly oblanceolate*, narrowed towards base and apex, the apex itself *normally obtuse*, but sometimes acute; leaflets placed on petiolules 2–8(–15) mm long, or sessile under certain habitat conditions; upper

[1] Up to 4 cm in Argentina according to Brücher (1957*b*).
[2] Up to 8-paired in Argentina according to Brücher (1957*b*).

1(–3) pairs of laterals *often decurrent on to the rhachis*, and sometimes decurrent even though petiolulate. *Terminal leaflets generally the same size or smaller than the upper laterals*, only occasionally larger. Interjected leaflets (0–)2–7(–18) pairs, but often placed irregularly, situated on rhachis and acroscopically and basiscopically on petiolules when these are well developed; *sometimes a large oblique interjected leaflet is situated at the lower angle between petiolule and rhachis*; interjected leaflets varying in size from less than 1 mm long and broad to 30×13 mm, orbicular, ovate or *elliptic-lanceolate*, narrowed towards each end and apically obtuse, normally sessile but occasionally shortly petiolulate and frequently slightly decurrent. All parts of leaf provided with rather short thick coarse hairs, frequent to dense, quite conspicuous on small leaves, less so on very well-developed ones; leaflet margins hairy and sometimes minutely denticulate, though not so markedly so as in *S. kurtzianum*. Pseudostipular leaflets generally present.

Inflorescence (2–)4–10(–30)-flowered, situated above the foliage or scarcely overtopping it. Peduncle once-forked, about (1–)2–10 cm long or perhaps longer in very well-developed plants, provided with a sparse to dense covering of short coarse spreading hairs. Pedicels in the flowering state (10–)15–20(–25) mm long, becoming longer in fruit; articulation sometimes placed below but generally at or above the centre, up to about two-thirds length above the base; pubescence as for peduncle but perhaps a little denser.

Calyx campanulate, 5–7 mm diam., with short lobes and *narrowly triangular to linear acumens* 1–2(–3) mm long, green to dark purple, moderately to densely covered with coarse short hairs.

Corolla pentagonal to sub-stellate (very occasionally almost rotate), (10–)15–20(–35) mm diam., the lobes 10–13 mm wide, 7–9 mm long of which 4–5 mm is the well-developed acumen; corolla colour *medium to pale violet* or purple (occasionally dark purple), sometimes darker in petal centres, sometimes much lighter and with pale acumens.

Anthers 4–6(–7) mm long; filaments 0·5–1·5 mm long.

Style straight, 7–10(–12) mm long, glabrous or clothed from base to centre with minute papillae; stigma globular to ovoid or clavate, 0·5–1 mm long, wider than style apex.

Berries globose to slightly ellipsoid, up to about 15 mm or even possibly to 20 mm. diam.

Chromosome number: $2n = 24$.

AFFINITIES

S. spegazzinii is probably related to *S. vidaurrei*, *S. gourlayi*, *S. oplocense*, and *S. kurtzianum* in Argentina, as well as to *S. pachytrichum* and *S. leptophyes* in Bolivia.

From *S. vidaurrei* it is distinguished by the better-dissected leaf with more numerous and generally petiolulate lateral leaflets and more numerous interjected leaflets. From *S. gourlayi* it differs in the denser, coarser pubescence, more numerous and narrower leaflets attenuated towards the base and apex, and by the form, number, and disposition of interjected leaflets. It may be distinguished from *S. oplocense* by the longer, narrower, better dissected leaf, the form of the anthers and the large stigma. It differs from *S. kurtzianum* in possessing a more dissected leaf, less coarsely hairy, higher pedicel articulation, and different colour pattern on the corolla.

When growing in very dry habitats *S. spegazzinii* may show some similarities to *S. infundibuliforme*, but can be distinguished from that species in possessing at least some leaflets narrowing to the base and not decurrent into the rhachis, and by normally possessing interjected leaflets, which are absent in *S. infundibuliforme*; furthermore, the typical wedge-shaped wings of *S. infundibuliforme* are absent in all but a few biotypes of *S. spegazzinii*.

S. spegazzinii is a widespread and highly variable species connected with the Bolivian *S. leptophyes* and *S. pachytrichum* probably by means of the intermediate species *S. vidaurrei*, *S. oplocense*, and *S. gourlayi*. It is recognizable by the long elliptic leaflets, narrowing towards each end, the tendency of the upper lateral leaflets to become decurrent on to the rhachis except in those plants with long petiolules (and even in these some decurrency can occur), the narrow interjected leaflets (in well-developed plants), coarse hairs, narrow triangular to linear calyx acumens, and medium purple substellate to pentagonal corolla.

SPECIMENS EXAMINED

ARGENTINA

Prov. SALTA:

Dept. Cachi, Valle Calchaquíes, Huasapampa. 8 Feb. 1943. *Castellanos* (s.n.) (BA 46997).

Dept. Cachi, entre Piedra Molino y Cachi. Alt. 3150 m? 22 Feb. 1956. 'Flor lila pálida con puntos blancos.' *Hjerting, Petersen & Rahn 317* (J.G.H., J.P.H.).

Dept. Cachi, entre Piedra Molino y Cachi. Alt. *c.* 3000 m? 22 Feb. 1956. 'Flor lila pálida. Xeromorphe.' *Hjerting, Petersen & Rahn 334* (C, J.G.H., J.P.H.). See Plates 140 and 141.

Dept. Cachi, 10 km de Payogasta hacia Salta. Alt. 2600 m. 23 Feb. 1956. *Hjerting, Petersen & Rahn 341* (J.G.H., J.P.H.).

Dept. Cachi, Cerro de Cachi. Jan. 1897. *C. Spegazzini* (s.n.) (LPS No. 14210—Syntype for *S. spegazzinii* Bitt.). See Plate 134.

Dept. Cafayate, Sierra de los Quilmes. 9 Jan. 1943. *Castellanos* (s.n.) (BA 46985, 46986).

Dept. San Carlos, Cachipampa to Amblayo. Alt. 3100 m. Apr. 1963. 'Fl. viol. Locus classicus.' *Brücher 1109* (BAA, CORD).

Dept. San Carlos, San Lucas. Alt. 2000 m. 14 Feb. 1959. 'Ufer d. Río San Lucas direkt im Sand unter Bäumen.' *Ross & Vervoorst 53* and *54* (Max.).

Dept. San Carlos, Amblayo—Cachipampa. Jan. 1897. *C. Spegazzini* (s.n.) (LPS No. 14211, Lectotype for *S. spegazzinii* Bitt.). See Fig. 38 and Plate 133.

Prov. TUCUMÁN:

Dept. Tafí, El Molle. 29 Jan. 1935. *Castellanos* (s.n.) (BA 14823).

Dept. Tafí, Amaichá, El Molle. 23 Dec. 1913. *Castillón 3191* (LIL).

Dept. Tafí, Infiernillo, Río Tafí. Alt. 2900 m. 20–27 Feb. 1952. *Petersen & Hjerting P–332b* (CPC: 2452. Tuber collection) (J.P.H.).

Dept. Tafí, Río Tafí (Infiernillo). Alt. 2800 m. 1 Jan. 1953. *Petersen & Hjerting 856b* (J.G.H., J.P.H.).

Dept. Tafí, Zanja de los Cardones. Alt. 2700 m. 13 Feb. 1959. 'Bachbett unter *Baccharis*, *Eupatorium*, etc.' *Ross & Vervoorst 30, 32, 35,* and *36* (Max.) See Plate 136.

Dept. Tafí, road from Santa María to Tafí del Valle, km 84, near Quebrada Honda. Alt. 3100 m. 13 Feb. 1959. 'Hochgrasflur, unter Felsen.' *Ross & Vervoorst 43* (p.p.) (Max.).

Dept. Tafí, Cuesta occidental de Lara. Alt. 3500–4000 m. 15 Feb. 1920. 'Flor color violeta obscuro.' *Schreiter 1215* (LIL).

Dept. Tafí, Los Cardones (Quebrada Amaichá). Alt. 2800 m. Dec. 1931. 'Flor violácea.' *Schreiter 7065* (LIL).

Dept. Trancas, Cuesta del Chorro. Alt. 3300 m. 1 Feb. 1927. 'Flor azul-lila.' *Schreiter 4839* (LIL).

Prov. CATAMARCA:

Dept. Andalgalá, El Candado. 20 Feb. 1915. 'Alta robusta, con flores blancas, azuladas o azules-coloradas; papa grande. Abundante.' *P. Jörgensen 1311* (p.p.) (LIL 62619).

Dept. Andalgalá, Ingenio del Arenal. Alt. 2500 m. 12 Feb. 1959. 'Bachbett unter Gesträuch.' *Ross & Vervoorst 16, 17, 18, 19, 20, 21, 22,* and *23* (Max.).

Dept. Andalgalá, in front of the Mina Capillitas on the other side of the valley. Alt. 2700 m. 12 Feb. 1959. 'Monte, trock. Forma. m. Kakteen.' *Ross & Vervoorst 25* (Max.).

Dept. Andalgalá, Capillitas. Mar. 187? *Schickendantz 226* (p.p.) (CORD).

Dept. Andalgalá, Cuesta La Negrilla, cerca de la Mina Capillitas. Alt. 3000–3100 m. 2 Feb. 1952. 'Fl. viol.' *Sleumer 2687* (p.p.) (LIL). *Sleumer 2688* (p.p.); H. 67 (LIL, US, J.G.H.).

Dept. Andalgalá, subida al Cerro Yutuyaco desde Capillitas. Alt. *c.* 3700 m. 3 Mar. 1952. 'Fl. pallide viol.' *Sleumer 2751* (LIL).

Dept. Andalgalá, near Mina Capillitas. Alt. *c.* 2700 m. Feb. 1959. Only living material, sent by a farmer through Dr. Vervoorst to Professor Ross. (E.B.S. No. 1395, Max.).

Dept. Belén, Corral Quemado. 7 Feb. 1930. 'La raíz lleva una papita del tamaño de una ciruela pequeña con granulaciones espaciadas.' *Cabrera* (s.n.) (LP 1379).

Dept. Belén, Sierra de Belén, Vayas altas. Alt. *c.* 2550 m. Jan. 1872. 'Planta ruderal; bei d. Alphütte.' *Lorentz 640* (CORD, GOET).

Dept. Belén, Sierra de Culampajá, El Puesto (Quinteros). Alt. 2700 m. 27 Jan. 1936. 'Fl. azul. "Papa Yuta".' *Peirano 10206* (LIL).

Dept. Belén, Barranca Larga. Alt. 2400 m. Feb. 1937. 'Flor violácea o azul.' *Schreiter 10274* and *10331* (LIL).

Dept. Belén, Las Mansas. Alt. 2300 m. Mar. 1938. 'Flor violácea.' *Schreiter 10594* (LIL).

Dept. Belén, Pozo de Piedra. Alt. 1900 m. *Sleumer 1* (CPC: 2464. No herbarium specimen).

Dept. Belén, Pozo de Piedra, a lo largo del camino, km 15–18. Alt. *c.* 1900–2000 m. 25 Jan.–2 Feb. 1952. 'Fl. violascentes. Fr. viridi-violascens. Lugares húmedos o asombreados.' *Sleumer & Vervoorst 2434* (LIL, US, J.G.H.). (Type for *S. sleumeri* Corr.)

Dept. Belén, Quebrada de los Potrerillos (El Rodeo), arriba de Granadillas. Alt. *c.* 2650 m. 26 Jan. 1952. 'Fl. violasc.' *Sleumer & Vervoorst 2489* (LIL, J.G.H.).

Dept. Belén, Quebrada de los Potrerillos (El Rodeo), arriba de Granadillas. Alt. *c.* 2700 m. 26 Jan. 1952. 'Fl. pallide violasc.' *Sleumer & Vervoorst 2490* (LIL, J.G.H.).

Dept. Belén, Quebrada del Río Blanco, Las Mesaditas, arriba de Granadillas. Alt. 2800 m. 28 Jan. 1952. 'Fl. violasc.' *Sleumer & Vervoorst 2534* (LIL, J.G.H.).

Dept. Santa María, Cerrillos. Mar. 1944. *T. Meyer 6873* (LIL).

Dept. Santa María, Sierra de Aconquija, Cerrillos. Alt. 3100 m. 16 Dec. 1933. 'Fl. azul.' *Peirano* (s.n.) (LIL 88694).

Dept. Santa María, Vuelta del Río. 24 Feb. 1948. *A. Reales 1002* (LIL).

Dept. Santa María, Caspinchango. 3 Feb. 1949. *A. Reales 1552* (LIL).

Dept. Santa María, Río Santa María. Alt. 2000 m. Dec. 1917. 'Flor celeste á violeta. Tiene tubérculos?' *Schreiter 358* (LIL, MVM—Osten 12402).

Dept. Santa María, Fuerte Quemado. Alt. 2000 m. Dec. 1917. 'Flor celeste a violeta. Tiene tubérculos?' *Schreiter 361* (LIL, MVM—Osten 12412).

Dept. Santa María, margen del Campo del Arenal cerca de Punta de Balasto, medanitos. Alt. *c.* 2400 m. 2 Mar. 1952. 'Fl. viol. pal.' *Sleumer 2664* (LIL, US, J.G.H.).

Dept. Tinogasta, Vallecito a Agua Negra. Alt. 2400 m. 7 Feb. 1930. 'Flor violácea.' *Schreiter 6299* (LIL).

Dept. Tinogasta, cerca de Rumi Rayana, sobre Río Chaschuil. Alt. 2630 m. 19 Mar. 1951. 'Entre rocas; con tubérculos pequeños.' *Vervoorst 3159* (LIL).

Prov. LA RIOJA:

Dept. Chilecito, Guanchín, El Bordo. 18 Apr. 1960. Seeds sent to Denmark. Herbarium specimen made in Copenhagen. *Reche 7* (J.G.H., J.P.H.).

Dept. Chilecito, Guanchín Viejo. 10–15 Jan. 1961. *Viirsoo 200* (J.P.H.).

Dept. Chilecito, Guanchín, Finca del Sr. Vilar. 10–15 Jan. 1961. *Viirsoo 201* (J.G.H., J.P.H.).

Dept. Famatina, Sierra Famatina, Los Corrales. Alt. 2000 m. Mar. 1963. 'Locus classicus.' *Brücher 1101* (BAA, CORD, SI).

Dept. Famatina, Sierra Famatina, entre 'Los Corrales' y 'Escalera de Famatina'. Alt. *c.* 2000 m. 10–12 Feb. 1906. 'In sepibus freq.; ad 1 m alto, fl. lilacinis.' *F. Kurtz 13445 b* (CORD—LECTOTYPE of *S. famatinae* Bitt. et Wittm.; GOET—drawing only). See Fig. 39 and Plate 135.

Dept. Famatina, Sierra Famatina, Agua Salado. 25 Feb. 1907. 'In umbrosis ad 1 m alt. Fl. violac.' *F. Kurtz 14466*[1] (CORD—ISOCOTYPE of *S. famatinae* Bitt. et Wittm.; Photograph of now destroyed Berlin specimen, made by Field Museum, Chicago). See Fig. 40.

Dept. Famatina, Las Gredas, entre Alto Carrizal y Corrales, 2 km abajo de Escalera de Famatina, en un campo de maíz. Alt. 1900 m. 7 Mar. 1956. *Petersen & Hjerting 1892* (J.G.H., J.P.H.).

Dept. Famatina, Corrales. Alt. 2050 m. 7 Mar. 1956. 'La flor (según dicen) blanca hasta lila; en un campo de maíz.' *Petersen & Hjerting 1893* (C, J.G.H., J.P.H.). See Plates 137 and 138.

Dept. Famatina, entre Corrales y Cuevas de Noroño. Alt. 2400 m. 8 Mar. 1956. *Petersen & Hjerting 1894* (J.G.H., J.P.H.).

Dept. Famatina, entre Cuevas de Noroño y Encrucijada. Alt. 2900 m. 8 Mar. 1956. 'En este lugar muchas papas diferentes (híbridos?).' *Petersen & Hjerting 1895* (C, J.G.H., J.P.H., Max. 927).

Dept. Famatina, entre Cuevas de Noroño y Encrucijada. Alt. 2900 m. 8 Mar. 1956. 'Hojas anchas.' *Petersen & Hjerting 1896a* (J.G.H., J.P.H.).

Dept. Famatina, abajo de La Encrucijada. Alt. 2950 m. 8 Mar. 1956. *Petersen & Hjerting 1900a* (C, J.G.H., J.P.H.).

Dept. Famatina, near Famatina, Playa de los Ramblones. Alt. *c.* 1500 m. 9 Apr. 1960. Only living material collected. *Reche* (s.n.) (E.B.S. No. 2078, Max.).

Dept. Famatina, Tamberías. 13 Apr. 1960. Tubers and seeds sent to Denmark. Herbarium specimen made in Copenhagen. *Reche 6* (J.G.H., J.P.H.).

Dept. Famatina, El Carrizal, cerca de la casa del Sr. Galíndez. 10–15 Jan. 1961. *Viirsoo 3* and *3a* (J.G.H., J.P.H.).

Prov. SAN JUAN:

Dept. Jáchal, Chepical y Quebrada de Gualcamayo. 11–13 Dec. 1957. 'Común en suelos arenosos sueltos o en la asociación de *Tessaria absinthioides*. Fls. azules!' Leg. Ruiz Leal and F. Roig, *Ruiz Leal 19368* and *19369*. (Ruiz Leal, J.G.H.). See Plate 139.

[1] One of the Córdoba sheets was originally numbered 14466*a* (see Fig. 40) and gave a general altitude range of 1400–4525 m.

FORMS INTERMEDIATE BETWEEN S. SPEGAZZINII AND S. VIDAURREI

Prov. JUJUY:

Dept. Tilcara, Garganta del Diablo. Alt. 3000 m. 19 Feb. 1956. 'De un lugar muy seco.' *Hjerting, Petersen & Rahn 264* (J.G.H., J.P.H.).

Dept. Tumbaya, Est. Volcán, quebrada occidental. Alt. 2200 m. 24 Jan. 1927. 'Flor morada obscura.' *Castillón 365* (LIL).

DISTRIBUTION

S. spegazzinii is widely distributed in the Andes of north-western Argentina (see Map 12). It has been collected in provs. Salta (depts. Cachi, Cafayate, San Carlos),[1] Tucumán (depts. Tafí, Trancas), Catamarca (depts. Andalgalá, Belén, Santa María, Tinogasta), La Rioja (depts. Chilecito, Famatina), and San Juan (dept. Jáchal). It thus extends for at least 600 km in a north-east–south-west direction from about 25–30° latitude south.

Forms apparently intermediate between *S. spegazzinii* and *S. vidaurrei* have been recorded for Jujuy (depts. Tilcara and Tumbaya). These need further study. No doubt *S. spegazzinii* occurs elsewhere in the drier regions of the Andes in north-west Argentina, and may even perhaps extend into Bolivia.

HABITAT

This species has been recorded for a wide altitude range, as follows: (1500–)1900–3100(–3700) m.

It is primarily a plant of dry places and occurs in the following phytogeographical formations of Cabrera (1953):

1. Western Subtropical province, upper part of Montano district, grassy meadows ('pajonales'). It is not very common here.
2. Prepuna province at higher altitudes with *Larrea*, giant cacti (*Trichocereus* spp.) and spiny shrubs. The species occurs most frequently in this province.
3. Monte province (including 'bosquecillos marginales', which are said by Cabrera to be enclaves of the Espinal province from Central Argentina), generally in the wetter parts, with *Prosopis*, *Larrea*, *Bougainvillea*, etc.
4. Puna province in the drier parts, grading down into drier prepuna and wetter 'pajonal' regions.

Although it seems to withstand dry conditions quite well, *S. spegazzinii* can also be found in damp shady places, though it never penetrates into the eastward-facing rain and cloud forests of the Aconquija mountains. It is frequently reported from sandy or stony soil, in stream beds or by the banks of streams and irrigation ditches, along roadsides and in hedges and bushy places. This species also penetrates into maize fields as a weed of cultivation, and Brücher (1957*b*) considers it a typical species of river valleys and irrigated fields, where it grows most luxuriously, attaining a height of 1 m or more. It thus shares with *S. chacoense* and certain other potato species a tendency to colonize ruderal habitats and to become a frequent invader of cultivated fields.

In more natural vegetation, *S. spegazzinii* is seen under the shade or protection of giant cacti, opuntias, *Prosopis*, *Baccharis*, *Eupatorium*, *Tessaria absinthioides* (in San Juan), *Bougainvillea spinosa*, *Larrea*, and many other shrubs and bushes.

At higher altitudes it occurs in meadows ('pajonales') or in open sandy and stony soil, often amongst dry stunted bushes. Frequently, also, it is found amongst grass in the protective shade of large rocks and walls.

[1] Also depts. Chicoana and Guachipas (see Appendix III).

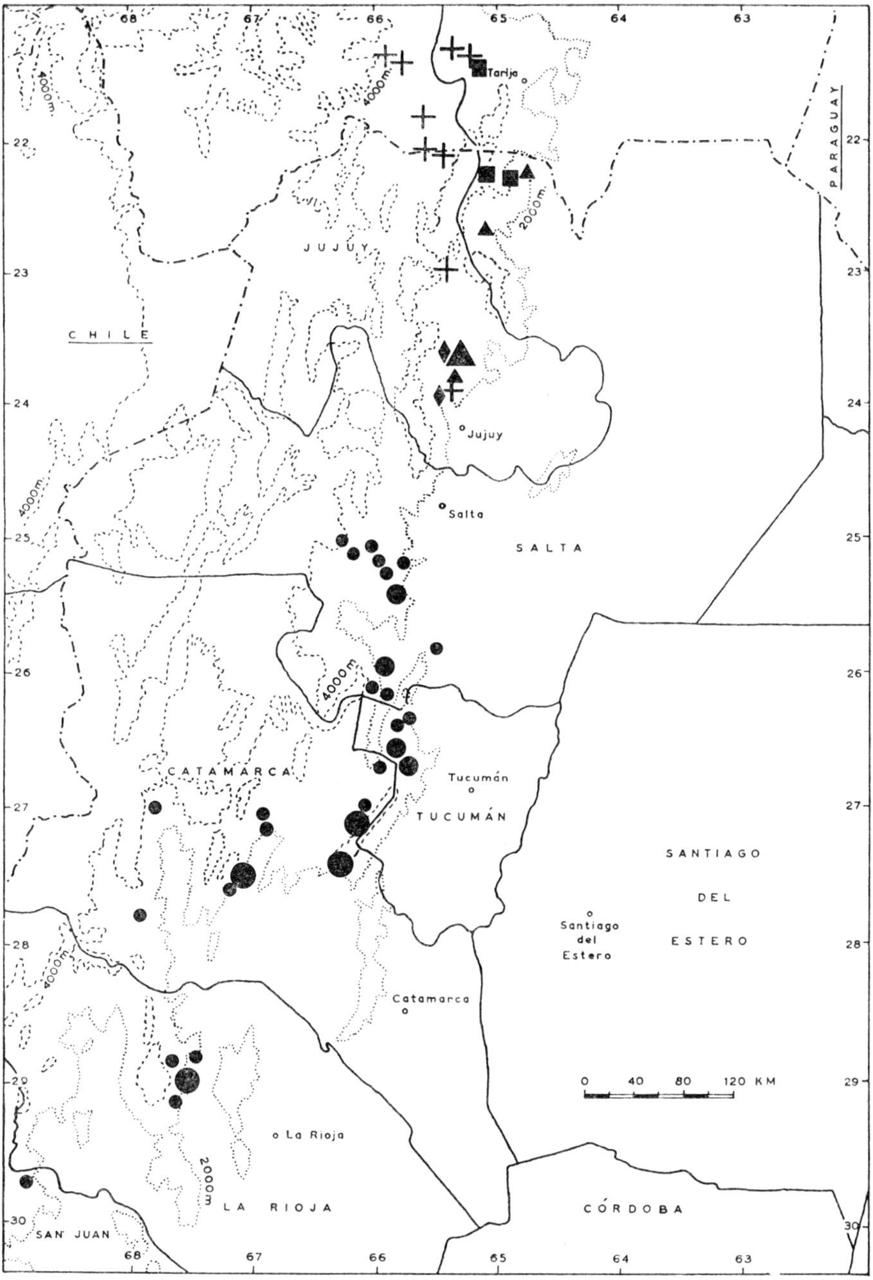

No. of Records

5 & over	3–4	1–2	
●	●	●	*Solanum spegazzinii* Bitt.
▲	▲	▲	*Solanum gourlayi* Hawkes
		+	*Solanum oplocense* Hawkes
		■	*Solanum vidaurrei* Cárd.
		♦	*Solanum spegazzini* Bitt./*Solanum vidaurrei* Cárd. intermediates

MAP 12. Distribution of *S. gourlayi* Hawkes, *S. oplocense* Hawkes, *S. vidaurrei* Cárd., and *S. spegazzinii* Bitt.

Brücher (loc. cit.) mentions that it grows in areas where the annual rainfall is about 300 mm, but it can no doubt spread into regions with a lower average annual rainfall by colonizing stream beds, ditches, cultivated fields, etc., or by vegetating only in those years when the rainfall is above average.

In the lower part of its range *S. spegazzinii* comes into contact with *S. chacoense* and in some places forms hybrids with it (also perhaps with *S. tarijense*). At the higher part it overlaps with *S. acaule* and hybridizes with that species also (both Argentine subspecies). Hybrids have also been seen between *S. spegazzinii* and *S. sanctae-rosae* in the zones of overlap, and even with *S. vernei* in a few places where the two species occur together, though they normally possess very distinct habitat requirements. Although the geographical distribution, altitude range, and ecological preferences of *S. spegazzinii* and *S. kurtzianum* overlap to a certain extent they have not generally been recorded from the same localities, and only a few hybrids have been reported.

TAXONOMIC NOTES

Bitter's type description of *S. spegazzinii* was published in 1913 and was based on two collections made by Carlos Spegazzini in prov. Salta in January 1897. The first was from Cerro de Cachi, whilst the second was gathered somewhere along the mule track from Amblayo to Cachipampa, a distince of some 30–35 km. (see Fig. 38 and Plates 133 and 134). We were fortunately able to examine both collections, thanks to the generosity of the Instituto Spegazzini. Both were of rather small delicate plants, 20 to 25 cm tall, with (3–)4(–5)-jugate leaf and 0–3 pairs of interjected leaflets. The primary lateral leaflets were lanceolate, sessile to subsessile, 15–20 × 5–7 mm, and the uppermost were generally shortly decurrent. The Cerro de Cachi collection had fewer and less conspicuous hairs than the Amblayo–Cachipampa one, but in neither was the pubescence very long and conspicuous. The general impression gained from these specimens is that they were collected in a very dry region and for this reason were considerably stunted. Unfortunately no altitude data are given by Spegazzini on the collection labels. Our own collections from the Amblayo to Cachipampa region were made at about 3000 m, however. (See Appendix III.)

No holotype was mentioned by Bitter and we therefore designate the larger left-hand specimen of Spegazzini from the Amblayo–Cachipampa collection (Inst. Spegazzini 14211) as LECTOTYPE for *S. spegazzinii* Bitt.

In the same publication, and immediately following the diagnosis of *S. spegazzinii*, Bitter described another species, *S. setulosistylum*. This was also based on a Carlos Spegazzini collection of January 1897, but made at Pampa Grande, about 80–100 km away from the type collections of *S. spegazzinii*. (See Fig. 21.) We have already demonstrated (p. 233) that *S. setulosistylum* must be regarded as a hybrid of *S. spegazzinii* and *S. chacoense* though Brücher (1964*a*) considers it to be a synonym of *S. chacoense*. It shows the influence of its *S. chacoense* parent in the more deeply stellate corolla and larger leaf, even though the leaf dissection is not unlike that of *S. spegazzinii*. Both species are to be found at Pampa Grande though we have not as yet found the hybrid. The pubescence, also, is intermediate between that of the two parents, and the type specimen matches very well certain hybrid forms between *S. chacoense* and *S. spegazzinii* collected by Sleumer and Vervoorst in prov. Catamarca, to which we have already referred. The fact that none of these hybrids shows the supposedly characteristic style setae of the Pampa Grande plants is apparently of no consequence. This character occurs in Argentina only in Pampa Grande and is there to be found in other species as well as *S. setulosistylum*. It is therefore not species specific.

A year later, in 1914, Bitter and Wittmack described a further species, *S. famatinae*, on the basis of two collections made by F. Kurtz (*13445b* and *14466*) in prov. La Rioja, Sierra Famatina, some 500 km distant from the type locality of *S. spegazzinii*.

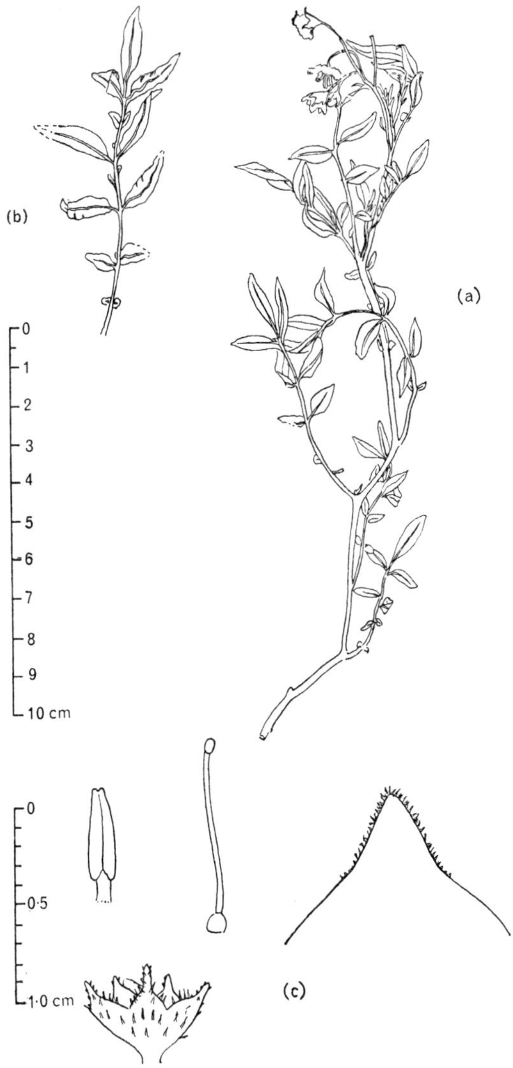

FIG. 38. *S. spegazzinii* Bitt. Material from Amblayo to Cachipampa, prov. Salta, Argentina, collected by C. Spegazzini (s.n.—Herb. La Plata—Spegazzini 14211). See also Plate 133.

(*a*) Isotype specimen, very close to the syntype (see Plate 134).

(*b*) Leaf from lectotype specimen with short petiolules and some interjected leaflets.

(*c*) Flower dissection from lectotype specimen. The corolla lobe may not be so steep-sided as shown in the drawing.

Compare with *Ross & Vervoorst 36* (Plate 136).

Both specimens were deposited in Berlin and were destroyed during the Second World War.

During our visit to Argentina in 1966 we were fortunate enough to find duplicate sheets of both these collections in the very fine Kurtz Type Herbarium at Córdoba. Bitter and Wittmack did not designate a type specimen, but since *Kurtz 13445b* was the first to be quoted in their description, and since Bitter made a drawing (now in Göttingen) of presumably the Berlin sheet of that collection, it is probable that they thought of it as the type.

Although we are here treating *S. famatinae* as a taxonomic synonym of *S. spegazzinii* we feel that it might be useful in the future to have a type specimen for reference purposes. Both specimens in the Córdoba herbarium agree well with Bitter and Wittmack's description, and in view of what has been said above *we designate* Kurtz 13445b *in Hb. Córdoba as* LECTOTYPE *of* ***S. famatinae*** *Bitt. & Wittm.* The Córdoba sheet of *Kurtz 14466* we designate as ISOCOTYPE. (See Plate 135 and Figs. 39 and 40.)

In a previous publication Hawkes (1956*b*) considered that *S. famatinae* might be a hybrid of *S. chacoense* × *S. kurtzianum*, since Bitter mentioned in the type description that the corolla was almost stellate ('substellata'), and the leaf of *Kurtz 14466* looks not unlike that of *S. chacoense*. Furthermore, the violet flower colour might have been inherited from the *S. kurtzianum* parent. At that time also, the better-developed and more typical specimens of *Kurtz 13445b* and *Kurtz 14466* were not available for study, nor had the living collections of Petersen and Hjerting in Sierra Famatina been made.

When these living collections and detailed herbarium specimens were examined it became quite clear that *S. famatinae* had nothing to do with *S. chacoense*, although it manifestly bears some similarity to *S. kurtzianum*. None of the seedling families shows any segregation for *S. chacoense* or *S. kurtzianum* characters and, indeed, they all remain remarkably uniform. Furthermore, *S. chacoense* is not to be seen in Sierra Famatina, though *S. kurtzianum* has been recorded in two or three places.

Brücher (1957*b*) also points out the non-hybrid nature of *S. famatinae* and gives useful notes on its habitat preference and distribution.

The decision to include *S. spegazzinii* and *S. famatinae* in the same species rests on a study of some sixty-six gatherings made by various collectors from a range of localities lying between Sierra Famatina in the south and Cerro de Cachi in the north (see Map 12) and including two from even further south, in prov. San Juan. It became quite clear during this study that the two species *S. spegazzinii* and *S. famatinae* represented two aspects of one and the same species, to which the name *S. spegazzinii* must be applied, for reasons of priority. The two species united here represent *not* the two ends of a geographical cline over a distance of some 500 km but rather two growth forms or ecads of one and the same species, between which innumerable intermediates could be seen. Kurtz's original specimens represent the growth of this species under conditions of adequate moisture and some shade (see field notes 'in shady places') and match very closely material of this species *grown under glass in Europe*! (See Plates 135, 137, and 138; Figs. 39 and 40.)

In material of this sort the habit is tall and lax, and the leaf is large and well dissected, with well-developed petiolules, hardly any decurrency, and numerous interjected leaflets springing from both rhachis and petiolules.

At the other extreme, the type collections of *S. spegazzinii* are of plants that we believe have developed with a lack of moisture, possibly in poor stony or sandy soil. Here the plants are small, with smaller leaves and fewer leaflets which are sessile, narrow, and generally decurrent; interjected leaflets are few and may often be entirely absent. (See Plates 133 and 134; Fig. 38.)

Interestingly enough, certain other collections may be seen (*Ross & Vervoorst 36*; see Plate 136) in which a range of leaf type occurs *on the same plant*, changing from

FIG. 39. *S. spegazzinii* Bitt. One of the original specimens on which Bitter and Wittmack's description of '*S. famatinae*' was based, redrawn from a tracing made by Bitter. (*Kurtz 13445b*—original in Herb. Berlin, now destroyed; Bitter's drawing in Herb. Göttingen). Collected at Escalera de Famatina, Sierra Famatina, prov. La Rioja, Argentina. This matches the Lectotype specimen (see Plate 135) very closely.

that typical of *S. spegazzinii* to that typical of *S. famatinae*. The 1966 collection of *Hawkes, Hjerting, Okada & Rahn 3425* also shows this range. Probably the rainfall improved through the growing season, thus inducing the change in leaf type during the growth of the individual.

Corroboration of this is forthcoming from Ross, who states that the plants of '*spegazzinii*' type are from dry soils and '*famatinae*' type from more humid situations. Our own observations in 1966 confirm this completely (see Appendix III). Although most of the collections cited above (pp. 389–91) are made without adequate habitat notes, those that possess these notes corroborate our thesis. We give below a table of habitat data, grouped into three categories of growth type, to illustrate this point (see Table 23).

It will be noted from the table that all plants of '*famatinae*' type (Group I) are from shaded and/or damp places, whilst those of '*spegazzinii*' type (Group 3) are from dry places, except perhaps a group of collections made by Ross and Vervoorst from a

FIG. 40. *S. spegazzinii* Bitt. Isocotype specimen of '*S. famatinae*' Bitt. & Wittm., collected by Kurtz (*No. 14466*—Herb. Córdoba) at Sierra Famatina, prov. La Rioja, Argentina. This matches very closely the lectotype specimen of Kurtz (*No. 13445b*—see Plate 135 and Fig. 39).

stream bed under *Baccharis* and *Eupatorium* (*Nos. 30, 32, 35, 36*). These, all taken from one restricted area, range from one morphological type to the other. They were made, according to Ross (verbal communication) to illustrate precisely this range, and probably represent plants from a single population, differing in the amount of moisture available and hence in their growth form. It might be objected at this point that such variation could be explained by postulating a hybrid swarm between the two species *S. famatinae* and *S. spegazzinii*. We shall demonstrate later, however, that tubers taken from '*spegazzinii*' type of plant will grow into '*famatinae*' plants when given adequate moisture and shading. Furthermore, we were able to see in our 1966

TABLE 23

Growth forms of S. spegazzinii

Locality	Collector and Number	Habitat note
Group 1. Plants with well-developed leaves, quite long petiolules and numerous interjected leaflets.		
'Famatinae' type		
Prov. La Rioja, Sierra Famatina	*Kurtz 13445b*	Frequent in hedges
	Kurtz 14466	In shady places
	P. & H. 1892, 1893	In a maize field
Prov. Catamarca, dept. Andalgalá	*Ross & Vervoorst 16, 17, 20*	Stream bed under bushes
Prov. Catamarca, dept. Belén	*Sleumer & Vervoorst 2434*	Damp or shady places
Prov. Tucumán, dept. Tafí	*Ross & Vervoorst 30*	Stream bed under *Baccharis, Eupatorium,* etc.
Prov. Salta, dept. San Carlos	*Ross & Vervoorst 53, 54*	Bank of Río San Lucas in sand under trees
Group 2. **Intermediate forms,** petiolules shorter and interjected leaflets fewer.		
Prov. Catamarca, dept. Andalgalá	*Ross & Vervoorst 18, 19, 21 23*	Stream bed under bushes
Prov. Tucumán, dept. Tafí	*Ross & Vervoorst 32, 36*	Stream bed under *Baccharis, Eupatorium,* etc.
Group 3. Plants with short stems, poorly developed leaves, sessile and often decurrent lateral leaflets and few or no interjected leaflets.		
'Spegazzinii' type		
Prov. Catamarca, dept. Tinogasta	*Vervoorst 3159*	Amongst rocks
Prov. Catamarca, dept. Andalgalá	*Ross & Vervoorst 25*	'Monte', dry formation with Cacti
Prov. Tucumán, dept. Tafí	*Ross & Vervoorst 43* (p.p.)	High grass meadows under rocks (probably dry microclimate)
	Ross & Vervoorst 35	Stream bed under *Baccharis, Eupatorium,* etc.

collections a range from one type to the other, even in the type locality for *S. spegazzinii* at Amblayo–Cachipampa (see Appendix III).

Another point arising from Table 23, which shows even more clearly if other collections with no habitat are considered also, is that there is no restriction of different morphological types to different parts of the geographical range of this species.

Apart from considerations of moisture it is highly likely that altitude and exposure also play an important role in conditioning the phenotype of *S. spegazzinii*. Thus, although the rainfall tends to increase with altitude the lower temperatures will prevent very luxurious growth; hence several high-altitude collections have a reasonable leaf dissection even though the stem is short and the leaf small. Eight collections with altitudes of 3000 m or over have been examined by us. Six of these possess a poor to intermediate growth similar to the type of *S. spegazzinii* and only two show reasonably good development of the *S. famatinae* type. Here again, our 1966 collections confirm this point (see Appendix III).

Although we ourselves are completely convinced of the conspecificity of *S. spegazzinii* and *S. famatinae*, it should be noted that Brücher (1957*b*) prefers to consider them as quite distinct. He maintains that the widespread species apparently equivalent to the taxon we are now discussing is to be called *S. famatinae*. Brücher incorrectly states that the habit and characteristic leaf form of this taxon remain almost unchanged no matter under what conditions of shading and soil-moisture content it is growing, even though he admits that the plants may vary in height from 10 cm to more than 1 m. We do not know whether Brücher has examined the type specimens of *S. spegazzinii*, and we are inclined to think that in 1957 he had no very clear concept of this species, since he says: 'In *S. spegazzinii* on the other hand, the terminal leaflet is excessively large, and the remaining leaflet pairs are mostly very irregular and of different size . . . *S. spegazzinii* has a similarity to *S. acaule* in its compact growth and broad leaves.' Unfortunately, Brücher gives no illustration of this material, but it rather sounds as though he had collected a form of *S. megistacrolobum* on that occasion. He did, however, collect *S. spegazzinii* from the 'type locality' in 1963, since we have seen specimens of his collection (*Brücher 1109*) from Córdoba and Fac. Agron., Buenos Aires. This material is rather similar to the H.P.R. collections from not too far away from the Amblayo–Cachipampa 'classical locality' (see below), agreeing with them in the rather long hairs and other features. Our own collections in 1966 from the 'type locality' matched Spegazzini's material very closely, however. Since Spegazzini merely stated on the label: 'Amblayo–Cachipampa', we cannot be certain of exactly where his collections were made; no altitude was given, either.

In his 1957 publication Brücher was comparing *S. spegazzinii* with *S. leptophyes*. Although these two species are, in our opinion, very similar, Brücher strenuously denies this. Since on that occasion he does not seem to have had true *S. spegazzinii* available, the fact that he sees enormous differences between his '*S. spegazzinii*' and *S. leptophyes* is perhaps not very surprising. Although he states that he has distributed herbarium material of '*S. spegazzinii*' to several institutes we have not been able to discover any of this pre-1957 material in the major herbaria of Europe, the U.S.A., or Argentina. The 1963 material is, of course, correctly identified *S. spegazzinii*.

In 1956 Hjerting, Petersen, and Rahn made three collections (*317*, *334* (see Plate 140), *341*) from the region between the two type localities of *S. spegazzinii*. None of these matches perfectly with the type specimens, and in general the 1956 collections show a longer and coarser pubescence and more rotate corolla than the type. Only a few specimens of each collection are close in leaf type, whilst in the larger plants there is a marked tendency to leaflet decurrency far greater than that of the type for *S. spegazzinii* itself. Yet such plants, when grown in Europe, show no decurrency at all (*H.P. & R. 334*, see Plate 141). So striking is the decurrency on the Argentine plants that we thought at first that they might be hybrids of *S. spegazzinii* with *S. infundibuliforme*. No signs of hybridity or of decurrency were shown in the seedling progenies, however, and we can only assume that the decurrency was conditioned by certain habitat factors which we are as yet unable to reproduce artificially. It is interesting to note, in passing, that extreme decurrency can be seen also on the Ruiz Leal and Roig collections from the other extreme of the range of this species, in prov. San Juan. (See Plate 139.)

Another curious feature of the H.P. & R. collections from prov. Salta is the more rotate corolla with pale or white acumens and often pale central petal streaks; this character is constant, and discernible also in hybrids. The rather coarse long pubescence is also genotypically controlled, and is relatively independent of habitat—a feature also noted by Brücher (1957*b*). On the other hand, most of the variation of pubescence density in other material of *S. spegazzinii* is probably due to habitat, since poorly developed plants of *S. spegazzinii* phenotype regularly exhibit dense pubescence, whilst those of *S. famatinae* phenotype do not. Despite the fact that these 1956 collections differed in several details from the type of *S. spegazzinii* the material from the 'type locality' that we collected in 1966 matched the type specimen very closely indeed. (See Appendix III.)

Another rather variable feature in *S. spegazzinii* is the position of the pedicel articulation. On the average it seems to be situated at the centre or up to two-thirds above the base of the pedicel. Yet in certain samples it seems rather low, just as we have seen in *S. oplocense*. Thus, Bitter, in his type description of *S. famatinae*, mentions the variation of the articulation from the centre down to one-third above the base. The low articulation, so characteristic of *S. kurtzianum*, is apparently rare in *S. spegazzinii*, and is only to be found on occasional plants where the inflorescence is not very well developed. This must have been the case with the material examined by Bitter; and indeed the low articulation of *Kurtz 13445b* can be seen on Bitter's drawing (Fig. 39) and on the lectotype specimen of *S. famatinae*, though not so clearly on the *S. spegazzinii* syntypes. The normal position, however, in well-developed plants seems to be median or up to two-thirds the length above the pedicel base.

We have already spoken of variation in the flower colour of *S. spegazzinii*; although in a large number of plants the flower seems to be a medium purple, it can vary, according to Brücher (1957*b*) from a deep blue to a light lilac colour; sometimes, also, the petal centres are darker in colour, though as we have seen with the Cachi plants, the reverse may sometimes occur.

In a general revision of the taxonomy of tuber-bearing Solanums, published in 1956, Hawkes placed *S. spegazzinii* and the Bolivian species *S. pachytrichum* as synonyms of the northern Bolivian species *S. leptophyes*, basing this on the delicate habit, narrow leaflets, slight decurrency and apparently similar corolla shape.

The exact locality and habitat of *S. leptophyes* are unknown but *S. pachytrichum* is from dry regions near Sucre, Bolivia, similar to those where *S. spegazzinii* is found in Argentina; the leaf dissection and pubescence of *S. pachytrichum* is very similar to that of the better-developed forms of *S. spegazzinii*, also. Unfortunately no living material has been available of *S. pachytrichum*.

Correll (1962) placed *S. spegazzinii* as a synonym of *S. leptophyes*, in accordance with the treatment proposed by Hawkes in 1956. There is no doubt that the two entities are closely related, but we have decided in the present treatment to separate them for reasons to be given below.

In the same work Correll separates *S. famatinae* from *S. spegazzinii* and ranks it as a synonym of *S. velascanum*. In our opinion *S. velascanum* is a taxonomic synonym of *S. kurtzianum* (see p. 412). We are certain that *S. famatinae* has nothing to do with *S. kurtzianum*, though admittedly the superficial resemblance between some of the specimens of these two taxa is quite strong. Correll himself thought that *S. velascanum* (including *S. famatinae*) might prove to be a hybrid of *S. kurtzianum* × *S. chacoense* and concludes: 'It is even more probable that with further study it will be found necessary to include this plant in *S. kurtzianum*.' With this we would tentatively agree, always assuming that the *S. famatinae* (= *S. spegazzinii*) part of Correll's concept of *S. velascanum* was not included.

A further species described by Correll (1961) is *S. sleumeri*, and this in our opinion is undoubtedly a synonym of *S. spegazzinii*, with the 'famatinae' phenotype (habitat: 'damp shady places'), developing under adequate conditions of shade and moisture. The species was based on a collection of Sleumer and Vervoorst (*2434*) which is part of a series of gatherings from prov. Catamarca, dept. Belén, Pozo de Piedra to Potrerillos, made by the same collectors (*Nos. 2489, 2490, 2534*). It would appear that Correll did not have the opportunity of examining the other three collections, though he did indeed examine one or two of the hybrids of *S. chacoense* × *S. spegazzinii* which we have cited previously (p. 235). These he classifies either as *S. setulosistylum* or as a new species, *S. puberulofructum*. We have already discussed the position of these two species when dealing with *S. chacoense* hybrids, so they need not be discussed in further detail at this point.

To sum up, then, we have maintained *S. spegazzinii* as a distinct species in the present treatment[1] for three reasons:

(*a*) Our knowledge of variation in Bolivian potatoes is insufficient for us to be absolutely certain that *S. leptophyes* should include *S. spegazzinii*; furthermore, not enough living material from Bolivia has been available for study up to now.

(*b*) In south Bolivia there is a series of forms apparently rather distinct both from *S. pachytrichum* from Central Bolivia and *S. spegazzinii* from Argentina; these forms have received the name *S. vidaurrei* and have already been referred to above (p. 384). No really good collections of the *S. leptophyes–S. pachytrichum–S. spegazzinii* complex have been collected from Sucre to Salta, a distance of about 650 km.

(*c*) Finally, there is a distribution gap between the northern part of the range of *S. spegazzinii* (about 25° S.) and the southern part of the range of *S. vidaurrei* (about 22° S.) with only two collections of possible intermediates from prov. Jujuy. Of these collections one from dept. Tumbaya, Volcán (*Castillón 365*), seems to be much closer to *S. spegazzinii* than the other from dept. Tilcara, Garganta del Diablo (*H.P. & R. 264*), which shows a greater similarity to *S. vidaurrei*. Whatever the final verdict may be about these two collections there can be no doubt that there is a partial or complete discontinuity in the region of Jujuy between *S. spegazzinii* and related forms to the north.

It seems probable that an ancestral species spread at one time all the way from La Paz to Sierra Famatina in the drier valleys of the eastern Andes and began to differentiate later into a series of separate units with restricted gene flow between them. One such unit is *S. spegazzinii* (including *S. famatinae*), another is *S. vidaurrei*, and a third is *S. leptophyes* (and *S. pachytrichum*).

Further careful collections of dried and living material in Bolivia and in provs. Jujuy and Salta in north-west Argentina may lead us to treat these units as geographical vicariads or subspecies. At the moment, however, information is lacking from Bolivia, and it seems safest to treat *S. spegazzinii* as a species distinct from *S. leptophyes*.

[1] This has been followed in the second edition (1963) of the 'Revision' by Hawkes, mentioned above.

NOTES ON LIVING PLANTS

Twelve collections of *S. spegazzinii* were grown by us, seven of them being from Sierra Famatina in La Rioja province (*P. & H. 1892, 1893, 1895, 1896a, 1900a; Reche 6, 7*); three were from Belén, Andalgalá, and Tinogasta respectively in Catamarca province (*Sleumer 1, 2688* (p.p.); *Vervoorst 3159*); one was from Cachi in Salta province (*H.P.R. 334*); and one was of unknown source (EBS: 510). The altitude range of these was from 1900 to 3100 m.

When grown in the field and glasshouse at Birmingham and in several parts of Denmark the uniformity of these different collections was high, except for the rather distinct *H.P.R. 334* from Salta.

The leaflet number ranged from 5 to 7 pairs of laterals and from 2 to 12 pairs of interjected leaflets. Not all leaflets were arranged exactly in pairs, and some plants exhibited a strong alternation of leaflets. Accessions varied slightly from each other in leaf dissection but there was also a fairly wide intraplant variation and one between different seedlings of any one family. In the good soil conditions of the experimental plots in England and Denmark the leaflets were rather broad, slightly more than twice as long as broad, sessile to shortly petiolulate, and with narrow decurrent wings running down on to the rhachis for a short distance.

Amongst the families of seedlings raised from the Petersen and Hjerting Sierra Famatina collections some showed a constantly narrower leaflet (*1893—see Plate 138; 1895*) than others (*1892, 1896a, 1900a*). The leaf and stem pubescence in the Catamarca collections was on the whole coarser than that of the La Rioja (Famatina) ones, but not so coarse and spreading as that of the Salta collection.

The seedling progeny raised from *Reche 7* was very variable and seemed to indicate segregation of characters derived possibly from a natural cross between *S. spegazzinii* and *S. kurtzianum*.

Glasshouse-grown material of all the above-mentioned seedling lines exhibited lateral and terminal leaflets that were narrower and more delicate, with longer petiolules and narrower interjected leaflets than those of the field seedlings. Such plants corresponded very closely to the lectotype sheet of '*S. famatinae*' (*Kurtz 13445b*) and Bitter's drawing, as well as the phenotype mentioned in Group I of Table 23.

Plants of *P. & H. 1893* were also grown in small pots under poor conditions of moisture and soil fertility and these gave phenotypes that closely matched the syntype specimens of *S. spegazzinii* (*Spegazzini* (s.n.) LPS 14210 and 14211) and the plants from drier habitats shown in Group 3 of Table 23.

These results reinforce very strongly our conclusions that the phenetic range of this species is a broad one, embracing both *S. spegazzinii* and '*S. famatinae*', and is due to differences in soil moisture and fertility and the relative humidity of the atmosphere. Some of these conditions have been reproduced by us and the resulting range of phenotypes satisfactorily matches most of those encountered in nature. We have not, however, been able to reproduce the extremely wide leaflet decurrency as shown in wild plants of *H.P.R. 334*, which disappeared completely when tubers and seeds of these plants were cultivated in Europe. (See Plates 140 and 141.) The extreme decurrency of the Ruiz Leal and Roig collections from San Juan is also most striking. Unfortunately we possess no living material of these but Ruiz Leal stated (verbal communication) that this collection was made from plants growing in a marsh, with *Typha* and other water-loving species. One of our 1966 collections (*H.H.R. 3427*) from a marshy locality, also shows a well-marked leaflet decurrency (see Appendix III).

High pedicel articulation is constant in every seedling family except those grown from *Vervoorst 3159*, where the articulation is median. The ovate sepals and narrow calyx acumens, which are 1–1·5(–3) mm long, are also reasonably constant. In general the substellate to pentagonal medium mauve corolla is maintained under cultivation though there are certain genotypically conditioned differences. Thus, the lobes are narrower in *P. & H. 1892*, *1893*, and *1896a*, and rather broader in *1895* and *1900a*. On the whole, and certainly in the Sierra Famatina collections, the medium to pale corolla possesses darker central petal streaks on the outer surface; in the Catamarca accessions, however, the centres and acumens are much paler (*Sleumer 1*; *Vervoorst 3159*), whilst in the Salta collection of *HPR 334* the acumens are quite white, whilst the rest of the corolla is red-purple below and paler above. There seems to be at least some evidence of a cline in corolla colour patterning, showing an increasing paling of petal centres and acumens from south to north. We cannot be certain of this, however, until a wider range of living collections has been examined.

On all the flowers analysed the style is clothed with quite long papillae on its lower third but no setae are present. It is very long in *P. & H. 1895* (to 12 mm) but rather shorter in other collections. The stigma is generally long and slender and more or less narrowly obconical in shape.

The berries are green, or green spotted with white, but no collections grown by us possess puberulent fruits.

A seedling progeny of *P. & H. 1896a* gave 5 out of 10 plants completely without stamens; in these aberrant plants the style was fasciated and in one seedling the corolla was aborted, also. The same phenomenon was observed in a progeny of seedlings from a *Sleumer 2688* (p.p.) intersib cross. This may be due to plasmon variations of the kind analysed by Koopmans (1952, 1954, 1955, 1959) and by Grun *et al.* (1962).

Ross (written communication) raised seedlings from berries obtained from *Ross & Vervoorst 16 to 23*, collected at Ingenio del Arenal, dept. Andalgalá, prov. Catamarca. The observed variation led him to conclude that the natural population may have been of hybrid origin. No other tuber-bearing *Solanum* species has yet been observed in that area, however. We made living collections from the same population in 1966 and are inclined to think that some of the variation is due to differences in light and soil moisture (see Appendix III).

The tuber yields of *S. spegazzinii* are, on average, rather poor and the individual tubers do not attain a great size. At Vandel, under artificial short days, different plants gave average tuber weights ranging from 2 to 5 grammes.

The high degree of eelworm resistance in *S. spegazzinii* has already been mentioned in Chapter 5. This species has also exhibited resistance to wart and *Verticillium* wilt.

CYTOLOGY

A somatic chromosome number of $2n = 24$ was reported by Brücher (1957*b*) from three different counts. This was verified by us in counts made during the same year from five accessions from La Rioja, one from Catamarca, and one from Salta province.

Little work has been published on meiosis in this species. Von Wangenheim (1957) observed normal pachytene pairing in an artificial hybrid of *S. vernei* × *S. spegazzinii* (as *S. famatinae*), followed by regular tetrad formation. No pairing disturbances accrued in the hybrid as compared with the parent species, and there seemed to be therefore no significant structural genomic differentiation between these species; nor for that matter could any major differences be seen between these and such widely differing species as *S. verrucosum*, *S. chacoense*, *S. goniocalyx*, and *S. microdontum*. He disagrees, therefore, with the thesis propounded by Gottschalk (1954, Gottschalk and Peters 1955, etc.) to the effect that large structural differences can be observed at pachytene between different species in the genus *Solanum*.

FERTILITY AND CROSSABILITY

S. spegazzinii is self-incompatible but all clones so far tested are fertile when crossed with each other. We have obtained crosses in various directions with *P. & H. 1892*,

1893, *1896a* and *1900a*, all from Sierra Famatina. All combinations gave abundant seed and the families raised were apparently fertile and uniform.

The hybrid of *Reche No. 6* (from Famatina) × *H.P.R. 334* (from Salta), which was grown in the experimental field at Taastrup, near Copenhagen, in 1963, showed a certain amount of variation, but most seedlings exhibited the long spreading hairs of the aberrant *H.P.R. 334* parent.

We have already described the artificial hybrids of *S. spegazzinii* (both normal and aberrant forms) with *S. chacoense* (p. 228), with *S. infundibuliforme* (p. 280), and with both subspecies of *S. vernei* (p. 368).

Crosses in both directions were made in 1957 between *S. spegazzinii* (*P. & H. 1892*, from Sierra Famatina) and *S. kurtzianum* (*Ruiz Leal 11906*, from prov. Mendoza, dept. Las Heras). Only a few seeds were produced from several berries and none of these germinated.

NATURAL HYBRIDS

(*a*) Natural hybrids have been recorded between ***S. spegazzinii*** and ***S. chacoense*** towards the lower part of the altitude range of the former species. They have received the names *S. setulosistylum* and *S. puberulofructum*. Since they appear to be fertile they can, and indeed do, show a wide range of variability (see pp. 233–6).

A possible natural hybrid has also been noted with ***S. tarijense*** in the far north of Argentina where the distribution areas of the two species come into contact (see p. 245).

At high altitudes occasional hybrids with ***S. acaule*** and its subsp. ***aemulans*** have been identified from herbarium specimens, but corroborative evidence from living collections is lacking so far (see pp. 268–9).

Hybrids with ***S. sanctae-rosae*** are not uncommon in several places and occur in regions of medium humidity at the humid end of the range of *S. spegazzinii* and the drier end of that of *S. sanctae-rosae* (see p. 310).

A rather doubtful hybrid of ***S. spegazzinii*** × ***S. microdontum*** subsp. ***giganto-phyllum*** has also been observed, but it is possible that this might involve ***S. vernei*** rather than *S. spegazzinii* (see p. 347).

Hybrids of ***S. spegazzinii*** with ***S. vernei*** subsp. ***vernei*** are also to be seen occasionally (see pp. 369–70).

(*b*) Finally, it is possible that in certain regions ***S. spegazzinii*** may hybridize with ***S. kurtzianum***, though we ourselves have been able to identify only two possible hybrid collections of this type. These seem to be roughly intermediate morphologically between the two species, though the articulation is high (*Castellanos* (s.n.) 1928) in both hybrids.

The two species overlap in the zone of (1500–)1900–2500(–3000) m, and they are both apparently well adapted to rather dry regions of the Prepuna phytogeographical zone. It is therefore at first sight rather surprising that hybrids are not seen more frequently.[1] On looking into the question more closely, however, it can be seen that the distributional overlap of the two species is rather narrow and that only in La Rioja province, and especially in Sierra Famatina, have both species been recorded with some frequency. Further south of this area *S. spegazzinii* has been recorded only once, whilst *S. kurtzianum* has been recorded only three times north of Sierra Famatina. (See Maps 12 and 13.)

Another point worth consideration is that *S. spegazzinii* is confined chiefly to the Prepuna province whilst *S. kurtzianum* is predominantly a species of the Monte province, even though each may be found in the province of the other on certain occasions. Possibly the two species require slightly different microhabitats, which would make it unlikely for them to grow in close proximity even though they exist in the same

[1] See hybrids of this type noted in Appendix III, however.

phytogeographical province. Finally, of course, there may be some sterility or other barrier preventing hybridization, as evidenced by the experimental hybrids mentioned above.

Both specimens identified by us as possible hybrids of *S. spegazzinii* × *S. kurtzianum* occur in localities from which the parents have not yet been collected although they are known in the same general areas. Obviously, more detailed collections and observations are needed here, as in so many other regions of Argentina, despite the fact that the country is one of the best-surveyed parts of South America so far as wild potato species are concerned.

It will be noted that *S. spegazzinii* apparently forms hybrids with a wide range of species, and at least seven or eight species and subspecies have been considered as possibly hybridizing with it. It thus exceeds even *S. chacoense* in its ability to form natural hybrids so far as numbers of species are concerned, but on the whole the quantities recorded of each hybrid are not very great. So far as we know, no large-scale introgression of another species into *S. spegazzinii* has taken place as has happened with *S. microdontum* and *S. chacoense*.

LOCALITIES OF NATURAL HYBRIDS

(*a*) With ***S. chacoense, S. tarijense, S. acaule, S. sanctae-rosae, S. microdontum,*** and ***S. vernei*** (see pp. 238, 246, 269–70, 312, 349, and 370–1)

(*b*) Probable hybrids of ***S. spegazzinii*** with ***S. kurtzianum***

Prov. LA RIOJA:

Dept. Chilecito, Sierra Famatina, El Rodado. 30 Jan. 1928. *Castellanos* (s.n.) (BA 28/343).

Dept. Chilecito, Sierra Famatina, Ciénaga de Cosme. 19 Jan. 1928. *Castellanos* (s.n.) (BA 28/355).

18. SOLANUM KURTZIANUM Bitter et Wittmack

(In Engler, *Bot. Jb.* **50**, Suppl., 548–50, 1914)

S. commersonii Dun., var. 2. *pilosiusculum* Hooker f. (*Flora Antarct.* **2**, 330, 1846)
S. commersonii Dun., var. 3. *glanduloso-pubescens* Hooker f. (ibid. 330, 1846)
S. velascanum Bitt. et Wittm. (in Engler, *Bot. Jb.* **50**, Suppl., 551–2, 1914)
S. macolae Buk. (*Revta argent. Agron.* **4**, 239–40, 1937)
(*S. juanense* Wittm., *nomen nudum*)

Herbaceous, tuberiferous, bearing round to oval tubers up to 2·5 cm long.

Stem below ground about 5–10 cm long, above ground (10–)20–35(–70) cm long, 1·5–2(–3) mm diam., ±upright, slightly zigzag and little branched, with generally rather short internodes and semi-rosette habit in the shorter plants, and with barely visible decurrent lines or very narrow wings (only occasionally with wings to 2 mm wide); stem clothed with sparse to moderately frequent coarse subappressed hairs.

Leaves often densely following, (7–)10–15(–22) cm long × (3·5–)5–8(–14) cm broad, 3–4(–5)-jugate; lateral leaflets broad-elliptic, ovate-elliptic, lanceolate or *narrowly oblong, apically obtuse* and very rarely acuminate, *basally rounded*, not very often attenuated into the petiolule; leaflets of uppermost pair *very oblique at the base, generally subsessile* and even slightly decurrent occasionally; other leaflets borne on (1–)3–6(–12) mm long petiolules; size of largest lateral leaflet (2·2–)3·5–6(–7·5) cm long × (0·7–)1·5–2·5(–3·8) cm broad; terminal leaflet ovate to obovate, generally slightly larger than the laterals, often gradually attenuated at the base, up to 7·5(–9) cm long × 3(–5·5) cm broad; interjected leaflets minute, not very frequent, 0–3(–5)-jugate, broadly ovate to rotund, obtuse, 2–4(–6) mm long, sessile or rarely subsessile, sometimes more or less decurrent on to the rhachis, which is sometimes very narrowly winged. Leaflets provided above with sparse (to frequent) rather short thick acute curved accumbent hairs and below with smaller hairs, chiefly on the veins; leaflet margins entire or denticulate, *provided with frequent stiff curved conspicuous hairs.* Pseudostipulate leaves well developed.

Inflorescence (2–)6–25(–30)-flowered; peduncle 3·5–10(–17) cm long, once or twice forked above; *pedicels short,* 9–15 mm long, *often articulated near the base* (1–4 mm above base) or sometimes to just below or above the centre; pedicels endowed, as the peduncles, with scattered rather stiff recurved hairs; pedicels often endowed with frequent minute stalked glands in the southern forms (prov. Mendoza) but sparsely glandular or completely without glands in the central and northern parts of its range (La Rioja, Catamarca).

Calyx campanulate, to about 5 mm long, with ovate lobes and spathulate, linear or triangular acumens which are straight or recurved, (0·5–)1–1·5(–3) mm long; pubescence of sparse to frequent thick short to long white acute accumbent hairs and a few glands in forms with glandular pedicels.

Corolla white or *white with a rich deep purple streak* on the centre of the lower surface of each petal, pentagonal to sub-stellate, 15–27 mm diam. (up to 37 mm in plants grown in Europe); free parts of petals (8–)10–12(–14) mm broad at the base, (5–)7·5–9 mm long, including the acumen which is 4–5 mm long and 4 mm broad at the base.

Filaments 0·5–1 mm long, glabrous; anthers about 4–6 mm long, 1·5–1·75 mm broad, cordate at the base.

Style 7–11 mm long, clothed on its lower $\frac{1}{3}$–$\frac{1}{2}$ with minute papillae; stigma thicker than style apex, ovoid to globose.[1]

Berries globular, about 1–1·5 cm diam., or perhaps larger.

Chromosome number: $2n = 24$.

AFFINITIES

S. kurtzianum is closely related to *S. spegazzinii*, from which it may be distinguished by the less well-dissected leaf with only 0–3(–5) pairs of interjected leaflets, rounded leaflet base and conspicuous marginal hairs. It differs from *S. spegazzinii* also in the generally low pedicel articulation, the coarse thick stem and inflorescence hairs and the white corolla with rich purple streaks externally on the petal centres.

From *S. chacoense*, to which it is also closely related and with which it occasionally forms hybrids, it is distinguishable by the lesser degree of leaf dissection, the coarse hairs on leaflet margins and other green parts, the obtuse and not acuminate leaflet apices, as also by the generally low pedicel articulation and the pentagonal corolla which is generally purple and white in colour.

From *S. maglia* it differs in the better-dissected leaf, very narrow stem wing, narrower oblong to elliptic leaflets which are not acuminate, distinct demarcation between anthers and filaments, and the generally purple-streaked corolla.

[1] Bitter states that it is barely emarginate at the apex but we have seen no material embodying this character.

S. kurtzianum probably forms an intermediate link between *S. spegazzinii*, on the one hand, and *S. maglia*, on the other. It also bears affinities to *S. chacoense* and to a lesser extent to *S. oplocense*. Its chief distinguishing features are the 3–5-jugate leaf with 0–3 pairs of interjected leaflets, the conspicuous coarse marginal hairs, the coarse hairs on stems and inflorescence, the short pedicels with often low articulation, and the white pentagonal to substellate corolla which frequently bears a rich purple radial streak on the centre of each petal on the outside surface only.

SPECIMENS EXAMINED

ARGENTINA

Prov. CATAMARCA:

Dept. Andalgalá, Río Pantanillos, entre Las Pampitas y Punta del Agua. Alt. 1800 m. 11 Feb. 1949. 'En tierra arenosa, muy seca.' *Petersen & Hjerting P–40* (J.G.H.) See Plate 145.

Dept. Belén, Quebrada de Belén. Alt. *c.* 1400–1500 m. 24 Jan. 1952. 'Fl. alb.' *Sleumer & Vervoorst 2357* (LIL, J.G.H.).

Dept. Belén, entre Pozo de Piedra y El Bajo, km 11. Alt. *c.* 1700 m. 3 Feb. 1952. 'Fl. alb. Sobre loess.' *Sleumer & Vervoorst 2645* (LIL, US, J.G.H.).

Prov. LA RIOJA:

Dept. Chilecito, Sierra de Famatina, Rodeo de Las Vacas. Alt. 3000–4000[1] m. Feb. 1913. 'Rocas graníticas.' *Flossdorf* (s.n.) (SI 15956).

Dept. Chilecito, Sierra de Famatina, Rodeo de Las Vacas. Alt. 3000–4000[1] m. Feb. 1913. 'Rocas graníticas.' *Flossdorf 57* (SI). (Probably the same collection as above.)

Dept. Chilecito, Sierra Famatina, cercanías de la 'Mina El Oro.' 23–25 Jan. 1879. *Hieronymus & Niederlein 415* (p.p.) (CORD, SI).

Dept. Famatina, Ruta 40 (km 680) yendo de Famatina a Tinogasta, entre Ángulos y Santa Cruz. 20 Mar. 1960. *A. T. Hunziker, Cocucci & Di Fulvo 15203* (CORD).

[Dept. Sanagasta?], Sierra Velasco, Yacuchi. Alt. 2000 m. Mar. 1963. 'Locus classicus.' *Brücher 1102* (BAA, CORD, SI, Ruiz Leal).

[Dept. Sanagasta?], Sierra Velasco, Cuesta del Sigú, Cuesta de la Puerta de Piedra. 8–11 Jan. 1879. *Hieronymus & Niederlein 68* (CORD).

[Dept. Sanagasta?], Sierra Velasco, Yacuchi. 25 Feb. 1908. 'Pl. herbacea tuberosa, fl. albis.' *F. Kurtz 15422* (CORD—LECTOTYPE; Photograph of the now destroyed Berlin type of *S. kurtzianum* Bitt. & Wittm., taken by Field Museum, Chicago. The following additional information is given: Alt. *c.* 2100 m. 'Here and there on bushy slopes'). See Plate 142.

[Dept. Sanagasta?], Sierra Velasco, Yacuchi. 25 Feb. 1908. 'Fl. albo-violascens.' *F. Kurtz 15380* (CORD—LECTOTYPE of *S. velascanum* Bitt. & Wittm.; Photograph of now destroyed Berlin type of *S. velascanum* Bitt. & Wittm., taken by Field Museum, Chicago. The following additional information is given: 'Near the ruined house "Yacuchi" among stones. Alt. *c.* 2100 m. Flowers from white to violet.') See Plate 143.

Prov. SAN JUAN:

Dept. Calingasta, salida de la Quebrada de Leoncito. *Echegaray* (s.n.) (CORD).

Dept. Zonda, Precordillera entre Barreales, Tontal y Retamito, Agua Pinto. 28 Feb. 1897. 'In sepibus freq. Fl. albi.' *F. Kurtz 9828* (CORD—ISOTYPE of '*S. juanense*' Wittm. *nom. nud.*). See Plate 144.

[1] These altitude data are certainly incorrect.

Prov. MENDOZA:

Dept. Capital (?), San Isidro, near the foot of the mountains at Mendoza. Mar. 1824. *Gillies* (s.n.) (K–Type for *S. commersonii* var. 3, *glanduloso-pubescens* Hooker f.).

Dept. Capital (?), foot of mountain at Chayado. Apr. 1824. *Gillies* (s.n.) (K—Type for *S. commersonii* Dun. var 2, *pilosiusculum* Hooker f.).

Dept. Las Heras, Melocotón, 50 km al S. de Villavicencio. Alt. 2000 m. 3 Mar. 1947. *Moyano* (s.n.) (LIL 199507, 199838).

Dept. Las Heras, Quebrada del Peral. Alt. 1800 m. 27 Mar. 1938. 'Muy abundante en las laderas de los cerros.' *Ruiz Leal 4990* (LIL, Ruiz Leal).

Dept. Las Heras, Cerro Alfalfal. Alt. 1900 m. 22 May 1938. *Ruiz Leal 5239* (Ruiz Leal).

Dept. Las Heras, cerca del Puesto de las Higueras. 12 May 1940. 'No escaso.' *Ruiz Leal 6746* (Ruiz Leal).

Dept. Las Heras, Quebrada del Toro. Alt. 2500 m. 18 Mar. 1945. 'En acarreos. Fl. blancas.' Leg. J. Semper, *Ruiz Leal 10133* (Ruiz Leal).

Dept. Las Heras, Quebrada del Potrero Puerta, en laderas con exposición al E. Abundante. 30 Mar. 1947. *Ruiz Leal 10747* (LIL, Ruiz Leal, J.G.H.).

Dept. Las Heras, La Crucesita. 25 Feb. 1949. 'Común. Estéril.' *Ruiz Leal 11892* (LIL, Ruiz Leal).

Dept. Las Heras, El Portezuelo (camino internacional a Chile). 30 Apr. 1949. *Ruiz Leal 11906* (H. 62) (LIL, Ruiz Leal). See Plate 147.

Dept. Las Heras, entre El Portezuelo y Villavicencio. 16 Dec. 1956. 'Protegido entre arbustos. Fl. blanca.' Leg. Ruiz Leal, and Roig, *Ruiz Leal* 18390 (Ruiz Leal, J.G.H.). See Plate 148.

Dept. Las Heras, Quebrada del Toro. Alt. 2500 m. 18 Mar. 1945. 'En la falda sud sobre terreno de acarreo. Hierba, 10–20 cm. Flor blanca. Escasa.' *Juan Semper 470* (LIL, SI).

Dept. Luján, Cacheuta. Jan. 1926. Leg. Sta. Negrete, *Hicken.* (s.n.) (SI 15949).

Dept. Luján, entre Vistalba y La Crucesita. Alt. 1000–1100 m. 17 Apr. 1938. 'Estéril.' *Ruiz Leal 5070* (LIL, Ruiz Leal).

Dept. Rivadavia, Huayquerías, Cañadón del Gringo Muerto, en las galerías de chañar. 13 Jan. 1957. Leg. Roig, and Ruiz Leal, *Roig 1273* and *1274* (J.G.H.).

Dept. Rivadavia, Huayquerías. 13 Jan. 1957. 'Formando colonias en el chañaral húmedo, denso. Fl. blanca con pigmentación violeta en la parte media inferior de los pétalos!' Leg. Ruiz Leal, and F. Roig, *Ruiz Leal 18435* (J.G.H., Ruiz Leal).

Dept. San Rafael, near San Rafael, Mármol Amarillo. 12 Dec. 1887. 'Fl. albi, tubera esculent.' *F. Kurtz 5497* (CORD).

Dept. San Rafael (Nevado). 1935. Leg. Orn. Carette, *Ruiz Leal 4043* (Ruiz Leal).

Dept. San Rafael, along the Río Diamante near San Rafael. Jan.–Feb. 1897. 'Fleurs blanches.' *Wilczek 240* (G, LAU).

Dept. Tunuyán, entre Los Arboles y Puesto del Manzano. 20 Mar. 1935. 'Al pie de arbustos sobre suelo humífero de mantillo arenoso.' *Ruiz Leal 3210* (Ruiz Leal).

Mendoza, locality unknown. Grown at Krasny Pakhar 1934 as No. 35–18. *Macola* (s.n.) (WIR No. 35–18). (Named *S. macolae* by Bukasov, though *not* the type.)

Collector and locality unknown. Grown from tubers presented by Professor S. M. Bukasov. CPC: 1296 (P.G.S. 0024. J.G.H. flower dissection). See Plate 146.

Probably ***S. kurtzianum***

Prov. MENDOZA:

Dept. Las Heras, Quebrada del Toro. Alt. *c.* 2500 m. 25 Feb. 1945. *Lourteig 817* (LIL).

Dept. Las Heras, Quebrada del Toro. Alt. 2000–2300 m. 7 Mar. 1937. 'En laderas, sobre tierra humífera, bajo arbustos.' *Ruiz Leal 4407* (LIL, Ruiz Leal, J.G.H.).

Dept. Las Heras, Quebrada Cajón de Minas (Salto). Alt. 2000 m. 24 Apr. 1938. *Ruiz Leal 5169* (LIL, Ruiz Leal).

DISTRIBUTION

S. kurtzianum grows in the drier valleys and hillsides of the western and north-western Argentine Andes (see Map 13). It has been recorded for the following provinces and departments:

Prov. Catamarca, depts. Andalgalá and Belén.
Prov. La Rioja, depts. Chilecito, Famatina, and Sanagasta.
Prov. San Juan, depts. Calingasta and Zonda.
Prov. Mendoza, depts. Capital, Las Heras, Luján, Rivadavia, San Rafael, and Tunuyán.

It probably occurs elsewhere in this area and has no doubt been insufficiently collected in provs. Catamarca, La Rioja, and San Juan for us to be very certain of its distribution in those provinces.

HABITAT

S. kurtzianum is probably better adapted to growth in dry regions than any other Argentine wild potato. The altitude range of this species is (1000–)1400–2500(–3000?) m. It has been recorded from very dry sandy soil, in loess, amongst stones and granitic rocks, on hillsides, road embankments, and bushy slopes, sometimes probably in the open but often protected by bushes, shrubs, or trees.

According to Bukasov (1958), *S. kurtzianum* grows in the arid *Larrea* steppe from Mendoza northwards. Our own observations in 1966 conform with this statement (see Appendix III).

In Mendoza at Las Huayquerías, according to Roig (1960), it occurs in the herbaceous layer of narrow dense gallery woods of 'chañar' (*Geoffraea decorticans*), together with other herbaceous species such as *Parietaria debilis*, *Castelia cuneato-ovata*, *Sonchus asper*, *Simsia dombeyana*, etc. Such woods apparently occur along the beds of temporary water courses.

This plant association is placed by Roig within the Monte phytogeographical province of Cabrera (1953) and others, and it seems clear from this and other distribution data that *S. kurtzianum* grows typically in the Monte province and in fact is practically the only wild potato to do so, apart from *S. chacoense* and *S. spegazzinii* which spread into it here and there, and *S. maglia* which may also occur on the border of this province. No doubt, however, *S. kurtzianum* grows best, or perhaps almost exclusively, in special parts of this province where there is more moisture or shading from other plants than could be thought of as typical for the province as a whole. It also seems to extend into the Prepuna province here and there. A few of the collections in Mendoza were probably gathered in the transition zone between the Monte and High Andean (and/or Puna) provinces.

S. kurtzianum is apparently better adapted to drier regions than *S. spegazzinii* and *S. chacoense*; it occurs generally at lower altitudes than *S. spegazzinii* and at higher ones than *S. chacoense*, but occasionally grows with one or other of these two species and is able to form hybrids with them when the overlap of distribution areas occurs.

TAXONOMIC NOTES

S. kurtzianum was described by Bitter and Wittmack from a single collection made by Kurtz (*No. 15422*) in prov. La Rioja, Sierra Velasco, Yacuchi, at about 2100 m

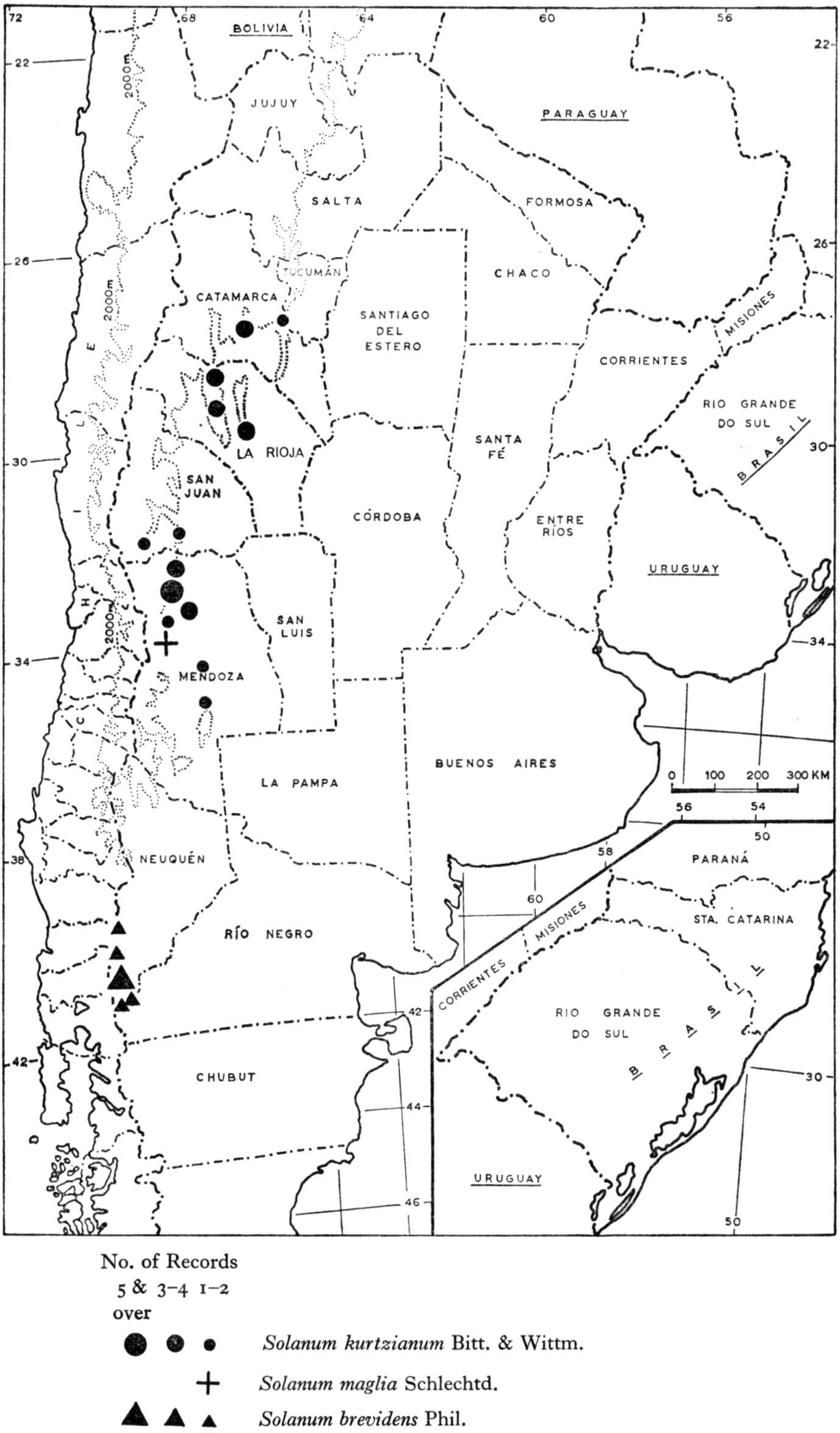

MAP 13. Distribution of *S. brevidens* Phil., *S. kurtzianum* Bitt. & Wittm., and *S. maglia* Schlechtd.

above sea level. In the same publication, following the description of *S. kurtzianum*, the same authors described a second species, *S. velascanum*, basing their description on another single collection of Kurtz (*No. 15380*) from approximately the same locality.

Both these type specimens were deposited in Berlin and destroyed during the last world war. Fortunately, isotype material of both was preserved in the Kurtz Type Herbarium at Córdoba, and we have thus been able to designate lectotypes, as follows:

F. Kurtz 15422 in Herb. Córdoba—LECTOTYPE for *S. kurtzianum* Bitt. et Wittm.
F. Kurtz 15380 in Herb. Córdoba—LECTOTYPE for *S. velascanum* Bitt. et Wittm.
(See Fig. 41 and Plates 142 and 143.)

It is clear from an examination of the lectotype specimens of *S. kurtzianum* and *S. velascanum* that the two species are basically similar, though differing somewhat in their dimensions. The lectotype of *S. kurtzianum* matches the type collection photograph very closely with its 4-jugate leaf, petiolulate to subsessile laterals, conspicuous marginal hairs, and pentagonal to substellate corolla. The articulation is low (to 4 mm above the base), there are no glands on the inflorescence and the calyx acumens range from 1 to 1·25 mm.

The lectotype of *S. velascanum* is a larger plant, 45 cm tall above ground, with 4–5-jugate leaves, conspicuous marginal hairs, no glands on the inflorescence, low pedicel articulation (3–7 mm above the base), calyx acumens linear to spathulate, to 2 mm long (to 5·5 mm in fruit) and rotate corolla. Both lectotypes possess narrowly winged stems and the gibberulous hairs that are always to be seen on living plants and are such a conspicuous feature of this species.

Even though the leaflets of *S. velascanum* are wider than those of *S. kurtzianum* all intermediates can be seen in material gathered by other collectors; furthermore, in details of hair type and distribution, leaflet form and leaf dissection, as well as in the typically low articulation the two specimens are remarkably similar.

The corolla form of these two specimens does not perhaps match so well. That of the *S. kurtzianum* lectotype is pentagonal, agreeing well with the other collections of that species to which we have already referred in the list of specimens examined. The corolla of *S. velascanum*, on the other hand, is said to be stellate, with the free part of each petal about 10 mm long × 3–4 mm broad at the base. Such a corolla lobe is completely wrong for *S. kurtzianum* and is more typical for *S. chacoense*; yet if one measures the corolla lobes visible in the photograph of the original type a length of 7 mm and breadth of 6 mm seem to be visible, which is possibly not so far removed from the proportions of the lobes in *S. kurtzianum*. Also, as we have seen, the lectotype specimen possesses a completely rotate corolla.

It is clear, then, that '*S. velascanum*' is merely a rather more vigorous and better-developed growth form of *S. kurtzianum*. It is not even a hybrid of *S. kurtzianum* with *S. chacoense*, as we had at first suspected when we considered the corolla form described by Bitter. In fact, the type specimen matches *S. kurtzianum* very well in all main features, and evidently was collected from more or less the same locality. We therefore class it as a taxonomic synonym of *S. kurtzianum*, since it follows that species in Bitter and Wittmack's treatment and is not as well known as *S. kurtzianum*.

Very few potato collections have been made from Sierra Velasco; apart from *S. kurtzianum* and *S. velascanum* (each a single collection) only two other collections have been obtained from that mountain range.[1] One is an older collection of Hieronymus and Niederlein. (*No. 68*—Herb. Córdoba), with rather small leaflets, low pedicel articulation, and no inflorescence glands. It comes fairly close to the type specimen of *S. kurtzianum*. The other is a collection by Brücher (*No. 1102*, in various herbaria), which also agrees very well with the type of *S. kurtzianum*.

[1] Collections of *S. chacoense* and *S. chacoense* × *S. kurtzianum* have already been noted (pp. 190 and 239). They are evidently from much lower altitudes.

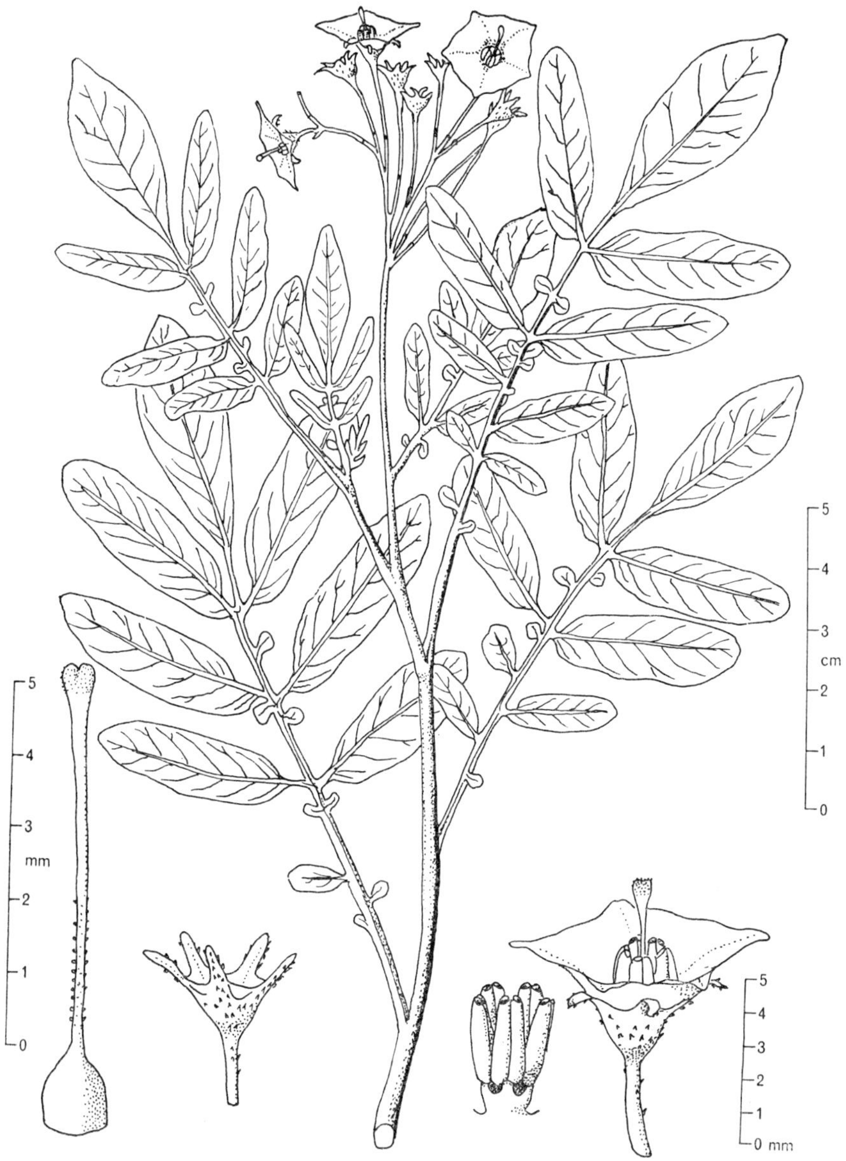

FIG. 41. *S. kurtzianum* Bitt. & Wittm. Copy of Bitter and Wittmack's figure of this species from Engler, *Bot. Jb.* **50**, Suppl., 549, 1914. Note the low pedicel articulation and linear calyx acumens.

Although the lectotype specimens of *S. kurtzianum* and *S. velascanum* vary from each other to some extent it must be realized that the variation of *S. kurtzianum* is necessarily far wider than can be encompassed in a single type specimen; our description and concept of this species are derived not only from the drawings and photographs referred to above but also from the complete range of material available to us, including the Mendoza specimens previously grouped under *S. macolae*.

The species *S. macolae* Buk. was described by the Russian potato breeder and taxonomist, S. M. Bukasov, in 1937 (= *S. mendozinum* Buk. *nom. nud*). Although the type description of *S. macolae* does not match that of *S. kurtzianum* exactly, one would hardly expect it to do so, since all potato species possess a wide range of infraspecific variability. From the living material that has been available for study, however, we are quite convinced that *S. macolae* should be placed as a taxonomic synonym of *S. kurtzianum*, since the characteristic leaf pubescence, low articulation and white and purple corolla (see Fig. 42 and Plate 146) are to be seen in both types and seem to us to be typical features of this species.

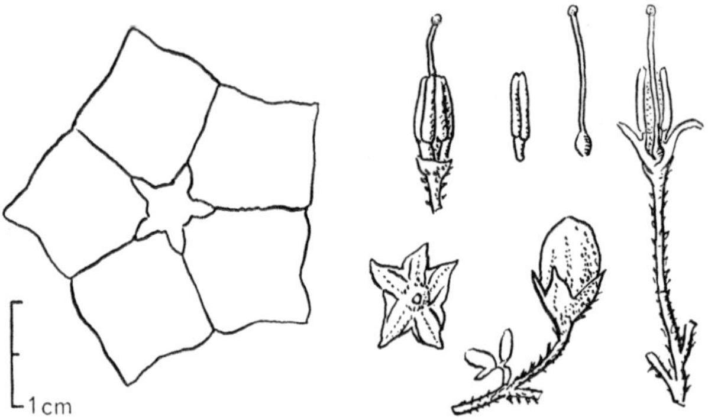

FIG. 42. *S. kurtzianum* Bitt. & Wittm. Flower dissection of '*S. macolae*' Buk., reproduced from Bukasov, *Vest. sots. Rasteniev.* No. 4, p. 11, 1940.

A further problem concerns the *nomen nudum*, *S. juanense* Wittmack. This, until recently, was known only as a rather poor Field Museum photograph of the destroyed Berlin specimen on which the name *S. juanense* had been written. We knew that this collection was made by F. Kurtz (*No. 9828*) in San Juan and thought it might be related to *S. kurtzianum*, if not that species itself.

We were extremely pleased to find in the Kurtz Type Herbarium at Córdoba an isotype specimen of *S. juanense* (*Kurtz 9828*) (see Plate 144). This agreed remarkably well with *S. kurtzianum* in the short basal internodes, 3–5-jugate leaves with 1–3-paired interjected leaflets, the thick marginal hairs of gibberulous type on leaflet margins and inflorescence branches, the presence of some glands on the inflorescence, the median to low articulation, typical calyx with 1·5 mm long acumens, and the pentagonal corolla, which was said by the collector to be white. We have not designated the specimen as a lectotype since *S. juanense* was never described, so far as we are aware. However, if in the future a lectotype is needed, then this specimen should certainly be used.

Brücher (1962) described a new species, *S. improvidum* from prov. Catamarca, of which we have not yet seen type or authenticated material. No specimens seem to be deposited in the Argentine herbaria, though possibly some might have been sent to European or American herbaria subsequent to the time when we received loaned material from them. The photographs (though not all the drawings) and the description given in his publication match the white-flowered forms of *S. kurtzianum* collected by us and by others (*A. T. Hunziker* et al. *15203*) fairly closely. We visited the type and one other locality cited by Brücher and found them to be typical dry monte scrub vegetation where *S. kurtzianum* normally occurs. No rain had fallen in 1966 by the time of our visit and no plants were growing. Nevertheless, we consider, on the whole, that *S. improvidum* may be a taxonomic synonym of *S. kurtzianum*; until we have seen authenticated material we cannot make a definite decision about it, however.[1]

Two of the varieties of *S. commersonii* described by J. D. Hooker in his *Flora Antarctica* in 1846 should also be grouped under *S. kurtzianum*. These are var. 2, *pilosiusculum* and var. 3, *glanduloso-pubescens*. They were collected by Gillies in prov. Mendoza, the one at Chayado [= Challao] (now within the city of Mendoza) and the other at San Isidro (dept. Capital). Neither is worth retaining as a variety of *S. kurtzianum*. No doubt they were collected from very dry regions since both are rather densely pubescent, and var. 3 possesses dense glandular hairs on the peduncle, pedicels, and calyx. We shall return to this point later (p. 416).

According to Brücher (1956*a*) *S. kurtzianum* should be included in series Commersoniana. Quite without any morphological comparisons and purely on supposed phytogeographical affinities he includes it in a Dry forest (Trocken wald) or Monte Oriental species group, even though on his map the area given does not in the slightest degree coincide with the Monte formation of Lorentz, to which he refers. Instead, he places it in an area coinciding with the Chaco formation of north-eastern Argentina and Paraguay together with '*S. chacoense, S. garciae, S. cordobense, S. millanii*, and *S. gibberulosum*'.

Furthermore, *S. kurtzianum* is separated from its undoubted synonym, *S. macolae*, in Mendoza and is removed some hundreds of miles eastwards from its known type locality in La Rioja province. Brücher thus unites parts of three species, *S. chacoense, S. kurtzianum*, and *S. commersonii* subsp. *malmeanum* into the so-called 'collective species' *S. chacoense*.

In Mendoza province Brücher designates an arid *Larrea* Steppe group of species, which is indeed the true Monte of Lorentz and others. Here he includes *S. mendozinum, S. rionegrinum*, and *S. macolae*. The first of these is merely a nomenclatural synonym for the third, which is in turn a taxonomic synonym of *S. kurtzianum*. The second species mentioned, *S. rionegrinum*, has nothing to do with the province of Río Negro in Argentina in which Brücher places it, but was in fact collected by the Río Negro in Uruguay and is a synonym of *S. commersonii* subsp. *commersonii* (see p. 167).

In his monograph on wild potatoes published in 1962 Correll places *S. kurtzianum* in series Transaequatorialia, together with all other Argentine species. We prefer to consider this series as included within series Tuberosa. He rightly includes *S. macolae* and Hooker's vars. *pilosiusculum* and *glanduloso-pubescens* as synonyms of *S. kurtzianum*.

Under *S. velascanum*, which he separates completely from *S. kurtzianum*, Correll groups *S. famatinae*, even though he concedes that vegetatively *S. velascanum* closely resembles *S. kurtzianum* and *S. chacoense*. We cannot agree that *S. velascanum* has much in common with *S. spegazzinii* (= *S. famatinae*), except for the somewhat similar corolla. As we have already pointed out, *S. velascanum* seems to be a good match with the type of *S. kurtzianum*, and comes well within the range of variability of that species.

S. kurtzianum seems to us to be a good species, with well-defined distribution area and ecological requirements. At the northern end of its range it perhaps hybridizes with *S. spegazzinii* and exchanges genes with it. Nevertheless the two species seem too distinct to be considered as two subspecies of a single species.

[1] See Appendix III, p. 492, footnote.

Rather characteristic features of *S. kurtzianum* are the short internodes and the somewhat zigzag stem. Young plants frequently form semi-rosettes and the juvenile leaves of these plants possess rather wider leaflets and an especially large and wide terminal leaflet in some cases, compared with the adult foliage. The coarse pubescence noted by Bitter and Wittmack is also most characteristic, and this is very easily visible on the leaflet edges. It also occurs on other parts of the leaf and often, though not invariably, on the stem, peduncles, pedicels, and calyx. We are not certain whether the slightly branched hairs on the corolla lobes mentioned by Bitter and Wittmack are constant, since they are apparently not visible on much of the dried material available to us. For this reason we have not mentioned this feature in our own description of *S. kurtzianum*.

The glandular pubescence mentioned by Hooker for his var. *glanduloso-pubescens* is very well marked both on the inflorescence (peduncles and pedicels) and on the petiole, rhachis, and petiolules of the leaf, as well as, to a less obvious extent, on the calyx and leaflets. Curiously enough it is also just as well marked in the other variety, *pilosiusculum*, although Hooker uses this character to differentiate the two varieties. Since glandular pubescence was not mentioned by Bitter in his type descriptions of *S. kurtzianum* and *S. velascanum* (and could not in any case be seen on the photographs) we thought that it might be worth while to examine a sample of the other specimens available to us, so as to assess its value in characterising the species as a whole. At the same time we scored the coarse non-glandular pubescence of the inflorescence and leaves as a standard of comparison.

The results are presented in Table 24 and give some indication of a cline showing increasing density of glandular pubescence from north to south. The glands are sparse to absent in material from Catamarca and La Rioja (cf. the type description from La Rioja) whilst they are very frequent on the inflorescences of seven out of nine collections from Mendoza. Since there is, so far as we know, no glandular species other than *S. kurtzianum*[1] to be found in Mendoza province we must assume that this characteristic is typical of that species in the south but that it becomes less and less frequent in the north. It would be of great interest to verify this by the scoring of collections or plants examined in the field in San Juan,[2] La Rioja, and Catamarca provinces, and it is to be hoped that such a project might one day be carried out. With the rather sparse material of this species at our disposal from the northern part of its range we can do no more than indicate that a cline of this type may be present. In so far as the coarse hairs are concerned they seem, on the whole, to be more abundant in the north than in the south. Some southern collections possess no coarse hairs at all on the inflorescence though others show a rather dense covering. On the leaves they range from moderate to very frequent in all collections.

The leaflet length is about three times the breadth, or even more in Bitter and Wittmack's type specimen. Yet in the type specimen of *S. velascanum* some leaflets are only 2½ or even only twice as long as broad. Leaflets with length/breadth ratios of 2·2 to 2·5 are mentioned in the type description of *S. macolae*, and this length/breadth ratio is perhaps more typical for *S. kurtzianum* than the dimensions of the type specimen. Several narrow-leaved collections from prov. Mendoza and elsewhere are also known, so it would seem that this range of leaflet length/breadth ratio is normal for the species and may depend partly on differences of genotype and partly on different environmental conditions.

The articulation is very low in what we have come to consider as typical *S. kurtzianum*; and in most collections it is situated well below the centre. In a few collections it varies from below to just about central. In two collections it rises to above the

[1] Some glands occur on *S. maglia*, however.

[2] Some glandular pubescence is evident in '*S. juanense*'.

centre. These are *Lourteig 817* from prov. Mendoza, dept. Las Heras, Quebrada del Toro, 2500 m, and *Ruiz Leal 4407* from the same locality but at 2000–2300 m. They differ also from typical *S. kurtzianum* in possessing a well-marked petiolule at the base of the distal pair of lateral leaflets on each leaf. We at first thought that these two collections might have been hybrids with *S. chacoense*, but there is no evidence for the occurrence of *S. chacoense* so far west and at such high altitudes in that region. It seems better to leave these as 'probably *S. kurtzianum*' until we have further collections from that area which may help to resolve the problem.

TABLE 24

Pubescence characters of S. kurtzianum *in different parts of its range*

Province	Collector and number	Glands		Coarse hairs	
		Inflor-escence	Leaves	Inflor-escence	Leaves
Catamarca	*P. & H. P–40*	+	+	++	+++
	Sleumer & Vervoorst 2357	+	—	++	++
	Sleumer & Vervoorst 2645	++	++	++	++
La Rioja	*Flossdorf 57*	+	+	+++	+++
Mendoza	*Gillies* (San Isidro)	+++	+++	+++	+++
	Moyano (s.n.)	++	++	++	++
	Ruiz Leal 10747	+++	++	—	++
	Ruiz Leal 18390	+++	+++	—	++
	Juan Semper 470	+++	+++	+++	+++
	Hicken (s.n.)	++	++	++	++
	Ruiz Leal 18435	+++	+++	—	++
	P. G. S. 0024	+++	++	++	++
	Ruiz Leal 4407 (Probably *S. kurtzianum*)	+++	++	++	++

Key: — absent
+ trace
++ moderate amount
+++ very frequent

A third doubtful collection (*Ruiz Leal 5169*) bears no flowers and possesses very wide and enlarged terminal leaflets. In this case, the material is not sufficiently mature for us to make a reliable determination.

The colour of the flower in *S. kurtzianum* is very characteristic and will be treated further in the section on living plants. Occasional white-flowered forms are known, and although we have insufficient data for an exact statement on their frequency it would seem that they were more often encountered in the northern part of the distribution area. Thus, they are noted for prov. Catamarca in two out of three collections, in La Rioja once only, in San Juan in one out of two collections, and in prov. Mendoza in five out of twenty-two collections. It is of interest to note that Kurtz's type collection of *S. kurtzianum* (*No. 15422*) is said to have white flowers, whilst in that of *S. velascanum* (*No. 15380*) the flowers are said to be whitish-violet (*albide violascens*). Our own collections in 1966 indicate that white-flowered forms are not uncommon in La Rioja province and may occur in mixtures with the white and purple flowered forms (see Appendix III).

NOTES ON LIVING PLANTS

When grown in the experimental field in Europe the stems of *S. kurtzianum* often develop short internodes, 1–3 cm long. In Leningrad, according to Bukasov (1937*b*), they may be up to 6 mm in diameter, even though such a dimension is not attained in South American material. The straight but narrow stem wing is also clearly observable on living plants.

Owing to the short internodes the leaves are frequently produced in a rosette in plants grown in the field, though this is not so in glasshouse-grown plants. Such differences in internode length can be seen amongst wild collections and is no doubt due to habitat differences which simulate the environments of experimental field and glasshouse. The leaves on material cultivated by us were normally 4–6-jugate with generally up to 1–3 pairs of interjected leaflets; thus the dissection was slightly greater than in South American material, which had only 3–4(–5)-jugate primary laterals with 0–2 pairs of interjecteds. In material sent from Germany (E.B.S. 42, sent from Russia) there were 4–6 pairs of interjected leaflets, so that it seems as though the Russian material of *S. kurtzianum* possesses a genotypically more dissected leaf than other material grown by us.

In all collections of *S. kurtzianum* the coarse marginal hairs, often set on the apices of small teeth, are very pronounced and characteristic.

The pedicel articulation is almost always in the lowest third but may occasionally ascend to a central position in some material.

The corolla shape varies from pentagonal to rotate, apparently according to growth conditions; in size it varies from 2·5 to 3·5 cm diam., the larger size being found in Bukasov's material of '*S. macolae*'. The purple streak on the centre of each petal, on the external surface only, is most characteristic, and is of a much more magenta colour than is normal for wild potato species. According to Dodds (1956) and Harborne (verbal communication) the pigment malvidin is found in the flowers of *S. kurtzianum* and in those of the Mexican species *S. hougasii* but not in any other wild potato species, though it occurs also in the tubers of the cultivar 'Congo'. Differences in the intensity of the colour and the width of the streak were seen by us in different seedlings; in some accessions the colour is absent altogether.

Perlova (1946) states that the berry is roundish and slightly shorter in the longitudinal than in the transverse axis.

The eelworm resistance of *S. kurtzianum* has been remarked on by several authors. It has also been shown to exhibit resistance to late blight, powdery mildew, *Verticillium* wilt, Colorado beetle, leaf hoppers, and viruses X, F, A, Y, and leaf-roll. (See Chapter 5.)

CYTOLOGY

A chromosome number of $2n = 24$ was reported by Bukasov (1937*b*) from counts made by E. V. Ivanovskaya. This was later confirmed by Gottschalk & Peters (1955) and by Gillies (1955). Our counts on various lines from 1951 onwards have also confirmed these results. Gottschalk & Peters (1955) in making a pachytene analysis stated that of 8 chromosomes analysed, 6 were similar to those in *S. tuberosum*, whilst 2 were distinct.

Meiotic analysis of *S. kurtzianum* × *S. microdontum* subsp. *gigantophyllum* colchicine-tetraploids by Swaminathan (1956) revealed a mean frequency of 1·19 quadrivalents, 0·095 trivalents, 21·025 bivalents, and 0·095 univalents per cell. This is a much higher bivalent ratio than that normally obtained from autotetraploids and may indicate a marked degree of genome differentiation between the two species. A genetical analysis also led that author to conclude that preferential pairing of chromosomes autosyndetically from each of the original parents must take place, as shown by the inheritance of stem wing and flower pigmentation. Nevertheless, tetrasomic inheritance was observed in respect of leaf dissection and it is concluded that the gene controlling this character must be carried on chromosomes that normally form quadrivalents in the tetraploid hybrid. From this one assumes that the bivalents and quadrivalents are not formed in a random manner and that cryptic structural differentiation of chromosomes has played an important role in the origin of differences between these two species.

FERTILITY AND CROSSABILITY

Most authors have found that *S. kurtzianum* is fully fertile but self-incompatible. Hybrids of *S. kurtzianum* × *S. microdontum* subsp. *gigantophyllum* were made by Swaminathan (1950, 1951, 1956), Malheiros-Gardé (1959*a*), and Grun (1961). Swaminathan reported regular bivalent pairing in these hybrids at MI with 85 per cent stainable pollen but only 69 per cent stainable pollen in the amphidiploid. Diploid and colchicine-tetraploid lines of *S. kurtzianum* were studied by Magoon *et al.* (1958) for leaflet length/breadth ratio and pollen stainability and diameter. Tetraploid *S. kurtzianum* displayed high fertility, according to Lebedeva (1959*a*, *b*).

Malheiros-Gardé (1959*a*) also made the crosses *S. kurtzianum* × *S. vernei* (and reciprocal), and *S. microdontum* subsp. *gigantophyllum* × *S. kurtzianum* (and reciprocal). Grun (1961) crossed *S. kurtzianum* with *S. tarijense*, *S. chacoense*, and *S. microdontum* subsp. *gigantophyllum* and all reciprocals. Kameraz (1957*b*) reported that *S. kurtzianum* was difficult to cross with *S. tuberosum*, though one cross had been achieved in breeding for Colorado Beetle resistance, using the variety Northern Star.

We ourselves have made a number of F_1 hybrids between *S. kurtzianum* and other diploid species. F_2 families have also been raised in some cases.

Hybrids have already been mentioned between *S. kurtzianum* and *S. chacoense* (p. 228), *S. tarijense* (p. 244), *S. acaule* subsp. *aemulans* (p. 265), and *S. vernei* (p. 368).

The F_1 hybrid (Family 121) of *S. maglia* (CPC: 2057, from prov. Mendoza, dept. San Carlos, Quebrada Alvarado) × *S. kurtzianum* (*Ruiz Leal 11906*, from prov. Mendoza, dept. Las Heras, El Portezuelo) showed the dominance of *S. maglia* in the large terminal leaflet, long inflorescence branches, characteristic stamens, and generally white corolla. Some flowers showed the purple streaks inherited from *S. kurtzianum* and in most the calyx acumens were poorly developed.

Three attempts at germinating intersib crosses were made, two of them unsuccessful. Only one F_2 family (Family 220) could be raised and of 30 seeds sown, only 10 germinated, giving 3 vigorous, 1 medium, 4 very poor, and 2 extremely unthrifty plants that hardly managed to struggle beyond the seed leaf stage. Thus the genetic breakdown shown in the F_2 and the unviability of the seeds deriving from the other two intersib crosses gives a good indication that *S. kurtzianum* and *S. maglia* are genetically most distinct from each other even though there are many points of similarity in their morphology.

NATURAL HYBRIDS

Natural hybrids of ***S. kurtzianum*** × ***S. chacoense*** have already been mentioned (p. 236) and have been noted by us for provs. Catamarca, La Rioja, and Mendoza.

There is some slight evidence of *S. chacoense* introgression in the *Lourteig 817* and *Ruiz Leal 4407* collections from Mendoza, Quebrada del Toro (see pp. 409–10 and 417). Natural hybrids of ***S. microdontum*** subsp. ***gigantophyllum*** by ***S. kurtzianum*** have already been described (p. 347). They are known so far only from prov. La Rioja, Sierra Famatina (see Appendix III).

Probable hybrids of ***S. spegazzinii*** with ***S. kurtzianum*** are discussed on p. 405. These are two collections from Sierra Famatina. We collected material in 1966 that seems to show influence of *S. spegazzinii,* also in Sierra Famatina (see Appendix III).

When we were reviewing our material of *S. kurtzianum* from Mendoza province we were at first inclined to think that some of the collections with broader leaflets were hybrids with *S. maglia*. After further examination these were seen to be merely broader-leaved forms or younger plants of *S. kurtzianum*, and we do not now believe we have seen any natural hybrids of *S. kurtzianum* with *S. maglia*. Apparently, so far as we have been able to see, the two species do not occur in the same localities in Mendoza province, or if they do, collections have not yet been made from the zones where they coincide.

LOCALITIES OF NATURAL HYBRIDS

1. With ***S. chacoense*** (see p. 239).
2. With ***S. microdontum*** (see p. 349).
3. With ***S. spegazzinii*** (see p. 406).

19. SOLANUM MAGLIA Schlechtendal

(*Hort. Halensis* **1**, 6, 1841)

S. collinum Dun. (in *DC Prodr.* **13**, 1, 36, 1852)

Plant large, bushy, with creeping elongated subterranean stolons (to about 1·5 m long), bearing bitter tubers up to 5 cm long, oval or occasionally irregularly fusiform in shape.

Stem robust, 30–80 cm high, green to dark violet, sparsely pubescent, provided with well-marked straight to undulate-crenulate wings which are *furnished marginally with large curved gibberulous hairs.*

Leaf 12–27(–35) cm long × 5–17(–25) cm broad; lateral leaflets (0–)2–3(–4)-jugate, decreasing in size rather rapidly from apex to base of leaf or in the 3–4-jugate leaves the uppermost two pairs quite large and more or less equal in size; laterals ovate-lanceolate, ovate, elliptic or oblong, 3–10(–13·5) × 1·5–5·5(–6·5) cm, or smaller in subsimple leaves (cf. *S. collinum*), sessile or shortly petiolulate (up to 5 mm long on basiscopic side), oblique at the base, the basiscopic side rounded, free or joined to the rhachis, and sometimes narrowly decurrent on to the rhachis, apically obtuse to acute and often acuminate; terminal leaflet longer and *much broader than the laterals*, broad-elliptic to broad-ovate or rhomboid, 5–11(–16) × 2·5–7(–10) cm, basally subcordate to broad-cuneate and slightly decurrent, apically obtuse or acute and acuminate; interjected leaflets 0(–3) pairs per leaf, present only in the 3–4-jugate leaves, sessile or shortly petiolulate, generally minute but occasionally to 1·5 cm long; leaf bright to deep green above, paler below, sparsely covered above with few-celled adpressed hairs which are denser and shorter below, *marginally minutely crenulate* with conspicuous curved gibberulous hairs; pseudostipular leaflets generally well developed, clasping the stem, semi-lunate, to 1·5 (or occasionally to 5) cm long.

Inflorescence generally many-flowered (5–)15–35-flowered; peduncle 4–20 cm long, once or twice forked, *the branches at length very long* (as long as the unbranched part); pedicels 1·5–2·5 cm long, generally articulated at about the centre but in some forms (cf. type for *S. collinum*) quite near the base, like the peduncles provided with many-celled acute slightly curved hairs and also generally with sparse (sometimes more frequent) stalked glands.

Calyx campanulate, 5–8 mm long; lobes 3–5 mm long including the 1–2(–5) mm long *linear acumens*; calyx clothed with frequent thick many-celled adpressed hairs and stalked glands.

Corolla *white, pentagonal to rotate*, 2–3·5 cm diam.; lobes 8–18 mm broad at the base × 5–9 mm long, including the 2–4 mm acumen.

Anthers in a *loose barrel-shaped column*, the apices not conniving, deep golden or yolk-yellow both inside and outside, *not clearly distinguished either in colour or form from the filaments* (cf. *S. calvescens*), *apically deeply emarginate, gradually narrowing at the base into the filaments*; *anther pores small or obsolete*; pollen scanty; filaments about 1–2 mm long, quite thick, rather pale yellow below but becoming darker above, *expanding gradually into anthers.*

Style 7–11 mm long, exserted 4–5 mm above the anthers, gently bent and slightly expanded towards the apex, covered for about $\frac{1}{4}$–$\frac{1}{2}$ way above the base with conspicuous spreading dense papillae; stigma generally *long obconical*, occasionally ± capitate, *apically clearly bilobed*, with two deep basal sinuses, about 0·5–1·5 mm long, not or only slightly thicker than style apex; ovary often oblong, sometimes slightly lobed above.

Berries rare, perhaps never normally developed on triploid plants.
Chromosome number: $2n = 24, 36$.

Note: The above description was made from both Chilean and Argentine collections. So far as is known no triploid biotypes are to be found in Argentina.

AFFINITIES

S. maglia may be distinguished from other potato species by the characteristic loose barrel-shaped anther column and deep yellow anthers which are not delimited sharply either in colour or form from the filaments (but see also *S. calvescens*). In its leaf pubescence it seems to show affinities with *S. kurtzianum*; the calyx of these two species is also rather similar. The form of the leaf is not unlike that of *S. weberbaueri* and *S. medians* from Peru, though this resemblance may not necessarily point to a close phylogenetic relationship. The slightly lobed ovary and long obconical bilobed stigma are also very characteristic features of *S. maglia*. *S. maglia* differs from *S. microdontum* in the gibberulous hairs, low pedicel articulation, and the absence of a sharp boundary between the anthers and filaments.

SPECIMENS EXAMINED

ARGENTINA

Prov. MENDOZA:

Dept. San Carlos, Quebrada de Alvarado. (Cultivado en Godoy Cruz.) 17 Nov. 1945. 'Tubérculos rojizos; flor blanca, ligeramente perfumada.' *Ruiz Leal 10470* (LIL, Ruiz Leal). See Plate 149.

Dept. San Carlos, Quebrada de Alvarado. Alt. 1500 m. *Schnack* (s.n.) (CPC: 2057) (Tubers sent by Sr. Alfonso Castronovo, Castelar; Accession no. 200/46). See Plate 150.

DISTRIBUTION

In Argentina *S. maglia* is known only from prov. Mendoza, dept. San Carlos (see Map 13). Its greatest distribution is in Chile, where it is to be found both on the coast and some distance inland, in provs. Santiago, Valparaíso (these are at the same latitude as Mendoza), Aconcagua, and Coquimbo.

HABITAT

In Argentina *S. maglia* has been found in the dry Andean Quebrada de Alvarado at about 1500 m altitude. This lies presumably in the Monte phytogeographical province, though according to Ruiz Leal (verbal communication) it really represents a transition zone between the Monte and High Andean provinces. He further states that *S. maglia* is here accompanied by plants such as *Pleurosorus*, *Maytenus*, *Ribes*, *Grindelia chiloensis*, and *Mulinum spinosum*. Its distribution in Argentina is not well known and it may have a wider geographical and altitudinal range than has hitherto been suspected. In Chile the species occurs from sea level to about 120 m (or possibly much higher, to 700 m, *fide* Correll 1962) and is to be found amongst maritime sands and rocks and on dry grassy or wooded hillsides, in full sun or partial shade, at least to about 35 km inland. Here again, little is known of its altitudinal limits in Chile, but since *S. maglia* is found at 1500 m in Argentina, only about 150 km away, one may suspect that it ascends higher in Chile than the previous records indicate, if the climate is not too wet. Along the coast of Chile we suspect that *S. maglia* may vegetate all the year round.

TAXONOMIC NOTES

S. maglia was described by Schlechtendal in 1841 from specimens collected by Poeppig (Pl. Chil. I. 72. Diar. 105) in 1828 on the seashore at Valparaíso, Chile. Since no holotype was mentioned by Schlechtendal we designate the Paris sheet of this collection as the LECTOTYPE for *S. maglia*. (See Fig. 43.) The name 'maglia' is an italianized version of the Araucanian word 'malla' or 'wild potato' which was first mentioned by Molina in 1782.[1]

In 1852 Dunal described a form of this species with less dissected leaf as *S. collinum*. This specimen was from a little inland, at Quillota, near Valparaíso, and had been gathered by Bertero (*No. 1328*) in 1829. (See Fig. 44.) Our studies of the variation in *S. maglia* and *S. collinum* show that the two species intergrade completely and that there is, furthermore, no geographical localization of the forms with less-dissected leaves in inland localities. *S. maglia* is, apart from the high Andean northern *S. infundibuliforme*, the only truly wild tuberiferous species known in Chile.

No suspicion of the occurrence of *S. maglia* in Argentina had been entertained until quite recently, when we discovered that it had been collected by Ruiz Leal[2] and by Schnack in prov. Mendoza. We should mention, however, that Bitter (1913) quotes Spegazzini (1902) as stating that it is to be found on the east side of the Andes at Carren-leofú (=Carrenleufú). This place is situated, so far as we can make out from our maps, at about 43½° S., 71½° W., where no authentic records of *S. maglia* have been found; it is possible that both Spegazzini and O. Kuntze (*Rev. gen. pl.* III, ii, 227) who recorded it for Patagonia, mistook escaped forms of *S. tuberosum* for *S. maglia*, and we ourselves have identified such forms of *S. tuberosum* from that region (see p. 435). Bitter, indeed, in 1913 described a so-called variety *chubutense* of *S. maglia* from specimens collected by Illin in Chubut province, but this was later shown by Hawkes (1956*a*) to be a variety of *S. tuberosum*, not of *S. maglia*.

It is very probable, therefore, that these southern occurrences of '*S. maglia*' refer in fact to escaped forms of *S. tuberosum*. It is certain, however, that genuine *S. maglia*

[1] Original Italian version: *Saggio Hist. Nat.* Bologna 1782; translated into German, 1786, into Spanish in 1788, and into English in 1809.

[2] Probably grown from tubers collected originally by Schnack.

FIG. 43. *S. maglia* Schlechtd. Lectotype specimen, collected by Poeppig (*I. 72*, *Diar. 105*—Herb. Paris) at Valparaíso, Chile. Note the well-dissected leaf with 3–4 pairs of lateral leaflets and the characteristic stamen and stigma.

is to be found in Mendoza province at 1500 m altitude, since, as we shall see later, tubers from the Schnack collection were grown at Cambridge, Birmingham, and Vandel, giving very typical *S. maglia* plants. (See Plate 150.)

Our hypothesis is that *S. maglia* was derived from *S. kurtzianum* in the Andes of Argentina, somehow crossed the 150 km of mountains that separate Mendoza from the coast of Chile, and in this latter region spread rapidly under conditions to which it seems better adapted than its original location. Although the altitude difference between the Argentine and Chilean parts of its range is some 1400 m this is not any greater than the total altitude range for many other wild potato species, as can be seen by referring to those described in the present work (see Fig. 2). An alternative hypothesis, put forward by Bitter (1912) and continued by Bukasov (1933*b*), is that *S. maglia* might be derived from the Peruvian coastal and mountain species *S. weberbaueri*, *S. medians*, and *S. hypacrarthrum*. If this were so, we should have to postulate a 'leap' of some 1500 km (1000 miles) over the Atacama desert for *S. maglia* to have reached its present separation from the nearest of these three species in south Peru. One might account for this by past climatic changes, but until we have some other confirmatory evidence it seems better to adhere provisionally to the first, and simpler, of the two hypotheses, namely, that the Chilean *S. maglia* was derived by a trans-Andean migration rather than by a western coastal one.

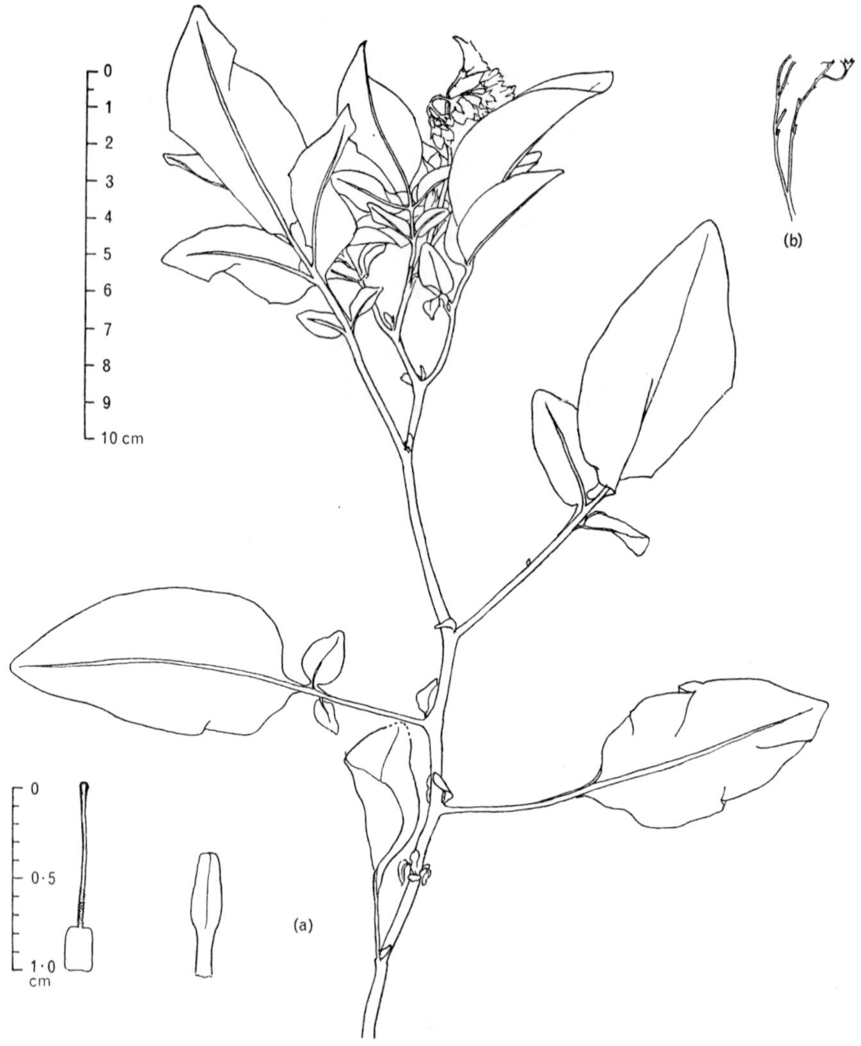

FIG. 44. *S. maglia* Schlechtd. (*a*) Type specimen of '*S. collinum*' Dun., collected by Bertero (*No. 1328*—Herb. DC., Geneva) at Quillota, Chile. This biotype shows a less-dissected leaf (0–2-jugate), than Poeppig's collection (3–4-jugate—see Fig. 43), which may be due partly to environmental conditions. All intermediates between these two extremes occur.

(*b*) Inflorescence from isotype specimen in Herb. Paris, showing low pedicel articulation.

NOTES ON LIVING PLANTS

The only living collections of *S. maglia* from Argentina are those grown from tubers collected by Schnack (CPC: 2057) in Quebrada Alvarado, Mendoza (see Plate 150).

The tubers are long oval, reddish brown, and sprout very slowly.[1] This slow sprouting does not seem to have been noted in Chilean collections of *S. maglia* and may perhaps be an adaptation to the special climatic conditions of Mendoza. Although the first leaves produced are simple, later in the season the plant develops 2–3-jugate leaves with 0–3 pairs of interjected leaflets. The lateral leaflets are oblique, sub-sessile, and tending towards decurrency. The stem is winged and moderately pubescent, and the pedicel articulation is $\pm$ median. Rather coarse hairs of various lengths are characteristic on all green parts. In all floral characters it agrees very well indeed with Chilean *S. maglia*. Living collections from Chile made by Hjerting and Rahn (*Nos. 528*, *532*, and *545*) differ very little from the Argentine material. It is interesting to note that the diploid *No. 532*, which from the sub-simple leaf would have been placed as *S. collinum*, gave plants in Vandel with 2–3-jugate leaves, though without interjected leaflets. Although certain of the differences in leaf dissection seem to be genetically controlled, there is no doubt that they are due also in part to differences of environment and the particular state of development reached by an individual plant when the specimen was gathered.

Chilean *S. maglia* has been collected many times in the living state, probably first by Caldcleugh in 1822 (see Sabine 1824) who considered it to be a form of *S. tuberosum*. It was grown at Kew for many years (see Baker 1884, Hooker 1884), was also collected by Verne at Quillota in 1911 (see Verne 1912), and by Reiche (see Wittmack 1914). These latter gatherings were grown by many investigators, including Wittmack and Bitter. All these workers stressed the specific distinctness of *S. maglia* from *S. tuberosum* and *S. commersonii*, though few found it of much value for breeding purposes.

S. maglia has exhibited resistance to wart, *Verticillium* wilt, common eelworm, horsenettle cyst and Osborne's cyst nematodes, leaf hoppers, flea beetles, and viruses X and A. Some of these results were obtained from Argentine and the rest from Chilean lines of this species (see Chapter 5).

CYTOLOGY

Rybin (1933) first reported on the cytology of *S. maglia*, finding a somatic number of $2n = 36$. Meiosis was extremely irregular, with plates of 15–19 chromosomes at second metaphase and one or two chromosomes lagging in the cytoplasm. Okuno (1951) confirmed these results, obtaining first metaphase plates with 12–18 bivalents and a corresponding number of univalents. Strangely enough, no multivalents were discovered, and Okuno considered that *S. maglia* was a natural $4x \times 2x$ hybrid.

Although *S. maglia* (triploid) is very infertile, Perlova (1940) obtained three seedlings from natural berries on plants grown in the Pamir mountains in south Russia. These were unbalanced diploids with 25 or 26 chromosomes. (See also below under 'fertility and crossability'.)

Until 1956 it was generally supposed that *S. maglia* was a triploid species, possibly of hybrid origin, maintaining itself entirely by vegetative reproduction. However, Hawkes (1956*b*) reported a diploid clone of this species (CPC: 2057) from Argentina and this has been used by us in numerous crosses. A collection of Hjerting and Rahn

[1] According to Montaldo and Sanz (1962) the Chilean forms have a very short dormancy period.

in Chile from prov. Valparaíso, Olmué, Granizo (*No. 532*), also showed a diploid chromosome complement. *S. maglia* thus apparently behaves in conformity with certain other 'triploid species' such as *S. commersonii*, *S. cardiophyllum*, and *S. medians*, in all of which diploid clones have been discovered, growing in the same region as the apparently more vigorous triploid ones. It is probable, therefore, that the triploid forms of *S. maglia*, as also of the other species mentioned above, are autotriploids formed by the union of unreduced gametes with normal reduced ones. The diploid form from Chile already referred to possessed a less-dissected leaf than the triploids and was referable to the *S. collinum* type. It is too soon to say, however, that all such *S. collinum* forms must be diploids until a wider cyto-taxonomic survey of *S. maglia* has been carried out.

FERTILITY AND CROSSABILITY

Both triploid and diploid forms of *S. maglia* are apparently rather male-sterile and produce scanty pollen. This is only to be expected for the triploids, but is rather surprising in the diploids. Juzepczuk (personal communication) states that *S. maglia* fails to set seed in the wild state and this is in general confirmed by Hjerting's observations in Chile though as yet we have no information about its fertility in Argentina. The diploid CPC: 2057 from Quebrada Alvarado, Mendoza, possesses scanty pollen and is practically male sterile, though it forms good seed in several cases when used as female parent.

Stout and Clarke (1924) record abundant pollen for triploid *S. maglia* but at least 90 per cent of empty grains. It is therefore almost completely pollen sterile. Irregular sized grains, mostly empty, are noted by Montaldo and Sanz (1962).

Perlova (1940, 1945*a*, 1945*b*) obtained more or less diploid seedlings (see previous section) from natural seed produced on triploid *S. maglia* grown in the Pamir mountains of south Russia. She considered that this was due to parthenogenesis and that the diploid strains so obtained (called by her, *S. pamiricum*) were similar to the original diploid strain of this autotriploid species. From an examination of Perlova's material we would like to suggest that an alternative explanation of these facts is possible. The seedlings of '*S. pamiricum*' are not uniform, and some possess pale blue flowers; the leaf dissection varies also from seedling to seedling. If one pollinates triploid with diploid species the only viable seedlings from this type of cross are found to have from 24 to 26 chromosomes, all those with a higher number of chromosomes being unviable (see von Oláh 1938). It is therefore likely that a natural pollination by some pale blue-flowered species has taken place on to the triploid *S. maglia* and that the so-called '*S. pamiricum*' is really a hybrid and not a true diploid of *S. maglia*.

We were able to obtain a small quantity of selfed seed from a diploid Chilean clone of *S. maglia* (*Hjerting & Rahn 532*) from which only two rather poor seedlings were obtained.

A cross was made between the Chilean and Argentine diploid forms of *S. maglia* (*H. & R. 532* × CPC: 2057), giving nine small fruits with rather few seeds in each. The reciprocal cross gave about the same result. The F_1 family was uniform, and was good *S. maglia*.

Using *S. maglia* as mother, many crosses have been reported, such as *S. maglia* × *S. tuberosum* (Sutton 1909, Perlova 1940), *S. maglia* × *S. chacoense* (Perlova 1945*b*), *S. maglia* × *S. kurtzianum* (Perlova 1945*b*), *S. maglia* × *S. edinense* (Paton 1909), *S. maglia* × *S. phureja* (Emme 1936). The chromosome numbers of these hybrids are not generally mentioned but Perlova (1940) records a count of $2n = 48$ for her *S. maglia* × *S. tuberosum* hybrids. This is in accordance with the numbers found by other workers in triploid × tetraploid crosses.

A hybrid of *S. maglia* and *S. chacoense* has already been mentioned (see p. 228). Hybrids between *S. maglia* and *S. microdontum* have also been described. They were rather variable but the characteristic *S. maglia* stamens were clearly dominant (see p. 344).

In an attempt to understand the relationship between *S. maglia* and *S. kurtzianum*, we crossed clones of these two species from the region of Mendoza (*S. maglia* (CPC: 2057) from Quebrada Alvarado and *S. kurtzianum* (*Ruiz Leal 11906*), from dept. Las Heras). The lack of thriftiness and vigour already mentioned in the F_2 family (see p. 419) gives a good indication that the two species *S. maglia* and *S. kurtzianum* are specifically distinct and that there are genetic barriers to free gene interchange between them.

In a similar type of cross between *S. maglia* and *S. medians* from Peru (both diploid) an F_1 family was raised and intersib crosses made. However, none of this seed germinated to give an F_2 generation.

Thus from these experimental results we may conclude that *S. maglia* is quite distinct from *S. medians* and from *S. kurtzianum*, not only morphologically, but also genetically.

NATURAL HYBRIDS

We have no exact information about natural hybrids of *S. maglia* in Argentina. It is certainly to be found growing in the same general region as *S. kurtzianum*, though the two species have not been recorded from the same locality and no natural hybrids have been seen by us. So far as we know, therefore, *S. maglia* does not form natural hybrids with other species, either in Argentina or in Chile.

Series TUBEROSA

CULTIVATED SPECIES

ONLY the indigenous cultivated potatoes are dealt with in the present work. The commonly grown commercial varieties that have been produced in recent times by plant-breeders in South America, Europe, the U.S.A., and elsewhere are omitted entirely, since their history is already fully documented in the plant-breeding and agricultural literature.

The indigenous cultivated potatoes of the area under survey are, so far as we are aware, entirely limited to Argentina, having presumably spread into that country southwards from Bolivia, and possibly also eastwards to a limited extent from southern Chile. By far the greatest number of varieties is to be found in the mountainous north-western provinces of Jujuy and Salta. A few collections have also been made in Tucumán, Catamarca, and Mendoza provinces. We shall deal further with these, and with the evidence for the cultivation of indigenous potato varieties in other provinces also, when the tetraploid varieties are being discussed (pp. 442–4).

Potato cultivations are chiefly to be seen in the Andean valleys but may also spread on to the upland plateaux. The altitude range is approximately 2500 to 3500 m, but extends from 3450 to 4000 m in the frost-resistant species, *S.* ×*juzepczukii*. No doubt some of the varieties were brought southwards from Bolivia a very long time ago by migrating tribes, though at what time and under what conditions it is impossible to say. Many varieties are to be found both in north Argentina and south Bolivia, and it seems highly likely that interchange of varieties takes place fairly frequently in the border provinces at the present day.

Most of the indigenous Argentine potato varieties are tetraploid and belong to *S. tuberosum* subsp. *andigena*. So far as we can tell no diploid cultivated species occur in Argentina, though Brücher (1963*b*) claims to have obtained samples of diploids from provs. Salta and Jujuy. Possibly one or two of our collections that were studied only in Argentina by Hjerting and Weisaeth might have been diploid, though on the whole we are inclined to think that they were not. Certainly we have not seen any variety that we would consider as belonging to *S. stenotomum*, though a few clones of the triploid *S.* ×*juzepczukii*[1] and the pentaploid *S.* ×*curtilobum* have been observed in cultivation.

It should be mentioned here that a group of red-skinned varieties, known as 'Papas Criollas', were commonly grown in certain medium-altitude valleys in Salta, Tucumán, and Catamarca provinces between 1500 and 2500 m, but may not be so common now. Most, or perhaps all, of these belong to *S. tuberosum* subsp. *tuberosum*, and since they have been grown for such a long time in the same area we shall deal with them shortly in this monograph. We shall also deal with what we consider to be the escaped forms of *S. tuberosum* subsp. *tuberosum* that are to be found in the southern provinces of Neuquén and Chubut.

In 1948 and 1949 one of the authors (J.P.H.), together with E. Petersen and J. Gundesen, made a collection of cultivated potatoes in north-west Argentina for the Instituto de Genética of the University of Tucumán. A total of 235 samples was obtained, and these were grown in the following summer at the potato-breeding station

[1] A triploid form, identified as *S.* × *chaucha* Juz. & Buk., was collected on our 1966 expedition (see Appendix III).

'El Suncho'. This station, which belongs to the Facultad de Agronomía, is situated at an altitude of 1700 m at 'Las Estancias', dept. Andalgalá, prov. Catamarca. The material was studied and classified into cultivars (= 'varieties') according to tuber and foliage characters. (See also Viirsoo 1954*a*, *b*.) Several of the varieties were obtained from the village of Humahuaca, prov. Jujuy, where a collection of varieties from this province was maintained in the Granja de la Escuela de Maestros Normales Regionales (Farm of the School for Rural Regional Teachers). A description of ten of the commonest of these varieties was published some twenty years ago by Ing. Agr. J. P. Gómez (1946).

A short description of most of the varieties collected by Hjerting, Petersen, and Gundesen is given below. The varieties have been grouped under species from observations made by one of us (J.P.H.) in the field in north-west Argentina and on some of the same varieties in Denmark (by J.P.H. and J.G.H.). Owing to the fact that the same variety was often collected with two or more distinct names, and conversely, more than one variety at times bore the same name, we have chosen the most commonly used name for each variety and have added the most usual synonyms in parentheses. It should be pointed out that the several collections grouped by us under one varietal name were not always completely identical in the foliage or form and size of the tubers. Nevertheless, they were so similar that it was thought best to group them as one variety or cultivar, which could thus be considered as containing two or more slightly different clones.

In this way we were able to classify the 235 collections into about fifty cultivars, though some of them were found in only a single locality.

In the ensuing account we first give a key for the identification of the cultivated species in Argentina. This is followed, for each species, by the name and synonyms, a short description, a list of the herbarium specimens examined, and then the tuber collections classified under cultivars, as mentioned above. Next come notes on distribution and finally a general discussion on the status of each species in Argentina and other matters of interest. With *S. tuberosum* subsp. *tuberosum* we shall deal first with the 'Papas Criollas' and then with the escaped or 'wild' material.

KEY TO CULTIVATED POTATO SPECIES IN ARGENTINA

1. Articulation of pedicel not higher than two-thirds of its length in fully developed flowers, generally situated near the centre; leaves often slightly arched at tip; corolla rotate to pentagonal **21. *S. tuberosum*** (p. 431)
   Articulation of pedicel high, 2–4(–6) mm below calyx base, always higher than three-quarters of pedicel length in fully developed flowers; leaves stiff, not arched at tip; corolla very rotate, with short lobes 2
2. Pedicels indistinctly articulated; corolla blue-violet, not more than 2·5 cm diam.; common peduncle short, 2–4 cm long **20. *S.*×*juzepczukii*** (p. 429)
   Pedicels distinctly articulated; corolla purple, 3–3·5 cm diam.; common peduncle longer, 5–10 cm or more long **22. *S.*×*curtilobum*** (p. 444)

TRIPLOID SPECIES

20. SOLANUM × JUZEPCZUKII Buk.

(*Proc. U.S.S.R. Congr. Genet.* **3**, 603–4, 1929)

Distinguished by the semi-rosette habit, at least when young, though occasionally with long stem when older, long straight leaves, short peduncle (2–4 cm long), straight pedicels with very high but indistinct articulation, small blue-violet corolla (to 2·5 cm diam.) with very short lobes and small acumens, and bitter, purple or white tubers.

HERBARIUM SPECIMENS EXAMINED

Prov. JUJUY:

Dept. Humahuaca, Sierra de Zenta. Alt. 4500 m.[1] Feb. 1931. 'Flor azul. 20 cm alto.' *Budín* (s.n.) (LIL 73581).

Dept. Tumbaya, El Moreno, cultivo del agente de policía. Alt. 3450 m. 'Casi silvestre; en seguida degenera.' Cultivado en la Est. Exp. 'El Suncho', Andalgalá, Prov. Catamarca, 1700 m. Herb. spec. taken 11 Feb. 1950. Vernacular name 'Luqui'. *Petersen & Hjerting Cult. 273* (J.G.H., J.P.H.).

Dept. Tumbaya, Angosto de San José de Chañi. Alt. 3550 m. Cultivado en la Est. Exp. 'El Suncho', Andalgalá, Prov. Catamarca, 1700 m. Herb. spec. taken 11 Feb. 1950. *Petersen & Hjerting Cult. 283* (J.G.H.).

Dept. Yavi, La Quiaca. Alt. 3450 m. Bought in the market and said to have been brought from Santa Victoria. Cultivado en la Est. Exp. 'El Suncho', Andalgalá, Prov. Catamarca, 1700 m. Herb. spec. taken 11 Feb. 1950. *Petersen & Hjerting Cult. 232* (J.G.H., J.P.H.).

TUBER COLLECTIONS GROWN IN ARGENTINA

Of the four herbarium collections of this species noted above, the last three were made from living tuber collections gathered in October 1948. Two of these (*Nos. 273* and *283*) may well belong to the same clone, since they were obtained in the same area. All three collections looked very similar under cultivation.

When grown in 'El Suncho' the yield was very poor indeed, probably due to the low altitude (1700 m) or to virus infection. The tubers were very small (15–25 grammes), many per plant, with white skin and very white flesh. The flowers were very small, blue-violet, and strongly rotate. The plant itself was low-growing, with a stem 40–50 cm high.

In Tumbaya *S.* × *juzepczukii* is known under the name 'Luqui' and this name is also used in most of Bolivia and in Puno, Peru (see Hawkes 1947). The Yavi collection was named 'Malcachi' which is a potato name also known in central to south Bolivia under the variant 'Malcacho', though not specifically applied there to this species. The variety was said to have been brought from Santa Victoria in Salta province. The tetraploid cultivar, 'Astilla' (see later, subsp. *andigena*), is also often grown under the name 'Malcachi' or 'Malgacha'.

According to Viirsoo (1954*a*) the variety Luqui showed resistance to frost, at least to 5° C below zero, when grown in Humahuaca. This is in accordance with the known frost resistance of this species, as pointed out by Bukasov (1932*a*). In line with the known sterility of *S.* × *juzepczukii*, Weiseth (unpubl.) observed that *P. & H. Cult. 232* possessed only 20 per cent stainable pollen. *P. & H. Cult. 273*, on the other hand, showed up to 65 per cent stainable pollen, though the grains were very small. Many workers have commented on the complete self-sterility of this species. It can sometimes be used successfully in crosses with other species, though with great difficulty.

DISTRIBUTION

S. × *juzepczukii* has so far been recorded in Argentina for prov. Jujuy, depts. Humahuaca and Tumbaya (and Yavi?), and prov. Salta, dept. Santa Victoria, where it is grown in small plots at altitudes of 3450–4000 m.

Its area of greatest concentration is around Lake Titicaca in south Peru and north Bolivia, though its general range is from central Peru to central Bolivia, and south-

[1] It is doubtful whether this collection could have been made at such a high altitude. Probably *c.* 4000 m would be more correct.

wards into north Argentina. When and how it extended its area into Argentina is impossible to say. It might have been brought by migrating tribes at a very early epoch, or, on the other hand, tubers might have been sent by lorry or train in comparatively recent times. Perhaps the species was more widespread in the past and has now been replaced by higher-yielding tetraploid potatoes. Even though it might be grown in south Bolivia no samples from cultivations in that region have been mentioned in the literature.

STATUS IN ARGENTINA

S. ×*juzepczukii* is a hybridogenic species, shown by Hawkes (1962*a*) to have arisen, probably spontaneously, as a cross between the tetraploid wild species, *S. acaule*, and the diploid cultivated one, *S. stenotomum*. This latter species is grown on the high plateaux from central Peru to central Bolivia and is not known in Argentina, so far as we know. Thus, unless the distribution area of *S. stenotomum* once extended to Argentina and has since been much reduced, *S.* ×*juzepczukii* could not have been formed in that country. A much more likely hypothesis is that it has been brought from Bolivia for cultivation in Argentina because of its well-known frost resistance, which enables it to be grown at higher altitudes than any other cultivated potato species, with the possible exception of *S.* × *curtilobum*.

Besides frost resistance *S.* ×*juzepczukii* is of interest for its resistance to wart, early blight, eelworm, and virus X, though not in every line tested. It has also exhibited high dry matter, again in certain clones. (See Chapter 5.)

TETRAPLOID SPECIES

21. SOLANUM TUBEROSUM L.

(*Sp. Pl.* **1**, 185, 1753).

S. esculentum Neck. (*Delic. Gallo-Belg.* **1**, 119, 1768)
S. aracatscha Bess., *nomen nudum* (*Cat. Pl. Hort. Crem.* 135, 1816)
S. sinense Blanco (*Flora Filip.* ed. i, 137, 1837)
S. parmentieri Molina, *nomen nudum* (ex *Walp. Rep.* **3**, 38, 1844–5)
S. tuberosum var. *chiloense* A. DC. (*Archs Sci. phys. nat.*) **15**, 437, 1886)
S. tuberosum var. *cultum* A. DC. (ibid.)
S. tuberosum var. *sabini* A. DC., *pro parte* (ibid.)
S. fonckii Phil., *nomen nudum*, ex Reiche (*Flora Chile* **5**, 351, 1910)
S. chiloense (A. DC.) Berthault (*Annls Sci. agron. fr.* **28**, 179, 1911)
S. cultum (A. DC.) Berthault (ibid.)
S. sabini (A. DC.) Berthault, *pro parte* (ibid.)
S. maglia Schlechtd., var. *guaytecarum* Bitt. (in *Reprium. nov. Spec. Regni veg.* **12**, 2, 1913). (= *S. fonckii* Phil.)
S. maglia Schlechtd., var. *chubutense* Bitt. (in ibid. 452, 1913)
S. maglia Schlechtd., subsp. *meridionale* Bitt. (in ibid. 453, 1913)
S. tuberosum L., '*sensu stricto*', Juz. et Buk. (*Trudy vses. Syezda Genet.* **3**, 609, 1929)
S. tuberosum L., var. *europaeum* Buk., *nomen nudum* (Suppl. 58 *Trudy prikl. Bot. Genet. Selek.* 96, 182, 1933)
S. molinae Juz. (*Izv. Akad. Nauk SSSR* Sér. Biol., No. 2, 308, 1937)
S. leptostigma Juz. (ibid. 310, 1937)
S. tuberosum L., var. *chubutense* (Bitt.) Hawkes (*Proc. Linn. Soc. Lond.* **166**, 130, 1956)
S. tuberosum L., var. *guaytecarum* (Bitt.) Hawkes (ibid.)
S. aquinas Buk., *nomen nudum* (*Züchter* **28**, 65–70, 1958)

Possibly also:

S. diemii Brücher (*Darwiniana* **13**, 108, 1964)
S. sanmartiniense Brücher (ibid. 111, 1964)

S. tuberosum is distinguished from the other species of cultivated potato by the often slightly arched leaves, leaflets always ovate to ovate-lanceolate, about twice as long as broad, pedicel articulation placed near the centre and short calyx lobes, arranged regularly. The corolla is rotate to almost pentagonal, with well-marked lobes, about half as long as broad.

A detailed and lengthy description was published by Hawkes (1956*a*, pp. 107–9) and will not be repeated in the present publication. Two distinct species had been recognized by Juzepczuk and Bukasov within this complex, namely, *S. tuberosum* (in the narrow sense) and *S. andigenum*.[1] These were grouped together by Hawkes (1956*a*) as two geographical subspecies of *S. tuberosum*, and the same treatment has been followed by us in the present monograph. Subsp. *tuberosum* occurs indigenously in the lowland areas of southern Chile and secondarily in Europe and elsewhere; subsp. *andigena* occurs in the higher parts of the Andes of South America and possibly secondarily in the mountains of Mexico and Central America.

In the area under review in the present monograph, subsp. *tuberosum* is cultivated in the temperate and subtropical regions of Argentina, Brazil, and Uruguay, at low altitudes, from sea level to 2000 m, or occasionally (in Argentina) to 2500 m(–3000 m). Subsp. *andigena*, on the other hand, is grown at higher altitudes and in Argentina is generally to be seen from 2500 to 3500 m. Between 2000 and 2500 m both subspecies may be cultivated in certain parts of South America (Hawkes 1951) and this also seems to be so in Argentina.

KEY TO SUBSPECIES

Leaf generally arched, set at wide angle to stem; leaflets not more than twice as long as broad, markedly acuminate, rather few; pedicels thickened at apex; corolla often with white acumens, often pale — Subsp. ***tuberosum***

Leaf only slightly arched, set at narrow angle to stem; leaflets rather more than twice as long as broad, obtuse to only slightly acuminate, generally frequent; pedicels not or barely thickened at apex; corolla rarely with white acumens, generally medium to dark-coloured, occasionally pure white — Subsp. ***andigena***

1. Subspecies *tuberosum*

Distinguished from subsp. *andigena* by the less dissected leaves, with wider leaflets, which are sessile to shortly petiolulate; the leaf as a whole is generally arched and set at a wider angle to the stem. Pedicel thickened above; corolla often white or pale coloured, frequently with white acumens and pale to white central petal streaks. Tubers formed under long days or under short days in the tropics or subtropics at altitudes from sea level up to 2000 m (occasionally to 2500 m (–3000 m)).

(*a*) Cultivated semi-indigenous forms

TUBER COLLECTIONS GROWN IN ARGENTINA

1. **Criolla Colorada** (synonyms: Papa Colorada, P. Criolla, P. Morada, P. Chaqueña, Papa de Amblayo, Colorana?, Almidonada Rosada, Mar del Plata). Grown in provs. Catamarca (dept. Andalgalá), Tucumán (dept. Trancas), Salta (depts. San

[1] Orthography revised to *andigena* by *Index Kewensis*.

Carlos and Chicoana), and Jujuy (dept. Humahuaca) from 1500 to 2500 m. One of the potato-growers stated that the potato had been brought from Chile by General Navarro more than seventy years ago and had been grown in the Andalgalá area since then. In 'El Suncho' the healthy plants yielded very well, but the variety was very susceptible to leaf-roll virus. The tuber flesh is yellow, the skin pink, and the eyes almost superficial and flat. The flower is pink to white and flowering is very late.

2. **Bocha** (de Trancas). Grown in provs. Tucumán (dept. Trancas) and Salta (dept. San Carlos) from 1500 to 3100 m. Tubers borne on long stolons, big, with reddish skin and white flesh with pink lines. Eyes deep. Flowers white. Very early variety.

3. **Bocha** (de Las Mesadas). Grown in Catamarca (dept. Andalgalá) and Salta (dept. San Carlos) at *c.* 1700 m. The tubers are similar to those of the above variety, but the plant is quite distinct.

4. **Bocha** (de Don Miguel). Found only at Las Mesadas, dept. Andalgalá, at 1650 m. Tubers very big, round, with a very characteristic luminous red colour round the eyes. The only one of the three 'Bocha' varieties which in the form of the tubers agrees with the name Bocha (= bowl). Very early variety.

DISTRIBUTION

These cultivated forms of subsp. *tuberosum* were collected in provs. Catamarca (dept. Andalgalá, Las Mesadas), Tucumán (depst. Trancas), Salta (depts. Chicoana and San Carlos), and Jujuy (dept. Humahuaca), where they were normally grown at altitudes of 1500 to 2500 m.

STATUS IN ARGENTINA

The general appearance of the plants and the colour of the tubers of this group of cultivars clearly indicate that they belong to subsp. *tuberosum* and not to subsp. *andigena*. Although many detailed inquiries were made amongst the growers it was never possible to discover the real origin of these potatoes, though it was clear that several of them had been grown in these areas for a very long time. The very name by which they are known, 'Papas Criollas' (= 'native born' potatoes but not of indigenous stock), reinforces the assumption that they are ancient cultivars in this area but not truly indigenous.

It is difficult to be certain whether these varieties are of European, North American, or Chilean origin. They might perhaps have been derived from potatoes first introduced into eastern Argentina (Buenos Aires province, for example), where potato cultivation is common, which were later brought to the north-west provinces, where early potato cultivation is carried out at low altitudes in almost frost-free areas during the winter. They might then have spread up into the mountain valleys of Trancas, Andalgalá, etc., where, by some process of semi-natural selection, series of well-adapted varieties were formed. Millán (1939) gives some information on five varieties that were grown in Argentina until 1936 and which were all imported from Europe or the U.S.A. One of these was called 'Bocha', another 'Chaqueña'. Perhaps several of these survived in the provinces of Tucumán and Catamarca, though it is unlikely that they could be the source of all the *S. tuberosum* cultivations in those provinces.

Another possibility is that the Papas Criollas might have come from Chile (perhaps brought by General Navarro, seventy years ago), where many varieties of the same subspecies are grown indigenously on the island of Chiloé and the surrounding mainland. It would be of considerable interest to compare these Argentine varieties with the Chilean ones, and it is to be hoped that further studies on this interesting group will be effected in the future, though they seem to be less frequently grown now (1966) than ten years ago.

(*b*) Escaped forms

HERBARIUM SPECIMENS EXAMINED

Prov. CHUBUT:

Región del Río Corcovado, entre Bolsón y Colonia 16 de Octubre. 1 Feb. 1901. *Illin 234* (BAF, BR, CORD, G, HBG, SI, UC. Type for *S. maglia* var. *chubutense*). See Fig. 45.

Fac. de Agronomía, Buenos Aires, cultivado. 24 Nov. 1926. 'Los tubérculos proceden del Lago Puelo, Chubut, donde se hallan espontanea, y es cultivado por algunos para consumo. Corola blanca con ligero tinte violeta-rosado. Anteros amarillos.' *Millán 577, 592* (BA).

Fac. de Agronomía, Buenos Aires, segundo año en cultivo. 1 Dec. 1927. Los tubérculos proceden del Chubut. *Millan 1027, 1028* (BA).

Prov. NEUQUÉN:

Parque Nacional Nahuel Huapi, Arrayanes Wald. Alt. 760 m. Feb. 1957. *Brücher 900* (SI—TYPE of '*S. diemii*' Brücher).

Collún-Có (San Martín de los Andes). Feb. 1957. 'Blüte weiss, Knolle mit Anthozyan.' *Brücher* (s.n.) (SI—TYPE of '*S. sanmartiniense*' Brücher).

Parque Nacional Nahuel Huapi, Península Quetrihué, en bosque de *Myrceugenella apiculata* de mucha sombra y húmedo. Alt. 800 m. 18 Jan. 1952. 'Plantas juveniles, estolones muy largos y poco enterrados.' *Diem 1997* (J.G.H.).

Parque Nacional Nahuel Huapi, Península Quetrihué, parte alta del bosque de *Myrceugenella apiculata*. Alt. 820 m. 28 Feb. 1952. 'Flor blanca, fondo levemente lila-rosado; 30–120 cm alto. Planta adulta, de un lugar con algo más luz.' *Diem 2050* (J.G.H.).

Parque Nacional Nahuel Huapi, Península Quetrihué, Los Arrayanes. Alt. *c.* 800 m. Plants grown in Birmingham (as H. 5) and Vandel (as *Hjerting No. 1650*) from original tubers sent by José Diem, July 1953. *Diem* (s.n.) (Herb. specimen made in Vandel, 1955. J.P.H.).

Parque Nacional Nahuel Huapi, Península Quetrihué, Estancia Quetrihué. Alt. 750 m. 16 Jan. 1951. *Sleumer 1520* (p.p.) (US, J.G.H.).

DISTRIBUTION

The forms of subsp. *tuberosum* that we have listed above and which we consider to be escapes have been found so far only in the provinces of Chubut and Neuquén, at altitudes of about 750 to 850 m above sea level. It is possible that they may have spread into Argentina naturally from Chile, or may represent relics of old cultivations left by the Indians, which, owing to the suitable climate, have in places become completely naturalized. Similar occurrences of subsp. *tuberosum* have been noted in Chile by Charles Darwin (Chonos Archipelago), Fonck (Chonos and Tres Montes; Guaytecas Archipelago), Juzepczuk (Chiloé), and several others (see Hawkes 1956*a*, p. 129, for collecting details of these Chilean forms).

STATUS IN ARGENTINA

A certain amount of controversy centres on these 'wild' forms of *S. tuberosum* subsp. *tuberosum* in Argentina and Chile. The Russian botanists Juzepczuk and Bukasov considered that the tetraploid wild potatoes of Chile, described by them under the names *S. leptostigma* and *S. molinae*, were genuinely and primitively wild, and represented certain relics of the ancestral stock from which *S. tuberosum* had been brought into domestication. Hawkes (1956*a*) pointed out that there appear to be no diploid tuber-bearing wild species in Chile from which the tetraploids could have been derived. It was postulated by him that the cultivated potatoes of Chile had been brought there by migrating tribes from Peru and Bolivia and were not derived *in situ* from wild plants. The tetraploid 'wild' species, *S. leptostigma* and *S. molinae*, could better be regarded as escapes from cultivation, and in this connexion it was of particular interest to note that they possessed a gene for red tuber colour which is not found in any other wild species but which occurs with some frequency in all cultivated ones (except, possibly, in *S. ajanhuiri* and *S.* ×*juzepczukii*). Paralleling these escaped Chilean forms of subsp. *tuberosum* is a series of forms of subsp. *andigena* which also seem to be escapes; these are also tetraploid and have been collected here and there in Mexico, Costa Rica, Colombia, Ecuador, Peru, and Bolivia. It is indeed true that none of the subsp. *andigena* escaped forms are found far from cultivated fields, and they generally persist as weeds of cultivation in potato and other fields. On the other hand, some of the subsp. *tuberosum* forms are reported to occur far removed from villages and settlements, and even to form part of the natural vegetation. It must not be forgotten, however, that these might be relics of potato cultivations of the wandering nomadic Indian tribes of the Patagonian area, who might not be expected to leave any trace of their activities apart from certain remnants of the plants they cultivated. Nevertheless, Millán (1942) mentions that a special kind of potato-storage silo of Araucanian type has been found at Llao-Llao, near Nahuel Huapi, Río Negro. This points perhaps to ancient potato cultivation in this region. The Andean tribes have lived in static agricultural village communities for many thousands of years, and even when the peoples have disappeared it is often quite possible to see traces of their villages and old agricultural terraces, as well as remnants of escaped potatoes here and there in suitable areas.

Fonck's collection *No. 102c* from the Chilean Guaytecas Archipelago and Darwin's *No. 194 bis* from the Chonos Archipelago, both of which were misidentified by Bitter as variants of *S. maglia*, are undoubtedly escapes or relics of old potato cultivations. It seems reasonably certain that Illin's and Millán's collections from Chubut and those of Diem, Brücher, and others from Neuquén are of a similar nature, as we shall show in more detail below.

We ourselves have examined eleven collections of wild potatoes from Chubut and Neuquén. Five of these were from Chubut, the oldest collection being that of Illin (*No. 234*), made in 1901, between Bolsón and Colonia 16 de Octubre. It was on this material that Bitter based his *S. maglia* var. *chubutense* (see Fig. 45), which is, in reality, a biotype of *S. tuberosum* subsp. *tuberosum*. Bitter grouped this and his Chilean var. *guaytecarum* as subsp. *meridionale* of *S. maglia*, but the name cannot be retained since both the taxa fall within the typical subspecies of *S. tuberosum*. Millán's four (or, more probably, two) collections were made in 1926 and 1927, the two from the latter year apparently representing a second year's growing of the two original collections. The specimens examined were of plants grown in Buenos Aires from collections made at Lago Puelo in Chubut, which lies only about 15 km from Bolsón, near where the Illin collection was made, and only about 150 km from the Isle of Chiloé in Chile, where potatoes have been grown for centuries. It is of interest to note also that Millán states on the herbarium labels of the two first-year collections that the potatoes are found growing spontaneously and are cultivated by some people for consumption. We have no doubts that all this material from Chubut should be classed as *S. tuberosum* subsp. *tuberosum*.

The other six collections from southern Argentina were from the Nahuel Huapi region of Neuquén. One of these was a tuber collection, kindly sent us by Sr. José Diem (Accession H. 5; *P. & H. 1650*). This possessed pale pink flowers with white acumens and central petal streaks, and pale pink tubers. Several chromosome counts

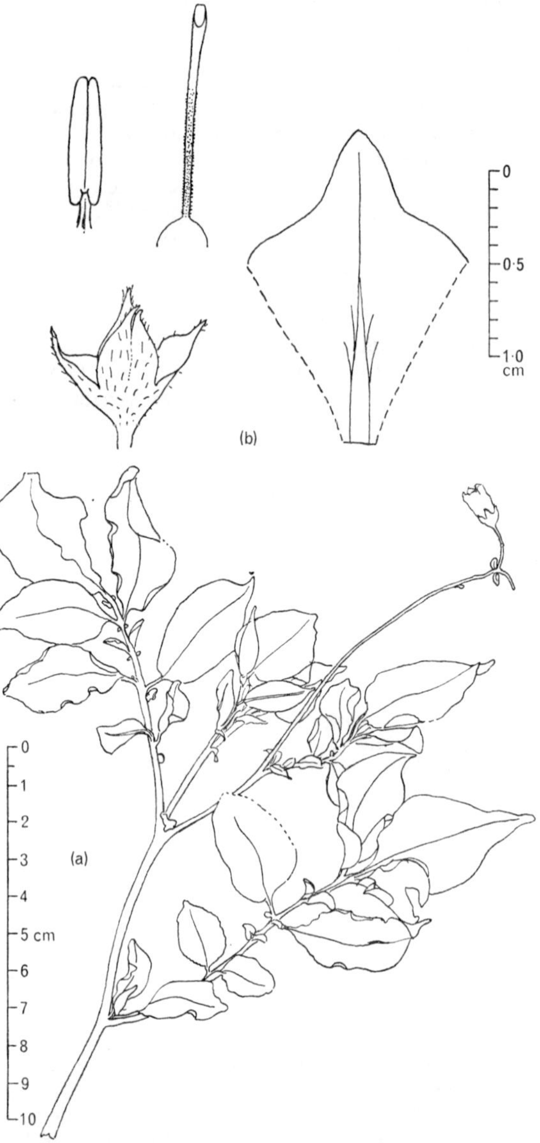

FIG. 45. *S. tuberosum* L. subsp. *tuberosum*. A naturalized or escaped form of this species collected by Illin in 1901 (*No. 234*) in Argentina, prov. Chubut. Bitter erroneously described this as var. *chubutense* of *S. maglia* Schlechtd.

(*a*) Whole plant (Herb. Geneva).

(*b*) Flower dissection (Herb. Darwinion). Note the typical anther of *S. tuberosum*, as opposed to *S. maglia*. (From Hawkes 1956*a*.)

were made on this material, both in Birmingham and Denmark. In all cases forty-eight somatic chromosomes were clearly counted. Our material is certainly to be grouped as a biotype of *S. tuberosum* subsp. *tuberosum*, and we consider it to be a remnant of some former cultivation that has become established as part of the natural vegetation. This material does not set berries, just as many European cultivars of that species, since the sterility is due to genetic causes, not to chromosome imbalance. The other collections from this area are known only as herbarium specimens, and were collected by Diem, Brücher, and Sleumer. When growing in shade the lateral leaflets are very much suppressed and the terminal enlarged (*Diem 1997*; *Sleumer 1520*, p.p.), but in places with more light the plant possesses the habit usual for this species, with well-dissected leaves and the terminal not much larger than the laterals (*Diem 2050*; Accession H. 5.).

We mentioned above that the Russian botanists described two Chilean tetraploid 'species', *S. leptostigma* and *S. molinae*. It is of some interest to note that the tubers of the former species are pink, thus demonstrating the presence of the gene for red pigmentation that is so often seen in *S. tuberosum* subsp. *tuberosum*. This colour is also to be seen in the collection of Diem (H. 5). The corolla of the Millán collections is white, tinged with pale rose-violet, whilst white, tinged 'rose-lilac' is mentioned by Diem for the flower colour of his *No. 2050*. In all these, therefore, the red gene is either certain or probable, lending strength to the hypothesis that they are escaped forms of *S. tuberosum*.

Several works on the 'wild' and cultivated potatoes of southern Argentina and Chile have been published by Brücher (1951, 1958, 1960*a*, 1960*b*, 1963*a*, 1963*b* 1963*d*, 1964*b*, 1964*c*). In these he agrees with Hawkes that the Chilean *S. leptostigma* and *S. molinae* cannot be regarded as ancestral forms of *S. tuberosum*. Both were acquired by Juzepczuk at second hand in Chiloé, and their geographical distribution on the island was not studied or their wild status established. He concurs that they may well be escapes and cites personal observations made in 1958 to corroborate this. Brücher mentions his searches in 1953, 1956, and 1958 in southern Argentina, in a region where there are no geographical barriers to a spread of Chilean species into Argentina (Chubut, Río Negro, Neuquén). Two wild taxa were collected by him, one in the Parque Nacional de Nahuel Huapi and the other in Collún-Có, near San Martín, both in prov. Neuquén. These were validly published in 1964 as new species, *S. diemii* and *S. sanmartiniense*, respectively. *S. diemii* was mentioned earlier by Brücher (1960*a*), though not with a valid description. In this publication he states that it forms part of the natural vegetation of evergreen Arrayan forest (*Myrceugenella*), and this information was also given by Diem in the field notes of his own two collections made in 1952, to which we have already referred (p. 434). The skin of the tubers was yellowish grey, according to Brücher, and the flowers were pale lilac. Brücher's plant was grown by José Diem in his garden, where it was said to have become 'spontaneous' and to have given good yields.

Brücher observed that the potato did not form fruits, and his chromosome counts indicated that the material was triploid, with thirty-six somatic chromosomes. He therefore concluded that it could not be an ancestral form of the tetraploid *S. tuberosum*, even though it was morphologically closely similar to it. In his 1964*b* paper the chromosome number of the other species, *S. sanmartiniense*, is given as $2n = 48$.

We have now examined the type material of these two species, which is deposited in Herb. Inst. Darwinion. There are no features (apart from the chromosome number of *S. diemii*) in them that would lead us to suppose that they were good species, specifically distinct from *S. tuberosum* subsp. *tuberosum*. In a footnote to his 1964 paper Brücher misquotes one of us (J.G.H. 1963) by saying that 'Hawkes expresses the opinion that *S. diemii* should be named *S. tuberosum* subsp. *tuberosum*'. What was actually stated (pp. 161–2) was: '*S. diemii*. An escaped form of *S. tuberosum* in the

Nahuel Huapi district of Argentina, which Brücher claims is a natural triploid. I have seen material from the same region which should be named *S. tuberosum* subsp. *tuberosum*.' This is not quite the same thing, but it seems probable to us now that in fact *S. diemii* should, indeed, be equated with *S. tuberosum* subsp. *tuberosum*.[1] The triploid nature of Brücher's specimen is most curious, and one can only assume that it is due to some kind of meiotic disturbance leading to the loss of a chromosome set. It would be most valuable to make further living collections in the Nahuel Huapi area to establish the extent of triploidy in the populations, since our own collections from the same place were undoubtedly tetraploid.

From an examination of Brücher's type specimen of *S. sanmartiniense* we are fairly certain that this also should be regarded as an escaped form of *S. tuberosum* subsp. *tuberosum*. The tubers are greyish rose[2] and the plant probably contains the same gene for red pigmentation (rather than purple) that is typically found in many cultivated varieties but not normally in wild potatoes, except for these doubtfully truly wild forms that we are at present discussing. The white corolla of *S. sanmartiniense* is of course not diagnostic in this respect.

Although Brücher agrees with Hawkes that the Chilean potatoes *S. molinae* and *S. leptostigma* are no more than escapes from cultivation, he does not apply the same reasoning to the Argentine 'species', *S. diemii* and *S. sanmartiniense*. One might perhaps with more justification argue that *S. diemii* was truly wild, because it seems to form part of the natural vegetation; but *S. sanmartiniense* has, in our view, every appearance of an escaped or weed potato, closely similar to the weed potatoes of the Andes in its mode of origin and status. And *S. diemii*, when viewed in context with all the other collections or accounts of wild potatoes in southern Argentina and Chile, must surely on balance be considered as a specially well-adapted escape and no more than that.

In order to clarify the comparison of the morphological features of these south Argentine and Chilean forms we have presented in the form of a table (Table 25) the features that we consider to be of particular significance. From this we can see that although there are some differences in flower and tuber colour, the leaflet number, both primary and interjected, varies very little. It will further be noted that all these taxa possess forty-eight somatic chromosomes, with the exception of the apparently triploid *S. diemii*.

Although many collectors have searched in southern Argentina and Chile no one has yet found any diploid wild species in that area from which the tetraploids under discussion might have been derived. They stand alone, with only the cultivated tetraploids of southern Chile to link on to. The diploid and triploid *S. maglia* can be discounted, since not only does it occur very much further north in a much drier zone, but it differs fundamentally in the morphology of its flowers from *S. tuberosum* and its supposed tetraploid progenitors in the south. It is interesting to note that Brücher (1960*b*, 1963*a*, 1963*b*, 1964*b*) found both triploid and diploid forms amongst the cultivated potatoes of southern Chile. The diploids are claimed to be forms of *S. stenotomum* brought by migrating tribes from the altiplano of Peru and Bolivia, and are not considered to be native.

To sum up this section we should like to reiterate the hypothesis put forward by Hawkes (1956*a*) that *S. tuberosum* was formed originally in the Peru-Bolivia area,

[1] According to Diem (written communication) Brücher's collections of *S. diemii* were made in the same place as those of Sleumer and Diem to which we have referred previously; we have had one of these in cultivation and have verified its tetraploid nature. We certainly have no doubts that it should be classed as *S. tuberosum* subsp. *tuberosum*, and we think it likely that *S. diemii* is part of the same colony, even though Brücher considers it to be triploid.

[2] The latin diagnosis (p. 111) states 'grisaceo-rosaceis', (greyish rose), though the Spanish text gives 'gris-morado' (grey purple), which is quite a different colour!

either as an autotetraploid from an already cultivated diploid species such as *S. stenotomum* or as an amphiploid hybrid between that species and the weed diploid *S. sparsipilum*.[1] Such a hybrid would appear to function cytologically as an autotetraploid, if, as seems likely, there was very little genome differentiation between the two diploid parents of this tetraploid derivative. Once formed, the tetraploid *S. tuberosum* would no doubt have spread rapidly because of its greater vigour and would eventually have been brought to southern Chile by migrating tribes. Since many of the Indian tribes of southern Chile and Patagonia were seminomadic it would be justifiable to assume that the 'escaped' forms of *S. tuberosum* might represent relics of their shifting cultivations existing in a greater or lesser equilibrium with the wild vegetation.

TABLE 25

Comparison of 'S. molinae', 'S. leptostigma', *Diem's potato (Hjerting* 1650), 'S. diemii' *and* 'S. sanmartiniense'

Characters	'*S. molinae*'	'*S. leptostigma*'	Diem's potato (*Hjerting 1650*)	'*S. diemii*'	'*S. sanmartiniense*'
Tuber colour	?	Pink	Pale pink	Yellowish (greyish-yellow)†	Greyish rose (grey-purple)†
Pairs of primary leaflets	2–4	3–4	3–4	3–4	2–3
Pairs of interjected leaflets	0–2	0–3‡	0–2	2–3§	Absent or 1–2; rather minute
Flower colour	Whitish, suffused with violet	White	Pale pink, with white acumens and central petal streaks	Lilac (lilac-purple with white streaks)†	White
Chromosome number	48	48	48	36	48

† Words in parentheses indicate Spanish description where this differs from the Latin.
‡ 0–3 pairs on type specimen but description mentions 'up to 8'.
§ 4–6 in cultivated plants (according to Brücher).

2. Subspecies *andigena* (Juz. et Buk.) Hawkes

(*Proc. Linn. Soc. Lond.* **166**, 130, 1956)

S. andigena[2] Juz. et Buk. (*Proc. U.S.S.R. Congr. Genet.* **3**, 609–10, 1929)
S. apurimacense Vargas (*Papas Sudperuanas* **2**, 58–59, 1956)
S. subandigena Hawkes (*Bull. imp. Bur. Pl. Breed. Genet.*, Camb. 128, 1944)

This subspecies may be distinguished from subsp. *tuberosum* by the generally more dissected leaves with narrower leaflets which are long-petiolulate, only slightly arched towards the tip and set at a narrower angle on the stem. Pedicel not or barely thickened above, corolla generally medium to dark pigmented, rarely paler towards the acumens, occasionally pure white.

Tubers formed under short days in the tropics and subtropics; not normally cultivated below 2000 m altitude. Not cultivated in Argentina at altitudes lower than (1900–)2500 m and extending in Bolivia, Peru, and elsewhere to 3800 m or higher.

[1] Not from *S. sparsipilum* alone, as Brücher (1964*b*, p. 108) misquotes.
[2] Originally published as *S. andigenum*. We follow here the corrected orthography cited in *Index Kewensis*, Suppls. IX and XI, 1947 and 1953.

HERBARIUM SPECIMENS EXAMINED

Prov. JUJUY:

Dept. Humahuaca, Humahuaca, Alt. 2930 m. 7 Feb. 1941. 'Cultivada junto con maíz. Pétalos intensamente violados.' *A. T. Hunziker 1292* (SI).

Dept. Tilcara, Tilcara, Alfarcito. Alt. 2650 m. 9 Feb. 1939. 'Papa Runa Blanca.' *Balls 5963* (P.G.S.).

Locality, etc. as for *5963*. 'Chacarera.' *Balls 5964* (P.G.S.).

Locality, etc. as for *5963*. 'Cuarentona.' *Balls 5965* (P.G.S., J.G.H.).

Locality, etc. as for *5963*. 'Papa Runa Morada.' *Balls 5966* (P.G.S., J.G.H.).

Locality, etc., as for *5963*. 'Papa Morada.' *Balls 5967* (P.G.S., J.G.H.).

Dept. Tilcara, Tilcara, Quebrada del Huasamayo. Alt. 2450 m. 9 Feb. 1939. *Balls 5976* (P.G.S., J.G.H.).

Dept. Tilcara, near Tilcara, Chorro valley, San Antonio. Alt. 2450 m. 13 Feb. 1939. 'Papa Runa.' *Balls 6054* (P.G.S., J.G.H.).

Locality, etc., as for *6054*. 'Cuarentona.' *Balls 6055* (P.G.S., J.G.H.).

Locality, etc., as for *6054*. *Balls 6056* (P.G.S., J.G.H.).

Dept. Yavi, La Quiaca, cultivated in the railway station garden. Alt. 3450 m. 19 Feb. 1939. *Balls 6079* (P.G.S., J.G.H.).

Locality, etc., as for *6079*. 'Papa Runa.' *Balls 6080* (P.G.S., J.G.H.).

Locality, etc., as for *6079*. 'Papa Morada.' *Balls 6081* (P.G.S., J.G.H.).

Dept. Yavi, La Quiaca, 'Manga Fiesta' market. Said to have been brought from Rodero Colorado, prov. Salta, dept. Iruya. Alt. 3450 m. Cultivated in the experimental station 'El Suncho', Andalgalá, prov. Catamarca, 1700 m. Herbarium specimen taken 1 Mar. 1950. 'Collareja.' *Petersen & Hjerting Cult. 258c* (J.G.H.).

Prov. SALTA:

Dept. La Viña, Coronel Moldes, de un cultivo en La Ollada (a más o menos 20 km de Coronel Moldes). Alt. *c.* 1200 m. [Probably the altitude of Coronel Moldes, not of La Ollada.] 2 Feb. 1941. 'En tierra humífera muy buena. Pétalos rosados.' *A. T. Hunziker 1282* (BA).

Prov. CATAMARCA:

Catamarca, no locality given. *Isolina Díaz Peña 125* (LP).

Prov. MENDOZA:

Dept. Maipú, Luriaga. 19 May 1937. 'Cultivada! N.V.: "Papa".' *Ruiz Leal 3456* (LIL, Ruiz Leal).

TUBER COLLECTIONS GROWN IN ARGENTINA (See also p. 429)

1. **Astilla** (Malgacha, Malcachi, Blanca). Salta and Jujuy up to the Bolivian border. Very widespread. The principal variety in the Valle de Calchaquíes and the Quebrada del Toro, 2300–3500 m. Foliage very vigorous, flower violet. Tubers long, with brownish skin and yellowish white flesh, which very quickly changes to violet when exposed. Deep eyes. Good flavour.

2. **Azul** or **Azulito** (Astra de Cabra, Luqui Morada). Jujuy (depts. Tumbaya and Humahuaca), 3400–3550 m. Plant low growing with many stems. Tubers with very dark violet skin and yellow eyes and brows. Flesh yellowish-white.

3. **Bola** (Papa del Cerro). Salta (depts. Cachi, Cafayate, Molinos, and Poma) and Tucumán (dept. Trancas), 2000–3050 m. Characterized by the small, round tubers with yellow skin and deep eyes. Flesh white; good flavour.

4. **Chacarera** (Cuarentona Blanca). Jujuy (depts. Humahuaca and Tilcara), 2900–3400 m. Plant with pale violet flowers but very rarely flowering. Tubers yellow, sometimes with violet spots, round with very deep eyes.

5. **Collareja** (Collareja Seca, C. Caspeada, C. Dulce, C. Blanca, Morada de Iruya, Runa Lisa). Jujuy (depts. Cochinoca, Humahuaca, Sta. Catalina, Tilcara, and Yavi) and Salta (depts. Iruya, Molinos, and Sta. Victoria), 2500–3800 m. A very widespread and esteemed variety from Humahuaca northwards; also known from Bolivia. Plant beautiful, with abundant celestial blue flowers. Tubers large, long oval to oval with violet to violet and yellow skin, with many very deep eyes. Flesh white. Taste very good. Yield normally very good even at 'El Suncho' (1700 m). According to information from several growers and Viirsoo (1954*a*), this variety is frost resistant.

6. **Colorana** (Tonca). Jujuy (depts. Humahuaca and Tumbaya), 2900–3550 m. Tubers drop-shaped with red skin and very yellow flesh; formed on very long stolons.

7. **Condorilla.** Jujuy (dept. Humahuaca), 2900 m. Flowers violet. Tubers long, curved, red and yellow, with very yellow flesh.

8. **Criolla Blanca** (Imilla, Blanca, Sallama Blanca, Colla). Tucumán (dept. Trancas) and Salta (depts. Caldera and Cachi), 1900–2600 m. The principal variety in the mountains of Trancas. Also known from Bolivia as 'Imilla'. Foliage yellowish green, very vigorous. Flowers violet with white sectors. Fruits very rare. Tubers oval to long with yellow skin and deep eyes; flesh white. Stolons long. Yield very good.

9. **Cuarentona** (Cuarentona Blanca, Ojosa, Caldeleña). Jujuy (depts. Tumbaya and Yavi) and Salta (depts. Chicoana, Iruya, Rosario de Lerma, and Sta. Victoria), 2700–3550 m. Widespread variety with violet flowers. Tubers roundish with red and violet marbled skin and deep eyes; long stolons. Flesh when boiled very yellow. Yields best in high altitudes. Said to be eaten with cheese.

10. **Cuarentona Amarilla.** Jujuy (depts. Tilcara? and Tumbaya), 2400?–3450 m. Characterized by the tubers, which have dark red skin and yellow flesh.

11. **Jueya.** Jujuy (dept. Tumbaya), 3450 m. Flowers dark rose; leaves crisped, on greenish-yellow stems. Tubers round, pink with yellow flesh.

12. **Manzana Rosada** (Manzana, M. Bola Rosada, M. Blanca). Jujuy (depts. Humahuaca, Tumbaya, and Yavi) and Salta (depts. Iruya and Rosario de Lerma), 2900–3450 m. The collections obtained under the above-mentioned names could probably all be included in the same variety. Tubers round with rosy (sometimes whitish) skin and yellowish white flesh. Brücher (1960*b*, 1963*a*) mentions a variety from Chiloé named Palmetta which he claims is identical with Papa Manzana from Jujuy. Both are said to be triploids.

13. **Misquila** (Cojón de Gallo). Jujuy (dept. Tumbaya) and Salta (dept. Rosario de Lerma), (2700)–3450 m. Flowers pink. Plant low with many stems, similar to the variety 'Azul'. Characterized by the tubers which are long and sometimes curved, dark red with white eyes and eyebrows. Flesh yellowish white.

14. **Morada** (Ojosa Negra). Jujuy (depts. Tumbaya and Yavi) and Salta (dept. Iruya), 3000–3500 m. The plant is very hairy, with dark violet stems. The tubers of this variety are the darkest of all the varieties we have seen, deep violet to almost black. The flesh is also often very deep violet, sometimes white. Perhaps more than one variety?

15. **Negra Overa** (Overa Negra, Ojosa Negra, Huareña). Jujuy (depts. Humahuaca and Yavi) and Salta (dept. Iruya), 2900–3500 m. Flowers celestial blue; leaves and stems yellowish green; plant tall. Tubers round, typically speckled ('overa') yellow and deep violet. Flesh white. Eyes very deep.

16. **Oca.** Salta (dept. Iruya), 2700–3500 m. Characterized by the dark rose-coloured flowers and the long, slender tubers, sometimes curved. Stem light green; leaves crisped. Tuber skin yellow, flesh whitish. Eyes almost flat.

17. **Oquilla.** Jujuy (dept. Humahuaca), 2900 m. Similar to Criolla Colorada, but tubers smaller and with deeper eyes.

18. **Overa Blanca** (Ojosa, O. Caspeada, O. Blanca, Overa, Negra Overa). Jujuy and Salta, 2500–3500 m. Under this name we have grouped several collections from all over the area, obtained under quite different names. Plant tall, with violet flowers. Tubers small, round, typically two-coloured, half yellow, half deep violet. Flesh white. Often several tubers on one stolon. Yield very low at low altitudes.

19. **Overa Negra** (Ojosa Negra, Chara Imilla Redonda Negra, Sayama Negra). Jujuy (depts. Humahuaca and Sta. Catalina) and Salta (dept. Sta. Victoria), 2900–3700 m. Flowers violet; stems dark; leaves thin, dark green. Tubers round, somewhat particoloured, but mostly dark violet. Flesh white.

20. **Runa** (Runa Blanca, R. Blanca Larga, R. Blanca Redonda, R. Morada Larga, R. Morada Bola, R. Negra). Jujuy and Salta, 2400–3650 m. This is the most common subsp. *andigena* variety in Argentina, cultivated in nearly all departments in Salta and Jujuy where potato-growing takes place. It was obtained under a series of names, but when grown under identical conditions in 'El Suncho', Catamarca, all samples were more or less the same, and we were not able to separate them into groups varying sufficiently to be given different varietal names. Flowers violet, stems weak. Tubers oval with brownish skin, which at once turns darker when placed in daylight ('Morada' ?) Eyes not very deep. Flesh white, flavour good. Very susceptible to *Phytophthora infestans*.

21. **Santa María.** Jujuy (dept. Humahuaca), 2900 m. Very interesting variety because of the tubers which are long and curved, with skin and flesh intense red, the reddest of all the varieties. The juice of the tubers is said to be used to dye cloth.

22. **Socombay** (Sacamba). Jujuy (dept. Humahuaca), 2900–3400 m. Quite different from all other varieties. Stems green and tender, almost procumbent. Tubers long, curved, with yellow smooth skin and yellow flesh; eyes almost flat.

23. **Tuni Chata Blanca** (Tuni Blanca Larga). Jujuy (depts. Humahuaca, Tilcara, and Tumbaya) and Salta (dept. Iruya), 2500–3450 m. Plant tall, with pale lilac flowers. Tubers oval, flat, with yellow skin and white flesh, almost with flat eyes. When placed in daylight the flesh turns red. The biggest tubers are less flat than the smaller ones.

24. **Tuni Morada Chata** (Tuni Chata Morada, T. Chata, T. Colorada). Jujuy (dept. Humahuaca) and Salta (depts. Iruya and Rosario de Lerma), 2700–3500 m. Similar in growth habit to the preceding, but tuber skin dark red and flesh white.

25. **Tunta.** Jujuy (dept. Yavi), (3450 m). Said to have been brought from Santa Victoria (Salta, 2700 m). Only found in one place. Tubers with yellow skin and very white flesh.

26. Four collections were obtained under the name of 'Cuarentona' and one as 'Ojosa Dulce', from different parts of Salta and Jujuy. They seem to belong to the same variety and were provisionally called 'Ojosa'. The variety is somewhat similar to Cuarentona, but the tubers have very dark violet skin and white flesh. Flowers violet; leaves very crisped.

27. A single collection obtained from Antofagasta de la Sierra in the north west of prov. Catamarca, at 3400 m, somewhat differing from the other collections. It was obtained by Dr. F. Vervoorst under the name of 'Papa Criolla'.

DISTRIBUTION

Subspecies *andigena* is chiefly grown in provs. Jujuy and Salta, though some varieties are occasionally cultivated in prov. Tucumán, and very occasionally in prov. Catamarca.

The chief potato-growing departments in prov. Jujuy are Humahuaca, Tilcara, Tumbaya, and Yavi, but samples have also been obtained from Santa Catalina and Cochinoca. In prov. Salta subsp. *andigena* varieties are cultivated chiefly in dept. Iruya, and to a lesser extent in Rosario de Lerma and Santa Victoria. One or two varieties have also been found in depts. Cafayate, Cachi, Caldera, Chicoana, Molinos, Poma, and San Carlos.

The general altitude range for the cultivation of subsp. *andigena* is 2500 to 3500 m. Some few cultivations have been observed as low as 1900 m and some as high as 3800 m.

It should be pointed out that this picture of the distribution is far from complete, and is constructed purely on the basis of the material collected or seen by us. Potato cultivation in these north-western provinces of Argentina is probably more widespread than we have indicated, and we think that many other varieties are to be found at places not visited by us. Nevertheless, we feel that it is of historical interest as well as of importance for plant-breeders and for those contemplating further collecting trips that the results, however incomplete, should be published now, rather than be delayed in the hope that further data will become available.

STATUS IN ARGENTINA

We have already mentioned that *S. tuberosum* subsp. *andigena* is almost certainly 'indigenous' to Argentina, by which we mean that it has probably been grown there for a very long period, at any rate in provs. Jujuy and Salta. Unfortunately, we have no documentary proof of this, or archaeological evidence either. Nevertheless, the highlands of those two provinces possess areas so well adapted to potato cultivation, that it is reasonable to suppose that potato agriculture spread at a fairly rapid rate into all the appropriate places shortly after the primary epoch of domestication, several millennia ago.

It seems fairly certain that there is a continuous interchange of varieties between northern Argentina and southern Bolivia, as evidenced by the presence of many identical varieties in these two areas and the similarity of native names for them. Some varieties, therefore, may represent recent introductions.

As we have already mentioned, subsp. *andigena* is rare in Tucumán and has only been recorded twice in Catamarca (herb. collection of Isolina Díaz Peña and tuber collection of F. Vervoorst). Millán (1945) talks of potatoes being grown in the high valleys of San Juan and Mendoza; these may possibly belong to subsp. *andigena*, though, unfortunately, the author does not give altitude data or detailed descriptions. The Ruiz Leal collection from Mendoza, dept. Maipú, Luriaga, is also of subsp. *andigena*, but is apparently from a low altitude, more or less 700 m, according to Ruiz Leal (verbal communication).

We have no data for potato cultivation in La Rioja province, but from its position and the high altitude of the western mountain regions one may hazard a guess that subsp. *andigena* might be found there also.

It is to be hoped that botanists and agricultural scientists will make a point of looking for cultivated as well as wild potatoes during the course of their travels in these remote western regions. If we find that subsp. *andigena* does indeed spread down in the form of ancient cultivations as far south as Mendoza it will be seen that there is a gap of only some 500–600 km, or possibly less, between it and the most northerly limit of ancient subsp. *tuberosum* cultivation in central Chile (see Hawkes 1956*a*, p. 127). This gap has been bridged by the wild species *S. maglia*, and it seems reasonable to suppose that the Chilean subspecies of *S. tuberosum* could have been derived from the original Andean stock by means of a gradual diffusion of potato agriculture down the eastern Andes to Mendoza and a crossing to the western side at about the same latitude. Under very different climatic conditions in southern Chile, with a mild oceanic climate and long summer day length, the potato evolved gradually into a distinct series of forms which can be considered as distinguishable from the Andean ones, and which thus merit subspecific distinction. Owing to the historical accident that forms similar to these were classed by Linnaeus as *S. tuberosum* in Europe in the eighteenth century the Andean ones cannot be considered as the typical subspecies even though they are undoubtedly more primitive and constitute the 'ancestral' fount of the species *S. tuberosum*.

S. tuberosum subsp. *andigena* has been shown to be of considerable interest to plant-breeders for qualities of resistance which can be incorporated into domestic varieties. The following resistance qualities have been noted in certain accessions (though not always in Argentine clones): wart, scab, powdery scab, early blight, *Verticillium* wilt, skin spot, *Fusarium* wilt, bacterial wilt, common eelworm, horsenettle cyst and Osborne's cyst nematode, flea beetle, viruses X, A, and leaf-roll, and frost resistance. High starch and dry matter have also been recorded. (See Chapter 5 for further details.)

PENTAPLOID SPECIES

22. SOLANUM × CURTILOBUM Juz. et Buk

(*Proc. U.S.S.R. Congr. Genet.* **3**, 609, 1929)

Distinguished by its semi-rosette habit, at least when young, though generally with a long stem when older, straight stiff leaves, common peduncle 5–10 cm long or more, high well-marked pedicel articulation, large purple circular corolla, 3·0–3·5 cm diam., with very short lobes and acumens, and bitter purple or white tubers.

HERBARIUM SPECIMEN EXAMINED

Prov. JUJUY:

Dept. Yavi, La Quiaca, comprado en el almacén de E. Farfán. Alt. 3450 m. Said to have been brought from Santa Victoria, prov. Salta. Cultivated in 'El Suncho' experimental station, Andalgalá, prov. Catamarca, at 1700 m. Herb. spec. taken 1 Mar. 1950. 'Collareja Blanca.' Was given the name 'Papa Odorata' because of the very pleasant flower scent. *Petersen & Hjerting Cult. 231* (J.G.H., J.P.H.).

TUBER COLLECTIONS GROWN IN ARGENTINA

Two collections of this pentaploid species were obtained, one called 'Collareja Dulce' (P. & H. Cult. 257), the other 'Collareja Blanca' (P. & H. Cult. 231). Since these samples have nothing to do with the variety 'Collareja', which belongs to *S. tuberosum* subsp. *andigena*, we have renamed them 'Papa Odorata', because of the very pleasant strong flower scent. The two collections were very similar and might have been derived from the same introduction from Bolivia at an earlier date. When grown at 'El Suncho', Catamarca province, the variety was very distinct from all the others, forming very low rosette plants at first, but later growing much higher. The stems were thick and dark violet and the flowers also were violet. Many fruits were formed at 'El Suncho', probably due to natural cross-pollination with the tetraploid varieties of *S. tuberosum* subsp. *andigena* that were cultivated around it. The skin of the tuber was violet with white eyes; flesh white.

The yield was poor at 'El Suncho', even though the plants looked very healthy. This was probably due to the fairly low altitude of 'El Suncho' (1700 m), compared with the altitude where this species is normally grown. This material withstood frosts of 5° C below zero, together with *S.* ×*juzepczukii* and *S. vernei*, according to Viirsoo (1954*a*).

DISTRIBUTION

S. × *curtilobum* is normally grown at very high altitudes (3400–4300 m) from central Peru to south Bolivia. Both of our collections were bought in La Quiaca on the border between Argentina and Bolivia. One (*P. & H. Cult. 231*) was said to have been brought from Santa Victoria, the other (*P. & H. Cult. 257*) from Rodero Colorado, dept. Iruya, both situated in the northern parts of prov. Salta, not too far distant from Bolivia. We have no knowledge as to whether they had been cultivated in those localities.

STATUS IN ARGENTINA

The presence of this cultivated pentaploid potato species has not been noted in Argentina before. We do not know whether it is an ancient cultigen there or has been imported recently. Judging from the facts given above it may well have been imported sporadically from Bolivia for consumption or for occasional cultivation only. We very much doubt whether it can be considered as 'indigenous' to Argentina.

S. × *curtilobum* is a complex hybrid, which was shown by Hawkes (1962*a*) to have arisen, probably spontaneously, from a cross between *S.* ×*juzepczukii* (which provided an unreduced gamete) and *S. tuberosum* subsp. *andigena*. It thus contains two sets of twelve chromosomes from the wild frost-resistant species *S. acaule*, one set from *S. stenotomum*, and two from *S. tuberosum* subsp. *andigena*.

The two immediate parents of *S.* × *curtilobum* are present in northern Argentina, but we have no evidence to show that *S.* × *curtilobum* was formed in that country. There is no evidence against that hypothesis either, other than the undisputed fact that this potato is very rare and is probably imported from time to time, as we have just mentioned.

Seeds from spontaneous berries formed in the experimental field at 'El Suncho' were grown at Birmingham. The chromosome number of the seedlings ranged from 53 to 54, and this would agree with the hypothesis that the *S.* × *curtilobum* parent provided gametes with 29 to 30 chromosomes, which were fertilized by twenty-four-chromosome gametes from subsp. *andigena*. The flowers of the seedlings also showed intermediate features between *S.* × *curtilobum* and *S. tuberosum* subsp. *andigena*.

Qualities of resistance in certain clones (mostly from Bolivia and Peru) of *S. curtilobum* have been recorded to the following diseases and pests: wart, powdery scab, early blight, *Fusarium* wilt, bacterial wilt, viruses X and A, and frost. High dry matter has also been seen in one clone. (For further details see Chapter 5.)

APPENDIX I

LIST OF HERBARIA FROM WHICH MATERIAL HAS BEEN CONSULTED

BA	Buenos Aires, Argentina: Museo Argentino de Ciencias Naturales 'Bernardino Rivadavia' e Instituto Nacional de Investigaciones de las Ciencias Naturales.
BAA	Buenos Aires, Argentina: Herbario de la Facultad de Agronomía y Veterinaria de la Universidad de Buenos Aires.
BAB	Buenos Aires, Argentina: Instituto de Botánica Agrícola del I.N.T.A., Castelar.
BAF	Buenos Aires, Argentina: Instituto de Botánica y Farmacología.
BG	Bergen, Norway: Universitetets Botaniske Museum.
BM	London, Great Britain: British Museum (Natural History).
BR	Bruxelles, Belgium: Jardin Botanique de l'État.
BREM	Bremen, Germany: Übersee-Museum.
BRLU	Bruxelles, Belgium: Laboratoire de Botanique, Systématique et de Phytogéographie.
C	Copenhagen, Denmark: Botanical Museum and Herbarium.
CGE	Cambridge, Great Britain: Botany School, University of Cambridge.
CORD	Córdoba, Argentina: Museo Botánico, Facultad de Ciencias E.F. y Naturales.
E	Edinburgh, Great Britain: Royal Botanic Garden.
F	Chicago, Illinois, U.S.A.: Chicago Natural History Museum.
FI	Firenze, Italy: Herbarium Universitatis Florentinae, Instituto Botanico.
G	Genève, Switzerland: Conservatoire et Jardin Botaniques.
GH	Cambridge, Massachusetts, U.S.A.: Gray Herbarium of Harvard University.
GOET	Göttingen, Germany: Systematisch-Geobotanisches Institut, Universität Göttingen.
GRO	Groningen, Netherlands: Department for Systematic Botany of the Botanical Laboratory.
H	Helsinki, Finland: Botanical Museum.
HBG	Hamburg, Germany: Staatsinstitut für allgemeine Botanik und Botanischer Garten.
JE	Jena, Germany: Institut für Spezielle Botanik und Herbarium Haussknecht.
K	Kew, Great Britain: The Herbarium and Library.
L	Leiden, Netherlands: Rijksherbarium.
LA	Los Angeles, California, U.S.A.: Herbarium, Department of Botany, University of California.
LAU	Lausanne, Switzerland: Musée Botanique Cantonal.

LD	Lund, Sweden: Botanical Museum.
LE	Leningrad, U.S.S.R.: Herbarium of the Komarov Botanical Institute of the Academy of Sciences of the U.S.S.R.
LIL	Tucumán, Argentina: Instituto Miguel Lillo.
LP	La Plata, Argentina: Museo de La Plata, División de Plantas Vasculares.
LPS	La Plata, Argentina: Instituto de Botánica C. Spegazzini.
LY	Lyon, France: Herbiers de la Faculté des Sciences de Lyon.
M	München, Germany: Botanische Staatssammlung.
MPU	Montpellier, France: Institut de Botanique, Université de Montpellier.
MVFA	Montevideo, Uruguay: Laboratorio de Botánica, Facultad de Agronomía.
MVM	Montevideo, Uruguay: Museo de Historia Natural.
NY	New York, U.S.A.: The New York Botanical Garden.
P	Paris, France: Muséum National d'Histoire Naturelle, Laboratoire de Phanérogamie.
S	Stockholm, Sweden: Botanical Department, Naturhistoriska Riksmuseum.
SI	San Isidro, Argentina: Instituto de Botánica Darwinion.
TCD	Dublin, Ireland: School of Botany, Trinity College.
TEX	Austin, Texas, U.S.A.: University of Texas Herbarium.
U	Utrecht, Netherlands: Botanical Museum and Herbarium.
UC	Berkeley, California, U.S.A.: Herbarium of the University of California.
UPS	Uppsala, Sweden: Institute of Systematic Botany, University of Uppsala.
US	Washington, D.C., U.S.A.: National Museum (Department of Botany).
W	Wien, Austria: Naturhistorisches Museum.
WIR	Leningrad, U.S.S.R.: Herbarium of Cultivated Plants, Institute of Plant Industry.
WU	Wien, Austria: Botanisches Institut und Botanischer Garten der Universität Wien.
Z	Zürich, Switzerland: Botanischer Garten und Institut für Systematische Botanik der Universität Zürich.

Also Special Collections at the Following Places

BAL	Estación Experimental de Balcarce, prov. Buenos Aires, Argentina.
P.G.S.	Commonwealth Potato Collections formerly at Potato Genetics Station, Cambridge, now at Scottish Plant Breeding Station, near Edinburgh.
Max.	Max-Planck-Institut für Züchtungsforschung, Köln-Vogelsang, Germany.
Ruiz Leal	Personal herbarium of Dr. A. Ruiz Leal, Mendoza, Argentina.
J.G.H.	Personal herbarium of J. G. Hawkes, at Birmingham University, England.
J.P.H.	Personal herbarium of J. P. Hjerting, at Copenhagen Botanic Gardens, Denmark.

APPENDIX II

A. SPECIES AND SUBSPECIES ABBREVIATIONS

(See Simmonds 1963)

acl	*S. acaule* Bitt. subsp. *acaule*
adg	*S. tuberosum* L. subsp. *andigena* (Juz. et Buk.) Hawkes
aem	*S. acaule* Bitt. subsp. *aemulans* (Bitt. et Wittm.) Hawkes et Hjerting
bal	*S. vernei* Bitt. et Wittm. subsp. *ballsii* (Hawkes) Hawkes et Hjerting
blv	*S. boliviense* Dun.
brd	*S. brevidens* Phil.
chc	*S. chacoense* Bitt. subsp. *chacoense*
clv	*S. calvescens* Bitt.
cmm	*S. commersonii* Dun. subsp. *commersonii*
cur	*S.* × *curtilobum* Juz. et Buk.
gig	*S. microdontum* Bitt. subsp. *gigantophyllum* (Bitt.) Hawkes et Hjerting
grl	*S. gourlayi* Hawkes
ifd	*S. infundibuliforme* Phil.
juz	*S.* × *juzepczukii* Buk.
ktz	*S. kurtzianum* Bitt. et Wittm.
mag	*S. maglia* Schlechtd.
mcd	*S. microdontum* Bitt. subsp. *microdontum*
mga	*S. megistacrolobum* Bitt.
mlm	*S. commersonii* Dun. subsp. *malmeanum* (Bitt.) Hawkes et Hjerting
mue	*S. chacoense* Bitt. subsp. *muelleri* (Bitt.) Hawkes et Hjerting
opl	*S. oplocense* Hawkes
sct	*S. sanctae-rosae* Hawkes
spg	*S. spegazzinii* Bitt.
tar	*S. tarijense* Hawkes
tbr	*S. tuberosum* L. subsp. *tuberosum*
vdr	*S. vidaurrei* Cárd.
vnt	*S. venturii* Hawkes et Hjerting
vrn	*S. vernei* Bitt. et Wittm. subsp. *vernei*

B. LIST OF COLLECTORS OF WILD AND CULTIVATED POTATOES IN ARGENTINA, BRAZIL, PARAGUAY, AND URUGUAY

Only those collectors who have made more than one collection are normally included in the list. Exceptions have been made for important collections in one or two cases. Note that the date refers only to collections of potatoes seen by us, not to total collections.

Collector	*Country*	*State or Province*	*Date*	*Number of collections*	*Species collected*
Abbiatti, D. & Claps, L.	Argentina	Jujuy, Salta	1945	4	chc, gig
Arechavaleta, J.	Uruguay		1885–1909	5	cmm
Argañarás, J. L.	Argentina	Sant. del Estero	1940–2	2	chc
Arrillaga	Uruguay		1956	2	cmm
Bâcle, C. H.	Argentina, Uruguay		1831–4	2	chc, cmm
Bailetti, E.	Argentina	Tucumán	1918–19	2	chc, gig
Balansa, B.	Paraguay		1875	1	chc
Balegno, B.	Argentina	Córdoba	1944–6	2	chc
Balls, E. K., Gourlay, W. B. & Hawkes, J. G.	,,	Jujuy, Salta	1939	37	acl, aem, bal, chc, gig, grl, ifd, mcd, mga, mcd × bal, acl × mga
Barba, R. de	,,	Neuquén	1946–7	5	brd
Berro, M. B.	Uruguay		1898–1914	12	cmm
Bertoni, M. S.	Argentina	Misiones	1944–5	3	chc, mlm
Birabén, M. Isabel H. Scott de & Birabén, M.	,,	Tucumán, Buenos Aires	1938–9	2	chc
Boock, O. J.	Brazil	Paraná	1959	1	mue
Brieger, F. G.	,,	Minas Gerais	1963	1	clv
Brizuela, A.	Argentina	Catamarca	1947	4	chc
Brizuela, J.	,,	,,	1950	3	chc
Bruch, C.	,,	Tucumán, Córdoba	1908–26	4	chc, gig, gig × vrn
Brücher, H.	,,	Jujuy, Salta, La Rioja, Neuquén	1957–63	7	chc, ktz, spg, tbr, acl × mga
Budín, E.	,,	Jujuy, Salta	1906–31	5	acl, juz

Collector	*Country*	*State or Province*	*Date*	*Number of collections*	*Species collected*
Burkart, A. E.	Argentina	Salta, Tucumán, Neuquén, La Pampa, Buenos Aires, Entre Ríos	1927–65	17	acl, brd, chc, cmm, gig, mlm, cmm × chc
Cabrera, A. L.	Uruguay, Argentina	Jujuy, Salta, Catamarca, Buenos Aires, Entre Ríos	1930–55	22	acl, chc, cmm, cmm × chc, ifd, mga, mlm, spg
Cabrera, A. L. & Schwabe, H.	Argentina	Salta	1949	3	acl, chc, ifd
Camañ, A.	,,	Buenos Aires	1958	1	cmm × chc ?
Castellanos, A.	,,	Jujuy, Salta, Tucumán, La Rioja, San Luis, Córdoba, Buenos Aires	1917–43	25	acl, chc, cmm, ifd, mga, sct, spg, acl × spg, ifd × mga, gig × ktz, spg × ktz ?
Castillón, L.	,,	Jujuy, Tucumán, Catamarca	1906–27	17	acl, bal, chc, gig, sct, spg, spg/vdr, vrn, chc × spg, vrn × spg
Chebataroff, J.	Uruguay		1939?	2	cmm
Commerson, P.	,,		1767	1	cmm
Correll, D. S., Dodds, K. S., Brücher, E. H. & Paxman, G. J.	,,	Jujuy, Tucumán	1960	11	gig, grl, ifd, mga, opl, sct, vrn ifd × mga, acl × ?
Courbon, A.	Uruguay		1852–6	5	cmm
Descole, H. R.	Argentina	Tucumán	1943	4	gig, sct, vrn
Diaz, R.	,,	,,	1933	3	acl, vnt, vrn
Dieckmann, J. G.	,,	Buenos Aires	1910–11	2	cmm
Diem, J.	,,	Neuquén	1952	4	brd, tbr
Dusén, P. K. H.	Brazil	Paraná	1914–16	2	mue
Fabris, H. A. *et al.*	Argentina	Jujuy, Buenos Aires	1959–65	4	cmm, mcd, gig, vnt
Felippone, F.	Uruguay		1922	10	cmm, mlm
Fiebrig, K.	Argentina	Salta	1903	1	mcd
Filipóvich, R.	,,	Salta, Catamarca	1948–9	3	chc
Flossdorf, A.	,,	La Rioja	1913	2	ktz
Fruchard	Uruguay		1867–76	2	cmm
García, P.	Argentina	Sant. del Estero	1944–5	2	chc
Garolera & Romero, E.	,,	Jujuy, Salta	1947	10	acl, chc, gig, mga, vrn, chc × gig, chc × ?
Gaudichaud-Beaupré, C.	Uruguay		1836	2	cmm
Giardelli, M. L.	Argentina	Córdoba	1935–9	2	chc

Gibert, E. J.	Uruguay		1858–69	5	cmm
Gillies, J.	Argentina	Mendoza, Buenos Aires	1820–4	4	cmm, ktz
Grondona, E.	,,	Buenos Aires	1958	10	cmm
Gülich, C. von	,,	Chaco	1883	2	chc, mlm
Hassler, E.	Paraguay		1903–16	11	chc, mlm
Hauman, L.	Argentina	Jujuy, Tucumán, Buenos Aires, Entre Ríos, Misiones	1906–23	12	acl, chc, cmm, ifd, mlm
Hawkes, J. G.	Brazil, Uruguay, Argentina	Río Grande do Sul, Buenos Aires	1958	18	cmm, mlm
Henry, T.	Uruguay		1935	5	chc, cmm
Herrera, J.	Argentina	Tucumán	1945	2	chc
Herter, W. G.	Uruguay		1907–47	8	cmm
Hicken, C. M.	Argentina	Mendoza, Córdoba, Buenos Aires	1899–1932	11	chc, cmm, ktz
Hieronymus, G. H. & Niederlein, G.	,,	La Rioja	1879	5	aem, ktz
Hjerting, J. P., Petersen, E. & Rahn, K.	,,	Jujuy, Salta, Tucumán	1956	65	acl, chc, gig, ifd, mga, sct, spg, vrn, chc × gig, chc × tar, gig × vrn, ifd × mga
Hjerting, J. P. & Rahn, K.	,,	Salta	1956	43	acl, bal, blv, chc, grl, mcd, mga, opl, tar, grl × vdr, mcd × bal, mcd × vrn, mga × blv, tar × gig, tar × ?
Hunziker, A. T.	,,	Jujuy, Salta, La Rioja, Córdoba, Buenos Aires	1938–56	15	adg, chc, cmm, gig, ktz, tar, chc × tar
Hunziker, J. H.	,,	Córdoba, Buenos Aires	1945–6	3	chc, cmm
Illin	,,	Chubut	1901	1	tbr
Jörgensen, P.	Paraguay, Argentina	Tucumán, Catamarca, Chaco, Formosa	1913–29	8	chc, gig, spg, gig × spg, vrn × spg
Kemp, F. J.	Paraguay, Uruguay		1954–60	2	chc, cmm
Knappe, P. T.	Argentina	Salta	1935?	2	chc
Krapovickas, A.	,,	Buenos Aires, Entre Ríos	1943–7	3	cmm, mlm
Kurtz, F.	,, Paraguay	Jujuy, La Rioja, San Juan, Mendoza	1885–1908	8	acl, aem, chc, ifd, ktz, mlm, spg

Collector	Country	State or Province	Date	Number of collections	Species collected
Legname, R. P. *et al.*	Argentina	Jujuy, Catamarca	1959–65	3	gig
Legrand, D.	Uruguay		1934–47	5	cmm
Lillo, M.	Argentina	Jujuy, Salta, Tucumán, Sant. del Estero	1888–1917	29	acl, chc, gig, sct, vnt, vrn
Lorentz, P. G.	,,	Entre Ríos, Tucumán, Catamarca	1872–7	5	cmm, spg, vrn
Lorentz, P. G. & Hieronymus, G. H.	,,	Salta, Tucumán	1873–4	8	acl, chc, gig, sct, vrn
Lourteig, A.	,,	Mendoza, Buenos Aires	1938–45	2	cmm, ktz?
Luna, F. E.	,,	Salta	1948–51	4	chc
Lunnasvaana, V.	,,	Misiones	1928	3	chc, mlm
Macola, T.	,,	Mendoza	1934?	1	ktz
Maldonado, R.	,,	San Luis, Río Negro, Sant. del Estero	1940–2	3	brd, chc
Martínez Crovetto, R.	,,	Buenos Aires, Entre Ríos	1942–3	2	cmm
Meyer, T.	,,	Jujuy, Salta, Tucumán, Catamarca, Corrientes, Misiones, Chaco	1933–51	24	chc, gig, ifd, mga, mlm, sct, spg, tar, vnt, vrn, chc × gig, sct × vnt, vnt × vrn
Millán, R. A.	,,	Tucumán, Córdoba, Buenos Aires, Misiones, Chubut	1926–41	13	chc, cmm, mlm, tbr
Millán, R. A. & Pujals, E. A.	,,	Buenos Aires, Santa Fé	1939	12	cmm, mlm
Monetti, L.	,,	Tucumán	1913–14	5	chc, gig, vrn
Montes, J. E.	Argentina, Paraguay	Misiones	1946–52	13	chc, mlm
Morel, I.	Argentina	Formosa	1947–9	9	chc
Mota da Costa, D.	Brazil	Río Grande do Sul	1955–9	8	cmm, mlm
Müller, F.	,,	Río Grande do Sul, Sta Catarina	1868	2	mue
Myndel Pedersen, T.	Paraguay, Argentina	Corrientes, Buenos Aires, Neuquén	1949–61	5	brd, chc, cmm, mlm
O'Donell, C. A.	Argentina	Jujuy, Tucumán	1941–7	4	acl, chc, gig
Olea, D.	,,	Tucumán	1933–45	5	acl, chc, vnt
Osten, C.	Uruguay, Argentina	Buenos Aires	1886–1926	13	cmm

Page, T. J.	Paraguay		1851	1	chc
Palacios, M., & Cuezzo A.	Brazil	Río Grande do Sul	1948	2	cmm
Parodi, L. R.	Argentina	Tucumán, Buenos Aires	1933–4	3	acl, chc, cmm
Peirano	,,	Tucumán, Catamarca	1933–6	3	chc, spg
Petersen, E. & Hjerting, J. P.	,,	Jujuy, Salta, Tucumán, Catamarca La Rioja, Corrientes, Misiones, Chaco, Sant. del Estero	1948–56	147	acl, adg, aem, bal, chc, cur, gig, grl, ifd, juz, ktz, mcd, mga, mlm, mue, opl, sct, spg, vnt, vrn, tar, aem × spg, chc × gig, chc × mcd, gig × vrn, sct × vnt, vrn × spg, vnt × vrn
Pierotti, S. A.	,,	Salta	1945–7	2	chc, mcd
Rambo, B.	Brazil	Río Grande do Sul	1945–57	7	cmm, mue
Ratera, E. L.	Argentina	Buenos Aires, Santa Fé	1935–43	4	chc
Reales, A.	,,	Catamarca	1948–9	3	chc, spg
Reche, J. O.	,,	La Rioja	1960	5	spg, gig × ktz
Regnell, A. F.	Brazil, Argentina	Minas Gerais, Río Grande do Sul; Jujuy	1845–1901*	8	acl, clv, cmm, ifd, mlm
Rocha, R.	Argentina	Salta, Tucumán	1948–9	2	chc
Rodríguez, D.?	,,	Tucumán, Chaco	1912–14	3	acl, chc, vrn
Rohmeder, W.	,,	Tucumán, Catamarca	1942–3	2	vrn, sct × vnt
Roig, F. A.	,,	Mendoza	1957	2	ktz
Rojas, T.	Paraguay, Argentina		1906/51	2	chc, mlm
Romero, E.	Argentina	Jujuy	1941	2	chc
Ross, H. & Vervoorst, F.	,,	Jujuy, Salta, Tucumán, Catamarca	1959	35	chc, gig, grl, ifd, mga, sct, spg, chc × gig, sct × spg, tar × spg
Ruiz Leal, A.	,,	Mendoza, San Luis, Córdoba, Neuquén	1935–57	26	adg, brd, chc, ktz, mag, spg, chc × ktz
Saint-Hilaire, A. P. de	Brazil	Río Grande do Sul	1816–21	2	cmm
Schenck, J. H. R.	,,	Sta Catarina	1886	1	mue
Schickendantz, F.	Argentina	Tucumán, Catamarca	1873–93	6	chc, gig, spg, sct?
Schreiter, R.	,,	Jujuy, Tucumán, Catamarca	1917–40	41	acl, chc, gig, ifd, sct, spg, vnt, vrn, mga × mcd, sct × vnt, vrn × gig, vrn × spg
Schulz, A. G.	,,	Chaco	1931–55	4	chc, mlm

* Later collections were made after his death by R. Fries and others.

Collector	*Country*	*State or Province*	*Date*	*Number of collections*	*Species collected*
Schwarz, G.	Argentina	Corrientes, Misiones	1944–9	5	mlm
Schwindt, E.	,,	Misiones	1949–50	7	chc, mlm
Sellow, F.	Brazil			5	cmm
Sleumer, H.	Argentina	Jujuy, Salta, Tucumán, Catamarca, Neuquén	1950–3	92	acl, aem, bal, brd, chc, gig, grl, ifd, mcd, mga, sct, spg, tar, tbr, vdr, vnt, vrn, acl × mga, chc × gig, chc × ktz, gig × bal, ifd × grl, mga × blv, sct × spg, sct × vnt, tar × sp.?
Sleumer, H. & Vervoorst, F.	,,	Salta, Catamarca	1952	29	acl, chc, gig, ktz, mga, sct, spg, vrn, chc × spg, sct × spg
Sparre, B.	,,	Tucumán	1949–52	13	acl, chc, gig, sct, vnt, sct × vnt, vnt × vrn
Spegazzini, C.	,,	Salta, Córdoba, Buenos Aires	1887–1921	12	chc, cmm, gig, spg, chc × spg
Stuckert, T.	,,	Tucumán, Córdoba, Chaco	1897–1911	9	chc, gig
Tweedie, J.	Uruguay, Argentina	Buenos Aires	1837	5	chc, cmm
Ule, E.	Brazil	Sta. Catarina	1888–9	2	cmm, mue
Venturi, S.	Argentina	Jujuy, Salta, Tucumán, Catamarca, Buenos Aires, Sant. del Estero	1902–29	27	acl, chc, cmm, gig, ifd, mga, sct, vnt, vrn, chc × gig, mga × mcd, sct × vnt
Vervoorst, F.	,,	Buenos Aires, Tucumán, Catamarca	1951–2	11	cmm, chc, sct, spg, vrn, chc × spg, chc × ktz, sct × spg
Viirsoo, E.	,,	La Rioja	1961	3	spg
Ybarrola, T. S.	,,	Corrientes	1944–5	5	chc, mlm

APPENDIX III

THE 1965–6 EXPEDITION TO ARGENTINA

WHILST we were preparing the manuscript for this book we began to feel very strongly that a final visit to Argentina would be necessary before the work could be said to have arrived at a reasonable state of completion. Many problems remained to be solved and several hypotheses needed to be tested by fresh observations and collections. For various reasons the visit had to be postponed until the 1965–6 season, by which time most of the text had been prepared and typed. We, therefore, decided to write a brief report of the expedition as an Appendix. The main text has been altered only as far as was necessary to bring it into line with our most recent findings, and we have added references to this Appendix where necessary. Obviously, this Appendix can give an account only of the most immediate results of the expedition, since several years' work are needed before the material can be evaluated reasonably well.

A pleasant but unforeseen result of this expedition was the discovery in the major Argentine botanical museums and institutes of a number of important herbarium collections which we had not seen before. The isotype collections in Córdoba of Kurtz's material were particularly valuable, since the types had been deposited in Berlin and were destroyed during the Second World War. There were also several recent collections in all the herbaria, which we had not seen. We have added to the main text all those which indicated an extension of the previously known ranges of a number of species, but unfortunately, time and space does not permit us to cite them all. We offer our apologies to those collectors whose material was seen too late and which has, therefore, not been included.

We were accompanied on this expedition by a Danish colleague, K. Rahn, of the Copenhagen Botanic Gardens. He had also collected with Hjerting and Petersen in 1955–6. Rahn was chiefly interested in the genus *Plantago*, but helped greatly with the potato collecting. Fortunately, wild potatoes and plantains are often found in the same areas, so that the two main interests of the expedition complemented each other. An Argentine colleague, Licenciado A. K. Okada, travelled and collected with us also, in January and early February.

As well as potatoes and plantains, we made general herbarium collections for Copenhagen and Kew. We also collected seeds and other living material for these and other botanic gardens. Gatherings of seeds and plants were made for special investigations of Commelinaceae, other genera in Solanaceae and *Solanum* sect. Morella.

We made a total of 1058 collections, including 220 potatoes, 280 plantains, 53 Commelinaceae, and 505 other collections. Living material totalled 599 collections, of which 155 were potatoes and 202 plantains. The living collections thus came to about three-fifths of the total.

The expedition began in November 1965 when Hjerting and Rahn made a few collections in Peru on their way out, before continuing to Argentina. They arrived at Buenos Aires at the end of the month and began to collect in Buenos Aires province early in December. They travelled west through that province to Mendoza, where Hawkes joined them on 1 January. Hawkes and Rahn had spent a few days previously in the Museum of Natural History, Santiago de Chile, working on type material of *Solanum* and *Plantago*, thanks to the kindness of the director, Professor Carlos Muñoz Pizarro.

We give a fairly detailed itinerary of our expedition in Table 26. The part of our expedition concerned with potatoes may be summarized as follows: We collected for a few days in Mendoza province with Professor A. Ruiz Leal and then travelled eastwards through San Luis to Córdoba. Here we were joined by Sr. Okada. After collecting in Córdoba province for a few days we went to Santa Fé and Entre Ríos, returning to Córdoba again. From Córdoba we drove westwards to San Juan and north to La Rioja (Sierra Famatina). Early in February Sr. Okada returned to Buenos Aires and we continued northwards through La Rioja and Catamarca to Tucumán. Our collections in Tucumán, southern Salta, and eastern Catamarca provinces lasted until the end of February. In March we continued northwards through Salta and Jujuy provinces to the Bolivian border. Finally, we retraced our steps, though with some variations in the route, through Jujuy, Salta, Tucumán, Catamarca, La Rioja, and Córdoba, returning to Buenos Aires on 30 March. From here, Hawkes went to Balcarce in Buenos Aires province from 1 to 3 April, to visit the potato-breeding group there. Finally, we sailed for Europe on 12 April on a Danish cargo ship, arriving towards the end of the month.

TABLE 26

Itinerary of the 1965–6 Expedition in Argentina (and Peru)

17 Nov. 1965	Hjerting and Rahn arrived at Lima.
18 and 19 Nov.	Worked at 'Museo de Historia Natural' by kind permission of the Director, Dr. Ramón Ferreyra.
20 Nov.	Lima to Loma de Chancay and return.
21 Nov.	Visit to Est. Exp. 'La Molina', with Ing. Agr. C. Ochoa.
23 Nov.	Lima to Buenos Aires by aeroplane.
24 Nov.–10 Dec.	In Buenos Aires, preparing the expedition and visiting Museums and Institutes.
11 Dec.	Buenos Aires to Maipú.
12 Dec.	Mar del Plata, Balcarce.
13 Dec.	Tandil, Juárez.
14 Dec.	Zubiaurre.
16 Dec.	Sierra de la Ventana.
18 Dec.	Peralta, Pigüé.
19 Dec.	General Acha in La Pampa province.
20 Dec.	Santa Rosa, Realicó to Unión in San Luis province.
21 Dec.	General Alvear, San Rafael, El Sosneado to Malargüe in Mendoza province.
22 Dec.	Malargüe to Atuel valley and return.
23 Dec.	Malargüe to Salado valley and return.
24 Dec.	Malargüe to San Rafael.
25 Dec.	San Rafael to Punta del Agua (Sierra del Nevado) and return.
26 Dec.	Arroyo Aguanda, Dique del Valle de Uco, Mendoza.
28 Dec.	Rahn left by aeroplane for Santiago, Chile.
1 Jan. 1966	Rahn and Hawkes arrived at Mendoza by aeroplane from Santiago.
2 Jan.	Mendoza to Quebrada de Alvarado and return.
4 Jan.	Mendoza to Villavicencio, Quebrada del Toro, and El Portezuelo, and return.
6 Jan.	Mendoza to San Luis.
7 Jan.	El Trapiche and Pancanta in Sierra de San Luis.
8 Jan.	La Carolina, La Toma, El Morro, La Toma.
9 Jan.	Villa del Carmen, Larca, Merlo.
10 Jan.	La Paz, Villa Dolores, Mina Clavero, Pampa de Achala, Ycho Cruz.
11 Jan.	Ycho Cruz to Córdoba.

14 Jan.	Córdoba to Río Ceballos, Salsipuedes, Sierra Chica, La Falda, Villa Allende, and return.
16 Jan.	Córdoba to San Francisco and Santa Fé.
17 Jan.	Crossed Río Paraná and thence to Villaguay.
18 Jan.	Colón and Concepción del Uruguay.
19 Jan.	Concordia, Paraná, Santa Fé.
20 Jan.	San José del Rincón, Esperanza, Rafaela, Córdoba.
22 Jan.	Río Ceballos, Sierra Chica, La Falda, Capilla del Monte.
23 Jan.	Ascended Cerro Uritorco, thence to La Cumbre.
24 Jan.	Cosquin, San José, Tanti, Sierra Grande, Los Gigantes, Salsacate (Tala Cañada).
25 Jan.	Salsacate to San Juan.
26 Jan.	San Juan to Valle de Zonda and return.
27 Jan.	San Juan westwards to Tamberías, Quebrada de Leoncito, and return.
28 Jan.	Talacasto, valley of Río Agua Negra, San José de Jáchal.
29 Jan.	Guandacol, Río Bermejo, Villa Unión, Chilecito.
30 Jan.	Chilecito to Famatina, Carrizal, Los Corrales, La Puntilla, and return.
31 Jan.	Chilecito to Guanchín Viejo and return.
1 Feb.	Chilecito to Nonogasta, La Puerta (Sierra Velasco), and return.
2 Feb.	Chilecito to La Florentina, along road to Mina de Oro and return.
3 Feb.	Chilecito to Ángulos, Campanas, Santa Cruz, and return.
4 Feb.	Carrizal, Los Corrales, La Encrucijada.
5 Feb.	Return from La Encrucijada to Chilecito.
7 Feb.	Olta, Olpas, Patquía, La Rioja.
8 Feb.	La Rioja to Dique Los Sauces and return.
9 Feb.	Catamarca, Cuesta de Totoral, Tucumán.
12 Feb.	Tucumán to Tapia, Raco, Río Raco, and return.
14 Feb.	Acheral, El Mollar, Tafí del Valle.
15 Feb.	Infiernillo, Quebrada Honda, Amaichá.
16 Feb.	Cafayate, San Lucas, Animaná, San Antonio, Cafayate.
17 Feb.	Santa María, Ingenio Arenal, Mina Capillitas, Andalgalá.
18 Feb.	Belén, Puerta de Belén, Pozo de Piedra, Belén.
19 Feb.	Andalgalá, 'El Suncho', El Alamito.
20 Feb.	Estancia 'Yunka Suma'.
21 Feb.	El Alamito to Tucumán.
23–26 Feb.	Tucumán to Estancia 'Las Pavas', Estancia 'Santa Rosa', and return.
1 Mar.	Ruiz de Los Llanos, Estancia 'Miraflores'.
2 Mar.	Estancia 'Miraflores' to Pampa Grande.
3 Mar.	At Pampa Grande and Cerro Cristal.
4 Mar.	Guachipas, La Vina, Quebrada de San Lorenzo, Salta.
5 Mar.	Carril, Escoipe, Cuesta del Obispo, Cachipampa, Payogasta, Cachi.
6 Mar.	Payogasta, Estancia 'Palermo Oeste', Cachipampa, towards Amblayo, Cuesta del Obispo, Salta.
7 Mar.	Camino de La Cornisa, Jujuy, Termas de Reyes, Jujuy.
8 Mar.	Volcán, Loma del Tambo, Abra Grande de Volcán, Maimará.
10 Mar.	Tilcara, Quebrada de Huasamayo, Garganta del Diablo, Tilcara, Purmamarca, Abra de Los Pibes, Tilcara.
11 Mar.	Humahuaca, Iturbe, Abra del Cóndor, Iturbe, Tres Cruces, Abra Pampa, La Quiaca.
12 Mar.	La Quiaca, Yavi, Rincón de Cajas, Abra de Lizoite and towards Santa Victoria; return to La Quiaca.
13 Mar.	Yavi, Rincón de Cajas, Abra de Lizoite, Patahuasi, Rodeopampa, La Huerta, Santa Victoria.
14 Mar.	Santa Victoria to San Felipe and return.
15 Mar.	Santa Victoria to La Quiaca.
16 Mar.	Tres Cruces, Esquinas Blancas, along railway by km 1313 to Iturbe, Humahuaca, Huacalera.
17 Mar.	Huacalera to Posta de Lozano.

TABLE 26—(*cont.*)

18 Mar.	Jujuy, camino de La Cornisa, Salta.
19 Mar.	Guachipas, Pampa Grande.
20 Mar.	Pampa Grande to Cerro Cristal and return.
21 Mar.	Pampa Grande to Tucumán.
24 Mar.	Catamarca, Quebrada de la Sébila, Aimogasta.
25 Mar.	San Blas, Pituil, Carrizal, Los Corrales, Famatina, La Puntilla, Chilecito.
26 Mar.	Chilecito to Guanchín Viejo, Las Talas, and return.
27 Mar.	Chamical, Olta, Cruz del Eje, La Cumbre, La Falda, Sierra Chica, Salsipuedes, Córdoba.
28 Mar.	Córdoba to Pampa de Achala and return.
29 Mar.	Alta Gracia, Río Cuarto, La Carlota.
30 Mar.	La Carlota to Buenos Aires.
1 Apr.	Buenos Aires to Mar del Plata, Balcarce.
2 Apr.	At Balcarce.
3 Apr.	Balcarce to Buenos Aires.
4–11 Apr.	Visits to Museums and Institutes in and around Buenos Aires (Castelar, San Isidro, La Plata, etc.).
12 Apr.	Sailed for Europe.

During the course of the expedition we travelled over 18 500 km (11 100 miles) by road in the 'Estanciera' (Estate Car) kindly loaned by the Kaiser Corporation of Argentina. We walked many scores of miles also, and travelled many more on horseback.

The weather on the first part of the expedition was excessively dry. No rain had yet fallen in Mendoza and San Juan by the time of our visit and we were too early for the rains in the lower, eastern part of Córdoba province. They were just beginning in Santa Fé and Entre Ríos provinces but potatoes had not yet started to develop. Hence we saw no living plants at all in Mendoza, San Juan, Santa Fé, and Entre Ríos, though in other more normal years potatoes should have been abundant by the time of our visit. Rainfall had been fairly satisfactory in the mountains of San Luis and Córdoba.

The rains began in La Rioja in the latter part of January, but by the time we arrived the plants were already beginning to dry up. In fact, in none of the drier areas of Argentina this year was precipitation at all adequate, even by the normal standards obtaining in those regions. It was thus always difficult to find material of species adapted to dry habitats, since in many of these places there had either been no rain or a completely inadequate amount. In many of the places we visited we knew from the evidence of previous herbarium collections that potatoes occurred there; yet this year we often drew a complete blank.

At the other extreme, places with a normally high rainfall had had an excess this year. Part of the 'Selva Tucumano-Boliviana', for example, had received so much rain that we actually saw plants of *S. microdontum* subsp. *gigantophyllum* with 'cell burst'. This is a phenomenon observed sometimes in the glasshouse on plants that have been excessively overwatered. Too much water is absorbed and areas of leaf tissue become powdery, owing to the swollen cells and burst tissues, whilst the leaves themselves become rolled up and distorted. Excessive rain and presumably heavy cloud seemed to have inhibited flowering in some areas, also.

Between these two extremes of precipitation we, of course, found some places that had received about the right amount of rain (Pampa Grande and Cuesta del Obispo, for example), though this was the exception, rather than the rule. This was unfortunate, since we had particularly hoped to be favoured by a 'good wet season' in order to collect adequate material in the dry regions of Mendoza, San Juan, Catamarca, and elsewhere. In this respect the botanists who live in Argentina are at a great advantage, since they stand a better chance of planning their collecting trips at short notice at times when the rainfall is of optimum quantity.

By and large, however, we were satisfied with the results of the expedition and have already been able to solve a number of problems. We hope that this book will act as an incentive to Argentine botanists and plant-breeders to solve other problems, which we were not able to deal with.

In the sections that follow we have presented our preliminary results under the species to which they apply and in the same order as in the main part of this book. We should, however, like to begin with a few general observations.

The phenomenon of species hybridization amongst wild potatoes in Argentina, and the nature and efficiency of the breeding barriers which keep them apart have been of constant interest to us. We have already remarked in various places that artificial hybrids are extremely easy to make in the experimental field, though of comparatively infrequent occurrence in nature. We have pointed out also that most species may be isolated from each other by means of ecological barriers, but we are not altogether satisfied that these are the only isolating mechanisms available.

In our 1966 expedition we were especially interested in making collections of hybrids. Our collections are no doubt 'overweighted' in this respect, that is to say, they do not represent a naturally occurring proportion of hybrid to non-hybrid material, but show bias in favour of hybrids. By thus observing and collecting hybrids, as well as the parent species, we hoped to gain some insight into the barriers to gene exchange in nature.

We were rather surprised to find that in many regions where hybrids had been found in other years we saw none at all or a very reduced amount. Thus, the Hjerting, Petersen, and Rahn collections of ten years ago and the Sleumer and Vervoorst ones of 1952 and 1953 included hybrid material from regions where in 1966 we observed the parent species only. In those years the rainfall seems to have been adequate for wild potato growth, whilst in 1966, as we have just seen, it was completely inadequate in most places.

This may point, not so much to a gradual decline of the hybrids over the intervening years (for we cannot be certain of this, since we have not sampled the areas over those years), but rather, that the species themselves are better adapted to their environment than the hybrids. The species seem able to survive under very adverse conditions, whilst the hybrids may only be able to do so when the conditions are near optimal. Such hybrids evidently might not possess such combinations of genes for perfect physiological adaptation to their environment, and a dry season would eliminate them. This idea runs contrary to Brücher's suggestion that hybrids are better adapted to certain areas than their parents. Perhaps in some years this might be so, but in 1966 it did not seem to be valid.

If our hypothesis were correct, then the fluctuations of the environmental conditions from season to season might be all that was necessary to prevent mass hybridization with a consequent complete blurring of the species boundaries. It would not, perhaps, prevent a limited amount of gene flow, since the introgressed forms of a species would presumably be selected for the same adaptive peak as the pure species. Or, alternatively, in some instances, as with *S. chacoense*, new adaptive peaks might have arisen to fit a range of biotypes to newly colonized mountain regions.

Our observations in 1966 showed that in places such as Pampa Grande, Cuesta del Obispo, Quebrada de San Lorenzo, and Estancia 'Yunka Suma', where hybrids had been plentiful in 1956, very few or none were observable in 1966. Introgressed forms were common, and were presumably as well adapted to their environment as the parental species. Hybrid segregates were largely absent. Since there seem to be no genetic barriers to species hybridization, and hybrids between diploid species are fully fertile under experimental conditions we can only assume that hybrids are also being formed in nature but at the same time eliminated by some other mechanism, such as the selective action of inclement seasons. It will be objected that there may be some other mechanism to prevent hybridization between species in nature which we do not yet comprehend. With this we agree, yet in 'good' years such as 1953 and 1956 hybrids seem to be plentiful enough.

One cannot always fall back on the hypothesis that the parents are adapted to two distinct habitats and the hybrid to neither, so that it cannot survive unless 'hybrid habitats' are provided, in the Andersonian sense. What intrigued us greatly in 1966 was that there seems to be an abundance of 'hybrid habitats', bushy scrubby places by roadsides and in previously felled and grazed forest areas where the natural vegetation no longer exists. In these highly modified habitats we saw plenty of examples of *both parental species* growing together in close proximity, but without any hybrids present. Surely one must assume here that each parental species has become adapted to these habitats by natural selection, but that because the hybrids would contain only portions of the genetic material of each parent their adaptation would be below the minimum required for survival. In other words, the adaptive mechanism is only fully effective when present in each parent in a more or less pure state. Evidently, such hybrid plants might be expected to survive in a 'good' year but to be eliminated in a 'bad' one (i.e. too wet or too dry).

It would be of very great interest to test this hypothesis by making careful observations for a period of five years or more in a number of carefully selected and easily accessible areas (such as the ones we mentioned above). One could mark out a well-defined series of fairly large plots and map and score the proportion of hybrid to non-hybrid plants, linking this information to meteorological data from the nearest recording stations. Even if our hypothesis were disproved, the experience of obtaining information and making careful readings over a period of years would add much useful data that might throw more light on the problem of species evolution in this group.

Another type of crossability barrier that might exist in nature between wild potato species is a difference in flowering time. For example, in Estancia 'Yunka Suma', the two species *S. microdontum* subsp. *gigantophyllum* and *S. vernei* subsp. *vernei* occur roughly in the same areas. Our observations in 1956 and earlier indicated that *S. vernei* grew in the open, on river terraces, or in partial shade only; *S. microdontum* was found in deep shade in the forests. In 1966, however, both species seemed to occur together both in shade and in the open, yet hybrids were few. Perhaps they had been eliminated by the mechanism we have postulated above; but it was interesting to note that by the time of our visit (20 February) *S. microdontum* was only just coming into flower, whilst *S. vernei* was already bearing large mature berries, and flowering had long since ceased. This mechanism would be an excellent one to prevent gene flow between species, yet we do not know whether it is normally operative. Flowering-time data from herbarium specimens are of no help in this case, so far as we can see. Here again, the making of careful observations over a period of years is the only way of finding out. In all probability there are several mechanisms at work to prevent widespread hybridization in nature, and only patient studies in the field will be able to show us little by little which are the most important.

DETAILED RESULTS OF THE 1965–6 EXPEDITION

1. SOLANUM BREVIDENS Phil.

The expedition did not visit the southern provinces of Río Negro and Neuquén, and consequently no collections of this species were made. Several herbarium collections were seen in Argentine herbaria, but they did not extend the known range appreciably, and consequently they have not been quoted in the text. An intriguing collection by Kurtz (*10081*) from Mendoza, Cordillera del Atuel, made actually by José M. Figueroa, was identified as *S. etuberosum* (Herb. Córdoba). The locality mentioned lies very close to the Chilean border, and it is possible that the collection was really made in Chile, not in Argentina. Professor Ruiz Leal has kindly agreed to look into this matter for us.

2. SOLANUM CALVESCENS Bitt.

No material of *S. calvescens* was collected, since this species is apparently endemic to the State of Minas Gerais in Brazil. However, whilst we were in Argentina a consignment was sent to Birmingham through the kindness of Professor F. G. Brieger, of Piracicaba, São Paulo. The collection data are as follows:

Estado de Minas Gerais, Alfenas, Fazenda da Flousta. Alt. 900 m. Nov. 1965. 'Entre pedras.' Leg. Cyro Paulino da Costa. *Brieger* (s.n.) (J.G.H.—Accession no. 775).

Alfenas lies some 100 km north-east of Caldas and this collection is of great interest, therefore, in showing an extension of the known range of *S. calvescens*. Like the Brieger collection mentioned previously (see p. 140), all the plants were triploid ($2n = 36$). Nevertheless, we still think that diploid clones of *S. calvescens* may exist in some part of its distribution area.

3. SOLANUM COMMERSONII Dun.

Our visit to the provinces of Santa Fé and Entre Ríos to examine and collect *S. commersonii* was wholly unsuccessful. Not only had the rains been delayed this year, but it seems as though subsp. *malmeanum* normally grows and flowers in the autumn and winter (May, June, etc.), rather than in the summer months (Jan., Feb.), when the temperatures are normally too high.

We had hoped to study the problem of the transition from subsp. *commersonii* in the south of Entre Ríos to subsp. *malmeanum* in the north by scoring and collecting in the zone where this takes place. We found no traces of wild potatoes, though we visited specific localities for this species known to Professor A. Burkart, who is writing a flora of the province. Later, at the Instituto Darwinion, of which Professor Burkart is Director, we saw his interesting transitional or intermediate material between the two subspecies (*Burkart Nos. 22408, 24241, 24251, 24255, 24257*; *Stella Soraru Nos. 7, 137, 140*). It is to be hoped that living collections will be made from the localities of these gatherings in due course.

A visit was made to Balcarce in Buenos Aires province where *S. commersonii* is known to occur. None was growing this year because of the very dry weather, but collections had been made previously by one of us in 1958 and the material had been shown to be all good diploid *S. commersonii* subsp. *commersonii*.

4. SOLANUM CHACOENSE Bitt.

During the time of our visit *S. chacoense* was growing and flowering in the mountains of San Luis, Córdoba, Tucumán, Salta, and Jujuy, as well as in the pampas of Buenos Aires province. The season was not well enough advanced in Mendoza or on the plains of Córdoba, Santa Fé, or Entre Ríos, though in any case *S. chacoense* is not very common in those regions.

We had hoped to score populations of this species in many parts of its distribution area, to assess more precisely its pattern of variability. We were not very successful with this project and only managed to score five populations, of which two were in San Luis, two in Córdoba, and one in Salta province. Everywhere else there were insufficient plants in flower to make scoring possible, since we needed a minimum of twenty plants for each. An analysis of these results will be published elsewhere, but meanwhile some general points can be made.

In San Luis and Córdoba *S. chacoense* shows a distinct regional pattern. The plants are vigorous, with a semi-rosette, straight stem wings which vary from 0·5 to 2·0 mm, gibbous hairs on leaf rhachis, stem, and peduncles, glabrescent or glabrous calyx with acumens 0·5–1·5 mm long, and the corolla with rather broad lobes $\pm$ as broad as long. In this region *S. chacoense* develops a broader corolla lobe than in any other. In a population scored at Larca (San Luis; *H.H.R. 3196*) some of the corollas were even pentagonal to rotate, yet the material agreed with *S. chacoense* in all other characters and there seemed to be no other species present with which it could have formed hybrids.

Before visiting San Luis in 1966 we had been strongly inclined to think that *S. chacoense* in that province was forming a series of introgressed hybrids with *S. kurtzianum*, since gibbous hairs are frequent in the latter species. After our expedition we were forced to change our views on this matter, since we saw no traces of *S. kurtzianum* there; furthermore, the characters of *S. chacoense* were much more constant than we expected, and it seemed rather unlikely that this regional variant was due to introgression. Whilst not discarding the introgression hypothesis entirely we should like to put forward an alternative one that seems more possible.

The *S. chacoense* populations in San Luis and Córdoba might perhaps represent the most primitive part of that species, lying closest to the white-flowered forms of *S. kurtzianum* in Catamarca and La Rioja provinces (*H.H.O.R. 3384*, for example). Although the San Luis and Córdoba forms of *S. chacoense* are varying in the direction of such forms of *S. kurtzianum*, very markedly in the corolla shape, for instance, this does not mean to say that the two groups of forms are indistinguishable. It is interesting to postulate, nevertheless, that *S. chacoense* might have been derived from an ancestral form in the north-west provinces which also gave rise to *S. kurtzianum*, and even perhaps to *S. maglia*. During the course of its spread southwards from somewhere in the region of Catamarca province, *S. chacoense* would have retained its primitive characters in San Luis and Córdoba; it might have gradually lost the gibbous hair type as it moved further from its point of origin, becoming adapted to growth in the Chaco of north-east Argentina and Paraguay. A secondary invasion into the highlands of Tucumán, Salta, etc., gave rise to the introgressed forms we have already described.

It is significant that in San Luis and Córdoba *S. chacoense* seems to form part of the natural Espinal and Chaco Serrano vegetation, though it also of course occurs as a weed of cultivated fields. In many other provinces, on the other hand, it seems to be entirely adventive, ruderal or viatical, or to have become established exclusively as a weed of cultivated fields. More needs to be known, however, about its status in north-east Argentina and Paraguay. We also need more information on the behaviour of experimental F_1 and F_2 hybrids between *S. chacoense* and *S. kurtzianum*, and further scorings of *S. chacoense* populations from all other parts of its distribution area should be made.

Is our knowledge of *S. chacoense* sufficient for a third subspecies to be distinguished, based on the San Luis and Córdoba forms? Bukasov, of course, described such a form as the species, *S. gibberulosum*, and Correll retained it as a *forma*. We are rather doubtful whether this should be done at present until more is known about these forms.

Up to 1966 we had not seen *S. chacoense* at altitudes of more than 2350 m. On the recent expedition we made one collection above Volcán in Jujuy province at 2580 m

(*H.H.R. 3764*) and one even at 2960 m (*H.H.R. 3783*). This is, of course, exceptional. Since both were growing in old cattle enclosures we are inclined to think that their presence at those altitudes could be accounted for by the cattle eating the berries of plants at lower altitudes and voiding the seeds at these higher elevations. Other than one collection at 2400 m, at Pie de la Cuesta del Obispo (*H.H.R. 3706*), the rest of our collections were made at altitudes below 2200 m.

Only eight hybrids of *S. chacoense* with other species were recorded. Three of these were with *S. microdontum*. This is an extremely low figure when one recalls the large numbers that have been seen in the past and the fact that we made a special search for hybrids of all types. Evidently the year was a bad one for hybrids, as we have already pointed out. Two were with *S. spegazzinii*, in the Pozo de Piedra area, and three were with *S. kurtzianum*, from Mendoza in one of the localities where Brücher had collected *S. ruiz-lealii*. This species, as we have already pointed out, is undoubtedly a hybrid of *S. chacoense* × *S. kurtzianum*.

No good hybrids of *S. chacoense* × *S. tarijense* were seen, though one collection (*H.H.R. 3703*) possessed a few glands which had probably been derived by introgression from the nearby *S. tarijense* (see under that species for further discussion).

Many of the collections of *S. chacoense* were highly introgressed with *S. microdontum* genes, especially in the areas of Pampa Grande and Cuesta del Obispo (both in prov. Salta). Only one collection of the Pampa Grande material possessed style setae, and the underground parts of this collection were infected with *Spongospora*. Much of the Tucumán and Salta material possessed the crisped stem wings of *S. microdontum*, but the calyx acumens were on the whole rather short.

We saw many plants of *S. chacoense* attacked by a large brown beetle ('bicho moro'—*Epicauta* sp.), despite the alleged insect resistance of this species. However, where *S. chacoense* was growing near fields of *S. tuberosum* the cultivated potato plants were much more damaged than the wild ones.

We looked, without success, for the hybrids of *S. chacoense* × *S. vernei* subsp. *ballsii* at Termas de Reyes, Jujuy province, where a possible hybrid of this type had been collected by Garolera and Romero. *S. chacoense* and *S. microdontum* were present, but no hybrids of any type, or *S. vernei* either. There is no reason, however, why *S. vernei* might not occur in the mountains above and have been washed down to the lower level where it formed a short-lived hybrid of the type postulated.

Hybrids of *S. chacoense* × *S. spegazzinii* (= *S.* × *setulosistylum*) were scarce, as we have already mentioned. We saw none at Pampa Grande, the type locality for this type of hybrid, though the parents were present. The collection made in cultivated fields at Pampa Grande (*H.H.R. 3645*) and of which we scored twenty-five plants, was also grown at Birmingham. Some plants possessed rotate-stellate corollas, whilst in others the corolla was completely rotate. The leaflets were much narrower in this material, and gibbous hairs were present also. It seems highly likely that this collection represents *S. chacoense*, introgressed with genes from *S. spegazzinii*, and thus not very far removed from *S.* × *setulosistylum*, which was first described from Pampa Grande. None of the plants scored or those grown in Birmingham possessed style setae but, as we have seen, this seems not to be a species-specific character. It was seen in one collection of *S. chacoense* and also in *S. microdontum*. No styles could be seen on the two flowers of *S. spegazzinii* collected at Pampa Grande, and we can only hope that it will be possible to assess the character at a later date on plants of this species grown from tubers.

Although seven hybrids of *S. chacoense* × *S. spegazzinii* had been collected from below Pozo de Piedra by Sleumer and Vervoorst in 1952 we found none at all there in 1966, and the soil was extremely dry. Nevertheless, two hybrids were found in the village itself (*H.H.R. 3531, 3532*), whilst *S. chacoense* was collected lower down (*H.H.R. 3523*) in an irrigation ditch. *S. spegazzinii* was also found in the same area.

We therefore have been able to confirm that the two parents are present, as well as a reduced number of hybrids. Allowing for the slight changes in locality due to fluctuating annual differences in rainfall, irrigation management, etc., we are well satisfied that the original collections of Sleumer and Vervoorst were as we have identified them and that the hybrids could have been produced in the same region.

Hybrids of *S. chacoense* × *S. kurtzianum* (*S.* × *ruiz-lealii*) were collected in Mendoza province (*H. & R. 3122, 3124*), in and by a cultivated field, not far from the Monte vegetation where the *S. kurtzianum* parent would be expected to occur. This is at or close to the type locality for *S.* × *ruiz-lealii*. Specimens grown from tubers in Europe showed clear evidence of their hybrid nature, with thin empty anthers, white stellate-pentagonal corollas, and the *S. chacoense* inflorescence type. We also made a collection in Catamarca province at Pozo de Piedra (*H.H.R. 3530*) of what we thought was this hybrid, but a plant raised from a tuber this year was clearly *S. spegazzinii*. We are still convinced that the herbarium specimen is of the hybrid, and can only conclude that the tuber came from some other plant whose long stolons had placed it by chance in the close proximity of the *S. chacoense* × *S. kurtzianum* hybrid. The plants were obtained in a sort of 'garden' or waste place behind the village stores, where probably a great deal of wild potato hybridization is taking place.

S. chacoense

Prov. JUJUY:

Dept. Capital, Termas de Reyes, side quebrada above hotel. 24° 10′ S. 65° 28′ W. Alt. 1750 m. 7 Mar. 1966. 'In a wall near hotel where path begins to enter the side quebrada.' *Hawkes, Hjerting & Rahn 3752* (J.G.H.).

Dept. Tumbaya, Volcán, base of Loma del Tambo. 23° 55′ S. 65° 27′ W. Alt. 2150 m. 9 Mar. 1966. 'Under bushes by path near cultivated area. No fls., fruits or tubers.' *Hawkes, Hjerting & Rahn 3757* (J.G.H.).

Dept. Tumbaya, Volcán, north-western slopes of cordillera on east side of Quebrada Grande, Loma del Tambo. 23° 55′ S. 65° 25′ W. Alt. 2580 m. 9 Mar. 1966. 'Growing in an old corral amongst large stones and bushes together with 3763 (*S. microdontum*) and 3765 (*S. chacoense* × *S. microdontum*). Rich soil. Fls. white.' *Hawkes, Hjerting & Rahn 3764* (K, J.G.H., J.P.H.).

Dept. Tumbaya, Volcán, western facing slopes on east side of Quebrada Grande, path from Volcán to Abra Grande de Volcán. 23° 55′ S. 65° 25′ W. Alt. 2960 m. 9 Mar. 1966. 'One plant only, near a large cattle enclosure amongst shrubs, mixed with 3780 (*S. vernei*). Highest altitude for this species yet recorded.' *Hawkes, Hjerting & Rahn 3783* (J.G.H.).

Prov. SALTA:

Dept. Chicoana, Quebrada de Escoipe, 28·8 km from Carril, 4 km above Mal Paso. 25° 09′ S. 65° 43′ W. Alt. 1700 m. 5 Mar. 1966. 'Growing on roadside scree, together with 3699 (*S. microdontum subsp. gigantophyllum*).' *Hawkes, Hjerting & Rahn 3700* (J.G.H., J.P.H.).

Dept. Chicoana, Quebrada de Escoipe, 2·2 km above Escoipe. 25° 11′ S. 65° 48′ W. Alt. 2075 m. 5 Mar. 1966. 'Prepuna zone, on side of road below the slopes on which *S. tarijense* (3702) was found. Some few glands, but mostly good *S. chacoense*.' *Hawkes, Hjerting & Rahn 3703* (K, J.G.H., J.P.H.).

Dept. Chicoana, 7·1 km above Escoipe, Pie de la Cuesta del Obispo. 25° 10′ S. 65° 50′ W. Alt. 2400 m. 5 Mar. 1966. 'Amongst scrub, *Adesmia*, etc. Some glands; warty spots on some berries.' *Hawkes, Hjerting & Rahn 3706* (J.G.H., J.P.H.).

Dept. Grl. M. M. de Güemes, camino de La Cornisa, road between La Caldera and El Carmen, at km 1648·9. *c.* 24° 30′ S. 65° 20′ W. Alt. 1480 m. 18 Mar. 1966. 'In rain forest by road; partial shade. Wide stem wings. No fruits, flowers or tubers.' *Hawkes, Hjerting & Rahn 3910* (J.G.H., J.P.H.).

Dept. Guachipas, route 9 from Ruiz de los Llanos to Pampa Grande at 41·6 km. *c.* 26° 00′ S. *c.* 65° 27′ W. Alt. 1100 m. 2 Mar. 1966. 'By side of road. Very large plant, 1 m tall; very dark green leaf. One berry, ±oval.' *Hawkes, Hjerting & Rahn 3629* (J.G.H., J.P.H.).

Dept. Guachipas, route 9 from Ruiz de los Llanos to Pampa Grande at km 69·5. *c.* 25° 54′ S. 65° 30′ W. Alt. 1650 m. 2 Mar. 1966. 'On roadside bank. Plants to 50 cm tall; no rosettes; stem wings broad; calyx pubescent; fls. white, stellate. No style setae.' *Hawkes, Hjerting & Rahn 3633* (K, J.G.H., J.P.H.). $2n = 24$.

Dept. Guachipas, route 9 from Ruiz de los Llanos to Pampa Grande at km 71. 25° 51′ S. 65° 30′ W. Alt. 1800 m. 2 Mar. 1966. 'By road amongst grass and bushes. Wide stem wings. No style setae.' *Hawkes, Hjerting & Rahn 3635* (J.G.H.).

Dept. Guachipas, Estancia Pampa Grande, near the lake. 25° 50′ S. 65° 30′ W. Alt. 1700 m. 2 Mar. 1966. 'West side of lake, by path. Under partial shade of trees and in bare ground.' *Hawkes, Hjerting & Rahn 3638* (J.G.H., J.P.H.).

Dept. Guachipas, Estancia Pampa Grande. 25° 50′ S. 65° 30′ W. Alt. 1700 m. 3 Mar. 1966. 'Paddock no. 15, near main road. Weed of maize field; very common in bottom of field. Fls. white, stellate; some tubers; no fruits.' *Hawkes, Hjerting & Rahn 3645* (K, J.G.H., J.P.H.). Introgressed with *S. spegazzinii*?

Dept. Guachipas, Estancia Pampa Grande. 25° 50′ S. 65° 30′ W. Alt. 1750 m. 3 Mar. 1966. 'Under shade of trees in conifer plantation above the lake. Fls. white, some with style setae. Some plants with blight, others with *Spongospora*.' *Hawkes, Hjerting & Rahn 3679* (J.G.H., J.P.H.).

Dept. Guachipas, 20·3 km from Pampa Grande on road to Salta (km 1480). 25° 42′ S. 65° 30′ W. Alt. 1920 m. 4 Mar. 1966. 'Amongst stones by road, growing with 3681 (*S. microd. gig.*). Fls. very large, stellate. Highly introgressed.' *Hawkes, Hjerting & Rahn 3682* (J.G.H., J.P.H.).

Prov. TUCUMÁN:

Dept. Monteros, road to Tafí del Valle at km 25 from Acheral, above Playa Larga. 27° 03′ S. 65° 39′ W. Alt. 850 m. 14 Feb. 1966. 'Amongst grass and other herbs by roadside; rain forest. No fls. Introgressed form.' *Hawkes, Hjerting & Rahn 3461* (J.G.H., J.P.H.).

Dept. Tafí, road from Tapia to Siambón, Raco, *c.* km 24, 'Huerta La Raqueña'. 26° 40′ S. 65° 26′ W. Alt. 1150 m. 12 Feb. 1966. 'Weed of field and by hedge. Some plants to 80 cm tall, no rosettes; others short and rosetted. Stem wings crisped. Calyx glabrous. Berries few so far, globular, green.' *Hawkes, Hjerting & Rahn 3432* (K, J.G.H., J.P.H.).

Dept. Tafí, path from Raco to San José de Chasquivil, valley of Río Raco. *c.* 26° 38′ S. 65° 28′ W. Alt. 1250–1350 m. 12 Feb. 1966. 'Amongst bushes by path in former rain forest. Fls. white, no berries; crinkled stem wings with some purple pigmentation.' *Hawkes, Hjerting & Rahn 3459* (J.G.H., J.P.H.).

Prov. CATAMARCA:

Dept. Belén, road from Puerta de Belén (San José) to Pozo de Piedra, 1·8 km along road. 27° 33′ S. 67° 02′ W. Alt. 1400 m. 18 Feb. 1966. 'Hedge by road, by irrigation ditch; in shade. Plants to 1 m tall. Fls. white, stellate; no fruits. Possibly all one clone.' *Hawkes, Hjerting & Rahn 3523* (K, J.G.H., J.P.H.).

Dept. Paclín, road from Catamarca to Tucumán, 7 km from the pass towards La Viña, El Duraznillo. 28° 04′ S. 65° 35′ W. Alt. 750 m. 9 Feb. 1966. 'Waste ground near road and stream. No gibbous hairs; purple crisped wings; semi-rosettes.' *Hawkes, Hjerting & Rahn 3429* (J.G.H., J.P.H.).

Prov. LA RIOJA:

Dept. Capital (on border with Catamarca), Quebrada de Sébila, 22·3 km from Chumbicha. *c.* 28° 46′ S. 66° 21′ W. Alt. 700 m. 24 Mar. 1966. 'Chaco-serrano woodland. By side of road in damp areas where rain water has accumulated.' *Hawkes, Hjerting & Rahn 3990* (J.G.H., J.P.H.).

Dept. Capital, as 3990, but 28·6 km from Chumbicha. *c.* 28° 44′ S. *c.* 66° 22′ W. Alt. 850 m. 24 Mar. 1966. 'Chaco-serrano woodland, damp places by side of road. Fls. white, stellate.' *Hawkes, Hjerting & Rahn 3991* (J.G.H., J.P.H.).

Prov. SAN LUIS:

Dept. Chacabuco, road from Villa del Carmen to Merlo, 10 km from Villa del Carmen, near turning to Estanzuela. 32° 50′ S. 65° 02′ W. Alt. 900 m. 9 Jan. 1966. 'In maize field by road. Many plants. Semi-rosettes; fls. white, broad stellate.' *Hawkes, Hjerting & Rahn 3191* (K, J.G.H., J.P.H.).

Dept. Chacabuco, road from Villa del Carmen to Merlo, 1 km along a side road to west, to Estanzuela. 32° 50′ S. 65° 03′ W. Alt. 900 m. 9 Jan. 1966. 'In tussocks of *Cortaderia* and other grasses in a palm grove by stream. Apparently natural vegetation. Plants to 50 cm, with buds and berries.' *Hawkes, Hjerting & Rahn 3194* (K, J.G.H., J.P.H.). $2n = 24$.

Dept. Chacabuco, between Larca and Merlo, 3·8 km south of Arroyo San Miguel, E. of Balcarce. 32° 34′ S. 65° 00′ W. Alt. 1000 m. 9 Jan. 1966. 'By side of road among grass and herbs, and in bushy places behind among shrubs and trees; also as a weed in cultivated field. Fls. white, varying from stellate to rotate-pentagonal.' *Hawkes, Hjerting & Rahn 3196* (K, J.G.H., J.P.H.).

Dept. General Pedernera, Cerro del Morro, south side, Estancia 'La Morena'. 33° 12′ S. 65° 25′ W. Alt. 1300 m. 8 Jan. 1966. 'Damp places on east-facing slope of a quebrada in shade of *Cortaderia*; rich soil, rotting leaves. No fls.' *Hawkes, Hjerting & Rahn 3186* (J.G.H., J.P.H.).

Dept. General Pedernera, same locality as 3186. 'Growing in old or dead *Cortaderia* tussocks in damp ground in quebrada by stream. Plants eaten by beetles.' *Hawkes, Hjerting & Rahn 3187* (J.G.H., J.P.H.).

Dept. General Pedernera, same locality as 3186. 'Growing in old *Cortaderia* tussocks in a damp quebrada. Young plants.' *Hawkes, Hjerting & Rahn 3188* (J.G.H., J.P.H.).

Dept. La Capital, road from San Luis to Trapiche, quebrada de Los Cóndores, in a side valley on west side of road. 33° 15′ S. 66° 13′ W. Alt. 850 m. 7 Jan. 1966. 'On S. facing slope in partial shade of trees in pockets of rich soil among stones.' *Hawkes, Hjerting & Rahn 3158* (J.G.H., J.P.H.). $2n = 24$.

Dept. La Capital, road from San Luis to Trapiche, above El Dique, 22 km from San Luis in north-east direction. 33° 15′ S. 66° 11′ W. Alt. 900 m. 7 Jan. 1966. 'Weed of maize field, rare; fls. stellate, white. 3–4 plants in corner of field.' *Hawkes, Hjerting & Rahn 3164* (K, J.G.H., J.P.H.).

Dept. Pringles, Valle de Pancanta, Hostería 'Las Verbenas', in garden. 32° 53′ S. 66° 07′ W. Alt. 1600 m. 8 Jan. 1966. 'Weed. Fls. white. Plants to 30 cm, not very abundant but hoed out.' *Hawkes, Hjerting & Rahn 3173* (K, J.G.H., J.P.H.). (Tubers sent later by Sr. Joaquín Suárez, *No. 4040*.) $2n = 24$.

Dept. San Martín, road from El Trapiche to La Toma, 10 km west of turning to Las Chacras. 32° 50′ S. 65° 55′ W. Alt. 1450 m. 8 Jan. 1966. 'Growing in a maize field, very eaten by insects, but cultivated potatoes nearby even more eaten. No fls.; large tubers.' *Hawkes, Hjerting & Rahn 3180* (J.G.H.). $2n = 24$.

Prov. CORDOBA:

Dept. Colón, 3·2 km from Río Ceballos towards Salsipuedes. 31° 10′ S. 64° 20′ W. Alt. 730 m. 14 Jan. 1966. 'In Monte bajo (or Espinal?). Shade of spiny bushes and trees. Appears to be part of natural vegetation. Plants in bud, some with berries.' *Hawkes, Hjerting, Okada & Rahn 3222* (J.G.H., J.P.H.).

Dept. Colón, 10 km from Salsipuedes on road to La Falda, above Almacén La Estancia, Sierra Chica. 31° 08′ S. 64° 21′ W. Alt. 1030 m. 14 Jan. 1966. 'Very abundant weed in cultivated field, no crop present. Fls. white, substellate, lobes 10 mm long, 14 mm wide.' *Hawkes, Hjerting, Okada & Rahn 3225* (K, J.G.H., J.P.H.). (Tubers collected 27 Mar. 1966 as *No. 4016*.)

Dept. Cruz del Eje, road from Tanti to Salsacate, Río Guasta, Sierra Grande. 31° 19′ S. 64° 52′ W. Alt. 1700 m. 24 Jan. 1966. 'In field border near river, much eaten by beetles.' *Hawkes, Hjerting, Okada & Rahn 3327* (J.G.H., J.P.H.).

Dept. Punilla, near Capilla del Monte, Sierra Chica, path up Cerro Uritorco. 30° 51′ S. 64° 30′ W. Alt. 1150 m. 23 Jan. 1966. 'In shade of *Lippia*, *Schinopsis marginata* (Quebracho serrano), etc., amongst *Puya* in rich soil among stones and rocks, steep slopes.' *Hawkes, Hjerting, Okada & Rahn 3277* (J.G.H., J.P.H.).

Dept. Punilla, same locality as 3277, but Alt. 1200 m. 'In shade of *Schinopsis marginata*, *Lippia*, etc. One plant in flower; fl. white, stellate.' *Hawkes, Hjerting, Okada & Rahn 3278* (J.G.H., J.P.H.).

Dept. Punilla, same locality as 3277, but Alt. 1680 m. 'Amongst rocks and shrubs. Fls. white, stellate.' *Hawkes, Hjerting, Okada & Rahn 3285* (J.G.H., J.P.H.).

Dept. Punilla, same locality as 3277, but Alt. 1550 m. 'Fls. white stellate.' *Hawkes, Hjerting, Okada & Rahn 3297* (K, J.G.H., J.P.H.).

Dept. Punilla, same locality as 3277, but Alt. 1100 m. 'Under shade of trees. Plants to 30 cm; fls. white, stellate.' *Hawkes, Hjerting, Okada & Rahn 3303* (K, J.G.H., J.P.H.). $2n = 24$.

Dept. Punilla, road from La Falda to Cosquín, 13 km north of Cosquín, Valle Hermoso. 31° 08′ S. 64° 29′ W. Alt. 850 m. 24 Jan. 1966. 'In a maize field and amongst grass just outside. Tall plants, to 1 m in maize field. Fls. substellate to pentagonal. Herbarium from plants outside field, berries from those inside it.' *Hawkes, Hjerting, Okada & Rahn 3305* (K, J.G.H., J.P.H.).

Dept. Punilla, road from Tanti to Salsacate, turning to Los Gigantes, Sierra Grande, La Esquina. 31° 26′ S. 64° 46′ W. Alt. 1800 m. 24 Jan. 1966. 'Damp rich soil by a stream. No fls.' *Hawkes, Hjerting, Okada & Rahn 3323* (J.G.H., J.P.H.).

Dept. San Javier, 7 km from route 148, 4·3 km east of La Paz. 32° 13′ S. 65° 07′ W. Alt. 650 m. 10 Jan. 1966. 'In half Monte, half Chaco scrub, probably Espinal; with *Prosopis*, *Larrea*, *Schinus*, (?) quebracho, etc., under shade of trees and shrubs in rich soil, apparently forming part of natural vegetation.' *Hawkes, Hjerting & Rahn 3198* (J.G.H., J.P.H.).

Dept. San Javier, between Villa Dolores and Mina Clavero, 0·5 km east of Las Rosas, by a stream, in hedge by road with bushes and shrubs. 31° 56′ S. 65° 03′ W. Alt. 750 m. 10 Jan. 1966. 'Fls. white, semi-stellate. Stem 60 cm. Two plants only.' *Hawkes, Hjerting & Rahn 3203* (J.G.H., J.P.H.).

Prov. BUENOS AIRES:

Castelar, Centro Nacional de Investigaciones Agropecuarias de I.N.T.A. 34° 39′ S. 58° 37′ W. Alt. *c.* 20 m. 9 Dec. 1965. 'In a fallow field, abundant colony of plants in flower.' *Hjerting & Rahn 3026* (J.G.H., J.P.H.) (seeds sent as *No. 4039*).

187·5 km from Buenos Aires on road to Mar del Plata, 9 km north of Sevigné. 36° 08′ S. 57° 48′ W. Alt. *c.* 30 m. 1 Apr. 1966. 'Road verge, among other herbs, very abundant locally. Plants without fls. or fruits.' *Hawkes, Hjerting, Okada & Rahn 4030* (J.G.H., J.P.H.).

Buenos Aires, Hort. Bot. Fac. Agron. y Vet. 5 Apr. 1966. 'Adventive. Sandy soil.' Tubers only, given by Professor O. Boelcke. *Hawkes, Hjerting & Rahn 4049.*

S. chacoense × *S. microdontum* (subsp. intermediate)

Prov. Jujuy, dept. Tumbaya, Volcán, north-western slopes of cordillera on east side of Quebrada Grande, Loma del Tambo. 23° 55′ S. 65° 25′ W. Alt. 2580 m. 9 Mar. 1966. 'Growing in an old corral amongst stones and bushes, mixed with 3764 (*S. chacoense*) and adjacent to 3763 (*S. microdontum*), in deep rich humus.' *Hawkes, Hjerting & Rahn 3765* (J.G.H., J.P.H.).

S. chacoense × *S. microdontum* subsp. *gigantophyllum*

Prov. Salta, dept. Chicoana, route 59 from Carril to Escoipe, Quebrada de Escoipe at *c.* 22 km from Carril. 25° 08′ S. 65° 38′ W. Alt. 1550 m. 5 Mar. 1966. 'By side of road in open ground. Plants very pubescent; fls. white.' *Hawkes, Hjerting & Rahn 3695* (J.G.H., J.P.H.).

S. chacoense (× *S. microdontum* subsp. *gigantophyllum*)

Prov. Salta, dept. Chicoana, route 59 from Carril to Escoipe, Quebrada de Escoipe at *c.* 26·2 km from Carril, 1·4 km above Mal Paso. 25° 09′ S. 65° 41′ W. Alt. 1650–1700 m. 5 Mar. 1966. 'Roadside banks and in woods under trees. All types of segregates. No pure ssp. *gigantophyllum* nearby.' *Hawkes, Hjerting & Rahn 3697* (J.G.H., J.P.H.). Material grown from seed at Birmingham showed F_2 segregation.

S. chacoense × *S. spegazzinii*

Prov. Catamarca, dept. Belén, Pozo de Piedra, lane behind El Almacén. 27° 30′ S. 67° 05′ W. Alt. 1950 m. 18 Feb. 1966. 'In hedge by cattle enclosure at side of lane. Fls. very pale violet.' *Hawkes, Hjerting & Rahn 3531* (J.G.H., J.P.H.).

Prov. Catamarca, dept. Belén, Pozo de Piedra, in 'garden' of El Almacén, amongst weeds and rubbish. 27° 30′ S. 67° 05′ W. Alt. 1950 m. 18 Feb. 1966. 'Plants in rosettes, bushy; fls. white.' *Hawkes, Hjerting & Rahn 3532* (K, J.G.H., J.P.H.).

S. chacoense × *S. kurtzianum*

Prov. Catamarca, dept. Belén, Pozo de Piedra, lane behind El Almacén. 27° 30′ S. 67° 05′ W. Alt. 1950 m. 18 Feb. 1966. 'Amongst shrubs by side of lane. Fls. very pale violet, almost white.' *Hawkes, Hjerting & Rahn 3530* (p.p.) (J.G.H., J.P.H.). *Note*: Plant grown from a tuber at Birmingham was *S. spegazzinii*. See text for discussion.

Prov. Mendoza, dept. San Rafael, Punta del Agua, E. of Cerro Nevado, arroyo. 35° 32′ S. 68° 04′ W. Alt. 750 m. 25 Dec. 1965. 'Amongst *Cortaderia* and *Salix*.' *Hjerting & Rahn 3122* (K, J.G.H., J.P.H.).

As 3122, but in field. *Hjerting & Rahn 3124* (J.G.H., J.P.H.).

5. SOLANUM TARIJENSE Hawkes

This species is very rare in Argentina, and we were able to make only two collections. The first was from Escoipe (*H.H.R. 3702*) in Salta province, where we had previously seen a hybrid with *S. chacoense*, collected by Hjerting, Petersen, and Rahn in 1956 (*No. 302*). This was growing by the roadside, and when we visited the same locality in 1966 we found *S. tarijense* on a dry hillside just above, in arid Prepuna vegetation. No good hybrids of *S. chacoense* × *S. tarijense* were seen on this occasion, though one or two plants of *S. chacoense* bore glands here and there, indicative of gene flow from the very glandular *S. tarijense*.

Curiously enough, plants of *H.H.R. 3702* of *S. tarijense*, when grown at Birmingham, showed purple interpetalar blotches. This presumably indicated some past hybridizations with a blue-flowered species, perhaps *S. spegazzinii*. This latter species is also to be found in Prepuna vegetation, as well as in other plant associations, so it could be involved in contributing a gene or genes for blue corolla blotch. Furthermore, we previously identified a possible hybrid of this combination (*S. tarijense* × *S. spegazzinii*), which was collected by Ross and Vervoorst near La Viña in prov. Salta, in 1959 (*No. 60*). The La Viña locality is some 35 km from Escoipe, and in similar climatic conditions. We visited it in 1966 but at this time no rain had fallen, and the soil was dust-dry. No wild potatoes were visible, therefore, though the hybrid had certainly been collected in 1959, and we are strongly inclined to think that *S. tarijense* is to be found at La Viña, also.

At Santa Victoria, in northern Salta province, *S. tarijense* had been collected by Sleumer in 1953 and by Hjerting and Rahn in 1956, together with hybrids and introgressed forms with a blue-flowered species. Again, there had been very little rain in the *S. tarijense* area in 1966, with the consequence that we saw only one small plant (not collected), in fields where it had previously been most abundant. We were given tubers from these fields and one of these sprouted in the summer of 1966 to give a typical white-flowered plant of *S. tarijense*. The other tubers, which had not sprouted by the time of writing, may not be *S. tarijense* but *S. tuberosum*.

We are no further forward in solving the problem of which blue-flowered species has hybridized with *S. tarijense* in the Santa Victoria area. We had suggested that it might be *S. vernei*, *S. gourlayi*, or *S. vidaurrei*, all of which occur in that region but at somewhat higher altitudes. It now occurs to us that *S. tuberosum* might also be added to that list. However, there is as yet no certain evidence, and more work is needed to solve this problem.

S. tarijense

Prov. SALTA:

Dept. Chicoana, 2·2 km above Escoipe on road from Carril to Cuesta del Obispo. 25° 11′ S. 65° 48′ W. Alt. 2075 m. 5 Mar. 1966. 'Dry stony slopes above road, with *Trichocereus*, spiny shrubs etc., Prepuna zone. Plants with glands and long hairs; large tubers; two berries white-verrucose.' *Hawkes, Hjerting & Rahn 3702* (J.G.H., J.P.H.). Evidence of introgression from a blue-flowered species, possibly *S. spegazzinii*.

Dept. Santa Victoria, Santa Victoria, potreros (cultivated fields) below the hospital on edge of town. 22° 15′ S. 64° 58′ W. Alt. *c.* 2300 m. 14 Mar. 1966. 'In maize fields. Plants died down. Tubers only.' *Hawkes, Hjerting & Rahn 3858*. There may be some tubers of *S. tuberosum* mixed with this collection.

6. SOLANUM ACAULE Bitt.

We made several collections of *S. acaule* subsp. *acaule* in Jujuy, Salta, and Tucumán provinces, and also saw possible subsp. *acaule* plants mixed with subsp. *aemulans* at La Encrucijada, Sierra Famatina, La Rioja province. Owing to the generally very dry conditions this year, even at high altitudes, many *S. acaule* collections were small and stunted. Thus at Abra del Cóndor (*No. 3808*) the plants were only a few centimetres in diameter. In that area *S. megistacrolobum* and the hybrid with *S. acaule* had been recorded previously but the plants this year were too underdeveloped for us to be able to say whether they were present in 1966 also. We made living tuber collections, however. No material of subsp. *acaule* or any other wild potatoes was seen at Abra de los Pibes, another area where collections had been made in the past.

On the other hand, material of subsp. *acaule* in the rich soil of cattle enclosures was well developed, and we made many fruiting collections. We also saw large plants by irrigation channels in the Iturbe region (*No. 3904*), growing together with *S. megistacrolobum*. No hybrids were found.

An interesting collection at Quebrada Honda in Tucumán province (*No. 3487*) showed *Spongospora* attack on its roots; it grew in close proximity to *S. sanctae-rosae*, and some plants, judging by the shape of the flowers, might have been hybrids.

Unfortunately, we were not able to get up to the Sierra de Los Quilmes near Cafayate, where Castellanos had collected *S. acaule* subsp. *acaule*, *S. spegazzinii* and the hybrids between them. This is an area of difficult access and it was impossible to hire horses at the time of our visit.

Subspecies *aemulans* was seen in abundance in the region of La Encrucijada, though it was already beginning to dry up by the time of our visit in early February. A few plants of *S. spegazzinii* were seen admixed with the subsp. *aemulans* but it was impossible to identify hybrids, owing to the stunted growth of the plants this year and the lack of flowering. A few flowers had been formed earlier, and had given rise to berries, from which seed collections were made.

Unfortunately, most of the mother tubers of subsp. *aemulans* developed secondary tubers, whose dormancy has been impossible to break, up to the time of writing. A few tubers produced normal plants, all with a somatic chromosome number of $2n = 48$, and rather less rotate corollas than those of subsp. *acaule*. In fact, some corollas were quite pentagonal. The pedicel articulation was well marked, and placed only about 1·5 mm below the calyx base.

In order to relate the phenotypes of plants collected in the wild to those of the same genetic material under cultivation in Europe we devised a 'tuber indexing' method (not, of course, to be confused with that used by virus workers). Within a collection we gave a letter to each mother tuber and the same letter to the plant which was developing from it. We could, of course, use only those mother tubers that were fairly hard and which had not been separated from their plants when they were being dug. A total of thirty-one tubers was indexed, but a rather small proportion of these survived to be planted in England. Tuber indexing was also carried out for collections of *S. spegazzinii* where suspected hybridization was taking place.

By this tuber-indexing method we hope to be able to relate the phenotypes of naturally occurring and experimentally grown material rather more exactly than we have done in the past. This method is perhaps not so necessary in a fairly uniform population, where hybridization or segregation is not occurring. But where there is great variation we hope to gain more insight into the processes at work by using this method.

S. acaule subsp. *acaule*

Prov. JUJUY:

Dept. Humahuaca, between Cóndor and Abra del Cóndor, 17 km from Iturbe on road to Iruya. 22° 53′ S. 65° 16′ W. Alt. 3875 m. 11 Mar. 1966. 'Rich soil amongst grass in damp stream bed; very small plants; no fls. or fruits; often protected by stones or *Stipa*. Might be *S. megistacrolobum* and *S. acaule* × *S. megistacrolobum* present also; plants too minute to identify with accuracy.' *Hawkes, Hjerting & Rahn 3808* (J.G.H., J.P.H.).

Dept. Humahuaca, between Iturbe and Tres Cruces, near km 1316 on railway. 22° 58′ S. 65° 27′ W. Alt. 3580 m. 16 Mar. 1966. 'Edge of irrigation channel; between railway line and river. Abundant by channel; many fruits.' *Hawkes, Hjerting & Rahn 3904* (J.G.H., J.P.H.).

Dept. Yavi, Rincón de Cajas to Santa Victoria road, 2·8 km above Rincón de Cajas. 22° 14′ S. 65° 17′ W. Alt. 4125 m. 12 Mar. 1966. 'In and around a stone corral, very rich soil. Fls. blue-violet, no visible articulation.' *Hawkes, Hjerting & Rahn 3813* (K, J.G.H., J.P.H.). See Plate 13.

Prov. SALTA:

Dept. Chicoana, road from Escoipe to Cachi, 15 km above Escoipe, Cuesta del Obispo. 25° 09′ S. 65° 52′ W. Alt. 2950 m. 5 Mar. 1966. 'In very short grazed grassland, in a damp flush.' *Hawkes, Hjerting & Rahn 3717* (J.G.H., J.P.H.).

Dept. Chicoana, same locality as 3717, but 21·4 km above Escoipe. Alt. 3200 m. 6 Mar. 1966. 'In grass by roadside.' Seeds and tubers only. *Hawkes, Hjerting & Rahn 3740.*

Dept. Santa Victoria, road from Yavi to Santa Victoria at 5·2 km east of Abra de Lizoite, bottom of a valley. 22° 10′ S. 65° 13′ W. Alt. 4400 m. 12 Mar. 1966. 'In stones by road and amongst short grass. Flat rosettes; also in a corral. Some plants with enlarged terminal leaflets on older leaves; berries yellow when ripe.' *Hawkes, Hjerting & Rahn 3814* (J.G.H., J.P.H.).

Dept. Santa Victoria, above Rodeopampa, 13 km from Santa Victoria on road to La Quiaca. 22° 15′ S. 65° 04′ W. Alt. 3200 m. 15 Mar. 1966. 'By roadside and on road itself. Berries. Puna formation.' *Hawkes, Hjerting & Rahn 3872* (J.G.H., J.P.H.).

Prov. TUCUMÁN:

Dept. Tafí, near Infiernillo, Quebrada Honda, above km 83 on road from Tafí del Valle to Cafayate, above the little house by the road. 26° 44′ S. 65° 44′ W. Alt. 3000–3270 m. 15 Feb. 1966. 'Growing in short grass. No fls. Flat rosettes; berries.' *Hawkes, Hjerting & Rahn 3482* (J.G.H.).

Dept. Tafí, Quebrada Honda, above km 83 on road from Tafí del Valle to Cafayate. 26° 44′ S. 65° 44′ W. Alt. 3550 m. 15 Feb. 1966. 'Place where horses and cattle shelter, under a very large rock; very rich soil; some plants attacked by *Spongospora subterranea* on roots, etc., canker form. Fls. dark purple, rather pentagonal. Part may be possible *S. acaule* × *S. sanctae-rosae* hybrids.' *Hawkes, Hjerting & Rahn 3487* (J.G.H., J.P.H.).

S. acaule subsp. *aemulans*

Prov. LA RIOJA:

Dept. Famatina, Sierra Famatina, La Encrucijada, just above, in a small valley. 28° 58′ S. 67° 42′ W. Alt. 3250–3300 m. 5 Feb. 1966. 'Dry shaly slope under protection of *Adesmia*, *Senecio*, etc. Some plants with fruits.' *Hawkes, Hjerting, Okada & Rahn 3396* (p.p.) (J.G.H., J.P.H.). (One specimen was *S. spegazzinii*.) $2n = 48$.

Dept. Famatina, Sierra Famatina, valley of Río Amarillo above La Encrucijada. 28° 58′ S. 67° 42′ W. Alt. 3250–3350 m. 5 Feb. 1966. 'Dry stony and sandy hillsides, occasionally on flat ground where cattle have stood.' *Hawkes, Hjerting, Okada & Rahn 3407* (p.p.) (J.G.H., J.P.H.). (Two specimens were *S. spegazzinii*.) $2n = 48$.

Dept. Famatina, same locality as 3407, but in a side quebrada leading off from La Encrucijada. Alt. 3450 m. 5 Feb. 1966. 'In sandy soil in sun, very dry; plants wilting.' *Hawkes, Hjerting, Okada & Rahn 3408* (p.p.) (J.G.H., J.P.H.). (Two specimens were *S. spegazzinii*.)

7. SOLANUM INFUNDIBULIFORME Phil.

There was very poor development of this dry Puna and Prepuna species in the dry season of 1966, and no hybrids were found in the regions where they had been recorded previously. For instance, we saw no hybrids with *S. megistacrolobum* between Humahuaca and Tres Cuces (see *H.P.R. 76*), though *S. infundibuliforme* itself was found near by (*H.H.R. 3811*). Nor did we find this same hybrid at Cajas (*H.P.R. 111*), though there we did find *S. megistacrolobum* (*H.H.R. 3823*). At Abra de los Pibes, where Sleumer had collected *S. infundibuliforme* × *S. gourlayi* and several other species, we found nothing whatsoever! We collected a possible hybrid of *S. infundibuliforme* × *S. megistacrolobum* between Iturbe and Tres Cruces (*No. 3905*) which, unfortunately, had no berries or tubers.

Our best collection of this species was made between Iturbe and Tres Cruces at km 1313·5 by the railway line, where it was growing abundantly and bearing many ripe fruits. Near by was *S. oplocense*, whilst one or two plants of *S. megistacrolobum* occurred in the *S. infundibuliforme* population.

It is quite evident that although *S. infundibuliforme* grows in dry regions and may be drought resistant up to a certain point, the dry season of 1966 was too arid for it, resulting in reduced and stunted growth, and a great decimation of hybrids.

S. infundibuliforme

Prov. JUJUY:

Dept. Humahuaca, between Cóndor and Abra del Cóndor, 15 km on road from Iturbe to Iruya. 22° 53′ S. 65° 16′ W. Alt. 3750 m. 11 Mar. 1966. 'In dry soil amongst bushes by road in stream bed; Puna formation, very dry. Two very small plants.' *Hawkes, Hjerting & Rahn 3802* (J.G.H.).

Dept. Humahuaca, road from Humahuaca to Tres Cruces at km 27 from Tres Cruces. *c.* 22° 59′ S. 65° 25′ W. Alt. 3850 m. 11 Mar. 1966. 'In disturbed soil and amongst stones by the roadside, not to be seen in open puna. Fls. mauve, white-edged. *S. acaule* also seen here, but not collected.' *Hawkes, Hjerting & Rahn 3811* (J.G.H., J.P.H.).

Dept. Humahuaca, between Iturbe and Tres Cruces, by railway at km 1313·5. 22° 57′ S. 65° 23′ W. Alt. 3550 m. 16 Mar. 1966. 'Dry shaly banks above railway. No fls. Mature berries frequent; berries purplish brown with whitish spots. (Possibly *S. megistacrolobum* here too.)' *Hawkes, Hjerting & Rahn 3906* (J.G.H., J.P.H.).

S. infundibuliforme × *S. megistacrolobum*?

Prov. JUJUY:

Dept. Humahuaca, between Iturbe and Tres Cruces, at km 1315·4 on railway line. 22° 58′ S. 65° 27′ W. Alt. 3570 m. 16 Mar. 1966. 'Growing on the ballast by the track. Small rosettes; no fls. or tubers.' *Hawkes, Hjerting & Rahn 3905* (J.G.H., J.P.H.).

8. SOLANUM MEGISTACROLOBUM Bitt.

Seven collections of this species were made, and it was observed in a number of other places as well. To the south of Pampa Grande in Salta province (25° 50′ S.) *S. megistacrolobum* disappears, and is replaced quite sharply by *S. sanctae-rosae*. No intergrading whatsoever seems to occur between these two species, even though they are geographical vicariads and apparently closely related.

We have mentioned that no hybrids of *S. megistacrolobum* with *S. acaule* were distinguished, though the two species were seen growing together. Neither did we see the hybrids of *S. megistacrolobum* × *S. microdontum* at Loma del Tambo, Volcán, which we had identified from herbarium specimens of Schreiter (*No. 2895*) and Venturi (*No. 9536*). This is a very large mountain ridge, and we might easily have missed both *S. megistacrolobum* and the hybrid. We did, of course, collect *S. microdontum* there, and we also collected material provisionally identified as *S. megistacrolobum* × *S. microdontum* subsp. *microdontum* between Santa Victoria and La Quiaca (*H.H.R.*

3881). Seedlings of this, when grown at Birmingham, showed long spreading hairs similar to those of *S. megistacrolobum* (*No. 3829*), and the leaf shape was also very close in both families. The flower colour has not yet been observed. Mature plants of *S. megistacrolobum* with ripe berries and completely simple leaves were collected at Pampa Grande in Salta province (*H.H.R. 3939*). We at first thought that these might be *S. sanctae-rosae*, but when we examined them further and raised a family of seedlings we found that they matched *S. megistacrolobum* very closely. Pampa Grande is thus the most southerly locality so far recorded for this species, and at a much lower altitude than normal. More work is needed on this collection and on the hybrids of *S. microdontum* subsp. *gigantophyllum* (*H.H.R. 3647, 3929*), which may possibly contain *S. megistacrolobum* germ plasm. (See p. 477.)

In 1956 Hjerting and Rahn had collected hybrids of *S. megistacrolobum* × *S. boliviense* above Rodeopampa (*No. 246*) and Sleumer had also gathered similar plants in 1953 (*No. 4040*). We found no hybrids of this type in 1966, though one of our collections from higher up, on the same road (*H.H.R. 3829*), looked as though it might be showing some influence from *S. boliviense*.

No hybrids of *S. megistacrolobum* × *S. gourlayi* or *S. oplocense* were found between Cóndor and Abra del Cóndor though they were collected by Petersen and Hjerting in 1952 (*P. & H. P–366*). For *S. megistacrolobum* also, our general impressions have been that 1966 was a year of purification, so far as species and hybrids of potatoes were concerned.

S. megistacrolobum

Prov. JUJUY:

Dept. Humahuaca, Abra del Cóndor, between Iturbe and Iruya, top of pass, 21·5 km from Iturbe. 22° 53′ S. 65° 15′ W. Alt. 4000 m. 11 Mar. 1966. 'Amongst a pile of large stones. Dry puna. No fruits; one poor flower; plants eaten down by animals.' *Hawkes, Hjerting & Rahn 3805* (J.G.H., J.P.H.).

Dept. Yavi, Rincón de Cajas, on road from Yavi to Santa Victoria. 22° 13′ S. 65° 18′ W. Alt. 4000 m. 12 Mar. 1966. 'Small plants growing under stones in a quarry above the police house.' *Hawkes, Hjerting & Rahn 3823* (J.G.H., J.P.H.).

Prov. SALTA:

Dept. Chicoana, road from Escoipe to Cachi, Cuesta del Obispo, 17·5 km above Escoipe. 25° 09′ S. 65° 52′ W. Alt. 3000 m. 5 Mar. 1966. 'In pajonales of *Stipa*, but ± open ground. Simple leaves, dark purple backs. One plant in bud.' *Hawkes, Hjerting & Rahn 3720* (J.G.H.).

Dept. Chicoana, same locality as 3720, but 21·4 km above Escoipe. Alt. 3200 m. 5 Mar. 1966. 'In grass by roadside.' *Hawkes, Hjerting & Rahn 3727* (J.G.H., J.P.H.).

Dept. Guachipas, Estancia Pampa Grande, western mountain range, between Cerro Cristal and Cerro Pirgua, on a ridge, north-facing slope. 25° 50′ S. 65° 34′ W. Alt. 2650–2670 m. 20 Mar. 1966. 'Stony grassy slope, dry soil. Ripe berries. Some plants still green, one with faded purple flower.' *Hawkes, Hjerting & Rahn 3939* (J.G.H., J.P.H.).

Dept. San Carlos, road from Cachipampa to Amblayo, 7·5 km from the turning. 25° 17′ S. 65° 54′ W. Alt. 3350 m. 6 Mar. 1966. 'In small *Verbena* bushes on puna, with *Stipa*, etc. Sandy soil. No fls.' *Hawkes, Hjerting & Rahn 3734* (J.G.H., J.P.H.).

Dept. Santa Victoria, road from Yavi to Santa Victoria at 16–16·7 km from Santa Victoria. 22° 15′ S. 65° 05′ W. Alt. 3450–3500 m. 13 Mar. 1966. 'Wet puna; by road amongst herbs and stones. Fls. deep purple; berries globular, slightly white-spotted.' *Hawkes, Hjerting & Rahn 3829* (J.G.H., J.P.H.).

S. megistacrolobum × *S. microdontum*?

Prov. SALTA:

Dept. Santa Victoria, 16 km from Santa Victoria on road to La Quiaca. 22° 15′ S. 65° 05′ W. Alt. 3450 m. 15 Mar. 1966. 'Amongst bushes in a small quebrada.' *Hawkes, Hjerting & Rahn 3881* (J.G.H., J.P.H.).

9. SOLANUM BOLIVIENSE Dun.

We saw no material of *S. boliviense* on the 1966 expedition. The original Argentine collections were made at Cuesta de San José, between Santa Victoria and Toldos, and we unfortunately had no time to visit that area. We have already mentioned that our collection *No. 3829* showed some influence of *S. boliviense* possibly. This will be verified by growing plants from seeds. No obvious hybrids of *S. megistacrolobum* × *S. boliviense* were seen, however.

10. SOLANUM SANCTAE-ROSAE Hawkes

We were pleased to see that *S. sanctae-rosae* was always manifestly quite distinct from its nearest relative, *S. megistacrolobum*. The latter grew as rosettes or with a straggling decumbent stem, whereas *S. sanctae-rosae* formed neat little bushy plants with erect stems and stiff leaves pointing obliquely upwards. The pointed, acuminate leaflets and rich purple corollas were also most characteristic.

We made four good collections at Quebrada Honda, near Infiernillo, above Tafí in Tucumán province. Here the plants were very typical, and only one possible hybrid with *S. acaule* was seen. (See *S. acaule*, subsp. *acaule*, *H.H.R. 3487*.)

Hybrids of *S. sanctae-rosae* × *S. venturii* were seen in Estancia Las Pavas (*H.H.R. 3590*, *3591*), a classic area for them, where Meyer, Venturi, and Petersen and Hjerting had also obtained similar collections at various times. We found no *S. sanctae-rosae* at Mina Capillitas, prov. Catamarca, despite the fact that it had been collected there by Sleumer in 1951; hybrids with *S. spegazzinii* from the same area were also not found, since in 1966 this region was extremely dry.

S. sanctae-rosae

Prov. TUCUMÁN:

Dept. Tafí, near Infiernillo, Quebrada Honda, above km 83 on road from Tafí del Valle to Cafayate, above the little house by the road. 26° 44′ S. 65° 44′ W. Alt. 3270 m. 15 Feb. 1966. 'In open ground amongst tussock grasses. Most plants not in flower; one with dark purple fls. Very common.' *Hawkes, Hjerting & Rahn 3483* (J.G.H., J.P.H.).

Dept. Tafí, near Infiernillo, Quebrada Honda, above km 83 on road from Tafí del Valle to Cafayate. 26° 44′ S. 65° 44′ W. Alt. 3300 m. 15 Feb. 1966. 'In open soil by side of path. No fls. or fruits; seen only in one place.' *Hawkes, Hjerting & Rahn 3485* (J.G.H., J.P.H.).

Dept. Tafí, same locality as 3483, but alt. 3550 m. 'Place where horses and cattle shelter, under a very large rock; very rich soil. Berries.' *Hawkes, Hjerting & Rahn 3488* (K, J.G.H., J.P.H.).

Dept. Tafí, same locality as 3483, but alt. 3200 m. 'Growing in short grass.' *Hawkes, Hjerting & Rahn 3496* (J.G.H.).

S. sanctae-rosae × *S. venturii*

Prov. TUCUMÁN:

Dept. Chicligasta, Quebrada del Río Las Pavas, between Puesto La Cascada and Las Cuevas. 27° 11′ S. 65° 58′ W. Alt. 2800–3000 m. 24 Feb. 1966. 'In Pajonales, amongst large stones. Plants 20–30 cm; fls. violet and violet with white or almost white.' *Hawkes, Hjerting & Rahn 3590* (J.G.H., J.P.H.).

Dept. Chicligasta, same locality as 3590. 'Pajonales, amongst large stones. No flowers. Mixed with plants of *S. venturii*.' *Hawkes, Hjerting & Rahn 3591* (J.G.H., J.P.H.).

11. SOLANUM MICRODONTUM Bitt.

We made six collections of subsp. *microdontum* in northern Salta, from 7 to 9·3 km above Santa Victoria on the road to La Quiaca. Most of the plants were in flower, and many berries were being produced. The plants were growing amongst bushes and in the open, not in woods or forests. Consequently, they were more well developed than usual, with tall stems, to 1 cm diameter and *straight* wings, up to 2·5 mm broad. In collection number *3863* some plants possessed dark pedicels and calyces, comparing closely with the Sleumer collections *4037* and *4038* from the same locality, made in 1953. Another Sleumer collection, *3752*, possessed the brown or dark calyx as well as a stiff upright growth and narrow leaves. We did not find plants with this combination of characters, but the leaves of some plants of our collection *3830* were very narrow. We suggested previously that the stiff growth and narrow leaves, as well as the brown calyx, might have been due to introgression with *S. boliviense*. We saw nothing in 1966 to prove or disprove that hypothesis. We have already mentioned that in 1966 no *S. boliviense* was seen.

The slightly mauve-tinted flowers and greenish-purple leaves and stems of collection *3869* indicated introgression, possibly from *S. vernei* subsp. *ballsii*. A possible hybrid with what we believe to be a new, undescribed species was also collected (*3879*).

Leaf dissection varied here, as it does in the type locality at Toldos. Three of our 1966 collections possessed simple leaves (*3863*, *3866*, *3871*), whilst others had lateral leaflets (*3868*, *3869*). In collection *3830* some leaves were simple and others 1-jugate.

All four collections of *S. microdontum* from Jujuy, dept. Tumbaya, Volcán (*3763*, *3771*, *3777*, *3781*) proved to be intermediate between subsp. *microdontum* and subsp. *gigantophyllum*. They possessed a wide and slightly crisped stem wing, and in general matched quite well the previously collected intermediates which we have already noted. The dimensions of these plants were greater than those of subsp. *microdontum*, at least in their leaves; and the crisped stem wings were well marked, though this character was not developed to such an extreme extent as in subsp. *gigantophyllum*.

No hybrids with *S. megistacrolobum* were seen in 1966 at Volcán, as we have already mentioned, though a well-marked hybrid with *S. chacoense*, occurring with both parents, has been noted. We also collected a hybrid with *S. vernei* subsp. *ballsii* (*No. 3782*) in a large cattle enclosure, where both parents were abundant. This was not completely intermediate between the two parents; its leaves were those of *S. microdontum*, but the flowers were white, tinted mauve. It probably represented a backcross or F_2 segregate, rather close to the *S. microdontum* parent. When one considers that both parent species were growing in great profusion very close together it seems at first sight astonishing that more hybrids are not present. However, the *S. vernei* seemed to have flowered much earlier than the *S. microdontum*, so that differences of flowering time may be an effective mechanism for isolation of these two species. Since no simple intermediate hybrids were present we may assume, perhaps, that when such hybrids are formed they are less well adapted than the parents to growth under extreme conditions, and so only back-crossed or introgressed types which possess the genetic background of one or the other parents, but not a mixture of both, are able to survive for long.

Subsp. *gigantophyllum* seems to be widespread, and abundant in those areas in which it has become established. We made twenty-nine collections, from Jujuy, Salta, Tucumán, Catamarca, and La Rioja provinces. The collections from La Rioja were particularly exciting, since *S. microdontum* had not previously been found so far south. This region of La Rioja, in the southern end of Sierra Famatina, is quite outside the selva Tucumáno-Boliviana, and the plants were growing here in damp valleys and amongst rocks and bushes, though occasionally under planted trees. It is of interest to note that *S. microdontum* has been seen in prov. Catamarca, dept. Belén, above Pozo de Piedra, also in regions where the selva does not occur, probably in similar situations to where it is found in La Rioja. We also saw it frequently in open bushy places in Tucumán and Salta provinces, generally in areas where the original forest cover had been cleared.

This subspecies conformed well to our previous concept of it, with thick stems, wide undulate and crisped stem wings and dense pubescence of shining hairs on all green parts.

Although *S. microdontum* normally grows in forests or in scrub and undergrowth, it is also to be found in open places by roadsides in areas of high rainfall. We saw it growing in a maize field at Pozo de Piedra, though this is exceptional, since the species does not seem to possess such 'weedy tendencies' as, for example, *S. chacoense*. In 1966 where it grew in forest or in deep shade it was not flowering very freely. Blossoming was starting by the middle of February, or in some cases a little earlier, whilst by mid March, most plants were bearing fruits. In the grassland areas above the forests ('pajonales') at about 2450 to 2700 m *S. microdontum* subsp. *gigantophyllum* seemed to be growing quite well. The plants were smaller than those in the forests, with short internodes and smaller leaves, but the crisped undulate stem wings were quite prominent.

Plants growing in the rain forests in many places, and especially in Quebrada de San Lorenzo (Salta province) were very stunted and generally without flowers (*3690*, *3693*). Several seemed to be suffering from excessive moisture and lack of light. No *S. chacoense* or hybrids were seen here this year. The same situation was evident at Camino de La Cornisa, between Salta and Jujuy, where all the subsp. *gigantophyllum* plants were very small, and few bore flowers. Only one *S. chacoense* plant was seen in 1966 but this was highly introgressed. No hybrids were present.

At Cuesta del Obispo (Salta) and Pampa Grande (Salta) conditions for growth seemed better. The plants were in flower by the beginning of March and some berries were forming. Abundant long setae were seen on the styles of one collection (*3634*) on the road from Ruíz de los Llanos to Pampa Grande. A few shorter setae were seen on some plants only at and near Ojo de Agua in the same region (*3680*, *3683*). Style setae had also been seen in this region in 1956 by Hjerting, Petersen, and Rahn, but only on *S. chacoense*. We also saw *S. chacoense* with style setae, and one of Spegazzini's plants of *S. microdontum* subsp. *gigantophyllum* exhibits them. They are, of course, known for *S.* × *setulosistylum*, a hybrid of *S. chacoense* × *S. spegazzinii*. We were not able to find setae on the styles of any other plants of *S. microdontum* in this or any other area, and the gene appears to be very localized, therefore.

It is interesting to note that although *S. microdontum* is not considered to be epiphytic, we observed one series of plants (*3537*) at Estancia 'Yunka Suma' growing on a branch of *Sambucus peruviana*.

Hybrids of *S. microdontum* subsp. *gigantophyllum* with *S. vernei* were seen at Pampa Grande (Salta province), growing with both parents, in the open grassy regions above the forests (*No. 3927*). They were also found in another place in the same region, together with subsp. *gigantophyllum* (*3940*). One collection was made, also, in prov. Tucumán, dept. Chicligasta (*3588*).

A hybrid population of subsp. *gigantophyllum* × *S. vernei* (probably intermediate between the two subspecies) was collected at Cuesta del Obispo. All intermediates between the two parents could be seen (*3711*), and the parents themselves were collected in the vicinity (gig: *3713*. vrn: *3714*).

A hybrid between *S. microdontum* (intermediate subspecies) and *S. vernei* subsp. *ballsii* was collected in prov. Jujuy, above Volcán (*3782*). A family of seedlings from this collection was raised in 1966 and showed clear genetic breakdown, with a range in morphology also from one parent to the other.

Perhaps our most interesting discovery was the hybrid of *S. microdontum* subsp. *gigantophyllum* × *S. kurtzianum*. This was found growing in the southern part of Sierra Famatina, in La Rioja province, not far from Chilecito. The hybrid occurred both at Guanchín Viejo and in the Mina de Oro valley, in areas where *S. kurtzianum* spread up through Monte vegetation into the Prepuna; here it came into contact with subsp. *gigantophyllum*, which descended from the scrub and grassland zone above. The few specimens we had seen previously of this hybrid had led us to suppose that it was a good species, for which we had proposed the name, *S. rechei*.

After a further examination of the *S. kurtzianum* material from the area mentioned we found to our surprise that much of it was involved in hybridization with *S. spegazzinii*, which has also been recorded in the same region. (See below.) We must, therefore, be prepared to concede that *S. rechei*, although mainly composed of *S. microdontum* × *S. kurtzianum* hybrids, may also contain some *S. spegazzinii* germplasm.

A very large population (which might have represented one or a very few clones) at Finca 'El Trapiche', on the road to Mina de Oro, had been sampled previously by Reche and material sent to Professor H. Ross (E.B.S. 2082). We took two tuber collections (*4008*, *4009*) and one herbarium and berry collection (*4007*) from these plants. The latter collection showed some virus mottle but the tubers were taken from parts of the population that seemed to be healthy.

The plants were quite abundant this year, but had been even more so in the year Reche collected them (personal communication). All the material of *S.* × *rechei* seemed to have produced remarkably large tubers, especially our collection *3373*.

At Pampa Grande, near the summit of Cerro Cristal, in Salta province, we made two collections of hybrids whose composition has not yet been determined with certainty (*3647*, *3929*). One of the parents is undoubtedly *S. microdontum* subsp. *gigantophyllum*, whilst the other may be *S. vernei* or *S. megistacrolobum*. Tuber material of *3647* when grown in Europe in 1966 gave plants smaller than subsp. *gigantophyllum*, though the vegetative characters were very close. The flowers, however, were large and showy, rather rotate, and much paler inside than outside. This material is still being studied.

Above Santa Victoria in Salta province we collected what appears to be a hybrid of subsp. *microdontum* with an as yet undescribed species. We shall mention this new species again below. (See p. 493.)

S. microdontum subsp. *microdontum*

Prov. SALTA:

Dept. Santa Victoria, road from La Quiaca to Santa Victoria, 7·2 km from Santa Victoria, between La Huerta and Rodeopampa. 22° 15′ S. 65° 02′ W. Alt. 2830 m. 13 Mar. 1966. 'Amongst bushes by side of road. Stem in largest plants to 10 mm diam.; wings straight, to 2·5 mm; fls. white; berries.' *Hawkes, Hjerting & Rahn 3830* (J.G.H., J.P.H.).

Dept. Santa Victoria, same locality as 3830, but 7 km from Santa Victoria and alt. 2725 m. 15 Mar. 1966. 'Below road, amongst bushes and stones. Stems branched; calyx and pedicels dark in some plants; fls. white; long style; many berries.' *Hawkes, Hjerting & Rahn 3863* (K, J.G.H., J.P.H.).

Dept. Santa Victoria, same locality as 3863, but 8·1 km from Santa Victoria and alt. 2850 m. 22° 15′ S. 65° 03′ W. 'By roadside; one plant; one berry.' *Hawkes, Hjerting & Rahn 3866* (J.G.H., J.P.H.).

Dept. Santa Victoria, same locality as 3863, but 9·3 km from Santa Victoria and alt. 2950 m. Just below Rodeopampa. 'Amongst bushes at point where river crosses road.' *Hawkes, Hjerting & Rahn 3868* (J.G.H., J.P.H.).

Dept. Santa Victoria, just below Rodeopampa, 9·3 km from Santa Victoria on road to La Quiaca. 22° 15′ S. 65° 03′ W. Alt. 2950 m. 15 Mar. 1966. 'In a wall by the road. Fls. slightly mauve-tinted; leaves and stems dark greenish-purple. Introgressed form.' *Hawkes, Hjerting & Rahn 3869* (K, J.G.H., J.P.H.).

Dept. Santa Victoria, same locality as 3869. By road. Plants very tall; leaves simple; fls. white.' *Hawkes, Hjerting & Rahn 3871* (J.G.H., J.P.H.).

S. microdontum subsp. *intermediate*

Prov. JUJUY:

Dept. Tumbaya, Volcán, north-western slopes of cordillera on east side of Quebrada Grande, Loma del Tambo. 23° 55′ S. 65° 25′ W. Alt. 2580 m. 9 Mar. 1966. 'Growing in an old corral amongst large stones and bushes, together with 3764 (*S. chacoense*) and *3765* (*S. chacoense* × *S. microdontum*). Rich soil. Fls. white.' *Hawkes, Hjerting & Rahn 3763* (J.G.H., J.P.H.).

Dept. Tumbaya, same locality as 3763, but alt. 2850 m. 'In pajonales.' *Hawkes, Hjerting & Rahn 3771* (J.G.H.).

Dept. Tumbaya, Volcán, western slopes of cordillera on east side of Quebrada Grande, path from Loma del Tambo to Abra Grande de Volcán. 23° 55′ S. 65° 25′ W. Alt. 3080 m. 9 Mar. 1966. 'Amongst bushes on a flat mesada where cattle have stood. Rich soil. Fls. and fruits.' *Hawkes, Hjerting & Rahn 3777* (J.G.H., J.P.H.).

Dept. Tumbaya, same locality as 3777, but alt. 2960 m. 'In a large cattle enclosure and amongst shrubs around it. Mixed with 3780 (*S. vernei*) and 3782 (*S. microdontum* × *S. vernei*). Fruits.' *Hawkes, Hjerting & Rahn 3781* (J.G.H., J.P.H.).

S. microdontum subsp. *gigantophyllum*

Prov. JUJUY:

Dept. Capital, Termas de Reyes, side quebrada above hotel. 24° 10′ S. 65° 28′ W. Alt. 1750–1850 m. 7 Mar. 1966. 'In scrub and amongst bushes and stones in bottom of quebrada. Damp humiferous soil.' *Hawkes, Hjerting & Rahn 3753* (J.G.H.).

Prov. SALTA:

Dept. Capital, Quebrada de San Lorenzo. 24° 45′ S. 65° 28′ W. Alt. 1500 m. 4 Mar. 1966. 'Under shade of *Phoebe porphyria*, *Alnus* etc. Rain forest; very rich damp soil. Lvs. simple, with cell burst, indicating too much rain; plants frequent but not well developed in general. Very few fls.' *Hawkes, Hjerting & Rahn 3690* (J.G.H.).

Dept. Capital, same locality as 3690, but alt. 1450 m. 'By car park at end of road, border of forest. Leaves lobed. No fls.' *Hawkes, Hjerting & Rahn 3693* (J.G.H.).

Dept. Chicoana, Quebrada de Escoipe, route 59 from Carril to Escoipe, 28·8 km from Carril and 4 km above Mal Paso. 25° 09′ S. 65° 43′ W. Alt. 1700 m. 5 Mar. 1966. 'Growing on roadside scree, also with 3700 (*S. chacoense*).' *Hawkes, Hjerting & Rahn 3699* (J.G.H., J.P.H.).

Dept. Chicoana, road from Carril to Cuesta del Obispo, Pie de la Cuesta, 7·1 km above Escoipe. 25° 10′ S. 65° 50′ W. Alt. 2400 m. 5 Mar. 1966. 'Amongst scrub by road—top of Prepuna, bottom of Pajonales. Leaves 2-jugate; poor stem wings. Introgressed?' *Hawkes, Hjerting & Rahn 3705* (J.G.H.).

Dept. Chicoana, same locality as 3705. 'Amongst scrub by road. Simple leaves.' *Hawkes, Hjerting & Rahn 3707* (J.G.H., J.P.H.).

Dept. Chicoana, road from Carril to Cachi, Cuesta del Obispo, 9·8 km above Escoipe. 25° 09′ S. 65° 52′ W. Alt. 2550 m. 5 Mar. 1966. 'Amongst bushes by side of road (+influence of *S. vernei*?).' *Hawkes, Hjerting & Rahn 3710* (J.G.H., J.P.H.).

Dept. Chicoana, same locality as 3710, but 11·6 km above Escoipe, 4·6 km above Pie de la Cuesta, where the road crosses a side quebrada on the right. 25° 09′ S. 65° 52′ W. Alt. 2700 m. 5 Mar. 1966. 'In the side quebrada, growing in Pajonales, above the mixture of *S. vernei*/*S. microdontum* (3711) and just above *S. venturii* (3712).' *Hawkes, Hjerting & Rahn 3713* (J.G.H., J.P.H.).

Dept. Guachipas, route 9 from Ruiz de los Llanos to Pampa Grande at km 71 from Ruiz de los Llanos. 25° 51′ S. 65° 30′ W. Alt. 1800 m. 2 Mar. 1966. 'By road amongst grass and bushes. Long STYLE SETAE!' *Hawkes, Hjerting & Rahn 3634* (J.G.H.).

Dept. Guachipas, Estancia Pampa Grande, Cerro Cristal. 25° 50′ S. 65° 35′ W. Alt. 2600 m. 3 Mar. 1966. 'Pajonales, open grassland. Fls. white. Some immature berries.' *Hawkes, Hjerting & Rahn 3663* (K, J.G.H., J.P.H.).

Dept. Guachipas, same locality as 3663, but alt. 2520 m. 'In Pajonales.' *Hawkes, Hjerting & Rahn 3667* (J.G.H., J.P.H.).

Dept. Guachipas, Estancia Pampa Grande, path to Cerro Cristal. 25° 50′ S. 65° 33′ W. Alt. 2380 m. 3 Mar. 1966. 'In *Alnus jorullensis* wood. Fls. white; styles glabrous.' *Hawkes, Hjerting & Rahn 3672* (J.G.H.).

Dept. Guachipas, same locality as 3672, but alt. 2200 m. 'Amongst low shrubs and grasses. Fls. white.' *Hawkes, Hjerting & Rahn 3677* (J.G.H.).

Dept. Guachipas, 19·2 km on road from Pampa Grande to Salta. 25° 42′ S. 65° 30′ W. Alt. 1900 m. 4 Mar. 1966. 'Amongst bushes by road. Fls. white. STYLE SETAE on some plants.' *Hawkes, Hjerting & Rahn 3680* (J.G.H., J.P.H.).

Dept. Guachipas, same locality as 3680, but 20·3 km from Pampa Grande (km 1480) and alt. 1920 m. 'Amongst stones and bushes by road, growing with 3682 (*S. chacoense*, introgressed). No style setae; berries.' *Hawkes, Hjerting & Rahn 3681* (J.G.H., J.P.H.).

Dept. Guachipas, 21·2 km from Pampa Grande on road to Salta, Ojo de Agua. 25° 42′ S. 65° 30′ W. Alt. 1950 m. 4 Mar. 1966. 'Amongst bushes by road. (*S. chacoense* nearby). Fls. and berries.' *Hawkes, Hjerting & Rahn 3683* (J.G.H., J.P.H.).

Dept. Guachipas, Estancia Pampa Grande, path to Cerro Cristal. 25° 50′ S. 65° 33′ W. Alt. 2300 m. 20 Mar. 1966. 'By path in *Alnus jorullensis* wood. Berries only.' *Hawkes, Hjerting & Rahn 3919*.

Dept. Guachipas, Estancia Pampa Grande, western mountain range, below Cerro Cristal. 25° 50′ S. 65° 34′ W. Alt. 2450 m. 20 Mar. 1966. 'Amongst bushes and tall grass in a small quebrada by path to Cerro Cristal. Growing with 3926 (*S. vernei*) and 3927 (*S. mcd. gig.* × *S. vernei*).' *Hawkes, Hjerting & Rahn 3928* (J.G.H.).

Dept. Guachipas, Estancia Pampa Grande, western mountain range, south-east slope of Cerro Cristal. 25° 50′ S. 65° 34′ W. Alt. 2500 m. 20 Mar. 1966. 'Pajonales vegetation, amongst tufts of *Stipa*; very common at this altitude.' *Hawkes, Hjerting & Rahn 3944* (J.G.H., J.P.H.).

Prov. TUCUMÁN:

Dept. Chicligasta, Quebrada del Río Las Pavas, near Puesto La Cascada. 27° 11′ S. 65° 57′ W. Alt. 2600 m. 24 Feb. 1966. 'Amongst rocks in Pajonales. Plant 70 cm high.' *Hawkes, Hjerting & Rahn 3573* (J.G.H., J.P.H.).

Dept. Famaillá, road to Tafí del Valle at km 36 from Acheral. 27° 01′ S. 65° 39′ W. Alt. 1350 m. 14 Feb. 1966. 'Amongst bushes and stones by roadside in rain forest. Plants to 1·75 m; stem to 1·5 cm diam., with large purple wings to 5 mm wide, crisped. Some fruits. Very abundant. Fls. white. Stem dark purple.' *Hawkes, Hjerting & Rahn 3464* (K, J.G.H., J.P.H.).

Dept. Monteros, road to Tafí del Valle at km 25 from Acheral, above Playa Larga. 27° 03′ S. 65° 39′ W. Alt. 850 m. 14 Feb. 1966. 'Amongst grass and other herbs by roadside; rain forest. Rare. No fls.' *Hawkes, Hjerting & Rahn 3462* (J.G.H.).

Dept. Tafí, path from Raco to San José de Chasquivil, valley of Río Raco *c.* 26° 38′ S. *c.* 65° 28′ W. Alt. 2000 m. 12 Feb. 1966. 'Edge of clearing on a ridge, always in full or partial shade. Fls. white, few so far; no fruits; stem purple.' *Hawkes, Hjerting & Rahn 3448* (K, J.G.H., J.P.H.).

Dept. Tafí, same locality as 3448, but alt. 1250–1350 m. 'Amongst bushes by path in former rain forest. Fls. white; no fruits.' *Hawkes, Hjerting & Rahn 3458* (J.G.H., J.P.H.).

Dept. Tafí, road to Tafí del Valle, at km 44 from Acheral, Duraznos Blancos. 26° 58′ S. 65° 39′ W. Alt. 1730 m. 14 Feb. 1966. 'Amongst bushes and herbs by road in rain forest zone. Fruits only.' *Hawkes, Hjerting & Rahn 3473.*

Prov. CATAMARCA:

Dept. Andalgalá, east slope of Nevados de Aconquija, Estancia 'Yunka Suma', Casa de Piedra to Los Queñoales. 27° 21′ S. 66° 02′ W. Alt. 2100 m. 20 Feb. 1966. 'Growing as an epiphyte on a dead branch of *Sambucus peruviana*. Some fls. and fruits. Tubers soft.' *Hawkes, Hjerting & Rahn 3537* (J.G.H., J.P.H.).

Dept. Belén, Pozo de Piedra, farm at end of road above el almacén. 27° 30′ S. 67° 05′ W. Alt. 2000 m. 18 Feb. 1966. 'Maize field in open-grassy borders, not amongst maize itself; medium damp climate. Plants to 40 cm; fls. white.' *Hawkes, Hjerting & Rahn 3528* (K, J.G.H., J.P.H.).

Prov. LA RIOJA:

Dept. Chilecito, Sierra Famatina, 21 km from Chilecito on road to Mina de Oro. 29° 07′ S. 67° 38′ W. In a quebrada to the north side of main valley. Alt. 2000–2050 m. 2 Feb. 1966. 'Amongst rocks, *Cortaderia*, *Cyperus*, etc. Lvs. simple, fls. white.' *Hawkes, Hjerting, Okada & Rahn 3372* (K, J.G.H., J.P.H.).

Dept. Chilecito, 23 km from Chilecito on road to Guanchín Viejo, finca of Domingo Rivero. 29° 12′ S. 67° 38′ W. Alt. 1800 m. 26 Mar. 1966. 'Near irrigation ditch, on other side.' *Hawkes, Hjerting & Rahn 4001* (K, J.G.H., J.P.H.).

S. microdontum subsp. *gigantophyllum* × *S. vernei* subsp. *vernei*

Prov. SALTA:

Dept. Guachipas, Estancia Pampa Grande, western mountain range, below Cerro Cristal. 25° 50′ S. 65° 34′ W. Alt. 2450 m. 20 Mar. 1966. 'Amongst bushes and tall grass in a small quebrada by path to Cerro Cristal. Growing with 3926 (*S. vernei*) and 3928 (*S. mcd. gig.*).' *Hawkes, Hjerting & Rahn 3927* (J.G.H., J.P.H.).

Dept. Guachipas, Estancia Pampa Grande, western mountain range, between Cerro Cristal and Cerro Pirgua, on a ridge, north-facing slope. 25° 50′ S. 65° 34′ W. Alt. 2650 m. 20 Mar. 1966. 'Rocky slope, amongst tall grasses. No fls. or berries. (*S. microdontum gig.* very abundant here, also).' *Hawkes, Hjerting & Rahn 3940* (J.G.H., J.P.H.).

Prov. TUCUMÁN:

Dept. Chicligasta, Quebrada del Río Las Pavas, between Puesto La Cascada and Las Cuevas. 27° 11′ S. 65° 58′ W. Alt. 2700 m. 24 Feb. 1966. 'Pajonales. Plants 40 cm high. No fls.' *Hawkes, Hjerting & Rahn 3588* (J.G.H., J.P.H.).

Hybrid population of *S. microdontum* subsp. *gigantophyllum* and *S. vernei* (intermediate subspecies)

Prov. SALTA:

Dept. Chicoana, road from Carril to Cachi, Cuesta del Obispo, 4·6 km above Pie de la Cuesta, 11·6 km above Escoipe, where the road crosses a side quebrada on the right. 25° 09′ S. 65° 52′ W. Alt. 2700 m. 5 Mar. 1966. 'Amongst scrub, in partial shade; range of leaf type from one species to the other.' *Hawkes, Hjerting & Rahn 3711* (J.G.H., J.P.H.).

S. microdontum subsp. intermediate × *S. vernei* subsp. *ballsii*

Prov. JUJUY:

Dept. Tumbaya, Volcán, western facing slopes on east side of Quebrada Grande, path from Volcán to Abra Grande de Volcán. 23° 55′ S. 65° 25′ W. Alt. 2960 m. 9 Mar. 1966. 'In a large cattle enclosure and amongst shrubs around it. Mixed with 3781 (*S. microdontum*) and 3780 (*S. vernei* subsp. *ballsii*). Plants with some lateral leaflets, white fls. tinted with pale purple, and some fruits.' *Hawkes, Hjerting & Rahn 3782* (K, J.G.H., J.P.H.).

S. microdontum subsp. *gigantophyllum* × *S. kurtzianum*(= *S. rechei*)

Prov. LA RIOJA:

Dept. Chilecito, Guanchín Viejo, near an irrigation channel at roadside, 600 m from the river crossing. 29° 12′ S. 67° 38′ W. Alt. 1800 m. 31 Jan. 1966. 'Damp very shady place by stream; wet humiferous soil. Fls. white; stem purple, wings very broad.' *Hawkes, Hjerting, Okada & Rahn 3357* (K, J.G.H., J.P.H.).

Dept. Chilecito, same locality as 3357, but 300 m further on towards the river crossing. 'Under trees and bushes. A mixed collection of hybrids and possibly pure *S. kurtzianum*.' *Hawkes, Hjerting, Okada & Rahn 3359* (J.G.H., J.P.H.).

Dept. Chilecito, same locality as 3357, but just above the river crossing. 'In grass and among bushes by road. No fls. *Hawkes, Hjerting, Okada & Rahn 3363* (J.G.H., J.P.H.).

Dept. Chilecito, same locality as 3357, but in a disused field by the road in full sun. 'Fls. white, stamens sterile (but berries seen later, in March).' *Hawkes, Hjerting, Okada & Rahn 3364* (K, J.G.H., J.P.H.).

Dept. Chilecito, Sierra Famatina, 21 km from Chilecito on road to Mina de Oro. 29° 07′ S. 67° 38′ W. Alt. 1950 m. 2 Feb. 1966. 'By a stream in grass, in partial shade of low shrubs. Fls. white.' *Hawkes, Hjerting, Okada & Rahn 3370* (J.G.H., J.P.H.).

Dept. Chilecito, same locality as 3370, but alt. 1900 m. 'In a hedge round an enclosure with walnut trees; also in enclosure itself, growing very abundantly in grass, in furrows with marrows. Stems purple, to 80 cm; fls. white.' *Hawkes, Hjerting, Okada & Rahn 3371* (K, J.G.H., J.P.H.). $2n = 24$.

Dept. Chilecito, same locality as 3370, but alt. 2000 m. 'In a quebrada to the north side of the main valley. Amongst bushes, rocks and *Cortaderia* in damp soil. Leaf soft. Tubers very large, to 5·5×4·5×3·5 cm.' *Hawkes, Hjerting, Okada & Rahn 3373* (K, J.G.H., J.P.H.).

Dept. Chilecito, 23 km from Chilecito on road to Guanchín Viejo. Finca of Domingo Rivero. 29° 12′ S. 67° 38′ W. Alt. 1800 m. 26 Mar. 1966. 'Seen and collected previously (No. 3357), mixed with two parents.' *Hawkes, Hjerting & Rahn 4002.* (J.G.H., J.P.H.).

Dept. Chilecito, road from Chilecito to Mina de Oro, finca 'El Trapiche', belonging to Sr. Antonio Vedia, above Los Talas, *c.* 10 km from Chilecito. 29° 08′ S. 67° 35′ W. Alt. *c.* 1500 m. 26 Mar. 1966. 'By an irrigation channel in rich damp soil among shrubs. Some plants 2 m high; some with mottle virus. Fls. white; stamens loose; a few berries.' *Hawkes, Hjerting & Rahn 4007* (K, J.G.H., J.P.H.).

Dept. Chilecito, same locality as 4007. 'A tuber collection, taken 20 m from 4007. Apparently no virused plants. Tubers only.' *Hawkes, Hjerting & Rahn 4008.*

Dept. Chilecito, same locality as 4007. 'Another tuber collection, taken 20 m in other direction from 4007. Apparently no virused plants. Tubers only.' *Hawkes, Hjerting & Rahn 4009.*

S. microdontum subsp. *gigantophyllum* × *S. vernei* or *S. megistacrolobum*

Prov. SALTA:

Dept. Guachipas, Estancia Pampa Grande, Cerro Cristal. 25° 50′ S. 65° 35′ W. Alt. 2690 m. 3 Mar. 1966. 'Rather dry N.-facing slope, amongst grass; occurring with 3646 (*S. spegazzinii*). Small plants; fls. dark purple.' *Hawkes, Hjerting & Rahn 3647* (J.G.H., J.P.H.).

Dept. Guachipas, Estancia Pampa Grande, western mountain range, Cerro Cristal, north face. 25° 50′ S. 65° 34′ W. Alt. 2650 m. 20 Mar. 1966. 'Amongst short grass in dry soil. No fls. or fruits.' *Hawkes, Hjerting & Rahn 3929* (J.G.H., J.P.H.).

S. microdontum subsp. *microdontum* × *Solanum sp. nov.*

Prov. SALTA:

Dept. Santa Victoria, above Rodeopampa, 14·7 km from Santa Victoria on road to La Quiaca. 22° 15′ S. 65° 05′ W. Alt. 3350 m. 15 Mar. 1966. 'Amongst bushes by road in quebrada crossing road. No fls.' *Hawkes, Hjerting & Rahn 3879* (J.G.H., J.P.H.).

12. SOLANUM VENTURII Hawkes et Hjerting

This species, which is to be found in the grasslands of the Montano district in very high rainfall areas, seems to possess the smallest tubers of all wild potatoes. Many plants collected had not developed tubers much larger than the head of a pin. Since the plants were only just beginning to flower in Tucumán and Catamarca provinces there were no berries, and we despaired of the tubers ever surviving the journey to Europe. Nevertheless, tubers of *3540* survived and gave rise to vigorous plants.

It is interesting to note that all four plants of *3540* were triploid, with thirty-six somatic chromosomes. The plants were large and vigorous, with dark purple stems. The leaves were fleshy, with the margins slightly rolled under. Probably diploid forms occur in the Aconquija mountains, where these triploids were collected. Brücher's material was diploid and was probably collected somewhere in that area.

On the whole, *S. venturii* seems to occur in a restricted ecological zone of high rainfall grassland ('pajonales') above the *Alnus jorullensis* forests. It generally grows in full sun and probably would not be able to withstand heavy grazing and trampling.

The Cuesta del Obispo material of *S. venturii* was found in much more easily accessible places than that from the Sierra de Aconquija, since it occurred quite close to the road. It was flowering and fruiting by the time of our visit. The Aconquija plants ($2n=36$), when grown at Birmingham, seemed to be closer morphologically to the type collections than the plants growing at Cuesta del Obispo which both possessed a somatic chromosome number of $2n = 24$.

A curious purple pigmentation of the leaf veins and of the cell sap of the short spreading 3–4-celled hairs on the leaf surface was rather striking in the Tucumán and Catamarca plants. This was not observed in the Salta ones.

We have no doubt at all that this is a good species, completely distinct from *S. microdontum*. We saw plants of these two species growing in the same localities but apparently no hybrids were formed.

As we have mentioned on p. 309, *S. venturii* frequently forms hybrids with *S. sanctae-rosae*, and there are zones of hybrids where pure *S. venturii* does not seem to occur. We were not able to make any new observations that would have led to a solution of this problem, owing to the extremely wet conditions this year in Quebrada del Río Las Pavas, which inhibited flowering and made travel excessively difficult. It would seem that our collection *3589* from that locality was somewhat introgressed. No. *3590* was the hybrid, whilst *3591* was a collection of non-flowering plants from the same area which probably were a mixture of the hybrid and *S. venturii*. No pure *S. sanctae-rosae* was collected there, though it is known to occur higher up.

S. venturii

Prov. SALTA:

Dept. Chicoana, road from Carril to Cachi, Cuesta del Obispo, 4·6 km above Pie de la Cuesta, 11·6 km above Escoipe, where the road crosses a side quebrada on the right. 25° 09′ S. 65° 52′ W. Alt. 2700 m. 5 Mar. 1966. 'Grass and scrub near road at entrance to side quebrada, edge of pajonales. Pure form, not hybridized.' *Hawkes, Hjerting & Rahn 3712* (J.G.H., J.P.H.).

Dept. Chicoana, road from Escoipe to Cachi, Cuesta del Obispo, 12·2 km above Escoipe. 25° 09′ S. 65° 52′ W. Alt. 2750 m. 5 Mar. 1966. 'In pajonales and under shade of shrubs in a small quebrada. One plant with fls.' *Hawkes, Hjerting & Rahn 3716* (J.G.H., J.P.H.).

Dept. Chicoana, road from Escoipe to Cachi, Cuesta del Obispo, 15·6 km from Piedra Molino, 11 km from Escoipe. 25° 09′ S. 65° 52′ W. Alt. *c.* 2700 m. 6 Mar. 1966. 'In a scree below the road, partial shade of bushes. Larger plants (contrast 3742), near a plant of *S. microdontum*. Plants to 60 cm. Fls. clear white. Abundant fls. and fruits.' *Hawkes, Hjerting & Rahn 3741* (J.G.H., J.P.H.). $2n = 24$.

Dept. Chicoana, same locality as 3741 and growing close to it. 'Same habitat. Smaller plants, further from the plant of *S. microdontum*. Abundant fls. and fruits. Fls. clear white.' *Hawkes, Hjerting & Rahn 3742* (J.G.H., J.P.H.). $2n = 24$.

Prov. TUCUMÁN:

Dept. Chicligasta, Quebrada del Río Las Pavas, between Puesto La Cascada and Las Cuevas. 27° 11′ S. 65° 58′ W. Alt. 2800–3000 m. 24 Feb. 1966. 'Pajonales, amongst big stones. Plants 15–20 cm.' *Hawkes, Hjerting & Rahn 3589* (J.G.H., J.P.H.).

Prov. CATAMARCA:[1]

Dept. Andalgalá, east slope of Nevados de Aconquija, Estancia 'Yunka Suma', Los Queñoales. 27° 21′ S. 66° 02′ W. Alt. 2450 m. 20 Feb. 1966. 'Pajonales, amongst tall grass, with scattered *Alnus* and *Polylepis*, in full light. Fls. white; leaf hairs with purple bases.' *Hawkes, Hjerting & Rahn 3539* (J.G.H., J.P.H.). (See Plate 4.)

[1] We have recently seen a collection from dept. Ambato, Sierra de Ambato, between El Rodeo and Cerro Manchado (*A. T. Hunziker 19177*), which is the most southerly locality yet known for this species.

Dept. Andalgalá, same locality as 3539. 'Growing on a moss-covered rock in full light or partial shade. Hairs purple; fls. white; tubers very minute. Long stolons.' *Hawkes, Hjerting & Rahn 3540* (J.G.H., J.P.H.). $2n = 36$.

Dept. Andalgalá, same locality as 3539. 'Pajonal, similar to 3539 but *c*. 200 m distant. Straggling plants among tall grasses, *Cosmos*, etc., in full sun. No tubers.' *Hawkes, Hjerting & Rahn 3541* (J.G.H., J.P.H.).

13. SOLANUM VERNEI Bitt. et Wittm.

We were able to make six collections of *S. vernei* subsp. *vernei*, two of the intermediate, and two of subsp. *ballsii*. The material from Pampa Grande in the south of Salta province represents a northward extension for the typical subspecies. Apparently the material further north of this, at Cuesta del Obispo and just north of Salta in dept. Caldera represents a series of intergrading forms between subsp. *vernei* and subsp. *ballsii*; further north, in prov. Jujuy and northern Salta, the typical subsp. *ballsii* can be found. From the morphology of the collections made in the field we could see that they conformed with this concept of the *S. vernei* variation pattern.

S. vernei was collected in forested areas (*3538*, *3542*), on bushy slopes (*3504*, *3721*, *3778*, *3780*, *3926*) and amongst grass or in fairly open ground (*3505*, *3593*, *3714*). It thus grows in a range of habitats rather similar to *S. microdontum*. Most of the collections were fruiting, rather than in flower, though we gathered some good flowering material at Infiernillo on plants up to 1·65 m high (*3504*).

As we have already noted, hybrids between *S. microdontum* and *S. vernei* were rare this year. This may be due, in part at least, to the difference in flowering time, which in 1966 was quite well marked.

Although *S. vernei* × *S. spegazzinii* hybrids were collected on three occasions at Infiernillo we saw none at all in 1966. However, our collection *3505* was from an apparently very dry river terrace ('mesada'), and might possess genes of *S. spegazzinii*. The leaflets were a little narrower than is usual in *S. vernei*, but even so, the plants were badly wilted. Curiously enough, we saw no *S. spegazzinii* in this area, though it has been collected there on other occasions.

S. vernei subsp. *vernei*

Prov. SALTA:

Dept. Guachipas, Estancia Pampa Grande, western mountain range, below Cerro Cristal. 25° 50′ S. 65° 34′ W. Alt. 2450 m. 20 Mar. 1966. 'Amongst bushes and tall grass in a small quebrada by path to Cerro Cristal. Growing with 3927 (*S. mcd. gig* × *S. vrn*) and 3928 (*S. mcd. gig.*).' *Hawkes, Hjerting & Rahn 3926* (J.G.H., J.P.H.).

Prov. TUCUMÁN:

Dept. Chicligasta, Quebrada del Río Las Pavas, between Puesto La Cascada and Las Cuevas. 27° 11′ S. 65° 58′ W. Alt. 2800–3000 m. 24 Feb. 1966 'Pajonales amongst stones, near *Polylepis*. Plants at least 50 cm. No fls.' *Hawkes, Hjerting & Rahn 3593* (J.G.H., J.P.H.).

Dept. Tafí, near Infiernillo, below km 83 on road from Tafí del Valle to Cafayate, in a quebrada of an affluent of Río Infiernillo. 26° 46′ S. 65° 45′ W. Alt. 2900 m. 15 Feb. 1966. 'Shaded valley, surrounded by shrubs, in rich soil, humid. Plants to 1·65 m high; others quite short, according to habitat. Rich purple fls.; many berries. Very common.' *Hawkes, Hjerting & Rahn 3504* (K, J.G.H., J.P.H.).

Dept. Tafí, same locality as 3504, but lower down. Alt. 2850 m. 'Dry mesada on side of quebrada. Full sun, in open. No fls.; some fruits; plants wilting.' *Hawkes, Hjerting & Rahn 3505* (J.G.H., J.P.H.).

Prov. CATAMARCA:

Dept. Andalgalá, east slope of Nevados de Aconquija, Estancia 'Yunka Suma', Casa de Piedra to Los Queñoales. 27° 21′ S. 66° 02′ W. Alt. 2270 m. 20 Feb. 1966. 'Growing in bushes by path, together with *S. microdontum*, in shade. *Alnus* woods. No fls.' *Hawkes, Hjerting & Rahn 3538* (J.G.H., J.P.H.).

Dept. Andalgalá, east slope of Nevados de Aconquija, Estancia 'Yunka Suma', Los Queñoales. 27° 21′ S. 66° 02′ W. Alt. 2450 m. 20 Feb. 1966. 'Bushy slope in half-shade of *Alnus*, etc. No fls. Plants to 1·5 m. Many berries.' *Hawkes, Hjerting & Rahn 3542* (J.G.H., J.P.H.).

S. vernei (intermediate subspecies)

Prov. SALTA:

Dept. Chicoana, road from Carril to Cachi, Cuesta del Obispo, 4·6 km above Pie de la Cuesta, 11·6 km above Escoipe, where the road crosses a side quebrada on the right. 25° 09′ S. 65° 52′ W. Alt. 2700 m. 5 Mar. 1966. 'By path, not far from 3712 (*S. venturii*) and 3713 (*S. mcd. gig.*); could be fairly pure.' *Hawkes, Hjerting & Rahn 3714* (J.G.H.).

Dept. Chicoana, road from Escoipe to Cachi, Cuesta del Obispo, 17·5 km above Escoipe. 25° 09′ S. 65° 52′ W. Alt. 3000 m. 5 Mar. 1966. 'Amongst bushes in Pajonal formation—possibly not typical (?+*gigantophyllum*).' *Hawkes, Hjerting & Rahn 3721* (J.G.H., J.P.H.).

S. vernei subsp. *ballsii*

Prov. JUJUY:

Dept. Tumbaya, Volcán, western slopes of cordillera on east side of Quebrada Grande, path from Loma del Tambo to Abra Grande de Volcán. 23° 55′ S. 65° 25′ W. Alt. 3175 m. 9 Mar. 1966. 'Amongst bushes, in partial shade. No fls. or fruits.' *Hawkes, Hjerting & Rahn 3778* (J.G.H., J.P.H.).

Dept. Tumbaya, Volcán, western facing slopes on east side of Quebrada Grande, path from Volcán to Abra Grande de Volcán. 23° 55′ S. 65° 25′ W. Alt. 2960 m. 9 Mar. 1966. 'In a large cattle enclosure and amongst shrubs around it. Mixed with 3781 (*S. microdontum*) and close to 3782 (*S. mcd.*×*S. vrn.*). Tall plants with fruits.' *Hawkes, Hjerting & Rahn 3780* (J.G.H., J.P.H.).

14. SOLANUM GOURLAYI Hawkes

Since this species grows in very dry regions in the high mountains of Salta and Jujuy, we were not surprised that very little of it was seen on the 1966 expedition.

We made three 'Probable *S. gourlayi*' collections, of which the one from Salta (*3870*) seems perhaps the most likely. The other two, from Jujuy province, match other collections of that species reasonably well so far as vegetative features are concerned. None of our collections bore flowers and the two Jujuy ones were rather stunted, because of lack of moisture. We were at first inclined to place them with *S. spegazzinii* but since no well-authenticated material of that species has been collected from so far north it will probably be better to leave them under *S. gourlayi* until they have received further study.

PROBABLY *S. gourlayi*

Prov. JUJUY:

Dept. Tilcara, near Tilcara, Garganta del Diablo, path from new road above Tilcara to Alfarcito. 23° 35′ S. 65° 22′ W. Alt. 2950 m. 10 Mar. 1966. 'In shale on steep bank above path. Shade. Very dry soil. Two small plants.' *Hawkes, Hjerting & Rahn 3791* (J.G.H.).

Dept. Tumbaya, Volcán, north-western slopes of cordillera on east side of Quebrada Grande, Loma del Tambo. 23° 55′ S. 65° 25′ W. Alt. 2800 m. 9 Mar. 1966. 'Amongst spiny bromeliads and in rock crevices. No fls., tubers or berries.' *Hawkes, Hjerting & Rahn 3770* (J.G.H., J.P.H.).

Prov. SALTA:

Dept. Santa Victoria, just below Rodeopampa, 9·3 km from Santa Victoria on road to La Quiaca. 22° 15′ S. 65° 03′ W. Alt. 2950 m. 15 Mar. 1966. 'In a wall by the road. No fls.' *Hawkes, Hjerting & Rahn 3870* (J.G.H.).

15. SOLANUM OPLOCENSE Hawkes

Only two collections of this species were made on the 1966 expedition, both from the Quebrada Grande, in Jujuy province.

The material from dept. Humahuaca (*3907*) was obtained in the same locality as *P. & H. 962* (see Plates 129 and 130), which was shown to be tetraploid. We found abundant fruiting plants in 1966 but none in flower. They were in exactly the same place by the railway at km 1313 to 1313·5, as in 1953 when Petersen and Hjerting collected them previously. Plants raised from seed at Copenhagen and Birmingham were fairly uniform and were also found to be tetraploid.

The other collection was from dept. Tumbaya. It had no fruits, flowers, or tubers, unfortunately, and the identification was made on leaves only. Both these collections agree quite well with *S. oplocense*, but *No. 3907* will, of course, be checked when the plants raised from seed come into flower.

S. oplocense

Prov. JUJUY:

Dept. Humahuaca, between Iturbe and Tres Cruces, by railway at km 1313–1313·5. 22° 57′ S. 65° 23′ W. Alt. 3530–3550 m. 16 Mar. 1966. 'On banks of rock and stones by railway; some plants with *3906* (*S. infundibuliforme*) at km 1313·5; others amongst bushes and in stony banks near railway (km 1313).' *Hawkes, Hjerting & Rahn 3907* (J.G.H., J.P.H.). $2n = 48$.

Dept. Tumbaya, Volcán, western facing slopes on east side of Quebrade Grande, path from Volcán to Abra Grande de Volcán 23° 55′ S. 65° 25′ W. Alt. 3100 m. 9 Mar. 1966. 'In open stony soil by path. No fls., fruits, or tubers.' *Hawkes, Hjerting & Rahn 3779* (J.G.H., J.P.H.).

16. SOLANUM VIDAURREI Cárd.

No material of this species was collected in 1966. Only two certain collections are known for Argentina, both of them in the Santa Victoria region. Unfortunately, we were not able to visit either of these localities and we were not lucky enough to encounter *S. vidaurrei* elsewhere.

17. SOLANUM SPEGAZZINII Bitt.

This is a widespread and very variable species which still requires further taxonomic investigation. We made seventeen collections of it on our 1966 expedition, including six from Salta province, one from Tucumán, three from Catamarca, and seven from

La Rioja. Most of the living collections were of tubers only, and unfortunately these mostly developed secondary tubers which have not yet sprouted. This strong secondary tuber dormancy, which develops under adverse conditions, may perhaps have been due to the selection pressure of the very dry conditions under which this species exists. A similar reaction can be seen in *S. kurtzianum.*

S. spegazzinii is a drought-resistant species of the dry Prepuna and Monte zones in the intermont basins and valleys behind the eastern mountain chains. Although it was seen in a number of places in 1966 it was evidently not so abundant as in 1959, when Ross and Vervoorst collected it. We looked at most of their localities and found that in some places, such as Mina Capillitas, *S. spegazzinii* was entirely absent. There was little at Pozo de Piedra in 1966, compared with 1952 when Sleumer and Vervoorst collected it there.

Fortunately for the collector, *S. spegazzinii* is a species with 'weedy tendencies', which often colonizes cultivated fields, and is seen as a weed of hedgerows, irrigation ditches, and roadsides, whilst at the same time it seems able to grow amongst dense vegetation by the rivers in really wet soil.

In the main part of the text we mentioned that we had been able to visit the 'type locality', as Brücher also did in 1963. The collection of Carlos Spegazzini that we designated as lectotype was made in prov. Salta between Amblayo and Cachipampa. Luckily, a motor road has recently been constructed in the valley of the Río Amblayo from Cachipampa as far down as a uranium mine near Amblayo. We made a collection from this valley, about half-way between the two localities mentioned (*3736*). One cannot, of course, speak of this as the type locality in a strict sense, since Spegazzini never defined it very precisely. Nevertheless, one imagines that he must have collected in approximately the location and altitude where we found this species. We made good collections of plants and tubers but found no flowering plants. Our plants not only agreed excellently with the type specimen of *S. spegazzinii*, especially those from the dry open sandy parts of a waterless river bed, but some of the larger plants growing amongst bushes in moister places varied towards the '*famatinae*' phenotype. This adds weight to our theory that the two 'species' are no more than two phenotypes produced by differences of soil moisture and exposure.

We also visited the type locality for '*S. famatinae*' in prov. La Rioja, dept. Famatina, between Los Corrales and Escalera de Famatina, where it was collected by Kurtz (*13445b*) in 1906. Typical '*famatinae*' plants were growing in the hedges, as Kurtz mentioned, and we made two collections from Los Corrales, just close by (*3344*, *3995*). These match the Kurtz collection quite closely and show the well-dissected leaf typical for plants in rich damp soils.

A collection from between Los Corrales and La Encrucijada (*3427*) was made in a marshy place and shows a wide leaflet decurrency, similar to that of the Ruiz Leal collections *19368* and *19369* from San Juan. It is of interest that Ruiz Leal's own collections were also made in a marshy locality (Ruiz Leal, personal communication).

It is difficult as yet to assess the variability of this highly variable species, which seems equally at home in very wet and in very dry habitats. We have an impression that material from Salta province possesses smaller leaves than that from Catamarca and La Rioja, but we cannot be clear about this until we have observed tuber and seed lines grown under comparable conditions. The Salta collections seem perhaps more highly variable than those from other provinces; thus, *No. 3509* is almost completely without interjected leaflets, whilst the other collections of this species generally possess a large number of them. The two Pampa Grande collections are of small plants (*3646*, *3930*) which are glabrescent on the upper leaf surfaces but the dwarfing may, of course, be due to exposure. During the 1956 expedition of Hjerting, Petersen, and Rahn a curious variant was found in the area between Cuesta del Obispo and Payogasta. This had very long hairs on all green parts and a curious colour pattern on the flowers. Such plants were not seen in 1966, since the area where they grew was extremely dry, when even the drought-resistant Larreas and Bulnesias were dying.

Our collections at Las Cuevas in Sierra Famatina, prov. La Rioja, on the route from Los Corrales to La Encrucijada (*3420, 3422, 3426*) were made in an open windswept valley at 2850 m, where there had been little rain this year. Consequently, the plants were very dwarfed and rosette-like in habit. We tuber-indexed these collections as far as possible (see explanation under *S. acaule* subsp. *aemulans*), in order to judge the amount of hybridization taking place. The plants in these three mass collections seemed to be mixtures of *S. spegazzinii*, *S. acaule* subsp. *aemulans*, and hybrids between them.[1] It will be remembered that Petersen and Hjerting in 1956 had collected possible hybrids of that type in the same region, but that no tubers survived. Collections of *S. acaule* subsp. *aemulans* (*3396, 3407, 3408*) contained an admixture of very stunted *S. spegazzinii* plants. We hope to assess the chromosome numbers of these accessions and to gain some insight into the amount of gene flow between the two species, as well as to arrive at a better understanding of subsp. *aemulans*.

Variation in the Ingenio Arenal material from dept. Andalgalá, Catamarca (*3518*) was perhaps greater than could be postulated for the pure species. Some plants had pale corollas, others very dark ones. In some the leaflets were narrow whilst in others they were broad and rather oblong. Some had smooth fruits, others puberulent ones (cf. '*S. puberulofructum*') and others white spotted ones. Ross and Vervoorst (personal communication) considered their collections from the same place to be due to possible hybridization (*Nos. 16–23*). If this is so, it would be necessary to find another possible parent for such hybrids. Curiously enough, a small family raised from seeds of *3518* in Birmingham during the autumn of 1966 showed great uniformity, but more work is still needed on this collection.

As we mentioned previously, we saw no hybrids of *S. chacoense* × *S. spegazzinii* (= *S.* × *setulosistylum*) at Pampa Grande, Salta province, though there was some influence of *S. spegazzinii* in collection *3645*. Whereas the *S. chacoense* occurred at 1700–1750 m, the *S. spegazzinii* collections were made much higher up, at 2650 to 2690 m. It should be pointed out that the Estancia extends up the valley behind the farm-house for a considerable distance, and we did not have time to investigate that area. The two species might be in contact there, or they might have come together in a favourable year or cycle of years to form hybrids, whilst withdrawing in unfavourable cycles of years. Such a possible yearly or cyclical fluctuation of weather, and hence of plant distribution must always be kept in mind when judging the geographical possibilities of hybridization between a pair of species.

Hybrids of the same type at Pozo de Piedra were very rare in 1966, whilst they seem to have been most abundant in 1952 when Sleumer and Vervoorst collected there. Again, the two parental species were not in contact in 1966, even though they were both present in the area.

Hybrids of *S. spegazzinii* × *S. kurtzianum* were not identified in the field. We have now cultivated most of the material we provisionally identified as *S. kurtzianum* from the Guanchín area of La Rioja province, and found it to be probably or certainly the hybrid of *S. spegazzinii* and *S. kurtzianum*. Plants from tubers showed intermediate characteristics between these two species, and plants raised from seed showed segregation of characters, and a range of variation from one parental species to the other. We have listed the lines grown by us in 1966 as hybrids of this type, whilst those not cultivated have been listed as 'probable hybrids'. Castellanos also collected two plants from the southern end of Sierra Famatina, probably not far from our hybrid collections, that we had previously identified as *S. spegazzinii* × *S. kurtzianum*, and it would seem that both these species have spread into the

[1] We have now identified hybrids of *S. acaule* subsp. *aemulans* × *S. spegazzinii* in these collections.

Prepuna scrub vegetation of the Guanchín area to form hybrids. Some hybridization has also been noted in the Pozo de Piedra region. Thus, our collection *3525* seems to be mostly *S. spegazzinii* × *S. kurtzianum*, with more influence from the latter parent. Some influence of this hybrid is to be seen in part of *3524* (*S. spegazzinii*), gathered from higher up in the same field. Evidently, *S. kurtzianum* was spreading up from a bushy quebrada into a field where *S. spegazzinii* was growing as a weed, and a zone of hybridization was being formed.

A very curious gathering (*3515*) from prov. Salta, dept. San Carlos, growing in a hedge and as a weed of a cultivated field behind, has been provisionally identified by us as a hybrid of *S. spegazzinii* × *S. tuberosum*. This possesses very long petiolules, broad leaflets, small calyx, and a corolla with pale central petal streaks. It is triploid ($2n = 36$). Probably subsp. *tuberosum* would have been involved here, because of the rather low altitude (2050 m).

S. spegazzinii

Prov. SALTA:

Dept. Cafayate, road from Amaichá to Cafayate, at 5·5 km south of Tolombón. 26° 15′ S. 65° 57′ W. Alt. 1650 m. 16 Feb. 1966. 'Under shade of *Atamisquea emarginata* and *Prosopis*, Monte with trees etc. Plants to 40 cm. Lvs. poorly dissected. Only one fl., pale mauve, all one colour. No fruits. Large colony, in one place only.' *Hawkes, Hjerting & Rahn 3509* (J.G.H., J.P.H.).

Dept. Chicoana, road from Escoipe to Cachi, Cuesta del Obispo, 17·5 km above Escoipe. 25° 09′ S. 65° 52′ W. Alt. 3000 m. 5 Mar. 1966. 'Amongst bushes in Pajonal formation, growing with 3721 (*S. vernei*). 2 plants only.' *Hawkes, Hjerting & Rahn 3722* (J.G.H.).

Dept. Guachipas, Estancia Pampa Grande, Cerro Cristal. 25° 50′ S. 65° 35′ W. Alt. 2690 m. 3 Mar. 1966. 'Rather dry N.-facing slope, among grass. Small plants; leaflets narrow, glabrous above; fls. dark purple.' *Hawkes, Hjerting & Rahn 3646* (J.G.H.).

Dept. Guachipas, Estancia Pampa Grande, western mountain range, near summit of Cerro Cristal. 25° 50′ S. 65° 34′ W. Alt. 2650–2680 m. 20 Mar. 1966. 'In dry grass on north-facing slope and amongst rocks right up to the summit. No fls.' *Hawkes, Hjerting & Rahn 3930* (J.G.H., J.P.H.).

Dept. San Carlos, San Lucas, banks of river. 25° 57′ S. 66° 00′ W. Alt. 2100 m. 16 Feb. 1966. 'In sand by banks of river, under shade of trees. Plants wilting, very dry; no fls. or fruits. Mother and daughter tubers.' *Hawkes, Hjerting & Rahn 3512* (J.G.H., J.P.H.).

Dept. San Carlos, road from Cachipampa to Amblayo at *c.* 18 km from the turning, valley of Río Amblayo. Locus classicus. 25° 22′ S. 65° 51′ W. Alt. 3050 m. 6 Mar. 1966 'In damp sand in river bed (no water), especially under protection of Compositae bushes (?*Baccharis*). No fls. Tubers. Range from *spg* to *fmt* leaf type.' *Hawkes, Hjerting & Rahn 3736* (J.G.H., J.P.H.).

Prov. TUCUMÁN:

Dept. Tafí, road from Tafí del Valle to Amaicha, at km 91·3. 26° 42′ S. 65° 47′ W. Alt. 2850 m. 15 Feb. 1966. 'Amongst thorny bushes in Prepuna zone. Plants small, frequently half eaten by animals, but larger under protection of spiny bushes. No fls. or fruits.' *Hawkes, Hjerting & Rahn 3507* (J.G.H., J.P.H.).

Prov. CATAMARCA:

Dept. Andalgalá, route 63, between Santa María and Andalgalá, Ingenio Arenal. 27° 14′ S. 66° 15′ W. Alt. 2800 m. 17 Feb. 1966. 'In stream bed in sand (dry) under

Baccharis and *Salix* and also in full sun. Plants from 5 to 70 cm according to position; hairs various; fls. almost white in sun plants; purple with darker central streak in shade plants; stunted plants *spg* type, others *fmt* type. Fruits white-spotted and warted in most plants, intermediate-sized ones with fine puberulence (= '*S. puberulofructum*'), but some not; never when young or nearly mature.' *Hawkes, Hjerting & Rahn 3518* (K, J.G.H., J.P.H.).

Dept. Belén, road from Puerta de Belén (San José) to Pozo de Piedra, 15·5 km along road. 27° 30′ S. 67° 05′ W. Alt. 1900 m. 18 Feb. 1966. 'In cultivated field of maize and beans; dry sandy soil. Bushy plants in open; leaflet edges smooth; fls. mauve.' *Hawkes, Hjerting & Rahn 3524* (J.G.H., J.P.H.).

Dept. Belén, Pozo de Piedra, lane behind El Almacén. 27° 30′ S. 67° 05′ W. Alt. 1950 m. 18 Feb. 1966. 'Amongst shrubs by side of lane. Fls. very pale violet, almost white.' *Hawkes, Hjerting and Rahn 3530* (p.p.) (J.G.H.). *Note*: Herbarium specimens were *S. chacoense* × *S. kurtzianum*. See text for discussion, p. 464.

Prov. LA RIOJA:

Dept. Famatina, 12 km from Carrizal, Los Corrales, foot of Sierra Famatina. 28° 49′ S. 67° 40′ W. Alt. 2100 m. 30 Jan. 1966. 'Maize field and damp places by river, amongst *Eupatorium*, etc. Plants 20–100 cm tall. Fls. pale mauve. Anthers yellow and brown. Seed collected as No. 3995.' *Hawkes, Hjerting, Okada & Rahn 3344* (K, J.G.H., J.P.H.).

Dept. Famatina, Sierra Famatina, valley of Río Amarillo, path from Los Corrales to La Encrucijada, Las Cuevas. 28° 53′ S. 67° 40′ W. Alt. 2850 m. 5 Feb. 1966. 'In open, amongst stones and small shrubs, full sun to partial shade. Very dry. Plants with substellate mauve or whitish fls.' *Hawkes, Hjerting, Okada & Rahn 3420* (J.G.H., J.P.H.).

Dept. Famatina, another collection from the same area as 3420, but *c.* 500 m distant. 'Less dry; plants better developed.' *Hawkes, Hjerting, Okada & Rahn 3422* (K, J.G.H., J.P.H.).

Dept. Famatina, Sierra Famatina, path from Los Corrales to La Encrucijada, Los Caracoles. 28° 52′ S. 67° 40′ W. Alt. 2650 m. 5 Feb. 1966. 'Dry stony banks by side of path in full sun. Plants varying in size. Fls. white to pale purple, probably according to dryness of soil.' *Hawkes, Hjerting, Okada & Rahn 3425* (K, J.G.H., J.P.H.).

Dept. Famatina, Sierra Famatina, valley of Río Amarillo, path from Los Corrales to La Encrucijada, Las Cuevas. 28° 53′ S. 67° 40′ W. Alt. 2850 m. 5 Feb. 1966. 'Same area as 3420 and 3422, taken half way between them, *c.* 250 m from each. Dry area; some plants with small pale fls.' *Hawkes, Hjerting, Okada & Rahn 3426* (J.G.H., J.P.H.).

Dept. Famatina, Sierra Famatina, path from Los Corrales to La Encrucijada, El Durazno. 28° 51′ S. 67° 40′ W. Alt. 2250 m. 5 Feb. 1966. 'Amongst tall bushes etc., in marshy place by path. Plants to 1 m tall. Fls. pale mauve to whitish.' *Hawkes, Hjerting, Okada & Rahn 3427* (J.G.H., J.P.H.).

Dept. Famatina, Sierra Famatina—at base, Los Corrales. 28° 49′ S. 67° 40′ W. Alt. 2100 m. 25 Mar. 1966. 'Amongst tall *Eupatorium* plants (2 m tall) in damp place by river (crossing fields from lane by house). Plants to 1 m tall, in partial shade. No fls.; berries. No. 3344 was from same locality.' *Hawkes, Hjerting & Rahn 3995* (J.G.H., J.P.H.).

Hybrids of *S. spegazzinii* × *S kurtzianum*

Prov. CATAMARCA:

Dept. Belén, road from Puerta de Belén (San José) to Pozo de Piedra, 15·5 km along road. 27° 30′ S. 67° 05′ W. Alt. 1900 m. 18 Feb. 1966. 'In a cultivated field of maize and beans, lower edge by a quebrada. Dry sandy soil. Leaflet edge rough-hairy; fls. white. Bushy plants in open.' *Hawkes, Hjerting & Rahn 3525* (J.G.H., J.P.H.).

Prov. LA RIOJA:

Dept. Chilecito, road from Chilecito to Guanchín Viejo, 19 km from Chilecito. 29° 12′ S. 67° 38′ W. Alt. 1650 m. 31 Jan. 1966. 'Scrub vegetation with cacti (Prepuna). Under shade of *Schinus* in rich loam. Plant with gibbous hairs and long calyx acumens. No fls.' *Hawkes, Hjerting, Okada & Rahn* 3349 (J.G.H., J.P.H.).

Dept. Chilecito, road from Chilecito to Guanchín Viejo, 21 km from Chilecito, Guanchín, Finca of Sr. Eugenio Silvano. 29° 12′ S. 67° 38′ W. Alt. 1700 m. 31 Jan. 1966. 'Prepuna scrub vegetation under shrubs in shade, rich soil. (See also *3353, 3997, 3998, 3999*.)' *Hawkes, Hjerting, Okada & Rahn 3352* (J.G.H., J.P.H.). Plants grown from tubers rather variable.

Dept. Chilecito, Guanchín Viejo to El Bordo. 29° 12′ S. 67° 38′ W. Alt. 1800 m. 31 Jan. 1966. 'In a grassy shrubby enclosure by road, extremely common. Fls. purple and white (4006 from same place).' *Hawkes, Hjerting, Okada & Rahn 3356* (K, J.G.H., J.P.H.). Plants grown from tubers very variable.

Dept. Chilecito, same locality as 3997 and same area as 3997, 3998, 3352, and 3353. *Hawkes, Hjerting & Rahn 3999* (J.G.H., J.P.H.) $2n = 24$.

Dept. Chilecito, 23 km from Chilecito on road to Guanchín Viejo, Finca of Domingo Rivero. 29° 12′ S. 67° 38′ W. Alt. 1800 m. 26 Mar. 1966. 'Field between ditch and road.' *Hawkes, Hjerting & Rahn 4003* (J.G.H., J.P.H.). Seedlings segregate for parental characters.

Dept. Chilecito, Guanchín Viejo, Finca of Horacio Alcayaga. 29° 12′ S. 67° 38′ W. Alt. 1800 m. 26 Mar. 1966. 'Same place from where 3356 was collected, but berries taken from plants in shade of trees. Berries only.' *Hawkes, Hjerting & Rahn 4006*. Seedlings segregate for parental characters.

Probable hybrids of *S. spegazzinii* × *S. kurtzianum*

Prov. LA RIOJA:

Dept. Chilecito, road from Chilecito to Guanchín Viejo, 21 km from Chilecito, Guanchín, Finca of Sr. Eugenio Silvano. 29° 12′ S. 67 38′ W. Alt. 1700 m. 31 Jan. 1966. 'Prepuna scrub vegetation under shrubs in shade, rich soil. A collection from near 3352.' *Hawkes, Hjerting, Okada & Rahn 3353* (J.G.H.).

Dept. Chilecito, Sierra Famatina, 21 km from Chilecito on road to Mina de Oro, in a quebrada to the north side of the main valley. 29° 07′ S. 67° 38′ W. Alt. 2050 m. 2 Feb. 1966. 'Growing in damp rich soil in shade of rock. One plant only; fl. pale mauve, darker petal centres.' *Hawkes, Hjerting, Okada & Rahn 3374* (J.G.H.).

Dept. Chilecito, Sierra Famatina, 13 km on road from Chilecito to Mina de Oro. 29° 08′ S. 67° 35′ W. Alt. 1650 m. 2 Feb. 1966. 'Under shade of bushes on steep stony slope in rich damp soil. No fls.' *Hawkes, Hjerting, Okada & Rahn 3382* (J.G.H., J.P.H.).

Dept. Chilecito, Guanchín, 21 km from Chilecito on road to Guanchín Viejo, Finca de Eugenio Silvano. 29° 12′ S. 67° 38′ W. Alt. 1700 m. 26 Mar. 1966. 'Under shade of bushes in Monte-Prepuna transition.' *Hawkes, Hjerting & Rahn 3997* (J.G.H.).

Dept. Chilecito, same locality as 3997, same exact spot as 3352 and same area as 3353. *Hawkes, Hjerting & Rahn 3998* (J.G.H.).

Probable hybrid of *S. spegazzinii* × *S. tuberosum*

Prov. SALTA:

Dept. San Carlos, San Antonio, 7 km west of Animaná. 26° 00′ S. 66° 00′ W. Alt. 2050 m. 16 Feb. 1966. 'Hedge by road amongst bushes and in open field behind. Dry sandy soil. Plants to 60 cm. Calyx short; corolla pale purple.' *Hawkes, Hjerting & Rahn 3515* (K, J.G.H., J.P.H.). $2n = 36$.

18. SOLANUM KURTZIANUM

The year 1966 was a disastrous one for *S. kurtzianum*, since there was practically no rain in the Monte scrub vegetation where this species normally grows. Thus, we were able to make only two collections of living plants in La Rioja province, whilst in Mendoza we made a tuber collection only because Professor Ruiz Leal showed us the exact place to dig. Three of our 'collections' are in the form of seeds taken from herbarium collections, with the permission of the original collectors.

In San Juan province we searched in a locality where *S. kurtzianum* had been found previously, but again, we had no success because of the drought. Similarly, other areas in this province which might have been suitable for *S. kurtzianum* growth in wetter years were too dry in 1966.

All our collections of what we thought to be this species in the Guanchín Viejo area, near Chilecito, in La Rioja province, were probably hybrids with *S. spegazzinii*. There were also, of course, hybrids with *S. microdontum* subsp. *gigantophyllum*. *S. kurtzianum* must be quite common there, but further field studies are required so that the exact pattern of hybridization and the areas of the parental species can be ascertained.

We did not manage to climb into the Yacuchi region of Sierra Velasco to look for topotype material of *S. kurtzianum*, but Brücher had done this a few years previously. His specimens matched the type very closely.

Two collections made in the Monte vegetation north of Famatina (*3383, 3384*) were found to be largely or completely the white-flowered form of *S. kurtzianum*. We suspect that it was this white-flowered form of the species which Brücher named '*S. improvidum*', but we have seen no authenticated material, though we have searched in all the herbaria and have written to the author of this species asking for specimens.[1]

We were able to observe that on the whole *S. kurtzianum* grew in the Monte and to some extent in the Prepuna phytogeographical provinces. It occasionally seems to come into contact with *S. spegazzinii*, and with *S. microdontum* subsp. *gigantophyllum*. Where it was forming hybrids with *S. chacoense* (*S.* × *ruiz-lealii*) the former was possibly growing in the fields, whilst *S. kurtzianum* was probably confined to the wild vegetation outside. In fact, *S. kurtzianum* does not seem to invade cultivated fields or act as a weed or ruderal plant in the same way as *S. chacoense* and *S. spegazzinii*.

S. kurtzianum

Prov. LA RIOJA:

Dept. Famatina, at km 680 on route 40 from Famatina to Tinogasta, between Ángulos and Santa Cruz. Original collection made by A. T. Hunziker, A. E. Cocucci, and E. Di Fulvo, 20 Mar. 1960, *No. 15203*. Seeds taken from herbarium specimen in Herb. Córdoba. Seeds only. *Hawkes, Hjerting, Okada & Rahn 3272*.

Dept. Famatina, between Campanas and Santa Cruz on route 40, at km 680, near road. 28° 32′ S. 67° 38′ W. Alt. 1600 m. 3 Feb. 1966. 'In typical Monte, with *Zuccagnia*, *Prosopis*, *Larrea cuneata*, *Acacia furcatispina*, etc. In damp place under bushes in rich

[1] Dr. Brücher recently kindly gave us an isotype specimen of *S. improvidum* (*Brücher 929*). We consider it to be a white-flowered form of *S. kurtzianum*.

soil in a little quebrada; not widely distributed. Some plants with pure white fls., others with purple petal streak.' *Hawkes, Hjerting, Okada & Rahn 3383* (J.G.H., J.P.H.).

Dept. Famatina, same locality as 3383, but at km 680·5. 'Damp stream hollow, in rich damp soil under shade of *Prosopis* and other Monte shrubs. Fls. all white; leaflets rather wide.' *Hawkes, Hjerting, Okada & Rahn 3384* (K, J.G.H., J.P.H.).

Prov. MENDOZA:

Dept. Las Heras, El Portezuelo, on the international road to Chile. 32° 32′ S. 68° 55′ W. Alt. 1500 m. 4 Jan. 1966. 'In typical Monte under *Larrea* and other bushes in rich soil. Plants dead, tubers not yet sprouting. Tubers only.' *Hawkes, Hjerting & Rahn 3152*.

Dept. Las Heras, entre Villavicencio y El Portezuelo. Berries taken from herbarium sheet of *Ruiz Leal No. 22621* in herb. Ruiz Leal. *Hawkes, Hjerting & Rahn 3154*.

Dept. Las Heras, Estancia Canota, Mina de Baritina. Original collection made 11 May 1963. 'Monte—dry slope. Berries to *c.* 1 cm diam. Berries taken from herbarium sheet of *Ruiz Leal No. 22696* in herb. Ruiz Leal.' *Hawkes, Hjerting & Rahn 3155*.

19. SOLANUM MAGLIA

Although we were taken to the original Argentine locality for this species by Professor Ruiz Leal we saw no plants and were not able to find tubers either, because of the drought. We were able to satisfy ourselves, however, that the area where it grew in Quebrada Alvarado seemed to possess a completely natural type of vegetation, with very little human influence. It represented, according to Ruiz Leal, a transition between Monte and High Andean vegetation types. Here there was apparently no *S. kurtzianum*, though it was said to occur at the bottom of the valley in Monte vegetation. We were also shown photographs of the plants and material under cultivation in Professor Ruiz Leal's laboratory.

Several other similar valleys are to be found in the vicinity of Quebrada Alvarado, and it is to be hoped that Professor Ruiz Leal will investigate them also, to see whether an extension of range for *S. maglia* can be established.

SOLANUM SPECIES NOVO

Four collections from above Rodeopampa on the road from Santa Victoria to La Quiaca in prov. Jujuy appear to belong to a hitherto undescribed species. The large purple flowers and rather poorly dissected leaves distinguish this material clearly from other Argentine species, but since it grows so close to the Bolivian border the species must surely occur in that country also. We shall not describe it here, but shall delay publication until we have studied it thoroughly and have ascertained that it is not an already described Bolivian species.

Prov. SALTA:

Dept. Santa Victoria, above Rodeopampa, 13 km from Santa Victoria on road to La Quiaca. 22° 15′ S. 65° 04′ W. Alt. 3200 m. 15 Mar. 1966. 'Roadside, in Puna. Fls. purple; berries. *S. acaule* and *S. megistacrolobum* found nearby.' *Hawkes, Hjerting & Rahn 3873* (J.G.H., J.P.H.). Some variation in seedling family grown in 1966.

Dept. Santa Victoria, same locality as 3873. 'Roadside, near 3872 (*S. acaule*) and 3873, in Puna, sheltered by bushes.' *Hawkes, Hjerting & Rahn 3874* (J.G.H.).

Dept. Santa Victoria, same locality as 3873, but about 50 m above it. 'Roadside; under large grass tussocks.' *Hawkes, Hjerting & Rahn 3875* (J.G.H., J.P.H.).

Dept. Santa Victoria, above Rodeopampa, 14·7 km from Santa Victoria on road to La Quiaca. 22° 15′ S. 65° 05′ W. Alt. 3350 m. 15 Mar. 1966. 'Amongst bushes and in shade of boulders in a quebrada crossing road at km 14·7. Fls. large, purple.' *Hawkes, Hjerting & Rahn 3878* (J.G.H., J.P.H.). Some variation in seedling family grown in 1966.

CULTIVATED SPECIES

Solanum × *juzepczukii*

This triploid cultivated species was given us by Sr. R. Schechaj, Director of 'El Suncho' Experimental Station in Catamarca province, under the name of *S. acaule*. It had been growing there for many years and probably represented part of the potato cultivar collection assembled in about 1950 by Hjerting and others. The chromosome number was $2n = 36$.

Prov. CATAMARCA

Dept. Andalgalá, road from Andalgalá to Concepción, Est. Exp. 'El Suncho'. 27° 27′ S. 66° 01′ W. Alt. 1750 m. 19 Feb. 1966. 'Growing as a weed in the plots, probably from a former introduction. Tubers only.' *Hawkes, Hjerting & Rahn 3534*. $2n = 36$.

Solanum × *chaucha*

This is one of eight tuber samples kindly sent to Buenos Aires by Sr. J. Fernández, from Mina Aguilar, prov. Jujuy. It bore no vernacular name. It possesses well-dissected leaves, large irregular calyx, mauve corolla and short style, with the stigma placed at anther pore level. This is a particularly interesting collection, since it represents the first record of *S.* × *chaucha* in Argentina, so far as we are aware. Chromosome number: $2n = 36$. Collection number *H.H.R. 4057*.

Solanum tuberosum subsp. *andigena*

The rest of the collection from Mina Aguilar, mentioned in the previous paragraph, has all been identified as *S. tuberosum* subsp. *andigena*. All seven lines were tetraploid, and possessed the usual leaf dissection and calyx and corolla form of this species. The collection numbers and vernacular names are as follows:

H.H.R. 4050 Papa Manzana Morada (= Collareja).
H.H.R. 4051 Papa Tuni (or Tune).
H.H.R. 4052 Papa Santa María.
H.H.R. 4053 Papa Oca. (*Note*: this has nothing to do with *Oxalis tuberosa*, to which this name is normally given.)
H.H.R. 4054 Papa Colorada.
H.H.R. 4055 Papa Runa.
H.H.R. 4056 Papa Azuleja.

BIBLIOGRAPHY

ADAMS, J. B. (1946) Aphid resistance in potatoes. *Am. Potato J.* **23**, 1–22.

AKELEY, R. V. and STEVENSON, F. J. (1943) Yield, specific gravity and starch content of tubers in a potato breeding program. Ibid. **20**, 203–17.

ANDERSON, E. (1949) *Introgressive hybridization.* New York.

—— (1954) *Plants, man and life.* London.

—— (1960) A semigraphical method for the analysis of complex problems. *Technometrics* **2**, 387–91.

ANON. (1945) Virus diseases. *Rep. Scott. Pl. Breed. Stn* 12–13.

—— (1946) Ibid. 17–18.

—— (1947) Ibid. 16–19.

—— (1949) Ibid. 19–22.

—— (1951*a*) Director's report. *Rep. Scott. Pl. Breed. Stn* 11–31.

—— (1951*b*) Appendix IV, Commonwealth potato collection. *21st Ann. Rep. Commonw. agric. Bur.* 28–31.

—— (1956) Director's report. *Rep. Scott. Pl. Breed. Stn* 5–20.

—— (1957) Ibid. 5–24.

—— (1958) Ibid. 5–30.

—— (1959) Ibid. 5–22.

—— (1960) Ibid. 5–23.

—— (1961) Ibid. 5–24.

—— (1962) *Rec. Scott. Pl. Breed. Stn* 5–27.

—— (1963) Ibid. 5–25.

—— (1964) Ibid. 5–30.

—— (1965) Ibid. 7–25.

ANSELMI LAUKIN, M. S. (1952) Estudio preliminar de la selección de papa con miras a la creación de un 'Instituto de la papa'. *Agronomía, Lima* **17**, No. 70, 57–71.

ANTIPOVA, L. K. (1950) (Potato resistance to *Epilachna.*) (*Fruit and vegetable gardens*) No. 8, 34–6.

AVANZI, M. G. (1949) Il numero cromosomico di specie e varietá di *Solanum* oriunde del Chile. *Caryologia* **2**, 111–12.

BAERECKE, M.-L. (1956) Ergebnisse der Resistenzzüchtung gegen das Blattrollvirus der Kartoffel. I. Mitteilung. *Z. Pflanzenz.* **36**, 395–412.

BAGNALL, R. H. (1960) Potato virus F latent in an imported variety and resistance to the virus in an interspecific potato hybrid. *Phytopathology* **50**, 460–4.

BAKER, J. G. (1884) A review of the tuber-bearing species of *Solanum. J. Linn. Soc. Lond.* **20**, 489–507.

BAZÁN DE SEGURA, C. (1953) Trabajos preliminares para la obtención de variedades resistentes al 'hielo', *Phytophthora infestans*, 1950–1. *Boln trim. Exp. agropec.* **2**, No. 2, 8–10.

—— and OCHOA, C. (1959) Búsqueda de fuentes de resistencia en papa al hongo *Verticillium albo-atrum. Infme mens. Estac. exp. agric. La Molina* **33**, No. 378, 1–7.

BEAMISH, K. I. (1955) Seed failure following hybridization between the hexaploid *Solanum demissum* and four diploid *Solanum* species. *Am. J. Bot.* **42**, 297–304.

—— COOPER, D. C., and HOUGAS, R. W. (1957) Induced tetraploid *Solanum* species and species hybrids: meiosis and use in breeding. Ibid. **44**, 305–10.

—— and SHEJKAL, J. (1955) The use of induced tetraploids in potato breeding. *Am. Potato J.* **32**, 386–7.

BENTHAM, G. and HOOKER, J. D. (1876) *Genera plantarum*. London.

BERTHAULT, P. (1911) Recherches botaniques sur les variétés cultivées du *Solanum tuberosum* et les espèces sauvages de *Solanum* tubérifères voisins. *Annls. Sci. agron. fr.* **2**, pp. 209.

BITTER, G. (1912) Solana nova vel minus cognita. *Reprium nov. Spec. Regni veg.* **10**, 529–65; **11**, 349–94, 431–73.

—— (1913) Ibid. **12**, 1–10, 49–90, 146–62, 433–67.

—— (1917) Solana africana, II. Engler, *Bot. Jb.* **54**, 499.

BLACK, W. (1945) Inheritance of resistance to blight (*Phytophthora infestans*) in potatoes: unbalanced segregations. *Proc. R. Soc. Edinb.* **62**, Sect. B, 171–81.

BLACK, W. and GALLEGLY, M. E. (1957) Screening of *Solanum* species for resistance to physiologic races of *Phytophthora infestans*. *Am. Potato J.* **34**, 273–81.

BOCZKOWSKA, M. (1951) Les pommes de terre résistantes au doryphore. *Revue hort.* **123**, 454–5.

BOYD, A. E. W. (1951) Susceptibility of *Solanum curtilobum* to *Spongospora subterranea* (Wallr.) Johnson. *Nature, Lond.* **167**, 412.

BRAUN, H. (1942) Biologische Spezialisierung bei *Synchytrium endobioticum* (Schilb.) Perc. *Z. Pflkrankh.* **52**, 481–6.

BRITTEN, J. and BOULGER, G. S. (1931) *A Biographical index of deceased British and Irish Botanists*. London.

BROILI, J. (1921) Arbeiten mit Wildbastarden von *Solanum*. *Mitt. biol. Reichanst. Ld-u. Forstw.* **21**, 154–5.

BRÜCHER, H. (1951) Probleme der Abstammung der Kulturkartoffel. *Naturw. Rdsch., Stuttg.* **8**, 345–50.

—— (1953) Über das natürliche Vorkommen von Hybriden zwischen *Solanum simplicifolium* und *Solanum subtilius* im Aconquija Gebirge. *Z. indukt. Abstamm.-u. VererbLehre* **85**, 12–19.

—— (1954) Cytologische und ökologische Beobachtungen an nordargentinischen *Solanum* Arten der Section *Tuberarium*. Teil I. Die Wildkartoffel-Arten des Aconquija-Gebirges. *Züchter* **24**, 281–95.

—— (1956*a*) Kritische Betrachtungen zur Nomenklatur argentinischer Wildkartoffeln. I. Die Serie *Commersoniana*. Ibid. **26**, 97–106.

—— (1956*b*) Critical observations on the taxonomy of Argentine wild potatoes. II. *S. vernei* Bitt. & Wittm. and its synonym *S. ballsii* Hawkes. *An. Dep. Invest. cient. Univ. nac. Cuyo* **2**, 1–13.

—— (1957*a*) Kritische Betrachtungen zur Nomenklatur argentinischer Wildkartoffeln. III. Die Serie *Cuneoalata*. *Züchter* **27**, 77–80.

—— (1957*b*) Kritische Betrachtungen zur Nomenklatur argentinischer Wildkartoffeln. IV. Die Serie *Tuberosa*. Ibid. 353–7.

—— (1958) Beiträge zur Abstammung der Kulturkartoffel. *Sber. dt. Akad. LandwWiss.* **7**, No. 8, pp. 44.

—— (1959*a*) Kritische Betrachtungen zur Nomenklatur argentinischer Wildkartoffeln. V. Die Serie *Acaulia*. *Züchter* **29**, 149–56.

—— (1959*b*) Kritische Betrachtungen zur Nomenklatur argentinischer Wildkartoffeln. VI. Die Serie *Alticola* (= *Megistacroloba*). Ibid. 257–64.

—— (1959*c*) La importancia de las variedades autóctonas y de las especies silvestres de 'Solanum' de la República Argentina para la fitogenética internacional. *Revta argent. Agron.* **26**, 53–6.

—— (1960*a*) Über das Vorkommen einer kultivierten triploiden 'Wildkartoffel' am Nahuel-Huapi-See in Südargentinien. *Züchter* **30**, 77–80.

—— (1960*b*) Problematisches zum Ursprung der Kulturkartoffel aus Chiloë. *Z. PflZücht.* **43**, 241–65.

Brücher, H. (1961) Primer hallazgo de *Heterodera rostochiensis* Woll. sobre papas silvestres. *Revta Fac. Cienc. agrar. Univ. nac. Cuyo* **8**, 7–18.
—— (1962) Nuevas especies de *Solanum* (Tuberarium) de la zona semiárida del N.W. Argentino. Ibid. **9**, No. 2, 7–14.
—— (1963*a*) Untersuchungen über die *Solanum* (Tuberarium) Cultivare der Insel Chiloë. *Z. Pflanzenz.* **49**, 7–54.
—— (1963*b*) Das südlichste Vorkommen diploider Kulturkartoffeln in Südamerika. *Qualitas Pl. Mater.* veg. **9**, 187–202.
—— (1963*c*) Aus der Urheimat unserer Kulturkartoffel. I. Über wilde Verwandte von *Solanum tuberosum*. *Kartoffelbau* **14**, 212–14.
—— (1963*d*) Aus der Urheimat unserer Kulturkartoffel. II. Wuchs die 'Urkartoffel' am Meeresstrand? *Kartoffelbau* **14**, 268–70.
—— (1964*a*) Kritische Betrachtungen zur Nomenklatur argentinischer Wildkartoffeln. VII. *Solanum setulosistylum* Bitter, eine seit 50 Jahren falsch interpretierte Species der Serie *Commersoniana*. *Züchter* **34**, 27–32.
—— (1964*b*) Dos nuevas especies de *Solanum* (sect. Tuberarium) del sur de la República Argentina. *Darwiniana* **13**, 104–14.
—— (1964*c*) Aus der Urheimat unserer Kulturkartoffel. III. Das andine Ursprungsgebiet der Kartoffeln. *Kartoffelbau* **15**, 160–2.
—— (1966) Kritische Betrachtungen zur Nomenklatur argentinischer Wildkartoffeln. VIII. *Solanum bijugum* und dessen synonyma. *Züchter* **36**, 103–6.
—— and Ross, H. (1953). La importancia de las especies tuberíferas de *Solanum* del noroeste Argentino como fuente de resistencia a las enfermedades. *Lilloa* **26**, 453–88.
Bukasov, S. M. (1932*a*) (Frost resistance in the potato.) *Trudȳ prikl. Bot. Genet. Selek.* Ser. 2, **3**, 287–97.
—— (1932*b*) (The results of 11 years' work with the potato, carried on at the Institute of Plant Industry.) Ibid. Ser. A, **4**, 67–70.
—— (1933*a*) (A revolution in the breeding of the potato.) (*Lenin Acad. Agric. Sci. Inst. Pl. Industr.*), Leningrad, pp. 44.
—— (1933*b*) (The potatoes of South America and their breeding possibilities.) *Trudȳ prikl Bot. Genet. Selek.* Suppl. 58, pp. 192.
—— (1936) The problems of potato breeding. *Am. Potato J.* **13**, 235–52.
—— (1937*a*) In Vavilov, N.I. (*Theoretical bases of plant breeding*) **3**, 1–76.
—— (1937*b*) Cuatro nuevas especies de *Solanum* de la Flora Argentina. *Revta argent. Agron.* **4**, 238–40.
—— (1938) (Interspecific hybridization in the potato.) *Izv. Akad. Nauk SSSR* 711–32.
—— (1939*a*) The origin of potato species. *Physis, B. Aires* **18**, 41–46.
—— (1939*b*) Interspecific hybridization in the potato. Ibid. 269–84.
—— (1940*a*) (New wild potato species from Argentina and Uruguay.) *Vest. sots. Rasteniev.* No. 4, 3–12.
—— (1940*b*) (The successes and failures of interspecific hybridization of the potato.) Ibid. No. 3, 39–48.
—— (1940*c*) Una nueva especie de *Solanum* del subgénero Tuberarium de la República Argentina. *Revta argent. Agron.* **7**, 363.
—— (1940*d*) (Classification of potato species on dry matter content.) *Vest. sots. Rasteniev.* No. 4, 144–5.
—— (1941*a*) The geography of the endemic potatoes of South America. *Revta argent. Agron.* **8**, 83–104.
—— (1941*b*) (The origin of species of potatoes.) *Vest. sots. Rasteniev.* No. 1, 157–64.

BUKASOV, S. M. (1951) (Breeding potatoes for resistance to Colorado beetle.) *Trudȳ prikl. Bot. Genet. Selek.* **29**, No. 1, 71–6.

——(1955*a*) (Production of wart-resistant potato varieties.) (*Fruit and vegetable gardens.*) No. 10, 90–93.

——(1955*b*) (Systematics of potato species.) *Problemȳ Bot.* **2**, 317–26.

—— (1958) Das Ausgangsmaterial für die Züchtung der Kartoffel – insbesondere gegen den Koloradokäfer resistenter Sorten. *Züchter* **28**, 65–70.

—— and KAMERAZ, A. J. (1948) (*Potato Breeding.*) Moscow and Leningrad.

—— ——(1959) (*Bases of Potato Breeding.*) Moscow.

—— and LEBEDEVA, N. A. (1955) (Breeding the potato for resistance to Colorado beetle (*Leptinotarsa decemlineata* Say.)) (*Colorado beetle and methods of controlling it.*) 118–33. Moscow.

—— and LECHNOVICZ, V. (1935) Importancia en la fitotecnia de las papas indígenas de la América del Sur. *Revta argent. Agron.* **2**, 173–83.

—— and NAUMOV, N. A. (1938) (*Wart resistant varieties of potato.*) Moscow.

CABRERA, A. L. (1953) Esquema fitogeográfico de la República Argentina. *Revta Mus. La Plata*, Nueva Serie **8**, Sección Botánica, 87–168.

CALDAS-AULETE, F. J. (1958) Dicionário contemporâneo da língua portugueza. Lisbõa. Brazilian edition by HAMÍLCAR GARCÍA.

CALDERONI, A. V., and INDUNI, C. J. (1952) In *Memoria de la primera reunión de papa realizada en la Estación Experimental de Balcarce*, 67–70.

DE CANDOLLE, A. (1852) (Auct. Dunal.) *Prodr. Syst. nat. Regni veg.* **13**, 1. Paris.

CÁRDENAS, M. (1956) New species of *Solanum* (Tuberarium–Hyperbasarthrum) from Bolivia. *Boln Soc. peru. Bot.* **5**, 9–45.

—— and HAWKES, J. G. (1945) New and little-known wild potato species from Bolivia and Peru. *J. Linn. Soc. Lond.* **53**, 91–108.

CARRIÈRE, E. A. (1883) Nouvelle espèce de pomme de terre. *Revue hort.* **55**, 496–500.

CASTRONOVO, A. (1950) Fuentes de resistencia a *Phytophthora infestans. Revta Invest. agric., B. Aires* **4**, 245–75.

CHARLES, A. W. (1956) The parasitism of *Alternaria solani* (Ell. and Mart.) Jones and Grout on certain species of *Solanum*, and their use as sources of disease resistance. *Abstr. Diss. Univ. Camb.* 1953–4, 4–5.

CHOUDHURI, H. C. (1943) Cytological studies in the genus *Solanum*. I. Wild and native cultivated 'diploid' potatoes. *Trans. R. Soc. Edinb.* **61**, 113–35.

——(1942–4) Cytological and genetical studies in the genus *Solanum*. II. Wild and cultivated 'diploid' potatoes. Ibid. 199–219.

——(1948) III. Genetical studies in wild and native cultivated potatoes. *Bull. bot. Soc. Beng.* **2**, 186–200.

CLARK, R. L. (1962) A study of the host range of the potato leaf-roll virus in some tuberous *Solanum* species. *Phytopathology* **52**, 360.

——(1963) Leafroll resistance in some tuberous solani under controlled aphid inoculations. *Am. Potato J.* **40**, 115–20.

COCKERHAM, G. (1943) Potato breeding for virus resistance. *Ann. appl. Biol.* **30**, 105–8.

——(1957) Experimental breeding in relation to virus resistance. *Proc. 3rd Conf. Potato Virus Diseases*, Lisse-Wageningen, 1957, 199–203.

COLE, C. S. and HOWARD, H. W. (1962) Further results from a field experiment on the effect of growing resistant potatoes on a potato root eelworm (*Heterodera rostochiensis*) population. *Nematologica* **7**, 57–61.

CORRELL, D. S. (1952) Section Tuberarium of the genus *Solanum* of North America and Central America. *U.S. Dept. Agric. Monogr.* No. 11, pp. 243. Washington.

CORRELL, D. S. (1961) New species and some nomenclatural changes in section Tuberarium of *Solanum*. *Wrightia* **2**, 169–97.

—— (1962) The potato and its wild relatives. *Tex. Res. Found. Bot. Studies* **4**, pp. 606.

CZAJKA, W. and VERVOORST, F. (1956) Die naturräumliche Gliederung Nordwest-Argentiniens. In *Petermanns Geographischen Mitteilungen*, Quartalsheft 2 and 3, 89–102, 196–208.

DAVY DE VIRVILLE, A. (1954) *Histoire de la Botanique en France*. Paris.

DEMIDOVICZ, A. F. (1938) (Breeding potatoes for drought resistance.) (*Trans. Kazan agric. Inst.*), 139–73.

DIEHL, R. (1934) État actuel de la maladie verruqueuse de la pomme de terre (la question des variétés). *Revue Path. vég. Ent. agric. Fr.* **21**, 1, 25–31.

DIONNE, L. A. (1963) Studies on the use of *Solanum acaule* as a bridge between *Solanum tuberosum* and species in the series Bulbocastana, Cardiophylla and Pinnatisecta. *Euphytica* **12**, 263–9.

—— and LAWRENCE, C. H. (1961) Early scab resistant derivatives of *Solanum chacoense* × *Solanum phureja*. *Am. Potato J.* **38**, 6–8.

DODDS, K. (1956) *47th Ann. Rep. John Innes Hortic. Inst.*

DOROZHKIN, N. A. (1936) (Powdery scab of potatoes.) (*White Russian Acad. Sci. Inst. Biol. Sci. Minsk* **5**.)

DREMLIUG, L. A. (1937) A frost resistant triple potato hybrid *Solanum acaule* × *S. tuberosum* (Fürstenkrone) × *S. tuberosum* (Centifolia). *Dokl. Akad. Nauk SSSR* **16**, 423–6.

DUNAL, M. F. (1813*a*) (in POIRET) *Encyclopédie méthodique botanique*, Suppl. 3.

—— (1813*b*) *Histoire des Solanum et des genres qui ont été confondus avec eux*. Paris, Strasbourg, Montpellier.

—— (1816) *Solanorum Generumque Affinium Synopsis*. Montpellier.

—— (1852) (in DE CANDOLLE) *Prodr. Syst. nat. Regni veg.* **13**, 1. Paris.

DUNNETT, J. M. (1957) Variation in pathogenicity of the potato root eelworm (*Heterodera rostochiensis* Woll.) and its significance in potato breeding. *Euphytica* **6**, 77–89.

—— (1960*a*) The role of *Solanum vernei* Bitt. et Wittm. in breeding for resistance to potato root eelworm (*Heterodera rostochiensis* Woll.). *Rep. Scott. Pl. Breed. Stn* 39–44.

—— (1960*b*) Potato breeders' strains of root eelworm (*Heterodera rostochiensis* Woll.). Rep. 5th int. Symp Pl. Nematology, 10–13 Aug. 1959, Uppsala. *Nematologica* Suppl. 2, 84–94.

EASTON, G. D. (1957) Virulence of potato virus Y isolates and immunity in the genus *Solanum*. *Diss. Abstr.* **17**, Pub. No. 24, 276, 2370.

ELLENBY, C. (1948) Resistance to the potato-root eelworm. *Nature, Lond.* **162**, 704.

—— (1952) Resistance to the potato root eelworm *Heterodera rostochiensis* Wollenweber. Ibid. **170**, 1016–17.

—— (1954) Tuber forming species and varieties of the genus *Solanum* tested for resistance to the potato root eelworm *Heterodera rostochiensis* Wollenweber. *Euphytica* **3**, 195–202.

EMME, E. K. (1936) (The genetics of the potato. I. Inheritance of corolla colour in 24-chromosome species of potato). *Biol. Zh.* **5**, 977–1000.

—— (1937) (Hybrids of *S. Emmeae* Juz. and *S. Commersonii* Dun.-species resistant to the Colorado beetle—genetical study II.) Ibid. **6**, 299–310.

—— (1938) (Studies on interspecific hybridization of tuber bearing potatoes section *Tuberarium* Bitter, genus *Solanum* L.) Ibid. **7**, 1093–4.

FIEDLER, H., and SCHREITER, J. (1959) Das Pachytän-Genom von *Solanum vernei*. *Z. VererbLehre* **90**, 62–5.

FILIPPOV, A. S. (1954) (The effect of low temperatures upon the winter hardiness of seedling potatoes.) *Agrobiologiya* No. 5, 43–9.

FIRBAS, H. (1962*a*) Zusammenhänge zwischen Trockenheit und Frostresistenz. *Z. Pflanzenz.* **48**, 29–35.

—— (1962*b*) Reaktion von Wildkartoffelarten unterschiedlicher Frostresistenz auf Kälteeinflüsse verschiedener Stärke und Dauer. Ibid. **48**, 101–5.

FIRBAS, H. and ROSS, H. (1961) Züchtung auf Frostresistenz bei der Kartoffel. I. Über die Frostresistenz des Laubes von Wildarten und Primitivformen der Kartoffel und ihre Beziehung zur Höhenlage des Artareals. Ibid. **45**, 259–99.

FOCKE, R. (1959) Ein Versuch zur Klärung der Vererbung der Nematodenresistenz (*Heterodera rostochiensis* Woll.) bei *Solanum vernei*. Tagber. dt. Akad. LandwWiss. Berl. 159–62.

FRIES, R. E. (1905) Zur Kenntnis der alpinen Flora im nördlichen Argentinien. *Nova Acta R. Soc. Scient. upsal.* Ser. 4, **1**, 1–205.

GELL, P. G. H., HAWKES, J. G., and WRIGHT, S. T. C. (1960). The application of immunological methods to the taxonomy of species within the genus *Solanum*. *Proc. R. Soc.* B **151**, 364–83.

GILLES, A. (1955) Recherches cytogénétiques sur les *Solanum* (Section *Tuberarium*) 1. Nombres chromosomiques et associations méiotiques. *Cellule* **57**, 7–31.

GOFFART, H. (1957) Fortschritte auf dem Gebiete der Züchtung nematodenresistenter Kartoffelsorten. *Kartoffelbau* **8**, 194–5.

—— (1962) Über das Auftreten aggressiver Biotypen der Kartoffelnematoden (*Heterodera rostochiensis* Woll.). *NachrBl. dt. PflSchutz-dienst., Stuttg.* **14**, 101–3.

—— and ROSS, H. (1954) Untersuchungen zur Frage der Resistenz von Wildarten der Kartoffel gegen den Kartoffelnematoden (*Heterodera rostochiensis* Wr.). *Züchter* **24**, 193–201.

GÓMEZ, J. P. (1946) *Labor desarrollada por las cátedras de botánica, trabajos agrícolas y granja e industrias rurales de la escuela de Maestros Normales Regionales de Humahuaca.* 1 y 2.

GOTTSCHALK, W. (1954) Die Chromosomenstruktur der Solanaceen unter Berücksichtigung phylogenetischer Fragestellungen. *Chromosoma* **6**, 539–626.

—— and PETERS, N. (1955) Die Chromosomenstruktur diploider Wildkartoffel-Arten und ihr Vergleich mit der Kulturkartoffel. Ein Beitrag zum Abstammungsproblem der Kartoffel. *Z. Pflanzenz.* **34**, 351–74.

GRISEBACH, A. (1879) *Symbolae ad floram argentinam.* Göttingen.

GRUN, P. (1961) Early stages in the formation of internal barriers to gene exchange between diploid species of *Solanum*. *Am. J. Bot.* **48**, 79–89.

—— AUBERTIN, MARY, and RADLOW, ANNE (1962) Multiple differentiation of plasmons of diploid species of *Solanum*. *Genetics* **47**, 1321–33.

HAMÍLCAR GARCÍA (1958) CALDAS-AULETE. Brazilian edition.

HANSEN, H. P. (1960) Tobacco mosaic virus carried in potato tubers. *Am. Potato J.* **37**, 95–101.

HARLAND, S. C. (1933) The genetical conception of the species. *Trop. Agric., Trin.* **11**, 51–3.

HASSLER, E. (1911) Ex Herbario Hassleriano: Novitates paraguarienses XI. *Reprium nov. Spec. Regni veg.* **11**, 115.

—— (1917) Le *Solanum chacoense* Bitter et les cellules à cristaux d'oxalate de chaux. *Annls Conserv. Jard. Bot., Genève.* **20**, 185–8.

HAWKES, J. G. (1941) Potato collecting expeditions in Mexico and South America. *Bull. imp. Bur. Pl. Breed. Genet., Camb.*, pp. 30.

—— (1944) Potato Collecting expeditions in Mexico and South America. II. Systematic classification of the collections. Ibid. pp. 142.

HAWKES, J. G. (1945) The indigenous American potatoes and their value in plant-breeding. Part I. Resistance to disease. Part II. Physiological properties, chemical composition, and breeding capabilities. *Emp. J. exp. Agric.* **13**, 11–40.
——(1947) Some observations on South American Potatoes. *Ann. appl. Biol.* **34**, 622–31.
——(1951) Cinco conferencias sobre el mejoramiento de la papa. *Instituto Nacional de Agricultura, Bol. Tec. No. 1.* Maracay, Venezuela, pp. 54.
——(1954) New *Solanum* species in sub-section *Hyperbasarthrum* Bitt. *Ann. Mag. nat. Hist.* Ser. 12, **7**, 689–710.
——(1956*a*) Taxonomic studies on the tuber-bearing Solanums. 1: *Solanum tuberosum* and the tetraploid species complex. *Proc. Linn. Soc. Lond.* **166**, 97–144.
——(1956*b*) A revision of the tuber-bearing Solanums. *Rep. Scott. Pl. Breed. Stn* 37–109.
——(1957) Potato breeding in Russia—July, 1956. *Euphytica* **6**, 38–44.
——(1958) (in KAPPERT, H., and RUDORF, W.) *Handbuch der Pflanzenzüchtung.* 2nd edn. Vol. III, Chapter 1. Potatoes: taxonomy, cytology and crossability, 1–43.
——(1962*a*) The origin of *Solanum juzepczukii* Buk. and *S. curtilobum* Juz. et Buk. *Z. Pflanzenz.* **47**, 1–14.
——(1962*b*) Introgression in certain wild potato species. *Euphytica* **11**, 26–35.
——(1963) A revision of the tuber-bearing Solanums (2nd edn.). *Rec. Scott. Pl. Breed. Stn* 76–181.
—— and HJERTING, J. P. (1960) Some wild potato species from Argentina. *Phyton*, Graz. **9**, 140–6.
HECKEL, E. D. (1904) Nuevas observaciones. *An. Dep. Ganad. Agric. Urug.* **7**, 259.
——(1905) Le *S. commersonii* ou pomme de terre aquatique. *Revue hort. Bouches Rhône*, 105.
——(1915) Sur le *S. Caldasii* Kunth. (*S. guaraniticum* Hassler) et sur la mutation gemmaire culturale de ses parties souterraines. *C. r. hebd. Seanc. Acad. Sci., Paris* **160**, 24, 54.
HIERONYMUS, J. (1882) *Plantae diaphoricae florae argentinae.* Buenos Aires.
HOFFMANN, G. M. (1958) Untersuchungen über die Resistenz von Wild-und Primitivkartoffeln gegen den Erreger des Kartoffelschorfes *Streptomyces scabies* Waksman und Henrici. *Züchter* **28**, 283–5.
HOLMBERG, E. L. (1898) *La Flora de la República Argentina, en Segundo Censo de la República Argentina* **1**, 385–474. Buenos Aires.
HOOKER, J. D. (1846) *Flora Antarctica* **2**.
——(1884) *Solanum maglia. Curtis's bot. Mag.* t. 6756.
HOSSEUS, C. C. (1921) Estudios comparativos sobre la vegetación de las provincias de La Rioja y de San Juan. *Boln acad. nac. Cienc. Córdoba* **26**, 5–160.
HOUGAS, R. W. (1953) *Inter-regional potato introduction station. Seed and tuber list.* Dept. of Genetics, Madison, Wisconsin (mimeographed).
——(1955) The inter-regional potato introduction project. *25th Ann. Rep. U.S.D.A. Nat. Potato Breed. Prog.*, 1954, 35–7.
—— ROSS, R. W., and BEAMISH, K. I. (1955). Hybridization of certain tuber-bearing *Solanum* species. *Am. Potato J.* **32**, 390.
HOWARD, H. W. (1955) Breeding potatoes for resistance to root eelworm (*Heterodera rostochiensis*). *Heredity* **9**, 150.
—— and SWAMINATHAN, M. S. (1952) Species differentiation in the section *Tuberarium* of *Solanum* with particular reference to the use of interspecific hybridization in breeding. *Euphytica* **1**, 20–8.
HROBRY, N. D. (1958) (Skin spot on potato varieties and species.) *Trudў príkl. Bot. Genet. Selek.* No. 1, 231–41.

HUDSON, M. A. (1961) The limitations of a cut leaf test for assessing the frost resistance of the tuber-bearing *Solanums*. *Euphytica* **10**, 169–79.

HUIJSMAN, C. A. (1955) Breeding for resistance to the potato root eelworm. II. Data on the inheritance of resistance in *andigenum-tuberosum* crosses obtained in 1954. Ibid. **4**, 133–40.

—— (1956) Breeding for resistance to the potato root eelworm in the Netherlands. *Nematologica* **1**, 94–9.

—— (1959) Nature and inheritance of the resistance to the potato root eelworm, *Heterodera rostochiensis* W. in *Solanum kurtzianum*. *Meded. LandbHoogesch. OpzoekStns Gent* **24**, 611–14.

—— (1960) Some data on the resistance against the potato root eelworm (*Heterodera rostochiensis W.*) in *Solanum kurtzianum*. *Euphytica* **9**, 185–90.

—— (1962) (Resistance to potato sickness and eating quality.) *Aardappelpootgoed* **19**, 43–4.

HUMBOLDT, F. H. A., BONPLAND, A., and KUNTH C. S. (1815–25) *Nova genera et species plantarum*. Paris (Vol. 3. 1818).

IVANOVSKAYA, E. V. (1939) Cytological study of *Solanum Millanii* Buk. et Lechn. *Dokl. Akad. Nauk SSSR* **24**, 389.

JONES, B. A., PLAISTED, R. L., and PETERSON, L. C. (1963) The behaviour of a chlorophyll mutant in *Solanum vernei* Bitt. et Wittm. *Am. Potato J.* **40**, 109–14.

JONES, F. G. W. (1954) First steps in breeding for resistance to potato root eelworm. *Ann. appl. Biol.* **41**, 348–53.

—— and PAWELSKA, K. (1963) The behaviour of populations of potato root eelworm (*Heterodera rostochiensis* Woll.) towards some resistant tuberous and other *Solanum* species. Ibid. **51**, 277–94.

JOSEFSSON, A. (1944) Torkresistens och rotmassa hos några potatissorter. *Sver. Utsädesfor. Tidskr.* **54**, 325–31.

JUZEPCZUK, S. W. (1937) (New species of the genus *Solanum* in the group *Tuberarium* Dun.) *Izv. Acad. Nauk SSSR* **2**, 295–331.

KAMERAZ, A. J. (1940) (Wild species as initial material in potato breeding.) *Vest. sots. Rasteniev.* No. 4, 13–30.

—— (1957*a*) (Wart resistance of interspecific potato hybrids.) *Vest. sel'. Nauk* No. 6, 35–42.

—— (1957*b*) (Breeding potatoes for resistance to Colorado beetle.) *Kartofel'* No. 1, 31–36.

—— (1961) (Breeding potatoes for resistance to Colorado beetle.) (*Proc. int. Conf. Stud. Colorado Beetle.*) Moscow, 1959, 174–82, from *Ref. Zh.* No. 8, abstr. 8G289.

—— (1964) (Breeding potatoes for resistance to blight and virus diseases.) (*Proc. All-Un. Inst. Pl. Industr.*) No. 2, 13–19.

KAWAKAMI, K. and MATSUBAYASHI, M. (1954) Studies on potato breeding by polyploid hybrids between related species. II. On the morphological characteristics of induced tetraploid plants in some species belonging to the series *Commersoniana*. (*Rep. Hyogo Univ. Agric., Ser. Agric.*) **1**, 131–6.

—— —— (1956) Studies on the species differentiation in the section *Tuberarium* of *Solanum*. II. Meiotic behaviour in the triploid F_1 hybrid from *S. longipedicellatum* × *S. chacoense*, with some regards to chromosome affinity between the parent species. (*Rep. Hyogo Univ. Agric., Ser. Agric.*) **2**, 143–8.

—— —— (1957) Studies on the species differentiation in the section *Tuberarium* of *Solanum*. IV. Cytological behaviour of the F_1 hybrids from *S. tuberosum* × *S. saltense* (*4x*), with special remarks on the polyploid nature of *S. tuberosum*. *Jap. J. Bot.* **16**, No. 1, 128–34.

—— —— (1958) Studies on the species differentiation in the section *Tuberarium* of *Solanum*. VI. Meiotic behaviour of amphiploids from *Solanum longipedicellatum* ×

S. chacoense and its bearing on genomic affinity between the parent species. (*Rep. Hyogo Univ. Agric., Ser. Agric.*) **3**, 124–30.

KAWAKAMI, K. and TAKAYAMA, A. (1955) Studies on the species differentiation in *Tuberarium*. I. Morphological analysis of *S. longipedicellatum* and *S. chacoense* and triploid hybrid between them. *Jap. J. Breed.* **5**, 69–74.

KING, CAPT. P. P. (1839) *Narrative of the voyages of the 'Adventure' and 'Beagle' between the years 1826 and 1836*, Vol. 1, 1839—First Expedition under the command of Captain P. P. King.

KNORR, L. C. (1948) Suscept range of the potato ring rot bacterium. *Am. Potato J.* **25**, 361–71.

KOOPMANS, A. (1951) Cytogenetic studies on *Solanum tuberosum* L. and some of its relatives. *Genetica* **25**, 193–337.

—— (1952) Changes in sex in the flowers of the hybrid *Solanum rybinii* × *S. chacoense*. Ibid. **26**, 359–80.

—— (1954) Changes in sex in the flowers of the hybrid *Solanum rybinii* × *S. chacoense*. II. Plasmatic influences upon gene action. Ibid. **27**, 273–85.

—— (1955) Changes in sex in the flowers of the hybrid *Solanum rybinii* × *S. chacoense*. III. Data about the reciprocal cross *Solanum chacoense* × *S. rybinii*. Ibid. 465–71.

—— (1959) Changes in sex in the flowers of the hybrid *Solanum rybinii* × *S. chacoense*. IV. Further data from the reciprocal cross *S. chacoense* × *S. rybinii*. Ibid. **30**, 384–90.

KOVALENKO, G. M. (1932) Hardy frost-resisting potato varieties. *Trudў prikl. Bot. Gen. Selek.* Ser. A, **3**, 127–30. Also *Am. Potato J.* **9**, 205–9.

KRUZHILIN, A. A. (1946) (New drought resistant varieties of potato.) *Sots. zern. Khoz.* No. 4, 41–45.

KUHN, R. and LÖW, I. (1954) Gluco-alcaloides de especies silvestres de *Solanum* y de *Lycopersicum* y su significación para la resistencia al escarabajo de la patata. *Ciencia, Méx.* **14**, 7–14.

KUNTZE, O. (1891–8) *Revisio generum plantarum vascularium omnium*. Leipzig.

LABERGERIE, J. (1907) Variation du *Solanum commersonii* et du *S. maglia*. *Bull. Séanc. Soc. natn. Agric. Fr.* **48**, 886–923.

LAMM, R., (1943) Notes on an octoploid *Solanum punae*. *Hereditas* **29**, 193–5.

—— (1945) Cytogenetic studies in *Solanum*, sect. *Tuberarium*. Ibid. **31**, 1–128.

—— (1953) Investigations on some tuber-bearing *Solanum* hybrids. Ibid. **39**, 97–112.

LANGENBUCH, R. (1952) Ist das Fehlen eines 'Frassstoffes' oder das Vorhandensein eines 'Vergällungsstoffes' die Ursache für die Resistenz der Wildkartoffel *Solanum chacoense* Bitt. gegenüber dem Kartoffelkäfer? *Z. PflKrankh.* **59**, 179–89.

LEBEDEVA, N. A. (1955) (The production of polyploids of some wild potato species.) *Bot. Zh. SSSR* **40**, 575–83.

—— (1959*a*) [Changes in the potato (*Solanum* spp.) accompanying polyploidy.] *Bot. Zh. SSSR* **44**, 1075–90.

—— (1959*b*) Das durch Polyploidie ausgelöste Varieren der Eigenschaften und Merkmale der Kartoffel. *Züchter* **29**, 107–17.

—— (1960) (Production of polyploids of the wild potato species *Solanum bulbocastanum* Dun., *S. cardiophyllum* Lindl. and *S. vernei* Bitt. et Wittm.) *Dokl. Akad. Nauk SSSR* **134**, 1211–13.

—— (1965) (Experimental polyploidy in breeding.) (*Potato and Vegetables*) **10**, No. 4, 20–4.

LECHNOVICZ, V. S. (1940) (The frost resistance of the new potato species *S. rionegrinum*.) *Vest. sots. Rasteniev.* No. 4, 31–2.

LESTER, R. (1965) Immunological studies on the tuber-bearing Solanums. I. Techniques and South American species. *Ann. Bot.* **29**, 609–24.

LILLO, M. (1919) Reseña fitogeográfica de la Provincia de Tucumán. *Primera Reunión Nacional Soc. Argent. Cienc. Nat., Tucumán*, 1916, 210–32.

LIPINSKY, S. (1957) (Breeding of potato varieties resistant to Colorado beetle. (*L. decemlineata* Say.)) *Postepy Nauk roln.* **4**, No. 6, 29–36.

LONGLEY, A. E. and CLARK, C. F. (1930) Chromosome behaviour and pollen production in the potato. *J. agric. Res.* **41**, 867–88.

LORENTZ, P. G. (1876) Cuadro de la vegetación de la República Argentina, in: R. NAPP. *La República Argentina*, 77–136.

MAGOON, M. L., HOUGAS, R. W., and COOPER, D. C. (1958*a*) Cytogenetic studies of South American diploid Solanums, section Tuberarium. *Am. Potato J.* **35**, 475–94.

—— —— —— (1958*b*) Cytogenetic studies of tetraploid hybrids in *Solanum* from hexaploid-diploid matings. *J. Hered.* **49**, 171–8.

—— COOPER, D. C., and HOUGAS, R. W. (1958) Cytogenetic studies of some diploid Solanums, section Tuberarium. *Am. J. Bot.* **45**, 207–21.

MAI, W. F. and PETERSON, L. C. (1952) Resistance of *Solanum ballsii* and *Solanum sucrense* to the golden nematode, *Heterodera rostochiensis* Wollenweber. *Science* **116**, 224–5.

MALHEIROS-GARDÉ, N. (1959*a*) Mechanisms of species isolation in tuberous *Solanum*. *Agronomia lusit.* **21**, 19–42.

—— (1959*b*) Fertilization and seed development in some crosses between diploid species of the genus *Solanum*. Ibid. 91–102.

MARTIUS, K. F. P. VON (1846) *Flora Brasiliensis* **10**. Munich.

—— (1906) Vitae Collectorum, in *Flora Brasiliensis* **1**, Part 1, 2–210. Munich.

MASTENBROEK, C. (1954) Nachtvorstresistentie in toekomstige aardappelrassen? *Meded. nat. Coöp. Aan-Verkoopver. Landb. Cent. Bur.* **38**, 75–6.

—— (1955) A note on resistance of *Solanum* species to powdery mildew. *Euphytica* **4**, 15–16.

—— (1956) Some experiences in breeding frost-tolerant potatoes. *Euphytica* **5**, 289–97.

MASUTANI, T. (1962*a*) (Cytogenetical studies of species related to the potato. I. Meiosis in diploid species.) (*Mem. Fac. Agric. Hokkaido Univ.*) **4**, 75–8.

—— (1962*b*) (Cytogenetical studies of species related to the potato. IV. The chromosome behaviour of amphidiploid plants of *Solanum laplaticum* Buk. × *S. stenotomum* Juz. et Buk.) Ibid. 84–7.

MATSUBAYASHI, M. (1955) Studies on the species differentiation in the section *Tuberarium* of *Solanum*. III. Behaviour of meiotic chromosomes in F_1 hybrid between *S. longipedicellatum* and *S. schickii* in relation to its parent species. (*Rep. Hyogo Univ. Agric., Ser. Agric.*) **2**, 25–31.

—— (1962) Studies on the species differentiation in *Solanum*, sect. *Tuberarium*, VIII. Genomic relationships between *S. demissum* and certain diploid *Solanum* species. (*Rep. Kihara Inst. biol. Res.*) No. 13, 57–68.

MILLÁN, R. (1939) Noticias acerca de las variedades de papa cuyo cultivo se radicó en la Argentina hasta 1936. *Revta argent. Agron.* **6**, 41–8.

—— (1942) Silo araucano para conservar papas. Ibid. **9**, 69–74.

—— (1945) La producción y el comercio exterior de papa en Argentina. Ibid. **12**, 1–11.

MOLINA, J. I. (1782) *Saggio Hist. Nat.* Bologna.

MONTALDO, A. (1950) Producción de semilla de papa. *Agricultura téc.* **10**, 32–42.

—— (1951*a*) Fitomejoramiento para resistencia a la nematosis de la papa. Ibid. **11**, 64–85.

—— (1951*b*) Fitomejoramiento para resistencia a la sarna polvorienta de la papa. Ibid. 140–8.

MONTALDO, A. and SANZ, C. (1962) Las especies de papas silvestres y cultivadas de Chile. Ibid. **22**, 66–152.

MORTON, C. V. (1944) Taxonomic studies of tropical American plants. Some South American species of *Solanum*. *Contr. U.S. natn. Herb.* **29**, 41–72.

MUJICA, R. F. (1939) La anguilulosis de la papa. *Boln Inf. No. 1. Serv. San. Veg. Chile, La Tierra* **1**, 4–10.

——(1942) Susceptibilidad de variedades de papas a la sarna polvorienta causada por la *Spongospora subterranea* (Wallr.) John. *Boln Dep. Sanid. veg., Santiago* **2**, No. 1, 17–19.

MÜLLER, K. O. and SELLKE, K. (1941) Über die Aussichten der Züchtung von 'käferfesten' Kartoffelsorten. *Mitt. biol. Reichsanst. Ld-u. Forstw.* **64**, 10–23.

MUÑOZ PIZARRO, C. (1960) *Las especies de plantas descritos por R. A. Philippi en el Siglo XIX*. Santiago.

MÜNTZING, A. (1933) Studies on meiosis in diploid and triploid *Solanum tuberosum* L. *Hereditas* **17**, 233–45.

NASSONOV V. A. and YAGER F. C. (1933) (Anatomical peculiarities of stems and leaves in hardy potato varieties.) *Trudȳ prikl. Bot. Genet. Selek.* Ser. A, **7**, 107–16.

NOVÁK, F. (1962) (The resistance of wild species of potato to wart (*S. endobioticum* (Schilb.) Perc.)). *Sb. čsl. Akad. zeměd. Věd.* Rada Rostlinná výroba **35**, 123–30.

NOVAKOVA, J. (1958) Czernani stonku u. brambor. Ibid. **31**, 1233–46.

OCHOA, C. M. (1960) Nuevas especies de *Solanum* (*Tuberarium* subsec. *Hyperbasarthrum*) del Perú. IV. *Agronomía, Lima* **27**, 363–72.

——(1962) *Los Solanum tuberíferos silvestres del Perú*. Lima, Peru.

OKUNO, S. (1951) Cytological studies on potatoes, with some remarks on genetical experiments. Part I. (*Jap. J. Genet.*) **26**, 79–103.

——(1952) Cytological studies on potatoes, with some remarks on genetical experiments. Part II. Ibid. **27**, 3–12.

OLÁH, L. VON (1938) Cytogenetische Untersuchungen in der Gattung *Solanum*, Sect. *Tuberarium*. III. *Solanum commersonii* Dun. und einige seiner Bastarde. *Z. indukt. Abstamm.-u. VererbLehre* **74**, 228–41.

OPPENHEIMER, H. C. (1933) Cytogenetische Untersuchungen an Bastarden knollentragender *Solanum*-Arten. I. *Solanum chacoense* Bitt. × *Solanum tuberosum* L. s. str. *Z. indukt. Abstamm.-u. VererbLehre* **65**, 72–98.

PAHARIA, K. D., PUSHKARNATH, and DESHMUKH, M. J. (1962) Resistance of potato varieties to charcoal rot. *Indian Potato J.* **4**, 84–7.

PAL, B. P. and PUSHKARNATH (1951) Potato breeding investigations in India. *Emp. J. exp. Agric.* **19**, 87–103.

PANDEY, K. K. (1960) Self-incompatibility in *Solanum megistacrolobum* Bitt. *Phyton, B. Aires*, **14**, 13–19.

——(1962) Incompatibility in some *Solanum* crosses. *Genetica* **22**, 24–30.

PARODI, L. R. (1961) Ciento cincuenta años de Botánica en la República Argentina *Boln Soc. argent. Bot.* **9**, 1–66.

PASFIELD OLIVER, S. (1909) *The life of Philibert Commerson*. London.

PATON, J. AIKMAN (1909) Notes on some hybrid tuberous Solanums. *Jl R. hort. Soc.* **35**, 53–5.

PENNELL, F. W. (1945) Historical sketch, in Verdoorn: Plants and plant sciences in Latin America, *Chronica bot.* **16**, 35–48.

PERLOVA, R. L. (1940) Spontaneous occurrence of diploid plants in the offspring of the triploid *Solanum maglia* Schlechtd. grown in the Pamirs. *Dokl. Akad. Nauk SSSR* **27**, 710–13.

——(1945*a*) (The self-fertility of South-American potatoes in Western Pamir.) *Izv. Akad. Nauk SSSR*, Sér. Biol. No. 4, 460–70.

PERLOVA, R. L. (1945*b*) Production of the original species of the Chile autotriploid *Solanum maglia* Schlechtd. at Pamir. *Dokl. Akad. Nauk SSSR* **48**, 56–8.

—— (1946) (Morphology of the berries as a taxonomic character of tuber-bearing species of *Solanum* L. (Section *Tuberarium* Bitt.).) *Bot. Zh. SSSR* **31**, 19–32.

—— (1951) (Biological characters of various potato species.) *Trudȳ glav. bot. Sada, Moscow* **2**, 217–63.

PHILIPPI, F. (1885) Reise nach der Provinz Tarapacá. *Verh. dt. wiss. Ver. Santiago Chile* **1**, 135–63.

PHILIPPI, R. A. (1873) Descripción de las nuevas plantas incorporadas últimamente en el herbario chileno. *An. Univ. Chile* **43**, 479–583.

—— (1891) *Verzeichniss der von Friedrich Philippi auf der Hochebene der Provinzen Antofagasta und Tarapacá gesammelten Pflanzen.* Leipzig.

PISSAREV, V. (1933) Kartoffelselektion auf Kälteresistenz. *Z. Zücht.* **18**, 582–94.

PLANCHON, L. (1909*a*) Mutation gemmaire du *Solanum commersonii. Annls Fac. Sci. Marseille* **18**, 1, 1–49.

—— (1909*b*) *S. commersonii* sauvage et muté. Ibid. 9, 167–203.

—— (1909*c*) Sur le *Solanum maglia* Schlechtd. Ibid. 10, 205–24.

PRAKKEN, R. and SWAMINATHAN, M. S. (1952) Cytological behaviour of some interspecific hybrids in the genus *Solanum*, sect. *Tuberarium. Genetica* **26**, 77–101.

PROKOSHEV, S. M. (1939) (Biochemistry of the potato.) (*Biochemistry of cultivated plants.*) **4**, 5–58.

—— and MATTISON, N. L. (1940) (Biochemical characteristics of new species of potatoes.) *Vest. sots. Rasteniev.* No. 4, 61–74.

PROPACH, H. (1937) Cytogenetische Untersuchungen in der Gattung *Solanum*, Sect. *Tuberarium*. II. Triploide und tetraploide Artbastarde. *Z. indukt. Abstamm.-u. VererbLehre* **73**, 143–54.

—— (1938*a*) Cytogenetische Untersuchungen in der Gattung *Solanum*, Sect. *Tuberarium*. IV. Tetraploide und sesquidiploide Artbastarde. Ibid. **74**, 376–87.

—— (1938*b*) Kreuzbarkeit von *Solanum*-Arten untereinander und mit Kulturkartoffeln und die Fertilität der Bastarde. *Forschungsdienst* **6**, 311–14.

—— (1940) Cytogenetische Untersuchungen in der Gattung *Solanum*, Sect. *Tuberarium*. V. Diploide Artbastarde. *Z. indukt. Abstamm.-u. VererbLehre* **78**, 115–27.

PUSHKARNATH (1953) Studies on sterility in potatoes. V. Genetics of self- and cross-incompatibility in *Solanum rybinii. Indian J. Genet. Pl. Breed.* **13**, 83–90.

RATERA, E. L. (1938*a*) Valor agrícola de algunas especies indígenas de *Solanum* (*Tuberarium*) de la República Argentina. *Revta Fac. Agron. Vet. Univ. B. Aires* **9**, 23–9.

—— (1938*b*) Determinación del número de cromosomas de varias especies de papas indígenas de la República Argentina. *Inst. Genét. Fac. Agron. Vet., Univ. B. Aires* **1**, No. 1, pp. 8.

—— (1940) Contribución al estudio del polen de papas. Observaciones en algunas variedades de papas cultivadas y en especies salvajes afines. Ibid. No. 4, pp. 19.

—— (1942) *Solanum chacoense* Bitter, espontáneo en los alrededores de Buenos Aires. *Agronomía, B. Aires* **31**, 56–60.

—— (1943*a*) Número de cromosomas de algunas Solanáceas Argentinas. *Revta Fac. Agron. Vet. Univ. B. Aires* **10**, 77–84.

—— (1943*b*) Observaciones sobre la biología floral de *Solanum chacoense* Bitt. Ibid. 451–7.

—— (1947*a*) Resistencia a las heladas de algunos *Solanum* (*Tuberarium*) argentinos. *An. Soc. cient. argent.* **143**, 258–63.

—— (1947*b*) Contribución al estudio del polen de papas. Observaciones en algunas variedades de papas cultivadas y en especies salvajes afines (Segunda contribución). *Revta Fac. Agron. Vet. Univ. B. Aires* **11**, 277–85.

RATERA, E. L. (1954) Número de cromosomas de algunos *Solanum* (*Tuberarium*) argentinos. Ibid. **13**, 470–3.

RAZUMOV, V. (1935) (Frost resistance of some potato spp.) *Trudӯ prikl. Bot. Genet. Selek.* Ser. 3, **6**, 221–6.

REDDICK, D. (1930) Frost-tolerant and blight-resistant potatoes. *Phytopathology* **20**, 987–91.

——(1939) Scab immunity. *Am. Potato J.* **16**, 71–6.

RENGGER, J. R. (1835) *Reise nach Paraguay*. Aarau.

ROIG, FIDEL A. (1956) Alteraciones florales en *Solanum subtilius*. *Revta argent. Agron.* **23**, 122–31.

——(1960) Bosquejo fitogeográfico de las provincias de Cuyo. *Fac. Cienc. Agrar., Mendoza, Publ.* 3, pp. 35.

ROSS, H. (1954*a*) Über die extreme Resistenz von *Solanum acaule* gegen das X-Virus. *Mitt. biol. BundAnst. Berl.* No. 80, 144–5.

——(1954*b*) Die Vererbung der 'Immunität' gegen das X-virus in tetraploidem *Solanum acaule. Atti del IX Congr. Intern. di Genetica, Bellagio, Italia, 1953; Caryologia* **6**, 1128–32.

——(1958*a*) Virusresistenzzüchtung an der Kartoffel. *Eur. Potato J.* **1**, No. 4, 1–19.

——(1958*b*) II. Ausgangsmaterial für die Züchtung; in KAPPERT, H., and RUDORF, W. (editors) *Handbuch der Pflanzenzüchtung*, 2nd edn, **3**, 43–59.

——(1960*a*) Über die Zugehörigkeit der knollentragenden *Solanum*-Arten zu den pflanzengeographischen Formationen Südamerikas und damit verbundene Resistenzfragen. *Z. Pflanzenz.* **43**, 217–40.

——(1960*b*) Über die Vererbung von Eigenschaften für Resistenz gegen das Y- und A-virus in *Solanum stoloniferum* und die mögliche Bedeutung für eine allgemeine Genetik der Virusresistenz in *Solanum* sect. *Tuberarium. Proc. 4th Conf. Potato Virus Dis.*, Braunschweig, 40–9.

——(1962) Über die Vererbung der Resistenz gegen den Kartoffelnematoden (*Heterodera rostochiensis* Woll.) in Kreuzungen von *Solanum famatinae* Bitt. et Wittm. mit *Solanum tuberosum* L. und mit *S. chacoense* Bitt. *Züchter* **32**, 74–80.

—— and BAERECKE, M.-L. (1950) III. Selection for resistance to mosaic virus (diseases) in wild species and in hybrids of wild species of potatoes. *Am. Potato J.* **27**, 275–84.

—— —— (1951) Über die Bedeutung der argentinischen *Solanum*-Arten *simplicifolium, vernei, berthaultii, acaule* und einiger Formen von *S. andigenum* für die Züchtung krankheitsresistenter Kartoffeln. *Z. Pflanzenz.* **30**, 280–91.

ROSS, R. W. and ROWE, P. R. (1965) Inventory of tuber-bearing *Solanum* species. *U.S.D.A. Bulletin 533*, pp. 73.

ROTHACKER, D. (1957*a*) Beiträge zur Resistenzzüchtung gegen den Kartoffelnematoden. I. Prüfung von Primitiv- und Wildkartoffeln auf das Verhalten gegenüber dem Kartoffelnematoden. *Züchter* **27**, 124–32.

——(1957*b*) Arbeiten zur Züchtung krebsresistenter Kartoffeln. I. Wild- und Primitivkartoffeln als Ausgangsmaterial für die Züchtung auf Krebsbiotypenresistenz. Ibid. 181–3.

——(1958) Beiträge zur Resistenzzüchtung gegen den Kartoffelnematoden (*Heterodera rostochiensis* Wollenweber). III. Untersuchungen über den Einfluss unterschiedlicher Kreuzungspartner auf die Ausbildung verschiedener Knolleneigenschaften bei Kartoffelkreuzungen, zugleich ein Beitrag zur Züchtungsmethodik. Ibid. **28**, 133–43.

ROTHACKER, D. (1959) Nematodenresistenzzüchtung auf der Basis von *Solanum vernei* Bitt. et Wittm. *TagBer. dt. Akad. LandwWiss. Berl.* No. 20, 135–7.

—— (1961) Die wilden und kultivierten mittel- und südamerikanischen Kartoffelspecies; in SCHICK, R. and KLINKOWSKI, M. (editors) *Die Kartoffel* **1**, 353–558.

—— and MÜLLER, W. A. (1960) Arbeiten zur Züchtung krebsresistenter Kartoffeln. II. Untersuchung kultivierter südamerikanischer Kartoffel-species auf ihr Verhalten gegenüber dem Krebsbiotyp G_1. *Züchter* **30**, 340–3.

—— and STELTER, H. (1957) Beiträge zur Resistenzzüchtung gegen den Kartoffelnematoden (*Heterodera rostochiensis* Wollenweber). II. Untersuchungen über die Vererbung der Nematodenresistenz bei den Arten *S. vernei* Bitt. et Wittm. und *S. tuberosum* L. subspecies *andigena* (Buk.) Hawkes. Ibid. **27**, 341–50.

—— —— (1961) *Solanum tarijense* Hawk.—eine weitere gegen den Kartoffelnematoden *Heterodera rostochiensis* (Wollenweber) resistente Species. *Naturwissenschaften* **48**, 742–3.

—— and WITT, I. K. (1959) Untersuchungen über das Verhalten zentral- und südamerikanischer Kartoffel-Species nach Infektion mit dem Rippenbräune-Stamm des Y-virus (RBV). *Züchter* **29**, 271–6.

RUDORF, W. (1958) The significance of wild species for potato breeding. *Eur. Potato J.* **1**, 10–20.

RYBIN, V. A. (1930) Karyologische Untersuchungen an einigen wilden und einheimischen kultivierten Kartoffeln Amerikas. *Z. indukt. Abstamm.-u. VererbLehre* **53**, 313–54.

—— (1933) (Cytological investigations of the South American cultivated and wild potatoes, and its significance for plant breeding.) *Trudȳ prikl. Bot. Genet. Selek.* Ser. 2, No. 2, 3–100.

RYDBERG, P. A. (1924) The section *Tuberarium* of the genus *Solanum* in Mexico and Central America. *Bull. Torrey bot. Club* **51**, 145–54; 167–76.

RZHAVITIN, V. N. (1936) (Testing potatoes for frost resistance by laboratory methods.) *Trudȳ prikl. Bot. Genet. Selek.* Ser. A, No. 20, 107–14.

SABINE, J. (1824) On the native country of the wild potato, with an account of its culture in the garden of the Horticultural Society. *Trans. hort. Soc. Lond.* **5**, 249–59.

SAINT-HILAIRE, A. DE (1822) Esquisse de mes voyages au Brésil et Paraguay. Reprinted by: *Chronica bot.* **10**, 1–61 (1946).

SANZ DE CORTÁZAR, C. (1955) Observaciones cromosomales en plantas chilenas. II Parte. *Agricultura téc.* **15**, 5–11.

SCHAPER, P. (1939) Arbeiten und Probleme zur züchterischen Bekämpfung des Kartoffelkäfers. *Z. Pflanzenz.* **23**, 239–322.

—— (1953) Beitrag zur Resistenz des *Solanum chacoense* (Bitt.) gegen den Kartoffelkäfer (*Leptinotarsa decemlineata* (Say.).). *Züchter* **23**, 115–21.

—— (1954) Resistenzzüchtung bei Hackfrüchten. (a) Kartoffel (*Solanum tuberosum* L.). *Saatgut-Wirt.* **6**, 11–12.

SCHECHAJ, R. (1954) Avaluación de las variedades de papa según su resistencia a *Phytophthora infestans* (Mont.) de Bary, en plantaciones de primavera y verano en la Estación Experimental 'El Suncho', Catamarca. *Revta agron. NO Argent.* **1**, 135–55.

SCHLECHTENDAL, D. F. L. (1841) *Hortus halensis*. Halle.

SCHREIBER, K. (1955) *Solanum*-Alkaloide IV. Über die Glycoalkaloide einiger Wild-Kartoffeln der Ser. *Commersoniana*. *Pharmazie* **10**, 379–86.

SCHROEDER, JUAN. (1928) El *Solanum commersonii* Dunal. *Boln Acad. nac. Cienc. Córdoba* **29**, 355–413.

SCHULTZ, E. S. *et al.* (1937) Recent developments in potato breeding for resistance to virus diseases. *Phytopathology* **27**, 190–7.

SCHWARTZ, M. and MÜLLER-BÖHME, H. (1938) Untersuchungen über die Kartoffelkäferwiderstandsfähigkeit von Kartoffelwildformen und von Kreuzungen solcher Wildformen mit Kulturformen. *Mitt. biol. Reichsanst. Ld.-u. Forstw.* **58,** 47–54.

SEITHE, A. (1962) Die Haararten der Gattung *Solanum L.* und ihre taxonomische Verwertung. Engler, *Bot. Jb.* **81,** 261–336.

SIMMONDS, N. W. (1963) Abbreviations of potato names. *Eur. Potato J.* **6,** 186–90.

SLEESMAN, J. P. (1940) Resistance in wild potatoes to attack by the potato leaf-hopper and the potato flea beetle. *Am. Potato J.* **17,** 9–12.

——(1960) The National potato-breeding program, 1959. *Bull. U.S. Dept. Agric.* 184–8.

——(1963) Abstracts of papers presented at the 50th Anniversary meeting of the Potato Association of America and 13th National Potato Utilization Conference, Perkins Inn, Riverhead, New York, 15–18 July, 1963. *Am. Potato J.* **40,** 177–9.

SMITH, H. B. (1927) Chromosome counts in the varieties of *S. tuberosum* and allied wild species. *Genetics* **12,** 84–92.

SPEGAZZINI, C. (1902) Nova addenda ad Floram patagonicam. *An. Soc. cient. argent.* **53,** 136–44.

STAPP, C. and HARTWICH, W. (1956) Zur Frage der Resistenzverschiedenheiten pflanzlicher Wirte gegenüber pathogenen Bakterien und ihre Ursachen. II. Mitteilung. Weitere Untersuchungen über *Erwinia phytophthora. Zentbl. Bakt. Parasitkde* Abt. 2, **109,** 23–5, 611–27.

—— and SPICHER, G. (1955) Zur Frage der Resistenzverschiedenheiten pflanzlicher Wirte gegenüber pathogenen Bakterien und ihre Ursachen. I. Untersuchungen mit *Erwinia phytophthora*, dem Erreger der Schwarzbeinigkeit und Knollennassfäule der Kartoffel. Ibid. **108,** 465–81.

STARCEVA, L. I. (1960) (The resistance of wild species and hybrids of potato to *Epilachna vigintioctomaculata* Mtsch. and diseases.) (*Symp. Wk. post-grad. jun. sci. Wkr. Inst. Pl. Industr.*) Leningrad, No. 4, Abst. 4G500, p. 49.

STEBBINS, G. L. (1945) The cytological analysis of species hybrids. *Bot. Rev.* **11,** 463–86.

STELTER, H. and ROTHACKER, D. (1965) Eine Bemerkunge nzu der Nematodenresistenz der Arten *S. multidissectum* Hawk., *S. kurtzianum* Bitt. et Wittm. und *S. juzepczukii* Buk. *Züchter* **35,** 180–6.

STELZNER, G. (1938) Künstliche Selektionsmethoden zur Züchtung frostharter Kartoffeln. Ibid. **10,** 271–5.

——(1943*a*) Wege zur züchterischen Nutzung des *Solanum chacoense* Bitt. in Hinblick auf die Züchtung käferresistenter Kartoffelsorten. Ibid. **15,** 33–8.

——(1943*b*) Über die Fertilitätsverhältnisse bei Bastardierungen zwischen der frostfesten Wildkartoffel *Solanum acaule* Bitt. und der Kulturkartoffel *Solanum tuberosum* L. Ibid. 143–4.

——(1950) Virusresistenz der Wildkartoffeln. *Z. Pflanzenz.* **29,** 135–58.

—— and TORKA, M. (1948) *Solanum macolae*, eine neue käferfeste Wildkartoffel. Weitere Untersuchungen zur Käferresistenz der Tuberarien. *Züchter* **19,** 68–9.

STOUT, A. B. and CLARK, C. F. (1924) Sterilities of wild and cultivated potatoes with reference to breeding from seed. *Bul.* 1195 *U.S. Dept. Agric.* pp. 32.

SUTTON, A. W. (1909) Notes on some wild forms and species of tuber-bearing Solanums. *J. Linn. Soc. Lond.* **38,** 446–51.

SWAMINATHAN, M. S. (1950) Einige Verfahren für die Verwendung wilder *Solanum*-Arten zu Zuchtzwecken. *Züchter* **20,** 358–60.

——(1951) Notes on induced polyploids in the tuber-bearing *Solanum* species and their crossability with *Solanum tuberosum. Am. Potato J.* **28,** 472–89.

——(1953) Studies on the inter-relationships between taxonomic series in the section *Tuberarium*, genus *Solanum*. I. *Commersoniana* and *Tuberosa*. Ibid. **30,** 271–81.

SWAMINATHAN, M. S. (1954) Nature of polyploidy in some 48-chromosome species of the genus *Solanum*, section *Tuberarium*. *Genetics* **39**, 59–76.

—— (1956) Disomic and tetrasomic inheritance in a *Solanum* hybrid. *Nature, Lond.* **178**, 599–600.

THIRUMALACHAR, M. J., and PUSHKARNATH (1953) Resistance of potato varieties to charcoal rot. *Am. Potato J.* **30**, 73–7.

THOMPSON, N. R. and OLIEN, C. R. (1964) Frost tolerance in *S. acaule* × *S. tuberosum* hybrids. Ibid. **41**, 305.

THUNG, T. H. (1947) Potato diseases and hybridization. *Phytopathology* **37**, 373–81.

TKACZENKO, P. I. (1938) (New frost resistant potato seedlings.) *Plodov. Khoz.* No. 4, 24–5.

—— (1941) (New interspecific hybrids with a wild potato species.) *Vest. Ovoshch. i Kartofel'* No. 2, 93–4.

TORKA, M. (1943) Die Widerstandsfähigkeit eines *Solanum chacoense*-Klons gegenüber dem Kartoffelkäfer (*Leptinotarsa decemlineata* Say.). *Züchter* **15**, 145–8.

—— (1949) Die Resistenz von *Solanum chacoense* Bitt. gegen *Leptinotarsa decemlineata* Say. und ihre Bedeutung für die Kartoffelzüchtung. *Z. Pflanzenz.* **28**, 63–78.

—— (1950) II. Breeding potatoes with resistance to the Colorado Beetle. *Am. Potato J.* **27**, 263–71.

—— (1951) Zur Selbststerilität von *Solanum chacoense* Bitt. *Z. Pflanzenz.* **30**, 309–14.

—— (1954) Die Käferresistenz der Ser. *Commersoniana* von *Solanum* Sect. *Tuberarium*. *Züchter* **24**, 138–9.

TOXOPEUS, H. J. (1958) Voorlopige resultaten van het onderzoek van de Wageningse Aardappel Collectie. De eerste tien jaren. *Meded. Sticht. PlVered.* No. 20, 30–7.

—— and HUIJSMAN, C. A. (1952) Genotypical background of resistance to *Heterodera rostochiensis* in *Solanum tuberosum*, var. *andigenum*. *Nature, Lond.* **170**, 1017–18.

—— —— (1953) Breeding for resistance to potato root eelworm. I. Preliminary data concerning the inheritance and the nature of resistance. *Euphytica* **2**, 180–6.

TROUVELOT, B. (1938) État actuel des recherches sur les solanées tubérifères résistantes au doryphore. *Revue Zool. agric. appl.* **37**, 177–80.

TURRILL, W. B. (1920) Botanical Exploration in Chile and Argentina. *Bull. misc. Inf. bot. Gdns. Kew* 57–66, 223–4.

UPCOTT, M. (1935) The cytology of triploid and tetraploid *Lycopersicon esculentum*. *J. Genet.* **31**, 1–19.

VARGAS, C. (1950) Algunos resultados de investigaciones con papas nativas peruanas. La primera asamblea latinoamericana de fitogenetistas. *Follets. misc. Secr. Agric. Ganad. Méx.* **3**, 187–200.

VAVILOVA, M. A. (1960) (Breeding potatoes for frost resistance under the conditions of the Kola peninsula.) (*Symp. Wk. Post-grad. jun. sci. Wkr. Inst. Pl. Industr.*) Leningrad, 98–109.

VERNE, C. (1912) Sur les *Solanum maglia* et *tuberosum* et sur les résultats d'expériences de mutations gemmaires culturales entreprises sur ces espèces sauvages. *C. r. hebd. Seánc. Acad. Sci., Paris* **155**, 505–9.

VESELOVSKY, I. A. and KUZNECOVA, O. A. (1963) (A new cold-resistant potato hybrid.) *Bot. Zh. SSSR* **48**, 564.

VIIRSOO, E. V. (1954*a*) Informe sobre la resistencia a las heladas de algunas especies argentinas del género *Solanum*. *Revta agron. NO. argent.* **1**, 83–6.

—— (1954*b*) Observaciones en papas andinas. Ibid. 87–98.

—— (1956) El valor amiláceo de las papas autóctonas de Bolivia y Peru. Ibid. **2**, 197–224.

—— (1963) Las papas sudamericanas y su resistencia a las heladas. Ibid. **4**, 97–109.

VILMORIN, R. DE (1929) Étude cytologique de *Solanum commersonii*. *Archs Anat. microsc.* **25**, 382–7.

—— and SIMONET, M. (1927) Variations du nombre des chromosomes chez quelques solanées. *C. r. hebd. Séanc. Acad. Sci., Paris* **184**, 164–6.

WALKER, R. I. (1955) Cytological and embryological studies in *Solanum*, section *Tuberarium*. *Bull. Torrey bot. Club* **82**, 87–101.

WALPERS, W. G. (1844–5) *Repertorium botanices systematicae* **3**. Leipzig.

WANGENHEIM, K.-H. VON (1954) Zur Ursache der Kreuzungsschwierigkeiten zwischen *Solanum tuberosum* L. und *S. acaule* Bitt. bzw. *S. stoloniferum* Schlechtd. et Bouché. *Z. Planzenz.* **34**, 7–48.

—— (1957) Das Pachytän und der weitere Ablauf der Meiose in diploiden *Solanum*-Arten und -Bastarden. *Chromosoma* **8**, 671–90.

WEBB, R. E., and BUCK, R. W. (1955) Reaction of some *Solanum* species to virus Y and *Verticillium albo-atrum*. *The National Potato-Breeding Program*. 1954. U.S. Dep. Agric., pp. 216.

—— and HOUGAS, R. W. (1959) Preliminary evaluation of *Solanum* species and species hybrids for resistance to disease. *Pl. Dis. Reptr* **43**, 144–51.

WEDDELL, H. A. (1855–7) *Chloris Andina*. Paris.

WEISÆTH, G. (1956*a*) Frostresistente potetarter fra Sør-Amerikas potetflora. *Norsk Landbr.* 491–3.

—— (1956*b*) Frostresistente potetarter. Ibid. 528–30.

WETTSTEIN, R. von (1895) (in ENGLER, A., and PRANTL, K.) *Die natürlichen Pflanzenfamilien* **4**, Part 3b, 4–38. Solanaceae. Leipzig.

WILLIAMS, T. D. (1956) The resistance of potatoes to root eelworm. *Nematologica* **1**, 88–93.

—— (1958) Potatoes resistant to root eelworm. *Proc. Linn. Soc. Lond.* **169**, 93–104.

WILTSHIRE, F. G. (1913) Bibliographical notes. LIV. The Botany of the Antarctic Voyage. *J. Bot. Lond.* **51**, 355–8.

WITTMACK, L. (1914) Einige neue *Solanum*-Arten aus der *Tuberarium*-Gruppe. Engler, *Bot. Jb.* **50**, Suppl., 539–55.

ZADINA, J. (1958) (Resistance of wild potatoes to *Spongospora* powdery scab (*S. subterranea* Johnson).) *Sb. čsl. Akad. zemed. Ved.*, Rada Rostlinna vyroba **31**, 1115–26.

ZAICZEVA, N. D. (1950) (Initial material for potato breeding.) *Selekts. Semenov.* No. 7, 31–7.

ZHUKOVSKY, P. M. (1959) (To the centres of origin of cultivated plants in Latin America.) *Bot. Zh. SSSR* **44**, 262–72, 722–37.

ZVEREVA, P. A. (1945) (New methods of breeding frost resistant potatoes.) *Dokl. vses. Akad. sel.-khoz. Nauk* Nos. 4–5, 31–3.

—— (1951) (New hybrids of potatoes able to withstand frost.) *Selekts. Semenov*, **4**, 35–7.

INDEX

[1] Synonyms are in italic type. Principal page references are in bold type.

PLATES

Wooded valley slopes (Montano district) with forests of *Alnus jorullensis* (aliso), *Podocarpus parlatorei* (pino) and other trees. In the far distance can be seen the high peaks of the Aconquija mountains. In this wooded region *S. vernei*, *S. microdontum* subsp. *gigantophyllum*, and (in the higher more open parts) *S. venturii* are to be found. Estancia 'Yunka Suma', dept. Andalgalá, prov. Catamarca.

Forests of the Montano district of Estancia 'Yunka Suma', where *S. vernei* and *S. microdontum* subsp. *gigantophyllum* are frequent. The trees are mostly *Alnus jorullensis* and *Sambucus peruviana*.

PLATE 3

S. microdontum subsp. *gigantophyllum* at Estancia 'Yunka Suma', dept. Andalgalá, prov. Catamarca, at 1550 m. This plant was growing amongst stones on the edge of the *Alnus* forest and cultivated fields in the Montano district (see *P. & H. P–26*).

S. venturii, growing in the open grassy meadows ('pajonales') at the upper limit of the Montano district, with occasional small trees and shrubs of *Polylepis australis* (queñoa). Estancia 'Yunka Suma', dept. Andalgalá, prov. Catamarca. Altitude 2450 m (*H.H.R. 3539*).

PLATE 5

Road in northern Misiones province at 350 m, in the Mixed Forest district of the Eastern Subtropical phytogeographical province. In this region *S. chacoense* subsp. *muelleri* was collected (*P. & H. P-66*).

Harvesting tubers of the wild potato species *S. chacoense* on the roof of a pig sty at Campo Largo, prov. Chaco. This is in the Chaco phytogeographical province at 100 m above sea level (*P. & H. P–78*). Photo: E. Petersen.

General view of the Quebrada Grande, prov. Jujuy, photographed from Loma del Tambo. The vegetation of the lower mountain slopes and valley bottom in un-irrigated parts belongs to the Prepuna province and is characterized by spiny shrubs and columnar cacti. This province gives way to the Puna province on the higher slopes.

PLATE 8

Prepuna vegetation in the Quebrada Grande, prov. Jujuy. *S. gourlayi* and *S. infundibuliforme* are to be found in certain parts of this phytogeographical province.

View of the lower Prepuna at 1300 m near Belén, prov. Catamarca. Note the typical vegetation of giant cacti (*Trichocereus* spp.), the bromeliad, *Dyckia velascana*, leguminous shrubs, etc. Photo: H. Sleumer.

PLATE 10

View of Monte vegetation in San Juan province, dept. Calingasta at Quebrada de Leoncito. In the distance are the snow-capped peaks separating Argentina from Chile. The plants in the foreground are *Larrea* (creosote bush). *S. kurtzianum*h as been found in this area.

PLATE 11

Llamas grazing on the Puna vegetation near Iturbe, prov. Jujuy, at about 3,400 m. This is typical dry puna, in which *S. megistracrolobum* and *S. infundibuliforme* were collected (*H.P. & R. 64, 75–80*). Photo: K. Rahn.

PLATE 12

Llamas grazing on wetter Puna vegetation, with *Stipa* grass dominant, on the road from La Quiaca to Santa Victoria, in Salta province, at about 4400 m. At this altitude the only wild potato species to be found is *S. acaule*.

Small plants of *S. acaule* growing at the base of a wall surrounding a corral or cattle enclosure. High Puna above Rincón de Cajas, dept. Yavi, prov. Jujuy, at 4125 m (*H.H.R. 3813*).

S. megistacrolobum growing in the Puna province in the region of San Gregorio between 3700 and 4000 m, dept. Tilcara, prov. Jujuy. This rosette form was at first thought to be a distinct species, *S. tilcarense*.

Nothofagus forest (Subantarctic province) near Ruca Malén in the Parque Nacional de Nahuel Huapi, at about 41° S. The non-tuberiferous species, *S. brevidens*, was collected in this region (*Böcher, Hjerting & Rahn 1711*). Photo: T. Böcher.

S. brevidens Phil. Plant grown at Birmingham from seed sent by Sr. J. Diem (*P. & H. 1890* = H. 610). This material matches very closely the specimens collected in Argentina and Chile (cf. Fig. 6 and Plate 17). Note the very short calyx acumens and corolla lobes.

S. brevidens Phil. A less pubescent form collected by Comber (*No. 942*) at Lake Huechulafquen, Argentina, which approaches var. *glabrescens* Hawkes. The pedicels above the articulation are glabrous but the leaf pubescence is dense, though rather short.

PLATE 18

S. calvescens Bitt. Syntype. Collected by S. E. Henschen (*Regnell III 1623*—herb. Uppsala) at Caldas, Minas Gerais, Brazil. Note the large terminal and few lateral leaflets (cf. *S. commersonii*), the well-developed petiolules and long style.

S. calvescens Bitt. Flower dissection of Topotype collection sent by Prof. F. G. Brieger in December 1963 and grown in Birmingham, 1964. Accession No. H. 653. Note the minute calyx acumens, broadly triangular corolla lobes and stamens with narrow anthers which are not sharply delimited from filaments.

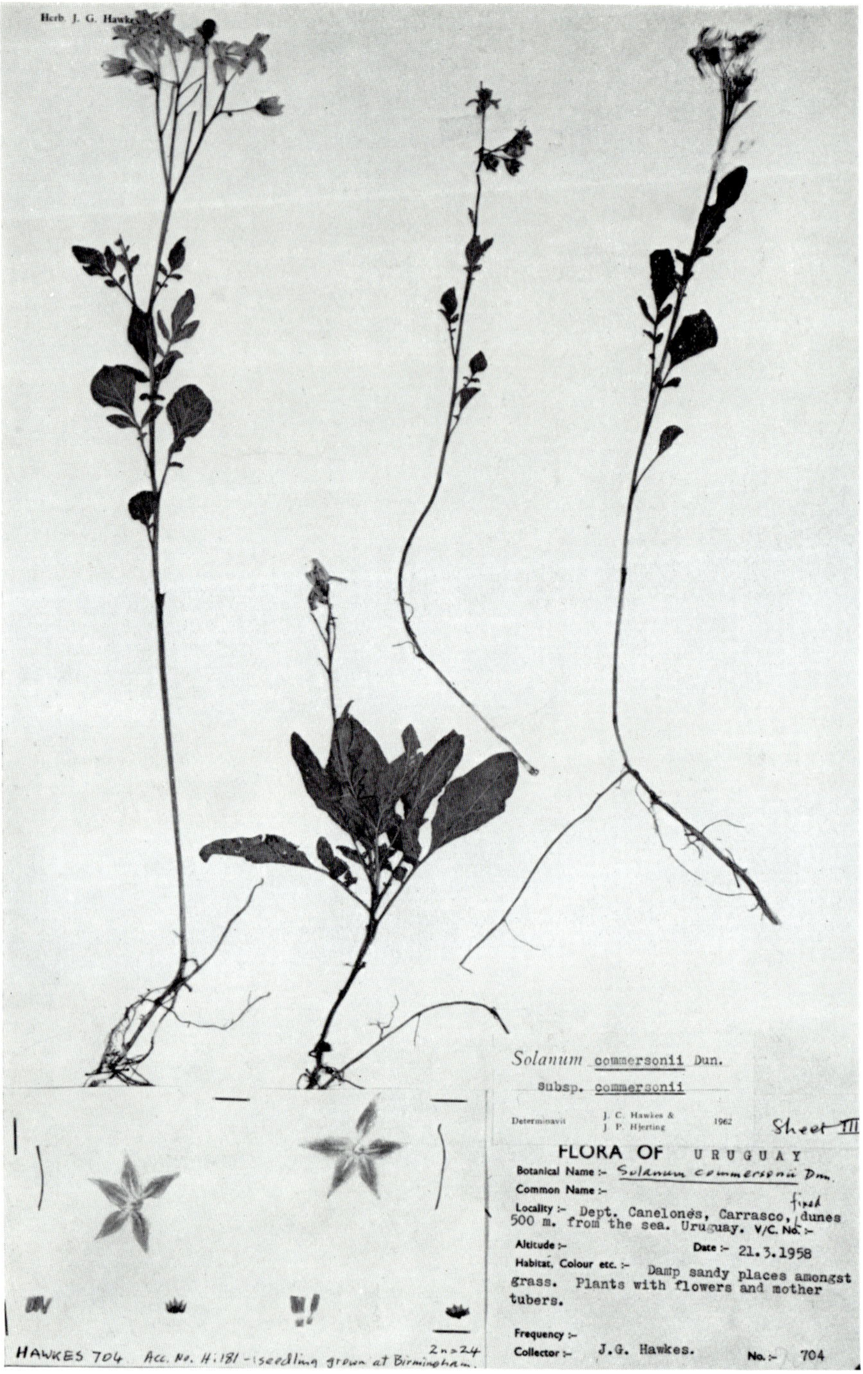

S. commersonii Dun. Typical growth of diploid material (*Hawkes 704*. $2n = 24$); collected in damp sandy places near the coast in Uruguay. Differences of habit (rosettes or long stemmed) are due to differences in the height of the surrounding vegetation. Note the sessile lateral leaflets and enlarged terminal, loose inflorescence, and large purple corolla.

S. commersonii Dun. Collected near Pelotas, Río Grande do Sul, Brazil (*Hawkes 717*). Typical habit and leaf form of material growing in good soil and open habitat. Note sessile, slightly decurrent lateral leaflets and enlarged terminal. Corolla as in Plate 20. General appearance of a diploid, though no counts made.

S. commersonii Dun. Collected on the coast near Montevideo, Uruguay (*Hawkes 701*). This is an anomalous triploid form with more upright growth and more petiolulate leaflets than those of the diploids (contrast with Plate 21).

S. commersonii Dun. Collected by B. Rambo (Accession No. H. 93) near Porto Alegre, Río Grande do Sul, Brazil. An anomalous triploid form; the present specimen was grown under glass in Birmingham but field-grown plants are of much the same form. This collection approaches very closely the 'conventional' or plant breeders' forms of *S. commersonii* which are also triploid. Note the well-marked petiolules and very reduced peduncle forks and compare with Plate 24.

S. commersonii Dun. The 'conventional' or plant breeders' triploid form of this species, considered by Bukasov to be *S. commersonii sensu stricto* (herb. Leningrad). The petiolulate leaflets and reduced peduncle branches are well marked.

S. commersonii Dun. Type specimen of '*S. henryi*' Buk. & Lechn., grown at Leningrad from material collected by T. Henry (Accession K. 189—herb. Leningrad, WIR) at La Estanzuela, Colonia, Uruguay. This diploid form may be compared with the diploid collections *Hawkes 705* & *706* from the same locality. Note the sessile leaflets and enlarged terminal, very similar to Commerson's type (Fig. 8).

S. commersonii Dun. Type specimen of '*S. mechonguense*' Buk., grown at Leningrad from material collected by R. Millán (*M.A. 31*—K. 1088—herb. Leningrad, WIR. $2n = 24$) at Mechongué, prov. Buenos Aires, Argentina. Compare with *Hawkes 708* from Balcarce in the same general region. There are more lateral leaflets than usual in this specimen.

S. commersonii Dun. Type specimen of '*S. mercedense*' Buk. (= *S. compactum* Buk., *nom. nud.*), grown at Leningrad from material collected by T. Henry (Accession K. 1071—herb. Leningrad, $2n = 24$, WIR), near Mercedes, dept. Soriano, Uruguay. This form of *S. commersonii* possesses fewer lateral leaflets and shorter calyx acumens than '*S. mechonguense*' (Plate 26).

S. commersonii Dun. Type specimen of '*S. sorianum*' Buk., grown at Leningrad from material collected near Mercedes, dept. Soriano, Uruguay, by T. Henry (Accession K. 1073—herb. Leningrad, WIR, $2n = 24$). This is said to be very similar to '*S. henryi*'.

S. commersonii Dun. Type specimen of '*S. rionegrinum*' Lechn., *nom. nud.*, grown at Leningrad from material collected by Rómulo Rubbo (Accession K. 1216, 2884—herb. Leningrad) at Naranjo, Río Negro, dept. Río Negro or Soriano, Uruguay. This seems highly likely to be a form of *S. commersonii* Dun.

PLATE 30

Probable hybrid of *S. commersonii* Dun. × *S. chacoense* Bitt., collected from Partido de La Plata, prov. Buenos Aires, Argentina, by A. Cabrera (*No. 5805*—herb. La Plata). This plant possesses the reduced peduncle branches, obtuse leaflet apex, and well-developed calyx acumen of *S. commersonii*, together with the long leaves, tall habit, and long petiolules of *S. chacoense*.

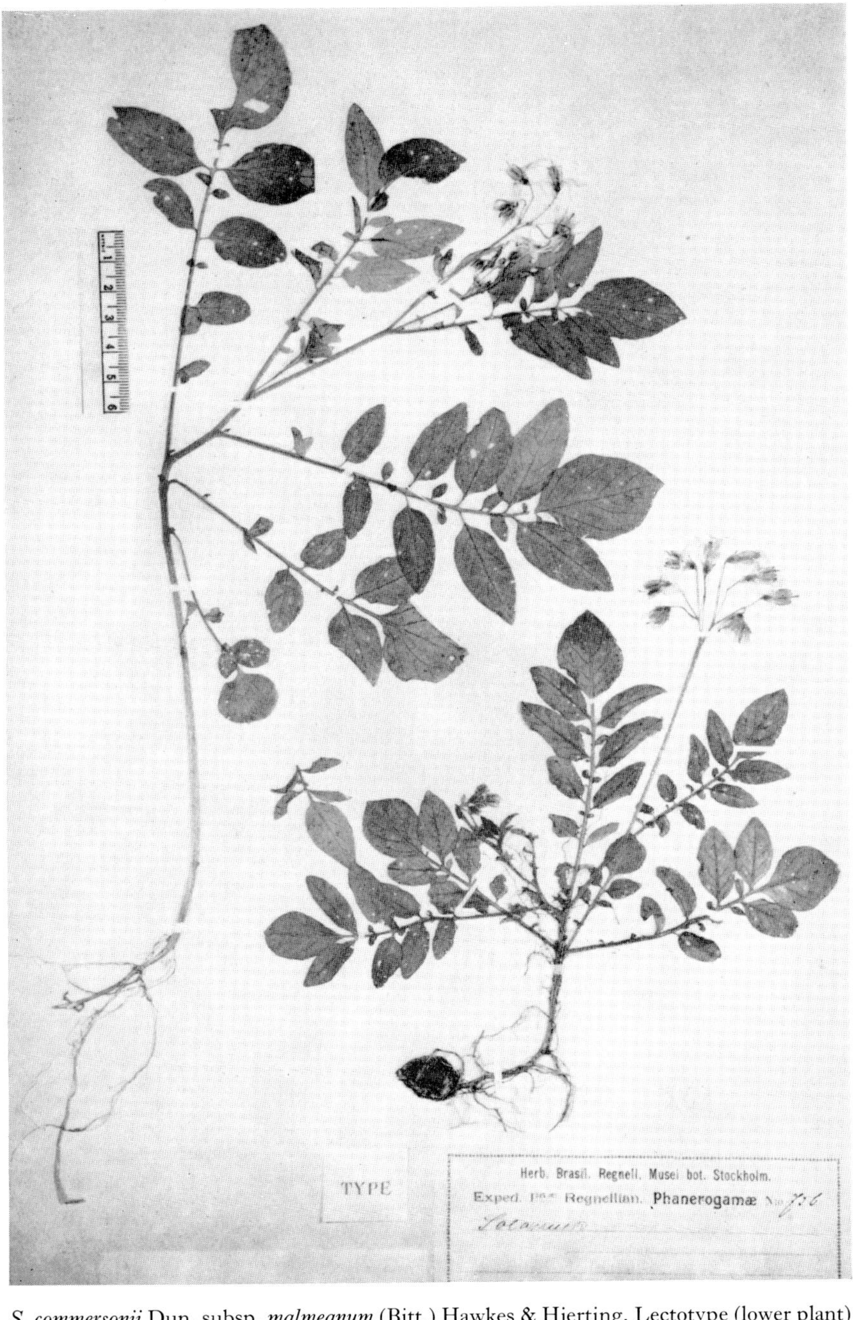

S. commersonii Dun. subsp. *malmeanum* (Bitt.) Hawkes & Hjerting. Lectotype (lower plant) and isotype of '*S. malmeanum*' Bitt., collected by G. A. Malme (*Regnell* 756—herb. Stockholm) at Ijuí, Río Grande do Sul, Brazil. Note the distinct petiolules, large number of lateral leaflets and semi-rosette habit of one plant and compare with Plate 32.

PLATE 32

S. commersonii Dun. subsp. *malmeanum* (Bitt.) Hawkes & Hjerting. A collection made by Petersen and Hjerting (*No. P–70*) in prov. Misiones, Argentina. This is triploid ($2n = 36$) and possesses fairly wide corolla lobes and a large number of petiolulate lateral leaflets.

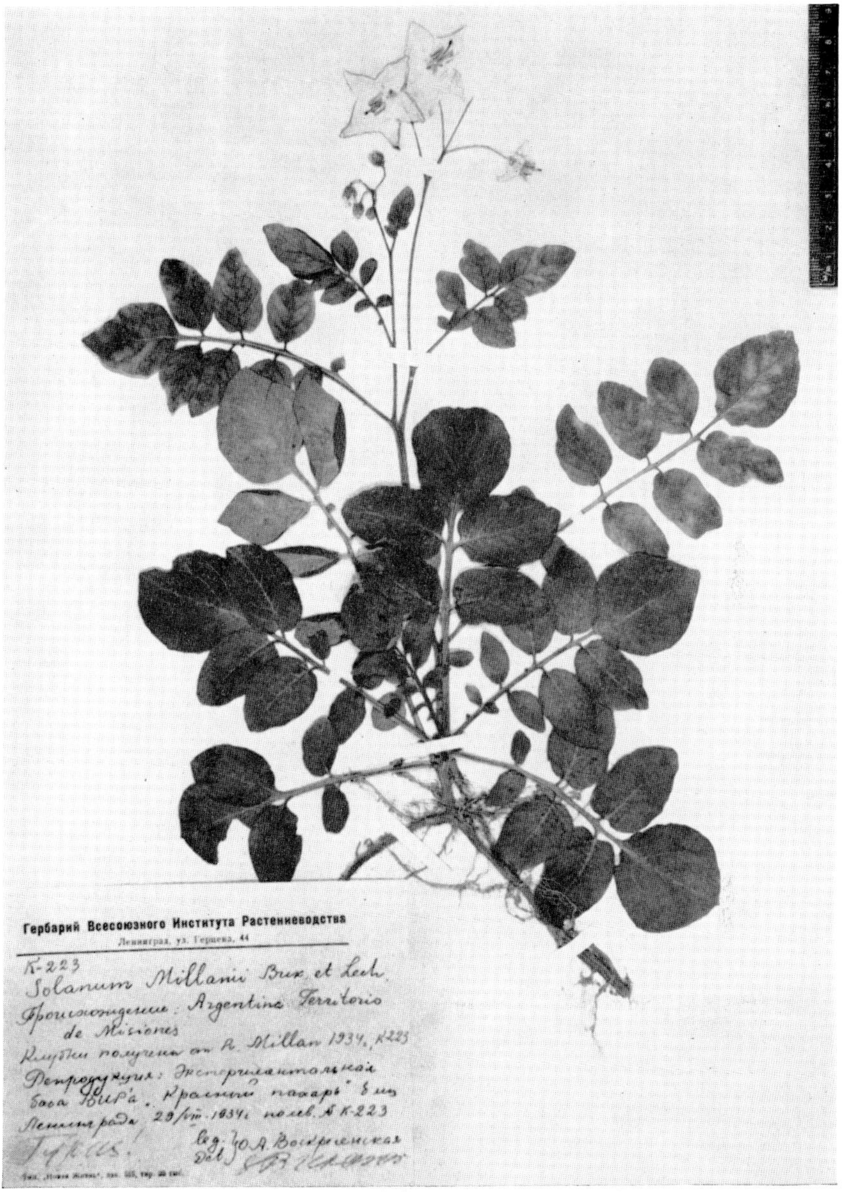

S. commersonii Dun. subsp. *malmeanum* (Bitt.) Hawkes & Hjerting. Type specimen of '*S. millanii*' Buk. & Lechn., grown at Leningrad from material collected by Millán (*No. 973*; accession K. 223—herb. Leningrad, WIR) at Loreto, prov. Misiones, Argentina. The original material was a mixture of diploid and triploid clones.

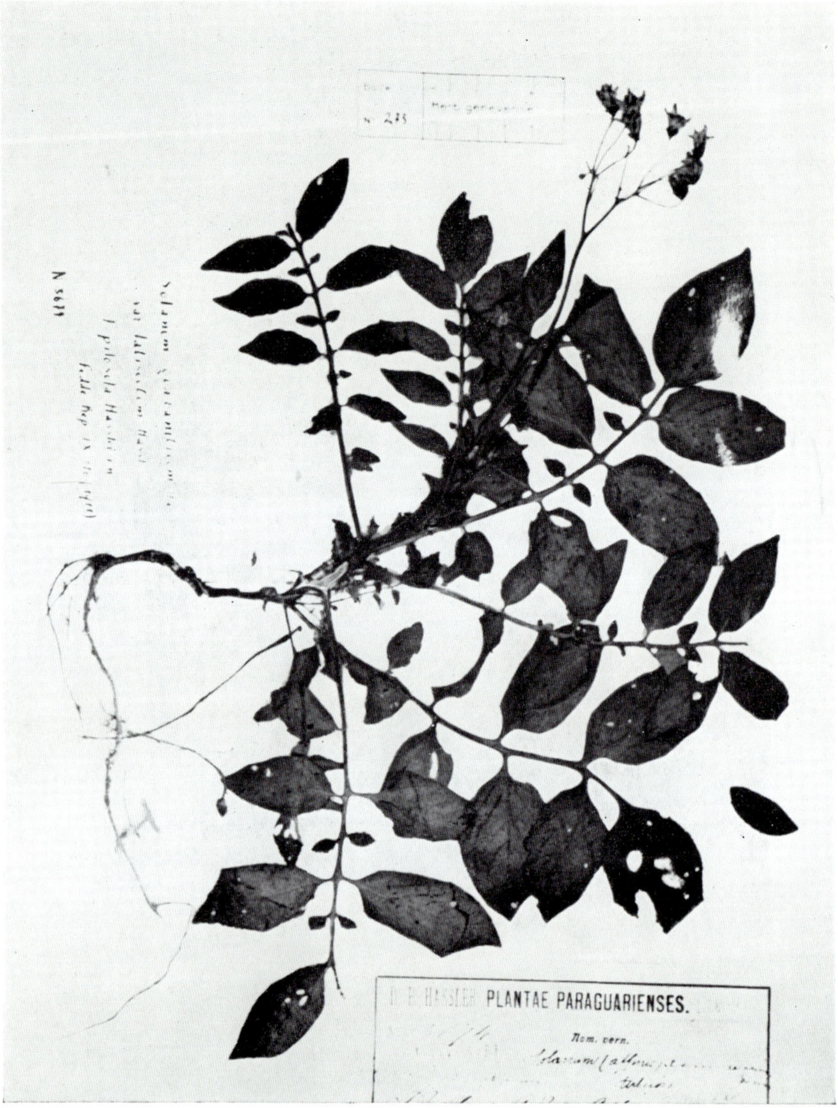

S. commersonii Dun. subsp. *malmeanum* (Bitt.) Hawkes & Hjerting. Type specimen of '*S. guaraniticum* Hassl. var. *latisectum* Hassl. f. *pilosulum*' Hassl. Collected by Hassler (*No. 3674*—herb. Geneva) at San Bernardino, Paraguay. Note the rosette habit and the shorter petiolules than in *S. chacoense*.

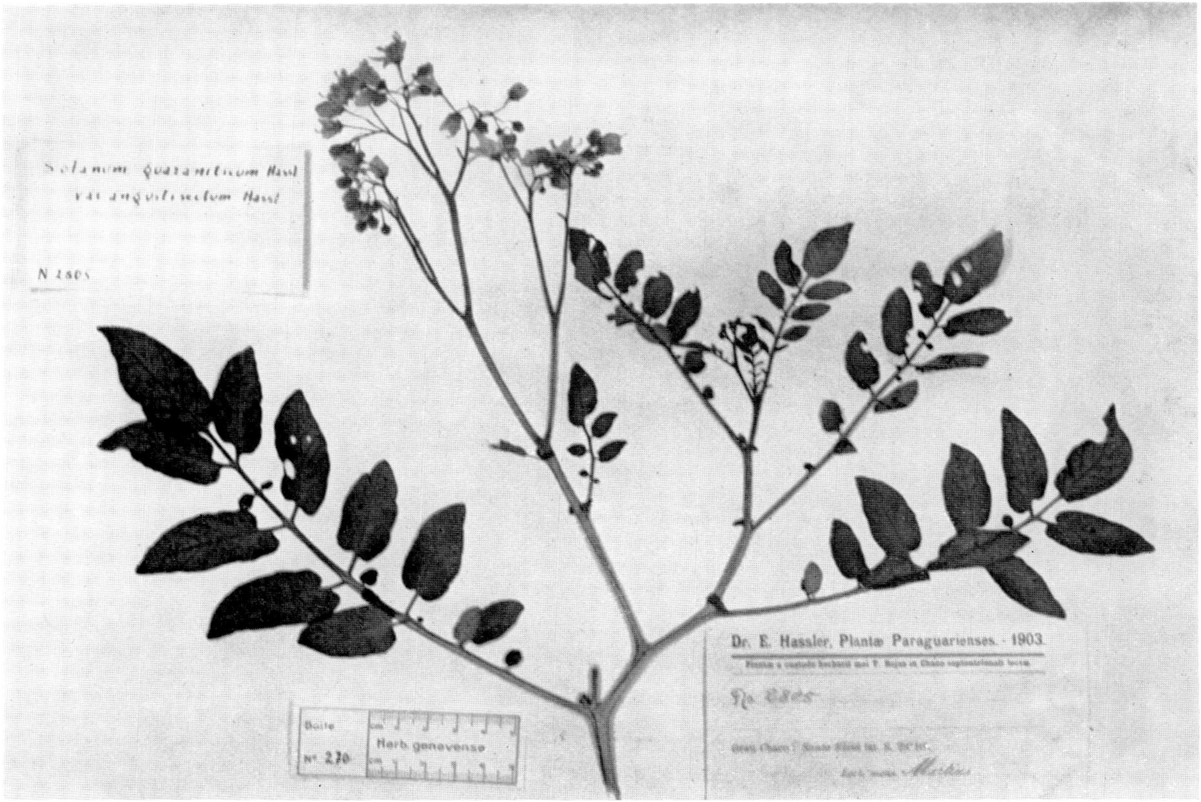

S. chacoense Bitt. Lectotype selected by Correll. One of Hassler's specimens of '*S. guaraniticum* Hassl. var. *angustisectum*' Hassl., collected by Hassler (*No. 2805*—herb. Geneva) at Santa Elisa, Paraguay (see also Fig. 15). Note the long petiolules and peduncles with several bifurcations, and contrast with Plate 34.

S. chacoense Bitt. Type for '*S. guaraniticum* Hassl. var. *latisectum* Hassl. f. *glabrescens*' Hassl. Collected by Hassler (*No. 326*—herb. Geneva) in the Cordilleras de Altos, Paraguay. This is a poorly developed form with wider leaflets (cf. Plate 37), but still showing the clearly marked petiolules typical of *S. chacoense*.

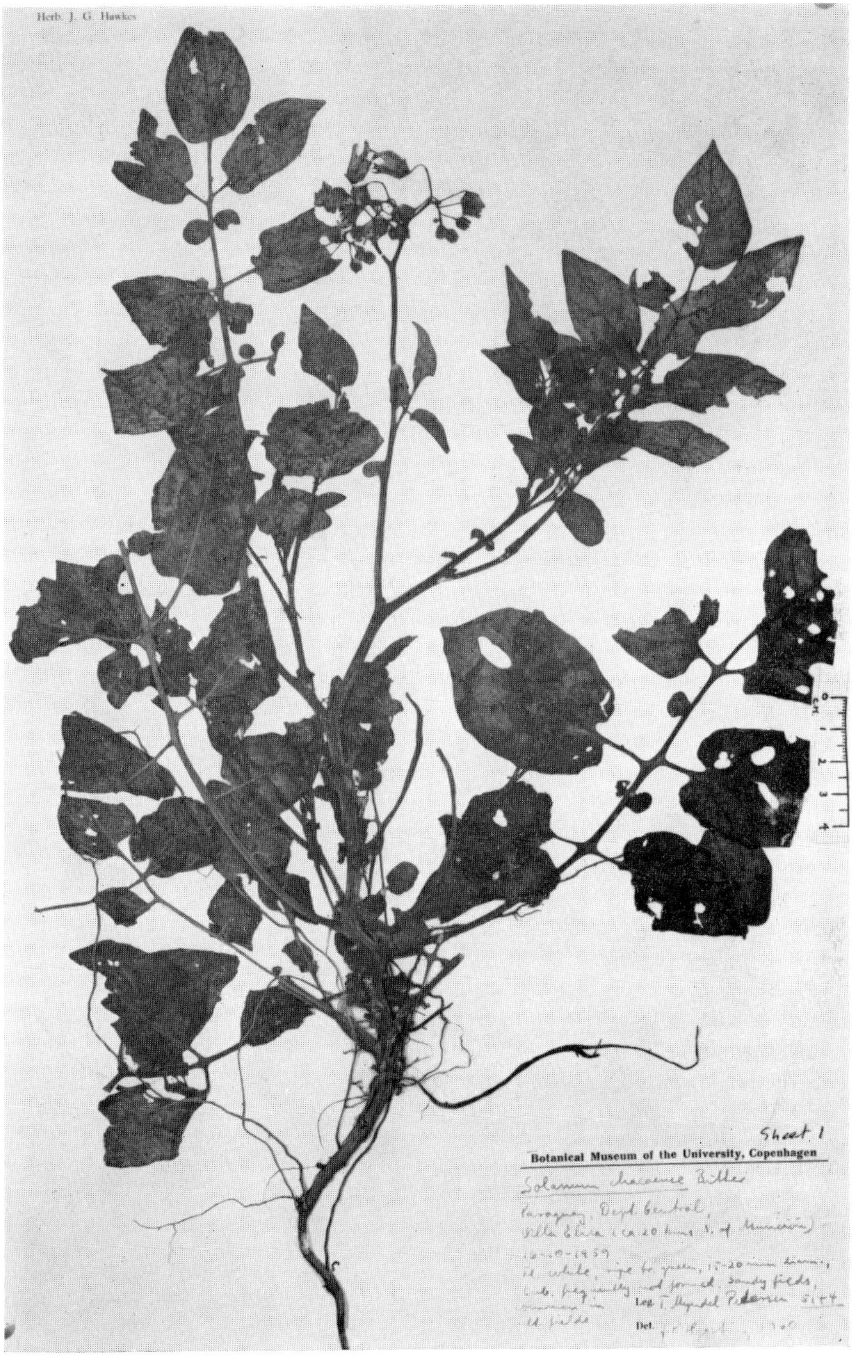

S. chacoense Bitt. A collection from Villa Elisa, near Asunción, Paraguay, collected by T. Myndel Pedersen (*No. 5144*), which matches very closely *Hassler 326* (see Plate 36) in leaflet dimensions. These forms with wider leaflets are not worth distinguishing from the typical form as they are connected to it by many intermediates.

S. chacoense Bitt. Type specimen of '*S. boergeri*' Buk., grown at Leningrad from material collected by T. Henry (accession K. 1070—herb. Leningrad, WIR) near Río Uruguay, Uruguay. This is a typical plains form of *S. chacoense* with long petiolules and very short calyx acumens, 0·5 mm long.

S. chacoense Bitt. Type specimen of '*S. laplaticum*' Buk., grown at Leningrad from material collected by R. Millán (*No. G–972*. Accession K. 597—herb. Leningrad, WIR) from the neighbourhood of the city of Buenos Aires, Argentina. This is undoubtedly a poor atypical stunted specimen of *S. chacoense*, possibly infected with virus. The long petiolules and short calyx acumens (0·3 mm) link it to that species and not to *S. commersonii* Dun.

PLATE 40

S. chacoense Bitt. Authenticated specimen of '*S. gibberulosum*' Juz. & Buk., grown at Leningrad (herb. J.G.H.), presumably from material of the type collection of Millán (*G–1183*) from Manfredi, Córdoba, Argentina. This is a regional variant of *S. chacoense* with vigorous habit and gibbous hairs. Correll classifies it as *S. chacoense* forma *gibberulosum*.

S. chacoense Bitt. Authenticated specimen of '*S. garciae*' Juz. & Buk., grown at Leningrad (herb. P.G.S.), presumably from the type collection of García (*Millán 1201*), collected near the city of Córdoba, Argentina. This is typical of plains *S. chacoense* and shows no gibbous hairs (contrast '*S. gibberulosum*').

S. chacoense Bitt. Type specimen of '*S. parodii*' Juz. & Buk., grown at Leningrad from material collected by R. Millán (*No. G–980*, accession K. 603—herb. Leningrad, WIR) at Río Chico, Tucumán, Argentina. The broad crisped stem wing shows that some *S. microdontum* introgression has taken place here.

S. chacoense Bitt. Authenticated specimen of '*S. schickii*' Juz. & Buk., grown at Leningrad (herb. Leningrad, WIR), presumably from the type material of Schick from Siambón, prov. Tucumán, Argentina.

S. chacoense Bitt. Type specimen of '*S. knappei*' Juz. & Buk., grown at Leningrad from material sent by P. T. Knappe (herb. Leningrad, WIR), and said to have been collected at Salta, Argentina. This specimen also shows *S. microdontum* introgression.

S. chacoense Bitt. Type specimen of '*S. jujuyense*' Hawkes. Collected by Balls, Gourlay and Hawkes (*Balls 5921*—herb. P.G.S.) from San Antonio, near Jujuy, Argentina. This material is highly introgressed, with crisped stem wings, no rosette, short petiolules, high articulation, pubescent calyx and 0·5–1 mm long calyx acumens.

PLATE 46

S. chacoense Bitt. Paratype specimen of '*S. saltense*' Hawkes, grown at Cambridge from material collected by Balls, Gourlay and Hawkes (*Balls 5936* p.p. CPC: 51b—herb. P.G.S.) near Salta, Argentina. This material shows evidence of strong introgression, possessing crisped stem wings, no rosettes, 1–3 mm long petiolules, above-centre articulation, 1 mm long calyx acumens and dense calyx pubescence.

S. chacoense Bitt. Type of '*S. emmeae*' Juz. & Buk., grown at Leningrad from material sent by P. T. Knappe (herb. Leningrad, LE), possibly collected near Tucumán, Argentina. Seems to be a plains form with semi-rosette habit, but nevertheless possesses a high pedicel articulation.

S. chacoense Bitt. subsp. *muelleri* (Bitt.) Hawkes & Hjerting. Typical material from N.E. Argentina, showing the narrow leaflets with oblique base and the fairly large corolla. Collected by Petersen and Hjerting (*No. P–66*; CPC: 2426) in prov. Misiones, Argentina.

S. chacoense Bitt. subsp. *muelleri* (Bitt.) Hawkes & Hjerting. Type specimen of '*S. muelleri* Bitt. f. *densipilosum*' Corr. A densely pubescent form collected by Smith, Reitz and Caldato (*No. 9516*—herb. US) at Chapecó, Santa Catarina, Brazil. Agrees well with typical subsp. *muelleri* in all features but pubescence.

PLATE 50

Natural hybrid of *S. chacoense* Bitt. × *S. microdontum* Bitt. subsp. *gigantophyllum* (Bitt.) Hawkes & Hjerting. Collected by Petersen and Hjerting (*No. P–32*) from dept. Andalgalá, prov. Catamarca, Argentina. Note the leaves intermediate in form between those of the two parents; stem wings well marked; corolla substellate. Compare with Plate 51.

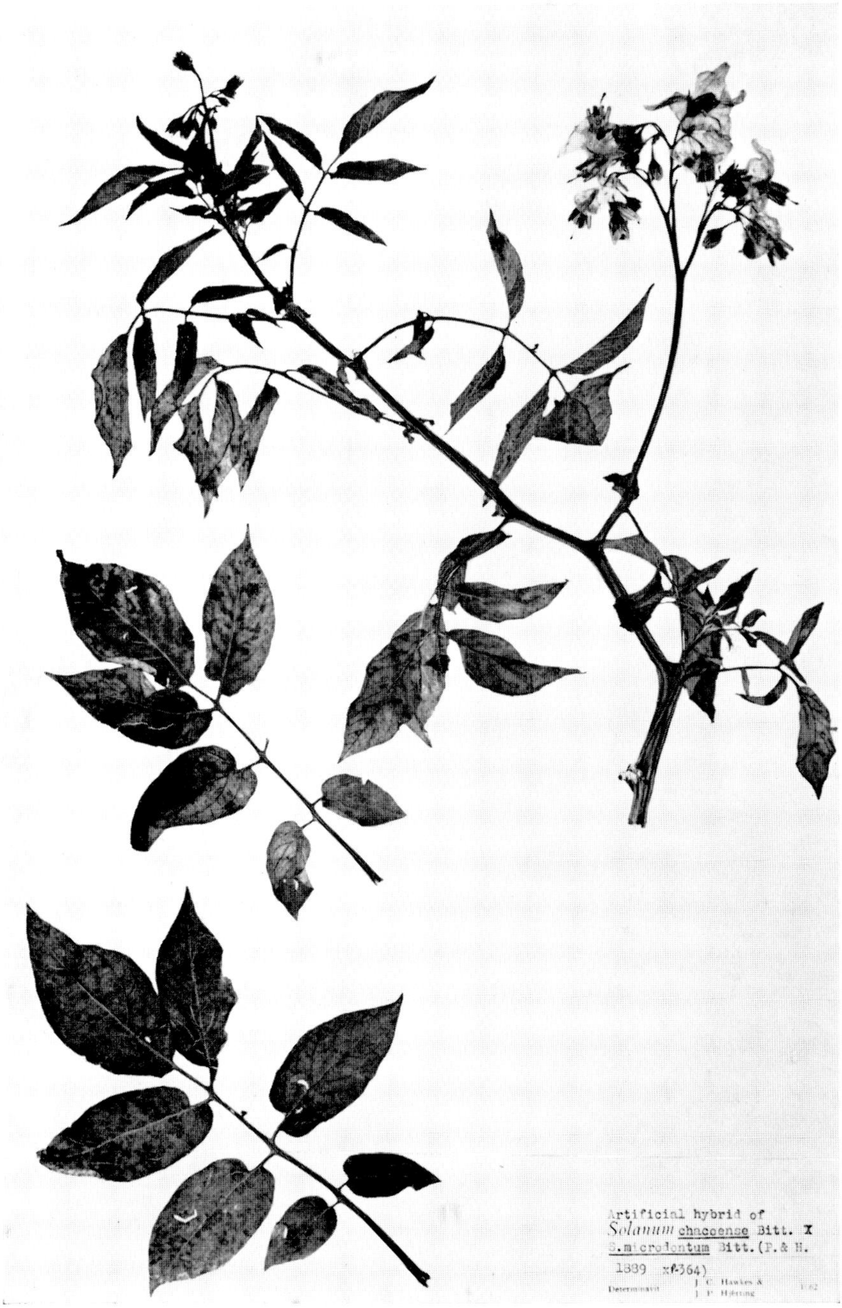

Artificial hybrid of *S. chacoense* Bitt. (*P. & H. 1889* from Tucumán) × *S. microdontum* Bitt. (*P. & H. P–364* from Salta—intermediate between two subspecies). This hybrid is approximately intermediate between the two parents but has a rather more robust stem and slightly wider leaflets than those of *P. & H. P–32* (see Plate 50).

Natural hybrid of *S. chacoense* Bitt. × *S. spegazzinii* Bitt. Collected by Sleumer and Vervoorst (*No. 2428*) at Pozo de Piedra, Belén, prov. Catamarca, Argentina. Leaf quite close to that of *S. spegazzinii*, but the large white sub-stellate corolla shows the strong influence of *S. chacoense* (see Fig. 21 and Plates 53, 54).

Natural hybrid of *S. chacoense* Bitt. × *S. spegazzinii* Bitt. Type collection of '*S. puberulo-fructum*' Corr., collected by Sleumer and Vervoorst (*No. 2435*) from Pozo de Piedra, Belén, prov. Catamarca, Argentina. This approaches close to *S. spegazzinii* and is one of a large number of interesting collections made in that region and showing considerable hybrid vigour (see Plates 52, 54, 55 and text pp. 235–6).

PLATE 54

Artificial hybrid of *S. chacoense* Bitt. (*H.P.R. 356* from prov. Salta, Argentina) × *S. spegazzinii* Bitt. (*H.P.R. 334* from prov. Salta, Argentina). This matches fairly well the natural hybrid (*Sleumer & Vervoorst 2428*) shown on Plate 52.

Natural hybrid of *S. chacoense* Bitt. × *S. kurtzianum* Bitt. & Wittm., collected by Vervoorst (*No. 3427* p.p.) in dept. Pomán, prov. Catamarca, Argentina. Note the very low articulation of the pedicels in flower and fruit and compare with Plate 56.

PLATE 56

Artificial hybrid of *S. chacoense* Bitt. (*P. & H. 1889*, from prov. Tucumán, Argentina) × *S. kurtzianum* Bitt. & Wittm. (*Ruiz Leal 11906* from dept. Las Heras, prov. Mendoza, Argentina). Note the similarities of leaf and inflorescence to Plate 55, though the pedicel articulation is not so low.

S. tarijense Hawkes. Topotype material grown at Birmingham from seed collected by M. Cárdenas at Tarija, Bolivia. (Accession No. H. 17.) This compares very closely with the wild material from Argentina collected by Sleumer (see Plate 58).

S. tarijense Hawkes. Collected by Sleumer (*No. 4023*) at Sta. Victoria, prov. Salta, Argentina. Compare with Plate 57.

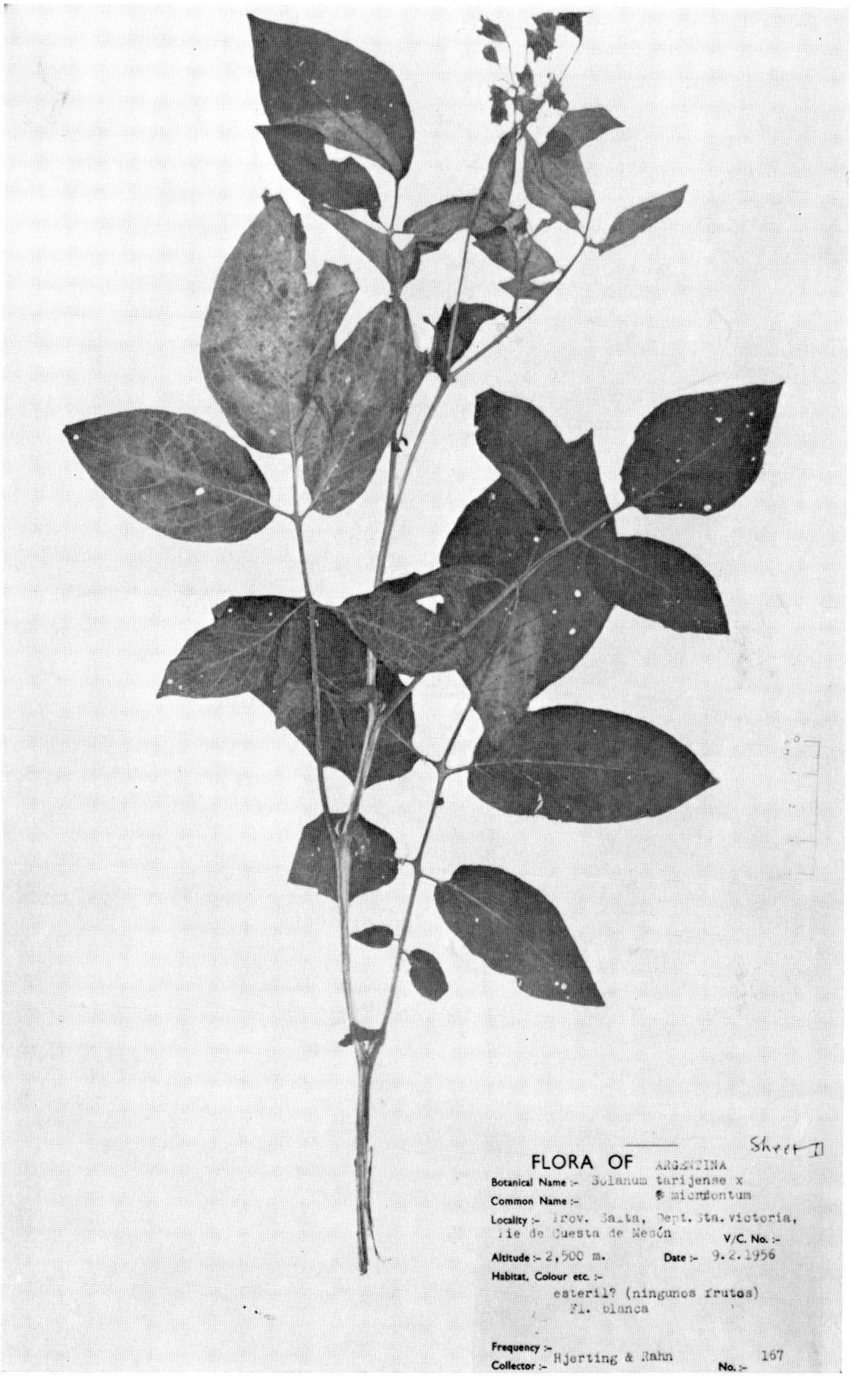

Natural hybrid of *S. tarijense* Hawkes × *S. microdontum* Bitt. collected by Hjerting and Rahn (*No. 167*) at Pie de Cuesta de Mesón in Prov. Salta, Argentina. Note the fewer broader leaflets and the wider terminal leaflet, showing the influence of *S. microdontum*.

Natural hybrid of *S. tarijense* Hawkes × blue-flowered species. Collected at Quebrada de San Felipe, prov. Salta, Argentina, by Sleumer (*No. 3786*). Note the pentagonal corolla which is pale blue in colour, showing that a species with dark blue rotate corolla was the other parent (possibly *S. vernei*, *S. vidaurrei*, or *S. gourlayi*).

S. acaule. Bitt. Lectotype specimen (upper left-hand plant) of *Fiebrig 3429* (herb. Stockholm), collected at Puna Patanca, south Bolivia, close to the Argentine border (see Map 7 and Fig. 22).

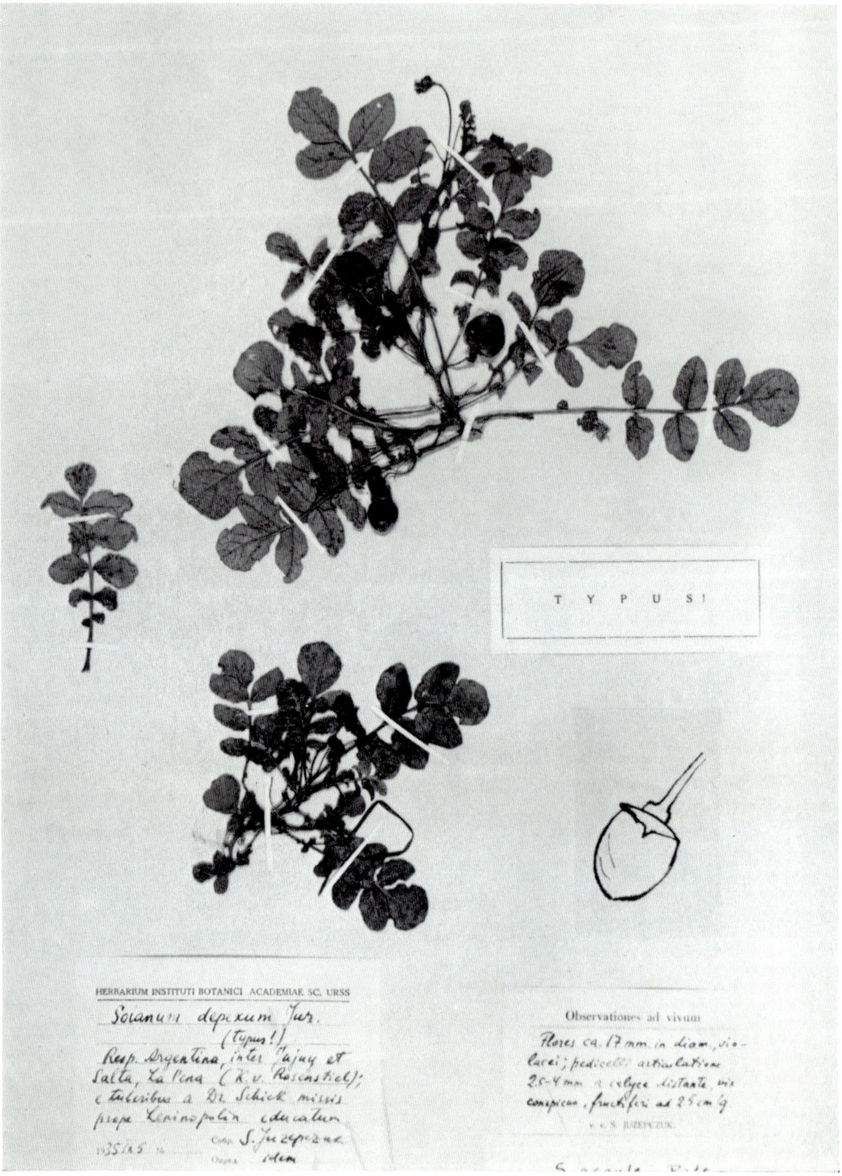

S. acaule Bitt. Type specimen of '*S. depexum*' Juz., collected by Rosenstiel at La Peña, between Jujuy and Salta, Argentina (herb. Leningrad, LE). This does not differ in any essential feature from the type material of *S. acaule*, except for the absence of interjected leaflets and the slightly visible articulation.

S. acaule Bitt. Type specimen of '*S. schreiteri*' Buk., *nom. nud.*, grown at Leningrad from material collected by Schick at San José, between Jujuy and Salta, Argentina (herb. Leningrad, WIR). This is very similar to the type specimen of '*S. depexum*', but does not possess an articulation. It does not differ materially from the type of *S. acaule*.

S. acaule Bitt. A very stunted form from extremely high altitude (4650 m), collected by Petersen and Hjerting (*No. P–361*) from above Mina Aguilar, prov. Jujuy, Argentina. Contrast this with Plate 65.

S. acaule Bitt. A collection with long stem and vigorous growth from a rather low altitude (3150 m) for this species at Mesón, prov. Salta, Argentina. Collected by Hjerting and Rahn (*No. 173*). Contrast this with Plate 64.

S. acaule Bitt. A vigorous plant collected by Petersen and Hjerting (*No. 1897*) from the southernmost part of its range, Sierra Famatina, prov. La Rioja, Argentina. Compare with Plate 67.

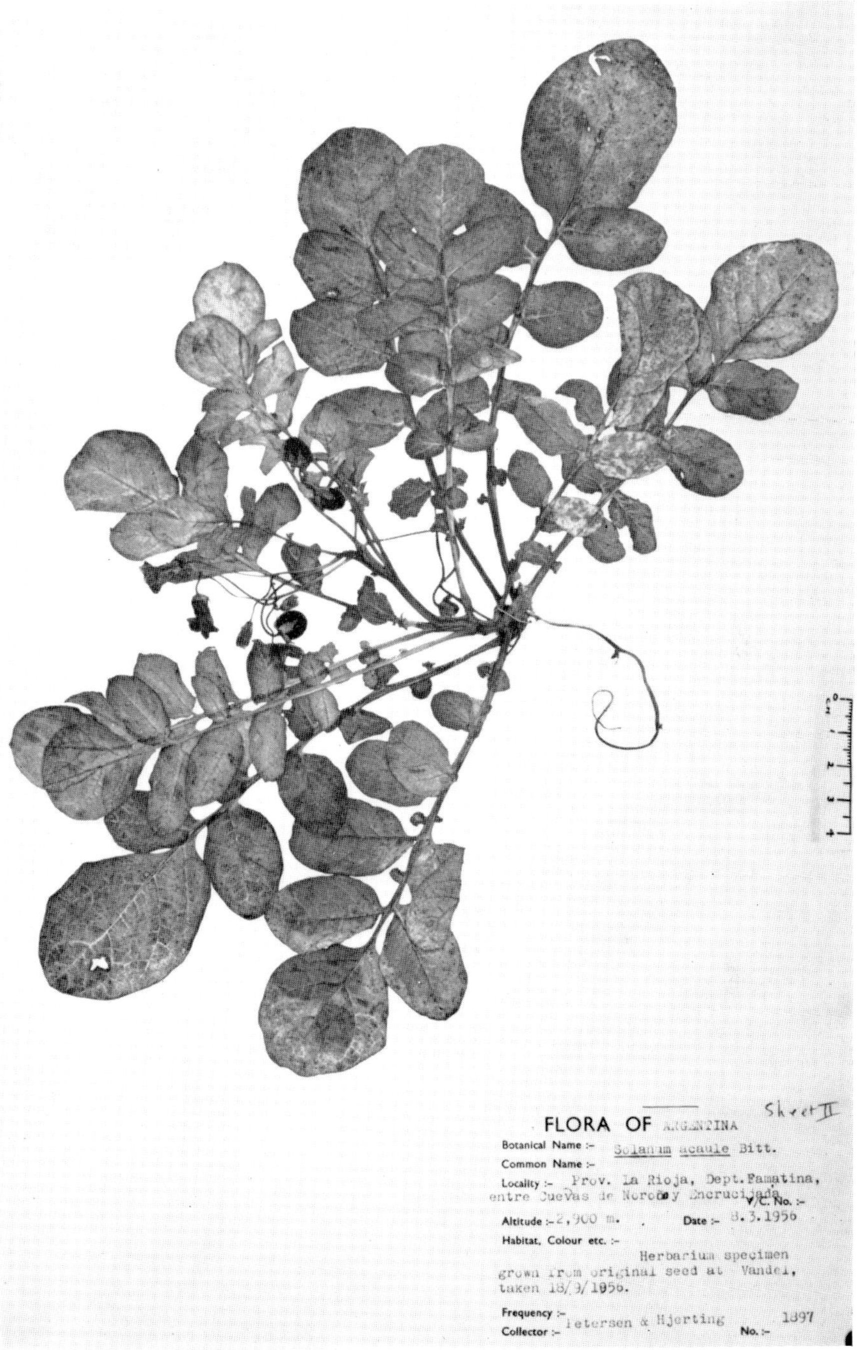

S. acaule Bitt. Plant of *Petersen & Hjerting 1897* (see Plate 66 for specimen from natural habitat) grown in the experimental field at Vandel, Denmark. The similarity in habit and leaf between plants from nature and from the experimental field is most striking. Contrast with subsp. *aemulans* from experimental field (Plate 69).

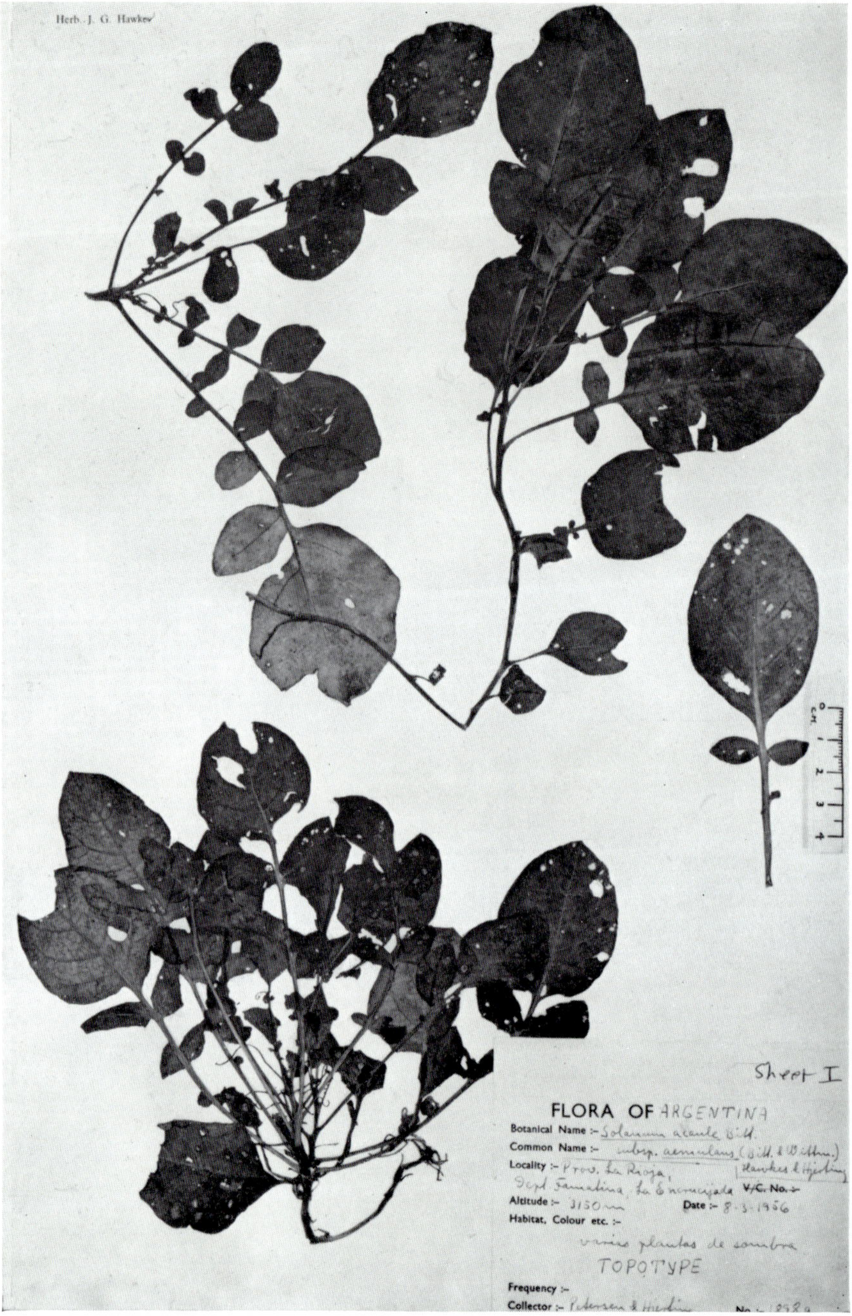

S. acaule Bitt. subsp. *aemulans* (Bitt. & Wittm.) Hawkes & Hjerting. Topotype collection made by Petersen and Hjerting (*No. 1898a*) at La Encrucijada, Sierra Famatina, prov. La Rioja, Argentina. Note the close similarity in leaf and habit between this and the isotype (see Fig. 23).

S. acaule Bitt. subsp. *aemulans* (Bitt. & Wittm.) Hawkes & Hjerting. Material of the topotype collection (*Petersen & Hjerting 1898a*) grown at Vandel, Denmark. Note the very close similarity of this specimen to Plate 68 and Fig. 23.

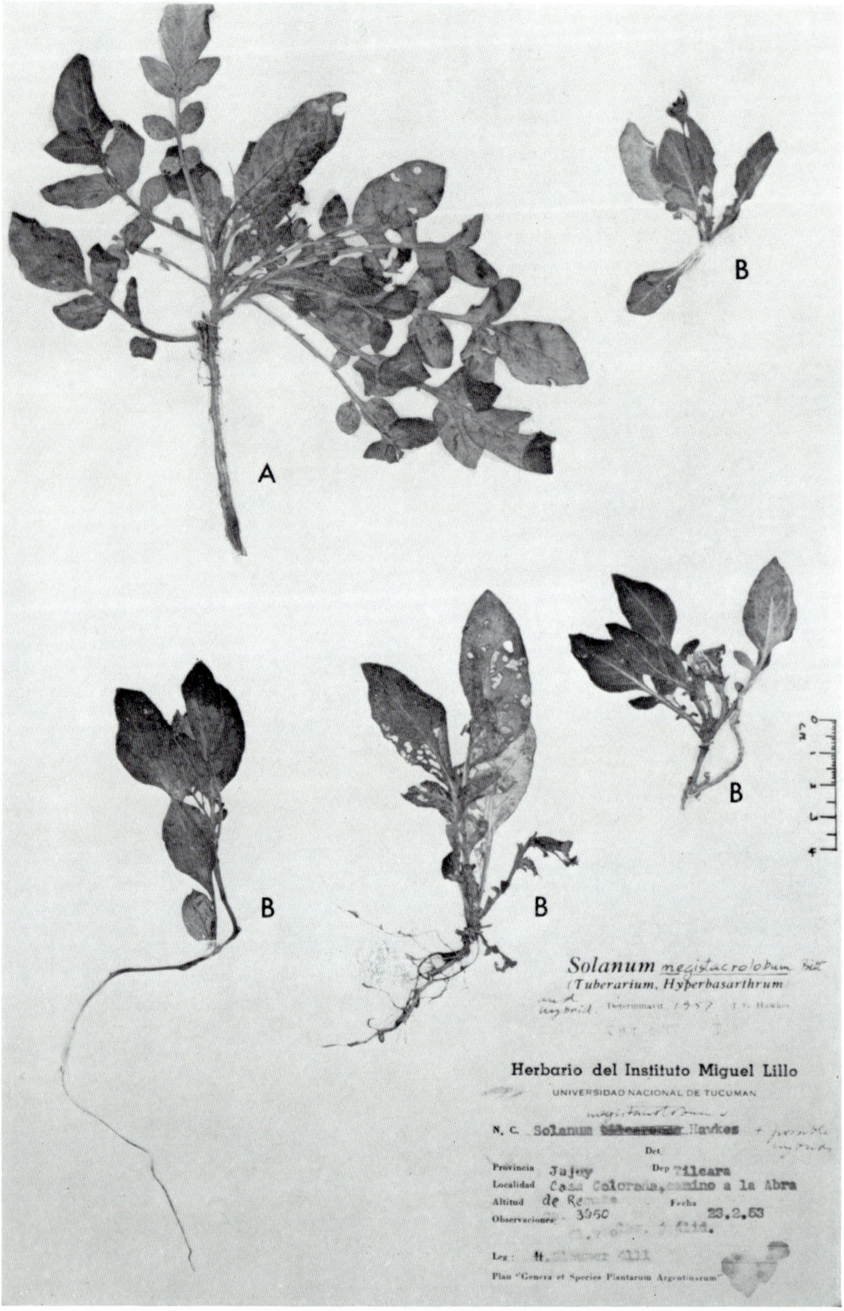

(A) Natural hybrid of *S. acaule* Bitt. × *S. megistacrolobum* Bitt.
(B) *S. megistacrolobum* Bitt.
This Sleumer collection (*No. 4111*) is from Casa Colorada, dept. Tilcara, prov. Jujuy, Argentina. The *S. acaule* parent was collected near by (*Sleumer 4114* p.p.).

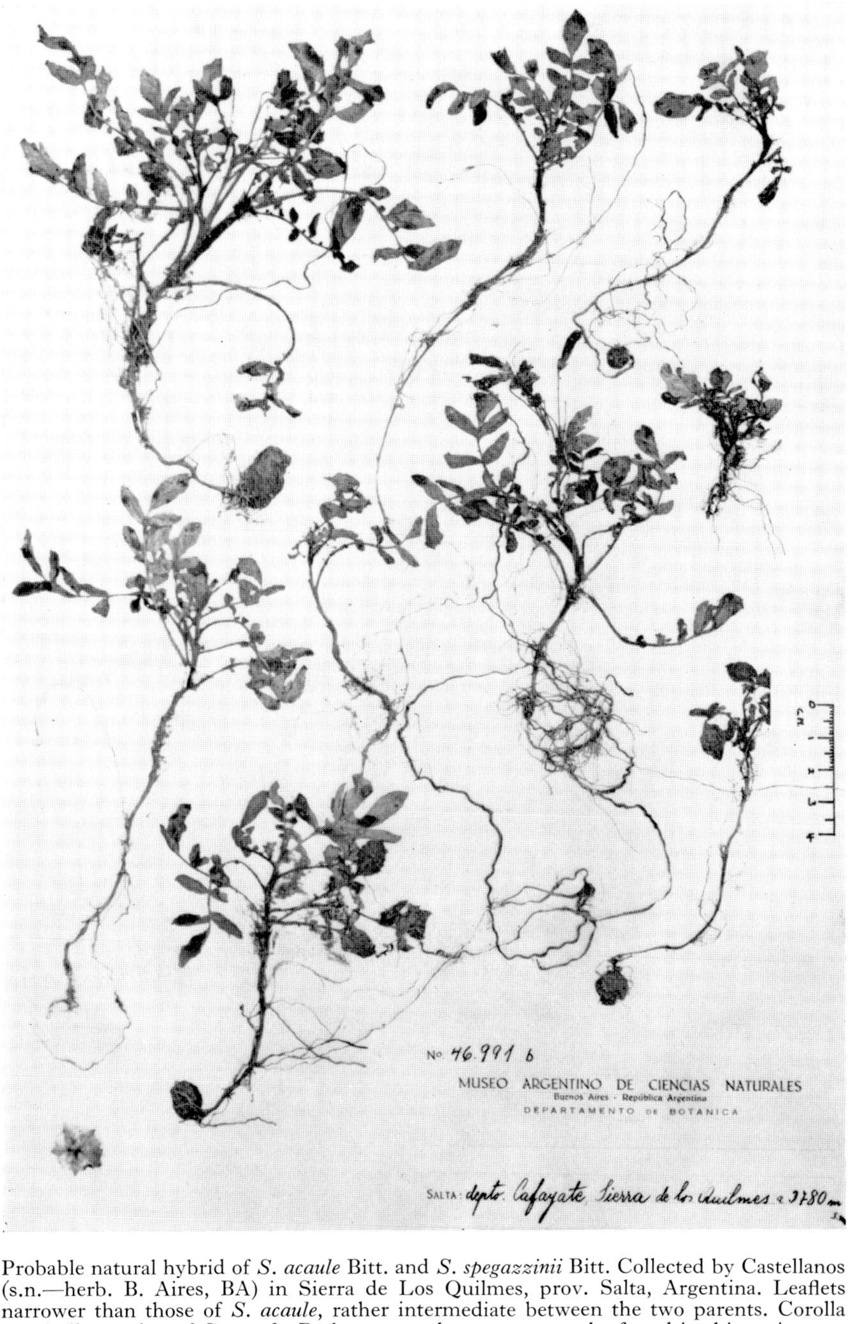

Probable natural hybrid of *S. acaule* Bitt. and *S. spegazzinii* Bitt. Collected by Castellanos (s.n.—herb. B. Aires, BA) in Sierra de Los Quilmes, prov. Salta, Argentina. Leaflets narrower than those of *S. acaule*, rather intermediate between the two parents. Corolla similar to that of *S. acaule*. Both presumed parents are to be found in this region.

PLATE 72

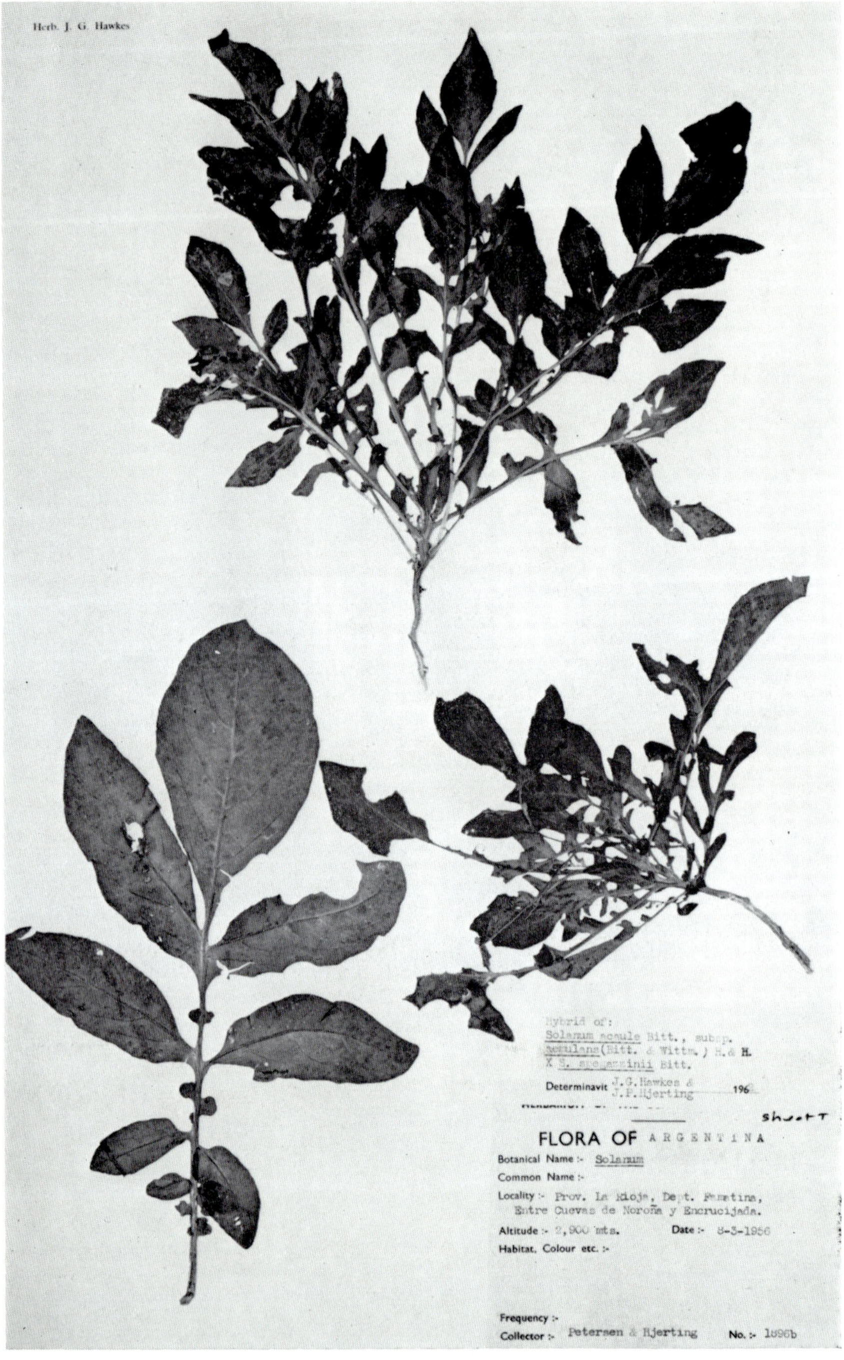

Probable natural hybrid between *S. acaule* Bitt. subsp. *aemulans* (Bitt. & Wittm.) Hawkes & Hjerting and *S. spegazzinii* Bitt. collected by Petersen and Hjerting (*No. 1896b*) at Sierra Famatina, prov. La Rioja, Argentina. Both parents are to be found in the vicinity. Note the strong leaflet decurrency and leaves intermediate between those of the two presumed parents.

S. infundibuliforme Phil. Lectotype specimen gathered by F. Philippi in Calcalhuay, prov. Tarapacá, Chile (now in Bolivia) (herb. Santiago).

S. infundibuliforme Phil. Another sheet of the Philippi collection from prov. Tarapacá, Chile (now in Bolivia). The left-hand plant was designated by Bitter as var. *angustepinnatum*, but is not worthy of varietal rank (formerly herb. Berlin).

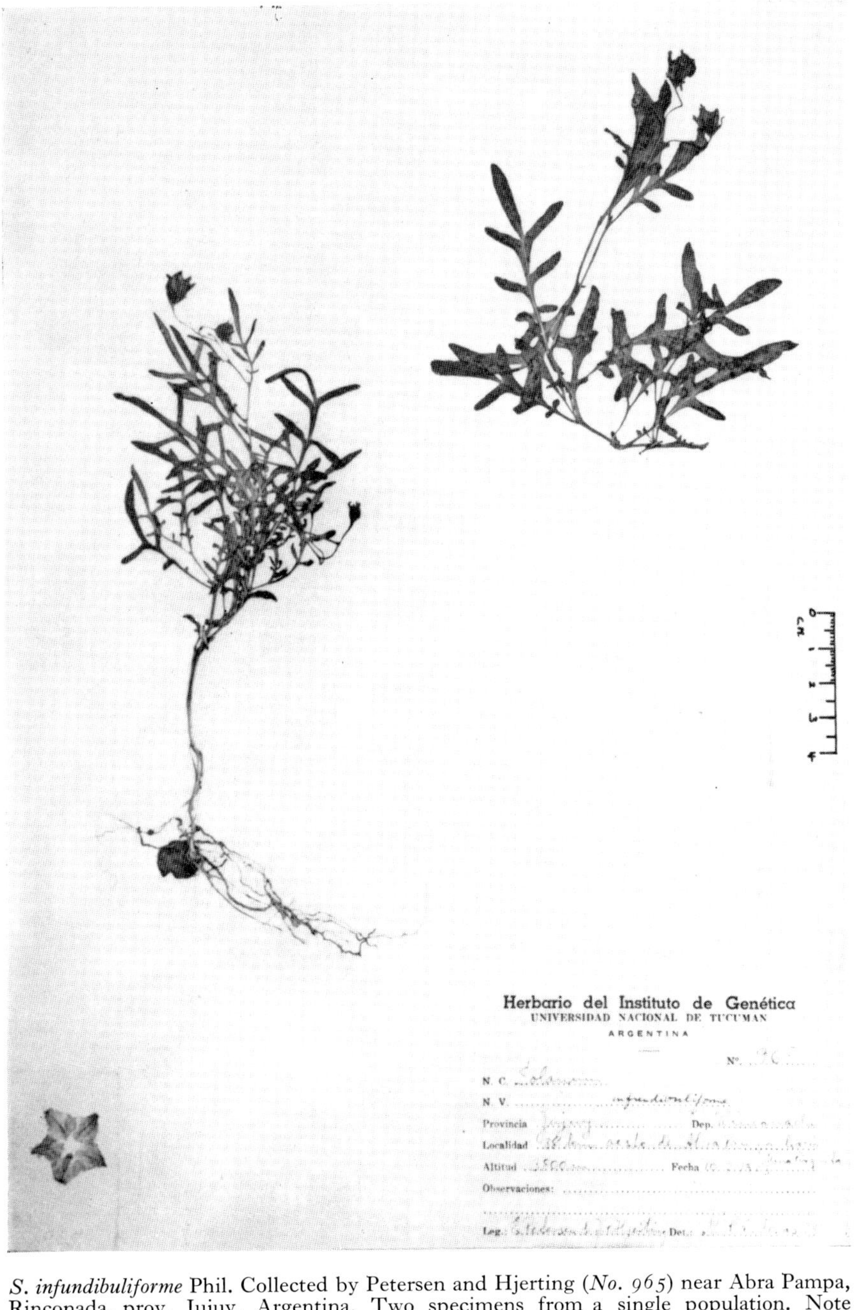

S. infundibuliforme Phil. Collected by Petersen and Hjerting (*No.* 965) near Abra Pampa, Rinconada, prov. Jujuy, Argentina. Two specimens from a single population. Note the difference in leaflet width and rhachis wing width, due probably to slight habitat differences. Compare with Plate 76.

S. infundibuliforme Phil. Seedlings raised in Vandel, Denmark, from *Petersen & Hjerting* 965. Compare with plants collected in the wild (Plate 75). In these Vandel plants a wide range of leaflet and wing width occurs, even on the same plant. The terminal leaflet exceeds the laterals in width on some leaves as can be seen in one of the Philippi specimens (see Plate 74). All seedlings were very uniform but showed a wide range of leaf type *on the same plant.*

(A) *S. infundibuliforme* Phil.

(B) Natural hybrid of *S. infundibuliforme* Phil. × *S. megistacrolobum* Bitt.

Collected in dept. Humahuaca, prov. Jujuy, Argentina, by Hjerting, Petersen and Rahn (*No. 76*). The variation in the two upper plants and the bottom right-hand one (A) comes well within *S. infundibuliforme* and is certainly due to habitat differences. The bottom left-hand plant (B) is a hybrid, with many features of *S. megistacrolobum*, such as the short lateral leaflets and very broad terminal.

(A) Natural hybrid of *S. infundibuliforme* Phil. × *S. megistacrolobum* Bitt.
(B) *S. megistacrolobum* Bitt.
Collected in dept. Tilcara, prov. Jujuy, Argentina, by Sleumer (*No. 4112*). Compare with Plate 77.

Artificial hybrid of *S. infundibuliforme* Phil. × *S. megistacrolobum* Bitt. (*Petersen & Hjerting 965* from Rinconada (see Plate 75) × *Petersen & Hjerting P–366* from Humahuaca, prov. Jujuy; both from Argentina). Compare with natural hybrids on Plates 77 and 78.

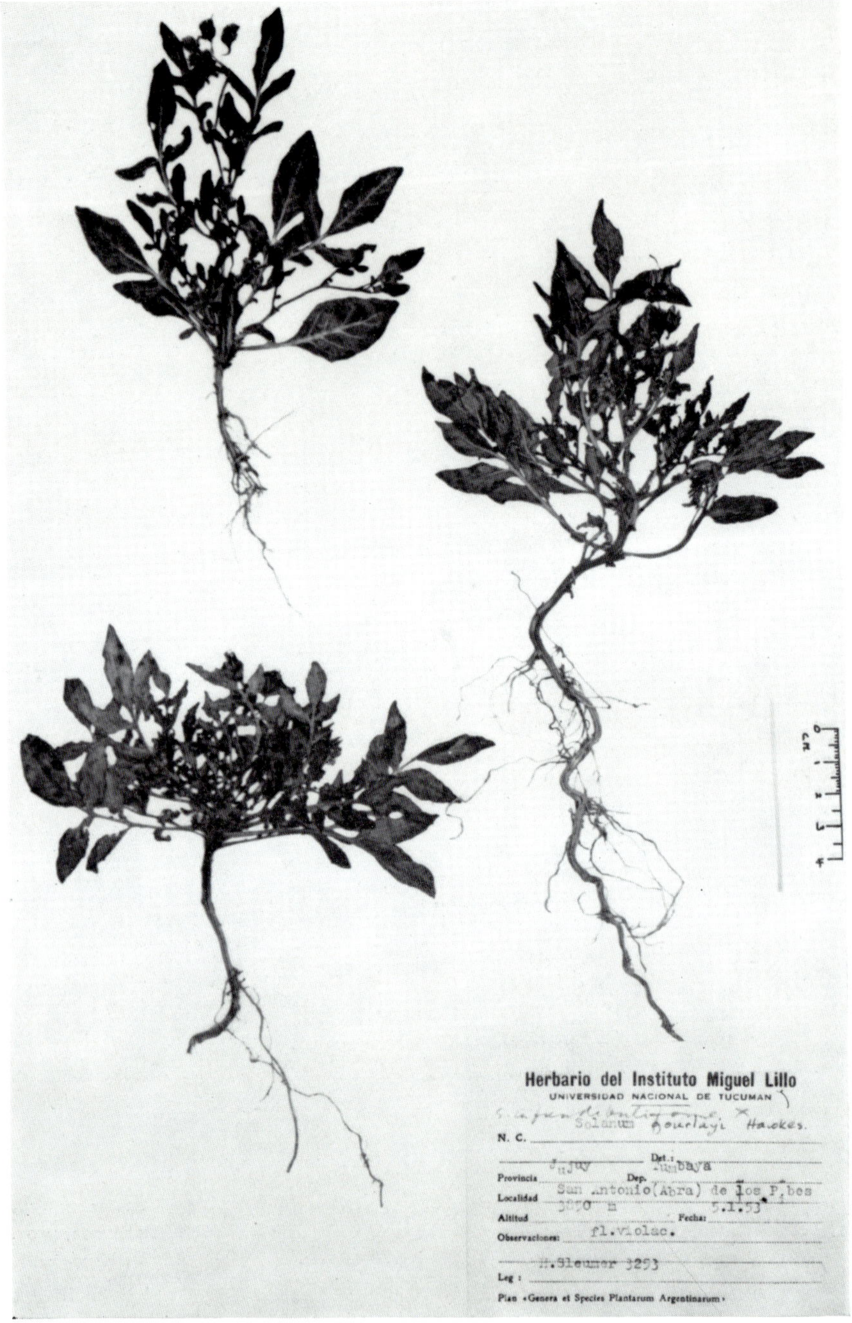

Natural hybrid of *S. infundibuliforme* Phil. × *S. gourlayi* Hawkes. Collected by Sleumer (*No. 3253* p.p.) at San Antonio de los Pibes, prov. Jujuy, Argentina. These plants appear to show a mixture of characters of both putative parents.

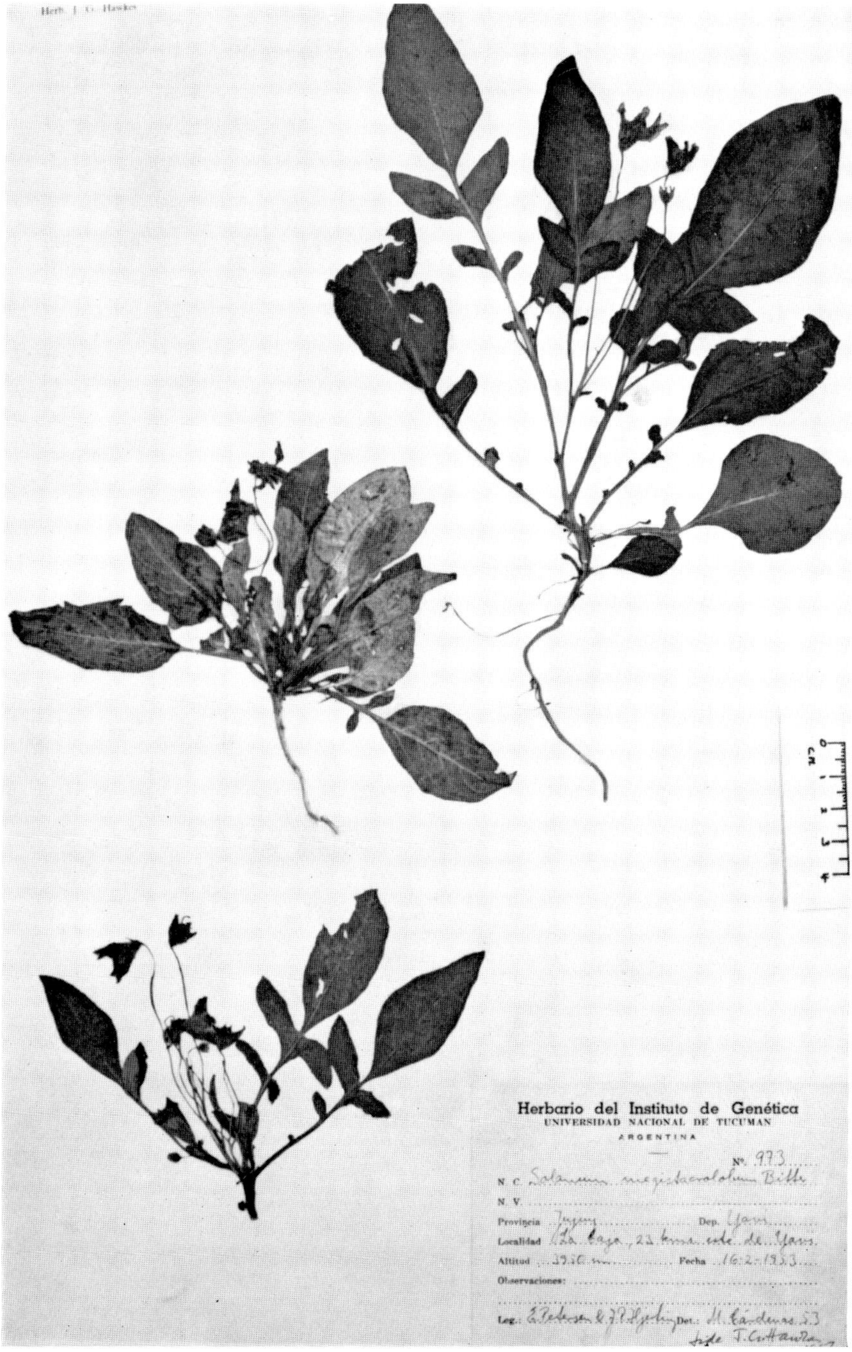

S. megistacrolobum Bitt. Collected by Petersen and Hjerting (*No. 973*) at Cajas, Yavi, prov. Jujuy, Argentina. A range of stem length and leaf dissection may be seen in these three plants, due no doubt to slight differences of habitat conditions (see also Plates 82 and 83).

PLATE 82

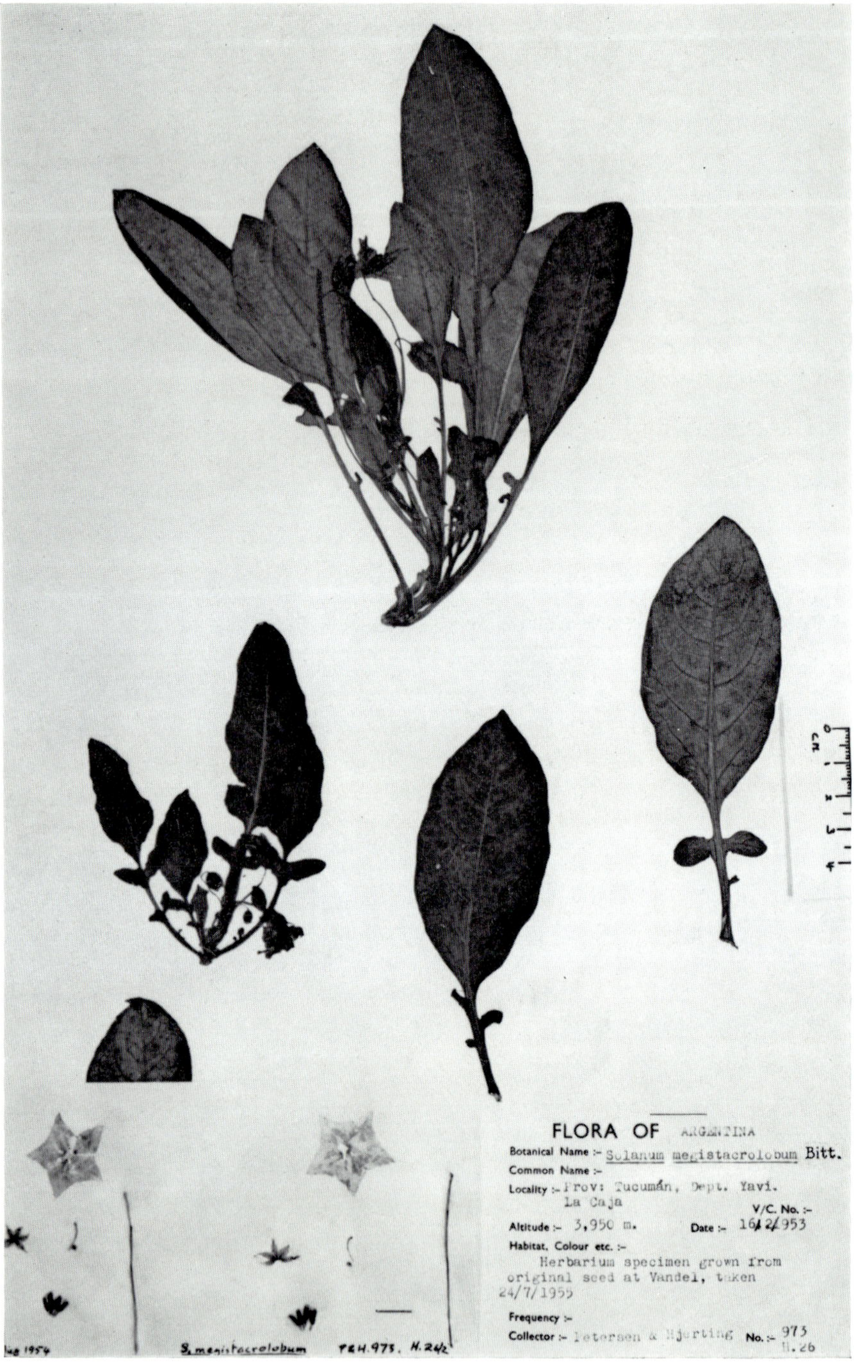

S. megistacrolobum Bitt. Material of *Petersen & Hjerting 973* grown in the experimental field at Vandel, Denmark. This is very similar in its habit to the two larger plants shown in Plate 81.

S. megistacrolobum Bitt. Collected by Hjerting, Petersen, and Rahn (*No. 77*) between Humahuaca and Tres Cruces, prov. Jujuy, Argentina. This is a shade form, rather similar to Fiebrig's type specimen (see Fig. 24).

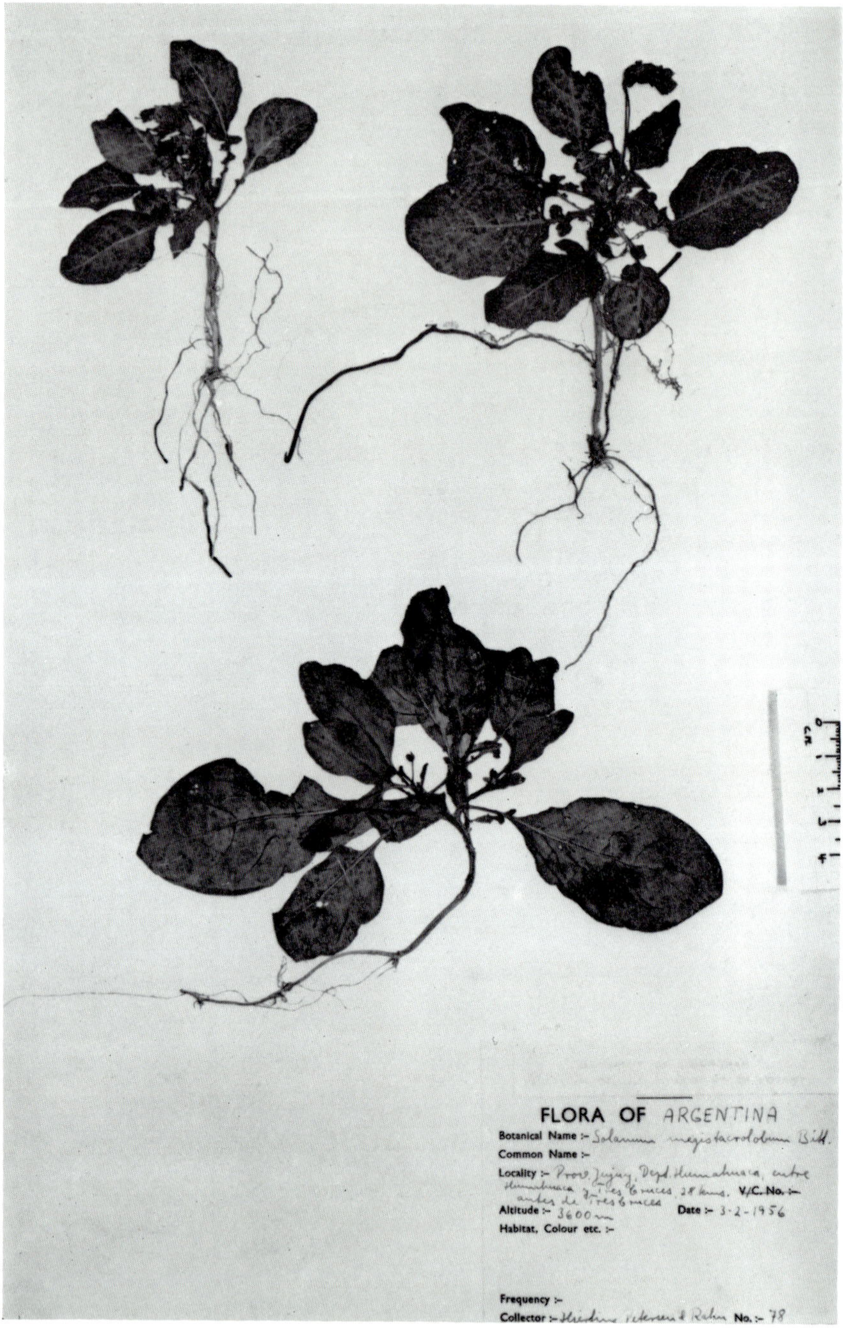

S. megistacrolobum Bitt. Specimen of *Hjerting*, *Petersen & Rahn 78*, collected very close to *H.P. & R. 77*, but in full sun (see Plate 83 for locality details). Note small stunted plants in rosette form and contrast with long stem of *H.P. & R. 77*.

PLATE 85

S. megistacrolobum Bitt. Specimen of *H.P. & R. 78* (Accession No. H. 103) grown under glass in Birmingham, England. Contrast with Plate 84 of genetically identical material in natural conditions in full sun. This matches well the shade material from natural conditions. (*H. P. & R. 77*, see Plate 83), and indicates the enormous range of variability to be seen in this species, connected with variations in habitat.

PLATE 86

Natural hybrid of *S. megistacrolobum* Bitt. × *S. boliviense* Dun., collected by Hjerting and Rahn (*No. 170*) at Cuesta de Mesón, Sta. Victoria, prov. Salta, Argentina. The left-hand plants are closer to *S. megistacrolobum* in habit, but the right-hand one is very similar to *S. boliviense*.

Natural hybrid of *S. megistacrolobum* Bitt. × *S. microdontum* Bitt. subsp. intermediate. Collected by Schreiter (*No. 2895*—herb. Lillo) at Volcán, prov. Jujuy, Argentina. The right-hand specimen, especially, shows a mixture of characters of the two putative parents.

S. boliviense Dun. Herbarium specimen collected near Sucre, Bolivia, by Balls, Gourlay, and Hawkes (*Balls 6190*—C.P.C. 97). Flower dissection and photo of pot plant grown at Cambridge, England, from material from the same locality (*Balls 6188*—C.P.C. 93). Both collections probably from type locality. Compare with Fig. 28—type specimen.

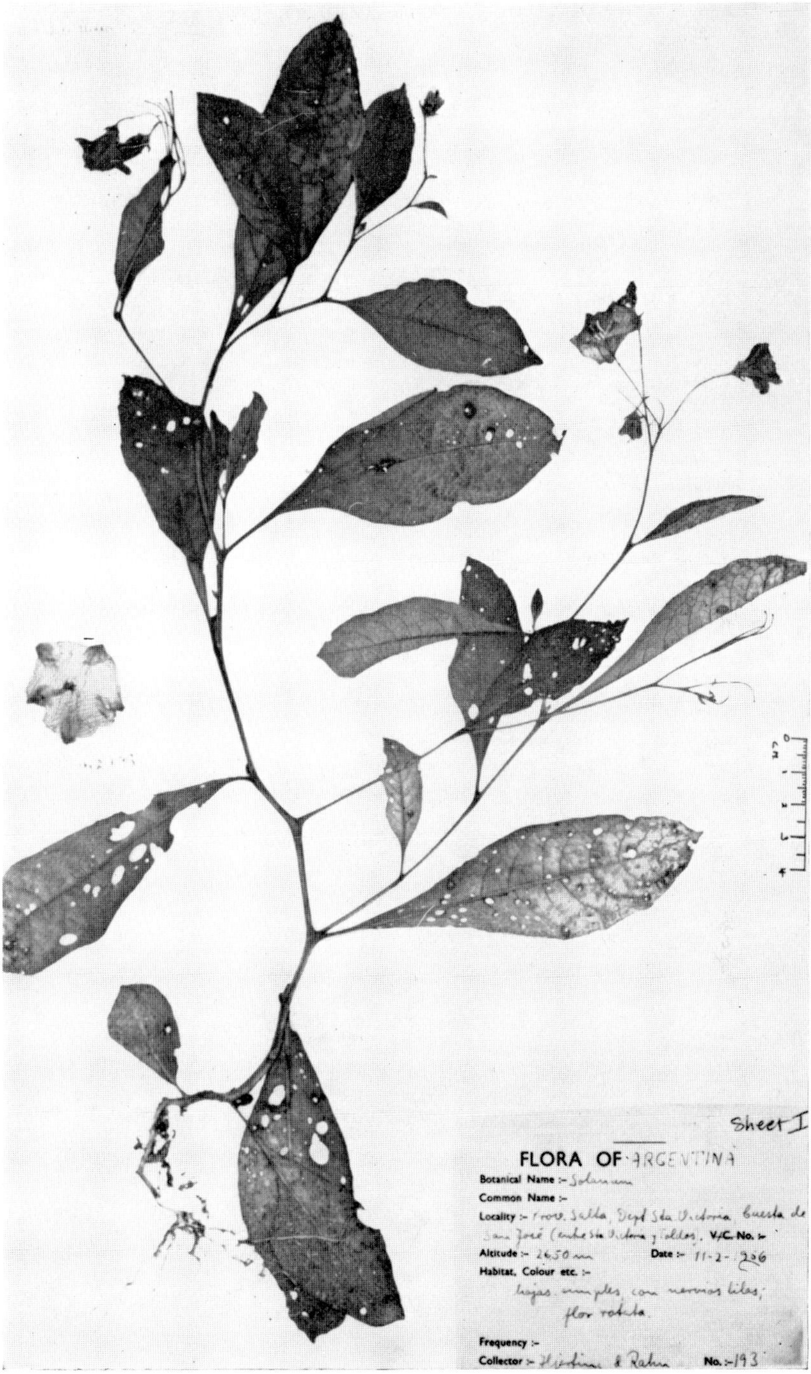

S. boliviense Dun. Collected by Hjerting and Rahn (*No. 193*) at Cuesta de San José, Sta. Victoria, prov. Salta, Argentina. This material matches the Bolivian probable topotype material (Plate 88) very closely in habit, leaf, and flower.

PLATE 90

S. sanctae-rosae Hawkes. Isotype, collected by Venturi (*No. 4737*—herb. Lillo) at Estancia Sta. Rosa, Chicligasta, prov. Tucumán, Argentina. Note variation in leaf shape between and within plants, and compare with Fig. 29(a), (b), and (c).

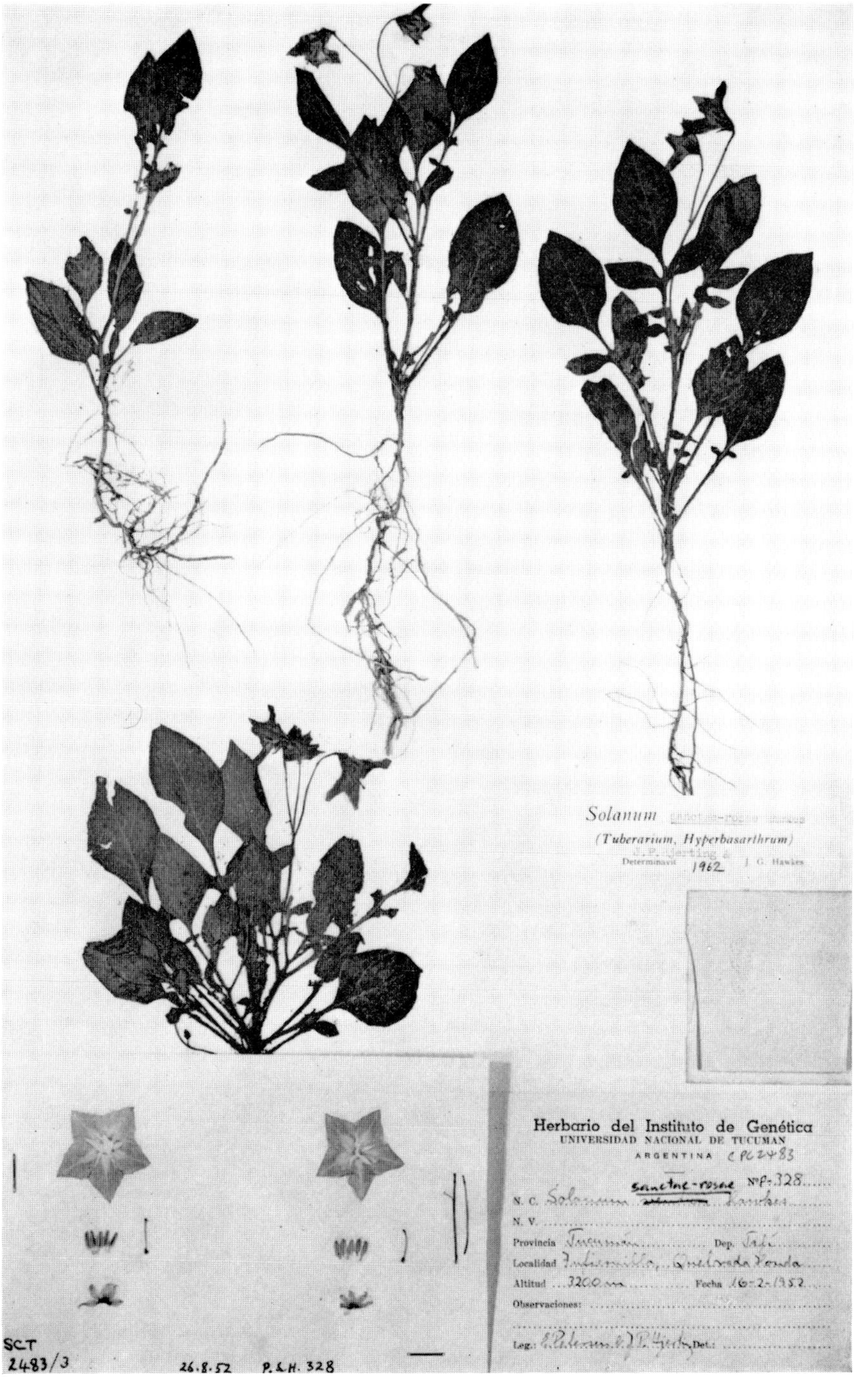

S. sanctae-rosae Hawkes. Two sheets from the same collection to illustrate the range of leaf dissection and habit. Collected by Petersen and Hjerting (*No. P–328*—CPC: 2483) at Infiernillo, dept. Tafí, prov. Tucumán, Argentina. Flower dissections made from plants of CPC: 2483 grown at Birmingham. Plate 92 is a shade form and matches closely the seedlings grown in England.

S. sanctae-rosae Hawkes. See legend to Plate 91.

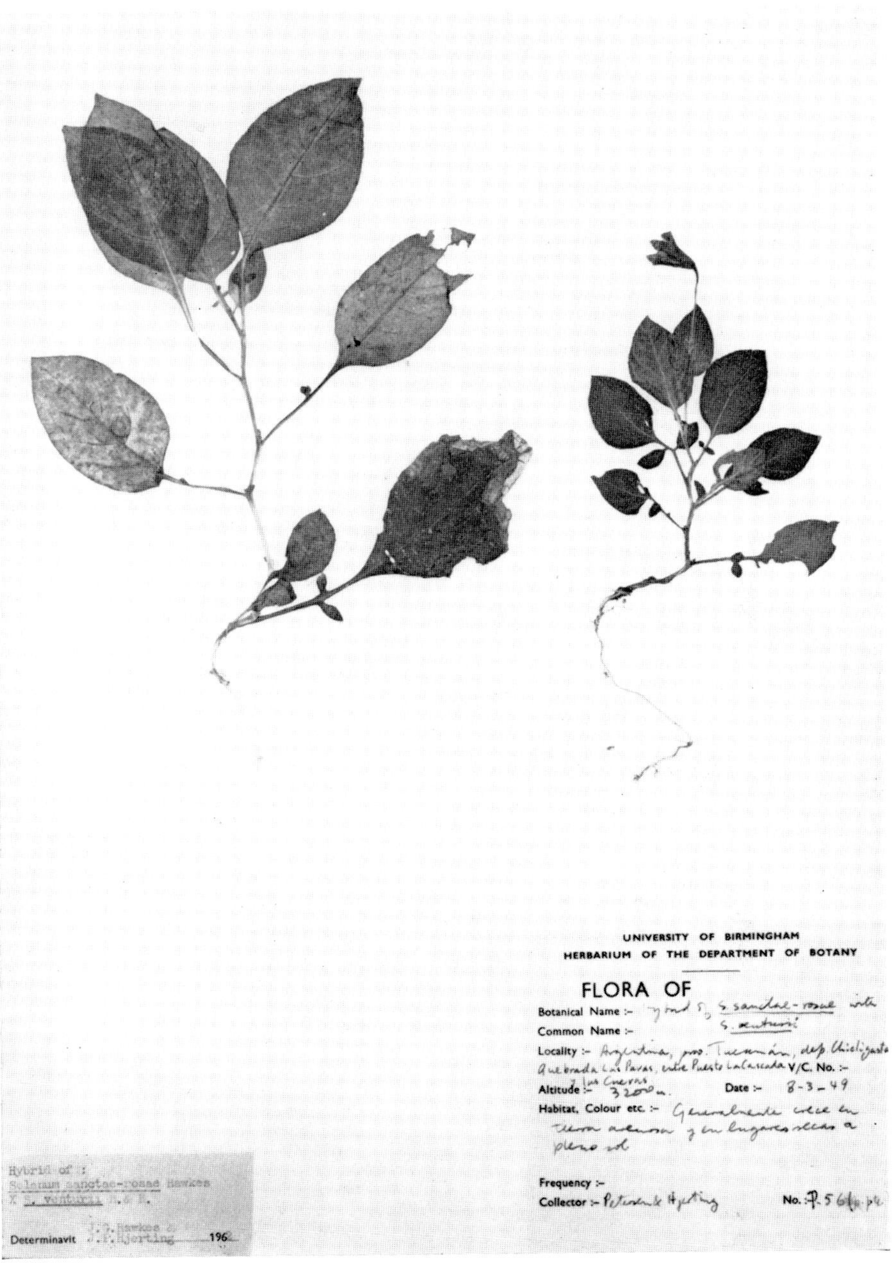

Natural hybrid of *S. sanctae-rosae* Hawkes × *S. venturii* Hawkes & Hjerting collected by Petersen and Hjerting (*No. P–56* p.p.) at Quebrada Las Pavas, dept. Chicligasta, prov. Tucumán, Argentina. Rather small plants from full sun.

PLATE 94

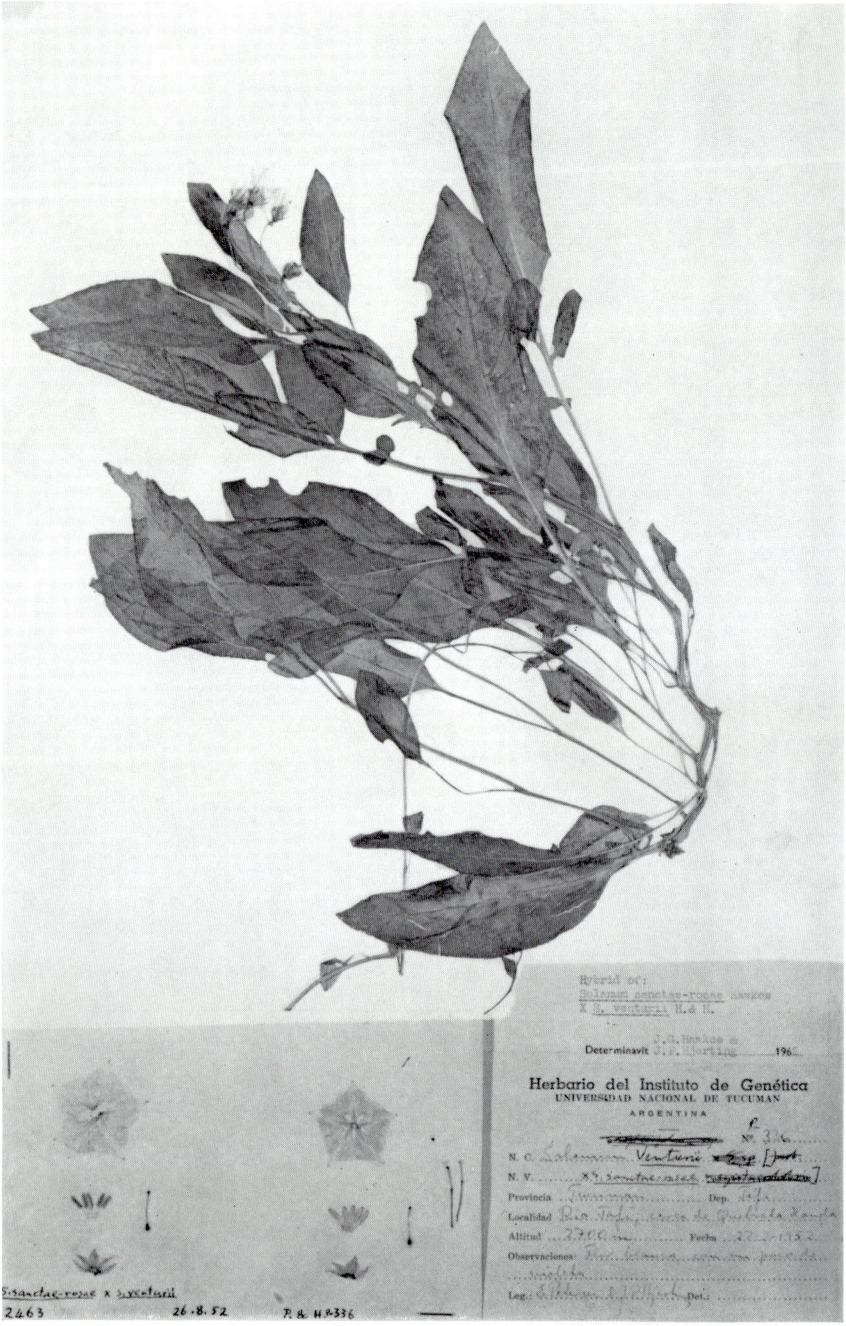

Natural hybrid of *S. sanctae-rosae* Hawkes × *S. venturii* Hawkes & Hjerting, collected by Petersen and Hjerting (*No. P–336*—CPC: 2463) from Quebrada Honda, Tafí, prov. Tucumán, Argentina. Flower dissection from plants grown at Cambridge. Corolla rotate, pale blue, shading to white near edges. Chromosome number: $2n = 24$.

S. sanctae-rosae Hawkes. Aberrant form, collected by Schickendantz (*No. 226*—photo of destroyed Berlin specimen) at Capillitas, prov. Catamarca, Argentina, and named '*S. catamarcae*' Bitt. but without a description.

PLATE 96

(A) Natural hybrid of *S. sanctae-rosae* Hawkes × *S. spegazzinii* Bitt.
(B) *S. sanctae-rosae* Hawkes.
Collected by Vervoorst (*No. 3482*) at Alto de La Cruz, dept. Pomán, prov. Catamarca, Argentina.

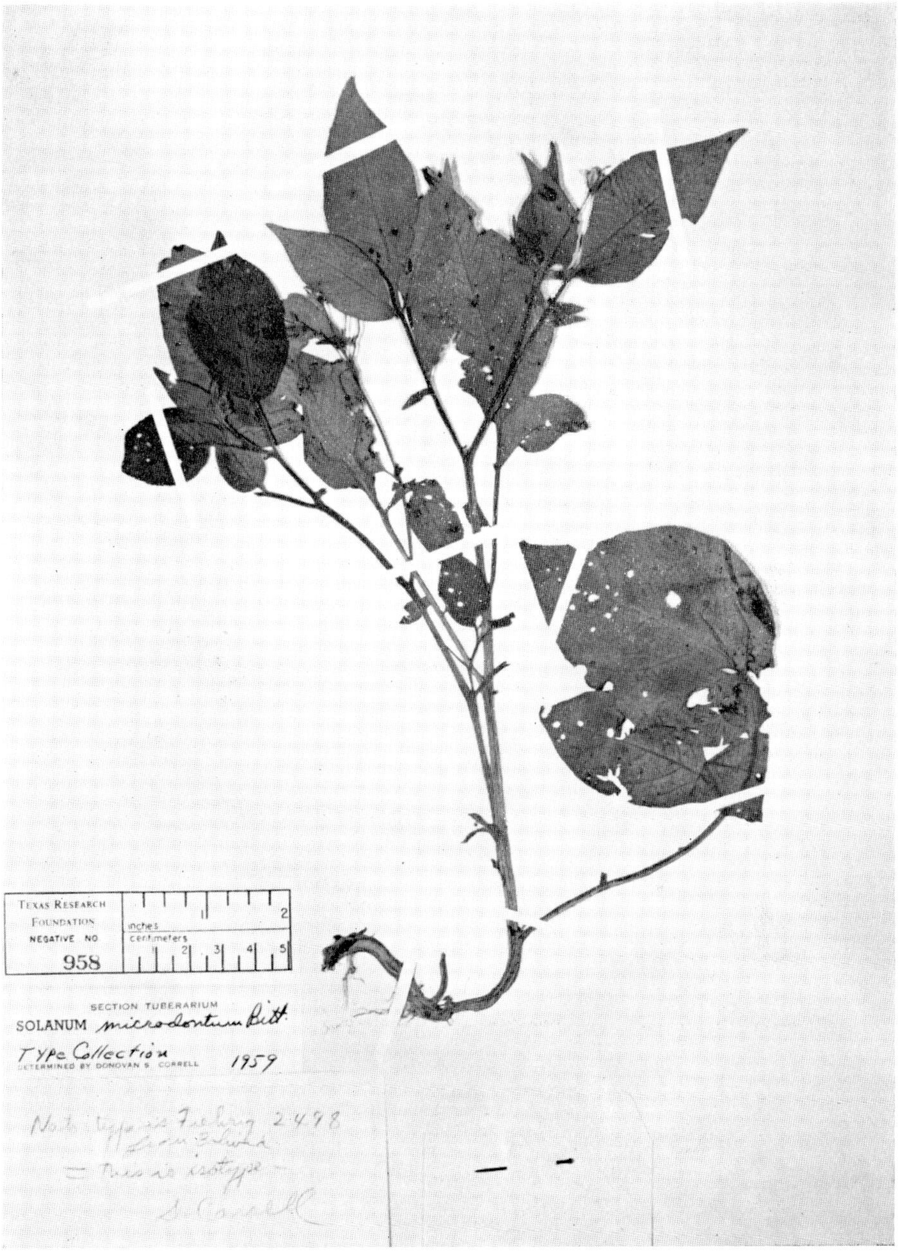

S. microdontum Bitt. Isotype. Collected by Fiebrig (*No. 2498*—herb. Göttingen) at Toldos, near Bermejo, prov. Salta, Argentina (formerly Bolivia). Leaves 1–2-jugate. Compare with the 0–1-jugate leaves of the lectotype (Fig. 30).

PLATE 98

S. microdontum Bitt. Type for '*S. bijugum*' Bitt. collected at Toldos, near Bermejo, prov. Salta, Argentina (formerly Bolivia) by Fiebrig (*No. 2253*—herb. Berlin. Specimen now destroyed). This is rather similar to the Göttingen sheet of *Fiebrig 2498* (see Plate 97) though the second pair of lateral leaflets is better developed.

S. microdontum Bitt. Topotype specimen collected by Hjerting and Rahn (*No. 226*) from Toldos, prov. Salta, Argentina. Leaf varies from 0–2-jugate. Matches very well the isotype from Göttingen (see Plate 97).

S. microdontum Bitt. Material of topotype grown in experimental field at Vandel, Denmark (*Hjerting & Rahn 226*—see Plate 99 for locality details). Note the close similarity to '*S. bijugum*' Bitt. (Plate 98).

S. microdontum Bitt. A well-marked bijugate form (cf. '*S. bijugum*', Plate 98) which was selected from a population with continuous variation in lateral leaflet development, from 0–2-jugate. Collected by Hjerting and Rahn (*No. 230*) from between Toldos and San José, prov. Salta, Argentina.

PLATE 102

S. microdontum Bitt. A curious and aberrant form with stiff erect habit and simple narrow leaves, possibly introgressed with *S. boliviense*. Collected by Sleumer (*No. 3752*) at Rodeopampa, prov. Salta, Argentina.

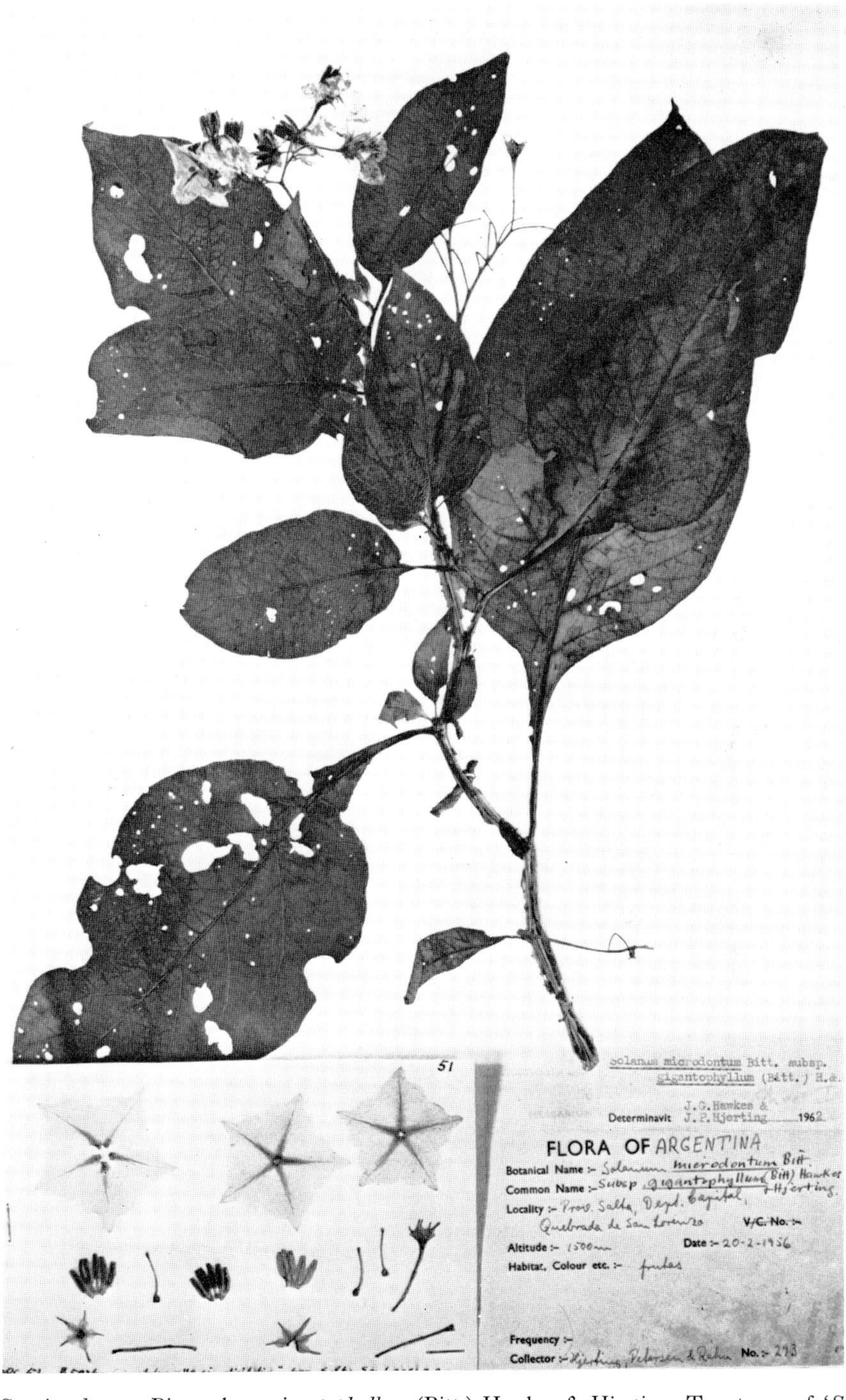

S. microdontum Bitt. subsp. *gigantophyllum* (Bitt.) Hawkes & Hjerting. Topotypes of '*S. simplicifolium*' Bitt. Herbarium specimen collected by Hjerting, Petersen, and Rahn (*No. 293*). Flower dissections from material collected by Balls, Gourlay, and Hawkes (*Balls 5936* p.p.—CPC: 51a); both from Quebrada de San Lorenzo, prov. Salta, Argentina. Both these biotypes are genotypically entire-leaved.

S. microdontum Bitt. subsp. *gigantophyllum* (Bitt.) Hawkes & Hjerting. Genotypically dissected-leaved biotype from Chicoana, Escoipe, prov. Salta, Argentina. Collected by Hjerting, Petersen, and Rahn (*No. 298*). Leaves 2-jugate.

S. microdontum Bitt. subsp. *gigantophyllum* (Bitt.) Hawkes & Hjerting. Material of *Hjerting, Petersen & Rahn 298* grown at Birmingham (see Plate 104 for locality details), showing the well-marked lateral leaflets that are genotypically determined.

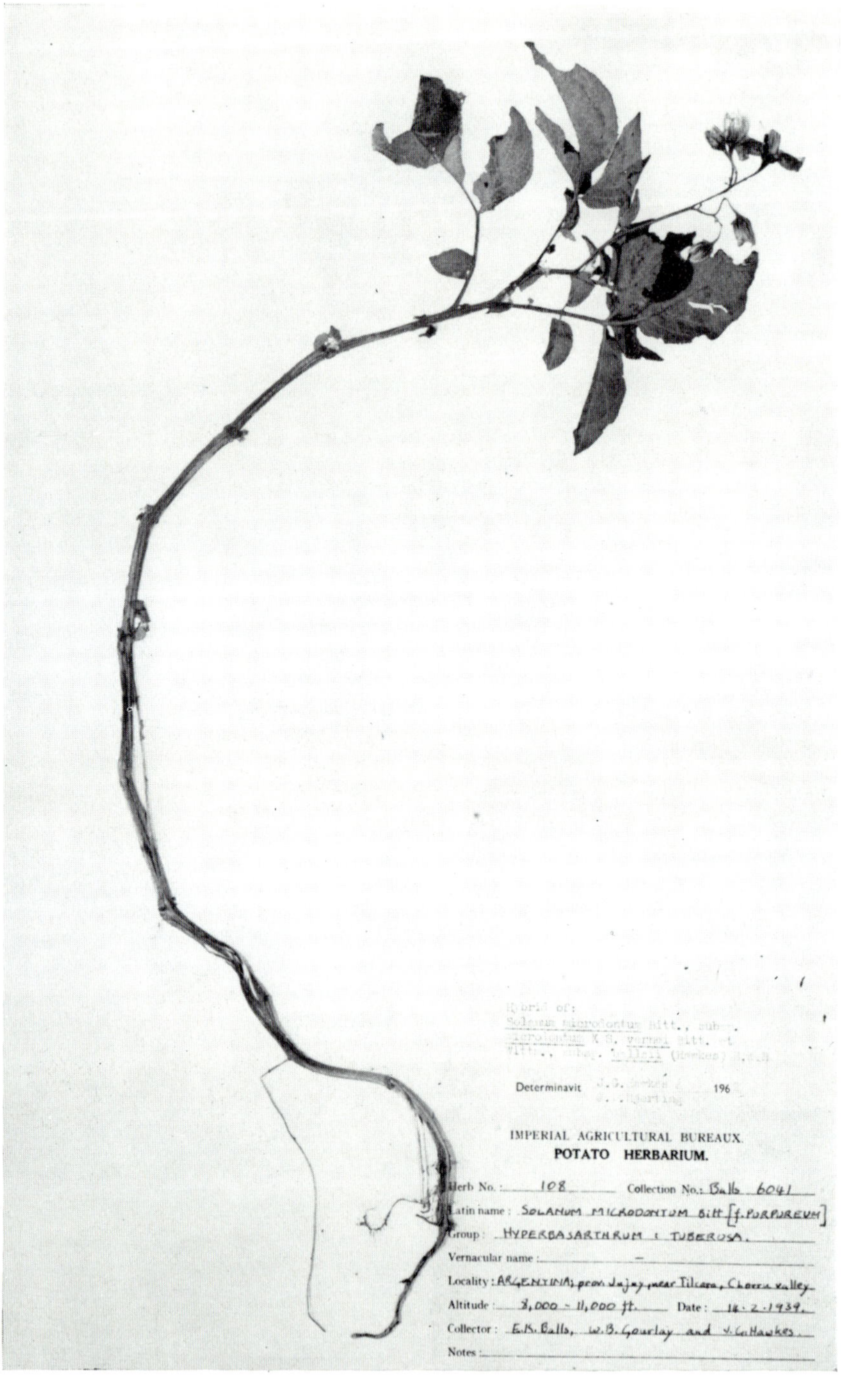

Natural hybrid of *S. microdontum* Bitt. × *S. vernei* Bitt. & Wittm. subsp. *ballsii* (Hawkes) Hawkes & Hjerting, collected by Balls, Gourlay, and Hawkes (*Balls 6041*—herb. P.G.S.) near Tilcara, prov. Jujuy, Argentina.

Artificial hybrid of *S. vernei* Bitt. & Wittm. subsp. *vernei* (E.B.S. 49/180, from the vicinity of Tucumán) × *S. microdontum* Bitt. (*Petersen & Hjerting P–364*, from dept. Iruya, prov. Salta, Argentina). Compare with Plate 108.

PLATE 108

Natural hybrid of *S. microdontum* Bitt. subsp. *gigantophyllum* (Bitt.) Hawkes & Hjerting × *S. vernei* Bitt. & Wittm. subsp. *vernei*. Collected by Petersen and Hjerting (*P–33*) at Estancia 'Yunka Suma', dept. Andalgalá, prov. Catamarca, Argentina. The flowers are pale purple and the leaves possess few laterals and enlarged terminals.

Natural hybrid of *S. microdontum* Bitt. subsp. *gigantophyllum* (Bitt.) Hawkes & Hjerting × *S. kurtzianum* Bitt. & Wittm. This was collected by Castellanos at Guanchín Viejo, at the southern end of Sierra Famatina, prov. La Rioja. We previously thought it was a good species, *S. rechei* Hawkes & Hjerting, but our 1966 expedition showed it to be a hybrid.

Natural hybrid of *S. microdontum* Bitt. subsp. *gigantophyllum* (Bitt.) Hawkes & Hjerting × *S. kurtzianum* Bitt. & Wittm. Collected by Reche in 1960. Herbarium specimen made by Ross from material grown at Cologne, Germany. (E.B.S. 2082.)

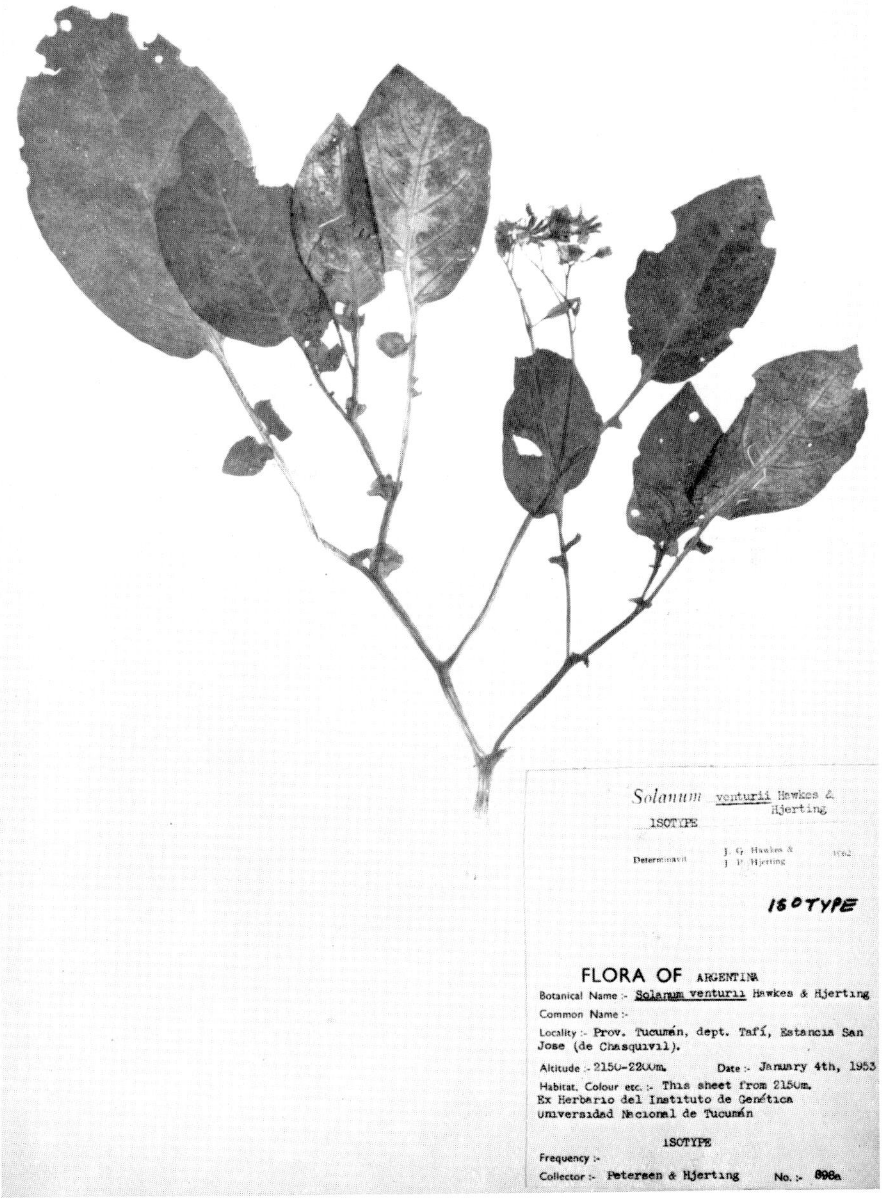

S. venturii Hawkes & Hjerting. Isotype, collected by Petersen and Hjerting (*No. 898a*—herb. J.G.H.) at 2150–2200 m, Estancia 'San José', dept. Tafí, prov. Tucumán, Argentina. Note the delicate habit, enlarged terminal leaflets, and very enlarged peduncle forks.

S. venturii Hawkes & Hjerting. Collected at Estancia 'Yunka Suma', dept. Andalgalá, prov. Catamarca, Argentina, alt. 2350–2800 m (*Petersen & Hjerting P–30*). Plants smaller than those of the type collection but still with well-marked stems and similar leaf dissection.

S. vernei Bitt. & Wittm. Original type specimen (formerly in herb. Berlin, now destroyed) of *Lorentz & Hieronymus 708*. Note the long leaves and well-marked petiolules, characteristic of the typical subspecies. See also Fig. 35.

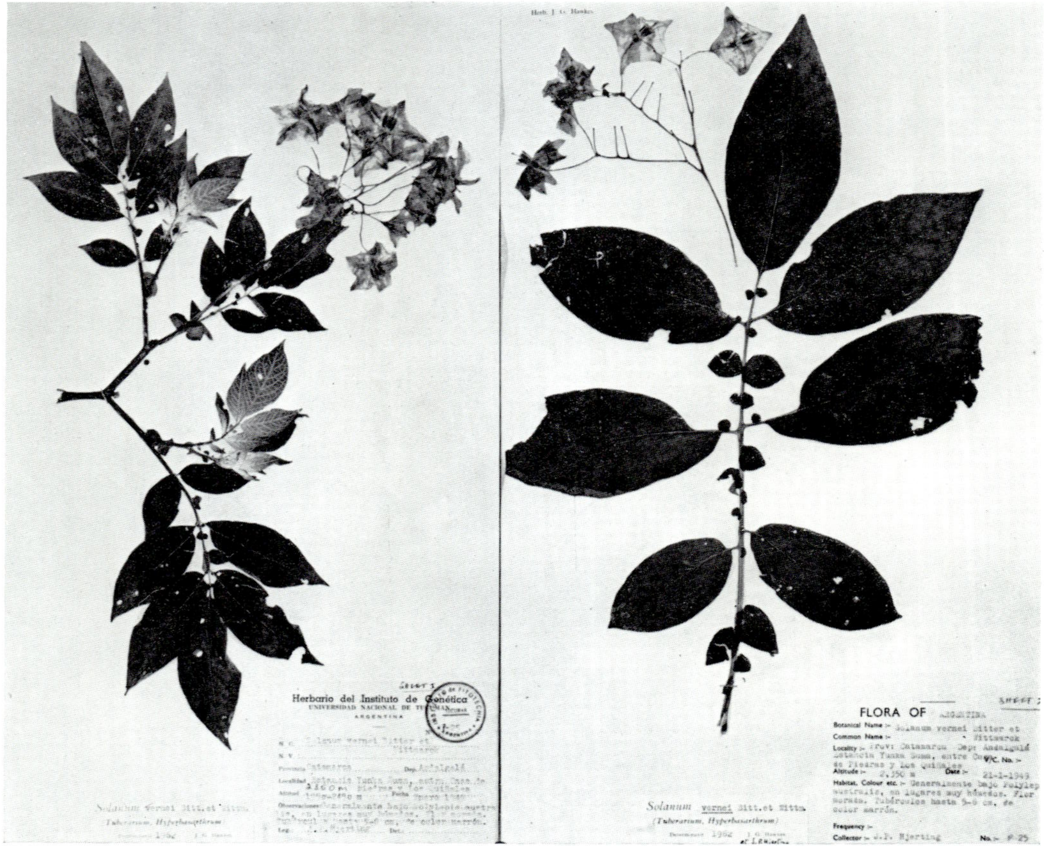

S. vernei Bitt. & Wittm., collected by Petersen and Hjerting (*No. P–25*) at Estancia 'Yunka Suma', dept. Andalgalá, prov. Catamarca, Argentina. Note the extremely large showy flowers and leaf white-felted below.

S. vernei Bitt. & Wittm. Tubers collected by Petersen and Hjerting (*No. P–25*); see Plate 114 for locality details. Tubers measure 3×4 to 7×8 cm.

S. vernei Bitt. & Wittm. subsp. *ballsii* (Hawkes) Hawkes & Hjerting. Isotype specimen of *Balls 6042* (p.p.) (CPC: 105) collected by Balls, Gourlay, and Hawkes, near Tilcara, prov. Jujuy, Argentina. Note the sessile lateral leaflets with auriculate base on the basiscopic side, and obtuse to subacute apex. See also Fig. 36.

Natural hybrid of *S. vernei* Bitt. & Wittm. × *S. spegazzinii* Bitt., collected by Petersen and Hjerting (*No. P–332a*) at Infiernillo, dept. Tafí, prov. Tucumán, Argentina. Leaflets narrower and with longer petiolules and smaller corolla than in *S. vernei*. Leaf type of *S. spegazzinii*. See Plate 118.

PLATE 118

Natural hybrid of *S. vernei* Bitt. & Wittm. × *S. spegazzinii* Bitt. One of the seedlings raised at Birmingham from a natural berry of *Petersen & Hjerting P–332a* (see Plate 117 for locality details). The seedlings showed a wide range of variation from *S. vernei* to *S. spegazzinii* and were undoubtedly F_2 segregates of this natural hybrid.

S. gourlayi Hawkes. Isotype specimen of *Balls 5979* collected by Balls, Gourlay, and Hawkes at San Gregorio, Tilcara, prov. Jujuy, Argentina. Note the wide leaflets and well-marked petiolules.

S. gourlayi Hawkes. Collected at Rodero Colorado, dept. Iruya, prov. Salta, Argentina, by Petersen and Hjerting (*No. P–365*). Note the wide range of phenotypes, and compare with Plate 121.

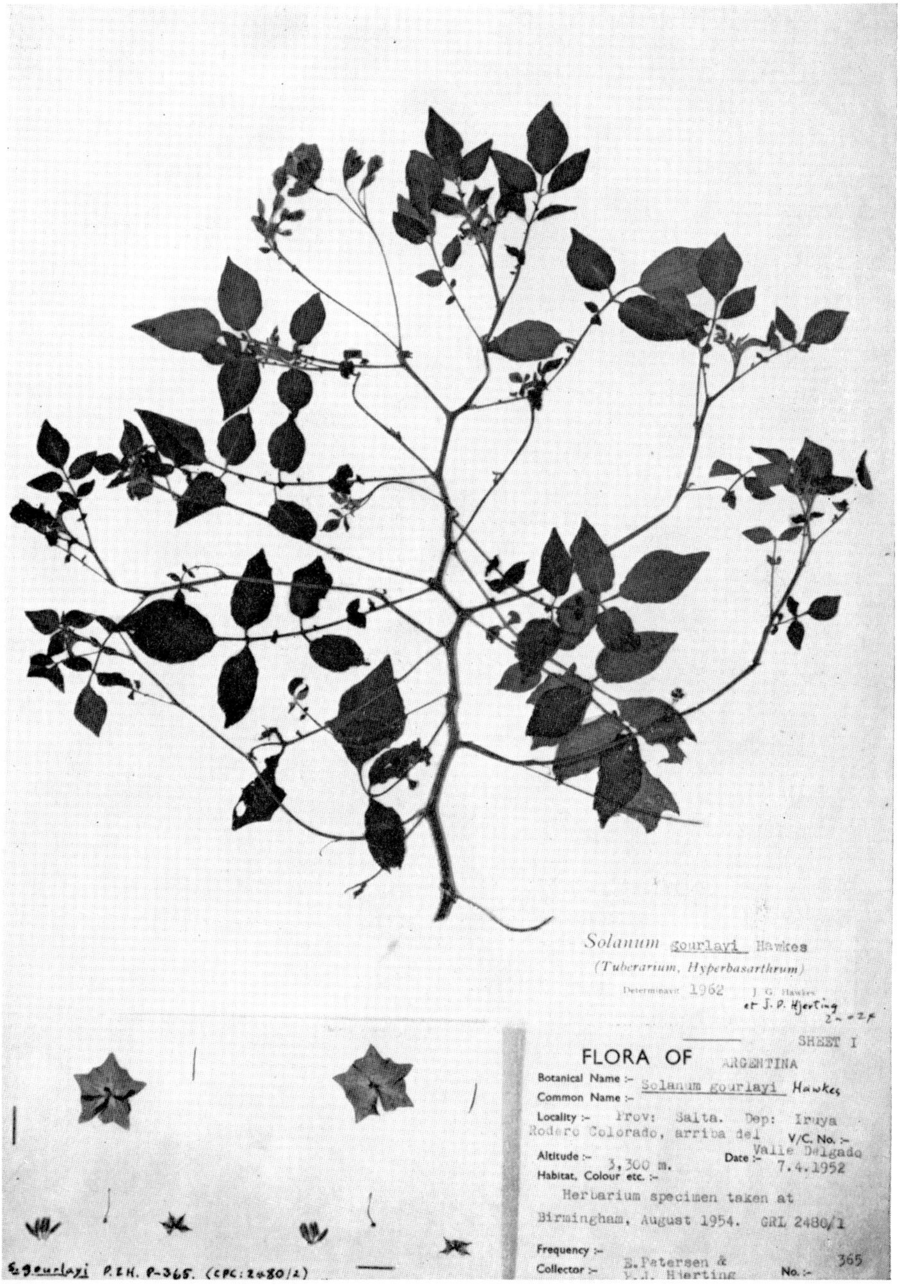

S. gourlayi Hawkes. Material of *Petersen & Hjerting P–365* (CPC: 2480/1—see Plate 120 for locality details) grown at Birmingham. Contrast with phenotype of identical material in Argentina. This comes closer to one of the isotype specimens (see Plate 119) and *Hjerting & Rahn 190* (see Plate 122).

S. gourlayi Hawkes. Collected by Hjerting and Rahn (*No. 190*) at Cuesta de San José, dept. Sta. Victoria, prov. Salta, Argentina. Note the very strong resemblance of this material to the Birmingham-grown *P. & H. P–365* (see Plate 121), and to a lesser extent to the isotype (Plate 119).

Probable natural hybrid of *S. gourlayi* Hawkes × *S. vidaurrei* Cárd. Collected by Hjerting and Rahn (*No. 254*) near Lizoite, dept. Santa Victoria, prov. Salta, Argentina. A range of leaf morphology from one parent to the other may be seen in this material.

S. oplocense Hawkes. Lectotype. Collected by Balls, Gourlay, and Hawkes (*Balls 6127*—herb. P.G.S.) between Oploca and Oro Ingenio, dept. Potosí, Bolivia.

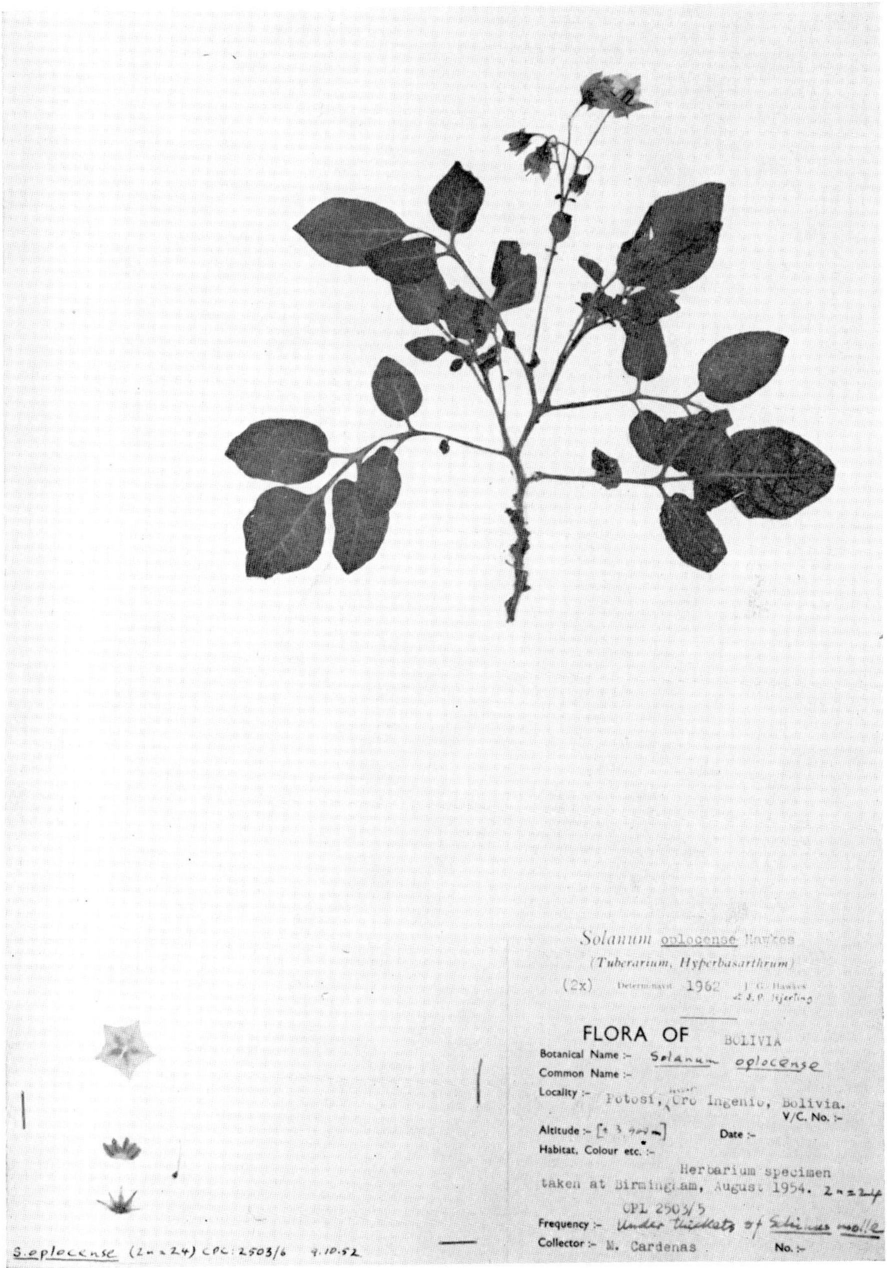

S. oplocense Hawkes. Topotype material collected by M. Cárdenas (CPC: 2503) and grown in Birmingham in 1954. $2n = 24$. Note few decurrent leaflets, long petiolules, and minute stigma.

S. oplocense Hawkes. Material collected by O. Braun (CPC: 2116) from the Salo district near Tupiza, dept. Potosí, Bolivia, and grown in Birmingham in 1954. $2n = 24$. Compare the leaf of this with those of Plates 125 and 128.

S. oplocense Hawkes. Isotype specimen of '*S. subandigenum* Hawkes var. *camarguense* Cárd.', collected by Cárdenas (*No. 5078*) near Camargo, dept. Chuquisaca, Bolivia. Compare with *Alandia & Ross 123* (Plate 128).

S. oplocense Hawkes. Specimen of '*S. camarguense*' or '*S. subandigenum* var. *camarguense*' collected by Alandia and Ross (*No. 123*—herb. Max) in Yavi, prov. Jujuy, Argentina. This is tetraploid, according to Ross. Note the close similarity between this specimen, *Cárdenas 5078* and CPC: 2116 (Plates 127 and 126).

S. oplocense Hawkes. Collected by Petersen and Hjerting (*No. 962*) between Iturbe and Tres Cruces, prov. Jujuy, Argentina. $2n = 48$. This is a smaller and more delicate biotype than the *Alandia & Ross 123* tetraploid shown in Plate 128.

S. oplocense Hawkes. Tetraploid material of *Petersen & Hjerting 962* grown in the experimental field at Birmingham. Compare with original collection—Plate 129.

S. vidaurrei Cárdenas. Isotype specimen of *Cárdenas 5075*, collected near Muyuquira, prov. Nor Cinti, dept. Chuquisaca, Bolivia. Note variation of leaf from petiolulate with 2 pairs of interjected leaflets to the more normal subsessile type with no interjected leaflets.

S. vidaurrei Cárdenas. Collection made by Sleumer (*No. 4059*) from Lizoite, dept. Sta. Victoria, prov. Salta, Argentina. Compare with Plate 131.

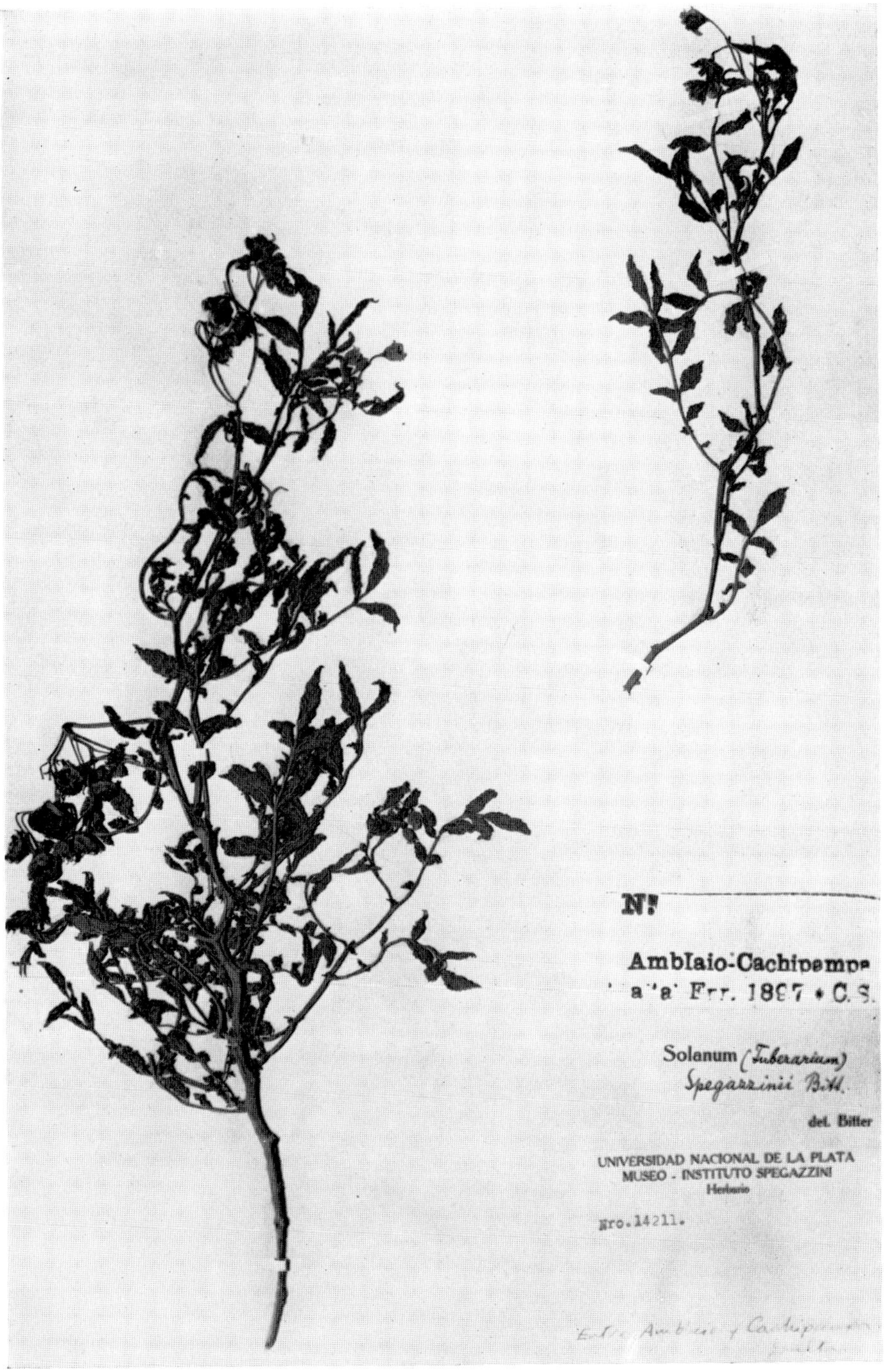

S. spegazzinii Bitt. Lectotype (left-hand plant) and isotype (right-hand plant) (see also Fig. 38), from Amblayo to Cachipampa, prov. Salta, Argentina, collected by C. Spegazzini (s.n.—herb. La Plata, Spegazzini 14211). Compare the larger left-hand plant with *Ross & Vervoorst 36* (Plate 136).

S. spegazzinii Bitt. Syntype specimen from Cerro de Cachi, prov. Salta, Argentina, collected by C. Spegazzini (s.n.—herb. La Plata, Spegazzini 14210). Note the sessile leaflets, slight decurrency in some upper laterals and few interjected leaflets. Compare with the lower leaves in *Ross & Vervoorst 36* (Plate 136).

S. spegazzinii Bitt. Duplicate of one of the original specimens on which Bitter and Wittmack's description of '*S. famatinae*' was made; collected by Kurtz (*No. 13445b*—herb. Córdoba) between Los Corrales and Escalera de Famatina, Sierra Famatina, prov. La Rioja, Argentina. Designated by us as Lectotype for '*S. famatinae*' Bitt. & Wittm. Note the well-dissected leaf and compare with Plates 137 and 138, and Figs. 39 and 40.

S. spegazzinii Bitt. Specimen collected by Ross and Vervoorst (*No. 36*—herb. Max Planck) at Zanja de los Cardones, dept. Tafí, prov. Tucumán, Argentina, in a fairly drys ituation and showing a clear transition from '*S. spegazzinii* type' leaf (see Plates 133 and 134 and Fig. 38) below, to the '*S. famatinae* type' leaf (see Plate 135 and Figs. 39 and 40) at the top of the plant. This specimen demonstrates very clearly the conspecificity of *S. spegazzinii* and *S. famatinae*.

S. spegazzinii Bitt. Specimen collected by Petersen and Hjerting (*No. 1893*) at Corrales, Sierra Famatina, prov. La Rioja, Argentina. This grew in a maize field under damp sheltered conditions and produced the '*S. famatinae*' phenotype (compare Plates 135, 138; Figs. 39, 40).

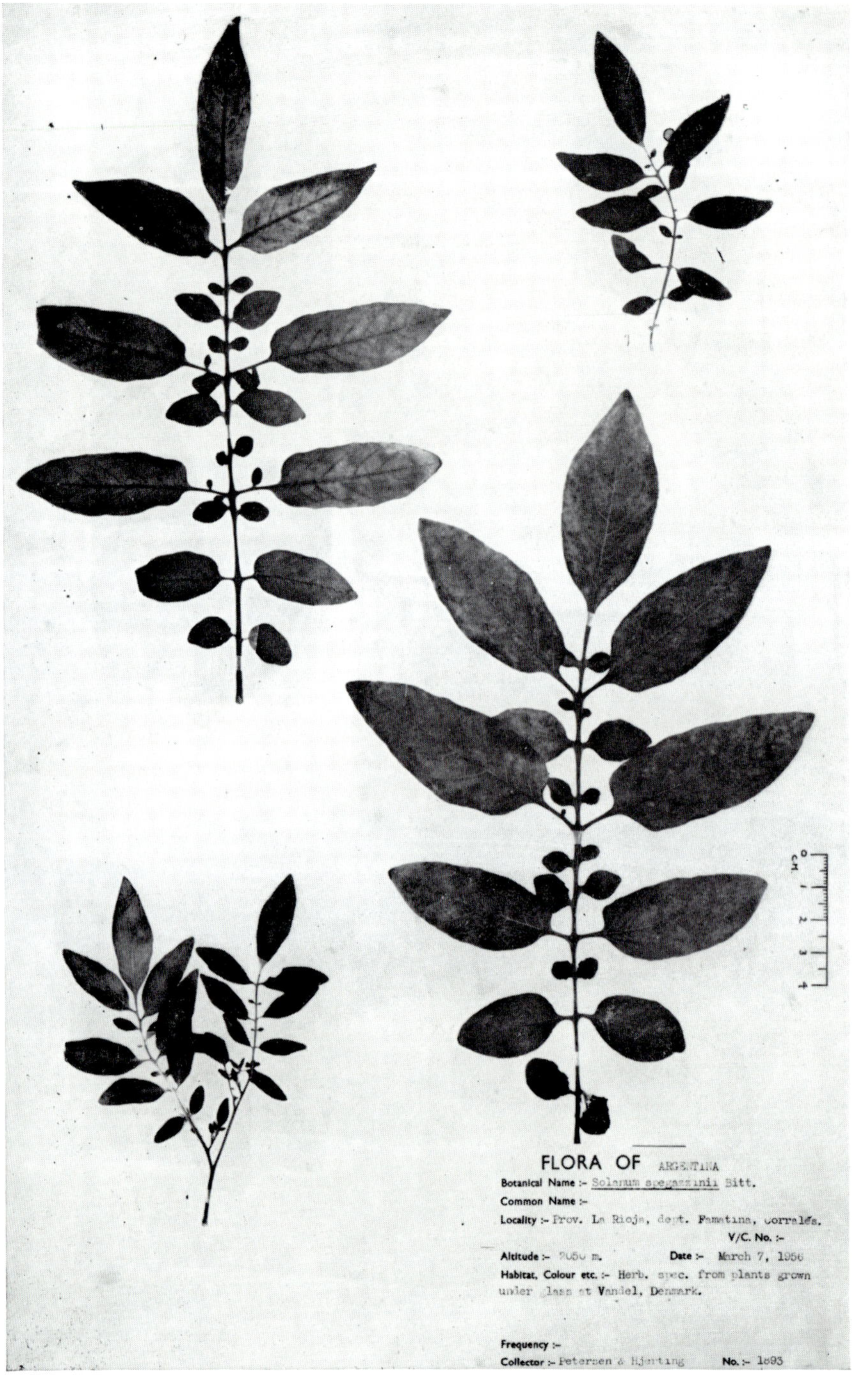

S. spegazzinii Bitt. Material grown under glass at Vandel, Denmark, of *P. & H. 1893* (see Plate 137). The leaf type matches those of plants growing in a maize field very closely.

S. spegazzinii Bitt. Specimens collected by Ruiz Leal and F. A. Roig (*No. 19369*) at Chepical, prov. San Juan, Argentina. The complete plant possesses the usual '*S. famatinae* type' leaf. The detached leaves from other plants of the same collection show extreme decurrency, which we have not been able to reproduce experimentally in this species. See also Plate 140.

S. spegazzinii Bitt. Collected by Hjerting, Petersen and Rahn (*No. 334*) from dept. Cachi, prov. Salta, Argentina. This material possesses much longer hairs than usual for *S. spegazzinii* and an extreme degree of decurrency on the lateral leaflets of all but the upper leaves (see Plate 141, however).

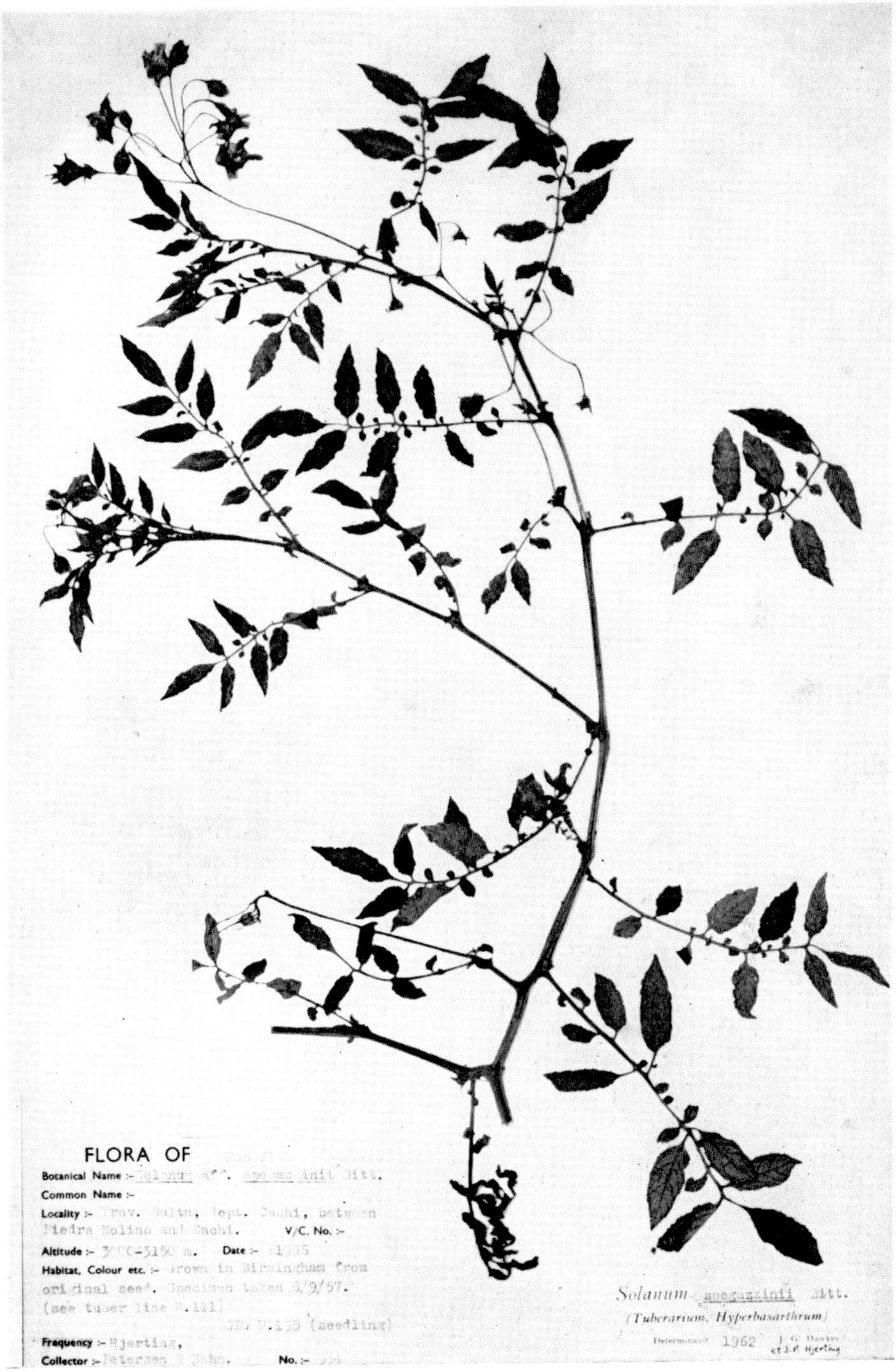

S. spegazzinii Bitt. Material grown from seeds of *H.P. & R. 334* (Accession No. H. 135) which contrasts strongly with the original specimen (see Plate 140). In this the leaflets are not at all decurrent and also possess long petiolules. The long pubescence is still to be seen in fresh specimens.

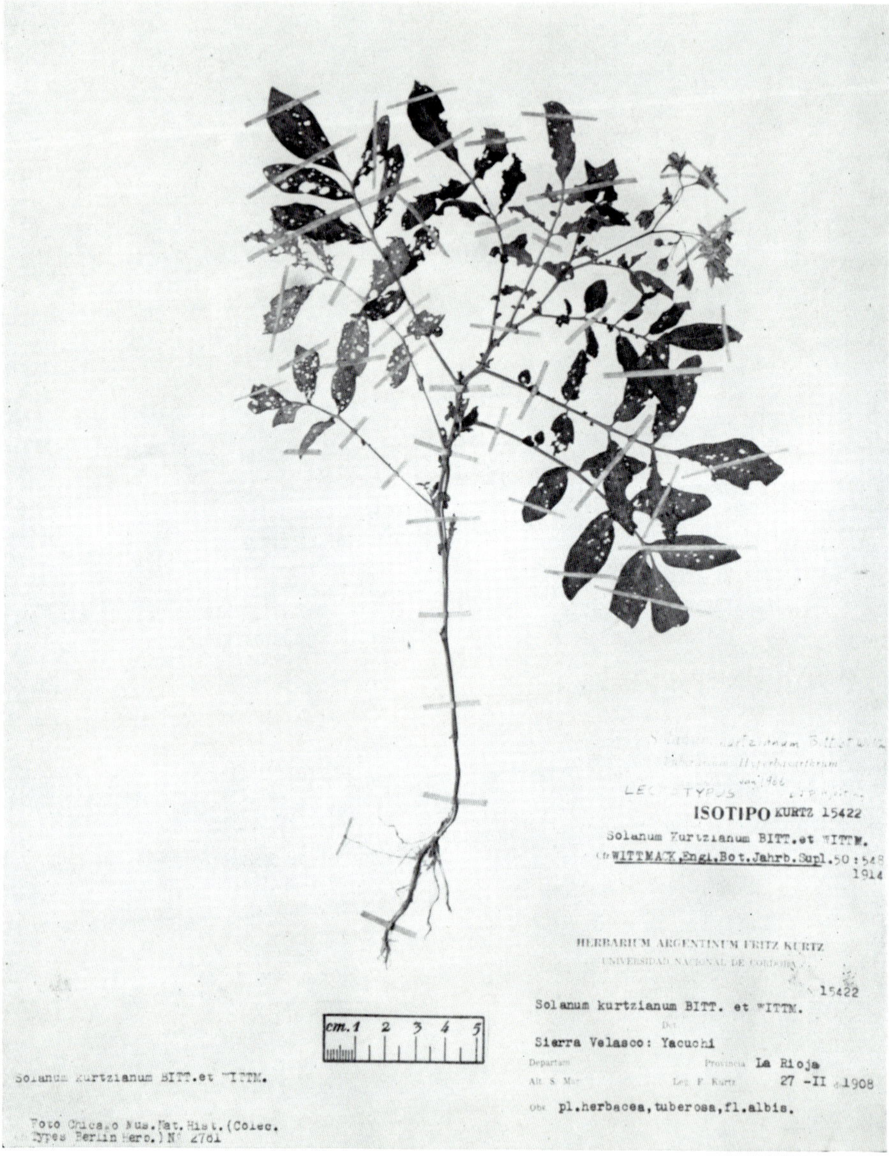

S. kurtzianum Bitt. & Wittm. Isotype specimen of *Kurtz 15422* from herb. Córdoba, now designated by us as Lectotype. This rather delicate specimen, collected in Sierra Velasco, prov. La Rioja, Argentina, matches the photograph of the now destroyed Berlin type specimen very closely. (Compare with Fig. 41.)

S. kurtzianum Bitt. & Wittm. Duplicate of the specimen on which Bitter and Wittmack based their description of '*S. velascanum*'. The original in herb. Berlin was destroyed. Collected by F. Kurtz (*No. 15380*—herb. Córdoba) in Sierra Velasco, prov. La Rioja, Argentina. Designated by us as Lectotype for '*S. velascanum*' Bitt. & Wittm. This is a larger plant than the type of *S. kurtzianum* (see Plate 142) but agrees with it in all essential details.

S. kurtzianum Bitt. & Wittm. Duplicate specimen from herb. Córdoba of '*S. juanense*' Wittm. *nom. nud.*, collected by F. Kurtz (*No. 9828*) in prov. San Juan, Argentina. The Berlin original was destroyed. Note the essential similarity in habit and leaf form to the other specimens of *S. kurtzianum* reproduced here and especially to Plate 148.

S. kurtzianum Bitt. & Wittm. Collected by Petersen and Hjerting (*No. P–40*) at Río Pantanillos, dept. Andalgalá, prov. Catamarca, Argentina. Note the low articulation and minutely serrate leaflet margins. This was obtained from the northern-most part of the geographical range of this species.

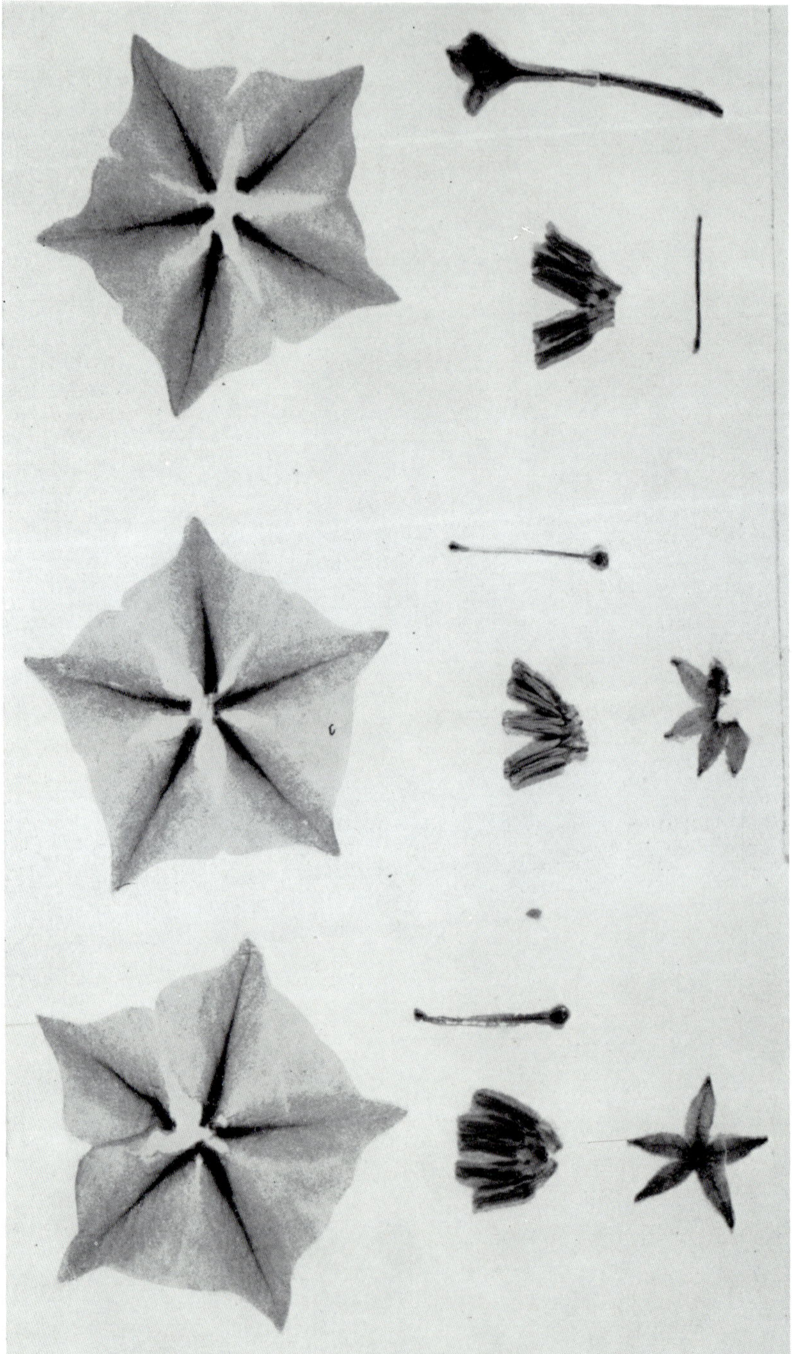

S. kurtzianum Bitt. & Wittm. Flower dissection of material grown at Cambridge (CPC: 1296) of the probable type collection of '*S. macolae*' Buk.

S. kurtzianum Bitt. & Wittm. Collected by Ruiz Leal (*No. 11906*—herb. Ruiz Leal) at El Portezuelo, dept. Las Heras, prov. Mendoza, Argentina. (A) original specimen. (B) flower dissection of seedling (H. 62) grown at Birmingham.

S. kurtzianum Bitt. & Wittm. Collected by Ruiz Leal and Roig (*No. 18390*—herb. Ruiz Leal) near El Portezuelo, dept. Las Heras, prov. Mendoza, Argentina. Note the broader leaflets, often to be seen in this species especially in the lower leaves.

S. maglia Schlechtd. Plant grown in Argentina from tubers collected by Ruiz Leal (*No. 10470*—herb. Lillo) at Quebrada de Alvarado, dept. San Carlos, prov. Mendoza, Argentina.

S. maglia Schlechtd. Grown in Cambridge and Birmingham from tubers collected by Schnack (s.n.—CPC: 2057) at Quebrada de Alvarado, dept. San Carlos, prov. Mendoza, Argentina. This material is fairly similar to Plate 149 and may perhaps be the same original collection.